W9-DCX-295

The swimming patterns of fish in a stream may be modeled by the diffusion equation. Learn how to use partial derivatives to determine how fish spread throughout a bay after being raised in a hatchery. (See the Extended Life Science Connection in Chapter 7 page 543.)

When a toxin enters the environment, its effects can be widespread. Learn how to use compartment models and differential equations to model the spread of pollutants in the environment. (See Exercise 18 in Section 9.6 on page 648.)

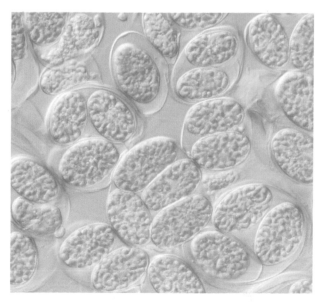

Axenic cultures are needed by biologists to generate culture of algae that are free of bacteria. Learn how to use calculus and the Poisson distribution to find the most efficient procedure to making axenic cultures. (See the Extended Life Science Connection in Chapter 10 on page 725.)

Calculus

FOR THE LIFE SCIENCES

Marvin L. Bittinger
Indiana University Purdue University Indianapolis

Neal Brand
University of North Texas

John Quintanilla
University of North Texas

PEARSON

Addison
Wesley

Boston San Francisco New York
London Toronto Sydney Tokyo Singapore Madrid
Mexico City Munich Paris Cape Town Hong Kong Montreal

Publisher:	Greg Tobin
Acquisitions Editor:	Carter Fenton
Project Editor:	Joanne Ha
Editorial Assistant:	Rachel Monaghan
Managing Editor:	Karen Wernholm
Production Supervisor:	Sheila Spinney
Cover Designer:	Susan Koski Zucker
Photo Researcher:	Geri Davis/The Davis Group, Inc.
Software Development:	Marty Wright and Mary Durnwald
Senior Marketing Manager:	Becky Anderson
Marketing Coordinator:	Maureen McLaughlin
Author Support Specialist:	Joe Vetere
Senior Manufacturing Buyer:	Evelyn Beaton
Production Coordination:	Elm Street Publishing Services, Inc.
Composition:	Beacon Publishing Services
Illustrations:	Network Graphics
Cover photo:	Salmon, Puget Sound © SuperStock, Inc.

For permission to use copyrighted material, grateful acknowledgment is made to the copyright holders on page 752, which is hereby made part of this copyright page.

Many of the designations used by manufacturers and sellers to distinguish their products are claimed as trademarks. Where those designations appear in this book, and Addison-Wesley was aware of a trademark claim, the designations have been printed in initial caps or all caps.

Library of Congress Cataloging-in-Publication Data

Bittinger, Marvin L.
    Calculus for the life sciences/Marvin Bittinger, Neal Brand, John Quintanilla.— 1st ed.
        p.   cm.
    Includes index.
    ISBN 0-321-27935-2
    1. Biomathematics. 2. Calculus. I. Brand, Neal E. II. Quintanilla, John, 1971-III. Title.

QH323.5.B576 2006
570'.1'515—dc22   2004062792

Copyright © 2006 Pearson Education, Inc. All rights reserved. No part of this publication may be reproduced, stored in a retrieval system, or transmitted, in any form or by any means, electronic, mechanical, photocopying, recording, or otherwise, without the prior written permission of the publisher. Printed in the United States of America. For information on obtaining permission for use of material in this work, please submit written request to Pearson Education, Inc., Rights and Contracts Department, 75 Arlington Street, Suite 300, Boston, MA 02116, fax your request to 617-848-7047, or e-mail at http://www.pearsoned.com/legal/permissions.htm.

9 10 11 12 V056 16 15 14

To Shari, Jeni, and Robin—NB

To Sandra and Sarah—JQ

Contents

v

6   Matrices   429

7   Functions of Several Variables   501

8   First-Order Differential Equations   545

9 Higher-Order and Systems of Differential Equations 597

10 Probability 653

Preface

This text is designed for either a two-semester sequence or a one-semester course in applied calculus for life science students. Students should be assured that calculus is indeed relevant to their future courses and careers in the life sciences. The topics in this book were chosen by surveying scientific journals and reflect the mathematical concepts used by geneticists, epidemiologists, environmental scientists, and other specialists in the life sciences. Applications are chosen from the contemporary literature in the life sciences, as opposed to physics or engineering, to show students the relevance of calculus to their own disciplines.

This text serves as excellent preparation for students who plan to take formal courses in differential equations, probability, or statistics. However, life science majors are often required to only take a calculus sequence to fulfill graduation requirements. As a result, this book has been written to provide mathematical tools necessary for these students.

A course in college algebra is a prerequisite for the text. The appendices and Chapter 1 provide a sufficient foundation to unify the diverse backgrounds of most students.

Intuitive Approach and Illustrative Applications

Although the word *intuitive* has many meanings and interpretations, its use here means "experienced-based." Throughout the text, when a particular concept is discussed, its presentation is designed so that the students' learning process is based on their earlier mathematical experience or on a new experience that is presented before the concept is formalized.

Intuition for mathematics is guided by illustrative applications. The applications in this text are chosen to highlight the importance of mathematical concepts in the life sciences. In addition, each chapter opener includes an application that serves as a preview of what students will learn in the chapter. An Index of Applications provides a quick reference for assigning problems.

Much of this book is based upon the tested and proven one-semester text *Calculus and Its Applications* by Marvin L. Bittinger. Reviews of algebra and functions may be found as appendixes. Chapter 1 begins by covering algebraic and trigonometric functions. The Technology Connections in Chapter 1 lay the foundation for the use of technology later in the book.

Chapter 2 begins a gradual approach to differentiation. The definition of the derivative in Chapter 2 is presented in the context of a discussion of average rates of change (see p. 95). This presentation is more accessible and realistic than the strictly geometric idea of slope.

Applications of differentiation are presented in Chapter 3. When maximum problems are introduced (see pp. 227–234), a function is derived that is to be maximized. Instead of forging ahead with the standard calculus solution, the student is asked to stop, make a table of function values, graph the function, and then estimate the maximum value. Also, relative minima and maxima (Sections 3.1–3.3) and absolute maxima and minima (Section 3.4) are covered in separate sections. Students will benefit by gradually building up to these topics as they consider graphing using calculus concepts.

Exponential and logarithmic functions are presented in Chapter 4. Both the definition of the number e and the derivative of e^x are explained in a biological context (see pp. 267–270). Population growth and decay models are introduced; these are covered more extensively in Chapters 8 and 9.

Chapter 5 covers integration and the Fundamental Theorem of Calculus. Antiderivatives are introduced first, followed by their connection to area and definite integrals. Applications of integration are given throughout the rest of the book.

Matrices are introduced in Chapter 6. Matrix multiplication, inversion, and the eigenvalues and eigenvectors of matrices are presented in a context appropriate to the life sciences. The exposition is motivated and illustrated by growth models using Leslie matrices. Difference equations are also solved by using eigenvalues and eigenvectors.

Multivariable calculus is covered in Chapter 7. Mathematical models using several variables are given as applications. Partial derivatives are defined in Section 7.2. These derivatives form the basis of partial differential equations, an important application of mathematics to the life sciences whose full treatment lies beyond the scope of an introductory text. Nevertheless, the Extended Life Science Connection at the end of the chapter interprets the solution of a common partial differential equation.

Chapter 8 covers first-order differential equations, one of the most important mathematical subjects for life science students to master. Compartment models describe how individual and systems of first-order differential equations arise in the life sciences (see pp. 557 and 617). Numerical solutions of differential equations are found by using spreadsheets.

Chapter 9 introduces higher-order and systems of differential equations. Systems of differential equations are first solved by reduction to higher-order differential equations. Comparisons between this reduction method and the eigenvalue/eigenvector method are emphasized. Nonlinear systems are also introduced with biological examples.

Probability is covered in Chapter 10. The first half of the chapter covers discrete notions of probability and their applications, especially to genetics and clinical medical testing (see pp. 666 and 688). Probabilities are represented as areas under a histogram, allowing a natural transition to probability for continuous distributions. This chapter provides the theoretical underpinning of statistics. In fact, important distributions that will be used in a future statistics class—including the normal, t-, and χ^2-distributions, are introduced in either the text or the exercises.

Pedagogical Features

Chapter Openers

Each chapter opener includes an application to motivate the chapter material. These applications also provide an intuitive introduction to a key calculus topic.

Section Objectives

As each new section begins, its objectives are clearly stated in the margin. These can be spotted easily by the student, and thus provide the answer to the typical question, "What material am I responsible for?"

Technology Connections

Though still optional, the use of technology reinforces the concepts presented throughout the text. For the topics that are computationally intensive (see Sections 3.6, 5.7, 8.5, and 9.6), technology greatly reduces the amount of effort required to solve exercises numerically.

Technology Connection features (101 in all) are included throughout the text. They illustrate the use of technology while students explore key ideas in calculus. When appropriate, art that simulates graphs and tables generated by a grapher is included as well. (In this text, we use the term *grapher* to refer to all graphing calculators and graphing software.)

There are four different types of Technology Connections.

- *Lesson/Teaching.* These provide students with an example, followed by exercises to work.
- *Checking.* These tell students how to verify a solution by using a grapher.
- *Exercises.* These are simply exercises for a student to solve using a grapher.
- *Exploration/Investigation.* These provide questions to guide students through an investigation.

Variety of Exercises

There are more than 4500 exercises in this text. The exercise sets include skill-based exercises, detailed art pieces, and extra graphs. They are also enhanced by the following features:

- *Synthesis Exercises.* Synthesis exercises are included at the end of most exercise sets and all Reviews and Chapter Tests. They require students to go beyond the immediate objectives of the section or chapter and are designed both to challenge students and to deepen their understanding (see pp. 295, 347, and 485).

- *Thinking and Writing Exercises.*   These exercises appear both in the exercise sets and in the synthesis section at the end of most exercise sets (see pp. 100, 407, and 457). They are denoted by the symbol **tw** . These ask students to explain mathematical concepts in their own words, thereby strengthening their understanding. The answers to these exercises are given in the *Instructor's Solutions Manual.*

- *Applications.*   A section of applied exercises is included in most exercise sets. Most of these applications are chosen from the life sciences and are motivated by current research or reflect current practices. Each exercise is accompanied by a brief description of its subject matter and often a citation to the scientific literature (see pp. 247, 276, and 442).

- *Technology Connection Exercises.*   These exercises appear in the exercise sets (see pp. 213, 312, and 380) and are indicated with the ⌐⌐ icon or under the heading "Technology Connection." These exercises also appear in the Reviews and the Chapter Tests. The Printed Test Bank supplement includes technology-based exercises as well. Exercises also appear in the Technology Connection feature (see pp. 201, 303, and 358). Answers to these exercises are given in the *Instructor's Solutions Manual.*

Tests and Reviews

- *Summary and Review.*   At the end of each chapter is a summary and review. These are designed to provide students with all the material they need for successful review. Answers for review exercises are located at the back of the book, together with section references so that students can easily find the correct material to restudy if they have difficulty with a particular exercise. In each summary and review, there is a list of Terms to Know, accompanied by page references to help students key in on important concepts (see p. 422).

- *Tests.*   Each chapter ends with a chapter test that includes synthesis and technology questions. The answers, with section references to the chapter tests, are at the back of the book. Six additional forms of each of the chapter tests and the final examination with answer keys appear, ready for classroom use, in the *Instructor's Manual/Printed Test Bank.*

Extended Life Science Connections

An Extended Life Science Connection ends each chapter. They are designed to demonstrate the importance of mathematical concepts to modern scientific research in fields such as ecology, genetics, paleontology, and epidemiology. The Extended Life Science Connections present multiple exercises that build on each other to form a conceptual framework for studying the given application. Group study is promoted by integrating concepts presented throughout the text. References to related literature are often provided.

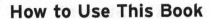

How to Use This Book

Two-Semester Sequence

This text may be used for a two-semester calculus sequence, introducing students to mathematical concepts that naturally arise in the life sciences. Instructors may decide to omit certain early topics in order to completely cover the later chapters in depth. If this approach is taken, any of the following topics may be omitted with little to no loss in continuity: Sections 3.6–3.7, 4.5, 5.7–5.8, 6.5, 7.4–7.5, and Bayes' Rule in Section 10.2.

One-Semester Course

Selections from this text may also be chosen for a one-semester calculus course. A standard calculus course may be based on the first five chapters. If little review from Chapter 1 is planned, then Chapter 7 may also be included. As shown below, instructors may also "fast-track" so that the course includes a substantial treatment of differential equations and/or probability.

Standard Calculus	Differential Equations	Probability
Chapters 1–5	Chapter 2	Chapter 2
Chapter 7	Sections 3.1–3.2	Sections 3.1–3.2
	Sections 4.1–4.4	Sections 4.1–4.2
	Sections 5.1–5.7	Sections 5.1–5.7
	Chapter 8	Chapter 10

Needless to say, instructors who teach one-semester courses culminating in either differential equations or statistics should use caution when selecting from the exercise sets. Time permitting, additional topics may be selected from the remaining chapters.

Supplements for the Student

Student's Solutions Manual (ISBN: 0-321-28605-7) This manual, written by Michael Butros, *Victor Valley Community College,* provides complete worked-out solutions for all odd-numbered exercises in the exercise sets (with the exception of the Thinking and Writing exercises).

Addison-Wesley Math Tutor Center The Addison-Wesley Math Tutor Center is staffed by qualified mathematics and statistics instructors who provide students with tutoring on examples and odd-numbered exercises from the textbook. Tutoring is available via toll-free telephone, toll-free fax, e-mail, and the Internet. Interactive, Web-based technology allows tutors and students to view and work through problems together in real time over the Internet. For more information, please visit our Website at www.aw-bc.com/tutorcenter or call us at 1-888-777-0463.

Supplements for the Instructor

Instructor's Solutions Manual (ISBN: 0-321-28604-9) Written by Michael Butros, *Victor Valley Community College,* this manual provides complete worked-out solutions to *all* exercises in the exercise sets.

Instructor's Manual/Printed Test Bank (ISBN: 0-321-38665-5) This manual, written by Mary Ann Teel, *University of North Texas,* contains six alternate test forms for each chapter test and six comprehensive final examinations with answers. The *IM/PTB* also has answers for the even-numbered exercises in the exercise sets. The answers to the odd-numbered exercises are at the back of the text.

TestGen with QuizMaster (ISBN: 0-321-28603-0) *TestGen* enables instructors to build, edit, print, and administer tests using a computerized bank of questions developed to cover all the objectives of the text. *TestGen* is algorithmically based, allowing instructors to create multiple equivalent versions of the same question or test with the click of a button. Instructors can also modify test bank questions or add new questions by using the built-in question editor, which allows users to create graphs, import graphics, and insert math notation, variable numbers, or text. Tests can be printed or administered online via the Internet or another network. *TestGen* comes packaged with *QuizMaster,* which allows students to take tests on a local area network. The software is available on a dual-platform Windows/Macintosh CD-ROM.

Acknowledgments

Finally, we wish to thank the following reviewers for their helpful comments and suggestions for this first edition of the text:

Stephen Aldrich, *Saint Mary's University of Minnesota*
Ward Canfield, *National Louis University*
John Daughtry, *East Carolina University*
Melvin Lax, *California State University, Long Beach*
Jeffrey Meyer, *Syracuse University*
Donald Mills, *Southern Illinois University*
J. B. Nation, *University of Hawaii*
Andreas Soemadi, *Kirkwood Community College*
Ali Zakeri, *California State University, Northridge*

M.L.B.
N.B.
J.Q.

Index of Applications

AGRICULTURE

Corn acreage, 691
Crop yield, 563
Ginseng roots, 306, 311
Harvesting, 572, 573
Maize leaf growth, 312
Maximum sustainable harvest, 333, 572
Plant thinning rule, 290
Soil organic matter, 633

ASTRONOMY

Angle of elevation of the Sun, 46
Satellite power, 323
Solar eclipse, 177
Solar radiation, 59, 527
Weight on Earth and the moon, 8

AUTOMOTIVE

Automobile accidents, minimizing, 226
Braking distance, 346
Engine emissions, 368, 377
Gas mileage, 99
Stopping distance on glare ice, 17, 131
Vanity license plates, 236

BIOLOGY

Algae population growth, 310, 360, 370
Animal weight gain, 736, 738
Axenic cultures, 725
Bacteria population growth, 267, 296, 309, 332, 333, 344, 360, 370, 394, 570, 571, 590, 592, 594, 605

Bacterium transfer, 692, 703, 709
Bee cells, 220, 227
Bird calls, 698, 703, 725
Bird population, 431, 432, 440, 441, 442, 443, 456, 470, 481, 485, 497, 499
   cerulean warblers, 429, 441, 443, 456, 461, 470, 485
   dinophilus gyrociliatus, 441, 470, 485
   ovenbirds, 434, 435, 446, 483
Cell cycle time, 705, 711, 724, 725
Competing species, 636, 637, 642, 643, 647, 648, 650, 651
DNA, 551
Deer population, 100, 113, 308
Dinosaurs, 39, 425
Evaluating habitats, 680, 690
Fairy shrimp population, 442, 443, 456, 470, 485
Fish growth, 503
Fish length, 720
Fish mass, 720
Fish population, 311, 642
Fox population, 494
Genetics, 663, 664, 668, 688, 717, 720
Growth of whales, 2, 113, 129
Growth rate for animals, 247
Growth rate of a fungus, 277, 309
Hardy-Weinberg Law, 677, 680, 682
Height
   estimating (anthropology), 18
   of a tree, 45
Home range of an animal, 38, 132
Homing pigeons, flights of, 233, 237
Honeybees, 47
Insect growth, 276, 295

Larvae and forest defoliation, 595
Limited predation, 643
Lotka-Volterra equations, 296, 644, 647
Mutations, 707, 708, 711
Otzi the Iceman, 321
Pea plant, 663, 664, 668, 676, 723
Photosynthesis, 93, 253
Planning an experiment, 309
Plant growth, 252, 423, 583
Plant mass, 495
Plant propagation, 489, 492, 493, 494
Polygenic traits, 678
Polymorphism, 260
Population growth, 499
Populations of organisms in a chemstat, 581
Predators and prey, 639, 642, 644, 645, 650, 651
Prey with inhibited growth, 642, 643, 647
Reynolds number, 294
Seagrass, 689, 712
Siberian Yuribei mammoth, 321
Species richness, 293, 535
Stocking fish, 543
Territorial area of an animal, 39, 131
Test cross, 668
Tree growth, 394, 424
Yeast growth, 594

BUSINESS

Advertising results, 130, 332
Candy sales, 357
Cost, minimizing, 236, 237, 257, 526
Franchise growth, 330
Postage function, 83, 113
Profit, maximizing, 235, 236, 526

xix

Index of Technology Connections

The following index lists, by topic, 67 optional Technology Connections. These features incorporate in-text comments, including guided explorations and exercises, that explain and encourage the use of technology to explore problem situations. The Technology Connection Features, along with the Technology Connection Exercises in the exercise sets, and the Extended Technology Applications, described in the Preface, bolster the optional integration of technology into the course.

1

Functions and Graphs

INTRODUCTION *This chapter reviews some common functions together with their graphs and applications, especially to the life sciences. Skills in using a graphing calculator or graphing software (henceforth referred to as "graphers") are also introduced in the optional Technology Connections.*

Those needing some algebra review or review of basic properties of functions might wish to study the appendices.

AN APPLICATION   Life Expectancy   Use the graph below to determine the average rate of change of life expectancy at birth.*

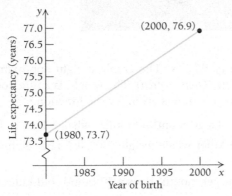

This problem appears as Exercise 57 in Exercise Set 1.1.

---

*National Center for Health Statistics.

1.1 Slope and Linear Functions

OBJECTIVES

- Graph horizontal and vertical lines.
- Graph linear functions.
- Find an equation of a line given the slope and a point on the line.
- Find an equation of a line given two points on the line.
- Solve applied problems involving proportions and other linear functions.

What Is Calculus?

What is calculus? Let's consider a simplified answer for now. We will develop the ideas more fully as the book progresses.

In algebra, we study functions and their graphs. For example, we can graph a function by plotting many points and then connecting the points to form a curve. We can also look at a graph and determine certain points on the graph. In calculus, we extend our study of functions by considering concepts such as the slope of the tangent line to the graph of a function at a specific point, the area below a function, and the highest or lowest point on the graph of a function. We will see how these concepts can be used to solve practical problems involving the rate at which a quantity changes and maximizing a quantity.

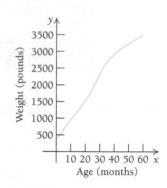

EXAMPLE 1 *Growth of Whales.* The graph at right approximates the weight, in pounds, of a killer whale (*Orcinus orca*). The weight is a function f of the age of the whale (in months). No equation is given for the function.[1]

a) What is the weight of a 20-month-old killer whale? That is, find $f(20)$.

b) At what age does a killer whale weigh 3000 lb? That is, find any inputs, x, for which $f(x) = 3000$.

c) How fast (in pounds per month) does a 20-month-old killer whale grow?

d) At what age does a killer whale grow at its fastest rate?

Solution

a) To estimate the weight of a 20-month-old killer whale, we locate 20 on the horizontal axis and move directly up until we reach the graph. Then we move across to the vertical axis. We estimate that value to be about 1500 lb—that is, $f(20) = 1500$.

---

[1]SeaWorld.

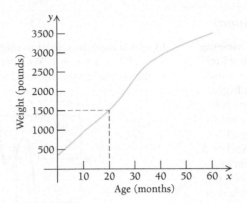
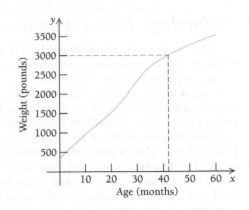

b) We locate 3000 on the vertical axis and move horizontally across to the graph and note that one input corresponds to 3000. We estimate that value to be about 42— that is, $f(42) = 3000$. The weight of a 42-month-old killer whale is approximately 3000 lb.

c) In algebra, questions like this are not usually considered. The *derivative* of a function is used to measure the rate of growth. We defer the answer to this type of question until Chapter 2.

d) Using algebra, we do not normally attempt to find the largest value of a function. We defer the answer to this type of question until Chapter 3, where we will discuss how to use the concept of derivative to find when a quantity is at its maximum.   ■

Appendix B gives a brief review of the basic properties of functions and graphs that are normally covered in algebra classes. If you have forgotten some of these concepts, it will be worth your time to review Appendix B before proceeding.

Calculators

Technology Connections are provided to show how technology can be used to help solve calculus problems and to clarify calculus concepts. The following Technology Connection shows how to use a calculator to graph a function. It also introduces the notation we will use to specify the graphing window.

Technology Connection

Introduction to the Use of a Graphing Calculator: Windows and Graphs

Viewing Windows

With this first of the optional Technology Connections, we begin to create graphs using a graphing calculator and computer graphing software, referred to simply as **graphers.** Most of the coverage will

refer to a TI-83 graphing calculator but in a somewhat generic manner, discussing features common to virtually all graphers. Although some reference to keystrokes will be mentioned, in general, exact details on keystrokes will be covered in the manual for your particular grapher.

(continued)

Technology Connection (continued)

One feature common to all graphers is the **viewing window.** This refers to the rectangular screen in which a graph appears. Windows are described by four numbers, [L, **R, B, T**], which represent the Left and **Right** endpoints of the x-axis and the **Bottom** and **Top** endpoints of the y-axis. A WINDOW feature can be used to set these dimensions. Below is a window setting of $[-20, 20, -5, 5]$ with axis scaling denoted as Xscl = 5 and Yscl = 1, which means that there are 5 units between tick marks on the x-axis and 1 unit between tick marks on the y-axis. Graphs are made up of black rectangular dots called **pixels.** Roughly speaking, the notation Xres = 1 is an indicator of the number of pixels used in making a graph.* We will usually leave it at 1 and not refer to it unless needed.

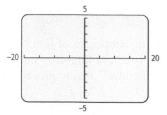

Axis scaling must be chosen with care, because tick marks become blurred and indistinguishable when too many appear. On some graphers, a setting of $[-10, 10, -10, 10]$, Xscl = 1, Yscl = 1, Xres = 1 is considered **standard.**

Graphs

The primary use for a grapher is to graph equations. For example, let's graph the equation $y = x^3 - 5x + 1$. The equation can be entered using the notation $y = x^3 - 5x + 1$. Some software uses Basic notation,

in which case the equation might be entered as $y = x^3 - 5*x + 1$. We obtain the following graph in the standard viewing window.

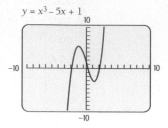

It is often necessary to change viewing windows in order to best reveal the curvature of a graph. For example, each of the following is a graph of $y = 3x^5 - 20x^3$, but with a different viewing window. Which do you think best displays the curvature of the graph?

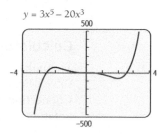

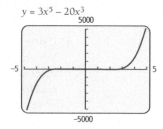

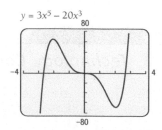

In general, choosing a window that best reveals a graph's characteristics involves some trial and error and, in some cases, some knowledge about the shape of that graph. We will learn more about the shape of graphs as we continue through the text.

---

*Xres sets pixel resolution at 1 through 8 for graphs of equations. At Xres = 1, equations are evaluated and graphed at each pixel on the x-axis. At Xres = 8, equations are evaluated and graphed at every eighth pixel on the x-axis. The resolution is better for smaller Xres values than for larger values.

To graph an equation like $3x + 5y = 10$, most graphers require that the equation be solved for y, that is, "$y = \ldots$" Thus we must rewrite and enter the equation as

$$y = \frac{-3x + 10}{5}, \quad \text{or} \quad y = -\frac{3}{5}x + 2.$$

Its graph is shown below in the standard window.

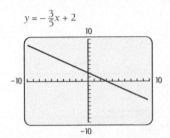

For the equation $x = y^2$, we would first obtain $y = \pm\sqrt{x}$ and then graph the individual equations $y_1 = \sqrt{x}$ and $y_2 = -\sqrt{x}$.

EXERCISES

Use a grapher to graph each of the following equations. Select the standard window $[-10, 10, -10, 10]$, with axis scaling Xscl $= 1$ and Yscl $= 1$.

1. $y = 2x - 2$ **2.** $y = -3x + 1$

3. $y = \frac{2}{5}x + 4$ **4.** $y = -\frac{3}{5}x - 1$

5. $y = 2.085x + 15.08$ **6.** $y = -\frac{4}{5}x + \frac{13}{7}$

7. $2x - 3y = 18$ **8.** $5y + 3x = 4$

9. $y = x^2$ **10.** $y = (x + 4)^2$

11. $y = 8 - x^2$ **12.** $y = 4 - 3x - x^2$

13. $y + 10 = 5x^2 - 3x$ **14.** $y - 2 = x^3$

15. $y = x^3 - 7x - 2$ **16.** $y = x^4 - 3x^2 + x$

17. $y = |x|$ (On most graphers, this is entered as $y = \text{abs}(x)$.)

18. $y = |x - 5|$

19. $y = |x| - 5$

20. $y = 9 - |x|$

Horizontal and Vertical Lines

In algebra, we learn to compute the slope of a line. This is a fundamental concept that underlies much of calculus. We therefore begin our study of calculus by reviewing properties of lines and linear functions.

Let's consider graphs of the equations $y = b$ and $x = a$.

EXAMPLE 2

a) Graph $y = 4$.

b) Decide whether the relation is a function.

Solution

a) The graph consists of all ordered pairs whose second coordinate is 4. For example, the points $(1, 4)$ and $(-2, 4)$ are on the graph.

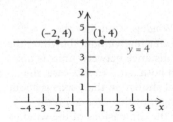

b) The vertical-line test holds. Thus this relation is a function. (See Appendix B to review the vertical-line test.)

EXAMPLE 3

a) Graph $x = -3$.

b) Decide whether the relation is a function.

Solution

a) The graph consists of all ordered pairs whose first coordinate is -3. For example, the points $(-3, 4)$ and $(-3, 1)$ are on the graph.

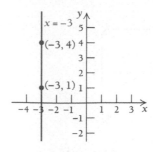

b) This relation is *not* a function because it fails the vertical-line test. The line itself meets the graph more than once—in fact, infinitely many times. ■

In general, we have the following.

THEOREM 1

The graph of $y = b$, or $f(x) = b$, a horizontal line, is the graph of a function.
The graph of $x = a$, a vertical line, is not the graph of a function.

Technology Connection

Visualizing Slope

Exploratory

Squaring a Viewing Window. Consider the $[-10, 10, -10, 10]$ viewing window shown on the right. Note that the distance between units is not visually the same on both axes. In this case, the length of the interval shown on the y-axis is about two-thirds of the length of the interval on the x-axis. If we change the dimensions of the window to $[-6, 6, -4, 4]$, we get a graph for which the units are visually about the same on both axes. Creating such a window is called **squaring the**

window. On a TI-83 grapher, there is a ZSQUARE feature for automatic window squaring. This feature alters the standard window dimensions to $[-15.1613, 15.1613, -10, 10]$.

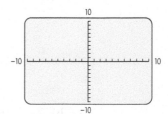

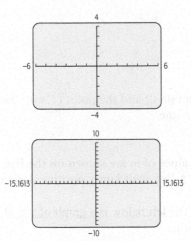

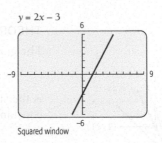

Squared window

Each of the following is a graph of the equation $y = 2x - 3$, but the viewing windows are different.

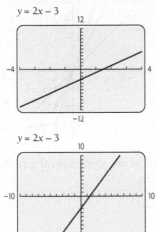

EXERCISES

Use a square viewing window for each of these exercises.

1. Graph $y = x + 1$, $y = 2x + 1$, $y = 3x + 1$, and $y = 10x + 1$. What do you think the graph of $y = 247x + 1$ will look like?

2. Graph $y = x$, $y = \frac{7}{8}x$, $y = 0.47x$, and $y = \frac{2}{31}x$. What do you think the graph of $y = 0.000018x$ will look like?

3. Graph $y = -x$, $y = -2x$, $y = -5x$, and $y = -10x$. What do you think the graph of $y = -247x$ will look like?

4. Graph $y = -x - 1$, $y = -\frac{3}{4}x - 1$, $y = -0.38x - 1$, and $y = -\frac{5}{32}x - 1$. What do you think the graph of $y = -0.000043x - 1$ will look like?

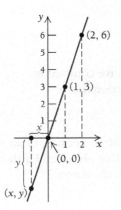

The Equation $y = mx$

Consider the following table of numbers and look for a pattern.

x	1	-1	$-\frac{1}{2}$	2	-2	3	-7	5
y	3	-3	$-\frac{3}{2}$	6	-6	9	-21	15

Note that the ratio of the bottom number to the top one is 3. That is,

$$\frac{y}{x} = 3, \quad \text{or} \quad y = 3x.$$

Ordered pairs from the table can be used to graph the equation $y = 3x$ (see the figure at left). Note that this is a function.

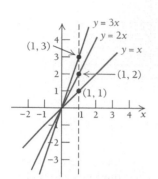

Lines of various slopes.

THEOREM 2

The graph of the function given by

$$y = mx \quad \text{or} \quad f(x) = mx$$

is the straight line through the origin $(0, 0)$ and the point $(1, m)$. The constant m is called the **slope** of the line.

Various graphs of $y = mx$ for positive values of m are shown on the left. Note that such graphs slant up from left to right. A line with larger positive slope rises faster than a line with smaller positive slope.

When $m = 0$, $y = 0x$, or $y = 0$. On the left below is a graph of $y = 0$. Note that this is both the x-axis and a horizontal line.

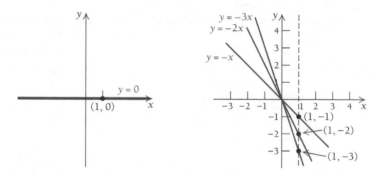

Graphs of $y = mx$ for negative values of m are shown on the right above. Note that such graphs slant down from left to right.

Direct Variation

There are many applications involving equations of the form $y = mx$, where m is some positive number. In such situations, we say that we have **direct variation**, and m (the slope) is called the **variation constant**, or **constant of proportionality.** Generally, only positive values of x and y are considered.

DEFINITION

The variable y **varies directly** as x if there is some positive constant m such that $y = mx$. We also say that y is **directly proportional** to x.

EXAMPLE 4 *Weight on Earth and the Moon.* The weight M, in pounds, of an object on the moon is directly proportional to the weight E of that object on Earth. An astronaut who weighs 180 lb on Earth will weigh 28.8 lb on the moon.

a) Find an equation of variation.

b) An astronaut weighs 19.2 lb on the moon. How much will the astronaut weigh on Earth?

Solution

a) Since $M = mE$, then $28.8 = m \cdot (180)$ so $0.16 = m$. Thus, $M = 0.16E$.

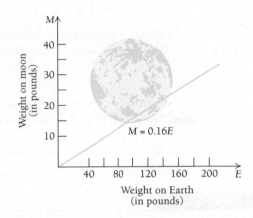

b) To find the weight on Earth of an astronaut who weighs 19.2 lb on the moon, we solve the equation of variation

$$19.2 = 0.16E \qquad \text{Substituting 19.2 for } M$$

and get

$$120 = E.$$

Thus an astronaut who weighs 19.2 lb on the moon weighs 120 lb on Earth. ■

The Equation $y = mx + b$

Compare the graphs of the equations

$$y = 3x \quad \text{and} \quad y = 3x - 2$$

(see the following figure). Note that the graph of $y = 3x - 2$ is a shift 2 units down from the graph of $y = 3x$, and that $y = 3x - 2$ has y-intercept $(0, -2)$. Note also that the graph of $y = 3x - 2$ is the graph of a function.

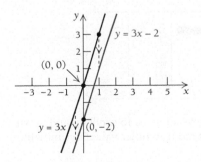

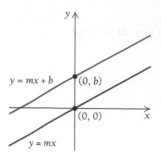

DEFINITION

A **linear function** is given by

$$y = mx + b \quad \text{or} \quad f(x) = mx + b$$

and has a graph that is the straight line parallel to $y = mx$ with y-intercept $(0, b)$. The constant m is called the **slope**. (See the figure at left.)

When $m = 0$, $y = 0x + b = b$, and we have what is known as a **constant function**. The graph of such a function is a horizontal line.

Technology Connection

Exploring b

One way to find ordered pairs of inputs and outputs of functions is to make use of the **TABLE** feature on a grapher. Let's consider the function $f(x) = 4x + 2$. We enter it into the grapher as $y_1 = 4x + 2$. To use the **TABLE** feature, we access the **TABLE SETUP** screen and choose the x-value at which the table will start and an increment for the x-value. For this function, let's set TblStart = 0.3 and ΔTbl = 1. This means that the table's x-values will start at 0.3 and increase by 1.

We also set Indpnt and Depend to Auto. The table that results when accessing the **TABLE** feature is shown below.

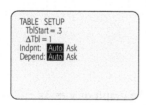

The scroll keys allow us to scroll up and down the table and extend it to other values not initially shown.

Let's use the **TABLE** feature and the graphing capabilities to explore the effect of b when we are graphing $f(x) = mx + b$.

EXERCISES

1. We look at functions of the form $y = x + b$.
   a) Plot $f(x) = x$ by entering $y_1 = x$. Then using the same viewing window, graph $y_2 = x + 3$ followed by $y_3 = x - 4$. Compare the graphs of y_2 and y_3 with that of y_1. Then without drawing them, compare the graphs of $y = x$ and $y = x - 5$.
   b) Use the **TABLE** feature to compare the values of y_1, y_2, and y_3 when $x = 0$. Then scroll through other values and see if you can determine a pattern.

2. We look at functions of the form $y = 3x + b$.
   a) Plot $f(x) = 3x$ by entering $y_1 = 3x$. Then using the same viewing window, graph $y_2 = 3x + 4$ followed by $y_3 = 3x - 2$. Compare the graphs of y_2 and y_3 with that of y_1. Then without drawing them, compare the graphs of $y = 3x$ and $y = 3x + 1$.
   b) Use the **TABLE** feature to compare the values of y_1, y_2, and y_3 when $x = 0$. Then scroll through other values and see if you can determine a pattern.

The Slope-Intercept Equation

Any nonvertical line is uniquely determined by its slope m and its y-intercept $(0, b)$. In other words, the slope describes the "slant" of the line, and the y-intercept is the point at which the line crosses the y-axis. Thus we have the following definition.

> **DEFINITION**
>
> $y = mx + b$ is called the **slope–intercept equation** of a line.

EXAMPLE 5   Find the slope and the y-intercept of $2x - 4y - 7 = 0$.

Solution   We solve for y:

$$-4y = -2x + 7$$
$$y = \tfrac{1}{2}x - \tfrac{7}{4}$$

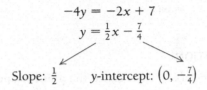

Slope: $\tfrac{1}{2}$ y-intercept: $\left(0, -\tfrac{7}{4}\right)$ ■

The Point-Slope Equation

Suppose that we know the slope of a line and some point on the line other than the y-intercept. We can still find an equation of the line.

EXAMPLE 6   Find an equation of the line with slope 3 containing the point $(-1, -5)$.

Solution   The slope is given as $m = 3$. From the slope–intercept equation, we have

$$y = 3x + b,$$

so we must determine b. Since $(-1, -5)$ is on the line, it follows that

$$-5 = 3(-1) + b,$$

so

$$-2 = b \quad \text{and} \quad y = 3x - 2.$$ ■

If a point (x_1, y_1) is on the line

$$y = mx + b, \tag{1}$$

it must follow that

$$y_1 = mx_1 + b. \tag{2}$$

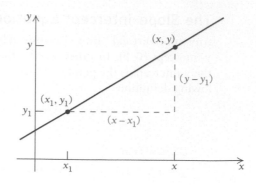

Subtracting equation (2) from equation (1) eliminates the b's, and we have

$$y - y_1 = (mx + b) - (mx_1 + b)$$
$$= mx + b - mx_1 - b$$
$$= mx - mx_1$$
$$= m(x - x_1).$$

DEFINITION

$y - y_1 = m(x - x_1)$ is called the **point–slope equation** of a line.

This definition allows us to write an equation of a line given its slope and the coordinates of *any* point on it.

EXAMPLE 7   Find an equation of the line with slope $\frac{2}{3}$ containing the point $(-1, -5)$.

Solution   Substituting in

$$y - y_1 = m(x - x_1),$$

we get

$$y - (-5) = \tfrac{2}{3}[x - (-1)]$$
$$y + 5 = \tfrac{2}{3}(x + 1)$$
$$y + 5 = \tfrac{2}{3}x + \tfrac{2}{3}$$
$$y = \tfrac{2}{3}x + \tfrac{2}{3} - 5$$
$$= \tfrac{2}{3}x + \tfrac{2}{3} - \tfrac{15}{3}$$
$$= \tfrac{2}{3}x - \tfrac{13}{3}.$$

Computing Slope

We now determine a method of computing the slope of a line when we know the co-ordinates of two of its points. Suppose that (x_1, y_1) and (x_2, y_2) are the coordinates of two different points, P_1 and P_2, respectively, on a line that is not vertical. Consider a right triangle with legs parallel to the axes, as shown in the following figure.

Which lines have the same slope?

The point P_3 with coordinates (x_2, y_1) is the third vertex of the triangle. As we move from P_1 to P_2, y changes from y_1 to y_2. The change in y is $y_2 - y_1$. Similarly, the change in x is $x_2 - x_1$. The ratio of these changes is the slope. To see this, consider the point–slope equation,

$$y - y_1 = m(x - x_1).$$

Since (x_2, y_2) is on the line, it must follow that

$$y_2 - y_1 = m(x_2 - x_1).$$

Since the line is not vertical, the two x-coordinates must be different, so $x_2 - x_1$ is nonzero and we can divide by it to get the following theorem.

THEOREM 3

$$m = \frac{y_2 - y_1}{x_2 - x_1} = \frac{\text{change in } y}{\text{change in } x} = \text{slope of line containing points } (x_1, y_1) \text{ and } (x_2, y_2)$$

EXAMPLE 8   Find the slope of the line containing the points $(-2, 6)$ and $(-4, 9)$.

Solution   We have

$$m = \frac{y_2 - y_1}{x_2 - x_1} = \frac{6 - 9}{-2 - (-4)} = \frac{-3}{2} = -\frac{3}{2}.$$

Note that it does not matter which point is taken first, so long as we subtract the coordinates in the same order. In this example, we can also find m as follows:

$$m = \frac{9 - 6}{-4 - (-2)} = \frac{3}{-2} = -\frac{3}{2}.$$ ∎

If a line is horizontal, the change in y for any two points is 0. Thus a horizontal line has slope 0. If a line is vertical, the change in x for any two points is 0. Thus the slope is *not defined* because we cannot divide by 0. A vertical line has no slope. Thus "0 slope" and "no slope" are two very distinct concepts.

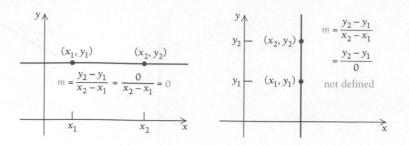

Applications of Linear Functions

Slope has many real-world applications. For example, numbers like 2%, 3%, and 6% are often used to represent the *grade* of a road, a measure of how steep a road on a hill or mountain is. A 3% grade $\left(3\% = \frac{3}{100}\right)$ means that for every horizontal distance of 100 ft, the road rises 3 ft, and a −3% grade means that for every horizontal distance of 100 ft, the road drops 3 ft. An athlete might change the grade of a treadmill during a workout. An escape ramp on an airliner might have a slope of about −0.6.

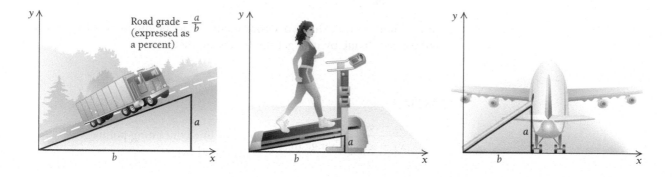

Road grade = $\frac{a}{b}$
(expressed as a percent)

Architects and carpenters use slope when designing and building stairs, ramps, or roof pitches. Another application occurs in hydrology. When a river flows, the strength or force of the river depends on how far the river falls vertically compared to how far it flows horizontally. Slope can also be considered as an **average rate of change.**

EXAMPLE 9   *Carbon Dioxide Concentration.* The concentration of carbon dioxide collected from air samples at Mauna Loa Observatory in Hawaii has increased steadily over the years, as shown in the graph on the next page. This concentration is measured in parts per million by volume (ppmv). Find the average rate of change of this concentration between 1960 and 2001.[2]

---

[2]C. D. Keeling and T. P. Whorf, "Atmospheric CO_2 records from sites in the SIO air sampling network." In *Trends: A Compendium of Data on Global Change.* Carbon Dioxide Information Analysis Center, Oak Ridge National Laboratory, U.S. Department of Energy, Oak Ridge, Tennessee (2002).

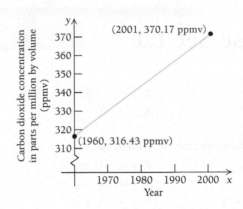

The Mauna Loa Observatory has measured the carbon dioxide concentration in the air above Hawaii since the 1950s.

Solution   First, we determine the coordinates of two points on the graph. In this case, they are given as (1960, 316.43 ppmv) and (2001, 370.17 ppmv). Then we compute the slope, or rate of change, as follows:

$$\text{Slope} = \text{Average rate of change} = \frac{\text{change in } y}{\text{change in } x}$$

$$= \frac{370.17 \text{ ppmv} - 316.43 \text{ ppmv}}{2001 - 1960} = 1.31 \frac{\text{ppmv}}{\text{yr}}$$

This result tells us that each year the carbon dioxide concentration has increased by about 1.31 parts per million by volume. A more detailed analysis of this increase will be discussed in the Extended Life Science Connection at the end of the chapter.

EXAMPLE 10   *Pressure and Temperature.* According to the ideal gas law, the pressure function $P(T)$ (in pounds per square inch, or psi) of a gas at temperature T (in degrees Kelvin) at constant volume is

$$P(T) = kT,$$

where k is a constant. Suppose the pressure in your car's tire is 36 psi in the afternoon, when the temperature is 27°C. What will be the pressure at night, when the temperature drops to 10°C?

Solution   To begin solving this problem, we convert 27°C into degrees Kelvin:[3]

$$T = 27 + 273 = 300.$$

We then solve for the constant k:

$$P(300) = k(300)$$
$$36 = 300k$$
$$0.12 = k.$$

At night, the temperature in degrees Kelvin will be $10 + 273 = 283$ degrees Kelvin. The pressure at this time will be

$$P(283) = (0.12)(283) = 33.96 \text{ psi}.$$

---

[3]To convert from degrees Celsius to degrees Kelvin, the true conversion is to add 273.15. We have rounded in this exercise.

Exercise Set 1.1

Graph.

1. $y = -4$
2. $y = -3.5$
3. $x = -4.5$
4. $x = 10$

Graph. Find the slope and the y-intercept.

5. $y = -3x$
6. $y = -0.5x$
7. $y = 0.5x$
8. $y = 3x$
9. $y = -2x + 3$
10. $y = -x + 4$
11. $y = -x - 2$
12. $y = -3x + 2$

Find the slope and the y-intercept.

13. $2x + y - 2 = 0$
14. $2x - y + 3 = 0$
15. $2x + 2y + 5 = 0$
16. $3x - 3y + 6 = 0$
17. $x = 2y + 8$
18. $x = -4y + 3$

Find an equation of the line:

19. with $m = -5$, containing $(1, -5)$.
20. with $m = 7$, containing $(1, 7)$.
21. with $m = -2$, containing $(2, 3)$.
22. with $m = -3$, containing $(5, -2)$.
23. with slope 2, containing $(3, 0)$.
24. with slope -5, containing $(5, 0)$.
25. with y-intercept $(0, -6)$ and slope $\frac{1}{2}$.
26. with y-intercept $(0, 7)$ and slope $\frac{4}{3}$.
27. with slope 0, containing $(2, 3)$.
28. with slope 0, containing $(4, 8)$.

Find the slope of the line containing the given pair of points, if it exists.

29. $(-4, -2)$ and $(-2, 1)$
30. $(-2, 1)$ and $(6, 3)$
31. $\left(\frac{2}{5}, \frac{1}{2}\right)$ and $\left(-3, \frac{4}{5}\right)$
32. $\left(-\frac{3}{4}, \frac{5}{8}\right)$ and $\left(-\frac{1}{2}, -\frac{3}{16}\right)$
33. $(3, -7)$ and $(3, -9)$
34. $(-4, 2)$ and $(-4, 10)$
35. $(2, 3)$ and $(-1, 3)$
36. $\left(-6, \frac{1}{2}\right)$ and $\left(-7, \frac{1}{2}\right)$
37. $(x, 3x)$ and $(x + h, 3(x + h))$
38. $(x, 4x)$ and $(x + h, 4(x + h))$

39. $(x, 2x + 3)$ and $(x + h, 2(x + h) + 3)$
40. $(x, 3x - 1)$ and $(x + h, 3(x + h) - 1)$
41.–52. Find an equation of the line containing the pair of points in each of Exercises 29–40.
53. Find the slope (or grade) of the treadmill.

54. Find the slope (or pitch) of the roof.

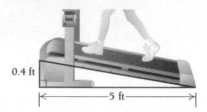

55. Find the slope (or head) of the river.

56. North Carolina state law requires that stairs have minimum treads (or width) of 9 in. and maximum risers (or height) of 8.25 in.[4] According to North Carolina law, what is the maximum grade of stairs in North Carolina?

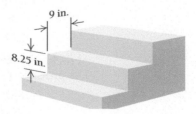

---

[4]North Carolina Office of the State Fire Marshall.

57. Find the average rate of change of life expectancy at birth.[5]

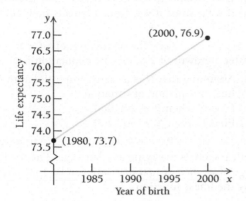

58. *Celsius and Fahrenheit.* If the temperature is C degrees Celsius, then the temperature is also F degrees Fahrenheit, where

$$F(C) = \frac{9}{5}C + 32.$$

a) Find $F(-10)$, $F(0)$, $F(10)$, and $F(40)$.
b) Suppose the outside temperature is 30 degrees Celsius. What is the temperature in degrees Fahrenheit?
c) What temperature is the same in both degrees Fahrenheit and in degrees Celsius?

59. *Energy Conservation.* The R-factor of home insulation is directly proportional to its thickness T.

a) Find an equation of variation if $R = 12.51$ when $T = 3$ in.
b) What is the R-factor for insulation that is 6 in. thick?

60. *Nerve Impulse Speed.* Impulses in nerve fibers travel at a speed of 293 ft/sec. The distance D traveled in t seconds is given by $D = 293t$. How long would it take an impulse to travel from the brain to the toes of a person who is 6 ft tall?

61. *Brain Weight.* The weight B of a human's brain is directly proportional to his or her body weight W.

a) It is known that a person who weighs 200 lb has a brain that weighs 5 lb. Find an equation of variation expressing B as a function of W.
b) Express the variation constant as a percent and interpret the resulting equation.
c) What is the weight of the brain of a person who weighs 120 lb?

[5]National Center for Health Statistics.

62. *Muscle Weight.* The weight M of the muscles in a human is directly proportional to his or her body weight W.

a) It is known that a person who weighs 200 lb has 80 lb of muscles. Find an equation of variation expressing M as a function of W.
b) Express the variation constant as a percent and interpret the resulting equation.
c) What is the muscle weight of a person who weighs 120 lb?

Muscle weight is directly proportional to body weight.

63. *Stopping Distance on Glare Ice.* The stopping distance (at some fixed speed) of regular tires on glare ice is given by

$$D(F) = 2F + 115,$$

where $D(F)$ is the stopping distance, in feet, when the air temperature is F, in degrees Fahrenheit.

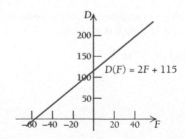

a) Find $D(0°)$, $D(-20°)$, $D(10°)$, and $D(32°)$.
tw b) Explain why the domain should be restricted to the interval $[-57.5°, 32°]$.

64. *Reaction Time.* While driving a car, you see a person suddenly cross the street unattended. Your brain registers the emergency and sends a signal to your foot to hit the brake. The car travels a distance D, in feet, during this time, where D is a function of the speed r, in miles per hour, that the car is traveling when you see the person. That reaction distance is a linear function given by

$$D(r) = \frac{11r + 5}{10}.$$

a) Find $D(5)$, $D(10)$, $D(20)$, $D(50)$, and $D(65)$.
b) Graph $D(r)$.
tw c) What is the domain of the function? Explain.

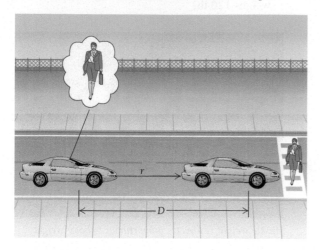

65. *Estimating Heights.* An anthropologist can use certain linear functions to estimate the height of a male or female, given the length of certain bones. The *humerus* is the bone from the elbow to the shoulder. Let x be the length of the humerus, in centimeters. Then the height, in centimeters, of a male with a humerus of length x is given by

$$M(x) = 2.89x + 70.64.$$

The height, in centimeters, of a female with a humerus of length x is given by

$$F(x) = 2.75x + 71.48.$$

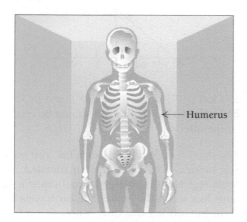

Humerus

A 26-cm humerus was uncovered in a ruins.

a) If we assume it was from a male, how tall was he?
b) If we assume it was from a female, how tall was she?

66. *Urban Population.* The population of a town is P. After a growth of 2%, its new population is N.

a) Assuming that N is directly proportional to P, find an equation of variation.
b) Find N when $P = 200,000$.
c) Find P when $N = 367,200$.

67. *Median Age of Women at First Marriage.* In general, our society is marrying at a later age. The median age of women at first marriage can be approximated by the linear function

$$A(t) = 0.08t + 19.7,$$

where $A(t)$ is the median age of women at first marriage t years after 1950. Thus, $A(0)$ is the median age of women at first marriage in the year 1950, $A(50)$ is the median age in 2000, and so on.

a) Find $A(0)$, $A(1)$, $A(10)$, $A(30)$, and $A(50)$.
b) What will be the median age of women at first marriage in 2010?
c) Graph $A(t)$.

SYNTHESIS

tw **68.** Explain and compare the situations in which you would use the slope–intercept equation rather than the point–slope equation.

 Technology Connection

69. Let $f(x) = x^2 + 3x + 1$.

a) Use the graphing window $[-0.01, 0.01, 0.99, 1.01]$ to graph the function $f(x)$ and the line $y = x + 1$ on the same screen.
b) Using the same graphing window, graph $f(x)$ and $y = mx + 1$ for different values of m. Try to make the line approximate the parabola as closely as you can in the window.
c) What line approximates the graph of $f(x)$ best in the graphing window?

1.2 Polynomial Functions

OBJECTIVES

■ Graph quadratic functions.
■ Solve polynomial equations.
■ Solve applied problems.

Linear functions are polynomial function of degree one. Here we study polynomials of higher degree, with particular emphasis on polynomials of degree two.

Quadratic Functions

DEFINITION

A **quadratic function** f is given by

$$f(x) = ax^2 + bx + c, \quad \text{where } a \neq 0.$$

Technology Connection

Graphs and Function Values

Consider the function $f(x) = x^2 - 2x - 3$ given in Example 1. To graph this function using a grapher, we just change the "$f(x) =$" notation to "$y =$" notation. We enter it into the grapher as $y_1 = x^2 - 2x - 3$ and graph it with the standard window.

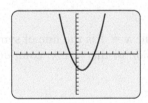

There are at least three ways in which to find function values on a grapher. The first is to use the TABLE feature, which we have already discussed. The second is to use the VALUE feature. We access the CALC menu and choose VALUE. To evaluate the function $f(x)$ at $x = -2$, that is, to find $f(-2)$, we enter -2 for x. The function value, or y-value, $y = 5$, appears together with a trace indicator showing the point $(-2, 5)$.

Function values can also be computed using the Y-VARS feature. Consult the manual for your particular grapher for details.

EXERCISES

1. Graph $f(x) = x^2 + 3x - 4$. Then find $f(-5)$, $f(-4.7)$, $f(11)$, and $f(2/3)$. (*Hint:* To find $f(11)$, be sure that the window dimensions for the x-values include $x = 11$.)

2. Graph $f(x) = 3.7 - x^2$. Then find $f(-5)$, $f(-4.7)$, $f(11)$, and $f(2/3)$.

3. Graph $f(x) = 4 - 1.2x - 3.4x^2$. Then find $f(-5)$, $f(-4.7)$, $f(11)$, and $f(2/3)$.

We can create hand-drawn graphs of quadratic functions using the following information.

The graph of a quadratic function $f(x) = ax^2 + bx + c$ is called a **parabola**.

a) It is always a cup-shaped curve, like those in Examples 1 and 2 that follow.

b) It has a turning point, a **vertex,** whose first coordinate is given by

$$x = -\frac{b}{2a}.$$

c) It has the vertical line $x = -b/2a$ as a line of symmetry (not part of the graph).

d) It opens up if $a > 0$ and opens down if $a < 0$.

EXAMPLE 1   Graph: $f(x) = x^2 - 2x - 3$.

Solution   Let's first find the vertex, or turning point. The x-coordinate of the vertex is

$$x = -\frac{b}{2a}$$

$$= -\frac{-2}{2(1)} = 1.$$

Substituting 1 for x in the equation, we find the second coordinate of the vertex:

$$y = f(1) = 1^2 - 2(1) - 3$$
$$= 1 - 2 - 3$$
$$= -4.$$

The vertex is $(1, -4)$. The vertical line $x = 1$ is the line of symmetry of the graph. We choose some x-values on each side of the vertex, compute y-values, plot the points, and graph the parabola:

$$f(x) = x^2 - 2x - 3.$$

x	$f(x)$	
1	-4	$\leftarrow$ Vertex
0	-3	
2	-3	
3	0	
4	5	
-1	0	
-2	5	

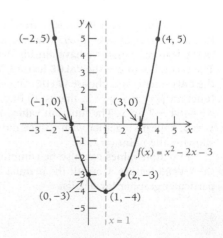

Technology Connection

EXERCISES

Using the procedure of Examples 1 and 2, graph each of the following functions. Use the **TABLE** feature to create an input–output table for each function. Then check the graph with a grapher.

1. $f(x) = x^2 - 6x + 4$

2. $f(x) = -2x^2 + 4x + 1$

EXAMPLE 2   Graph: $f(x) = -2x^2 + 10x - 7$.

Solution Let's first find the vertex, or turning point. The x-coordinate of the vertex is

$$x = -\frac{b}{2a}$$

$$= -\frac{10}{2(-2)} = \frac{5}{2}.$$

Substituting $\frac{5}{2}$ for x in the equation, we find the second coordinate of the vertex:

$$y = f\left(\tfrac{5}{2}\right) = -2\left(\tfrac{5}{2}\right)^2 + 10\left(\tfrac{5}{2}\right) - 7$$

$$= -2\left(\tfrac{25}{4}\right) + 25 - 7$$

$$= \tfrac{11}{2}.$$

The vertex is $\left(\tfrac{5}{2}, \tfrac{11}{2}\right)$, and the line of symmetry is $x = \tfrac{5}{2}$. We choose some x-values on each side of the vertex, compute y-values, plot the points, and graph the parabola:

$$f(x) = -2x^2 + 10x - 7.$$

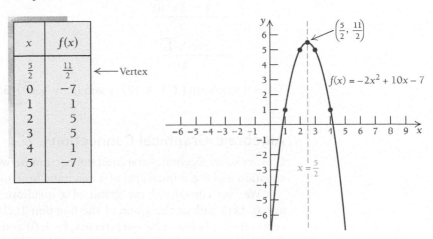

x	$f(x)$
$\frac{5}{2}$	$\frac{11}{2}$
0	-7
1	1
2	5
3	5
4	1
5	-7

First coordinates of points at which a quadratic function intersects the x-axis (x-intercepts), if they exist, can be found by solving the quadratic equation $ax^2 + bx + c = 0$. If real-number solutions exist, they can be found using the *quadratic formula*.

THEOREM 4

The Quadratic Formula

The solutions of any quadratic equation $ax^2 + bx + c = 0$, $a \neq 0$, are given by

$$x = \frac{-b \pm \sqrt{b^2 - 4ac}}{2a}.$$

When solving a quadratic equation, $ax^2 + bx + c = 0$, $a \neq 0$, first try to factor and then use the Principle of Zero Products (see Appendix A). When factoring is not possible or seems difficult, try the quadratic formula. It will always give the solutions.

EXAMPLE 3   Solve: $3x^2 - 4x = 2$.

Solution   We first rewrite this equation in the standard form $ax^2 + bx + c = 0$, and then determine a, b, and c:

$$3x^2 - 4x - 2 = 0,$$
$$a = 3, \quad b = -4, \quad c = -2.$$

We then use the quadratic formula:

$$x = \frac{-b \pm \sqrt{b^2 - 4ac}}{2a}$$
$$= \frac{-(-4) \pm \sqrt{(-4)^2 - 4(3)(-2)}}{2 \cdot 3}$$
$$= \frac{4 \pm \sqrt{40}}{6}$$
$$= \frac{4 \pm 2\sqrt{10}}{6}$$
$$= \frac{2 \pm \sqrt{10}}{3}.$$

The solutions are $\left(2 + \sqrt{10}\right)/3$ and $\left(2 - \sqrt{10}\right)/3$.   ■

Algebraic–Graphical Connection

Let's make an algebraic–graphical connection between the solutions of a quadratic equation and the x-intercepts of a quadratic function.

We just considered the graph of a quadratic function $f(x) = ax^2 + bx + c$, $a \neq 0$. Let's look at the graph of the function $f(x) = x^2 + 6x + 8$ and its x-intercepts, shown below. The **x-intercepts,** $(-4, 0)$ and $(-2, 0)$, are the points at which the graph crosses the x-axis. These pairs are also the points of intersection of the graphs of $f(x) = x^2 + 6x + 8$ and $g(x) = 0$ (the x-axis).

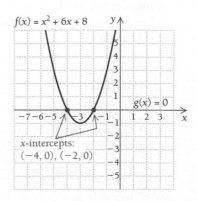

Now let's consider solving the quadratic equation $x^2 + 6x + 8 = 0$. We use factoring although we could use the quadratic formula instead.

$$x^2 + 6x + 8 = 0$$
$$(x + 4)(x + 2) = 0 \qquad \text{Factoring}$$
$$x + 4 = 0 \quad or \quad x + 2 = 0 \qquad \text{Principle of Zero Products}$$
$$x = -4 \quad or \qquad x = -2.$$

We see that the solutions of $0 = x^2 + 6x + 8$, -4 and -2, are the first coordinates of the x-intercepts, $(-4, 0)$ and $(-2, 0)$, of the graph of $f(x) = x^2 + 6x + 8$.

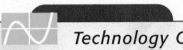

Technology Connection

The TRACE Feature

The trace feature can be used to determine coordinates of points on a graph. When this feature is activated, a cursor appears on the graph, and the coordinates of that point are displayed. The left and right cursor keys are used to move the cursor to other points on the graph.

Let's use this feature to find the solution to $x^2 - 6x + 7 = 0$. We first graph the function $f(x) = x^2 - 6x + 7$ using the window $[0, 5, -4, 10]$. We select **TRACE**.

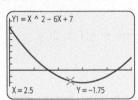

The coordinates at the bottom indicate that the cursor is at the point with coordinates $(2.5, -1.75)$. By using the left and right cursor keys, we can obtain the coordinates of other points. Watch the value of y as you move the cursor to the left. The cursor passes through the points $(1.5957447, -0.028067)$ and then $(1.5425532, 0.1241512)$. Somewhere between these two points the graph crosses the line $y = 0$. The x-coordinate of this point is a solution to $x^2 - 6x + 7 = 0$. To approximate this solution more accurately, we graph using the window $[1.5425532, 1.5957447, -0.028067, 0.1241512]$.

Again, we select **TRACE** and look for the point where $y = 0$.

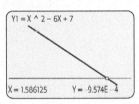

We see that one solution is approximately 1.586.

EXERCISES

1. Use the **TRACE** feature to approximate the other solution to $x^2 - 6x + 7 = 0$.

2. Use the **TRACE** feature to approximate both solutions to $x^2 - x - 1 = 0$.

Complex Numbers

If $b^2 - 4ac < 0$, then there are no real-number solutions of the quadratic equation $ax^2 + bx + c = 0$. However, there are solutions in an expanded number system called the **complex numbers.** To use complex numbers, we formally define

$$i = \sqrt{-1}.$$

Squaring both sides, we obtain $i^2 = -1$. If $b^2 - 4ac < 0$, then the solutions of the quadratic equation will involve i.

EXAMPLE 4   Solve $x^2 + 25 = 0$.

Solution   This quadratic equation may be solved directly:

$$x^2 + 25 = 0$$
$$x^2 = -25$$
$$x = \pm\sqrt{-25}$$
$$x = \pm\sqrt{25}\sqrt{-1} = \pm 5i.$$

The solutions are $5i$ and $-5i$. This may be verified by substitution:

$$(5i)^2 + 25 = 5^2 i^2 + 25 = (25)(-1) + 25 = 0,$$
$$(-5i)^2 + 25 = (-5)^2 i^2 + 25 = (25)(-1) + 25 = 0.$$

EXAMPLE 5   Solve $x^2 + 2x + 5 = 0$.

Solution   We see that $a = 1, b = 2$, and $c = 5$. Using the quadratic formula, we find

$$x = \frac{-b \pm \sqrt{b^2 - 4ac}}{2a}$$
$$= \frac{-(2) \pm \sqrt{(2)^2 - 4(1)(5)}}{2(1)}$$
$$= \frac{-2 \pm \sqrt{-16}}{2}$$
$$= \frac{-2 \pm \sqrt{16}\sqrt{-1}}{2} = \frac{-2 \pm 4i}{2} = -1 \pm 2i.$$

The solutions are $-1 + 2i$ and $-1 - 2i$.

Complex numbers will be used when we study differential equations in Chapter 9.

Polynomial Functions

Linear and quadratic functions are part of a general class of *polynomial functions*.

DEFINITION

A **polynomial function** f is given by

$$f(x) = a_n x^n + a_{n-1}x^{n-1} + \cdots + a_2 x^2 + a_1 x + a_0,$$

where n is a nonnegative integer and $a_n, a_{n-1}, \ldots, a_1, a_0$ are real numbers, called the **coefficients** of the polynomial.

The **degree** of a polynomial is the highest power of x. For example, a polynomial of degree one has the form $f(x) = mx + b$ with $m \neq 0$. The following are examples of polynomial functions:

$$f(x) = -5, \qquad \text{(A constant function, degree zero)}$$

$$f(x) = 4x + 3, \qquad \text{(A linear function, degree one)}$$

$$f(x) = -x^2 + 2x + 3, \qquad \text{(A quadratic function, degree two)}$$

$$f(x) = 2x^3 - 4x^2 + x + 1. \qquad \text{(A cubic function, degree three)}$$

In general, creating graphs of polynomial functions other than linear and quadratic functions is difficult unless we use a grapher. We use calculus to sketch such graphs in Chapter 3. Some **power functions,** such as

$$f(x) = ax^n,$$

are relatively easy to graph.

EXAMPLE 6   Using the same set of axes, graph $f(x) = x^2$ and $g(x) = x^3$.

Solution   We set up a table of values, plot the points, and then draw the graphs.

x	x^2	x^3
-2	4	-8
-1	1	-1
$-\frac{1}{2}$	$\frac{1}{4}$	$-\frac{1}{8}$
0	0	0
$\frac{1}{2}$	$\frac{1}{4}$	$\frac{1}{8}$
1	1	1
2	4	8

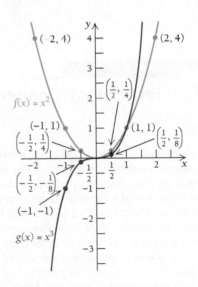

We saw that the graph of a polynomial of degree one is a straight line and the graph of a polynomial of degree two is a parabola. As the degree of a polynomial increases, the complexity of its graph also tends to increase. For example, a line has a constant slope, while a parabola has a positive slope on one side of its vertex and a negative slope on the other side. As we will see in Chapter 3, a polynomial of degree n has at most $n - 1$ turning points where the sign of the slope changes. Some typical polynomials of degree three (cubic) and degree four (quartic) are shown in the figures on the following page.

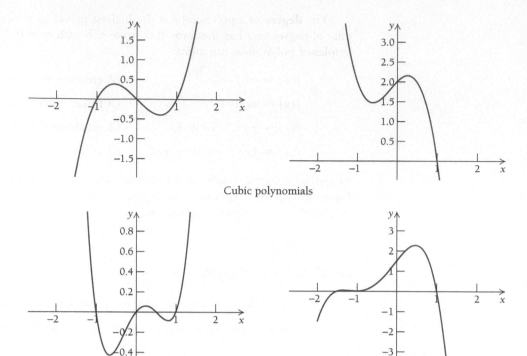

Cubic polynomials

Quartic polynomials

Technology Connection

Regression

In this Technology Connection, we investigate fitting data to a polynomial function. For real data, typically the data does not perfectly follow a polynomial function of a specified degree. In order to approximate the data with a polynomial function, we use a technique called *regression*. In Chapter 7 we will learn the formula and investigate the derivation for **linear regression.** Here we will investigate how to use a grapher to perform regression.

Let's use regression to fit the data:

x	y
0	2.7
1	4.4
2	5.4
3	6.7
4	8
5	9.3
6	10.6

First, we enter the x data into list L1 and the y data into list L2. Press the STAT key and select EDIT. Enter the data as shown in the screen below.

L1	L2	L3 2
0	2.7	———
1	4.4	
2	5.4	
3	6.7	
4	8	
5	9.3	
6	10.6	

L2(7) = 10.6

After entering the data, we again press the STAT key. Under the CALC menu we select LINREG(ax+b). The screen below shows the result:

LinReg
y = ax+b
a = 1.289285714
b = 2.860714286

To graph the data and the regression line together, we press the | STAT PLOT | key and turn PLOT 1 on. We then plot the graph of $y = 1.2893x + 2.8607$ with a window of $[0, 6, 0, 11]$.

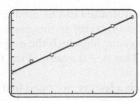

The TI-83 is able to fit data to other functions. Use **QuadReg, CubicReg,** or **QuartReg** to fit data to a polynomial of degree two, three, or four, respectively.

EXERCISES

Fit the data to a polynomial of degree n.

1. $n = 1$,

x	y
0	1.06
1	0.63
2	0.24
3	−0.09
4	−0.48
5	−0.81
6	−1.23

2. $n = 2$,

x	y
0	5.3
1	2.3
2	0.4
3	−0.6
4	−0.7
5	0.1
6	1.9

3. $n = 3$,

x	y
0	−3.1
1	0.6
2	−2.1
3	−5.2
4	−2.8
5	11.4
6	42.1

Solving Polynomial Equations

The quadratic formula provides a method for solving a quadratic equation. Although there are similar formulas for third- and fourth-degree polynomials, they are so complicated that they are generally of little use. The situation is even worse for polynomials of degree higher than four. There is not even a complicated algebraic formula to solve these equations. To algebraically solve higher-degree polynomials, we rely on factoring.

EXAMPLE 7   Solve the polynomial equation $x^3 + x^2 - 4x = 4$.

Solution   We first move all the terms to the left side and then we factor.

$$x^3 + x^2 - 4x = 4$$
$$x^3 + x^2 - 4x - 4 = 0 \qquad \text{Subtract 4 from both sides.}$$
$$x^2(x + 1) - 4(x + 1) = 0 \qquad \text{Factor out } x^2 \text{ and } -4.$$
$$(x^2 - 4)(x + 1) = 0 \qquad \text{Factor out } (x + 1).$$
$$(x + 2)(x - 2)(x + 1) = 0. \qquad \text{Factor difference of squares.}$$

Since the product of the three terms $(x + 2)$, $(x - 2)$, and $(x + 1)$ is 0, one of the terms must be zero. We have three solutions:

$$x + 2 = 0 \qquad x - 2 = 0 \qquad x + 1 = 0$$
$$x = -2 \qquad x = 2 \qquad x = -1.$$

There are many polynomial equations where factoring is not feasible. In these cases, we can use a grapher to approximate the solutions.

Technology Connection

Solving Polynomial Equations

The INTERSECT Feature

Consider solving the equation

$$x^3 = 3x + 1.$$

Doing so amounts to finding the x-coordinates of the point(s) of intersection of the graphs of the two functions

$$f(x) = x^3 \quad \text{and} \quad g(x) = 3x + 1.$$

We enter the functions as

$$y_1 = x^3 \quad \text{and} \quad y_2 = 3x + 1$$

and then graph. We use a $[-3, 3, -5, 8]$ window to see the curvature and possible points of intersection.

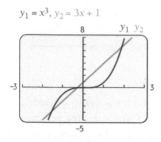

There appear to be at least three points of intersection. Using the INTERSECT feature in the CALC menu, we see that the point of intersection on the left is about $(-1.53, -3.60)$.

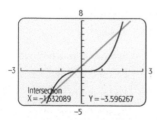

In a similar manner, we find the other points of intersection to be about $(-0.35, -0.04)$ and $(1.88, 6.64)$. The solutions of $x^3 = 3x + 1$ are the x-coordinates of these points:

$$-1.53, -0.35, \quad \text{and} \quad 1.88.$$

The ZERO Feature

A **ZERO**, or **ROOT**, feature can be used to solve an equation. The word "zero" in this context refers to an input, or x-value, for which the output of a function is 0. That is, c is a **zero** of the function f if $f(c) = 0$.

To use such a feature, we must first get a 0 on one side of the equation. Thus, to solve $x^3 = 3x + 1$, we consider $x^3 - 3x - 1 = 0$ by subtracting $3x$ and then 1. Graphing $y = x^3 - 3x - 1$ and using the ZERO feature to find the zero on the left, we obtain a screen like the following.

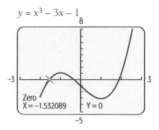

We see that $x^3 - 3x - 1 = 0$ when $x \approx -1.53$, so -1.53 is an approximate solution of the equation $x^3 = 3x + 1$. Proceeding in a similar manner, we can approximate the other solutions as -0.35 and 1.88.

EXERCISES

Using the **INTERSECT** feature, solve the equation.

1. $x^2 = 10 - 3x$ **2.** $2x + 24 = x^2$

3. $x^3 = 3x - 2$ **4.** $x^4 - 2x^2 = 0$

Using the **ZERO** feature, solve the equation.

5. $0.4x^2 = 280x$
(*Hint:* Use $[-200, 800, -100{,}000, 200{,}000]$.)

6. $\frac{1}{3}x^3 - \frac{1}{2}x^2 = 2x - 1$

7. $x^2 = 0.1x^4 + 0.4$

8. $0 = 2x^4 - 4x^2 + 2$

9. $x^4 + x^3 = 4x^2 + 2x - 4$

10. $11x^2 - 9x - 18 = x^4 - x^3$

Find the zeros of the function.

11. $f(x) = 3x^2 - 4x - 2$

12. $f(x) = -x^3 + 6x^2 + 5$

Exercise Set 1.2

Using the same set of axes, graph the pair of equations.

1. $y = \frac{1}{2}x^2$ and $y = -\frac{1}{2}x^2$

2. $y = \frac{1}{4}x^2$ and $y = -\frac{1}{4}x^2$

3. $y = x^2$ and $y = (x - 1)^2$

4. $y = x^2$ and $y = (x - 3)^2$

5. $y = x^2$ and $y = (x + 1)^2$

6. $y = x^2$ and $y = (x + 3)^2$

7. $y = x^3$ and $y = x^3 + 1$

8. $y = x^3$ and $y = x^3 - 1$

For each of the following, state whether the graph of the function is a parabola. If the graph is a parabola, then find the parabola's vertex.

9. $y = x^2 + 4x - 7$

10. $y = x^3 - 2x + 3$

11. $y = 2x^4 - 4x^2 - 3$

12. $y = 3x^2 - 6x$

Graph.

13. $y = x^2 - 4x + 3$

14. $y = x^2 - 6x + 5$

15. $y = -x^2 + 2x - 1$

16. $y = -x^2 - x + 6$

17. $y = 2x^2 + 4x - 7$

18. $y = 3x^2 - 9x + 2$

19. $y = \frac{1}{2}x^2 + 3x - 5$

20. $y = \frac{1}{3}x^2 + 4x - 2$

Solve.

21. $x^2 - 2x = 2$

22. $x^2 - 2x + 1 = 5$

23. $3y^2 + 8y + 2 = 0$

24. $2p^2 - 5p = 1$

Solve. Some of your answers may involve i.

25. $x^2 - 2x + 10 = 0$

26. $x^2 + 6x + 10 = 0$

27. $x^2 + 6x = 1$

28. $x^2 + 4x = 3$

29. $x^2 + 4x + 8 = 0$

30. $x^2 + 10x + 27 = 0$

31. $4x^2 = 4x + 1$

32. $-4x^2 = 4x - 1$

APPLICATIONS

33. *Fruit Stacking.* The number of oranges stacked in a pyramid is approximated by the function

$$f(x) = \frac{1}{6}x^3 + \frac{1}{2}x^2 + \frac{1}{3}x,$$

where $f(x)$ is the number of oranges and x is the number of layers. Find the number of oranges when the number of layers is 7, 10, and 12.

34. *NBA Payrolls.* The average payroll P (in millions of dollars) for teams in the National Basketball Association (NBA) can be approximated by

$$P = 4.8565 + 0.2841x + 0.1784x^2,$$

where x is the number of years since the 1985–86 season.

a) Estimate the average payroll for the 2009–10 season.

b) Use the quadratic formula to predict when the average NBA payroll will be $100 million.

35. *Baseball Ticket Prices.* The average ticket price for a major league baseball game can be modeled by the function

$$p(x) = 9.41 - 0.19x + 0.09x^2,$$

where x is the number of years after 1990.[6] Use the quadratic formula to predict when the average price of a ticket will be $50.

36. *Target Weight.* The target weight w for an adult man of medium build is

$$w(h) = 0.0728h^2 - 6.986h + 289,$$

where h is in inches (for $62 \leq h \leq 76$) and w is in pounds.[7]

a) Find the target weight of an adult man of medium build who is 6 ft (72 in.) tall.

b) If a man of medium build has achieved his target weight of 170 lb, how tall is he?

SYNTHESIS

37. Let $f(x) = x^3 - x^2$.

tw a) For very large values of x, is x^3 or x^2 larger? Explain.

tw b) Use (a) to describe what $f(x)$ looks like when x is very large.

c) Use a grapher to plot the graph of $f(x)$ for $100 \leq x \leq 200$. Does the graph confirm (b)?

38. Let $f(x) = x^4 - 10x^3 + 3x^2 - 2x + 7$.

tw a) For very large values of x, is x^4 or $|-10x^3 + 3x^2 - 2x + 7|$ larger? Explain.

tw b) Use (a) to describe what $f(x)$ looks like when x is very large.

---

[6]Major League Baseball.

[7]Metropolitan Life Insurance Company.

c) Use a grapher to plot the graph of $f(x)$ for $100 \le x \le 200$. Does the graph confirm (b)?

39. Let $f(x) = x^2 + x$.

 a) For values of x very close to 0, is x^2 or x larger? Explain.

 b) Use (a) to describe what $f(x)$ looks like when x is very close to 0.

 c) Use a grapher to plot the graph of $f(x)$ for $-0.01 \le x \le 0.01$. Does the graph confirm (b)?

40. Let $f(x) = x^3 + 2x$.

 a) For values of x very close to 0, is x^3 or $2x$ larger? Explain.

 b) Use (a) to describe what $f(x)$ looks like when x is very close to 0.

 c) Use a grapher to plot the graph of $f(x)$ for $-0.01 \le x \le 0.01$. Does the graph confirm (b)?

 Technology Connection

Use the **ZERO** feature or the **INTERSECT** feature to approximate the zeros of the function to three decimal places.

41. $f(x) = x^3 - x$ (Also use algebra to find the zeros for the function.)

42. $f(x) = x^3 - 2x^2 - 2$

43. $f(x) = 2x^3 - x^2 - 14x - 10$

44. $f(x) = x^4 + 4x^3 - 6x^2 - 160x + 300$

45. $f(x) = x^8 + 8x^7 - 28x^6 - 56x^5 + 70x^4 + 56x^3 - 28x^2 - 8x + 1$

Use the **REGRESSION** feature to fit a polynomial of degree n to the data.

46. $n = 1$,

x	y
-3	-30.1
-2	-22.4
-1	-13.8
0	-5.2
1	3.1
2	11.5
3	18.7

47. $n = 1$,

x	y
2	3.5
3	3.2
4	2.8
5	2.7
6	2.4
7	2.2
8	1.7

48. $n = 2$,

x	y
-2	-0.25
-1	-1.58
0	-0.73
1	2.35
2	6.51
3	15.5
4	22.47

49. $n = 2$,

x	y
-2	-19.2
-1	-24.1
0	-27.3
1	-29.9
2	-30.2
3	-27.1
4	-23.4

50. $n = 3$,

x	y
1	27.5
2	1.3
3	-25.1
4	-45.0
5	-58.4
6	-67.3

51. $n = 4$,

x	y
0	-210
1	-75.1
2	-0.23
3	18.4
4	-13.5
5	-77.1
6	-156.2
7	-220.5
8	-242.1
9	-182.3

1.3 Rational and Radical Functions

OBJECTIVES

■ Graph functions and solve applied problems.
■ Manipulate radical expressions and rational exponents.
■ Graph certain rational functions.

Many functions other than polynomials have applications in science. In this section, we investigate some common algebraic functions that will arise in the remainder of this book.

Rational Functions

DEFINITION
Functions given by the quotient, or ratio, of two polynomials are called **rational functions.**

The following are examples of rational functions:

$$f(x) = \frac{x^2 - 9}{x - 3},$$

$$g(x) = \frac{x^2 + 3x + 8}{x + 4},$$

$$h(x) = \frac{x - 3}{x^2 - x - 2}.$$

The domain of a rational function is restricted to those input values that do not result in division by zero. Thus for f above, the domain consists of all real numbers except 3. To determine the domain of h, we set the denominator equal to 0 and solve:

$$x^2 - x - 2 = 0$$

$$(x + 1)(x - 2) = 0$$

$$x = -1 \quad or \quad x = 2.$$

Therefore, -1 and 2 are not in the domain. The domain of h consists of all real numbers except -1 and 2.

One important class of rational functions is given by $f(x) = k/x$, where k is constant.

EXAMPLE 1   Graph: $f(x) = 1/x$.

Solution   We make a table of values, plot the points, and then draw the graph.

x	$f(x)$
-3	$-\frac{1}{3}$
-2	$-\frac{1}{2}$
-1	-1
$-\frac{1}{2}$	-2
$-\frac{1}{4}$	-4
$\frac{1}{4}$	4
$\frac{1}{2}$	2
1	1
2	$\frac{1}{2}$
3	$\frac{1}{3}$

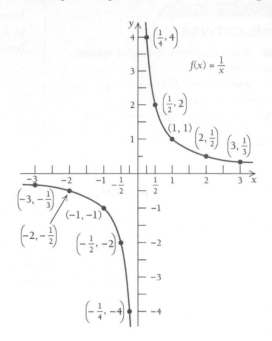

Technology Connection

Graphs of Rational Functions

Consider the rational function given by

$$f(x) = \frac{2x + 1}{x - 3}.$$

Let's consider its graph, two versions of which are shown below.

CONNECTED Mode

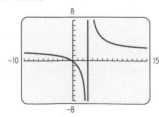

DOT Mode

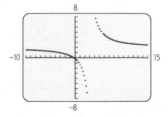

Here we see a disadvantage of the grapher. **CONNECTED** mode can lead to an *incorrect* graph. Because, in **CONNECTED** mode, a grapher connects plotted points with line segments, it connects branches of the graph, making it appear as though the vertical line $x = 3$ is part of the graph.

On the other hand, in **DOT** mode, a grapher simply plots dots representing coordinates of points.

If you have a choice in plotting functions given by rational expressions, use DOT mode.

EXERCISES
Graph each of the following functions using DOT mode.

1. $f(x) = \dfrac{4}{x - 2}$

2. $f(x) = \dfrac{x}{x + 2}$

3. $f(x) = \dfrac{x^2 - 1}{x^2 + x - 6}$

4. $f(x) = \dfrac{x^2 - 4}{x - 1}$

5. $f(x) = \dfrac{10}{x^2 + 4}$

6. $f(x) = \dfrac{8}{x^2 - 4}$

7. $f(x) = \dfrac{2x + 3}{3x^2 + 7x - 6}$

8. $f(x) = \dfrac{2x^3}{x^2 + 1}$

9. $f(x) = \dfrac{x^2 - 9}{x - 3}$.

In Example 1, note that 0 is not in the domain of the function because it would yield a denominator of zero. The function is decreasing over the intervals $(-\infty, 0)$ and $(0, \infty)$. The function $f(x) = 1/x$ is an example of **inverse variation.**

DEFINITION

y **varies inversely** as x if there is some positive number k such that $y = k/x$. We also say that y is **inversely proportional** to x.

EXAMPLE 2 *Pressure and Volume.*   Under the ideal gas law, if the temperature of a gas remains constant, then the pressure P of a gas is inversely proportional to its volume V. The pressure is measured in pounds per square inch (psi), and the volume is measured in cubic inches.

The volume of air in a bicycle pump is 4 cubic inches and has a pressure of 15 psi. The air is then compressed to a volume of 1.5 cubic inches. What is the new pressure of the air after compression?

Solution   We know that $P = k/V$, so $15 = k/4$ and $k = 60$. Thus,

$$P = \frac{60}{V}.$$

We substitute 1.5 for V and compute P:

$$P = \frac{60}{1.5} = 40 \text{ psi.}$$

The graphs of most rational functions are rather complicated and are best dealt with using the tools of calculus that we will develop in Chapters 2 and 3. However, some rational functions simplify, allowing us to more easily determine their graphs. We illustrate with one example.

EXAMPLE 3   Graph: $f(x) = \dfrac{x^2 - 9}{x - 3}$.

Solution   The domain of $f(x)$ consists of all real numbers except 3. To determine the graph, we factor the numerator and simplify.

$$f(x) = \frac{x^2 - 9}{x - 3}$$

$$= \frac{(x + 3)(x - 3)}{x - 3}$$

$$= \frac{x + 3}{1} \cdot \frac{x - 3}{x - 3}$$

$$= x + 3$$

The function $x + 3$ has domain of all x including 3. Otherwise the two functions $f(x)$ and $x + 3$ are identical. The graph of $y = x + 3$ is a line with slope 1 and y-intercept 3. The graph of $f(x)$ is the same line, except the point $(3, 6)$ is deleted as shown in the graph below by placing a hollow dot at $(3, 6)$.

x	1	0	2	-3	4	-1	-2
y	4	3	5	0	7	2	1

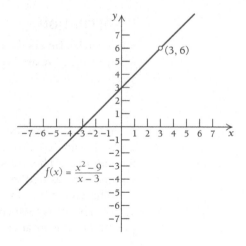

$f(x) = \dfrac{x^2 - 9}{x - 3}$

Technology Connection

EXERCISES

Graph.

1. $f(x) = |x|$
2. $f(x) = |x^2 - 4|$

Absolute-Value Functions

The following is an example of an **absolute-value function** and its graph. The absolute value of a number is its distance from 0 on the number line. We denote the absolute value of a number x as $|x|$.

EXAMPLE 4   Graph: $f(x) = |x|$.

Solution   We make a table of values, plot the points, and then draw the graph.

x	-3	-2	-1	0	1	2	3
$f(x)$	3	2	1	0	1	2	3

We can think of this *function as being defined piecewise* by considering the definition of absolute value:

$$f(x) = |x| = \begin{cases} x, & \text{if } x \geq 0, \\ -x, & \text{if } x < 0. \end{cases}$$

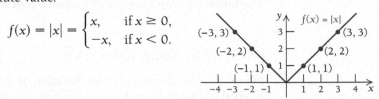

Technology Connection

Graphing Functions Defined Piecewise

Graphing functions defined piecewise on a grapher generally involves the use of inequality symbols. On the TI-83, the **TEST** menu is used. The absolute-value function defined piecewise in Example 4 is entered as follows.

```
Plot1 Plot2 Plot3
\Y1 = X(X ≥ 0) + -X(X < 0)
\Y2 =
\Y3 =
\Y4 =
\Y5 =
\Y6 =
```

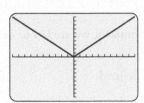

The function

$$f(x) = \begin{cases} 4, & \text{for } x \leq 0 \\ 4 - x^2, & \text{for } 0 < x \leq 2 \\ 2x - 6, & \text{for } x > 2 \end{cases}$$

may be graphed by entering the following.

```
Plot1 Plot2 Plot3
\Y1 = 4(X ≤ 0) + (4 – X ^ 2)(X > 0)
(X ≤ 2) + (2X – 6)(X > 2)
\Y2 =
\Y3 =
\Y4 =
\Y5 =
```

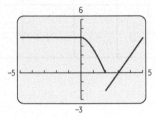

EXERCISES
Graph.

1. $f(x) = \begin{cases} -x - 2, & \text{for } x < -2 \\ x^2, & \text{for } -2 \leq x < 2 \\ x + 6, & \text{for } x \geq 2 \end{cases}$

2. $f(x) = \begin{cases} x - 1, & \text{for } x < -1 \\ x + 1, & \text{for } -1 \leq x \leq 1 \\ 0, & \text{for } x > 1 \end{cases}$

3. $f(x) = \begin{cases} 1/x, & \text{for } x \leq -1 \\ 0, & \text{for } -1 < x < 3 \\ -1/x, & \text{for } x \geq 3 \end{cases}$

Square-Root Functions

The following is an example of a **square-root function** and its graph.

EXAMPLE 5   Graph: $f(x) = -\sqrt{x}$.

Solution   The domain of this function is just the nonnegative numbers—the interval $[0, \infty)$. You can find approximate values of square roots on your calculator. We set up a table of values, plot the points, and then draw the graph.

x	0	1	2	3	4	5
$f(x)$, or $-\sqrt{x}$	0	-1	-1.4	-1.7	-2	-2.2

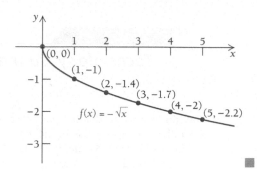

Power Functions with Rational Exponents

We are motivated to define rational exponents so that the laws of exponents still hold (Appendix A). For example, if the laws of exponents are to hold, we would have

$$a^{1/2} \cdot a^{1/2} = a^{1/2+1/2} = a^1 = a.$$

Thus we are led to define $a^{1/2}$ as $\sqrt{a}$. Similarly, we are led to define $a^{1/3}$ as the cube root of a, $\sqrt[3]{a}$. In general,

$$a^{1/n} = \sqrt[n]{a}, \quad \text{provided } \sqrt[n]{a} \text{ is defined.}$$

Again, if the laws of exponents are to hold, we would have

$$\sqrt[n]{a^m} = (a^m)^{1/n} = (a^{1/n})^m = a^{m/n}.$$

An expression $a^{-m/n}$ is defined by

$$a^{-m/n} = \frac{1}{a^{m/n}} = \frac{1}{\sqrt[n]{a^m}}.$$

EXAMPLE 6   Convert to rational exponents.

a) $\sqrt[3]{x^2} = x^{2/3}$

b) $\sqrt[4]{y} = y^{1/4}$

c) $\dfrac{1}{\sqrt[3]{b^5}} = \dfrac{1}{b^{5/3}} = b^{-5/3}$

d) $\dfrac{1}{\sqrt{x}} = \dfrac{1}{x^{1/2}} = x^{-1/2}$

e) $\sqrt{x^8} = x^{8/2}$, or x^4

EXAMPLE 7   Convert to radical notation.

a) $x^{1/3} = \sqrt[3]{x}$

b) $t^{6/7} = \sqrt[7]{t^6}$

c) $x^{-2/3} = \dfrac{1}{x^{2/3}} = \dfrac{1}{\sqrt[3]{x^2}}$

d) $e^{-1/4} = \dfrac{1}{e^{1/4}} = \dfrac{1}{\sqrt[4]{e}}$ ■

EXAMPLE 8   Simplify.

a) $8^{5/3} = (8^{1/3})^5 = \left(\sqrt[3]{8}\right)^5 = 2^5 = 32$

b) $81^{3/4} = (81^{1/4})^3 = \left(\sqrt[4]{81}\right)^3 = 3^3 = 27$ ■

Technology Connection

Graphing Radical Functions

Graphing functions defined by radical expressions involves approximating roots. Since the square root of a negative number is not a real number, y-values may not exist for some x-values. For example, y-values for the graph of $f(x) = \sqrt{x} - 1$ do not exist for x-values that are less than 1 because square roots of negative numbers would result.

We must enter $y = \sqrt{x} - 1$ using parentheses around the radicand as $y_1 = \sqrt{(x - 1)}$. Some graphers supply the left parenthesis automatically.

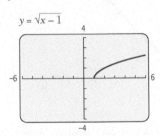

$y = \sqrt{x - 1}$

Similarly, y-values for the graph of $f(x) = \sqrt{2} - x$ do not exist for x-values that are greater than 2.

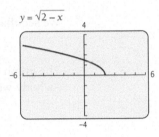

$y = \sqrt{2} - x$

EXERCISES

Graph each of the following functions. Then use the **TABLE** and **TRACE** features to determine the domain and the range of each function. The **MATH** menu contains $\boxed{\sqrt[3]{}}$ and $\boxed{\sqrt[x]{}}$ keys to enter kth roots.

1. $f(x) = \sqrt{x}$ **2.** $g(x) = \sqrt{x} + 2$
3. $f(x) = \sqrt[3]{x}$ **4.** $f(x) = \sqrt[3]{x} - 2$
5. $f(x) = \sqrt[4]{x} - 1$ **6.** $F(x) = \sqrt[5]{6} - x$
7. $g(x) = 5 - \sqrt{x} + 3$ **8.** $f(x) = 4 - \sqrt[3]{x}$
9. $f(x) = x^{2/3}$ **10.** $g(x) = x^{1.41}$

Use the **GRAPH** and **TABLE** features to determine whether each of the following is correct.

11. $\sqrt{x + 4} = \sqrt{x} + 2$
12. $\sqrt{25x} = 5\sqrt{x}$

The function $f(x) = \sqrt{x}$, may be rewritten as $f(x) = x^{1/2}$, or $f(x) = x^{0.5}$. The power functions

$$f(x) = ax^k, \quad k \text{ rational},$$

do occur in applications.

EXAMPLE 9 *Home Range.*   The *home range* of an animal is defined as the region to which the animal confines its movements. It has been hypothesized in statistical studies* that the area H of that region can be approximated by the function

$$H = W^{1.41},$$

where W is the weight of the animal. Graph the function.

Solution   We can approximate function values using a power key $\boxed{y^x}$ on a calculator.

W	0	10	20	30	40	50
H	0	26	68	121	182	249

We see that

$$H = W^{1.41} = W^{141/100} = \sqrt[100]{W^{141}}.$$

The graph is shown below. Note that the function values increase from left to right. As body weight increases, the area over which the animal moves increases.

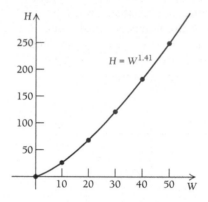

$$H = W^{1.41}$$

*J. M. Emlen, *Ecology: An Evolutionary Approach*, p. 200 (Addison-Wesley, Reading, MA, 1973).

Exercise Set 1.3

Using the same set of axes, graph the pair of equations.

1. $y = |x|$ and $y = |x + 3|$

2. $y = |x|$ and $y = |x + 1|$

3. $y = \sqrt{x}$ and $y = \sqrt{x + 1}$

4. $y = \sqrt{x}$ and $y = \sqrt{x - 2}$

Graph.

5. $y = \dfrac{2}{x}$

6. $y = \dfrac{3}{x}$

7. $y = \dfrac{-2}{x}$

8. $y = \dfrac{-3}{x}$

9. $y = \dfrac{1}{x^2}$ 10. $y = \dfrac{1}{x-1}$

11. $y = \sqrt[3]{x}$ 12. $y = \dfrac{1}{|x|}$

13. $f(x) = \dfrac{x^2-9}{x+3}$ 14. $g(x) = \dfrac{x^2-4}{x-2}$

15. $f(x) = \dfrac{x^2-1}{x-1}$ 16. $g(x) = \dfrac{x^2-25}{x+5}$

Convert to expressions with rational exponents.

17. $\sqrt{x^3}$ 18. $\sqrt{x^5}$

19. $\sqrt[5]{a^3}$ 20. $\sqrt[4]{b^2}$

21. $\sqrt[7]{t}$ 22. $\sqrt[8]{c}$

23. $\dfrac{1}{\sqrt[3]{t^4}}$ 24. $\dfrac{1}{\sqrt[5]{b^6}}$

25. $\dfrac{1}{\sqrt{t}}$ 26. $\dfrac{1}{\sqrt{m}}$

27. $\dfrac{1}{\sqrt{x^2+7}}$ 28. $\sqrt{x^3+4}$

Convert to radical notation.

29. $x^{1/5}$ 30. $t^{1/7}$

31. $y^{2/3}$ 32. $t^{2/5}$

33. $t^{-2/5}$ 34. $y^{-2/3}$

35. $b^{-1/3}$ 36. $b^{-1/5}$

37. $e^{-17/6}$ 38. $m^{-19/6}$

39. $(x^2-3)^{-1/2}$ 40. $(y^2+7)^{-1/4}$

41. $\dfrac{1}{t^{2/3}}$ 42. $\dfrac{1}{w^{-4/5}}$

Simplify.

43. $9^{3/2}$ 44. $16^{5/2}$ 45. $64^{2/3}$

46. $8^{2/3}$ 47. $16^{3/4}$ 48. $25^{5/2}$

Determine the domain of the function.

49. $f(x) = \dfrac{x^2-25}{x-5}$ 50. $f(x) = \dfrac{x^2-4}{x+2}$

51. $f(x) = \dfrac{x^3}{x^2-5x+6}$ 52. $f(x) = \dfrac{x^4+7}{x^2+6x+5}$

53. $f(x) = \sqrt{5x+4}$ 54. $f(x) = \sqrt{2x-6}$

APPLICATIONS

55. *Territorial Area.* **Refer to Example 9.** The *territorial area* of an animal is defined as its defended region, or exclusive region. For example, a lion has a cer-tain region over which it is ruler. The area T of that region can be approximated by the power function

$$T = W^{1.31},$$

where W is the weight of the animal. Complete the table of approximate function values and graph the function.

W	0	10	20	30	40	50	100	150
T	0	20						

56. *Zipf's Law.* According to Zipf's Law, the number of cities with a population greater than S is inversely proportional to S. In 2000, there were 48 U.S. cities with a population greater than 350,000. Estimate the number of U.S. cities with a population greater than 200,000.[8]

57. *Body Surface Area.* A person whose mass is 75 kg has surface area approximated by

$$f(h) = 0.144h^{1/2},$$

where $f(h)$ is measured in square meters and h is the person's height in centimeters.[9]

a) Find the approximate surface area of a person whose mass is 75 kg and whose height is 180 cm.
b) Find the approximate surface area of a person whose mass is 75 kg and whose height is 170 cm.
c) Graph the function $f(h)$ for $0 \le h \le 200$.

58. *Dinosaurs.* The body mass y (in kilograms) of a theropod dinosaur may be approximated by the function

$$y = 0.73x^{3.63},$$

where x is the total length of the dinosaur (in meters).[10]

a) Find the body mass of *Coelophysis bauri*, which has a total length of 2.7 m.
b) Find the body mass of *Sinraptor dongi*, which has a total length of 7 m.

---

[8]U.S. Bureau of the Census.
[9]U.S. Oncology.
[10]F. Seebacher, "A New Method to Calculate Allometric Length-Mass Relationships of Dinosaurs," *Journal of Vertebrate Pale-ontology,* Vol. 21, pp. 51–60 (2001).

c) Suppose a therapod has a body mass of 5000 kg. Find its total length.

Capillaries. The velocity of blood in a blood vessel is inversely proportional to the cross-sectional area of the blood vessel. This relationship is called the *law of continuity.*[11]

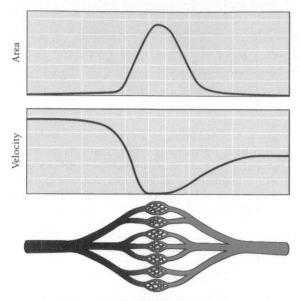

59. Suppose, in an adult male, blood leaves the aorta at 30 cm/sec, and the cross-sectional area of the aorta is 3 cm². Given that blood travels in the capillaries at 0.026 cm/sec, find the total cross-sectional area of his capillaries.[12]

60. Suppose, in an adult female, blood leaves the aorta at 28 cm/sec, and the cross-sectional area of the aorta is 2.8 cm². Given that blood travels in the capillaries at 0.025 cm/sec, find the total cross-sectional area of her capillaries.

SYNTHESIS

Solve.

61. $x + 7 + \dfrac{9}{x} = 0$ (*Hint:* Multiply both sides by x.)

62. $1 - \dfrac{1}{w} = \dfrac{1}{w^2}$

---

[11] N. A. Campbell and J. B. Reece, *Biology* (Benjamin Cummings, New York, 2002).

[12] Notice this is the total cross-sectional area of all capillaries, not the cross-sectional area of a single capillary.

63. *Pollution Control.* Pollution control has become a very important concern in all countries. If controls are not put in place, it has been predicted that the function

$$P = 1000t^{5/4} + 14{,}000$$

will describe the average pollution, in particles of pollution per cubic centimeter, in most cities at time t, in years, where $t = 0$ corresponds to 1970 and $t = 37$ corresponds to 2007.

a) Predict the pollution in 2007, 2010, and 2020.
b) Graph the function over the interval $[0, 50]$.

tw 64. At most, how many y-intercepts can a function have? Explain.

tw 65. Explain the difference between a rational function and a polynomial function. Is every polynomial function a rational function?

 Technology Connection

Use the **ZERO** feature or the **INTERSECT** feature to approximate the zeros of the function to three decimal places.

66. $f(x) = \dfrac{1}{2}(|x - 4| + |x - 7|) - 4$

67. $f(x) = \sqrt{7 - x^2}$

68. $f(x) = |x + 1| + |x - 2| - 5$

69. $f(x) = |x + 1| + |x - 2|$

70. $f(x) = |x + 1| + |x - 2| - 3$

1.4 Trigonometric Functions

OBJECTIVES

■ Find angle measures in degrees and radians.

■ Compute trigonometric functions from right triangles.

■ Solve right triangles.

■ Apply trigonometric identities.

Angles and Rotations

We now introduce the trigonometric functions. As we will see, these functions will be useful for modeling phenomena that repeat with a predictable cycle.

We first consider a rotating ray, with its endpoint at the origin of an *xy*-plane. The ray starts in position along the positive half of the *x*-axis. A counterclockwise rotation is called *positive,* and a clockwise rotation is called *negative.* Note that the rotating ray and the positive half of the *x*-axis form an **angle.** Thus we often speak of "rotations" and "angles" interchangeably.

The rotating ray is often called the *terminal side* of the angle, and the positive half of the *x*-axis is called the *initial side.* Notice that the terminal side may lie on an axis or in the first, second, third, or fourth quadrant, as illustrated in the figure.

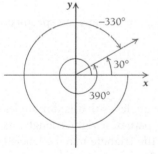

Measures of Rotations or Angles

The size, or *measure,* of an angle, or rotation, may be given in degrees. A complete revolution has a measure of 360°, half a revolution has a measure of 180°, and so on. Notice that an angle does not have to correspond to an angle of a triangle. For example, world-class figure skaters will attempt jumps with three (360° × 3 = 1080°) or even four (360° × 4 = 1440°) complete rotations in competition.

The angles with measures 30°, 390°, and −330° have the same terminal side in the first quadrant. Even though these angles share the same terminal side, they are different angles since the rotation taken by the ray to reach its terminal side is different for these three angles. These angles are called **coterminal** since they share the same terminal side. Since these angles differ by some number of full rotations, they differ by a multiple of 360°:

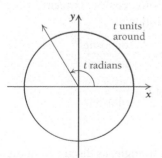

$$390° = 30° + 360°(1) = −330° + 360°(2).$$

> **CAUTION** Angles may share the same terminal side and still have different measures.

Measurement in Radians

The length of an object is usually measured in units like inches, feet, centimeters, meters, and so on. Likewise, angles are usually measured using either **degrees** or **radians.** It will turn out that the radian measure will be very useful in calculus.

Consider a circle centered at the origin with radius 1. We refer to it as a *unit circle.* The measure of an angle in *radians* is the distance around the circumference of this circle traveled by the rotating ray.

Since the circumference of the circle is $2\pi \cdot 1$, or 2π, a complete rotation (360°) has a measure of 2π radians. Half of this (180°) is π radians, and one-fourth (90°) is $\pi/2$ radians.

In general, we can convert from degrees to radians and vice versa using the following formula.

THEOREM 5

$$\frac{\text{Radian measure}}{\pi} = \frac{\text{Degree measure}}{180°}$$

EXAMPLE 1

a) Convert 270° to radians.

b) Convert $-3\pi/4$ radians into degrees.

Solution

a) Let x be the measure of 270° expressed in radians. Then

$$\frac{x}{\pi} = \frac{270°}{180°}, \quad \text{or} \quad x = \frac{3\pi}{2}.$$

b) Let y be the measure of $-3\pi/4$ expressed in degrees. Then

$$\frac{-3\pi/4}{\pi} = \frac{y}{180°}, \quad \text{or} \quad y = -135°.$$

In this book, if an angle is given but neither degrees nor radians are specified, then radian measure is assumed.

Trigonometric Functions of Acute Angles

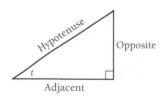

We now define the six trigonometric functions for *acute* angles—that is, for angles between 0 and $\pi/2$ radians. Any such angle t may be part of a right triangle, as illustrated in the margin. Notice that the three sides of the triangle may be labeled as the *hypotenuse,* the leg *opposite* the angle t, and the leg *adjacent* to the angle t. Ratios of two of the three sides are used to define the trigonometric functions.

DEFINITION

Let t be an acute angle of a right triangle. Then the **sine, cosine, tangent, cotangent, secant,** and **cosecant** of t are defined as follows:

$$\sin t = \frac{\text{opposite}}{\text{hypotenuse}}, \quad \cos t = \frac{\text{adjacent}}{\text{hypotenuse}}, \quad \tan t = \frac{\text{opposite}}{\text{adjacent}},$$

$$\csc t = \frac{\text{hypotenuse}}{\text{opposite}}, \quad \sec t = \frac{\text{hypotenuse}}{\text{adjacent}}, \quad \cot t = \frac{\text{adjacent}}{\text{opposite}}.$$

These definitions are independent on the size of the triangle, as discussed in the exercises.

The trigonometric functions of $\pi/6$, $\pi/3$, and $\pi/4$ may be computed by using certain right triangles along with these definitions.

EXAMPLE 2 Compute $\sin(\pi/6)$, $\cos(\pi/6)$, and $\tan(\pi/6)$.

Solution First, we convert $\pi/6$ into degrees:

$$\frac{\pi/6}{\pi} = \frac{\text{Degree measure}}{180°}, \quad \text{or} \quad \text{Degree measure} = 30°.$$

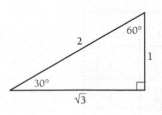

Next, we form a right triangle that contains an angle of 30°. The other acute angle has measure $90° - 30° = 60°$. We recall from geometry that a 30–60–90 right triangle has side lengths proportional to 1, $\sqrt{3}$, and 2, as shown in the figure. Therefore,[13]

$$\sin 30° = \sin \frac{\pi}{6} = \frac{1}{2}, \quad \cos 30° = \cos \frac{\pi}{6} = \frac{\sqrt{3}}{2}, \quad \text{and}$$

$$\tan 30° = \tan \frac{\pi}{6} = \frac{1}{\sqrt{3}}.$$

EXAMPLE 3 Compute $\sin(\pi/3)$, $\cos(\pi/3)$, $\tan(\pi/3)$.

Solution First, we convert $\pi/3$ into degrees:

$$\frac{\pi/3}{\pi} = \frac{\text{Degree measure}}{180°}, \quad \text{or} \quad \text{Degree measure} = 60°.$$

Next, we form a right triangle that contains an angle of 60°. Notice that this is the same triangle used in the previous example, except that the roles of opposite and adjacent sides have been reversed. Therefore,

$$\sin 60° = \sin \frac{\pi}{3} = \frac{\sqrt{3}}{2}, \quad \cos 60° = \cos \frac{\pi}{3} = \frac{1}{2}, \quad \text{and}$$

$$\tan 60° = \tan \frac{\pi}{3} = \frac{\sqrt{3}}{1} = \sqrt{3}.$$

EXAMPLE 4 Compute $\sin(\pi/4)$, $\cos(\pi/4)$, and $\tan(\pi/4)$.

Solution First, we convert $\pi/4$ into degrees:

$$\frac{\pi/4}{\pi} = \frac{\text{Degree measure}}{180°}, \quad \text{or} \quad \text{Degree measure} = 45°.$$

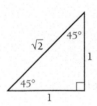

Next, we form a right triangle that contains an angle of 45°. The other acute angle has measure $90° - 45° = 45°$. We recall from geometry that, in a 45–45–90 right triangle, the legs have the same length and the hypotenuse is $\sqrt{2}$ times as long as each leg. Therefore,

$$\sin 45° = \sin \frac{\pi}{4} = \frac{1}{\sqrt{2}}, \quad \cos 45° = \cos \frac{\pi}{4} = \frac{1}{\sqrt{2}}, \quad \text{and}$$

$$\tan 45° = \tan \frac{\pi}{4} = \frac{1}{1} = 1.$$

---

[13]Though we do not do so here, fractions with square roots in the denominator may be *rationalized* as follows:

$$\frac{1}{\sqrt{3}} = \frac{1}{\sqrt{3}} \cdot \frac{\sqrt{3}}{\sqrt{3}} = \frac{\sqrt{3}}{3}.$$

The values found in the previous three examples are shown below and should be committed to memory.

t, radians	t, degrees	$\sin t$	$\cos t$	$\tan t$
$\pi/6$	$30°$	$1/2$	$\sqrt{3}/2$	$1/\sqrt{3}$
$\pi/4$	$45°$	$1/\sqrt{2}$	$1/\sqrt{2}$	1
$\pi/3$	$60°$	$\sqrt{3}/2$	$1/2$	$\sqrt{3}$

For these and other angles, approximations of the trigonometric functions can be found on your calculator.

CAUTION   If the calculator mode is not set appropriately for radians or degrees, then it will give incorrect answers.

Trigonometric Identities

Since ratios involving the three sides of a right triangle are used to define the six trigonometric functions, there are many interrelationships between them. Some of these interrelationships—called *trigonometric identities*—are listed in the following theorem.[14] These identities will be used when we study the derivatives and integrals of trigonometric functions.

CAUTION   We use the shorthand notation $\sin^2 t$ to represent $(\sin t)^2$.

THEOREM 6

The following are trigonometric identities:

Reciprocal: $\dfrac{1}{\sin t} = \csc t$ $\dfrac{1}{\cos t} = \sec t$ $\dfrac{1}{\tan t} = \cot t$

Ratio: $\dfrac{\sin t}{\cos t} = \tan t$ $\dfrac{\cos t}{\sin t} = \cot t$

Pythagorean: $\sin^2 t + \cos^2 t = 1$

Sum:

$$\sin(s + t) = \sin s \cos t + \cos s \sin t$$
$$\cos(s + t) = \cos s \cos t - \sin s \sin t$$

---

[14]Technically, these identities are true as long as both sides of each identity are defined.

Technology Connection

Radians and Degrees

Most calculators allow computation of trigonometric functions in both radians and degrees. This is done by using the **MODE** button and choosing radians or degrees, as appropriate. Notice that very different answers are obtained in radian mode and in degree mode.

$\sin(\pi/6) = 0.5$,
in radian mode, but

$\sin(\pi/6)° \approx 0.009138$.
in degree mode

EXERCISE

Use a calculator to verify the values of the trigonometric functions shown in the table above. For each angle, use both radian mode and degree mode.

Proof   We prove the Pythagorean Identity here; the rest are left as exercises. To begin, notice that the Pythagorean theorem for a right triangle states that

$$(\text{opposite})^2 + (\text{adjacent})^2 = (\text{hypotenuse})^2.$$

Dividing both sides of this equation by $(\text{hypotenuse})^2$, we find

$$\frac{(\text{opposite})^2}{(\text{hypotenuse})^2} + \frac{(\text{adjacent})^2}{(\text{hypotenuse})^2} = 1$$

$$\left(\frac{\text{opposite}}{\text{hypotenuse}}\right)^2 + \left(\frac{\text{adjacent}}{\text{hypotenuse}}\right)^2 = 1$$

$$(\sin t)^2 + (\cos t)^2 = 1$$

$$\sin^2 t + \cos^2 t = 1. \qquad \blacksquare$$

EXAMPLE 5   Use a calculator to find $\cot(\pi/5)$.

Solution   Your calculator may not have a button for cotangents. However, this can be found using a Reciprocal Identity:

$$\cot \frac{\pi}{5} = \frac{1}{\tan(\pi/5)} \approx \frac{1}{0.726542528} \approx 1.37638192. \qquad \blacksquare$$

CAUTION   If your calculator has a button for $\boxed{\text{TAN}^{-1}}$ or $\boxed{\text{INV}}$ $\boxed{\text{TAN}}$, this is not equivalent to cotangent.

EXAMPLE 6   Find $\sin 75°$ without using a calculator.

Solution   Using the Sum Identity, we find that

$$\sin 75° = \sin(30° + 45°) = \sin 30° \cos 45° + \cos 30° \sin 45°$$

$$= \frac{1}{2} \cdot \frac{1}{\sqrt{2}} + \frac{\sqrt{3}}{2} \cdot \frac{1}{\sqrt{2}}$$

$$= \frac{1 + \sqrt{3}}{2\sqrt{2}}. \qquad \blacksquare$$

Solving Right Triangles

The trigonometric functions can be used to find the sides of right triangles.

EXAMPLE 7   A surveyor stands 60 ft away from a tree, holding a device 5 ft above the ground. The angle of elevation, as shown in the figure, from the device to the top of the tree is 33°. Find the height of the tree.

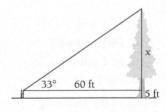

Solution   Using the tangent function, we find that

$$\tan 33° = \frac{x}{60}$$

$$x = 60 \tan 33° \approx 38.96.$$

The height of the tree is $h = x + 5$, or approximately 44 ft. $\qquad \blacksquare$

Technology Connection

Determining Angles

Let's determine the acute angle whose sine is 0.2. On the TI-83 and many other graphers, this is done using the $\boxed{\text{SIN}^{-1}}$ key. If the grapher is in degree mode, then the answer is approximately 11.537°. If the calculator is in radian mode, then the answer is approximately 0.20136 radians. The $\boxed{\text{COS}^{-1}}$ or $\boxed{\text{TAN}^{-1}}$ key can be used to find an angle if we know its cosine or tangent, respectively.

EXERCISES

Use a grapher to approximate the acute angle in degrees.

1. Find t if $\sin t = 0.12$.

2. Find t if $\cos t = 0.73$.

3. Find t if $\tan t = 1.24$.

Use a grapher to approximate the acute angle in radians.

4. Find t if $\sin t = 0.85$.

5. Find t if $\cos t = 0.62$.

6. Find t if $\tan t = 0.45$.

At times we wish to know an angle in a right triangle when we are given the lengths of two sides. We can find the angle by using the definition of a trigonometric function.

EXAMPLE 8 Find the angle of elevation of the sun if a building 25 ft high casts a

a) 25-ft shadow.

b) 10-ft shadow.

Solution

a) We can see from the figure that $\tan t = \dfrac{25}{25} = 1$. The angle must be 45° since $\tan 45° = 1$.

b) In this case, $\tan t = \dfrac{25}{10} = 2.5$. We have not seen any special angle whose tangent is 2.5. To find the angle we use the $\boxed{\text{TAN}^{-1}}$ key on a calculator to find $t \approx 68.20°$.

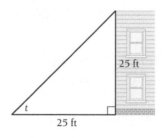
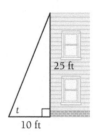

Exercise Set 1.4

In Exercises 1–6, convert from degrees into radians. Draw a picture of each angle on the xy-plane.

1. 120°
2. 150°
3. 240°
4. 300°
5. 540°
6. −450°

In Exercises 7–12, convert from radians into degrees. Draw a picture of each angle on the xy-plane.

7. $3\pi/4$
8. $7\pi/6$
9. $3\pi/2$
10. 3π
11. $-\pi/3$
12. $-11\pi/15$

Determine if the following pairs of angles are coterminal.

13. 15° and 395°
14. 225° and −135°
15. 107° and −107°
16. 140° and 440°
17. $\pi/2$ and $3\pi/2$
18. $\pi/2$ and $-3\pi/2$
19. $7\pi/6$ and $-5\pi/6$
20. $3\pi/4$ and $-\pi/4$

Use a calculator to find the values of the following trigonometric functions.

21. $\sin 34°$
22. $\sin 82°$
23. $\cos 12°$
24. $\cos 41°$

25. tan 5°

26. tan 68°

27. cot 34°

28. cot 56°

29. sec 23°

30. csc 72°

31. $\sin(\pi/5)$

32. $\cos(2\pi/5)$

33. $\tan(\pi/7)$

34. $\cot(3\pi/11)$

35. $\sec(3\pi/8)$

36. $\csc(4\pi/13)$

37. $\sin(2.3)$

38. $\cos(0.81)$

Use a calculator to find the degree measure of an acute angle whose trigonometric function is given.

39. $\sin t = 0.45$

40. $\sin t = 0.87$

41. $\cos t = 0.34$

42. $\cos t = 0.72$

43. $\tan t = 2.34$

44. $\tan t = 0.84$

Use a calculator to find the radian measure of an acute angle whose trigonometric function is given.

45. $\sin t = 0.59$

46. $\sin t = 0.26$

47. $\cos t = 0.60$

48. $\cos t = 0.78$

49. $\tan t = 0.11$

50. $\tan t = 1.26$

Solve for the missing side x.

51.

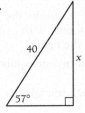

52.

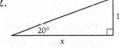

53.

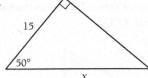

54.

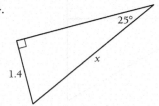

Solve for the missing angle t. Express your answer in degrees.

55.

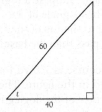

56.

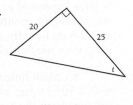

57.

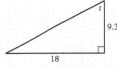

58.

59. Use a Sum Identity to find cos 75°.

APPLICATIONS

60. *Honeybees.* Honeybees communicate the location of food sources to other bees in a hive through an elaborate dance.[15] Through such a dance, the hive learns that a food source is 200 m away at an angle 20° north of the sun, which is rising due east. Find the x- and y-coordinates of the food source. (Think of east as the positive x-direction and north as the positive y-direction.)

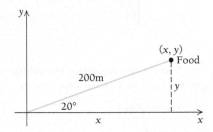

61. *Grade of a Road.* On a 5-mi stretch of highway, the road decreases in elevation at an angle of 4°. How much lower is a car after traveling on this part of the highway? (Remember that there are 5280 feet in a mile.)

[15]Karl von Frisch, *The Dance Language and Orientation of Bees,* Harvard University Press, 1971.

62. *Grade of a Road.* The tangent of a road's angle of elevation *t* is called the *grade* of the road; the grade is often expressed as a percentage. Suppose a highway through a mountain pass has a grade of 5% and is 6 mi long from the base to the top of the pass. How much higher is the pass than the base?

63. *Golf.* A certain hole on a golf course is 330 yd long with a 40° dogleg, as illustrated in the figure. The distance from the tee to the center of the dogleg is 180 yd, while the distance from the center of the dogleg to the green is 150 yd.

a) Find *x*.
b) Find *y*.
c) Find *z*, the straight-line tee-to-green distance.

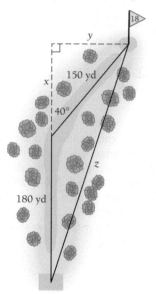

Blood Velocity. Ultrasound measures the velocity of blood (in cm/s) through a blood vessel using

$$v = \frac{77,000d \sec t}{f}.$$

In this formula, f is the emitted ultrasound beam frequency, d is the Doppler shift (or the difference between the emitted and received beam frequencies), and t is the angle between the ultrasound beam and the blood vessel.[16]

64. Suppose $f = 5,000,000$ Hz, $d = 200$ Hz, and $t = 60°$. Determine the blood velocity through the vessel.

65. Suppose $f = 4,000,000$ Hz, $d = 100$ Hz, and $t = 65°$. Determine the blood velocity through the vessel.

SYNTHESIS

66. *Washington Monument.* While standing in the Mall in Washington, D.C., a tourist observes the angle of elevation to the top of the Washington Monument to be 67°. After moving 1012 ft farther away from the Washington Monument, the angle of elevation changes to 24°.

a) Use the small triangle to find *x* in terms of *h*.
b) Use the large triangle to find the height of the Washington Monument.

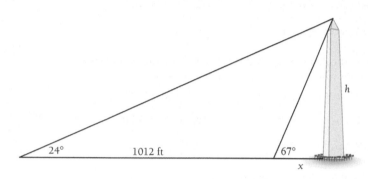

67. *Proportions of 30–60–90 Triangles.* Consider the adjacent 30–60–90 right triangles, each with hypotenuse of length 2, shown in the figure.

a) Explain why the two triangles form one equilateral triangle.
b) Explain why the short leg of each triangle has length 1.
c) Use the Pythagorean theorem to find the length of the long leg.
d) Explain how this figure gives the trigonometric functions of $\pi/6$ and $\pi/3$.

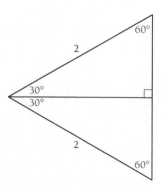

---

[16]Triton Technology, Inc.

68. *Proportions of 45–45–90 Triangles.* **Consider the** 45–45–90 right triangle shown in the figure, with a leg of length 1.

a) Explain why the other leg also has length 1.

b) Use the Pythagorean theorem to find the length of the hypotenuse.

c) Explain how this figure gives the trigonometric functions of $\pi/4$.

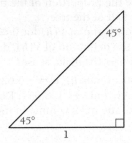

69. Refer to the figure below.

a) Use the small right triangle to show that $\tan t = 5/7$.

b) Use the large right triangle to show that $\tan t = 10/14$.

tw c) Why don't the trigonometric functions depend on the size of the triangle?

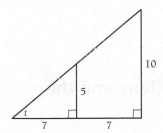

Trigonometric Identities. **Many interrelationships** between the trigonometric functions can be found, as shown in the following exercises.

70. *Reciprocal.* **Use the definitions of the trigonometric** functions to derive the Reciprocal Identities.

71. *Ratio.* **Use the definitions of the trigonometric** functions to derive the Ratio Identities.

72. *Sum Identity for Sine.* **Refer to the figure below to** answer these questions.[17]

a) Show that $u = \sin t$ and $v = \cos t$.

b) Use geometry to show that $r = s$.

c) Show that $w = \sin s \cos t$.

d) Show that $x = \cos s \sin t$.

e) Conclude that $\sin(s + t) = (w + x)/1 = \sin s \cos t + \cos s \sin t$.

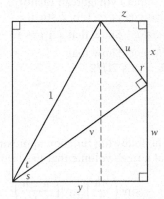

73. *Sum Identity for Cosine.* **Use the figure from the** previous exercise.

a) Show that $u = \sin t$ and $v = \cos t$.

b) Use geometry to show that $r = s$.

c) Show that $y = \cos s \cos t$.

d) Show that $z = \sin s \sin t$.

e) Conclude that $\cos(s + t) = (y - z)/1 = \cos s \cos t - \sin s \sin t$.

74. *Cofunction.*

a) Use the figure to show that
$$\cos\left(\frac{\pi}{2} - t\right) = \sin t.$$

b) Show that $\sin\left(\frac{\pi}{2} - t\right) = \cos t.$

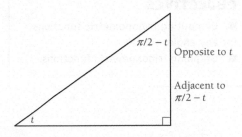

75. *Pythagorean.* **Show that** $1 + \tan^2 t = \sec^2 t$. (*Hint:* Begin with the Pythagorean Identity of Theorem 6 and divide both sides by $\cos^2 t$.)

76. *Pythagorean.* **Show that** $1 + \cot^2 t = \csc^2 t$.

77. *Double-Angle for Sine.* **Show that** $\sin 2t = 2 \sin t \cos t$. (*Hint:* Let $s = t$ and use a Sum Identity.)

---

[17]R. B. Nelson, *Proofs Without Words II* (Mathematical Association of America, Washington, DC, 2000).

78. *Double-Angle for Cosine.*

a) Show that $\cos 2t = \cos^2 t - \sin^2 t$.

b) Show that $\cos 2t = 2\cos^2 t - 1$. (*Hint:* Use part (a) and a Pythagorean Identity.)

c) Show that $\cos 2t = 1 - 2\sin^2 t$.

79. *Power Reduction.* Show that $\sin^2 t = (1 - \cos 2t)/2$.

80. *Power Reduction.* Show that
$$\cos^2 t = (1 + \cos 2t)/2.$$

 Technology Connection.

Tree Volume. The following function approximates the proportion V of a tree's volume that lies below height h:

$$V(h) = \sin^p\left(\frac{\pi h}{2}\right) \sin^q\left(\frac{\pi\sqrt{h}}{2}\right)$$
$$\times \sin^r\left(\frac{\pi\sqrt[3]{h}}{2}\right) \sin^s\left(\frac{\pi\sqrt[4]{h}}{2}\right).$$

In this equation, the exponents p, q, r, and s are constants and h is the proportion of the total height of the tree. For example, $h = 0$ corresponds to the bottom of the tree, $h = 1$ corresponds to the top, and $h = 1/2$ corresponds to halfway up the tree.[18]

81. a) Compute $V(0)$ and $V(1)$.

tw b) Explain your answers to part (a) in terms of tree volume.

82. For *Corymbia gummifera,* $p = -3.728$, $q = 48.646$, $r = -123.208$, and $s = 86.629$.

a) Determine the proportion of the tree's volume in the lower half of the tree.

b) Use a grapher to plot $V(h)$ for $0 \le h \le 1$.

tw c) Based on the definition of $V(h)$, does your answer to part (b) make sense?

83. For *Eucalyptus viminalis,* $p = -5.621$, $q = 74.831$, $r = -195.644$, and $s = 138.959$. Determine the proportion of the tree's volume in the lower half of the tree.

---

[18]H. Bi, "Trigonometric variable-form taper equations for Australian eucalyptus," *Forest Science,* Vol. 46, pp. 397–409 (2000).

1.5 Trigonometric Functions and the Unit Circle

OBJECTIVES

■ Computing trigonometric functions for any angle.

■ Graphing trigonometric functions.

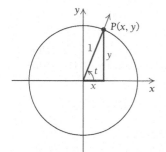

The Unit Circle

We have already discussed the evaluation of trigonometric functions for acute angles, that is, angles with measures between 0 and $\pi/2$ radians. In this section, we extend the definition of these functions to include angles that are not necessarily acute.

Consider the unit circle and an acute angle t as shown in the figure. To find the x- and y-coordinates of P, we use the trigonometric functions and the red triangle:

$$\cos t = \frac{\text{adjacent}}{\text{hypotenuse}} = \frac{x}{1} = x \quad \text{and} \quad \sin t = \frac{\text{opposite}}{\text{hypotenuse}} = \frac{y}{1} = y.$$

We also notice that the Pythagorean theorem states that $x^2 + y^2 = 1$; this is the equation of the unit circle.

We use the above observations about $\sin t$ and $\cos t$ to define trigonometric functions for any angle, whether acute or not.

DEFINITION

Let (x, y) be the point where the terminal side of an angle with measure t intersects the unit circle. Then we define

$$\sin t = y, \qquad \cos t = x, \qquad \tan t = \frac{y}{x},$$

$$\csc t = \frac{1}{y}, \qquad \sec t = \frac{1}{x}, \qquad \cot t = \frac{x}{y}.$$

However, $\tan t$, $\cot t$, $\sec t$, and $\csc t$ are *undefined* if division by zero occurs.

Though we do not prove them here, the identities found in Theorem 6 and the exercises of Section 1.4 still apply with this extended definition.

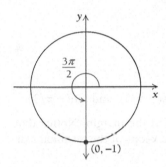

EXAMPLE 1   Use the unit circle to find $\sin(3\pi/2)$, $\cos(3\pi/2)$, and $\tan(3\pi/2)$.

Solution   We refer to the figure in the margin. An angle of measure $3\pi/2$ radians corresponds to three-quarters of a full rotation since $(3\pi/2)/(2\pi) = 3/4$. The terminal side of $t = 3\pi/2$ intersects the unit circle at the point $(0, -1)$. Therefore,

$$\sin\frac{3\pi}{2} = -1 \quad \text{and} \quad \cos\frac{3\pi}{2} = 0.$$

However, $\tan(3\pi/2)$ is undefined since division by zero is not permitted.   ■

For many angles, the trigonometric functions may be computed by using certain reflections around the unit circle, as discussed in the next few examples.

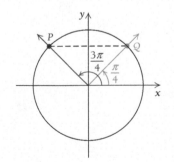

EXAMPLE 2   Use the unit circle to find $\sin(3\pi/4)$, $\cos(3\pi/4)$, and $\tan(3\pi/4)$.

Solution   To solve this problem, we refer to the figure in the margin. Notice that the terminal side for $t = 3\pi/4$ is the reflection about the y-axis of the terminal side of the angle $\pi/4$. The coordinates of the point Q are

$$\left(\cos\frac{\pi}{4}, \sin\frac{\pi}{4} \right) = \left(\frac{1}{\sqrt{2}}, \frac{1}{\sqrt{2}} \right).$$

Therefore, the coordinates of the point P may be obtained by reflecting Q across the y-axis. We conclude that

$$\sin\underbrace{\frac{3\pi}{4} = \frac{1}{\sqrt{2}}}_{y\text{-coordinate } P} \quad \text{and} \quad \cos\underbrace{\frac{3\pi}{4} = -\frac{1}{\sqrt{2}}}_{x\text{-coordinate } P}.$$

Using a Ratio Identity, we find that

$$\tan\frac{3\pi}{4} = \frac{\sin(3\pi/4)}{\cos(3\pi/4)} = \frac{1/\sqrt{2}}{-1/\sqrt{2}} = -1.$$ ∎

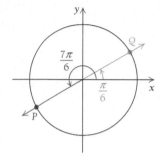

EXAMPLE 3   Use the unit circle to find $\sin(7\pi/6)$, $\cos(7\pi/6)$, and $\tan(7\pi/6)$.

Solution   To solve this problem, we refer to the figure in the margin. Notice that the terminal side for $t = 7\pi/6$ is in line with but directly opposite from the angle $\pi/6$. The coordinates of the point Q are

$$\left(\cos\frac{\pi}{6}, \sin\frac{\pi}{6}\right) = \left(\frac{\sqrt{3}}{2}, \frac{1}{2}\right).$$

Therefore, the coordinates of the point P may be obtained by negating both of the x- and y-coordinates of Q. We conclude that

$$\underset{\text{y-coordinate } P}{\sin\frac{7\pi}{6} = -\frac{1}{2}} \quad \text{and} \quad \underset{\text{x-coordinate } P}{\cos\frac{7\pi}{6} = -\frac{\sqrt{3}}{2}}.$$

Using a Ratio Identity, we find that

$$\tan\frac{7\pi}{6} = \frac{\sin(7\pi/6)}{\cos(7\pi/6)} = \frac{-1/2}{-\sqrt{3}/2} = \frac{1}{\sqrt{3}}.$$ ∎

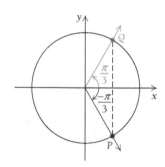

EXAMPLE 4   Use the unit circle to find $\sin(-\pi/3)$, $\cos(-\pi/3)$, and $\tan(-\pi/3)$.

Solution   To solve this problem, we refer to the figure in the margin. Notice that the terminal side for $t = -\pi/3$ is the reflection about the x-axis of the terminal side of the angle $\pi/3$. The coordinates of the point Q are

$$\left(\cos\frac{\pi}{3}, \sin\frac{\pi}{3}\right) = \left(\frac{1}{2}, \frac{\sqrt{3}}{2}\right).$$

Therefore, the coordinates of the point P may be obtained by reflecting Q across the x-axis. We conclude that

$$\underset{\text{y-coordinate of } P}{\sin\left(-\frac{\pi}{3}\right) = -\frac{\sqrt{3}}{2}} \quad \text{and} \quad \underset{\text{x-coordinate of } P}{\cos\left(-\frac{\pi}{3}\right) = \frac{1}{2}}.$$

Using a Ratio Identity, we find that

$$\tan\left(-\frac{\pi}{3}\right) = \frac{\sin(-\pi/3)}{\cos(-\pi/3)} = \frac{-\sqrt{3}/2}{1/2} = -\sqrt{3}.$$ ∎

Graphing sin t and cos t

Note that $\sin t$ and $\cos t$ are functions of t defined for all real numbers t. Some values of $\sin t$ and $\cos t$ are listed in the table in the margin on the next page. You may wish to verify these values by using the unit circle. These points are plotted on the following two figures to show their graphs. Both graphs are examples of **sinusoids.**

Angle, t	$\sin t$	$\cos t$
-2π	0	1
$-3\pi/2$	1	0
$-\pi$	0	-1
$-\pi/2$	-1	0
0	0	1
$\pi/6$	$1/2$	$\sqrt{3}/2$
$\pi/4$	$1/\sqrt{2}$	$1/\sqrt{2}$
$\pi/3$	$\sqrt{3}/2$	$1/2$
$\pi/2$	1	0
π	0	-1
$3\pi/2$	-1	0
2π	0	1
$5\pi/2$	1	0
3π	0	-1
$7\pi/2$	-1	0
4π	0	1

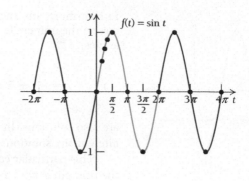

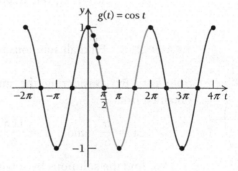

Note in particular that $\sin t = 0$ has solutions $t = 0$, $\pm\pi$, $\pm 2\pi, \ldots$ Also, the equation $\cos t = 0$ has solutions $t = \pm\pi/2, \pm3\pi/2, \pm5\pi/2, \ldots$

Both graphs "repeat" themselves. More specifically, the sections in blue are repeated infinitely often to the left and to the right. Both of these blue sections take input values on an interval with length 2π. For this reason, we say that 2π is the **period** of these graphs.

For both $f(t) = \sin t$ and $g(t) = \cos t$, the maximum is 1, while the minimum is -1. (This makes sense because $\cos t$ and $\sin t$ correspond to the coordinates of a point on the unit circle.) The distance of the maximum and minimum from the **mid-line,** or the t-axis, is 1. For this reason, we say that the **amplitude** of both $\sin t$ and $\cos t$ is equal to 1.

In the graph of $f(t) = \sin t$, the function is at the mid-line and is increasing when $t = 0$. However, for $g(t) = \cos t$, the function is at the maximum when $t = 0$.

Trigonometric Equations

A trigonometric equation such as $\sin t = -1/2$ may be solved by using a unit circle. As shown in the margin, the angle $t = 7\pi/6$ intersects the unit circle at $P(-\sqrt{3}/2, -1/2)$, and the angle $t = 11\pi/6$ intersects the unit circle at $Q(\sqrt{3}/2, -1/2)$. From the definition of $\sin t$, we observe

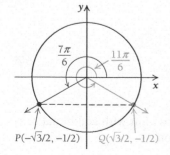

$$\sin\left(\frac{7\pi}{6}\right) = -\frac{1}{2} \quad \text{and} \quad \sin\left(\frac{11\pi}{6}\right) = -\frac{1}{2}.$$

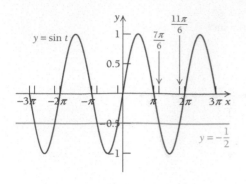

Furthermore, any angle coterminal with either $7\pi/6$ or $11\pi/6$ will intersect the unit circle at either P or Q. Therefore, for any integer n, the angles

$$t = \frac{7\pi}{6} + 2\pi n \quad \text{and} \quad t = \frac{11\pi}{6} + 2\pi n$$

are also solutions. In conclusion, the equation $\sin t = -1/2$ has *infinitely* many solutions, as can be seen in the graph in the margin.

This particular equation for $\sin t$ was solved by finding points on the unit circle with a given y-coordinate. Similar equations involving the cosine function may be solved by finding points on the unit circle with given x-coordinates.

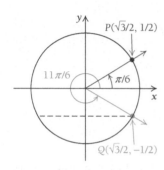

EXAMPLE 5 Find all solutions to $\cos 3t = \dfrac{\sqrt{3}}{2}$.

Solution By considering the unit circle, we know that two solutions can be obtained from

$$3t = \frac{\pi}{6} \quad \text{and} \quad 3t = \frac{11\pi}{6}.$$

We find the solutions by solving

$$3t = \frac{\pi}{6} + 2\pi n \quad \text{and} \quad 3t = \frac{11\pi}{6} + 2\pi n,$$

where n is an integer. Solving for t gives the solutions

$$t = \frac{\pi}{18} + \frac{2\pi}{3}n \quad \text{and} \quad t = \frac{11\pi}{18} + \frac{2\pi}{3}n.$$

EXAMPLE 6 Solve $2\cos^2 t + 3\cos t - 2 = 0$.

Solution We solve this equation by factoring.

$$2\cos^2 t + 3\cos t - 2 = 0$$
$$(2\cos t - 1)(\cos t + 2) = 0.$$

Since $\cos t \neq -2$ for any t, the solutions come from $\cos t = \frac{1}{2}$. By considering the unit circle, for each integer n,

$$t = \frac{\pi}{3} + 2\pi n \quad \text{and} \quad t = \frac{5\pi}{3} + 2\pi n$$

are solutions.

Amplitude, Period, and Mid-Line

Other sinusoids are functions based on either $\sin t$ or $\cos t$ but with additional constants. To see this, let's consider the graph of the function

$$h(t) = 2\sin 3t + 4$$

for $0 \leq t \leq 2\pi$. The table below shows the values of $h(t)$ for various values of t. To compute this table, we first multiply t by 3, apply the sine function, multiply by 2, and then add 4. For example,

$$h(0) = 2 \sin(3[0]) + 4 = 2 \sin 0 + 4 = 2(0) + 4 = 4,$$

$$h\left(\frac{\pi}{6}\right) = 2 \sin\left(3\left[\frac{\pi}{6}\right]\right) + 4 = 2 \sin\frac{\pi}{2} + 4 = 2(1) + 4 = 6,$$

and so on. These points are shown on the following graph.

> **CAUTION** To compute an expression such as $2 \sin 0 + 4$, adding 4 is the last step, not the first step.

t	$h(t) = 2 \sin 3t + 4$
0	4
$\pi/6$	6
$\pi/3$	4
$\pi/2$	2
$2\pi/3$	4
$5\pi/6$	6
π	4
$7\pi/6$	2
$4\pi/3$	4
$3\pi/2$	6
$5\pi/3$	4
$11\pi/6$	2
2π	4

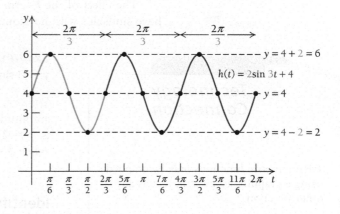

We notice that the graph of $h(t) = 2 \sin 3t + 4$ has the same repeating behavior as seen before. However, the amplitude, period, and mid-line have changed. Could we make a sketch of this graph without directly substituting values of t?

First, we see from the graph of h that the mid-line is the horizontal line $y = 4$, corresponding to the $+4$ in the definition of h. Second, the maximum is 6, and the minimum is 2. The amplitude is thus $6 - 4 = 4 - 2 = 2$. This amplitude corresponds to the coefficient 2 in the definition of h. Not surprisingly, the average of the maximum and minimum is the mid-line: $(6 + 2)/2 = 4$.

Third, we see that three cycles of h occur between 0 and 2π, whereas only one cycle occurs in the same interval for $\sin t$. The length of any one cycle is therefore one-third of the period for $\sin t$. In other words, the period is $p = 2\pi/3$. The number 3 is the coefficient within the sine function in the definition of h.

From these observations, we may infer the following theorem. The proof of this theorem is described in the exercises.

THEOREM 7

Given either the function $y = a \sin bt + k$ or $y = a \cos bt + k$, where a and b are both positive:

a) The mid-line is the horizontal line $y = k$.

b) The amplitude is a, so the maximum and minimum are $k + a$ and $k - a$, respectively.

c) The period is $p = 2\pi/b$.

d) For $y = a \cos bt + k$, the graph is at its maximum when $t = 0$. For $y = a \sin bt + k$, the graph is at its mid-line and is increasing when $t = 0$.

The effect of the k-term is to shift the sinusoid vertically. It is also possible to have sinusoids with horizontal shifts, as discussed in the exercises.

Technology Connection

Graphing Sinusoids

Trigonometric functions may be graphed using a graphing calculator, just as any other function.

EXERCISES

Use the window $[-2, 2, -2, 4]$ to sketch the graphs of the following functions. Be sure the grapher is in radian mode.

1. $y = \cos(\pi x)$

2. $y = \cos(\pi x) + 1$

3. $y = \cos(2\pi x) + 1$

4. $y = 3 \cos(2\pi x) + 1$

5. $y = 3 \cos(2\pi [x - 0.5]) + 1$

How does each change in the function affect the graph?

EXAMPLE 7 Find the amplitude, period, and mid-line of $y = 3 \sin 4t - 1$. Also, find the maximum and minimum.

Solution For this function, $a = 3$, $b = 4$, and $k = -1$. So the amplitude is 3, the mid-line is the line $y = -1$, and the period is $p = 2\pi/4 = \pi/2$. Also, the maximum and minimum are $-1 + 3 = 2$ and $-1 - 3 = -4$, respectively. ∎

Identifying Trigonometric Functions from Their Graphs

EXAMPLE 8

Identify a, b, and k so that the graph below is of the function

$$f(t) = a \sin bt + k.$$

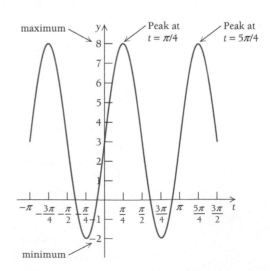

Solution

a) The mid-line is obtained by averaging the maximum and minimum: $k = (-2 + 8)/2 = 3$. The graph is at the mid-line at $t = 0$ (in other words, the point $(0, 3)$ is on the graph) and is increasing at $t = 0$. So, it is appropriate to use the model $f(t) = a \sin bt + k$ for this graph.

b) The amplitude may be obtained by subtracting the mid-line k from the maximum: $a = 8 - 3 = 5$.

c) The period may be obtained by finding the distance between successive peaks: $p = 5\pi/4 - \pi/4 = \pi$. Since $p = 2\pi/b$,

$$\frac{2\pi}{b} = \pi, \quad \text{or} \quad b = 2.$$

In summary,

$$f(t) = 5 \sin 2t + 3. \qquad \blacksquare$$

Frequency

If t is measured in seconds, the **frequency** is defined to be $f = b/(2\pi)$, or the reciprocal of the period. The frequency is measured in cycles per second, or Hertz (Hz).

EXAMPLE 9   *Sea Waves.* A buoy oscillates as waves go past. It moves a total of 2.5 m from its lowest point to its highest point, and it returns to its lowest point every 8 sec. In the absence of waves, the buoy would be exactly at sea level. Model the buoy's motion with a trigonometric function given that it is at its highest point when $t = 0$. Also, find the frequency of the buoy's motion.

Solution   Since the buoy is at its highest point when $t = 0$, it will be convenient to model its motion using

$$g(t) = a \cos bt + k.$$

Without the waves, the buoy would be at sea level, and so we set $k = 0$.
   The motion has a period of 8 sec, and so we can solve for b:

$$8 = \frac{2\pi}{b}, \quad \text{so} \quad b = \frac{2\pi}{8} = \frac{\pi}{4}.$$

The difference between the highest and lowest points is 2.5 m, and so

$$2a = 2.5, \quad \text{or} \quad a = \frac{2.5}{2} = 1.25.$$

The equation of motion is therefore $g(t) = 1.25 \cos(\pi t/4)$.
   Finally, the frequency is given by

$$f = \frac{b}{2\pi} = \frac{\pi/4}{2\pi} = \frac{1}{8} \text{ Hz.}$$

This may also have been found by taking the reciprocal of the period. $\qquad \blacksquare$

Exercise Set 1.5

Sketch the following angles.

1. $5\pi/4$

2. $-5\pi/6$

3. $-\pi$

4. 2π

5. $13\pi/6$

6. $-7\pi/4$

Use a unit circle to compute the following trigonometric functions.

7. $\cos(9\pi/2)$

8. $\sin(5\pi/4)$

9. $\sin(-5\pi/6)$

10. $\cos(-5\pi/4)$

11. $\cos 5\pi$

12. $\sin 6\pi$

13. $\tan(-4\pi/3)$

14. $\tan(-7\pi/3)$

Use a calculator to evaluate the following trigonometric functions.

15. $\cos 125°$

16. $\sin 164°$

17. $\tan(-220°)$

18. $\cos(-253°)$

19. $\sec 286°$

20. $\csc 312°$

21. $\sin(1.2\pi)$

22. $\tan(-2.3\pi)$

23. $\cos(-1.91)$

24. $\sin(-2.04)$

Find all solutions of the given equation.

25. $\sin t = \dfrac{1}{2}$

26. $\sin t = -1$

27. $\sin 2t = 0$

28. $2 \sin\left(t + \dfrac{\pi}{3}\right) = -\sqrt{3}$

29. $\cos\left(3t + \dfrac{\pi}{4}\right) = -\dfrac{1}{2}$

30. $\cos(2t) = 0$

31. $\cos(3t) = 1$

32. $2 \cos\left(\dfrac{t}{2}\right) = -\sqrt{3}$

33. $2 \sin^2 t - 5 \sin t - 3 = 0$

34. $\cos^2 x + 5 \cos x = 6$

35. $\cos^2 x + 5 \cos x = -6$

36. $\sin^2 t - 2 \sin t - 3 = 0$

For the following functions, find the amplitude, period, and mid-line. Also, find the maximum and minimum.

37. $y = 2 \sin 2t + 4$

38. $y = 3 \cos 2t - 3$

39. $y = 5 \cos(t/2) + 1$

40. $y = 3 \sin(t/3) + 2$

41. $y = \dfrac{1}{2} \sin(3t) - 3$

42. $y = \dfrac{1}{2} \cos(4t) + 2$

43. $y = 4 \sin(\pi t) + 2$

44. $y = 3 \cos(3\pi t) - 2$

For each of the following graphs, determine if the function should be modeled by either $y = a \sin bt + k$ or $y = a \cos bt + k$. Then find a, b, and k.

45.

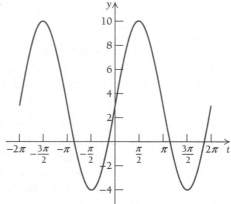

46.

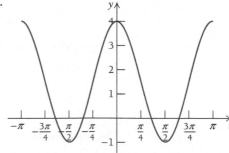

47.

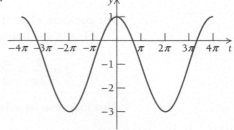

48.

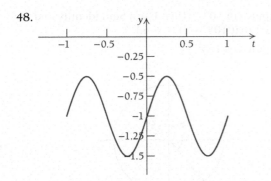

APPLICATIONS

Solar Radiation. The annual radiation (in megajoules per square centimeter) for certain land areas of the northern hemisphere may be modeled with the equation[19]

$$R = 0.339 + 0.808 \cos l \cos s - 0.196 \sin l \sin s - 0.482 \cos a \sin s.$$

In this equation, l is the latitude (between 30° and 60°) and s is the slope of the ground (between 0° and 60°). Also, a is the aspect, or the direction that the slope faces. For a slope facing due north, $a = 0°$, and for a slope facing south, $a = 180°$. For a slope facing either east or west, $a = 90°$.

49. Find the annual radiation of north-facing land at 40° north latitude with a 30° slope.

50. Find the annual radiation of south-facing land at 30° north latitude with a 20° slope.

51. Find the annual radiation of southeast-facing land at 50° north latitude with a 55° slope.

52. Find the annual radiation of flat land at 50° north latitude.

Lung Capacity. As we breathe, our lungs increase and decrease in volume. The volume of air that we inhale and exhale with each breath is called *tidal volume*. The maximum possible tidal volume is called the *vital capacity*, normally approached during strenuous physical activity. Even at vital capacity, the lungs are never completely drained of air; the minimum volume of the lungs is called the *residual volume*.[20]

53. Suppose a man watching television breathes once every 5 sec. His average lung capacity is 2500 mL, and his tidal volume is 500 mL. Express the volume of his lungs using the model $V(t) = a \cos bt + k$, where time 0 corresponds to the lungs at their largest capacity.

54. A woman undergoes her ordinary strenuous workout, breathing once every 2 sec. Her tidal volume is 3400 mL, and her residual volume is 1100 mL. Express the volume of her lungs using the model $V(t) = a \cos bt + k$, where time 0 corresponds to the lungs at their largest capacity.

tw 55. Explain why a periodic model like the cosine function may be reasonable for describing lung capacity.

56. *Body Temperature.* In a laboratory experiment, the body temperature T of rats was measured.[21] |The body temperatures of the rats varied between 35.33°C and 36.87°C during the course of the day. Assuming that the peak body temperature occurred at $t = 0$, model the body temperature with a function of the form $T(t) = a \cos bt + k$.

Sound Waves. The pitch of a sound wave is measured by its frequency. Humans can hear sounds in the range from 20 to 20,000 Hz, while dogs can hear sounds as high as 40,000 Hz. The loudness of the sound is determined by the amplitude.[22]

57. The note A above middle C on a piano generates a sound modeled by the function $g(t) = 4 \sin(880\pi t)$, where t is in seconds. Find the frequency of A above middle C.

58. The note A below middle C on a piano generates a sound modeled by the function $g(t) = 4 \sin(440\pi t)$, where t is in seconds. Find the frequency of A below middle C.

59. *Blood Pressure.* During a period of controlled breathing, the systolic blood pressure p of a volunteer averaged 143 mmHg with an amplitude of 5.3 mmHg and a frequency of 0.172 Hz. Assuming that the blood pressure was highest when $t = 0$, find a model $p(t) = a \cos bt + k$ for blood pressure as a function of time.

60. *Blood Pressure.* During an episode of sleep apnea, the systolic blood pressure averaged 137 mmHg with an amplitude of 6.7 mmHg and a frequency of 0.079 Hz. Assuming that the blood pressure was highest when $t = 0$, find a model

[19]B. McCune and D. Keon, "Equations for potential annual direct incident radiation and heat load," *Journal of Vegetation Science,* Vol. 13, pp. 603–606 (2002).

[20]G. J. Borden, K. S. Harris, and L. J. Raphael, *Speech Science Primer,* 4th ed. (Lippincott Williams & Wilkins, Philadelphia, 2003).

[21]H. Takeuchi, A. Enzo, and H. Minamitani, "Circadian rhythm changes in heart rate variability during chronic sound stress," *Medical and Biological Engineering and Computing,* Vol. 39, pp. 113–117 (2001).

[22]N. A. Campbell and J. B. Reece, *Biology* (Benjamin Cummings, New York, 2002).

$p(t) = a \cos bt + k$ for blood pressure as a function of time.[23]

SYNTHESIS

Using a calculator, find the x- and y-coordinates of the following points on the unit circle.

61.

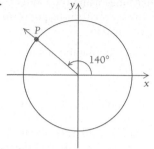

62.

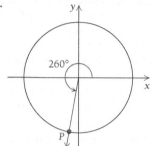

63.

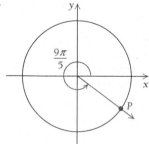

64.

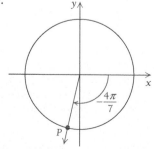

65. Compute $\sin 105°$. (*Hint:* Use a Sum Identity and the fact that $105° = 45° + 60°$.)

66. Compute $\cos 165°$.

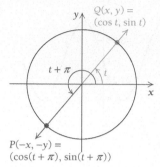

67. *Half-Rotations.*

    a) Use the figure to explain why
       $\sin(t + \pi) = -\sin t$ and $\cos(t + \pi) = -\cos t$.

    b) Rederive the results of part (a) using Sum Identities.

    c) Show that $\tan(t + \pi) = \tan t$.

68. *Amplitude and Mid-Line.* Consider the function $g(t) = a \sin bt + k$, where a and b are positive.

    a) Show that the maximum and minimum of $g(t)$ are $k + a$ and $k - a$, respectively.

    b) Show that the mid-line is the line $y = k$.

    c) Show that the amplitude of $g(t)$ is a.

69. *Period.*

tw a) Use a unit circle to explain why
      $\sin t = \sin(t + 2\pi)$ for all numbers t.

    b) Let $g(t) = a \sin bt + k$. Show that
      $g(t + 2\pi/b) = g(t)$.

tw c) Why does the result of part (b) imply that the period of $g(t)$ is $2\pi/b$?

Frequency Detection in the Ear. Basilar fibers in the ear detect sound, and they vary in length, tension, and density throughout the basilar membrane. A fiber is affected most by sound frequencies near the *fundamental frequency* f of the fiber, which is approximately

$$f = \frac{1}{2L} \sqrt{\frac{T}{d}}.$$

In this formula, L is the length of the fiber, T is the tension of the fiber, and d is the density of the fiber.[24]

tw **70.** At the apex of the basilar membrane, the fibers are long, flexible, and wide. At the apex, the basilar fibers are most affected by what kind of sound frequencies?

[23]M. Javorka, I. Žila, K. Javorka, and A. Čalkovská, "Do the oscillations of cardiovascular parameters persist during voluntary apnea in humans?" *Physiological Research,* Vol. 51, pp. 227–238 (2002).

[24]N. A. Campbell and J. B. Reece, *Biology* (Benjamin Cummings, New York, 2002).

Fibers of
sensory
neurons

Oval
window   Perilymph

Vestibular
canal

Stapes

Round
window

Tympanic   Basilar   Cochlear
canal   membrane   duct

Relative
lengths of   Base   Apex
basilar fibers
within
different
regions of
basilar
membrane

Hz   Hz   Hz   Hz
20,000   1,500   500   20
(High   (Low
notes)   notes)

tw 71. At the base of the basilar membrane, the fibers are short, stiff, and narrow. At the base, the basilar fibers are most affected by what kind of sound frequencies?

Notes on a Piano. A key n notes (including half-notes) above the A above middle C generates a sound wave proportional to

$$y = \sin(880 \cdot 2^{n/12} \pi t),$$

where t is in seconds. Use this equation to solve the following exercises.

72. Find the frequency of the A above middle C ($n = 0$).

73. Find the frequency of middle C ($n = -9$).

74. Find the note with a frequency of $2 \cdot 440 = 880$ Hz. *Note:* There are 12 notes in an octave (including half-notes).

75. Find the note with a frequency of $4 \cdot 440 = 1760$ Hz.

76. Find the (approximate) note with a frequency of $3 \cdot 440 = 1320$ Hz.

77. Find the (approximate) note with a frequency of $5 \cdot 440 = 2200$ Hz.[25]

[25]If you translate your answers to Exercises 76 and 77 downward by one and two octaves (12 and 24 notes), respectively, you have two of the three notes of a *major chord*. The third note of this particular major chord is A above middle C.

    Technology Connection

78. *Graphing a Trigonometric Identity.*
   a) Use a grapher to sketch the graph of $y = -\sin^2 t$. Use the graphing window $[-2\pi, 2\pi, -1, 1]$.
   b) The graph in part (a) should be periodic. Use a cosine model of the form $y = a \cos bt + k$ to model its graph.
   tw c) Use Exercise 79 of Section 1.4 to explain your results.

79. *Graphing a Trigonometric Identity.*
   a) Use a grapher to sketch the graph of $y = \cos^2 t$. Use the graphing window $[-2\pi, 2\pi, -1, 1]$.
   b) The graph in part (a) should be periodic. Use a cosine model of the form $y = a \cos bt + k$ to model its graph.
   tw c) Use Exercise 80 of Section 1.4 to explain your results.

80. *Graphing a Trigonometric Identity.*
   a) Use a grapher to sketch the graph of $y = 2 \sin t \cos t$. Use the graphing window $[-2\pi, 2\pi, -1, 1]$.
   b) The graph in part (a) should be periodic. Use a sine model of the form $y = a \sin bt + k$ to model its graph.
   tw c) Use Exercise 77 of Section 1.4 to explain your results.

Horizontal Shifts. The functions $y = a \sin(b[t - h]) + k$ and $y = a \cos(b[t - h]) + k$ incorporate a horizontal shift h. This amount is sometimes called the *phase shift*.

Graph the following functions on $[-2\pi, 2\pi]$.

81. a) $y = 3 \sin t + 4$
   b) $y = 3 \sin(t - \pi/4) + 4$
   tw c) How does the horizontal shift change the graph?

82. a) $y = 2 \cos 2t - 1$
   b) $y = 2 \cos(2[t + \pi/6]) - 1$
   tw c) How does the horizontal shift change the graph?

Oscillations with Trends. Graph the following functions on the interval $0 \le t \le 10$.

83. $f(t) = t + \sin 2t$

84. $f(t) = 5t + 3 \sin 4t$

85. $f(t) = t^2/10 + 3 + 4 \sin 5t$

Chapter 1   Summary and Review

Terms to Know

Review Exercises

These review exercises are for test preparation. They can also be used as a lengthened practice test. Answers are at the back of the book. The answers also contain bracketed section references, which tell you where to restudy if your answer is incorrect.

1. *Live Births to Women of Age x.* **The following graph relates the number of live births B per 1000 women of age x.**

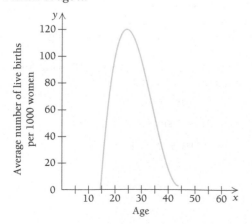

Use the graph to answer the following.

a) What is the incidence of live births to women of age 20?

b) For what ages is the incidence of live births approximately 100 per 1000 women?

A function f is given by $f(x) = 2x^2 - x + 3$. Find each of the following.

2. $f(-2)$ 3. $f(1 + h)$ 4. $f(0)$

A function f is given by $f(x) = (1 - x)^2$. Find each of the following.

5. $f(-5)$ 6. $f(2 - h)$ 7. $f(4)$

Graph.

8. $f(x) = 2x^2 + 3x - 1$ 9. $y = 3x^2 - 6x + 1$

10. $y = |x + 1|$ 11. $f(x) = (x - 2)^2$

12. $f(x) = \dfrac{x^2 - 16}{x + 4}$

13. The graph of $f(x)$ is given below.
   a) Determine $f(2)$.
   b) Find all x-values such that $f(x) = 2$.

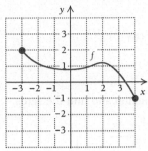

Graph.

14. $x = -2$ 15. $y = 4 - 2x$

16. Write an equation of the line containing the points $(4, -2)$ and $(-7, 5)$.

17. Write an equation of the line with slope 8, containing the point $\left(\frac{1}{2}, 11\right)$.

18. For the linear equation $y = 3 - \frac{1}{6}x$, find the slope and the y-intercept.

Solve.

19. $x^2 + 5x + 4 = 0$ 20. $x^2 - 7x + 12 = 0$
21. $x^2 + 2x = 8$ 22. $x^2 + 6x = 20$
23. $x^3 + 3x^2 - x - 3 = 0$
24. $x^4 + 2x^3 - x - 2 = 0$

Find the average rate of change.

25.

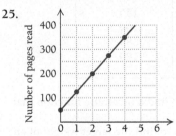

Number of days spent reading

26.

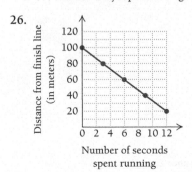

Number of seconds spent running

27. *Muscle Weight and Body Weight.* The weight M of muscles in a human is directly proportional to body weight W. It is known that a person who weighs 150 lb has 60 lb of muscles. What is the muscle weight of a person who weighs 210 lb?

28. Solve for x: $5x^2 - x = 7$.

29. Convert to radical notation: $y^{1/6}$.

30. Convert to rational exponents: $\sqrt[20]{x^3}$.

31. Simplify: $27^{2/3}$.

32. Graph:

$$f(x) = \begin{cases} x - 5, & \text{for } x > 2, \\ x + 3, & \text{for } x \leq 2. \end{cases}$$

33. *Study Time and Grades.* A math instructor asked her students to keep track of how much time each spent studying the chapter on percent notation in her basic mathematics course. She collected the information together with test scores from that chapter's test. The data are listed in the following table.

Study Time, x (in hours)	Test Grade, G (in percent)
9	74
11	94
13	81
15	86
16	87
17	81
21	87
23	92

   a) Using the data points $(9, 74)$ and $(23, 92)$, find a linear function that fits the data.
   b) Use the function to predict the test scores of a student who studies for 18 hr and for 25 hr.

Determine the following exactly.

34. $\sin(2\pi/3)$

35. $\cos(-\pi)$

36. $\tan(7\pi/4)$

37. Solve for the missing side x.

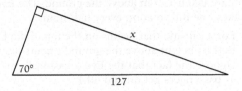

Find all t that solve the equation.

38. $\sin t = 1$

39. $\tan t = \sqrt{3}$

40. $4 \cos 2t = 8$

41. $12 \cos^2\left(2t - \dfrac{\pi}{4}\right) = 9$

42. $2 \sin^2 t + 7 \sin t - 4 = 0$

For the following functions, find the amplitude, period, and mid-line. Also, find the maximum and minimum.

43. $y = 2 \sin(t/3) - 4$

44. $y = \dfrac{1}{2} \cos(2\pi t) + 3$

For each of the following graphs, determine if the function should be modeled by either $y = a \sin bt + k$ or $y = a \cos bt + k$. Then find a, b, and k.

45.

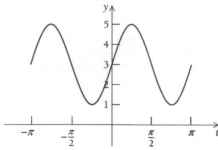

46.

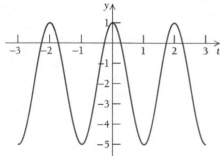

47. *London Eye.* The London Eye is a large observation wheel that allows tourists to see downtown London. Tourists enter from a platform 1 m off the ground and are taken 135 m above the ground. The Eye makes one full rotation every half-hour.

   a) For a capsule that begins on the top of the Eye, find its height above the ground t hours later. (Ignore the fact that the Eye occasionally stops as passengers get on and off.)

   b) How high above the ground will the capsule be 10 min after reaching the bottom?

SYNTHESIS

48. Simplify: $(64^{5/3})^{-1/2}$.

 Technology Connection

Graph the function and find the zeros.

49. $f(x) = x^3 - 4x$

50. $f(x) = \sqrt[3]{|9 - x^2|} - 1$

51. Approximate the points of intersection of the graphs of the two functions in Exercises 49 and 50.

52. *Study Time and Grades.* Use the data in Exercise 33.

   a) Use the **REGRESSION** feature to fit a linear function to the data.

   b) Use the function to predict the test scores of a student who studies for 18 hr and for 25 hr.

tw c) Compare your answers to those found in Exercise 33.

53. *Target Weight.* The following chart shows the target weight w (in pounds) for an adult man of large build of various heights h (in inches).[26]

Height, h (inches)	Weight, w (pounds)
63	146.5
66	155
69	165.5
72	184.5
75	189

   a) Use regression to find a quadratic function that fits the data.

   b) Use the function to find the target weight of an adult man of large build who is 5 ft 7 in. (67 in.) tall.

---

[26]Metropolitan Life Insurance Company.

Chapter 1 Test

1. *Time Spent on Home Computer.* The following graph relates the average number of minutes spent per month A on a home computer to a person's age x.

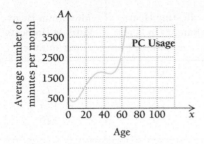

Source: Media Matrix; The PC Meter Company

Use the graph to answer the following.

a) What is the average use, in minutes per month, of persons of age 20?

b) For what ages is the average use approximately 3000 min per month?

2. A function is given by $f(x) = x^2 + 2$. Find

a) $f(-3)$.

b) $f(x + h)$.

3. A function is given by $f(x) = 2x^2 + 3$. Find

a) $f(-2)$.

b) $f(x + h)$.

4. What are the slope and the y-intercept of $y = -3x + 2$?

5. Find an equation of the line with slope $\frac{1}{4}$, containing the point $(8, -5)$.

6. Find the slope of the line containing the points $(2, -5)$ and $(-3, 10)$.

Find the average rate of change.

7.

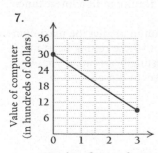

Number of years of use

8.

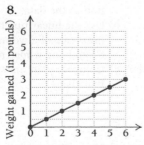

Number of bags of feed used

9. *Body Fluids.* The weight F of fluids in a human is directly proportional to body weight W. It is known that a person who weighs 180 lb has 120 lb of fluids. Find an equation of variation expressing F as a function of W.

10. For the following graph of function f, determine (a) $f(1)$ and (b) all x-values such that $f(x) = 4$.

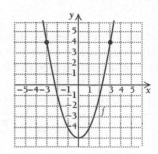

11. Solve for x: $x^2 + 4x - 2 = 0$.

12. Graph: $f(x) = 4/x$.

13. Convert to rational exponents: $1/\sqrt{t}$.

14. Convert to radical notation: $t^{-3/5}$.

15. Graph: $f(x) = \dfrac{x^2 - 1}{x + 1}$.

Determine the following exactly.

16. $\sin(11\pi/6)$

17. $\cos(-3\pi/4)$

18. $\tan \pi$

19. Solve for the missing side x.

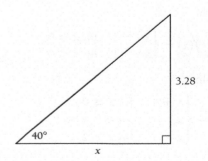

Find all values of t that satisfy the equation.

20. $\tan^2 t = 3$

21. $\cos^2 t = 2$

22. $2\sin^3(2t) - 3\sin^2(2t) - 2\sin(2t) = 0$

For the following functions, find the amplitude, period, and mid-line. Also, find the maximum and minimum.

23. $y = 4\cos(2t) + 4$

24. $y = 6\sin(t/3) - 10$

For each of the following graphs, determine if the function should be modeled by either $y = a\sin bt + k$ or $y = a\cos bt + k$. Then find a, b, and k.

25.

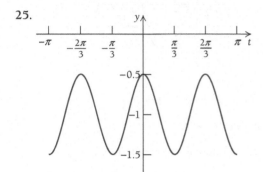

26.

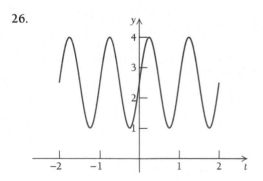

27. Graph:
$$f(x) = \begin{cases} x^2 + 2, & \text{for } x \geq 0, \\ x^2 - 2, & \text{for } x < 0. \end{cases}$$

28. Graph:
$$f(x) = \begin{cases} x + 1, & \text{for } x \leq 0 \\ x, & \text{for } 0 < x \leq 2 \\ x - 1, & \text{for } x > 2. \end{cases}$$

29. *Maximum Heart Rate.* A person exercising should not exceed a maximum heart rate, which depends on his or her gender, age, and resting heart rate. The following table shows data relating resting heart rate and maximum heart rate for a 20-yr-old woman.

Resting Heart Rate, r (in beats per minute)	Maximum Heart Rate, M (in beats per minute)
50	170
60	172
70	174
80	176

Source: American Heart Association

a) Using the data points (50, 170) and (80, 176), find a linear function that fits the data.
b) Use the linear function to predict the maximum heart rate of a woman whose resting heart rate is 62 and one whose resting heart rate is 75.

SYNTHESIS

30. Solve for x: $3x + \dfrac{8}{x} = 1$.

 Technology Connection

Graph the function and find the zeros.

31. $f(x) = x^3 - 9x^2 + 27x + 50$

32. $f(x) = \sqrt[3]{|4 - x^2|} + 1$

33. Approximate the points of intersection of the graphs of the two functions in Exercises 31 and 32.

34. *Maximum Heart Rate.* Use the data in Exercise 29.
a) Use the **REGRESSION** feature to fit a linear function to the data.
b) Use the linear function to predict the maximum heart rate of a woman whose resting heart rate is 62 and one whose resting heart rate is 75.
tw c) Compare your answers to those found in Exercise 29.

35. *Time Spent on Home Computer.* The following data relate the average number of minutes spent per month A on a home computer to a person's age x.

Age (in years)	Average Use (in minutes per month)
6.5	363
14.5	645
21	1377
29.5	1727
39.5	1696
49.5	2052
55	2299

Source: Media Matrix; The PC Meter Company

a) Use the **REGRESSION** feature to fit linear, quadratic, cubic, and quartic functions to the data.

b) Plot the data and graph each function on the same set of axes.

tw c) Decide which function best fits the data and explain your result.

tw d) Compare the result of part (a) with the answer found in Exercise 1.

Extended Life Science Connection

CARBON DIOXIDE CONCENTRATIONS

Extended Life Science Applications occur at the end of each chapter. They are designed to consider certain applications in greater depth, make use of grapher skills, and allow for possible group or collaborative learning.
Ecologists continue to be concerned with the rate that atmospheric carbon dioxide (CO_2) concentration is increasing. This increase in carbon dioxide emissions could lead to the phenomenon of global warming, or the trend of average global temperatures to rise over the years.

The world's longest continuous record of atmospheric carbon dioxide concentrations is located at Mauna Loa, Hawaii. This site is located over 11,000 ft above sea level in a barren lava field of an active volcano. This location is unaffected by both local vegetation and human activities, and so ecologists believe the data collected from this site are representative of the atmosphere in the middle troposphere (see photo on page 15).

Carbon dioxide levels between 1959 and 2001 are graphed at right and listed on page 68. We see an overall trend toward an increase in carbon dioxide levels.

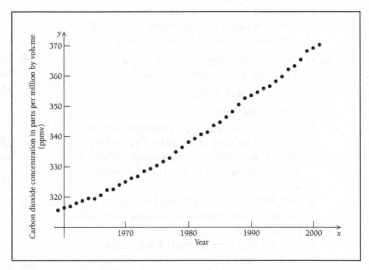

EXERCISES

1. a) Use a grapher that performs linear regression. Consider all the data to find a linear function that fits the data.

b) Graph the linear function.

c) Use the function to estimate the carbon dioxide concentration in January 1990 and in January 2000. Do these estimates appear reasonable?

Year	Carbon dioxide concentration in January (in parts per million by volume)	Year	Carbon dioxide concentration in January (in parts per million by volume)	Year	Carbon dioxide concentration in January (in parts per million by volume)
0. 1959	315.58	15. 1974	329.35	30. 1989	352.76
1. 1960	316.43	16. 1975	330.40	31. 1990	353.66
2. 1961	316.89	17. 1976	331.74	32. 1991	354.72
3. 1962	317.94	18. 1977	332.92	33. 1992	355.98
4. 1963	318.74	19. 1978	334.97	34. 1993	356.70
5. 1964	319.57	20. 1979	336.23	35. 1994	358.36
6. 1965	319.44	21. 1980	338.01	36. 1995	359.96
7. 1966	320.62	22. 1981	339.23	37. 1996	362.05
8. 1967	322.33	23. 1982	340.75	38. 1997	363.18
9. 1668	322.57	24. 1983	341.37	39. 1998	365.32
10. 1969	324.00	25. 1984	343.70	40. 1999	368.15
11. 1970	325.06	26. 1985	344.97	41. 2000	369.08
12. 1971	326.17	27. 1986	346.29	42. 2001	370.17
13. 1972	326.77	28. 1987	348.02		
14. 1973	328.54	29. 1988	350.43		

d) Use the function to predict the carbon dioxide concentration in January 2010 and January 2050.

e) Use the function to predict when the carbon dioxide concentration will reach 500 parts per million (by volume).

2. a) Use quadratic regression to find a quadratic polynomial function

$$y = ax^2 + bx + c$$

that fits the data.

b) Graph the quadratic polynomial function.

c) Use the function to estimate the carbon dioxide concentration in January 1990 and in January 2000. Do these estimates appear reasonable?

d) Use the function to predict the carbon dioxide concentration in January 2010 and January 2050.

e) Use the function to predict when the carbon dioxide concentration will reach 500 parts per million (by volume).

3. a) Consider all the data to find a cubic polynomial function

$$y = ax^3 + bx^2 + cx + d$$

that fits the data.

b) Graph the cubic polynomial function.

c) Use the function to estimate the carbon dioxide concentration in January 1990 and in January 2000. Do these estimates appear reasonable?

d) Use the function to predict the carbon dioxide concentration in January 2010 and January 2050.

e) Use the function to predict when the carbon dioxide concentration will reach 500 parts per million (by volume).

4. By examining monthly data, atmospheric carbon dioxide concentrations can be approximated by

$$C(t) = 312.7 + 0.74t + 0.01188t^2 - 0.5407 \sin(2\pi t),$$

where t is the number of years since January 1957.

a) Graph $y = 312.7 + 0.74t + 0.01188t^2$ for $0 \le t \le 50$.

tw b) What does the graph in part (a) represent?

c) Graph $y = -0.5407 \sin(2\pi t)$ for $45 \le t \le 50$.

tw d) What does the graph in part (c) represent?

e) Graph $y = C(t)$ for $45 \le t \le 50$.

tw f) Explain the behavior of the graph in part (e).

tw **5.** Discuss the pros and cons of each type of function for modeling the past and future carbon dioxide concentrations.

2

Differentiation

AN APPLICATION   The median weight w of a boy whose age t is between 0 and 36 mo can be approximated by the function

$$w(t) = 0.000758t^3 - 0.0596t^2 + 1.82t + 8.15,$$

where t is measured in months and w is measured in pounds. Use this function to approximate the rate of change of the boy's weight with respect to time on his first birthday.

This problem appears as Exercise 8 in Exercise Set 2.6.

INTRODUCTION   *With this chapter, we begin our study of calculus. The first concepts we consider are those of limits and continuity. Then we apply those concepts to establishing the first of the two main building blocks of calculus: differentiation.*

Differentiation is a process that takes a formula for a function and derives a formula for another function, called a derivative, *that allows us to find the slope of the tangent line to a curve at a point. We also find that a derivative can represent an instantaneous rate of change. Throughout the chapter, we will learn various techniques for finding derivatives.*

2.1 Limits and Continuity: Numerically and Graphically

OBJECTIVES

■ Find limits of functions, if they exist, using numerical or graphical methods.

■ Determine the continuity of a function from its graph.

■ Determine continuity of a function at a point.

In this section, we give an intuitive (meaning "based on prior and present experience") treatment of two important concepts: *limits* and *continuity*.

Limits

Kinetic energy is the energy of motion. The kinetic energy of molecules is measured by temperature. If all the kinetic energy is removed from the molecules of a lab sample, then the temperature of that sample is defined to be absolute zero. Absolute zero is equivalent to $-273°C$ (Celsius), or $0°K$ (Kelvin). It is impossible for scientists to remove all the kinetic energy from a sample. Nevertheless, they can attain temperatures very close to absolute zero by repeatedly removing enough energy to reduce the temperature by half, as measured in degrees Kelvin. So a sample's temperature could be reduced from $6°K$, to $3°K$, to $1\frac{1}{2}°K$, and so on. Note that the temperature could never be zero, but the temperature could be made closer and closer to absolute zero. We say that the **limit** of the temperature is zero.

One important aspect of the study of calculus is the analysis of how function values, or outputs, change when the inputs change. Basic to this study is the notion of limit. Suppose the inputs get closer and closer to some number. If the corresponding outputs get closer and closer to a number, then that number is called a *limit*.

For example, scientists extrapolate properties of matter at absolute zero by observing properties of matter as the temperature gets close to absolute zero. They have observed that as the temperature of a wire gets very close to absolute zero, the resistance of the wire gets close to zero. That is, the limit of the resistance is 0 as the temperature gets closer and closer to absolute zero.

Consider the function f given by

$$f(x) = 2x + 3.$$

Suppose we select input numbers x closer and closer to the number 4, and look at the output numbers $2x + 3$. Study the input–output table and the graph that follow.

In the table and the graph, we see that as input numbers approach 4, but do not equal 4, from the left, output numbers approach 11.

As input numbers approach 4, but do not equal 4, from the right, output numbers approach 11. Thus we say:

As *x approaches* 4 from either side, $2x + 3$ *approaches* 11.

An arrow, $\rightarrow$, is often used for the wording "approaches from either side." Thus the statement above can be written:

As $x \rightarrow 4$, $2x + 3 \rightarrow 11$.

Limit Numerically

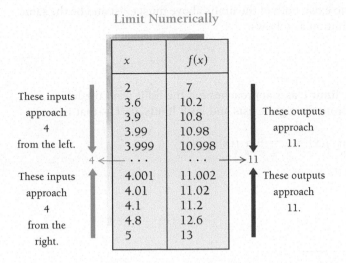

These inputs approach 4 from the left.

These outputs approach 11.

These inputs approach 4 from the right.

These outputs approach 11.

Limit Graphically

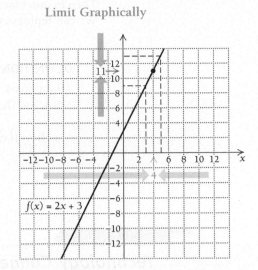

The number 11 is said to be the *limit* of $2x + 3$ as x approaches 4 from either side, but does not equal 4. We can abbreviate this statement as follows:

$$\lim_{x \to 4} (2x + 3) = 11.$$

This is read, "The limit, as x approaches 4 from either side, of $2x + 3$ is 11."

DEFINITION

A function f has the **limit** l as x approaches a from either side, written

$$\lim_{x \to a} f(x) = L,$$

if *all* values $f(x)$ are close to L for values of x that are arbitrarily close, but not equal, to a.

The phrase "from either side" used in the preceding definition is very important. For example, as the time approaches midnight on New Year's Eve 2006 from the left, the limit of the year is 2006 since at any time before midnight, the year is still 2006. But if we think of time going backward and approach the same New Year's Eve from the right, the limit of the year would be 2007 since at any time after midnight, the year would be 2007.

There will be times when we wish to consider only the limit from one side or the other. We use the notation

$$\lim_{x \to a^+} f(x)$$

to indicate the limit from the right and

$$\lim_{x \to a^-} f(x)$$

to indicate the limit from the left.

Then in order for a limit to exist, both of the limits above must exist and be the same. We can rephrase the definition as follows.

DEFINITION

A function f has the **limit** L as x approaches a if the limit from the left exists and the limit from the right exists and both limits are L—that is,

$$\lim_{x \to a^+} f(x) = \lim_{x \to a^-} f(x) = L = \lim_{x \to a} f(x).$$

Technology Connection

Finding Limits Using the TABLE and TRACE Features

Exploratory

Consider the function $f(x) = \dfrac{x^2 - 9}{x - 3}$. Let's use the TABLE feature to complete the following table. Note that the inputs do not have the same increment from one to the next, but do approach 3 from either the left or the right. We set up a table with Indpnt in ASK mode. Then we enter the inputs shown and use the corresponding outputs to complete the table.

x	$f(x)$
2	5
2.9	5.9
2.99	5.99
2.999	5.999
$3 \leftarrow$	$\rightarrow$?
3.001	6.001
3.01	6.01
3.1	6.1
4	7

Now, we set the table in AUTO mode and starting (TblStart) with a number near 3, we make tables for some increments (ΔTbl) like 0.1, 0.01, and so on, and like -0.1, -0.01, and so on, to determine $\lim_{x \to 3} f(x)$.

TABLE SETUP
TblStart = 2.97
ΔTbl = .01 ▮
Indpnt: **Auto** Ask
Depend: **Auto** Ask

X	Y1
2.97	5.97
2.98	5.98
2.99	5.99
3	ERROR
3.01	6.01
3.02	6.02
3.03	6.03

X = 2.97

Note that we get an error when we try to compute $f(3)$. This is because 3 is not in the domain of $f(x) = \dfrac{x^2 - 9}{x - 3}$. When we investigate a limit as x approaches 3, what is important is the value of $f(x)$ when x is near but not equal to 3.

Now, using the TRACE feature with the graph, we move the cursor from left to right so that the x-coordinate approaches 3 from the left. We may need to make several window changes to see what happens. For example, let's try $[2.3, 3.4, 4, 10]$. Next, we move the cursor from right to left so that the x-coordinate approaches 3 from the right. In general, the TRACE feature is not an efficient way to find limits, but it will help you to visualize the limit process in this early stage of your learning.

With what we have observed using the TABLE and TRACE features, let's complete the following:

$$\lim_{x \to 3^+} f(x) = \underline{\quad 6 \quad} \quad \text{and} \quad \lim_{x \to 3^-} f(x) = \underline{\quad 6 \quad}.$$

Thus,

$$\lim_{x \to 3} f(x) = \underline{\quad 6 \quad}.$$

EXERCISES

Use the **TABLE** and **TRACE** features, making your own tables, to find each of the following.

1. $\lim\limits_{x \to 2} \dfrac{x^3 - 8}{x^2 - 4}$

2. $\lim\limits_{x \to -1} \dfrac{x^2 + 3x + 2}{x^3 + 3x^2 - 2x - 4}$

3. $\lim\limits_{x \to 0} \dfrac{\sin x}{x}$

4. $\lim\limits_{x \to 0} \dfrac{1 - \cos x}{x}$

EXAMPLE 1   Consider the function $f(x) = \dfrac{x^3 - 8}{x - 2}$. Graph the function and find the limit $\lim\limits_{x \to 2} f(x)$.

Solution   We check the limit both numerically, with an input–output table, and graphically.

Limit Numerically

$x \to 2^-, (x < 2)$	$f(x)$
1	7
1.9	11.41
1.99	11.94
1.999	11.994
1.9999	11.999

$x \to 2^+, (x > 2)$	$f(x)$
2.0001	12.001
2.001	12.006
2.01	12.06
2.1	12.61
3	19

Limit Graphically

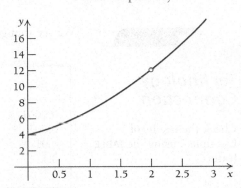

As x approaches 2 from either the left or right, the value of $f(x)$ approaches 12. That is,

$$\lim_{x \to 2} f(x) = 12.$$

Since substituting $x = 2$ into the formula for $f(x)$ entails dividing by zero, $x = 2$ is not in the domain of f. It is important to distinguish between the value of the function and the limit of the function. In this case, the limit of the function exists at $x = 2$, even though the function is not defined at $x = 2$.   ∎

EXAMPLE 2   Consider the function $g(x) = \dfrac{|x|}{x}$. Determine $\lim\limits_{x \to 0} g(x)$.

Solution   We first make a table for $g(x)$. For example, $g(-0.1) = \dfrac{|-0.1|}{-0.1} = -1$ and $g(0.1) = \dfrac{|0.1|}{0.1} = 1$.

Limit Numerically

$x \to 0^-, (x < 0)$	$g(x)$
-1	-1
-0.1	-1
-0.01	-1
-0.001	-1
-0.0001	-1

Limit Graphically

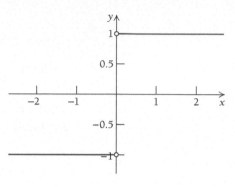

We see that $g(x) = -1$ whenever $x < 0$.

$x \to 0^+, (x > 0)$	$g(x)$
0.0001	1
0.001	1
0.01	1
0.1	1
1	1

Technology Connection

Check the results of Example 2 using the TABLE feature.

We see that $g(x) = 1$ whenever $x > 0$. In this case,

$$\lim_{x \to 0^-} g(x) = -1$$

and $$\lim_{x \to 0^+} g(x) = 1.$$

Since the left- and right-hand limits are different, we say that $\lim_{x \to 0} g(x)$ does not exist.   ■

The "Wall" Method

Let's look at the limits found in Examples 1 and 2 in another way that might make the concept more meaningful. For Example 1, we draw a "wall," or vertical line, at $x = 2$. This is shown in blue in the graph below. We follow the curve from the left with a pencil until we hit the wall and mark the location with an $\times$, assuming that it can be determined. Then we follow the curve from the right until we hit the wall and mark that location with an $\times$. If the locations are the same, we have a limit. Thus, $\lim\limits_{x \to 2} f(x) = 12$.

From Example 2, the graph of $g(x) = \dfrac{|x|}{x}$ with the wall at $x = 0$ is shown below. Since the two $\times$s are at different locations on the wall, $\lim\limits_{x \to 0} g(x)$ does not exist.

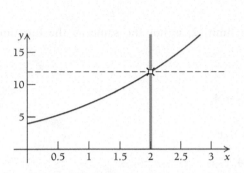
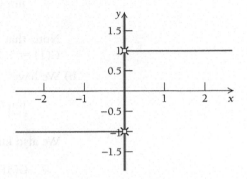

CAUTION   The limit at a number a *does not depend* on the function value at a even if that function value, $f(a)$, exists. That is, whether or not a limit exists at a has *nothing* to do with the function value $f(a)$.

EXAMPLE 3   Consider the function G defined as follows:

$$G(x) = \begin{cases} 5, & \text{for } x = 1, \\ \dfrac{x^2 - 1}{x - 1}, & \text{for } x \neq 1. \end{cases}$$

Graph the function and find each of the following limits, if they exist.

a) $\lim\limits_{x \to 1} G(x)$ b) $\lim\limits_{x \to 3} G(x)$

Solution   The graph follows.

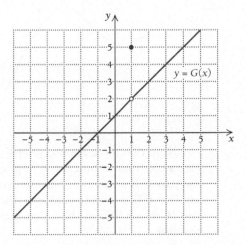

a) As inputs x approach 1 from the left, outputs $G(x)$ approach 2, so the limit from the left is 2. As inputs x approach 1 from the right, outputs $G(x)$ approach 2, so the limit from the right is 2. Since the limit from the left, 2, is the same as the limit from the right, 2, we have

$$\lim_{x \to 1} G(x) = 2.$$

Note that the limit, 2, is not the same as the function value at 1, which is $G(1) = 5$.

b) We have

$$\lim_{x \to 3} G(x) = 4.$$

We also know that

$$G(3) = 4.$$

In this case, the function value and the limit are the same.   ■

After working through Example 3, we might ask, "When can we substitute to find a limit?" The answer lies in the following development of the concept of continuity.

Continuity

When the limit of a function is the same as its function value, it satisfies a condition called **continuity at a point.**

If, at each point on an interval, the value of a function is the same as its limit at that point, then we say the function is **continuous** on that interval.

The following are the graphs of functions that are *continuous* over the whole real line $(-\infty, \infty)$.

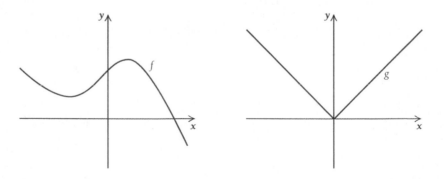

Note that there are no "jumps" or holes in the graph. We can generally recognize a continuous function if its graph can be traced without lifting the pencil from the paper. If a function has one point at which it fails to be continuous, then we say that it is *not continuous* over the whole real line. The graphs of functions F, G, and H, which follow, show that these functions are *not* continuous over the whole real line.

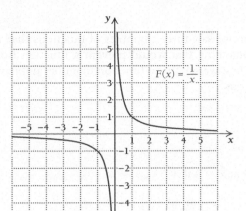

$F(x) = \dfrac{1}{x}$

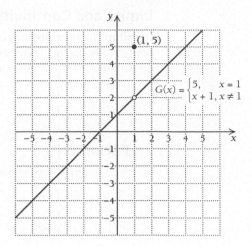

$(1, 5)$

$G(x) = \begin{cases} 5, & x = 1 \\ x + 1, & x \neq 1 \end{cases}$

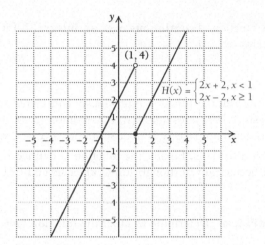

$(1, 4)$

$H(x) = \begin{cases} 2x + 2, & x < 1 \\ 2x - 2, & x \geq 1 \end{cases}$

A continuous curve.

In each case, the graph *cannot* be traced without lifting the pencil from the paper. However, each case represents a different situation. Let's discuss why each case fails to be continuous over the whole real line.

The function F fails to be continuous over the whole real line $(-\infty, \infty)$. Since F is not defined at $x = 0$, the point $x = 0$ is not part of the domain, so $F(0)$ does not exist and there is no point $(0, F(0))$ on the graph. Thus there is no point to trace at $x = 0$. However, F is continuous over the intervals $(-\infty, 0)$ and $(0, \infty)$.

The function G is not continuous over the whole real line since it is not continuous at $x = 1$. As x approaches 1, $G(x)$ seems to approach 2. However, at $x = 1$, $G(x)$ *jumps* up to 5. Thus G is discontinuous at $x = 1$.

The function H is not continuous over the whole real line since it is not continuous at $x = 1$. Let's trace the graph of H starting to the left of $x = 1$. As x approaches 1, $H(x)$ approaches 4. However, at $x = 1$, $H(x)$ *drops* down to 0; and just to the right of $x = 1$, $H(x)$ is close to 0. Since the limit from the left is 4 and the limit from the right is 0, we see that $\lim\limits_{x \to 1} f(x)$ does not exist. Thus $H(x)$ is discontinuous at $x = 1$.

Limits and Continuity

Recall in the previous discussion we saw three examples of functions that were not continuous. Each illustrated something that would prevent a function from being continuous. In the definition below, we say that a function that does not have one of these three problems is continuous.

DEFINITION

A function f is **continuous** at $x = a$ if:

a) $f(a)$ exists,   (The output $f(a)$ exists.)

b) $\lim\limits_{x \to a} f(x)$ exists, and   (The limit as $x \to a$ exists.)

c) $\lim\limits_{x \to a} f(x) = f(a)$.   (The limit is the same as the output.)

A function is **continuous over an interval** I if it is continuous at each point in I.

Note that the function F previously considered on page 77 is not continuous at $x = 0$ since $F(0)$ does not exist, so condition (a) of the definition fails. The function H considered on page 77 is not continuous at $x = 1$ since $\lim\limits_{x \to 1} H(x)$ does not exist, so condition (b) fails. Finally, the function G defined on page 77 is not continuous at $x = 1$ because, even though $\lim\limits_{x \to 1} G(x)$ exists and $G(1)$ exists, the two are not equal.

EXAMPLE 4   Determine whether the function given by

$$f(x) = 2x + 3$$

is continuous at $x = 4$.

Solution   This function is continuous at 4 because:

a) $f(4)$ exists,   ($f(4) = 11$)

b) $\lim\limits_{x \to 4} f(x)$ exists,   and $[\lim\limits_{x \to 4} f(x) = 11$ (as shown on page 71)],

c) $\lim\limits_{x \to 4} f(x) = 11 = f(4)$.

In fact, $f(x) = 2x + 3$ is continuous at any point on the real line.   ∎

Some common examples of functions that are continuous over all the real numbers include $f(x) = c, f(x) = x, f(x) = \sin x$, and $f(x) = \cos x$. It would be impossible to list all continuous functions, so instead we list a few basic properties from which we can determine the continuity of many functions.

CONTINUITY PRINCIPLES

The following continuity principles, which we will not prove, allow us to determine whether a function is continuous.

C1. If $f(x)$ is continuous, then so is $cf(x)$, for any constant c.

C2. If $f(x)$ and $g(x)$ are continuous, then so are $f(x) + g(x)$ and $f(x) - g(x)$.

C3. If $f(x)$ and $g(x)$ are continuous, then so is $g(x) \cdot f(x)$.

C4. If $f(x)$ and $g(x)$ are continuous, then so is $\dfrac{g(x)}{f(x)}$ for any value of x where $f(x) \neq 0$.

C5. If $f(x)$ and $g(x)$ are continuous, then so is the composition $(f \circ g)(x) = f(g(x))$, as long as the composition is defined.

EXAMPLE 5   Provide an argument to show the function is continuous.

a) $f(x) = x^2 - 3x + 2$

b) $s(x) = \sin(2x + 1)$

c) $h(x) = \dfrac{x^2 - 3x + 2}{x - 1}$ for $x \neq 1$

Solution

a) We first note that x and the constant function 2 are continuous. By property C1, $3x$ is continuous and by property C3, $x \cdot x = x^2$ is continuous. Property C2 then implies that $f(x) = x^2 - 3x + 2$ is continuous, since each of the individual functions x^2, $3x$, and 2 are continuous.

b) We know that x is a continuous function. It follows from C1 that $2x$ is continuous. Since $2x$ and 1 are continuous, property C2 says that $2x + 1$ is continuous. If we let $g(x) = 2x + 1$ and $f(x) = \sin x$, then we may write $s(x)$ as $(f \circ g)(x) = f(g(x)) = \sin(2x + 1)$. By C5, $s(x)$ is continuous, since both $f(x)$ and $g(x)$ are continuous.

c) The function $x^2 - 3x + 2$ is continuous by part (a). The denominator, $x - 1$, is continuous by C2, since both x and the constant 1 are continuous. By C4, the quotient $h(x) = \dfrac{x^2 - 3x + 2}{x - 1}$ is continuous for all x except where $x - 1 = 0$. That is, $h(x)$ is continuous for all $x \neq 1$.   ∎

In a similar fashion, we can show that any polynomial is continuous for all real numbers. Furthermore, we can show that all rational functions, radical functions, and trigonometric functions are continuous over their domains.

Other Limits

There are two special limits that we will need when we compute derivatives in Section 2.5. The first is

$$\lim_{x \to 0} \frac{\sin x}{x}.$$

If the function $f(x) = \dfrac{\sin x}{x}$ were continuous at 0, then it would be simple to compute the limit at 0; we would just substitute $x = 0$. However, since the denominator is x, $f(x)$ is not even defined at $x = 0$. We investigate this limit numerically and graphically.

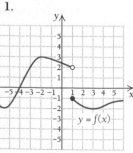

Technology Connection

Cosine Limit

Use the **TABLE** function and the graph to determine

$$\lim_{x \to 0} \frac{\cos x - 1}{x}.$$

Explain why we cannot use the continuity properties to determine this limit.

Limit Numerically

x	$\dfrac{\sin x}{x}$
-0.5	0.95885
-0.4	0.97355
-0.3	0.98507
-0.2	0.99335
-0.1	0.99833
-0.01	0.99998
0.01	0.99998
0.1	0.99833
0.2	0.99335
0.3	0.98507
0.4	0.97355
0.5	0.95885

Limit Graphically

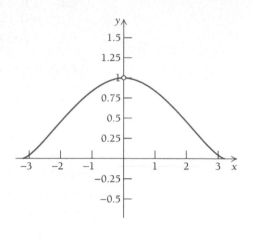

We can see from both the table and the graph that

$$\lim_{x \to 0} \frac{\sin x}{x} = 1.$$

The other trigonometric limit needed in Section 2.5 is

$$\lim_{x \to 0} \frac{\cos x - 1}{x} = 0.$$

This limit is investigated in the Technology Connection located in the margin, and it is investigated algebraically in Section 2.2.

Exercise Set 2.1

Determine whether each of the following is continuous.

1.

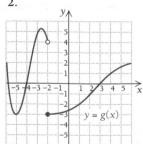

$y = f(x)$

2.

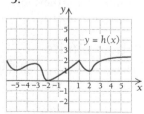

$y = g(x)$

3.

$y = h(x)$

4.

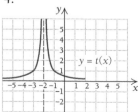

$y = t(x)$

Use the graphs and functions in Exercises 1–4 to answer each of the following.

5. a) Find $\lim\limits_{x \to 1^+} f(x)$, $\lim\limits_{x \to 1^-} f(x)$, and $\lim\limits_{x \to 1} f(x)$.

   b) Find $f(1)$.
   c) Is f continuous at $x = 1$?
   d) Find $\lim\limits_{x \to -2} f(x)$.
   e) Find $f(-2)$.
   f) Is f continuous at $x = -2$?

6. a) Find $\lim\limits_{x \to 1^+} g(x)$, $\lim\limits_{x \to 1^-} g(x)$, and $\lim\limits_{x \to 1} g(x)$.

   b) Find $g(1)$.
   c) Is g continuous at $x = 1$?
   d) Find $\lim\limits_{x \to -2} g(x)$.
   e) Find $g(-2)$.
   f) Is g continuous at $x = -2$?

7. a) Find $\lim\limits_{x \to 1} h(x)$.

   b) Find $h(1)$.
   c) Is h continuous at $x = 1$?
   d) Find $\lim\limits_{x \to -2} h(x)$.
   e) Find $h(-2)$.
   f) Is h continuous at $x = -2$?

8. a) Find $\lim\limits_{x \to 1} t(x)$.

   b) Find $t(1)$.
   c) Is t continuous at $x = 1$?
   d) Find $\lim\limits_{x \to -2} t(x)$.
   e) Find $t(-2)$.
   f) Is t continuous at $x = -2$?

In Exercises 9–12, use the graphs to find the limits and answer the related questions.

9. Consider $f(x) = \begin{cases} 4 - x, & \text{for } x \neq 1, \\ 2, & \text{for } x = 1. \end{cases}$

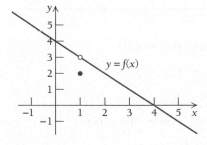

   a) Find $\lim\limits_{x \to 1^+} f(x)$.

   b) Find $\lim\limits_{x \to 1^-} f(x)$.

   c) Find $\lim\limits_{x \to 1} f(x)$.

d) Find $f(1)$.
e) Is f continuous at $x = 1$?
f) Is f continuous at $x = 2$?

10. Consider $f(x) = \begin{cases} 4 - x^2, & \text{for } x \neq -2, \\ 3, & \text{for } x = -2. \end{cases}$

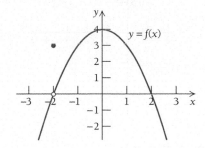

   a) Find $\lim\limits_{x \to -2^+} f(x)$.

   b) Find $\lim\limits_{x \to -2^-} f(x)$.

   c) Find $\lim\limits_{x \to -2} f(x)$.

   d) Find $f(-2)$.
   e) Is f continuous at $x = -2$?
   f) Is f continuous at $x = 1$?

11. Refer to the graph of f below to determine whether each statement is true or false.

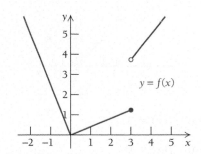

   a) $\lim\limits_{x \to 0^+} f(x) = f(0)$

   b) $\lim\limits_{x \to 0^-} f(x) = f(0)$

   c) $\lim\limits_{x \to 0^+} f(x) = \lim\limits_{x \to 0^-} f(x)$

   d) $\lim\limits_{x \to 3^+} f(x) = \lim\limits_{x \to 3^-} f(x)$

   e) $\lim\limits_{x \to 0} f(x)$ exists.

   f) $\lim\limits_{x \to 3} f(x)$ exists.

   g) f is continuous at $x = 0$.

   h) f is continuous at $x = 3$.

12. Refer to the graph of g below to determine whether each statement is true or false.

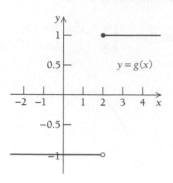

a) $\lim\limits_{x \to 2^+} g(x) = g(2)$

b) $\lim\limits_{x \to 2^-} g(x) = g(2)$

c) $\lim\limits_{x \to 2^+} g(x) = \lim\limits_{x \to 2^-} g(x)$

d) $\lim\limits_{x \to 2} g(x)$ exists.

e) g is continuous at $x = 2$.

13. Refer to the graph of f below to determine whether each statement is true or false.

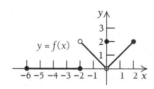

a) $\lim\limits_{x \to -2^+} f(x) = 1$

b) $\lim\limits_{x \to -2^-} f(x) = 0$

c) $\lim\limits_{x \to -2^-} f(x) = \lim\limits_{x \to -2^+} f(x)$

d) $\lim\limits_{x \to -2} f(x)$ exists.

e) $\lim\limits_{x \to -2} f(x) = 2$

f) $\lim\limits_{x \to 0} f(x) = 0$

g) $f(0) = 2$

h) f is continuous at $x = -2$.

i) f is continuous at $x = 0$.

j) f is continuous at $x = -1$.

14. Refer to the graph of f below to determine whether each statement is true or false.

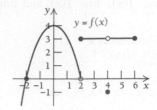

a) $\lim\limits_{x \to 2^-} f(x) = 3$

b) $\lim\limits_{x \to 2^+} f(x) = 0$

c) $\lim\limits_{x \to 2^-} f(x) = \lim\limits_{x \to 2^+} f(x)$

d) $\lim\limits_{x \to 2} f(x)$ exists.

e) $\lim\limits_{x \to 4} f(x)$ exists.

f) $\lim\limits_{x \to 4} f(x) = f(4)$

g) f is continuous at $x = 4$.

h) f is continuous at $x = 0$.

i) $\lim\limits_{x \to 3} f(x) = \lim\limits_{x \to 5} f(x)$

j) f is continuous at $x = 2$.

15. Refer to the graph of f below to determine whether each statement is true or false.

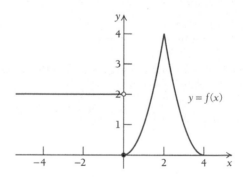

a) $\lim\limits_{x \to 0^+} f(x) = f(0)$

b) $\lim\limits_{x \to 0^-} f(x) = f(0)$

c) $\lim\limits_{x \to 0^+} f(x) = \lim\limits_{x \to 0^-} f(x)$

d) $\lim\limits_{x \to 2^+} f(x) = \lim\limits_{x \to 2^-} f(x)$

e) $\lim\limits_{x \to 0} f(x)$ exists.

f) $\lim\limits_{x \to 2} f(x)$ exists.

g) f is continuous at $x = 0$.

h) f is continuous at $x = 2$.

16. Refer to the graph of g below to determine whether each statement is true or false.

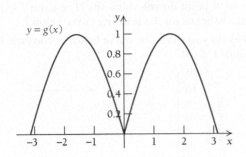

$y = g(x)$

a) $\lim_{x \to 0^+} g(x) = g(0)$

b) $\lim_{x \to 0^-} g(x) = g(0)$

c) $\lim_{x \to 0^+} g(x) = \lim_{x \to 0^-} g(x)$

d) $\lim_{x \to 0} g(x)$ exists.

e) g is continuous at $x = 0$.

APPLICATIONS

The Postage Function. Postal rates are $0.37 for the first ounce and $0.23 for each additional ounce (or fraction thereof). Formally speaking, if x is the weight of a letter in ounces, then p(x) is the cost of mailing the letter, where

$$p(x) = \$0.37, \qquad \text{if } 0 < x \le 1,$$
$$p(x) = \$0.60, \qquad \text{if } 1 < x \le 2,$$
$$p(x) = \$0.83, \qquad \text{if } 2 < x \le 3,$$

and so on, up to 13 oz (at which point postal cost also depends on distance). The graph of p is shown below.

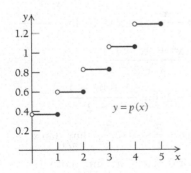

$y = p(x)$

17. Is p continuous at 1? at 1.5? at 2? at 2.01?

18. Is p continuous at 2.99? at 3? at 3.04? at 4?

Using the graph of the postage function, find each of the following limits, if it exists.

19. $\lim_{x \to 1^-} p(x), \lim_{x \to 1^+} p(x), \lim_{x \to 1} p(x)$

20. $\lim_{x \to 2^-} p(x), \lim_{x \to 2^+} p(x), \lim_{x \to 2} p(x)$

21. $\lim_{x \to 2.6^-} p(x), \lim_{x \to 2.6^+} p(x), \lim_{x \to 2.6} p(x)$

22. $\lim_{x \to 3} p(x)$

23. $\lim_{x \to 3.4} p(x)$

Taxicab Fares. In New York City, taxicabs charge passengers $2.00 for entering a cab and then $0.30 for each one-fifth of a mile (or fraction thereof) that the cab travels. (There are additional charges for slow traffic and idle times, but these are not considered in this problem.) Formally speaking, if x is the distance traveled in miles, then C(x) is the cost of the taxi fare, where

$$C(x) = \$2.00, \qquad \text{if } x = 0,$$
$$C(x) = \$2.30, \qquad \text{if } 0 < x < 0.2,$$
$$C(x) = \$2.60, \qquad \text{if } 0.2 \le x < 0.4,$$
$$C(x) = \$2.90, \qquad \text{if } 0.4 \le x < 0.6,$$

and so on. The graph of C is shown below.

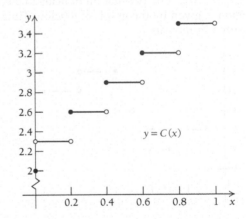

$y = C(x)$

24. Is C continuous at 0.1? at 0.2? at 0.25? at 0.267?

25. Is C continuous at 2.3? at 2.5? at 2.6? at 3.0?

Using the graph of the taxicab fare function, find each of the following limits, if it exists.

26. $\lim_{x \to 1/4^-} C(x), \lim_{x \to 1/4^+} C(x), \lim_{x \to 1/4} C(x)$

27. $\lim_{x \to 0.2^-} C(x), \lim_{x \to 0.2^+} C(x), \lim_{x \to 0.2} C(x)$

28. $\lim_{x \to 0.6^-} C(x), \lim_{x \to 0.6^+} C(x), \lim_{x \to 0.6} C(x)$

29. $\lim_{x \to 0.5} C(x)$

30. $\lim_{x \to 0.4} C(x)$

Population Growth. In a certain habitat, the deer population as a function of time (measured in years) is given in the graph of p below.

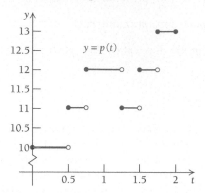

31. Identify each point where the population function is discontinuous.

32. At each point where the function is not continuous, identify an event that might have occurred in the population to cause the discontinuity.

33. Find $\lim\limits_{t \to 1.5^+} p(t)$.

34. Find $\lim\limits_{t \to 1.5^-} p(t)$.

Population Growth. The population of bears in a certain region is given by the graph of p below. Time t is measured in months.

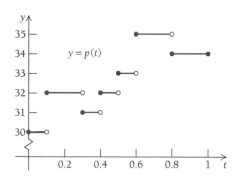

35. Identify each point where the population function is discontinuous.

36. At each point where the function is not continuous, identify an event that might have occurred in the population to cause the discontinuity.

37. Find $\lim\limits_{t \to 0.6^+} p(t)$.

38. Find $\lim\limits_{t \to 0.6^-} p(t)$.

A Learning Curve. In psychology one often takes a certain amount of time t to learn a task. Suppose that the goal is to do a task perfectly and that you are practicing the ability to master it. After a certain time period, what

is known to psychologists as an "I've got it!" experience occurs, and you are able to perform the task perfectly.

tw **39.** At what point do you think the "I've got it!" experience happens on the learning curve below?

tw **40.** Why do you think the curve below is constant for inputs $t \geq 20$?

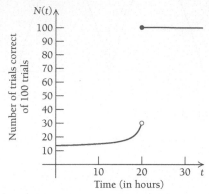

Using the graph above, find each of the following limits, if it exists.

41. $\lim\limits_{t \to 20^+} N(t), \ \lim\limits_{t \to 20^-} N(t), \ \lim\limits_{t \to 20} N(t)$

42. $\lim\limits_{t \to 30^-} N(t), \ \lim\limits_{t \to 30^+} N(t), \ \lim\limits_{t \to 30} N(t)$

43. Is N continuous at 20? at 30?

44. Is N continuous at 10? at 26?

SYNTHESIS

tw **45.** Discuss three ways in which a function may not be continuous at a point a. Draw graphs to illustrate your discussion.

Use the Continuity Properties C1–C5 to justify that the function is continuous. Then give the limit using the fact that the function is continuous.

46. $f(x) = x^2 + 5x - 5; \ \lim\limits_{x \to 3} f(x)$

47. $f(x) = 3x^3 + 2x^2 - 9x + 4; \ \lim\limits_{x \to 1} f(x)$

48. $g(x) = \dfrac{x}{x - 1}, \text{ for } x \neq 1; \ \lim\limits_{x \to -1} \dfrac{x}{x - 1}$

49. $g(x) = \dfrac{x^2 + 9x - 7}{x + 2},$

for $x \neq -2; \ \lim\limits_{x \to 1} \dfrac{x^2 + 9x - 7}{x + 2}$

50. $\tan x, \text{ for } -\dfrac{\pi}{2} < x < \dfrac{\pi}{2}; \ \lim\limits_{x \to \pi/4} \tan x$

51. $\cot x, \text{ for } 0 < x < \pi; \ \lim\limits_{x \to \pi/3} \cot x$

52. $\sec x, \text{ for } -\dfrac{\pi}{2} < x < \dfrac{\pi}{2}; \ \lim\limits_{x \to \pi/6} \sec x$

53. $\csc x, \text{ for } 0 < x < \pi; \ \lim\limits_{x \to \pi/4} \csc (x)$

54. $f(x) = \sqrt{x^2 + 2x + 4}$ (*Hint: f* is a composition of functions); $\lim\limits_{x \to 3} f(x)$.

55. $f(x) = \sqrt{\sin x}$, for $0 < x < \pi$; $\lim\limits_{x \to \pi/3} \sqrt{\sin x}$

56. $g(x) = \sin^2 x$; $\lim\limits_{x \to \pi/4} \sin^2 x$

57. $g(x) = \cos(2x + 3\pi)$; $\lim\limits_{x \to \pi/6} \cos(2x + 3\pi)$

 Technology Connection

In Exercises 58–63, use a grapher to determine the limits. Make a table for each and draw the graph.

58. $\lim\limits_{x \to 5} \dfrac{x^2 + 3x - 40}{x^2 + 4x - 45}$

59. $\lim\limits_{x \to -1} \dfrac{x^3 + x^2 - x - 1}{x^3 + 1}$

60. $\lim\limits_{x \to 0} \dfrac{1 - \cos x}{x}$

61. $\lim\limits_{h \to 0} \dfrac{\tan h}{h}$

62. $\lim\limits_{x \to 0} \dfrac{\sin x - x}{x^3}$

63. $\lim\limits_{h \to 0} \dfrac{\csc h - 1}{h}$

2.2 Limits: Algebraically

Using Limit Principles

OBJECTIVE

■ Find limits using algebraic methods.

If a function is continuous at a, we can substitute to find the limit.

EXAMPLE 1 Find $\lim\limits_{x \to 0} \sqrt{x^2 - 3x + 2}$.

Solution Using the Continuity Principles, we have shown that polynomials such as $x^2 - 3x + 2$ are continuous for all values of x. When we restrict x to values for which $x^2 - 3x + 2$ is nonnegative, it follows from Principle C5 that $\sqrt{x^2 - 3x + 2}$ is continuous. Since $x^2 - 3x + 2$ is nonnegative when $x = 0$, we can substitute to find the limit:

$$\lim\limits_{x \to 0} \sqrt{x^2 - 3x + 2} = \sqrt{0^2 - 3 \cdot 0 + 2}$$
$$= \sqrt{2}. \qquad ■$$

EXAMPLE 2 Find $\lim\limits_{x \to 2} \sin(x^2 - 4)$.

Solution By the Continuity Principles, we know that $x^2 - 4$ is continuous. Therefore, the composition $\sin(x^2 - 4)$ is also continuous since the sine function is continuous. To find the limit, we simply substitute $x = 2$.

$$\lim\limits_{x \to 2} \sin(x^2 - 4) = \sin(2^2 - 4) = \sin(0) = 0. \qquad ■$$

Using the fact that many of the usual functions from algebra and trigonometry are continuous, we can compute many limits by simply evaluating the function at the point in question. It is also possible to use Limit Principles to compute limits. These principles can be used when we are uncertain of the continuity of the function.

LIMIT PRINCIPLES

Let c be a constant and suppose that $\lim_{x \to a} f(x) = L$ and $\lim_{x \to a} g(x) = M$.

L1. $\lim_{x \to a} cf(x) = cL$.

L2. $\lim_{x \to a} (f(x) + g(x)) = L + M$ and $\lim_{x \to a} (f(x) - g(x)) = L - M$.
(The limit of a sum or difference is the sum or difference of the limits.)

L3. $\lim_{x \to a} (f(x) \cdot g(x)) = L \cdot M$.
(The limit of a product is the product of the limits.)

L4. $\lim_{x \to a} \dfrac{g(x)}{f(x)} = \dfrac{M}{L}$, provided that $L \neq 0$.
(The limit of a quotient is the quotient of the limits.)

L5. If the function $H(x)$ is continuous at L, then
$\lim_{x \to a} H(f(x)) = H(\lim_{x \to a} f(x)) = H(L)$.

EXAMPLE 3   Find

$$\lim_{x \to 2} \sqrt{2x + 12}.$$

Solution   We first observe that the function $2x + 12$ is continuous, so

$$\lim_{x \to 2} (2x + 12) = 2 \cdot 2 + 12 = 16.$$

We also know that $\sqrt{x}$ is a continuous function for $x > 0$. We use Limit Principle L5 with $f(x) = 2x + 12$ and $H(x) = \sqrt{x}$ to conclude that

$$\lim_{x \to 2} \sqrt{2x + 12} = \sqrt{\lim_{x \to 2} (2x + 12)} = \sqrt{16} = 4. \qquad \blacksquare$$

EXAMPLE 4   Find

$$\lim_{x \to -3} \frac{x^2 - 9}{x + 3}.$$

Solution   Consider the graph of the rational function $r(x) = (x^2 - 9)/(x + 3)$, shown to the right. We can see that the limit is -6. But let's also examine the limit using the Limit Principles.

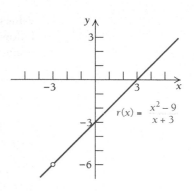

The function $(x^2 - 9)/(x + 3)$ is not continuous at $x = -3$. We use some algebraic simplification and then some Limit Principles:

$$\lim_{x \to -3} \frac{x^2 - 9}{x + 3} = \lim_{x \to -3} \frac{(x + 3)(x - 3)}{x + 3}$$

$$= \lim_{x \to -3} (x - 3) \qquad \text{Simplifying, assuming } x \neq -3$$

$$= \lim_{x \to -3} x - \lim_{x \to -3} 3 \qquad \text{By L2}$$

$$= -3 - 3 = -6. \qquad \blacksquare$$

It is important to keep in mind in Example 4 that the simplification

$$\frac{x^2 - 9}{x + 3} = x - 3, \quad x \neq -3,$$

can be done only for x-values *not* equal to -3. This is one reason why the definition of the limit $\lim_{x \to 3} f(x)$ on page 71 considers inputs x close to but not equal to -3.

In Example 4, the limit could be found after algebraic simplification. When computing limits involving trigonometric functions, it is frequently helpful to use trigonometric identities and to remember that $\lim_{x \to 0} \dfrac{\sin x}{x} = 1$.

Technology Connection

EXERCISES

Find each limit, if it exists, using the TABLE feature.

1. $\lim_{x \to -2} (x^4 - 5x^3 + x^2 - 7)$

2. $\lim_{x \to 1} \sqrt{x^2 + 3x + 4}$

3. $\lim_{x \to 3} \dfrac{x - 3}{x^2 - 9}$

EXAMPLE 5   Compute $\lim_{x \to 0} \dfrac{1 - \cos x}{x}$ algebraically.

Solution   Since the denominator of $\dfrac{1 - \cos x}{x}$ is approaching zero, we cannot use Limit Principle L4. Instead, we rewrite the function algebraically.

$$\frac{1 - \cos x}{x} = \frac{1 - \cos x}{x} \cdot \frac{1 + \cos x}{1 + \cos x} \qquad \text{Multiply by 1.}$$

$$= \frac{1 - \cos^2 x}{x(1 + \cos x)}$$

$$= \frac{\sin^2 x}{x(1 + \cos x)} \qquad \text{Pythagorean Identity}$$

$$= \frac{\sin x}{x} \cdot \frac{\sin x}{1 + \cos x}.$$

We can now compute the limit.

$$\lim_{x \to 0} \frac{1 - \cos x}{x} = \lim_{x \to 0} \left(\frac{\sin x}{x} \cdot \frac{\sin x}{1 + \cos x} \right)$$

$$= \lim_{x \to 0} \frac{\sin x}{x} \cdot \lim_{x \to 0} \frac{\sin x}{1 + \cos x} \qquad \text{L3}$$

$$= 1 \cdot \lim_{x \to 0} \frac{\sin x}{1 + \cos x}$$

$$= \frac{\lim_{x \to 0} \sin x}{\lim_{x \to 0} (1 + \cos x)} \qquad \text{L4}$$

$$= \frac{0}{2}$$

$$= 0.$$

In Section 2.3, we will encounter expressions with two variables, x and h. Our interest is in those limits where x is fixed as a constant and h approaches zero.

EXAMPLE 6   Find $\lim_{h \to 0} (3x^2 + 3xh + h^2)$.

Solution   We treat x as a constant since we are interested only in the way in which the expression varies when h approaches 0. We use the Limit Principles to find that

$$\lim_{h \to 0} (3x^2 + 3xh + h^2) = 3x^2 + 3x(0) + 0^2$$

$$= 3x^2.$$

The student can check any limit about which there is uncertainty by using an input–output table. The following is a table for this limit.

h	$3x^2 + 3xh + h^2$	
1	$3x^2 + 3x \cdot 1 + 1^2,$	or $3x^2 + 3x + 1$
0.8	$3x^2 + 3x(0.8) + (0.8)^2,$	or $3x^2 + 2.4x + 0.64$
0.5	$3x^2 + 3x(0.5) + (0.5)^2,$	or $3x^2 + 1.5x + 0.25$
0.1	$3x^2 + 3x(0.1) + (0.1)^2,$	or $3x^2 + 0.3x + 0.01$
0.01	$3x^2 + 3x(0.01) + (0.01)^2,$	or $3x^2 + 0.03x + 0.0001$
0.001	$3x^2 + 3x(0.001) + (0.001)^2,$	or $3x^2 + 0.003x + 0.000001$

From the pattern in the table, it appears that

$$\lim_{h \to 0} (3x^2 + 3xh + h^2) = 3x^2.$$ ■

Denominator Approaching Zero

In Examples 4 and 5, we computed limits where the denominator was approaching 0. We cannot rely on Limit Principle L4 directly to compute these limits; first, we must simplify algebraically. In Example 4, we were able to factor and cancel $x + 3$, allowing us to compute the limit. We now investigate what happens if the denominator does not cancel.

EXAMPLE 7   Compute the limit

$$\lim_{x \to 4} \frac{x + 3}{x - 4}.$$

Solution   As x approaches 4, the numerator approaches $4 + 3 = 7$, and the denominator approaches $4 - 4 = 0$. Therefore if x is very close to 0, the value of the function $\frac{x + 3}{x - 4}$ is a number close to 7 divided by a number close to zero. The table gives input and output values for this function.

x	$\dfrac{x+3}{x-4}$
3.9	−69
3.99	−699
3.999	−6999
4.001	7001
4.01	701
4.1	71

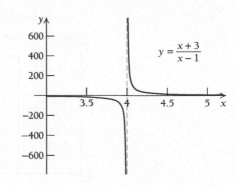

We can see that as x approaches 4 from either the left or the right, the value of $\dfrac{x+3}{x-4}$ does not approach any fixed value. Therefore $\lim\limits_{x\to4}\dfrac{x+3}{x-4}$ does not exist. ■

Based on this example, we give one more Limit Principle.

L6. If $\lim\limits_{x\to a} f(x) \neq 0$ and $\lim\limits_{x\to a} g(x) = 0$, then $\lim\limits_{x\to a}\dfrac{f(x)}{g(x)}$ does not exist.

EXAMPLE 8   Find

$$\lim_{x\to0} \cot x.$$

Solution   To compute the limit, we use a trigonometric identity to rewrite $\cot x$ in terms of $\sin x$ and $\cos x$.

$$\lim_{x\to0} \cot x = \lim_{x\to0} \frac{\cos x}{\sin x}$$

Since

$$\lim_{x\to0} \cos x = 1 \neq 0$$

and

$$\lim_{x\to0} \sin x = 0,$$

Limit Principle L6 states that $\lim\limits_{x\to0}\dfrac{\cos x}{\sin x} = \lim\limits_{x\to0} \cot x$ does not exist. ■

Summary

The following is a summary of the three methods we can use to determine a limit. Consider

$$\lim_{x\to2} f(x),$$

where

$$f(x) = \frac{x^2 - 4}{x - 2}.$$

Limit Numerically

$x \to 2^-$	$f(x)$
1	3
1.5	3.5
1.8	3.8
1.9	3.9
1.99	3.99
1.999	3.999

$x \to 2^+$	$f(x)$
2.8	4.8
2.2	4.2
2.1	4.1
2.01	4.01
2.001	4.001
2.0001	4.0001

Limit Graphically

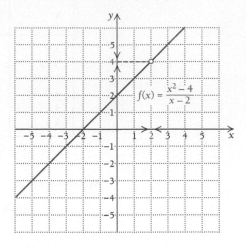

$$f(x) = \frac{x^2 - 4}{x - 2}$$

We see from either the input–output tables on the left or the graph on the right that

$$\lim_{x \to 2^-} f(x) = 4 \quad \text{and} \quad \lim_{x \to 2^+} f(x) = 4,$$

so

$$\lim_{x \to 2} f(x) = 4.$$

Limit Algebraically

The function is not continuous at $x = 2$. We use some algebraic simplification and then some Limit Principles:

$$\lim_{x \to 2} \frac{x^2 - 4}{x - 2} = \lim_{x \to 2} \frac{(x + 2)(x - 2)}{x - 2}$$

$$= \lim_{x \to 2} (x + 2) \qquad \text{Simplifying, assuming } x \neq 2$$

$$= \lim_{x \to 2} x + \lim_{x \to 2} 2 \qquad \text{By L2}$$

$$= 2 + 2$$

$$= 4.$$

We may also say that $\lim_{x \to 2} (x + 2) = 4$ since the polynomial $x + 2$ is continuous.

Exercise Set 2.2

Find the limit using the algebraic method. Verify using the numerical or graphical method.

1. $\lim\limits_{x \to 1} (x^2 - 3)$

2. $\lim\limits_{x \to 1} (x^2 + 4)$

3. $\lim\limits_{x \to 0} \dfrac{3}{x}$

4. $\lim\limits_{x \to 0} \dfrac{-4}{x}$

5. $\lim\limits_{x \to 3} (2x + 5)$

6. $\lim\limits_{x \to 4} (5 - 3x)$

7. $\lim\limits_{x \to -5} \dfrac{x^2 - 25}{x + 5}$

8. $\lim\limits_{x \to -4} \dfrac{x^2 - 16}{x + 4}$

9. $\lim\limits_{x \to -2} \dfrac{5}{x}$

10. $\lim\limits_{x \to -5} \dfrac{-2}{x}$

11. $\lim\limits_{x \to 2} \dfrac{x^2 + x - 6}{x - 2}$

12. $\lim\limits_{x \to -4} \dfrac{x^2 - x - 20}{x + 4}$

Find the limit. Use the algebraic method.

13. $\lim\limits_{x \to 5} \sqrt[3]{x^2 - 17}$

14. $\lim\limits_{x \to 2} \sqrt{x^2 + 5}$

15. $\lim\limits_{x \to \pi/4} (x + \sin x)$

16. $\lim\limits_{x \to \pi/6} (\cos x + \tan x)$

17. $\lim\limits_{x \to 0} \dfrac{1 + \sin x}{1 - \sin x}$

18. $\lim\limits_{x \to 0} \dfrac{1 + \cos x}{\cos x}$

19. $\lim\limits_{x \to 2} \dfrac{1}{x - 2}$

20. $\lim\limits_{x \to 1} \dfrac{1}{(x - 1)^2}$

21. $\lim\limits_{x \to 2} \dfrac{3x^2 - 4x + 2}{7x^2 - 5x + 3}$

22. $\lim\limits_{x \to -1} \dfrac{4x^2 + 5x - 7}{3x^2 - 2x + 1}$

23. $\lim\limits_{x \to 2} \dfrac{x^2 + x - 6}{x^2 - 4}$

24. $\lim\limits_{x \to 4} \dfrac{x^2 - 16}{x^2 - x - 12}$

25. $\lim\limits_{h \to 0} (6x^2 + 6xh + 2h^2)$

26. $\lim\limits_{h \to 0} (10x + 5h)$

27. $\lim\limits_{h \to 0} \dfrac{-2x - h}{x^2(x + h)^2}$

28. $\lim\limits_{h \to 0} \dfrac{-5}{x(x + h)}$

29. $\lim\limits_{x \to 0} \dfrac{\tan x}{x}$

30. $\lim\limits_{x \to 0} x \csc x$

SYNTHESIS

Find the limit, if it exists.

31. $\lim\limits_{h \to 0} \dfrac{\sin x \sin h}{h}$

32. $\lim\limits_{h \to 0} \dfrac{\sin x \cos h - \sin x}{h}$

33. $\lim\limits_{x \to 0} \dfrac{x^2 + 3x}{x^2 - 2x^4}$

34. $\lim\limits_{x \to 0} \dfrac{x^2 - 2x^4}{x^2 + 3x}$

35. $\lim\limits_{x \to 0^+} \dfrac{x\sqrt{x}}{x + x^2}$

36. $\lim\limits_{x \to 0} \dfrac{x + x^2}{x\sqrt{x}}$

37. $\lim\limits_{x \to 2} \dfrac{x - 2}{x^2 - x - 2}$

38. $\lim\limits_{x \to -1} \dfrac{x^2 - 1}{x + 1}$

39. $\lim\limits_{x \to 3} \dfrac{x^2 - 9}{2x - 6}$

40. $\lim\limits_{x \to -2} \dfrac{3x^2 + 5x - 2}{x^2 - 3x - 10}$

Technology Connection

Further Use of the TABLE Feature. In Section 2.1, we discussed how to use the TABLE feature to find limits. Consider

$$\lim\limits_{x \to 0} \dfrac{\sqrt{1 + x} - 1}{x}.$$

Input–output tables for this function are shown below. The table on the left uses TblStart $= -1$ and ΔTbl $= 0.5$. By using smaller and smaller step values and beginning closer to 0, we can refine the table and obtain a better estimate of the limit. On the right is an input–output table with TblStart $= -0.03$ and ΔTbl $= 0.01$.

x	y		x	y
-1	1		-0.03	0.503807
-0.5	0.585786		-0.02	0.502525
0	ERROR		-0.01	0.501256
0.5	0.449490		0	ERROR
1	0.414214		0.01	0.498756
1.5	0.387426		0.02	0.497525
2	0.366025		0.03	0.496305

It appears that the limit is 0.5. We can verify this by graphing

$$y = \frac{\sqrt{1 + x} - 1}{x}$$

and tracing the curve near $x = 0$, zooming in on that portion of the curve.

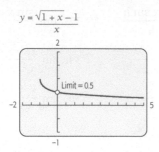

$y = \frac{\sqrt{1+x}-1}{x}$

We see that

$$\lim_{x \to 0} \frac{\sqrt{1 + x} - 1}{x} = 0.5.$$

This can be verified algebraically. (*Hint*: Multiply by 1, using $\dfrac{\sqrt{1 + x} + 1}{\sqrt{1 + x} + 1}$.)

In Exercises 41–48, find the limit. Use the TABLE feature and start with ΔTbl $= 0.1$. Then move to 0.01, 0.001, and 0.0001. When you think you know the limit, graph and use the TRACE feature to verify your assertion. Then try to verify algebraically.

41. $\displaystyle\lim_{a \to -2} \frac{a^2 - 4}{\sqrt{a^2 + 5} - 3}$

42. $\displaystyle\lim_{x \to 1} \frac{\sqrt{x} - 1}{x - 1}$

43. $\displaystyle\lim_{x \to 0} \frac{\sqrt{3 - x} - \sqrt{3}}{x}$

44. $\displaystyle\lim_{x \to 0} \frac{\sqrt{4 + x} - \sqrt{4 - x}}{x}$

45. $\displaystyle\lim_{x \to 1} \frac{x - \sqrt[4]{x}}{x - 1}$

46. $\displaystyle\lim_{x \to 0} \frac{\sqrt{7 + 2x} - \sqrt{7}}{x}$

47. $\displaystyle\lim_{x \to 4} \frac{2 - \sqrt{x}}{4 - x}$

48. $\displaystyle\lim_{x \to 0} \frac{7 - \sqrt{49 - x^2}}{x}$

2.3 Average Rates of Change

OBJECTIVES

- Compute an average rate of change.
- Find a simplified difference quotient.

Let's say that a car travels 110 mi in 2 hr. Its *average rate of change* (*speed*) is 110 mi/2 hr, or 55 mi/hr (55 mph). On the other hand, suppose that you are on the freeway and you begin accelerating. Glancing at the speedometer, you see that at that *instant* your *instantaneous rate of change* is 55 mph. These are two quite different concepts. The first you are probably familiar with. The second involves ideas of limits and calculus. To understand *instantaneous rate of change,* we first use this section to develop a solid understanding of *average rate of change*.

Photosynthesis is the conversion of light energy to chemical energy that is stored in glucose or other organic compounds.[1] In the process, oxygen is produced. The following graph approximates the amount of oxygen a plant produces by photosynthesis. Time 0 is taken to be 8 A.M. The rate of photosynthesis depends on a number of factors, including the amount of light and temperature.

---

[1]N. A. Campbell and J. B. Reece, *Biology*, 6th ed. (Benjamin Cummings, New York 2002).

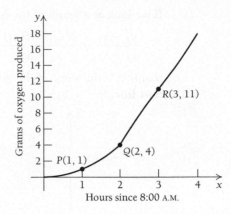

EXAMPLE 1 *Photosynthesis.* How much oxygen was produced from 9 A.M. until 10 A.M.?

Solution At 9 A.M., 1 g of oxygen had been produced. At 10 A.M., 4 g of oxygen had been produced. In the hour from 9 A.M. until 10 A.M., the amount produced was

$$4 \text{ g} - 1 \text{ g} = 3 \text{ g}.$$

Note that 3 is the slope of the line from P to Q. ■

EXAMPLE 2 *Photosynthesis.* What was the average number of grams of oxygen produced per hour between 9 A.M. and 11 A.M.?

Solution We have

$$\frac{11 \text{ g} - 1 \text{ g}}{11 \text{ A.M.} - 9 \text{ A.M.}} = \frac{10 \text{ g}}{2 \text{ hr}}$$

$$= 5 \frac{\text{g}}{\text{hr}} \text{ (grams per hour).}$$ ■

Let's consider a function $y = f(x)$ and two inputs x_1 and x_2. The *change in input,* or the *change in x,* is

$$x_2 - x_1.$$

The *change in output,* or the *change in y,* is

$$y_2 - y_1,$$

where $y_1 = f(x_1)$ and $y_2 = f(x_2)$.

DEFINITION

The **average rate of change of y with respect to x**, as x changes from x_1 to x_2, is the ratio of the change in output to the change in input:

$$\frac{y_2 - y_1}{x_2 - x_1}, \quad \text{where } x_2 \neq x_1.$$

If we look at a graph of the function, we see that

$$\frac{y_2 - y_1}{x_2 - x_1} = \frac{f(x_2) - f(x_1)}{x_2 - x_1}$$

and that this is the slope of the line from $P(x_1, y_1)$ to $Q(x_2, y_2)$. The line $\overleftrightarrow{PQ}$ is called a **secant line.**

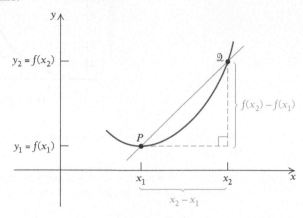

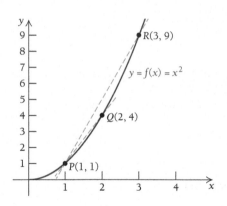

EXAMPLE 3   For $y = f(x) = x^2$, find the average rate of change as:

a) x changes from 1 to 3.

b) x changes from 1 to 2.

c) x changes from 2 to 3.

Solution   The graph at the left is not necessary to the computations, but gives us a look at two of the secant lines whose slopes are being computed.

a) When $x_1 = 1$,

$$y_1 = f(x_1) = f(1) = 1^2 = 1;$$

and when $x_2 = 3$,

$$y_2 = f(x_2) = f(3) = 3^2 = 9.$$

The average rate of change is

$$\frac{y_2 - y_1}{x_2 - x_1} = \frac{f(x_2) - f(x_1)}{x_2 - x_1}$$

$$= \frac{9 - 1}{3 - 1}$$

$$= \frac{8}{2} = 4.$$

b) When $x_1 = 1$,

$$y_1 = f(x_1) = f(1) = 1^2 = 1;$$

and when $x_2 = 2$,

$$y_2 = f(x_2) = f(2) = 2^2 = 4.$$

The average rate of change is

$$\frac{4 - 1}{2 - 1} = \frac{3}{1} = 3.$$

c) When $x_1 = 2$,

$$y_1 = f(x_1) = f(2) = 2^2 = 4;$$

and when $x_2 = 3$,

$$y_2 = f(x_2) = f(3) = 3^2 = 9.$$

The average rate of change is

$$\frac{9 - 4}{3 - 2} = \frac{5}{1} = 5.$$

For a linear function, the average rates of change are the same for any choice of x_1 and x_2; that is, they are equal to the slope m of the line. As we saw in Example 3, a function that is not linear has average rates of change that vary with the choice of x_1 and x_2.

Difference Quotients as Average Rates of Change

We now use a different notation for average rates of change by eliminating the subscripts. Instead of x_1, we will write simply x.

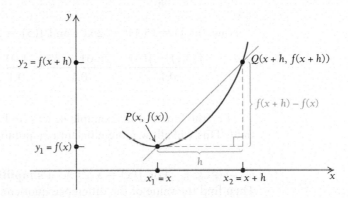

To get from x_1, or x, to x_2, we move a distance h. Thus, $x_2 = x + h$. Then the average rate of change, also called a **difference quotient,** is given by

$$\frac{y_2 - y_1}{x_2 - x_1} = \frac{f(x_2) - f(x_1)}{x_2 - x_1} = \frac{f(x + h) - f(x)}{(x + h) - x} = \frac{f(x + h) - f(x)}{h}.$$

DEFINITION

The average rate of change of f with respect to x is also called the **difference quotient.** It is given by

$$\frac{f(x + h) - f(x)}{h}, \quad \text{where } h \neq 0.$$

The difference quotient is equal to the slope of the line from a point $P(x, f(x))$ to a point $Q(x + h, f(x + h))$.

CAUTION Keep in mind that, in general, $f(x + h) \neq f(x) + h$. You should check this using a function like $f(x) = x^2$.

Technology Connection

EXERCISES

Use the **TABLE** feature to show that $f(x + h) \neq f(x) + h$ for each of the following functions.

1. $f(x) = x^2$; let $x = 6$ and $h = 2$.

2. $f(x) = x^3 - 2x^2 + 4$; let $x = 6$ and $h = 0.1$.

EXAMPLE 4   For $f(x) = x^2$, find the difference quotient when:

a) $x = 5$ and $h = 3$.

b) $x = 5$ and $h = 0.1$.

Solution

a) We substitute $x = 5$ and $h = 3$ into the formula:

$$\frac{f(x + h) - f(x)}{h} = \frac{f(5 + 3) - f(5)}{3} = \frac{f(8) - f(5)}{3}.$$

Now $f(8) = 8^2 = 64$ and $f(5) = 5^2 = 25$, and we have

$$\frac{f(8) - f(5)}{3} = \frac{64 - 25}{3} = \frac{39}{3} = 13.$$

The difference quotient is 13. It is also the slope of the line from $(5, 25)$ to $(8, 64)$.

b) We substitute $x = 5$ and $h = 0.1$ into the formula:

$$\frac{f(x + h) - f(x)}{h} = \frac{f(5 + 0.1) - f(5)}{0.1} = \frac{f(5.1) - f(5)}{0.1}.$$

Now $f(5.1) = (5.1)^2 = 26.01$ and $f(5) = 25$, and we have

$$\frac{f(5.1) - f(5)}{0.1} = \frac{26.01 - 25}{0.1} = \frac{1.01}{0.1} = 10.1.$$ ◼

For the function in Example 4, let's find a general form of the difference quotient. This will allow more efficient computations.

EXAMPLE 5   For $f(x) = x^2$, find a **simplified** form of the **difference quotient.** Then find the value of the difference quotient when $x = 5$ and $h = 0.1$.

Solution   We have

$$f(x) = x^2,$$

so

$$f(x + h) = (x + h)^2 = x^2 + 2xh + h^2.$$

Then

$$f(x + h) - f(x) = (x^2 + 2xh + h^2) - x^2 = 2xh + h^2.$$

Thus,

$$\frac{f(x + h) - f(x)}{h} = \frac{2xh + h^2}{h} = \frac{h(2x + h)}{h} = 2x + h, \quad h \neq 0.$$

It is important to note that a difference quotient is defined only when $h \neq 0$. The simplification above is valid only for nonzero values of h.

When $x = 5$ and $h = 0.1$,

$$\frac{f(x + h) - f(x)}{h} = 2x + h = 2 \cdot 5 + 0.1 = 10 + 0.1 = 10.1.$$ ◼

EXAMPLE 6   For $f(x) = x^3$, find a simplified form of the difference quotient.

Solution   Now $f(x) = x^3$, so

$$f(x + h) = (x + h)^3$$
$$= x^3 + 3x^2h + 3xh^2 + h^3.$$

(This is shown in Appendix A.) Then

$$f(x + h) - f(x) = (x^3 + 3x^2h + 3xh^2 + h^3) - x^3$$
$$= 3x^2h + 3xh^2 + h^3.$$

Thus,

$$\frac{f(x + h) - f(x)}{h} = \frac{3x^2h + 3xh^2 + h^3}{h}$$
$$= \frac{h(3x^2 + 3xh + h^2)}{h}$$
$$= 3x^2 + 3xh + h^2, \quad h \neq 0.$$

Again, this is true *only* for $h \neq 0$.   ■

EXAMPLE 7   For $f(x) = 3/x$, find a simplified form of the difference quotient.

Solution   Now

$$f(x) = \frac{3}{x},$$

so

$$f(x + h) = \frac{3}{x + h}.$$

Then

$$f(x + h) - f(x) = \frac{3}{x + h} - \frac{3}{x}$$

$$= \frac{3}{x + h} \cdot \frac{x}{x} - \frac{3}{x} \cdot \frac{x + h}{x + h}$$

Here we are multiplying by 1 to get a common denominator.

$$= \frac{3x - 3(x + h)}{x(x + h)}$$
$$= \frac{3x - 3x - 3h}{x(x + h)}$$
$$= \frac{-3h}{x(x + h)}.$$

Thus,

$$\frac{f(x + h) - f(x)}{h} = \frac{\dfrac{-3h}{x(x + h)}}{h}$$

$$= \frac{-3h}{x(x + h)} \cdot \frac{1}{h} = \frac{-3}{x(x + h)}, \quad h \neq 0.$$

This is true *only* for $h \neq 0$.   ■

Exercise Set 2.3

For the functions in each of Exercises 1–12, (a) find a simplified form of the difference quotient and (b) complete the following table.

x	h	$\dfrac{f(x + h) - f(x)}{h}$
4	2	
4	1	
4	0.1	
4	0.01	

1. $f(x) = 7x^2$

2. $f(x) = 5x^2$

3. $f(x) = -7x^2$

4. $f(x) = -5x^2$

5. $f(x) = 7x^3$

6. $f(x) = 5x^3$

7. $f(x) = \dfrac{5}{x}$

8. $f(x) = \dfrac{4}{x}$

9. $f(x) = -2x + 5$

10. $f(x) = 2x + 3$

11. $f(x) = x^2 - x$

12. $f(x) = x^2 + x$

APPLICATIONS

13. *Growth of a Baby.* The median weight of boys is given in the graph below. Use the graph to estimate:[2]

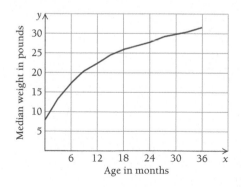

Median weight in pounds vs. Age in months

a) The average growth rate of a typical boy during his first year of life. (Your answer should be in pounds per month.)

b) The average growth rate of a typical boy during his second year of life.

c) The average growth rate of a typical boy during his third year of life.

d) The average growth rate of a typical boy during his first 3 yr of life.

e) When does the graph indicate that a boy's growth rate is greatest during his first 3 yr of life?

14. *Growth of a Baby.* Use the graph in Exercise 13 to estimate:

a) The average growth rate of a typical boy during his first 9 mo of life. (Your answer should be in pounds per month.)

b) The average growth rate of a typical boy during his first 6 mo of life.

c) The average growth rate of a typical boy during his first 3 mo of life.

tw d) Based on your answers in parts (a)–(c) and the graph, estimate what the average growth rate of a typical boy should be the first few weeks of his life.

---

[2]Centers for Disease Control. Developed by the National Center for Health Statistics in collaboration with the National Center for Chronic Disease Prevention and Health Promotion (2000).

15. *Growth of a Baby.* Use the following graph to estimate:

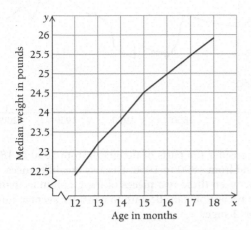

a) The average growth rate of a typical boy between ages 12 mo and 18 mo. (Your answer should be in pounds per month.)

b) The average growth rate of a typical boy between ages 12 mo and 14 mo.

c) The average growth rate of a typical boy between ages 12 mo and 13 mo.

tw d) Based on your answers in parts (a)–(c) and the graph, estimate the average growth rate of a typical boy when he is 12 mo old.

16. *Temperature During an Illness.* The temperature T, in degrees Fahrenheit, of a patient during an illness is shown in the following graph, where t is the time, in days.

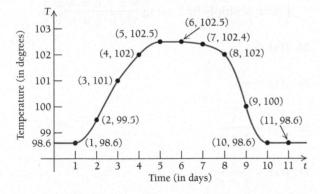

a) Find the average rate of change of T as t changes from 1 to 10. Using this rate of change, would you know that the person was sick?

b) Find the rate of change of T with respect to t, as t changes from 1 to 2; from 2 to 3; from 3 to 4; from 4 to 5; from 5 to 6; from 6 to 7; from 7 to 8; from 8 to 9; from 9 to 10; from 10 to 11.

c) When do you think the temperature began to rise? reached its peak? began to subside? was back to normal?

tw d) Explain your answers to part (c).

17. *Memory.* The total number of words $M(t)$ that a person can memorize in time t, in minutes, is shown in the following graph.

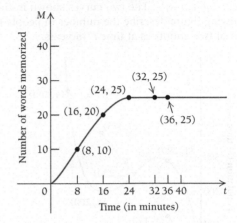

a) Find the average rate of change of M as t changes from 0 to 8; from 8 to 16; from 16 to 24; from 24 to 32; from 32 to 36.

tw b) Why do the average rates of change become 0 after 24 min?

18. *Average Velocity.* A car is at a distance s, in miles, from its starting point in t hours, given by

$$s(t) = 10t^2.$$

a) Find $s(2)$ and $s(5)$.

b) Find $s(5) - s(2)$. What does this represent?

c) Find the average rate of change of distance with respect to time as t changes from $t_1 = 2$ to $t_2 = 5$. This is known as **average velocity,** or **speed.**

19. *Average Velocity.* An object is dropped from a certain height. It is known that it will fall a distance s, in feet, in t seconds, given by

$$s(t) = 16t^2.$$

a) How far will the object fall in 3 sec?

b) How far will the object fall in 5 sec?

c) What is the average rate of change of distance with respect to time during the period from 3 to 5 sec? This is also the *average velocity.*

20. *Gas Mileage.* At the beginning of a trip, the odometer on a car reads 30,680 and the car has a full tank of gas. At the end of the trip, the odometer reads 30,970. It takes 15 gal of gas to refill the tank.

a) What is the average rate of change of the number of miles with respect to the number of gallons?

b) What is the average rate of consumption (that is, the rate of change of the number of miles with respect to the number of gallons)?

21. *Population Growth.* The two curves shown in the following figure describe the number of people in each of two countries at time *t*, in years.

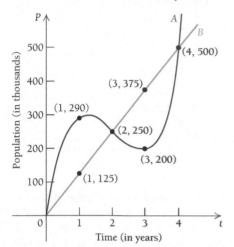

a) Find the average rate of change of each population (the number of people in the population) with respect to time *t* as *t* changes from 0 to 4. This is often called an **average growth rate.**

tw b) If the calculation in part (a) were the only one made, would we detect the fact that the populations were growing differently? Explain.

c) Find the average rates of change of each population as *t* changes from 0 to 1; from 1 to 2; from 2 to 3; from 3 to 4.

tw d) For which population does the statement "the population grew by 125 million each year" convey the least information about what really took place? Explain.

SYNTHESIS

Deer Population. The deer population in California from 1800 to 2000 is approximated in the graph shown below. Use this graph to answer Exercises 22 and 23.[3]

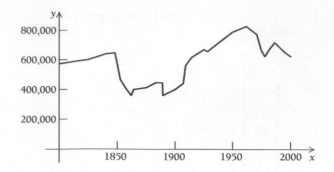

tw **22.** Consider the parts of the graph from 1850 to 1860 and from 1890 to 1960. Discuss the differences between these two pieces of the graph in as many ways as you can. Be sure to consider average rates of change.

tw **23.** Pick out pieces of the graph where the slopes and shapes are similar and pieces where the slopes and shapes are different. Explain the differences and the similarities. Can you identify historical occurrences that correspond to when the graph changes?

Find the simplified difference quotient.

24. $f(x) = mx + b$

25. $f(x) = ax^2 + bx + c$

26. $f(x) = ax^3 + bx^2$

27. $f(x) = \sqrt{x}$

$\left(\text{Hint: Multiply by 1 using } \dfrac{\sqrt{x + h} + \sqrt{x}}{\sqrt{x + h} + \sqrt{x}}. \right)$

28. $f(x) = x^4$

29. $f(x) = \dfrac{1}{x^2}$

30. $f(x) = \dfrac{1}{1 - x}$

31. $f(x) = \dfrac{x}{1 + x}$

32. $f(x) = \sqrt{3 - 2x}$

33. $f(x) = \dfrac{1}{\sqrt{x}}$

34. $f(x) = \dfrac{2x}{x - 1}$

---

[3]California Department of Fish and Game.

2.4 Differentiation Using Limits of Difference Quotients

OBJECTIVES

■ Find derivatives and values of derivatives.

■ Find equations of tangent lines.

We will see in Sections 2.4–2.6 that an instantaneous rate of change is given by the slope of a tangent line to the graph of a function. Here we see how to take the limit of a simplified difference quotient as h approaches 0 to define the slope of a tangent line.

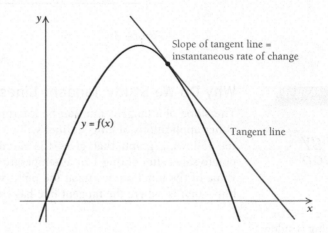

Slope of tangent line = instantaneous rate of change

$y = f(x)$

Tangent line

Tangent Lines

A line tangent to a circle is a line that touches the circle exactly once.

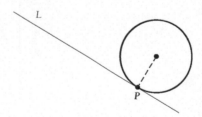

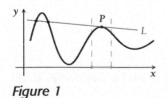

Figure 1

This definition becomes unworkable with other curves. For example, consider the curve shown in Fig. 1. Line L touches the curve at point P but meets the curve at other places as well. It is considered a tangent line, but "touching at one point" cannot be its definition.

Note in Fig. 1 that over a small interval containing P, line L does touch the curve exactly once. This is still not a suitable definition of a *tangent line* because it allows a line like M in Fig. 2 to be a tangent line, which we will not accept.

In calculus, we think of a tangent line as a line that just "skims" the function at some point, regardless of how many times it intersects the function. In Fig. 3, the line T_2 just skims the function, making it a tangent line even though it crosses the function. On the other hand, the line L_1 intersects the function only once but doesn't skim the function. Consequently, L_1 is not a tangent line.

Figure 2

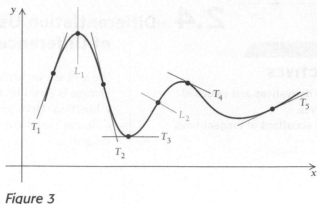

Figure 3

Technology Connection

Exploratory

Graph $y_1 = 3x^5 - 20x^3$ with the viewing window $[-3,3,-80,80]$, with Xscl = 1 and Yscl = 10. Then also graph the lines $y_2 = -7x - 10$, $y_3 = -30x + 13$, and $y_4 = -45x + 28$. Which appears to be tangent to the graph of y_1 at $(1, -17)$? It might be helpful to zoom in near the point $(1, -17)$. Use smaller viewing windows to try to refine your guess.

Why Do We Study Tangent Lines?

The slope of a tangent line can be interpreted as an instantaneous rate of change. Other applications of tangent lines will be discussed in Chapter 3. For now, look at the following graph that gives the number of bacteria in a sample of leftover potato salad after sitting 1 hr at temperature T. Note that the largest (or maximum) value of the function occurs at the point where the graph has a horizontal tangent line—that is, where the tangent line has slope 0.

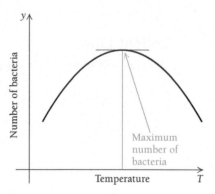

Differentiation Using Limits

We will define *tangent line* in such a way that it makes sense for *any* curve. To do this, we use the notion of limit.

We obtain the line tangent to the curve at point P by considering secant lines through P and neighboring points Q_1, Q_2, and so on. As the points Q approach P, the secant lines approach the tangent line. Each secant line has a slope. The slopes m_1, m_2, m_3, and so on, of the secant lines approach the slope m of the tangent line. In fact, we *define* the **tangent line** as the line that contains the point P and has slope m, where m is the limit of the slopes of the secant lines as the points Q approach P.

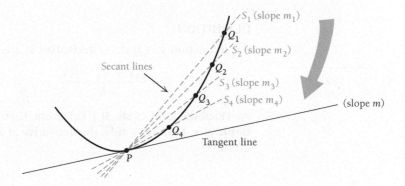

How might we calculate the limit m? Suppose that in Fig. 4 P has coordinates $(x, f(x))$. Then the first coordinate of Q is x plus some number h, or $x + h$. The coordinates of Q are $(x + h, f(x + h))$.

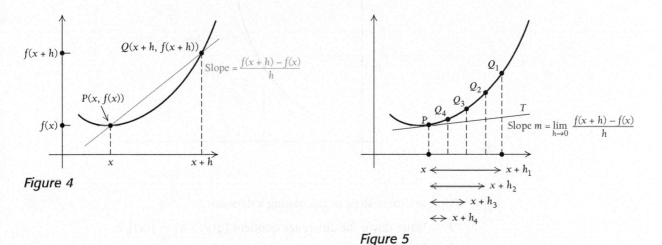

Figure 4

Figure 5

From Section 2.3, we know that the slope of the secant line $\overleftrightarrow{PQ}$ is given by the difference quotient

$$\frac{f(x + h) - f(x)}{h}.$$

Now, as we see in Fig. 5, as the points Q approach P, $x + h$ approaches x. That is, h approaches 0. Thus we have the following.

The slope of the tangent line $= m = \lim_{h \to 0} \dfrac{f(x + h) - f(x)}{h}$.

The formal definition of the *derivative of a function f* can now be given. We will designate the derivative at x as $f'(x)$, rather than $m(x)$.

"Nothing in this world is so
powerful as an idea whose
time has come."
                          Victor Hugo

DEFINITION

For a function $y = f(x)$, its **derivative** at x is the function f' defined by

$$f'(x) = \lim_{h \to 0} \frac{f(x + h) - f(x)}{h},$$

provided the limit exists. If $f'(x)$ exists, then we say that f is **differentiable** at x. In other words, the derivative at x is the slope of the tangent line at x.

This is the basic definition of *differential calculus*.

Let's now calculate some formulas for derivatives. That is, given a formula for a function f, we will be trying to find a formula for f'. This process is called **differentiation.**

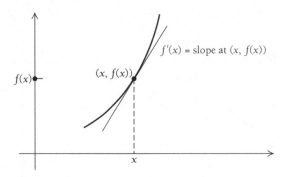

There are three steps in calculating a derivative.

1.   Write down the difference quotient $[f(x + h) - f(x)]/h$.
2.   Simplify the difference quotient.
3.   Find the limit as h approaches 0.

EXAMPLE 1   For $f(x) = 3x - 4$, find $f'(x)$.

Solution   We follow the steps above. Thus:

1. $\dfrac{f(x + h) - f(x)}{h} = \dfrac{[3(x + h) - 4] - (3x - 4)}{h}$;

2. $\dfrac{f(x + h) - f(x)}{h} = \dfrac{3x + 3h - 4 - 3x + 4}{h}$

$$= \frac{3h}{h} = 3, \quad h \neq 0;$$

3. $\lim\limits_{h \to 0} \dfrac{f(x + h) - f(x)}{h} = \lim\limits_{h \to 0} 3 = 3,$

since 3 is a constant.

Thus if $f(x) = 3x - 4$, then $f'(x) = 3$.

A general formula for the derivative of a linear function

$$f(x) = mx + b$$

is $$f'(x) = m.$$

The formula could be verified in a manner similar to that used in Example 1.

Consider the graph of $f(x) = x^2$ that follows. Tangent lines are drawn at various points on the graph. Estimate the slope of each line and complete the table. Can you guess a formula for $f'(x)$?

Lines	x	$f'(x)$
L_1	-1	
L_2	$-\frac{1}{2}$	
L_3	0	
L_4	$\frac{1}{2}$	
L_5	1	

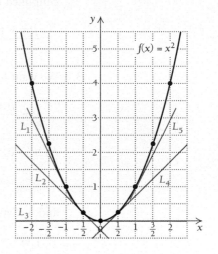

Now let's see if we can find a formula for the derivative of

$$f(x) = x^2.$$

We first find the slope of the tangent line at $x = 4$. That would be $f'(4)$. Then we find the general formula for $f'(x)$.

EXAMPLE 2   For $f(x) = x^2$, find $f'(4)$.

Solution   We have

1. $\dfrac{f(4 + h) - f(4)}{h} = \dfrac{(4 + h)^2 - 4^2}{h};$

2. $\dfrac{f(4 + h) - f(4)}{h} = \dfrac{16 + 8h + h^2 - 16}{h}$

   $= \dfrac{8h + h^2}{h}$

   $= \dfrac{h(8 + h)}{h}$

   $= 8 + h, \quad h \neq 0;$

3. $\displaystyle\lim_{h \to 0} \dfrac{f(4 + h) - f(4)}{h} = \lim_{h \to 0} (8 + h) = 8.$

Thus, $f'(4) = 8.$

EXAMPLE 3   For $f(x) = x^2$, find (the general formula) $f'(x)$.

Solution

1.   We have

$$\frac{f(x + h) - f(x)}{h} = \frac{(x + h)^2 - x^2}{h}.$$

2.   In Example 5 of Section 2.3, we showed how this difference quotient can be simplified to

$$\frac{f(x + h) - f(x)}{h} = 2x + h.$$

3.   We want to find

$$\lim_{h \to 0} \frac{f(x + h) - f(x)}{h} = \lim_{h \to 0} (2x + h).$$

As $h \to 0$, we see that $2x + h \to 2x$. Thus,

$$\lim_{h \to 0} (2x + h) = 2x,$$

and we have

$$f'(x) = 2x.$$ ■

   We can check the result of Example 3 in Example 2. We know that a general formula is $f'(x) = 2x$. Thus, $f'(4) = 2(4) = 8$, as we found in Example 2. This formula also tells us, for example, that at $x = -3$, the curve has a tangent line whose slope is

$$f'(-3) = 2(-3) = -6.$$

We can also say:

- The tangent line to the curve at the point $(-3, 9)$ has slope -6.
- The tangent line to the curve at the point $(4, 16)$ has slope 8.

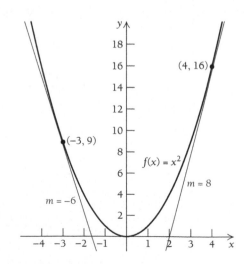

EXAMPLE 4   For $f(x) = x^3$, find $f'(x)$. Then find $f'(-1)$ and $f'(1.5)$.

Solution

1. We have

$$\frac{f(x + h) - f(x)}{h} = \frac{(x + h)^3 - x^3}{h}.$$

2. In Example 6 of Section 2.3, we showed how this difference quotient can be simplified to

$$\frac{f(x + h) - f(x)}{h} = 3x^2 + 3xh + h^2.$$

3. We then have

$$\lim_{h \to 0} \frac{f(x + h) - f(x)}{h} = \lim_{h \to 0} (3x^2 + 3xh + h^2) = 3x^2.$$

(An input–output table for this is shown in Example 6 of Section 2.2.) Thus, for $f(x) = x^3$, we have $f'(x) = 3x^2$. Then

$$f'(-1) = 3(-1)^2 = 3 \quad \text{and} \quad f'(1.5) = 3(1.5)^2 = 6.75.$$

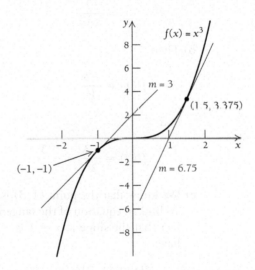

You should know that Examples 1–4 and Example 5, which follows, involve a somewhat lengthy process, but in Section 2.5 we will begin to develop some faster techniques. It is very important in this section, however, to fully understand the concept of a derivative.

EXAMPLE 5   For $f(x) = 3/x$:

a) Find $f'(x)$.

b) Find $f'(1)$ and $f'(2)$.

c) Find an equation of the tangent line to the curve at the point $(1, 3)$.

d) Find an equation of the tangent line to the curve at the point $\left(2, \frac{3}{2}\right)$.

Solution

a) **1.** We have

$$\frac{f(x + h) - f(x)}{h} = \frac{[3/(x + h)] - (3/x)}{h}.$$

2. In Example 7 of Section 2.3, we showed that this difference quotient can be simplified to

$$\frac{f(x + h) - f(x)}{h} = \frac{-3}{x(x + h)}.$$

3. We want to find

$$\lim_{h \to 0} \frac{f(x + h) - f(x)}{h} = \lim_{h \to 0} \frac{-3}{x(x + h)}.$$

As $h \to 0$, $x + h \to x$, so we have

$$f'(x) = \lim_{h \to 0} \frac{-3}{x(x + h)}$$

$$= \frac{-3}{x^2}.$$

b) Then

$$f'(1) = \frac{-3}{1^2} = -3$$

and

$$f'(2) = \frac{-3}{2^2} = -\frac{3}{4}.$$

c) We know that the point $(1, 3)$ is on the graph of the function because $f(1) = 3$. To find an equation of the tangent line to the curve at the point $(1, 3)$, we use the fact that the slope at $x = 1$ is -3, as we found in the preceding work. Now we have

    Point: $(1, 3)$,

    Slope: -3.

We substitute into the point–slope equation (see Section 1.1):

$$y - y_1 = m(x - x_1)$$
$$y - 3 = -3(x - 1)$$
$$y = -3x + 3 + 3$$
$$= -3x + 6.$$

The equation of the tangent line to the curve at $x = 1$ is $y = -3x + 6$ (see the following figure).

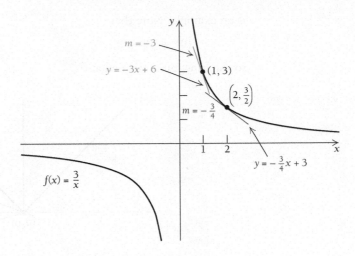

d) To find an equation of the tangent line to the curve at $x = 2$, we use the fact that the slope at $x = 2$ is $-\frac{3}{4}$, as we found in part (b). Now we have

Point: $\left(2, \frac{3}{2}\right)$,

Slope: $-\frac{3}{4}$.

We substitute into the point–slope equation:

$$y - y_1 = m(x - x_1)$$
$$y - \tfrac{3}{2} = -\tfrac{3}{4}(x - 2)$$
$$y = -\tfrac{3}{4}x + \tfrac{3}{2} + \tfrac{3}{2}$$
$$= -\tfrac{3}{4}x + 3.$$

The equation of the tangent line to the curve at $x = 2$ is

$$y = -\tfrac{3}{4}x + 3.$$

Note that because $f(0)$ does not exist, we cannot evaluate the difference quotient

$$\frac{f(0 + h) - f(0)}{h}.$$

Thus, $f'(0)$ does not exist. We say that "f is not differentiable at 0." ∎

Technology Connection

Curves and Tangent Lines

EXERCISES

1. For $f(x) = 3/x$, find $f'(x)$, $f'(-2)$, and $f'\left(-\frac{1}{2}\right)$.

2. Find an equation of the tangent line to the curve $f(x) = 3/x$ at $\left(-2, -\frac{3}{2}\right)$ and an equation of the tangent line to the curve at $\left(-\frac{1}{2}, -6\right)$. Then graph the curve $f(x) = 3/x$ and both tangent lines. Use different viewing windows near the points of tangency.

When a function is not defined at a point, it is not differentiable at that point. Also, if a function is discontinuous at a point, it is not differentiable at that point.

It can happen that a function f is defined and continuous at a point but that its derivative f' is not defined. The function f given by

$$f(x) = |x|$$

is an example. Note that

$$f(0) = |0| = 0,$$

so the function is defined at 0.

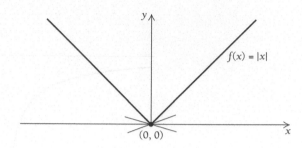

Suppose we try to draw a tangent line at $(0, 0)$. A function like this with a corner (not smooth) would seem to have many tangent lines at $(0, 0)$, and thus many slopes. In a situation like this, the derivative does not exist. Let's try to calculate the derivative to see why.

Since

$$f'(x) = \lim_{h \to 0} \frac{|x + h| - |x|}{h},$$

at $x = 0$ we have

$$f'(0) = \lim_{h \to 0} \frac{|0 + h| - |0|}{h} = \lim_{h \to 0} \frac{|h|}{h}.$$

As we saw in Example 2 of Section 2.1, $\lim_{h \to 0} \dfrac{|h|}{h}$ does not exist.

Graphical Interpretation of Differentiable Functions

If a function has a "corner" or "cusp," it will not have a derivative at that point.

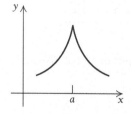
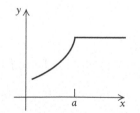
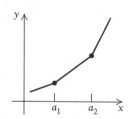

A function may also fail to be differentiable at a point by having a vertical tangent at that point. For example, the function shown below has a vertical tangent at point a. Recall that since the slope of a vertical line is undefined, there is no derivative at such a point.

Technology Connection

Vertical Tangent Lines

Use the graphing window $[-1, 1, -1, 1]$ to graph the function $f(x) = \sqrt[3]{x}$. On the TI-83, the cube root function can be found on the MATH menu. What line appears to be tangent to the graph at the point $(0, 0)$? Graph $f(x) = \sqrt[n]{x^m} = x^{m/n}$ for various values of n and m. Try to guess for which values of $\frac{m}{n}$ the tangent line to the graph is vertical.

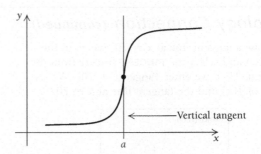

The function $f(x) = |x|$ illustrates the fact that although a function may be continuous at each point in an interval I, it may not be differentiable at each point in I. That is, continuity does not imply differentiability. On the other hand, if we know that a function is differentiable at each point in an interval I, then it is continuous over I. That is, if $f'(a)$ exists, then f is continuous at a. The function $f(x) = x^2$ is an example of a function that is differentiable over the interval $(-\infty, \infty)$ and is thus continuous everywhere. Also, if a function is discontinuous at some point a, then it is not differentiable at a. Thus, when we know that a function is differentiable over an interval, it is *smooth* in the sense that there are no "sharp points," "corners," or "breaks" in the graph.

Technology Connection

Numerical Differentiation and Drawing Tangent Lines

Graphers have the capability of finding slopes of tangent lines, that is, of taking the derivative at a specific x-value. Let's consider the function

$$f(x) = x(100 - x)$$

and find the value of the derivative at $x = 70$.

Calculating the Derivative

Select nDeriv (numerical derivative) from the MATH menu and enter the function, the variable, and the value at which the derivative is to be evaluated. When we enter nDeriv($x(100 - x)$, x, 70), the grapher returns -40.

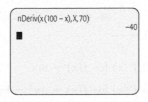

We see that the derivative of the function at $x = 70$ is -40. That is, the slope of the tangent line to the curve at $x = 70$ is -40.

Drawing the Tangent Line

We first graph the function using the viewing window $[-10, 100, -10, 3000]$, with Xscl $= 10$ and Yscl $= 1000$.

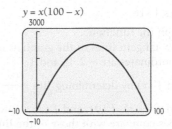

(continued)

Technology Connection (*continued*)

To draw a tangent line at $x = 70$, we go to the home screen and select the **TANGENT** feature from the **DRAW** menu. Then we enter Tangent(Y_1, 70). We see the graph of $f(x)$ and the tangent line at $x = 70$.

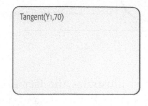

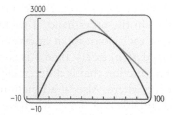

EXERCISES

For each of the following functions, evaluate the derivative at the given point. Then draw the graph and the tangent line.

1. $f(x) = x(100 - x)$;
$x = 20, x = 37, x = 50, x = 90$

2. $f(x) = -\frac{1}{3}x^3 + 6x^2 - 11x - 50$;
$x = -5, x = 0, x = 7, x = 12, x = 15$

3. $f(x) = 6x^2 - x^3$;
$x = -2, x = 0, x = 2, x = 4, x = 6.3$

4. $f(x) = x\sqrt{4 - x^2}$;
$x = -2, x = -1.3, \ x = -0.5, x = 0, x = 1,$
$x = 2$

tw 5. For the function in Exercise 4, try to draw a tangent line at $x = 3$ and estimate the derivative. What goes wrong? Explain.

CAUTION Some calculators will give answers for derivatives even though they do not exist. For example, $f(x) = |x|$ does not have a derivative at $x = 0$, but some calculators will give 0 for the answer. Try entering nDeriv(abs(X), X, 0) to see what your calculator does.

Exercise Set 2.4

In Exercises 1–16:

 a) Graph the function.
 b) Draw tangent lines to the graph at points whose x-coordinates are -2, 0, and 1.
 c) Find $f'(x)$ by determining $\lim\limits_{h \to 0} \dfrac{f(x + h) - f(x)}{h}$.
 d) Find $f'(-2)$, $f'(0)$, and $f'(1)$. How do these slopes compare with those of the lines you drew in part (b)?

1. $f(x) = 5x^2$
2. $f(x) = 7x^2$
3. $f(x) = -5x^2$
4. $f(x) = -7x^2$
5. $f(x) = x^3$
6. $f(x) = -x^3$
7. $f(x) = 2x + 3$
8. $f(x) = -2x + 5$

9. $f(x) = -4x$
10. $f(x) = \frac{1}{2}x$
11. $f(x) = x^2 + x$
12. $f(x) = x^2 - x$
13. $f(x) = 2x^2 + 3x - 2$
14. $f(x) = 5x^2 - 2x + 7$
15. $f(x) = \dfrac{1}{x}$
16. $f(x) = \dfrac{5}{x}$

17. Find $f'(x)$ for $f(x) = mx$.
18. Find $f'(x)$ for $f(x) = ax^2 + bx + c$.

19. Find an equation of the tangent line to the graph of $f(x) = x^2$ at the point $(3, 9)$, at $(-1, 1)$, and at $(10, 100)$. See Example 3.

20. Find an equation of the tangent line to the graph of $f(x) = x^3$ at the point $(-2, -8)$, at $(0, 0)$, and at $(4, 64)$. See Example 4.

21. Find an equation of the tangent line to the graph of $f(x) = 5/x$ at the point $(1, 5)$, at $(-1, -5)$, and at $(100, 0.05)$. See Exercise 16.

22. Find an equation of the tangent line to the graph of $f(x) = 2/x$ at the point $(-1, -2)$, at $(2, 1)$, and at $\left(10, \frac{1}{5}\right)$.

23. Find an equation of the tangent line to the graph of $f(x) = 4 - x^2$ at the point $(-1, 3)$, at $(0, 4)$, and at $(5, -21)$.

24. Find an equation of the tangent line to the graph of $f(x) = x^2 - 2x$ at the point $(-2, 8)$, at $(1, -1)$, and at $(4, 8)$.

List the points in the graph at which each function is not differentiable.

25.

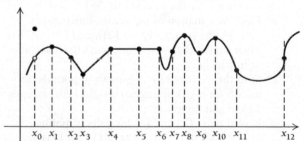

26.

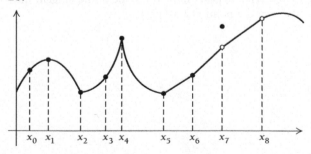

27. *The Postage Function.* Consider the postage function defined on page 83. At what values is the function not differentiable?

28. *The Taxicab Function.* Consider the taxicab function defined on page 83. At what values is the function not differentiable?

29. *Baseball Ticket Prices.* Consider the model for major league baseball average ticket prices in Exercise 35 of Exercise Set 1.2. At what values is the function not differentiable?

tw 30. *Deer Population.* Using the graph below to model the population of deer in California, in which years is the function not differentiable? Explain.

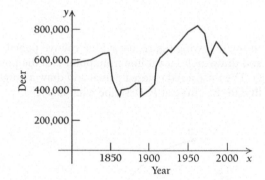

31. *Growth of Whales.* The graph below approximates the weight, in pounds, of a killer whale (*Orcinus orca*).[4] The age is given in months. Find all months where the weight function is not differentiable.

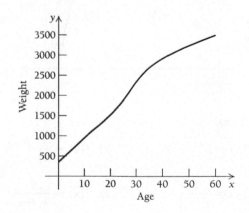

[4]SeaWorld.

SYNTHESIS

tw **32.** Which of the following appear to be tangent lines? Try to explain why or why not.

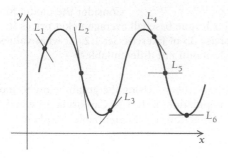

tw **33.** In the following figure, use a blue colored pencil and draw each secant line from point P to the points Q. Then use a red colored pencil and draw a tangent line to the curve at P. Describe what happens.

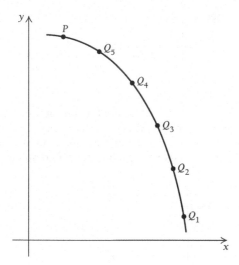

Find $f'(x)$.

34. $f(x) = x^4$ **35.** $f(x) = \dfrac{1}{x^2}$

36. $f(x) = \dfrac{1}{1-x}$ **37.** $f(x) = \dfrac{x}{1+x}$

38. $f(x) = \sqrt{x}$

$$\left(\text{Multiply by 1, using } \frac{\sqrt{x+h}+\sqrt{x}}{\sqrt{x+h}+\sqrt{x}}. \right)$$

39. $f(x) = \dfrac{1}{\sqrt{x}}$ **40.** $f(x) = \dfrac{3x}{x+5}$

41. Consider the function f given by

$$f(x) = \frac{x^2 - 9}{x + 3}.$$

For what values is this function not differentiable?

 Technology Connection

42.–47. Use a grapher to do the numerical differentiation and draw the tangent lines in each of Exercises 19–24.

48. *Growth of a Baby.* The median weight w of a girl whose age t is between 0 and 36 mo can be approximated by the function

$$w(t) = 0.0006t^3 - 0.0484t^2 + 1.61t + 7.6,$$

where t is measured in months and w is measured in pounds. Use this approximation to make the following computations for a girl with median weight.[5]

Note: Some graphers use only the variables x and y, so you may need to change the variables when entering the function.

a) Graph w over the interval $[0, 36]$.
b) Find the equation of the secant line passing through the points $(12, w(12))$, and $(36, w(36))$. Then sketch the secant line using the same axes as in part (a).
c) Find the average rate of growth in pounds per month for a girl of median weight between ages 12 mo and 36 mo.
d) Repeat parts (b) and (c) for pairs of points $(12, w(12))$, and $(24, w(24))$; $(12, w(12))$ and $(18, w(18))$; $(12, w(12))$ and $(15, w(15))$.
e) What appears to be the slope of the tangent line at the point $(12, w(12))$?

---

[5]Centers for Disease Control. Developed by the National Center for Health Statistics in collaboration with the National Center for Chronic Disease Prevention and Health Promotion (2000).

2.5 Differentiation Techniques: Introduction

OBJECTIVES

■ Differentiate using the Power Rule.
■ Differentiate using the
Sum-Difference Rule.
■ Differentiate the sine and cosine
functions.

Leibniz's Notation

When y is a function of x, we will also designate the derivative, $f'(x)$, as*

$$\frac{dy}{dx},$$

which is read "the derivative of y with respect to x." This notation was invented by the German mathematician Leibniz. For example, if $y = x^2$, then

$$\frac{dy}{dx} = 2x.$$

We can also write

$$\frac{d}{dx} f(x)$$

to denote the derivative of f with respect to x. For example,

$$\frac{d}{dx} x^2 = 2x.$$

The value of dy/dx when $x = 5$ can be denoted by

$$\left.\frac{dy}{dx}\right|_{x=5}.$$

Thus for $dy/dx = 2x$,

$$\left.\frac{dy}{dx}\right|_{x=5} = 2 \cdot 5, \quad \text{or} \quad 10.$$

In general, for $y = f(x)$,

$$\left.\frac{dy}{dx}\right|_{x=a} = f'(a).$$

Historical Note: The German mathematician and philosopher Gottfried Wilhelm von Leibniz (1646–1716) and the English mathematician, philosopher, and physicist Sir Isaac Newton (1642–1727) are both credited with the invention of the calculus, though each made the invention independently of the other. Newton used the dot notation $\dot{y}$ for dy/dt, where y is a function of time, and this notation is still used, though it is not as common as Leibniz's notation.

> **CAUTION** The symbol $\frac{dy}{dx}$ does not represent a fraction; it is a traditional way to denote the derivative. It does not mean dy divided by dx, nor does it mean $\frac{y}{x}$ by canceling out d. Think of $\frac{dy}{dx}$ as a single entity.

The Power Rule

In the remainder of this section, we will develop rules and techniques for efficient differentiation.

Look for a pattern in the table on the following page, which contains functions and derivatives that we have found in previous work.

---

*The notation $D_x y$ is also used.

Function	Derivative
x^2	$2x^1$
x^3	$3x^2$
x^4	$4x^3$
$\dfrac{1}{x} = x^{-1}$	$-1 \cdot x^{-2} = \dfrac{-1}{x^2}$
$\dfrac{1}{x^2} = x^{-2}$	$-2 \cdot x^{-3} = \dfrac{-2}{x^3}$

Perhaps you have discovered the following theorem.

THEOREM 1

The Power Rule

For any real number k,

$$\frac{d}{dx}x^k = k \cdot x^{k-1}.$$

We proved this theorem for the cases $k = 2, 3$, and -1 in Examples 3 and 4 and Exercise 15, respectively, of Section 2.4. We will not prove the other cases in this text, though the proofs are similar. Note that this rule holds no matter what the exponent. That is, to differentiate x^k, we write the exponent k as the coefficient, followed by x with an exponent 1 less than k.

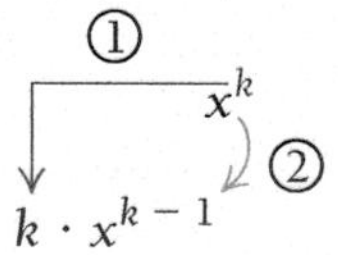

① Write the exponent as the coefficient.

$$\begin{array}{c} ① \\ x^k \\ \downarrow \quad ② \\ k \cdot x^{k-1} \end{array}$$

② Subtract 1 from the exponent.

EXAMPLE 1 $\dfrac{d}{dx}x^5 = 5x^4$

EXAMPLE 2 $\dfrac{d}{dx}x = 1 \cdot x^{1-1} = 1 \cdot x^0 = 1$

EXAMPLE 3 $\dfrac{d}{dx}x^{-4} = -4 \cdot x^{-4-1} = -4x^{-5}, \quad \text{or } -\dfrac{4}{x^5}$

The **Power Rule** also allows us to differentiate $\sqrt{x}$. To do so, it helps to first convert to an expression with a rational exponent.

EXAMPLE 4 $\dfrac{d}{dx}\sqrt{x} = \dfrac{d}{dx}x^{1/2} = \dfrac{1}{2} \cdot x^{(1/2)-1} = \dfrac{1}{2}x^{-1/2}, \quad \text{or } \dfrac{1}{2\sqrt{x}}$

EXAMPLE 5 $\dfrac{d}{dx}x^{-2/3} = -\dfrac{2}{3}x^{(-2/3)-1}$

$\qquad\qquad\qquad = -\dfrac{2}{3}x^{-5/3}, \quad \text{or} -\dfrac{2}{3}\dfrac{1}{x^{5/3}}, \quad \text{or} -\dfrac{2}{3\sqrt[3]{x^5}}$ ∎

Technology Connection

More on Numerical Differentiation and Tangent Lines

Consider $f(x) = x\sqrt{4 - x^2}$, graphed below.

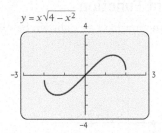

To find the value of dy/dx at a point, we select dy/dx from the **CALC** menu.

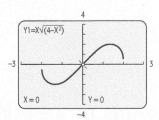

We see the graph and the cursor at the point $(0, 0)$. We key in the desired x-value or use the arrow keys to move the cursor to the desired point. We then press $\boxed{\text{ENTER}}$ to obtain the value of the derivative at the given x-value.

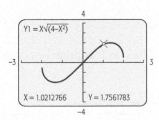

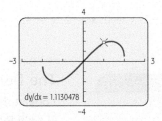

If the arrow keys were used to position the cursor when the derivative was found, we can use the **TANGENT** feature from the **DRAW** menu to draw the tangent line at the point where the derivative was found. This can be done directly from the Graph screen. The equation of the tangent line will also be displayed.

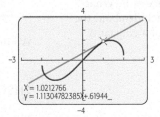

If an x-value was keyed in when the derivative was found, we can also use the **TANGENT** feature to draw the tangent lines. In this case, we must access the **TANGENT** feature from the home screen and supply the name of the function and the x-coordinate of the point of tangency. The equation of the tangent line will not be displayed when this is done.

(continued)

Technology Connection *(continued)*

EXERCISES

For each of the following functions, find the derivative and draw the tangent line at the given point. (*Hint:* When selecting the viewing window, choose x-dimensions that include the given points.)

1. $f(x) = x(200 - x)$;
$x = 24, x = 138, x = 150, x = 190$

2. $f(x) = -\frac{1}{3}x^3 + 6x^2 - 11x - 50$;
$x = -5, x = 0, x = 7, x = 12, x = 15$

3. $f(x) = 6x^2 - x^3$;
$x = -2, x = 0, x = 2, x = 4, x = 6.3$

4. *Median Weight.* The median weight of a girl whose age x is between 0 and 36 mo can be approximated by the function $f(x)$ given below, where x is the age of the girl measured in months and $f(x)$ is the girl's weight measured in pounds.[6]

$$f(x) = 0.0006x^3 - 0.0484x^2 + 1.61x + 7.6;$$
$$x = 0, x = 12, x = 24, x = 36$$

---

[6]Centers for Disease Control. Developed by the National Center for Health Statistics in collaboration with the National Center for Chronic Disease Prevention and Health Promotion (2000).

Technology Connection

Exploratory

Graph the constant function $y = -3$. Then find the derivative of this function at $x = -6$, $x = 0$, and $x = 8$. What do you conclude about the derivative of a constant function?

The Derivative of a Constant Function

Look at the graph of the constant function $F(x) = c$ shown below. What is the slope at each point on the graph?

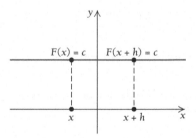

We now have the following.

THEOREM 2

The derivative of a constant function is 0. That is, $\dfrac{d}{dx}c = 0$.

Proof: Let F be the function given by $F(x) = c$. Then

$$\frac{F(x + h) - F(x)}{h} = \frac{c - c}{h}$$

$$= \frac{0}{h} = 0.$$

The difference quotient is always 0. Thus, as h approaches 0, the limit of the difference quotient approaches 0, so

$$F'(x) = \lim_{h \to 0} \frac{F(x + h) - F(x)}{h} = \lim_{h \to 0} 0 = 0.$$

■

Regardless of the value of a constant, its derivative is 0. For example, $\frac{d}{dx}5 = 0$ and $\frac{d}{dx}(3.6)^{2.82}\sin(2.39) = 0$ since both 5 and $(3.6)^{2.82}\sin(2.39)$ are constants.

The Derivative of a Constant Times a Function

Now let's consider differentiating functions such as

$$f(x) = 5x^2 \quad \text{and} \quad g(x) = -7x^4.$$

Note that we already know how to differentiate x^2 and x^4. Let's again look for a pattern in the results of Exercise Set 2.3.

Function	Derivative
$5x^2$	$10x$
$4x^{-1}$	$-4x^{-2}$
$-7x^2$	$-14x$
$5x^3$	$15x^2$

Perhaps you have discovered the following.

THEOREM 3

The derivative of a constant times a function is the constant times the derivative of the function. Using derivative notation, we can write this as

$$\frac{d}{dx}[c \cdot f(x)] = c \cdot \frac{d}{dx}f(x).$$

Proof: Let g be the function given by $g(x) = cf(x)$. Then

$$\frac{g(x + h) - g(x)}{h} = \frac{cf(x + h) - cf(x)}{h}$$

$$= c\left[\frac{f(x + h) - f(x)}{h}\right].$$

As h approaches 0, the limit of the preceding expression is the same as c times $f'(x)$. Thus,

$$g'(x) = \lim_{h \to 0}\frac{g(x + h) - g(x)}{h} = \lim_{h \to 0}c\left[\frac{f(x + h) - f(x)}{h}\right]$$

$$= c\lim_{h \to 0}\frac{f(x + h) - f(x)}{h} = cf'(x).$$

Combining this rule with the Power Rule allows us to find many derivatives.

EXAMPLE 6 $\dfrac{d}{dx}5x^4 = 5\dfrac{d}{dx}x^4 = 5 \cdot 4 \cdot x^{4-1} = 20x^3$

EXAMPLE 7 $\dfrac{d}{dx}(-9x) = -9\dfrac{d}{dx}x = -9 \cdot 1 = -9$

With practice you will be able to differentiate many such functions in one step.

EXAMPLE 8 $\dfrac{d}{dx}\dfrac{-4}{x^2} = \dfrac{d}{dx}(-4x^{-2}) = -4 \cdot \dfrac{d}{dx}x^{-2}$

$$= -4(-2)x^{-2-1}$$

$$= 8x^{-3}, \quad \text{or} \quad \dfrac{8}{x^3}$$

EXAMPLE 9 $\dfrac{d}{dx}(-x^{0.7}) = -1 \cdot \dfrac{d}{dx}x^{0.7}$

$$= -1 \cdot 0.7 \cdot x^{0.7-1}$$

$$= -0.7x^{-0.3}$$

The Derivative of a Sum or a Difference

In Exercise 11 of Exercise Set 2.4, you found that the derivative of

$$f(x) = x^2 + x$$

is

$$f'(x) = 2x + 1.$$

Note that the derivative of x^2 is $2x$, the derivative of x is 1, and the sum of these derivatives is $f'(x)$. This illustrates the following.

Technology Connection

Exploratory

Enter the following functions into a grapher:

$f(x) = x(100 - x)$,

$g(x) = x\sqrt{100 - x^2}$.

Then enter a third function as the sum of the above two: $Y_3 = Y_1 + Y_2$. Find the derivative of each of the three functions at $x = 6$ using the grapher's numerical differentiation feature. (See the Technology Connection on p. 111.)

Compare your answers. How do you think you can find the derivative of a sum?

> **THEOREM 4**
>
> The Sum-Difference Rule
>
> *Sum.* The derivative of a sum is the sum of the derivatives:
>
> $$\dfrac{d}{dx}[f(x) + g(x)] = \dfrac{d}{dx}f(x) + \dfrac{d}{dx}g(x).$$
>
> *Difference.* The derivative of a difference is the difference of the derivatives:
>
> $$\dfrac{d}{dx}[f(x) - g(x)] = \dfrac{d}{dx}f(x) - \dfrac{d}{dx}g(x).$$

Proof: For the Sum Rule, the proof is based on Limit Principle L2. Let S be the function defined by $S(x) = f(x) + g(x)$. Then

$$\dfrac{S(x + h) - S(x)}{h} = \dfrac{[f(x + h) + g(x + h)] - [f(x) + g(x)]}{h}$$

$$= \dfrac{f(x + h) - f(x)}{h} + \dfrac{g(x + h) - g(x)}{h}.$$

As h approaches 0, the two terms on the right approach $f'(x)$ and $g'(x)$, respectively, so their sum approaches $f'(x) + g'(x)$.

Thus,

$$S'(x) = \lim_{h \to 0} \frac{S(x + h) - S(x)}{h}$$

$$= \lim_{h \to 0} \left[\frac{f(x + h) - f(x)}{h} + \frac{g(x + h) - g(x)}{h} \right]$$

$$= \lim_{h \to 0} \frac{f(x + h) - f(x)}{h} + \lim_{h \to 0} \frac{g(x + h) - g(x)}{h} \qquad \text{By L2}$$

$$= f'(x) + g'(x). \qquad \blacksquare$$

The proof of the Difference Rule is similar.

Any function that is a sum or a difference of several terms can be differentiated term by term.

EXAMPLE 10 $\dfrac{d}{dx}(3x + 7) = \dfrac{d}{dx}(3x) + \dfrac{d}{dx}(7)$

$$= 3\frac{d}{dx}(x) + \frac{d}{dx}(7)$$

$$= 3 \cdot 1 + 0$$

$$= 3 \qquad \blacksquare$$

EXAMPLE 11 $\dfrac{d}{dx}(5x^3 - 3x^2) = \dfrac{d}{dx}(5x^3) - \dfrac{d}{dx}(3x^2)$

$$= 5\frac{d}{dx}x^3 - 3\frac{d}{dx}x^2$$

$$= 5 \cdot 3x^2 - 3 \cdot 2x$$

$$= 15x^2 - 6x \qquad \blacksquare$$

EXAMPLE 12 $\dfrac{d}{dx}\left(24x - \sqrt{x} + \dfrac{2}{x} \right) = \dfrac{d}{dx}(24x) - \dfrac{d}{dx}\left(\sqrt{x} \right) + \dfrac{d}{dx}\left(\dfrac{2}{x} \right)$

$$= 24 \cdot \frac{d}{dx}x - \frac{d}{dx}x^{1/2} + 2 \cdot \frac{d}{dx}x^{-1}$$

$$= 24 \cdot 1 - \frac{1}{2}x^{(1/2)-1} + 2(-1)x^{-1-1}$$

$$= 24 - \frac{1}{2}x^{-1/2} - 2x^{-2}$$

$$= 24 - \frac{1}{2\sqrt{x}} - \frac{2}{x^2} \qquad \blacksquare$$

CAUTION   The derivative of $f(x) + c$, a function plus a constant, is just the derivative of the function $f'(x)$. The derivative of $c \cdot f(x)$, a function times a constant, is the constant times the derivative $c \cdot f'(x)$. That is, for a product the constant is retained, but for a sum it is not.

Derivative of the Sine and Cosine Functions

We now investigate the derivatives of the sine and cosine functions. To do so, we use the definition of the derivative:

$$\frac{d}{dx} f(x) = \lim_{h \to 0} \frac{f(x + h) - f(x)}{h}.$$

If $f(x) = \sin x$, we have

$$\frac{d}{dx} \sin x = \lim_{h \to 0} \frac{\sin(x + h) - \sin x}{h}$$

$$= \lim_{h \to 0} \frac{\sin x \cos h + \cos x \sin h - \sin x}{h} \qquad \text{Using the Sum Identity from Section 1.4}$$

$$= \lim_{h \to 0} \left[\frac{\sin x \cos h - \sin x}{h} + \frac{\cos x \sin h}{h} \right]$$

$$= \lim_{h \to 0} \left[\sin x \left(\frac{\cos h - 1}{h} \right) + \cos x \left(\frac{\sin h}{h} \right) \right]$$

$$= \sin x \left(\lim_{h \to 0} \frac{\cos h - 1}{h} \right) + \cos x \left(\lim_{h \to 0} \frac{\sin h}{h} \right). \qquad \text{Using Limit Principles from Section 2.2}$$

On page 80, we saw that

$$\lim_{h \to 0} \frac{\sin h}{h} = 1 \quad \text{and} \quad \lim_{h \to 0} \frac{\cos h - 1}{h} = 0.$$

Therefore,

$$\frac{d}{dx} \sin x = \sin x \cdot (0) + \cos x \cdot (1)$$

$$= \cos x.$$

The derivation of the derivative of $\cos x$ may be found in the exercises. We state the results for $\sin x$ and $\cos x$ in the following theorem.

THEOREM 5

$$\frac{d}{dx} \sin x = \cos x \quad \text{and} \quad \frac{d}{dx} \cos x = -\sin x.$$

We illustrate this theorem with some examples.

EXAMPLE 13

$$\frac{d}{dx} (4 \cos x + 3) = \frac{d}{dx} (4 \cos x) + \frac{d}{dx} 3$$

$$= 4 \frac{d}{dx} \cos x + 0$$

$$= 4(-\sin x)$$

$$= -4 \sin x.$$

EXAMPLE 14 Find the derivative of $f(x) = 2 \sin x - 3 \cos x + x - 1$, and find an equation for the tangent line at the point $(0, -4)$.

Solution We first find the derivative.

$$\frac{d}{dx}(2 \sin x - 3 \cos x + x - 1) = \frac{d}{dx}(2 \sin x) - \frac{d}{dx}(3 \cos x) + \frac{d}{dx}x - \frac{d}{dx}1$$

$$= 2\frac{d}{dx}\sin x - 3\frac{d}{dx}\cos x + 1 - 0$$

$$= 2 \cos x - 3(-\sin x) + 1$$

$$= 2 \cos x + 3 \sin x + 1.$$

The derivative is $f'(x) = 2 \cos x + 3 \sin x + 1$.

We next use the derivative to find the tangent line. We know the point $(0, -4)$ is on the graph of $f(x)$ since $f(0) = 2 \sin 0 - 3 \cos 0 + 0 - 1 = -4$. The slope of the tangent line at the point $(0, -4)$ is

$$m = f'(0) = 2 \cos 0 + 3 \sin 0 + 1 = 3.$$

The slope–intercept form for the equation of a line gives:

$$y - (-4) = 3(x - 0)$$
$$y + 4 = 3x$$
$$y = 3x - 4.$$

Exercise Set 2.5

Find $\dfrac{dy}{dx}$.

1. $y = x^7$

2. $y = x^8$

3. $y = 3x^2 + 3$

4. $y = 5x^4 - 8$

5. $y = x^3(4)$

6. $y = (\sin x)(10)$

7. $y = 3x^{2/3} + 1$

8. $y = -2\sqrt{x}$

9. $y = \sqrt[4]{x^3}$

10. $y = (\sqrt[7]{x})^4$

11. $y = 4 \sin x$

12. $y = 3 \cos x$

13. $y = \sin x + \dfrac{1}{x}$

14. $y = \sqrt{x} - \dfrac{1}{\sqrt{x}}$

15. $y = (2x + 1)^2$

16. $y = (3x - 2)^2$

Find $f'(x)$.

17. $f(x) = 0.25x^{3.2}$

18. $f(x) = 0.32x^{12.5}$

19. $f(x) = 10 \sin x - 12 \cos x$

20. $f(x) = \sqrt{5} \sin x$

21. $f(x) = \sqrt[3]{9} \cos x$

22. $f(x) = \dfrac{2}{x} - \dfrac{x}{2}$

23. $f(x) = \dfrac{5}{x} + \dfrac{x}{5}$

24. $f(x) = 3x^{-2/3} + x^{3/4} + 14x^{6/5} + \dfrac{8}{x^3}$

25. $f(x) = -2x^{1/2} - 4x^{1/4} + 6 + 2x^{-1/4} - 7x^{-1/2}$

Find $f'(x)$. Some algebraic simplification is needed before differentiating.

26. $f(x) = x^3\sqrt{x}$

27. $f(x) = x^3\sqrt[5]{x^3} - x^2\sqrt[5]{x^3}$

28. $f(x) = \dfrac{2x + 1}{x}$

29. $f(x) = \dfrac{3x - 4}{x^2}$

30. $f(x) = \dfrac{x^3 + 2x^2 + 3x + 4}{x^2}$

31. $f(x) = \dfrac{4x^2 + 3x - 2}{x}$

Find the derivatives of the functions.

32. $g(x) = \dfrac{2}{x} - \dfrac{4}{x^2} + \dfrac{6}{x^3}$

33. $p(x) = 3\left(\dfrac{2}{x^4} - \dfrac{2}{x^3}\right)$

34. $r(x) = 4(3 \cos x + 2 \sin x) - 5$

35. $s(x) = \sqrt{2}(3 \cos x - 2 \sin x)$

36. $f(x) = \dfrac{3 \sin x}{2} - \dfrac{5 \cos x}{8}$

37. $q(x) = 8\left(\dfrac{\sqrt{5} \sin x}{3} - \dfrac{\sqrt[3]{5} \cos x}{7}\right)$

38. $W(x) = \dfrac{x^2 + 2x - 3 + 4x \cos x}{3x}$ (Simplify before differentiating.)

39. $U(x) = \dfrac{\sqrt{x} \sin x - 2x + 3}{\sqrt{x}}$ (Simplify before differentiating.)

Find an equation of the tangent line at the indicated point.

40. $f(x) = 2 \sin x; \ (0, 0)$

41. $f(x) = 3 \cos x + 1; \ (0, 4)$

42. $f(x) = x^{3/2} + x^{1/2}; \ (4, 10)$

43. $f(x) = x^{4/3} + 2x^{1/3}; \ (8, 20)$

SYNTHESIS

Find the derivative $\dfrac{dy}{dx}$. Some algebraic simplification is necessary before differentiation.

44. $y = \sqrt{\dfrac{\sqrt{x}}{\sqrt[3]{x}}}$ **45.** $y = \sqrt[3]{\dfrac{\sqrt[3]{x^2}}{\sqrt{x^3}}}$

46. $y = 4(\sin^2 x + \cos^2 x) \sin x$

47. $y = 3(1 + \tan^2 x) \cos^3 x$

48. $y = \sin\left(x + \dfrac{\pi}{4}\right)$ (*Hint:* Use the Sum Identity for sine.)

49. $y = \cos\left(x + \dfrac{\pi}{6}\right)$

Find an equation of the tangent line at the indicated point.

50. $y = \left(x + \dfrac{1}{x}\right)^2; \ \left(2, \dfrac{25}{4}\right)$

51. $y = \left(\sqrt{x} - \dfrac{1}{\sqrt{x}}\right)^2; \ (1, 0)$

52. Prove the Difference Rule (part of Theorem 4).

53. Prove $\dfrac{d}{dx} \cos x = -\sin x$.

tw 54. Sketch graphs of the sine and cosine functions on the same axes. Draw tangent lines to the sine function and discuss how the slope of these lines corresponds to the graph of the cosine function.

tw 55. Write a paragraph comparing the Power Rule, the Sum–Difference Rule, and the rules for differentiating a constant times a function.

tw 56. Write a short biographical paper on the lives of Leibniz and/or Newton. Emphasize the contributions they made to many areas of science and society.

 Technology Connection

Graph each of the following. Draw tangent lines at various points. Estimate those values of x at which the tangent line is horizontal.

57. $f(x) = 1.6x^3 - 2.3x - 3.7$

58. $f(x) = 10.2x^4 - 6.9x^3$

59. $f(x) = \dfrac{5x^2 + 8x - 3}{3x^2 + 2}$

60. $f(x) = \dfrac{\sin x - 1}{\cos x} + \dfrac{\cos x - 1}{\sin x}, \ 0 < x < \dfrac{\pi}{2}$

2.6 Instantaneous Rates of Change

OBJECTIVES

■ Given a formula for distance, find velocity and acceleration.

■ Find instantaneous rates of change.

Now we are ready to move from average rates of change to instantaneous rates of change. In Section 2.3, we found that for a function f:

$$\text{Average rate of change} = \text{Simplified difference quotient}$$
$$= \frac{f(x + h) - f(x)}{h}.$$

If we let h approach 0, we find the instantaneous rate of change.

DEFINITION

For any function f,

$$\text{Instantaneous rate of change} = f'(x) = \lim_{h \to 0} \frac{f(x + h) - f(x)}{h}.$$

In Sections 2.4 and 2.5, we focused on this limit as the slope of the tangent line to a curve, finding some fast ways to compute this limit. Now we focus on this limit as an instantaneous rate of change.

Let's say that a car travels 108 mi in 2 hr. Its *average* **speed** (or *average velocity*) is 108 mi/2 hr, or 54 mi/hr. This is the *average rate of change* of distance with respect to time. At various times during the trip, however, the speedometer did not read 54. Thus we say that 54 is the *average*. A snapshot of the speedometer taken at any instant would indicate *instantaneous* speed, or **instantaneous rate of change.**

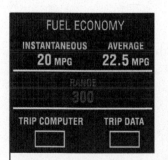

Some automobiles describe fuel economy by giving average *miles per gallon* and *instantaneous miles per gallon. Next time you see such a display, think of the instantaneous mpg part of the display as a derivative.*

Average rates of change are given by difference quotients. If distance s is a function of time t and h is the duration of the trip, then the average rate of change of distance with respect to time, called **average velocity,** is given by

$$\text{Average velocity} = \frac{\text{Difference in distance}}{\text{Difference in time}} = \frac{s(t + h) - s(t)}{h}.$$

Instantaneous rates of change are found by letting h approach 0. Thus,

$$\text{Instantaneous velocity} = \lim_{h \to 0} \frac{s(t + h) - s(t)}{h} = s'(t).$$

EXAMPLE 1   *Velocity.* A cheetah can attain a speed of over 50 mph $\approx$ 73 ft/sec in 3 sec.[7] Suppose a cheetah runs in such a way that distance s (in feet) from the starting point is a function of time t (in seconds) as follows:

$$s(t) = 12.5t^2$$

for $0 \le t \le 3$.

a) Find the average velocity between times $t = 1$ and $t = 3$.

b) Find the (instantaneous) velocity when $t = 3$.

Solution

a) From $t = 1$ to $t = 3$, $h = 2$, so

$$\frac{\text{Difference in feet}}{\text{Difference in seconds}} = \frac{s(t + h) - s(t)}{h} = \frac{s(1 + 2) - s(1)}{2}$$

$$= \frac{s(3) - s(1)}{2} = \frac{12.5 \cdot 3^2 - 12.5 \cdot 1^2}{2} = 50 \, \frac{\text{ft}}{\text{sec}}.$$

b) The instantaneous velocity is given by $s'(t)$. Using techniques learned in Section 2.5, we see that $s'(t) = 25t$ and $s'(3) = 25 \cdot 3 = 75 \, \dfrac{\text{ft}}{\text{sec}}$. ■

We generally use the letter v for velocity. We have the following definition.

DEFINITION

$$\textbf{Velocity} = v(t) = \lim_{h \to 0} \frac{s(t + h) - s(t)}{h} = s'(t)$$

The rate of change of velocity is called **acceleration.** We generally use the letter a for acceleration. Thus the following definition applies.

DEFINITION

$$\textbf{Acceleration} = a(t) = v'(t)$$

[7]AfriCat UK: Cheetah information, http://www.africatuk.org/cheetah.htm.

EXAMPLE 2 *Distance, Velocity, and Acceleration.* For $s(t) = 10t^2$, find $v(t)$ and $a(t)$, where s is the distance from the starting point, in miles, and t is in hours. Then find the distance, velocity, and acceleration when $t = 4$ hr.

Solution We have

$$v(t) = s'(t) = 20t,$$

$$a(t) = v'(t) = 20.$$

Then $s(4) = 10(4)^2 = 160$ mi,

$$v(4) = 20(4) = 80 \text{ mi/hr}, \quad \text{and}$$

$$a(4) = 20 \text{ mi/hr}^2.$$

If this distance function applies to a vehicle, then at time $t = 4$ hr, the distance is 160 mi, the velocity, or instantaneous speed, is 80 mi/hr, and the acceleration is 20 miles per hour per hour, which we abbreviate as 20 mi/hr^2.   ■

EXAMPLE 3 *Pendulum Motion.* The motion of a swinging pendulum is given by

$$s(t) = 28 \sin t,$$

where t is the time measured in seconds and $s(t)$ is the position of the pendulum along a line measured in millimeters.

a) Find the velocity of the pendulum at time t.
b) Find the times when the pendulum's velocity is 14 mm/sec.
c) Find the acceleration of the pendulum at time t.
d) Find the times when the pendulum has acceleration 0.

Solution

a) We find the velocity by computing the derivative of the position function:

$$v(t) = s'(t) = 28 \cos t.$$

b) To find when the pendulum's velocity is 14 mm/sec, we set the derivative to 14:

$$28 \cos t = 14$$

$$\cos t = \frac{1}{2}$$

$$t = \frac{\pi}{3} + 2\pi n \quad \text{and} \quad t = \frac{5\pi}{3} + 2\pi n$$

All solutions differ from either $\frac{\pi}{3}$ or $\frac{5\pi}{3}$ by an integer multiple of 2π. This makes sense because a pendulum periodically repeats its motion.

c) The acceleration is given by the derivative of the velocity:

$$a(t) = v'(t) = \frac{d}{dt}(28 \cos t) = -28 \sin t.$$

d) To find when the pendulum has acceleration 0, we set $a(t)$ equal to 0.

$$-28 \sin t = 0$$
$$\sin t = 0$$
$$t = n\pi.$$

All solutions are multiples of π.   ■

In general, derivatives give instantaneous rates of change. Instead of saying "the derivative of y with respect to x," we will sometimes say "the *rate of change of y with respect to x.*"

EXAMPLE 4   *Volume of a Cancer Tumor.* The spherical volume V of a cancer tumor is given by

$$V(r) = \tfrac{4}{3}\pi r^3,$$

where r is the radius of the tumor, in centimeters.

a) Find the rate of change of the volume with respect to the radius.

b) Find the rate of change of the volume at $r = 1.2$ cm.

Solution

a) $\dfrac{dV}{dr} = V'(r) = 3 \cdot \dfrac{4}{3} \cdot \pi r^2 = 4\pi r^2.$

(This turns out to be the surface area.)

b) $V'(1.2) = 4\pi(1.2)^2 = 5.76\pi \approx 18\,\dfrac{\text{cm}^3}{\text{cm}} = 18 \text{ cm}^2.$ ■

EXAMPLE 5   *Population Growth.* The initial population in a bacteria colony is 10,000. After t hours, the colony has grown to a number $P(t)$ given by

$$P(t) = 10{,}000(1 + 0.86t + t^2).$$

a) Find the rate of change of the population P with respect to time t. This is also known as the **growth rate.**

b) Find the number of bacteria present after 5 hr. Also, find the growth rate when $t = 5$.

c) At what time is the growth rate 200,000 bacteria per hour?

Solution

a) Note that $P(t) = 10{,}000 + 8600t + 10{,}000t^2$. Then

$$P'(t) = 8600 + 20{,}000t.$$

b) The number of bacteria present when $t = 5$ is given by

$$P(5) = 10{,}000 + 8600 \cdot 5 + 10{,}000 \cdot 5^2 = 303{,}000.$$

The growth rate when $t = 5$ is given by

$$P'(5) = 8600 + 20{,}000 \cdot 5 = 108{,}600\,\frac{\text{bacteria}}{\text{hr}}.$$

Thus at $t = 5$, there are 303,000 bacteria present, and the colony is growing at the rate of 108,600 bacteria per hour.

c) We set the growth rate to 200,000 bacteria per hour and solve for t.

$$8600 + 20{,}000t = 200{,}000$$
$$20{,}000t = 191{,}400$$
$$t = \frac{191{,}400}{20{,}000} = 9.57.$$

After 9.57 hr, the growth rate is 200,000 bacteria per hour. ■

EXAMPLE 6 *Whale Growth.* The growth of a killer whale (*Orcinus orca*) can be modeled by the function

$$f(t) = 412 + 44.8t + 1.1t^2 - 0.0167t^3,$$

where $0 \leq t \leq 60$ is the time in months after birth and $f(t)$ is the weight in pounds.[8]

a) Find the rate of change of the weight of the whale.

b) Find the weight of the whale at 40 mo.

c) Find the rate of change of the weight of the whale at 40 mo.

d) At what time does the whale stop growing?

Solution

a) The rate of change is given by the derivative:

$$f'(t) = 44.8 + 2.2t - 0.0501t^2.$$

b) At 40 mo, the whale weighs

$$f(40) = 412 + 44.8(40) + 1.1(40)^2 - 0.0167(40)^3 = 2895.2 \text{ lb.}$$

c) At 40 mo, the rate of change of the whale's weight is

$$f'(40) = 44.8 + 2.2(40) - 0.0501(40)^2 = 52.64 \, \frac{\text{lb}}{\text{mo}}.$$

d) The whale stops growing when the rate of growth is 0. We set $f'(t)$ equal to 0:

$$44.8 + 2.2t - 0.0501t^2 = 0$$

To solve this equation, we can use the quadratic formula. In this case, $a = -0.0501$, $b = 2.2$, and $c = 44.8$.

$$x = \frac{-b \pm \sqrt{b^2 - 4ac}}{2a}$$

$$= \frac{-2.2 \pm \sqrt{(2.2)^2 - 4(-0.0501)(44.8)}}{2(-0.0501)}$$

$$\approx 21.96 \pm 37.10.$$

There are two solutions to this equation, $t = 21.96 - 37.1 = -15.14$ and $t = 21.96 + 37.1 = 59.06$. For this problem, only the positive answer makes sense. At age approximately 59.06 mo, the whale's growth rate is 0. ■

---

[8]SeaWorld.

Exercise Set 2.6

1. Given

$$s(t) = t^3 + t,$$

where s is in feet and t is in seconds, find each of the following.

a) $v(t)$ b) $a(t)$
c) The velocity and acceleration when $t = 4$ sec

2. Given

$$s(t) = 3t + 10,$$

where s is in miles and t is in hours, find each of the following.

a) $v(t)$ b) $a(t)$
c) The velocity and acceleration when $t = 2$ hr. When the distance function is given by a linear function, we have *uniform motion*.

tw d) What does uniform motion mean in terms of velocity and acceleration?

3. Given

$$s(t) = -10t^2 + 2t + 5,$$

where s is in meters and t is in seconds, find the following.

a) $v(t)$
b) $a(t)$
c) The velocity and acceleration when $t = 1$
d) The time when the velocity is 1 m/sec.

4. Given

$$s(t) = t^2 - \frac{1}{2}t + 3,$$

where s is in meters and t is in seconds, find the following.

a) $v(t)$
b) $a(t)$
c) The velocity and acceleration when $t = 2$
d) The time when the velocity is 5 m/sec.

5. Given

$$s(t) = 5t + 2\sin t,$$

where s is in meters and t is in seconds, find the following.

a) $v(t)$
b) $a(t)$

c) The velocity and acceleration when $t = \dfrac{\pi}{4}$
d) All times when the velocity is 3 m/sec.

6. Given

$$s(t) = 3t - \cos t,$$

where s is in millimeters and t is in seconds, find the following.

a) $v(t)$
b) $a(t)$

c) The velocity and acceleration when $t = \dfrac{\pi}{3}$

d) All times when the velocity is 2 mm/sec.

7. *Advertising.* A firm estimates that it will sell N units of a product after spending a dollars on advertising, where

$$N(a) = -a^2 + 300a + 6$$

and a is in thousands of dollars.

a) What is the rate of change of the number of units sold with respect to the amount spent on advertising?
b) How many units will be sold after spending $10,000 on advertising?
c) What is the rate of change at $a = 10$?

tw d) Explain the meaning of your answers to parts (a) and (c).

8. *Growth of a Baby.* The median weight of a boy whose age t is between 0 and 36 mo can be approximated by the function

$$w(t) = 8.15 + 1.82t - 0.0596t^2 \\ + 0.000758t^3,$$

where t is measured in months and w is measured in pounds. Use this approximation to make the following computations for a boy with median weight.[9]

a) The rate of change of weight with respect to time
b) The weight of the boy at birth
c) The rate of change of the weight of the boy at birth
d) The weight of the boy on his first birthday
e) The rate of change of the boy's weight with respect to time on his first birthday
f) The average rate of change of the boy's weight during his first year of life
g) A time when the instantaneous rate of change is the same as the average rate of change over the first year

9. *Growth of a Baby.* The median weight of a girl whose age t is between 0 and 36 mo can be approximated by the function

$$w(t) = 7.6 + 1.61t - 0.0484t^2 + 0.0006t^3,$$

where t is measured in months and w is measured in pounds. Use this approximation to make the following computations for a girl with median weight.[10]

a) The rate of change of weight with respect to time
b) The weight of the girl at birth
c) The rate of change of the weight of the girl at birth
d) The weight of the girl on her first birthday
e) The rate of change of the girl's weight with respect to time on her first birthday
f) The average rate of change of the girl's weight during her first year of life
g) A time when the instantaneous rate of change is the same as the average rate of change over the first year

10. *Stopping Distance on Glare Ice.* The stopping distance on glare ice (at some fixed speed) of regular tires is given by a linear function of the air temperature F,

$$D(F) = 2F + 115,$$

where $D(F)$ is the stopping distance, in feet, when the air temperature is F, in degrees Fahrenheit.

a) Find the rate of change of the stopping distance D with respect to the air temperature F.
tw b) Explain the meaning of your answer to part (a).

11. *Healing Wound.* The circumference C, in centimeters, of a healing wound is given by

$$C(r) = 2\pi r,$$

where r is the radius, in centimeters.

a) Find the rate of change of the circumference with respect to the radius.
tw b) Explain the meaning of your answer to part (a).

12. *Healing Wound.* The circular area A, in square centimeters, of a healing wound is given by

$$A(r) = \pi r^2,$$

where r is the radius, in centimeters.

a) Find the rate of change of the area with respect to the radius.
tw b) Explain the meaning of your answer to part (a).

13. *Temperature During an Illness.* The temperature T of a person during an illness is given by

$$T(t) = -0.1t^2 + 1.2t + 98.6,$$

where T is the temperature, in degrees Fahrenheit, at time t, in days.

a) Find the rate of change of the temperature with respect to time.
b) Find the temperature at $t = 1.5$ days.
c) Find the rate of change at $t = 1.5$ days.
tw d) Why would the sign of $T'(t)$ be significant to a doctor?

14. *Blood Pressure.* For a certain dosage of x cubic centimeters (cc) of a drug, the resulting blood pressure B is approximated by

$$B(x) = 0.05x^2 - 0.3x^3.$$

a) Find the rate of change of the blood pressure with respect to the dosage. Such a rate of change is often called the *sensitivity*.
tw b) Explain the meaning of your answer to part (a).

15. *Territorial Area.* The territorial area T of an animal is defined as its defended, or exclusive, region. The area T of that region can be approximated using the animal's body weight W by

$$T = W^{1.31}$$

---

[9],[10]Centers for Disease Control. Developed by the National Center for Health Statistics in collaboration with the National Center for Chronic Disease Prevention and Health Promotion (2000).

(see Section 1.3).

a) Find dT/dW.

tw b) Explain the meaning of your answer to part (a).

These bears may be determining territorial area.

16. *Home Range.* The home range H of an animal is defined as the region to which the animal confines its movements. The area of that region can be approximated using the animal's body weight W by

$$H = W^{1.41}$$

(see Section 1.3).

a) Find dH/dW.

tw b) Explain the meaning of your answer to part (a).

17. *Sensitivity.* The reaction R of the body to a dose Q of medication is often represented by the general function

$$R(Q) = Q^2\left(\frac{k}{2} - \frac{Q}{3}\right),$$

where k is a constant and R is in millimeters, if the reaction is a change in blood pressure, or in degrees of temperature, if the reaction is a change in temperature. The rate of change dR/dQ is defined to be the *sensitivity*.

a) Find a formula for the sensitivity.

tw b) Explain the meaning of your answer to part (a).

18. *Population Growth Rate.* The population of a city grows from an initial size of 100,000 to an amount P given by

$$P(t) = 100,000 + 2000t^2,$$

where t is in years.

a) Find the growth rate.

b) Find the number of people in the city after 10 yr (at $t = 10$).

c) Find the growth rate at $t = 10$.

tw d) Explain the meaning of your answers to parts (a) and (c).

19. *Median Age of Women at First Marriage.* The median age of women at first marriage can be approximated by the linear function

$$A(t) = 0.08t + 19.7,$$

where $A(t)$ is the median age of women at first marriage t years after 1950.

a) Find the rate of change of the median age A with respect to time t.

tw b) Explain the meaning of your answer to part (a).

20. *View to the Horizon.* The view V, or distance, in miles that one can see to the horizon from a height h, in feet, is given by

$$V = 1.22\sqrt{h}.$$

a) Find the rate of change of V with respect to h.

b) How far can one see to the horizon from an airplane window from a height of 40,000 ft?

c) Find the rate of change at $h = 40,000$.

tw d) Explain the meaning of your answers to parts (a) and (c).

SYNTHESIS

tw 21. Explain and compare average rate of change and instantaneous rate of change.

tw 22. Discuss as many interpretations as you can of the derivative of a function at a point x.

 Technology Connection

In Exercises 23–28, graph the position function $s(t)$. Then graph the velocity and acceleration functions.

23. $s(t) = 3t^2 - 2t + 7$

24. $s(t) = -32t^2 + 30t + 5$

25. $s(t) = 23.7 - 15.8t + 0.023t^2 + 0.0046t^3$

26. $s(t) = 0.12 - 0.032t - 0.0065t^2 + 0.00051t^3$

27. a) $s(t) = 2\cos t + 3\sin t$

tw b) How do the acceleration and the position functions compare?

28. a) $s(t) = 3\cos t - 2\sin t$

tw b) How do the acceleration and the position functions compare?

2.7   Differentiation Techniques: The Product and Quotient Rules

OBJECTIVES

■ Differentiate using the Product and Quotient Rules.

The Product Rule

The derivative of a sum is the sum of the derivatives, but the derivative of a product is *not* the product of the derivatives. To see this, consider x^2 and x^5. The product is x^7, and the derivative of this product is $7x^6$. The individual derivatives are $2x$ and $5x^4$, and the product of these derivatives is $10x^5$, which is not $7x^6$.

CAUTION

$$\frac{d}{dx}[f(x)g(x)] \neq f'(x)g'(x).$$

The following is the rule for finding the derivative of a product.

THEOREM 5

The Product Rule

Suppose that $F(x) = f(x) \cdot g(x)$, where $f(x)$ and $g(x)$ are differentiable. Then

$$F'(x) = \frac{d}{dx}[f(x) \cdot g(x)] = f(x) \cdot \left[\frac{d}{dx}g(x)\right] + \left[\frac{d}{dx}f(x)\right] \cdot g(x).$$

Proof: We compute the derivative $F'(x)$:

$$F'(x) = \lim_{h \to 0} \frac{F(x+h) - F(x)}{h}$$

$$= \lim_{h \to 0} \frac{f(x+h)g(x+h) - f(x)g(x)}{h}.$$

In order to compute this limit, we introduce a term and its additive inverse to the numerator. The idea is to make it factor with both terms in the numerator.

$$\frac{f(x+h)g(x+h) - f(x)g(x)}{h}$$

$$= \frac{f(x+h)g(x+h) - f(x+h)g(x) + f(x+h)g(x) - f(x)g(x)}{h}$$

$$= \frac{f(x+h)[g(x+h) - g(x)] + [f(x+h) - f(x)]g(x)}{h}$$

$$= f(x+h)\frac{g(x+h) - g(x)}{h} + \frac{f(x+h) - f(x)}{h}g(x).$$

We can now simplify the difference quotient:

$$F'(x) = \lim_{h \to 0} \left[f(x+h) \frac{g(x+h) - g(x)}{h} + \frac{f(x+h) - f(x)}{h} g(x) \right]$$

$$= \lim_{h \to 0} f(x+h) \frac{g(x+h) - g(x)}{h} + \lim_{h \to 0} \frac{f(x+h) - f(x)}{h} g(x)$$

$$= \lim_{h \to 0} f(x+h) \lim_{h \to 0} \frac{g(x+h) - g(x)}{h} + \left[\lim_{h \to 0} \frac{f(x+h) - f(x)}{h} \right] g(x)$$

$$= f(x)g'(x) + f'(x)g(x). \qquad \text{By continuity of } f \qquad \blacksquare$$

EXAMPLE 1   Find $\dfrac{d}{dx}[(x^4 - 2x^3 - 7)(3x^2 - 5x)]$.

Solution   We have

$$\frac{d}{dx}[(x^4 - 2x^3 - 7)(3x^2 - 5x)] = (x^4 - 2x^3 - 7)\frac{d}{dx}(3x^2 - 5x)$$

$$+ \left(\frac{d}{dx}(x^4 - 2x^3 - 7) \right)(3x^2 - 5)$$

$$= (x^4 - 2x^3 - 7)(6x - 5)$$

$$+ (4x^3 - 6x^2)(3x^2 - 5x).$$

Note that we could have multiplied the polynomials and then differentiated, avoiding the use of the Product Rule, but this would have been more work. $\blacksquare$

For many products, it is not possible to multiply the terms and use the differentiation formulas we learned in Section 2.6. In these cases, the Product Rule is essential.

EXAMPLE 2   Let $f(x) = x \cos x$. Find $f'(x)$.

Solution   We use the Product Rule.

$$f'(x) = x \cdot \frac{d}{dx} \cos x + \left(\frac{d}{dx} x \right) \cos x$$

$$= x(-\sin x) + 1 \cos x$$

$$= -x \sin x + \cos x. \qquad \blacksquare$$

The Quotient Rule

The derivative of a quotient is *not* the quotient of the derivatives. To see why, consider x^5 and x^2. The quotient x^5/x^2 is x^3, and the derivative of this quotient is $3x^2$. The individual derivatives are $5x^4$ and $2x$, and the quotient of these derivatives is $(5/2)x^3$, which is not $3x^2$.

CAUTION

$$\frac{d}{dx} \frac{N(x)}{D(x)} \neq \frac{N'(x)}{D'(x)}$$

The rule for differentiating quotients is as follows.

THEOREM 6

The Quotient Rule

If $N(x)$ and $D(x)$ are differentiable and

$$Q(x) = \frac{N(x)}{D(x)},$$

then

$$Q'(x) = \frac{D(x) \cdot N'(x) - D'(x) \cdot N(x)}{[D(x)]^2}.$$

Proof: We compute the derivative as the limit of the difference quotient:

$$Q'(x) = \lim_{h \to 0} \frac{Q(x + h) - Q(x)}{h}$$

$$= \lim_{h \to 0} \frac{\dfrac{N(x + h)}{D(x + h)} - \dfrac{N(x)}{D(x)}}{h}$$

$$= \lim_{h \to 0} \frac{\dfrac{D(x)N(x + h) - D(x + h)N(x)}{D(x + h)D(x)}}{h} \qquad \text{Subtracting the fractions}$$

$$= \lim_{h \to 0} \frac{1}{D(x + h)D(x)} \frac{D(x)N(x + h) - D(x + h)N(x)}{h}$$

$$= \frac{1}{[D(x)]^2} \lim_{h \to 0} \frac{D(x)N(x + h) - D(x + h)N(x)}{h}. \qquad \text{By continuity of } D$$

In the numerator, we subtract and add a term that will factor with both terms.

$$\frac{D(x)N(x + h) - D(x + h)N(x)}{h}$$

$$= \frac{D(x)N(x + h) - D(x)N(x) + D(x)N(x) - D(x + h)N(x)}{h}$$

$$= D(x)\frac{N(x + h) - N(x)}{h} - \frac{D(x + h) - D(x)}{h}N(x).$$

We can now complete the computation of the derivative:

$$Q'(x) = \frac{1}{[D(x)]^2} \lim_{h \to 0} \frac{D(x)N(x + h) - D(x + h)N(x)}{h}$$

$$= \frac{1}{[D(x)]^2} \lim_{h \to 0} \left[D(x)\frac{N(x + h) - N(x)}{h} - \frac{D(x + h) - D(x)}{h}N(x) \right]$$

$$= \frac{1}{[D(x)]^2} \left[D(x) \lim_{h \to 0} \frac{N(x + h) - N(x)}{h} \right.$$

$$\left. - \lim_{h \to 0} \frac{D(x + h) - D(x)}{h}N(x) \right]$$

$$= \frac{1}{[D(x)]^2} [D(x)N'(x) - D'(x)N(x)]$$

$$= \frac{D(x)N'(x) - D'(x)N(x)}{[D(x)]^2}.$$

EXAMPLE 3   Differentiate: $f(x) = \dfrac{x^2 - 3x}{x - 1}$.

Solution   We have

$$f'(x) = \frac{(x - 1)(2x - 3) - 1(x^2 - 3x)}{(x - 1)^2}$$

$$= \frac{2x^2 - 5x + 3 - x^2 + 3x}{(x - 1)^2}$$

$$= \frac{x^2 - 2x + 3}{(x - 1)^2}.$$

It is possible but not necessary to multiply out $(x - 1)^2$ to get $x^2 - 2x + 1$.   ■

Technology Connection

Checking Derivatives Graphically

Consider the derivative in Example 3. To check the answer, first we enter the function:

$$y_1 = \frac{x^2 - 3x}{x - 1}.$$

Then we enter the derivative:

$$y_2 = \frac{x^2 - 2x + 3}{(x - 1)^2}.$$

For the third function, we enter

$$y_3 = \text{nDeriv}(Y_1, x, x).$$

Next, we deselect y_1 and graph y_2 and y_3. We use different graph styles so that we see each graph as it appears on the screen.

$y_2 = \frac{x^2 - 2x + 3}{(x - 1)^2}$, $y_3 = \text{nDeriv}(Y_1, x, x)$

Since the graphs appear to coincide, it appears that $y_2 = y_3$ and we have a check of the result. This is considered a partial check, however, because the graphs might not coincide at a point that is not in the viewing window.

We can also use a table to check that $y_2 = y_3$.

X	Y2	Y3
5.97	1.081	1.081
5.98	1.0806	1.0806
5.99	1.0803	1.0803
6	1.08	1.08
6.01	1.0797	1.0797
6.02	1.0794	1.0794
6.03	1.079	1.079
X = 5.97		

Suppose that we had incorrectly calculated the derivative to be $y_2 = (x^2 - 2x - 8)/(x - 1)^2$. We see in the following that the graphs do not agree:

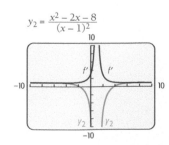

$y_2 = \frac{x^2 - 2x - 8}{(x - 1)^2}$

(continued)

EXERCISES

1. For the function

$$f(x) = \frac{x^2 - 4x}{x + 2},$$

use graphs and tables to determine which of the following seems to be the correct derivative.

a) $f'(x) = \dfrac{-x^2 - 4x - 8}{(x + 2)^2}$

b) $f'(x) = \dfrac{x^2 - 4x + 8}{(x^2 - 4x)^2}$

c) $f'(x) = \dfrac{x^2 + 4x - 8}{(x + 2)^2}$

2.-3. Check the results of Examples 1–2 using a grapher.

EXAMPLE 4   Differentiate: $g(t) = \dfrac{t \sin t + t^3}{\cos t}$.

Solution   Using the Quotient Rule:

$$g'(t) = \frac{(\cos t)(\sin t + t \cos t + 3t^2) - (-\sin t)(t \sin t + t^3)}{\cos^2 t}$$

$$= \frac{(\cos t \sin t + t \cos^2 t + 3t^2 \cos t) + (t \sin^2 t + t^3 \sin t)}{\cos^2 t}$$

$$= \frac{\cos t \sin t + 3t^2 \cos t + t^3 \sin t + t(\cos^2 t + \sin^2 t)}{\cos^2 t}$$

$$= \frac{\cos t \sin t + 3t^2 \cos t + t^3 \sin t + t}{\cos^2 t}. \qquad \text{Using a Pythagorean Identity} \quad ■$$

EXAMPLE 5   Differentiate: $r(x) = \dfrac{3x^2 + 2x - 9}{5}$.

Solution   The Quotient Rule could be used, but it is not required; $r'(x)$ can be found more easily if the fraction is first simplified.

$$r(x) = \frac{3x^2 + 2x - 9}{5} = \frac{3}{5}x^2 + \frac{2}{5}x - \frac{9}{5}.$$

Now the derivative can be computed.

$$r'(x) = \frac{3}{5} \cdot 2x + \frac{2}{5} = \frac{6}{5}x + \frac{2}{5}. \qquad\qquad ■$$

Derivatives of Trigonometric Functions

In Section 2.5 we learned how to compute the derivative of the sine and cosine functions. With the Quotient Rule, we are now able to compute the derivative of the other four trigonometric functions.

EXAMPLE 6   Compute.

a) $\dfrac{d}{dx} \tan x$

b) $\dfrac{d}{dx} \sec x$

Solution

a) We rewrite the tangent function in terms of the sine and cosine functions and then differentiate.

$$\frac{d}{dx} \tan x = \frac{d}{dx} \frac{\sin x}{\cos x}$$

$$= \frac{\cos x \left(\dfrac{d}{dx} \sin x \right) - \left(\dfrac{d}{dx} \cos x \right) \sin x}{\cos^2 x}$$

$$= \frac{\cos x \cos x - (-\sin x) \sin x}{\cos^2 x}$$

$$= \frac{\cos^2 x + \sin^2 x}{\cos^2 x}$$

$$= \frac{1}{\cos^2 x} \qquad \text{Pythagorean Identity}$$

$$= \sec^2 x. \qquad \text{Reciprocal Identity}$$

b) We use a Reciprocal Identity to rewrite $\sec x$.

$$\frac{d}{dx} \sec x = \frac{d}{dx} \frac{1}{\cos x}$$

$$= \frac{\cos x \left(\dfrac{d}{dx} 1 \right) - \left(\dfrac{d}{dx} \cos x \right) \cdot 1}{\cos^2 x}$$

$$= \frac{0 + \sin x}{\cos^2 x}$$

$$= \frac{1}{\cos x} \frac{\sin x}{\cos x}$$

$$= \sec x \tan x. \qquad \blacksquare$$

The derivatives of $\csc x$ and $\cot x$ can also be computed using the Quotient Rule. We summarize the derivatives of the trigonometric functions in Theorem 7.

THEOREM 7

$$\frac{d}{dx} \sin x = \cos x \qquad \frac{d}{dx} \tan x = \sec^2 x \qquad \frac{d}{dx} \sec x = \sec x \tan x$$

$$\frac{d}{dx} \cos x = -\sin x \qquad \frac{d}{dx} \cot x = -\csc^2 x \qquad \frac{d}{dx} \csc x = -\csc x \cot x$$

Theorem 7 can be used in conjunction with the rules of differentiation to compute derivatives of a variety of functions.

EXAMPLE 7   Compute the derivative.

a) $\dfrac{d}{dx}(x \tan x)$

b) $\dfrac{d}{dx}\left(\dfrac{\sec x + \csc x}{x}\right)$

Solution

a) We use the Product Rule.

$$\frac{d}{dx}(x \tan x) = \left(x \frac{d}{dx}\tan x\right) + \left(\frac{d}{dx}x\right)\tan x$$
$$= x \sec^2 x + 1 \cdot \tan x$$
$$= x \sec^2 x + \tan x.$$

b) We use the Quotient Rule.

$$\frac{d}{dx}\left(\frac{\sec x + \csc x}{x}\right) = \frac{x\dfrac{d}{dx}(\sec x + \csc x) - \left(\dfrac{d}{dx}x\right)(\sec x + \csc x)}{x^2}$$

$$= \frac{x(\sec x \tan x - \csc x \cot x) - (\sec x + \csc x)}{x^2}$$

$$= \frac{x \sec x \tan x - x \csc x \cot x - \sec x - \csc x}{x^2}.$$ ∎

Exercise Set 2.7

Differentiate.

1. $y = x^3 \cdot x^8$, two ways

2. $y = x^4 \cdot x^9$, two ways

3. $y = x\sqrt{x}$, two ways

4. $y = x^2\sqrt[3]{x^2}$, two ways

5. $y = \dfrac{x^8}{x^5}$, two ways

6. $y = \dfrac{x^5}{x^8}$, two ways

7. $f(x) = (x + 5)(x - 5)$, two ways

8. $f(x) = \dfrac{x^2 - 9}{x + 3}$, two ways

9. $y = (8x^5 - 3x^3 + 2)(3x^4 - 3\sqrt{x})$

10. $y = (7x^5 + 3x^3 - 50)(9x^8 - 7x\sqrt{x})$

11. $f(x) = (\sqrt{x} - \sqrt[3]{x})(2x + 3)$

12. $f(x) = (\sqrt[3]{x} + 2x)\tan x$

13. $g(x) = \sqrt{x}\tan x$

14. $g(x) = \sqrt[3]{x}\sec x$

15. $f(t) = (2t + 3)^2$ (*Hint:* Use algebra before differentiating.)

16. $r(t) = (5t - 4)^2$ (*Hint:* Use algebra before differentiating.)

17. $g(x) = (0.02x^2 + 1.3x - 11.7)(4.1x + 11.3)$

18. $g(x) = (3.12x^2 + 10.2x - 5.01)$
    $\times (2.9x^2 + 4.3x - 2.1)$

19. $g(x) = \sec x \csc x$

20. $p(x) = \cot x \csc x$

21. $q(x) = \dfrac{\sin x}{1 + \cos x}$

22. $q(x) = \dfrac{1 - \cos x}{\sin x}$

23. $s(t) = \tan^2 t$ (*Hint:* Use algebra before differentiating.)

24. $s(t) = \sec^2 t$ (*Hint:* Use algebra before differentiating.)

25. $y = \left(x + \dfrac{2}{x}\right)(x^2 - 3)$

26. $y = (4x^3 - x^2)\left(x - \dfrac{5}{x}\right)$

27. $q(x) = \dfrac{3x^2 - 6x + 4}{5}$

28. $q(x) = \dfrac{4x^3 + 2x^2 - 5x + 13}{3}$

29. $y = \dfrac{x^2 + 3x - 4}{2x - 1}$

30. $y = \dfrac{x^2 - 4x + 3}{x^2 - 1}$

31. $w = \dfrac{3t - 1}{t^2 - 2t + 6}$

32. $q = \dfrac{t^2 + 7t - 1}{2t^2 - 3t - 7}$

33. $f(x) = \dfrac{x}{x^{-1} + 1}$

34. $f(x) = \dfrac{x^{-1}}{x + x^{-1}}$

35. $y = \dfrac{\tan t}{1 + \sec t}$

36. $y = \dfrac{\cot t}{1 + \csc t}$

37. $w = \dfrac{\tan x + x \sin x}{\sqrt{x}}$

38. $w = \dfrac{\cos x - x^2 \csc x}{\sqrt{x}}$

39. $y = \dfrac{1 + \sqrt{t}}{1 - \sqrt{t}}$

40. $y = \dfrac{\dfrac{2}{3x} - 1}{\dfrac{3}{x^2} + 5}$

41. $f(t) = t \sin t \tan t$

42. $g(t) = t(\csc t)(1 + \cos t)$

43.–84. Use a grapher to check the results of Exercises 1–42.

85. Let $f(x) = \dfrac{x}{x + 1}$ and $g(x) = \dfrac{-1}{x + 1}$.

   a) Compute $f'(x)$.
   b) Compute $g'(x)$.
 tw c) Compare your answers in parts (a) and (b) and explain.

86. Let $f(x) = \dfrac{x^2}{x^2 - 1}$ and $g(x) = \dfrac{1}{x^2 - 1}$.

   a) Compute $f'(x)$.
   b) Compute $g'(x)$.
 tw c) Compare your answers in parts (a) and (b) and explain.

87. Let $f(x) = \sin^2 x + \cos^2 x$.

   a) Compute $f'(x)$. Don't simplify $f(x)$ before differentiating, but simplify your answer.
 tw b) Explain your answer.

88. Let $f(x) = \tan^2 x$ and $g(x) = \sec^2 x$.

   a) Compute $f'(x)$.
   b) Compute $g'(x)$.
 tw c) Compare your answers to parts (a) and (b) and explain.

89. Find an equation of the tangent line to the graph of $y = 8/(x^2 + 4)$ at the point $(0, 2)$ and at $(-2, 1)$.

90. Find an equation of the tangent line to the graph of $y = 4x/(1 + x^2)$ at the point $(0, 0)$ and at $(-1, -2)$.

91. Find an equation of the tangent line to the graph of $y = \dfrac{\sqrt{x}}{x + 1}$ at the point where $x = 1$.

92. Find an equation of the tangent line to the graph of $y = \dfrac{x^2 + 3}{x - 1}$ at the point where $x = 2$.

93. Find an equation of the tangent line to the graph of $y = x \sin x$ at the point $\left(\dfrac{\pi}{4}, \dfrac{\sqrt{2}\pi}{8}\right)$.

94. Find an equation of the tangent line to the graph of $y = x \tan x$ at the point $\left(\dfrac{\pi}{4}, \dfrac{\pi}{4}\right)$.

APPLICATIONS

95. *Temperature During an Illness.* The temperature T of a person during an illness is given by

$$T(t) = \dfrac{4t}{t^2 + 1} + 98.6,$$

where T is the temperature, in degrees Fahrenheit, at time t, in hours.

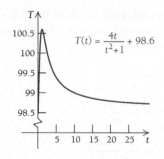

$$T(t) = \frac{4t}{t^2+1} + 98.6$$

a) Find the rate of change of the temperature with respect to time.
b) Find the temperature at $t = 2$ hr.
c) Find the rate of change at $t = 2$ hr.

96. *Population Growth.* The population P, in thousands, of a small city is given by

$$P(t) = 10 + \frac{50t}{2t^2 + 9},$$

where t is the time, in years.

a) Find the growth rate.
b) Find the population after 8 yr.
c) Find the growth rate at $t = 12$ yr.

97. *Lighthouse.* A lighthouse illuminates a straight wall as indicated in the figure. The light rotates at the rate of 1 radian per second, so the angle t in the figure is also the time measured in seconds. The distance $s(t)$ is measured in feet.

a) Find a formula for the function $s(t)$.
b) Find the rate at which the light moves across the wall.
c) Find all times when the rate is 200 ft/sec.

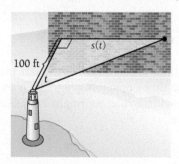

98. *Pendulum.* The position of a pendulum that is slowing due to friction is given by

$$s(t) = \frac{3 \cos t}{\sqrt{t} + 1},$$

where t is time measured in seconds and the position $s(t)$ is measured in inches.

a) Find the velocity function for the pendulum.
b) Find the velocity of the pendulum at time $t = 1$.
c) Find the velocity of the pendulum at time $t = \dfrac{\pi}{3}$.

SYNTHESIS

Differentiate.

99. $g(x) = \dfrac{x^2 + 1}{x^2 - 1} \tan x$

100. $g(t) = \dfrac{t^3 - 1}{t^3 + 1} \csc t$

101. $s(t) = \dfrac{\tan t}{t \cos t}$

102. $f(x) = \dfrac{x^2 \sin x - x \cos x}{x^2 \sin x + x \cos x}$

103. $g(x) = \dfrac{x \sin x \cos x - \tan x}{x + \cos x}$

104. $f(x) = \dfrac{\sqrt{x} \sin x - x\sqrt{x} \cos x}{x^2 + 2x + 3}$

105. In this exercise we will develop a rule for finding the derivative of a product of more than two functions.

a) Compute $\dfrac{d}{dx} (x - 1)(x - 2)(x - 3)$ without multiplying it out. (*Hint:* Think of the function as a product of two functions, $(x - 1)[(x - 2)(x - 3)]$.)

b) Compute $\dfrac{d}{dx} (2x + 1)(3x - 5)(-x + 3)$.

tw c) Based on your calculations in parts (a) and (b), develop a rule that will give the derivative of a product of three functions.

tw d) Guess a rule and describe it carefully in words for a product of more than three functions. Try your rule on $f(x) = x(x + 1)(2x + 3)(-x + 1)$ and then check your answer by multiplying out some of the factors before you differentiate.

 Technology Connection

For the function in each of Exercises 106–111, graph f and f'. Then estimate points at which the tangent line is horizontal.

106. $f(x) = x^2(x - 2)(x + 2)$

107. $f(x) = \left(x + \dfrac{2}{x}\right)(x^2 - 3)$

108. $f(x) = \dfrac{x^3 - 1}{x^2 + 1}$

109. $f(x) = \dfrac{0.3x}{0.04 + x^2}$

110. $f(x) = \dfrac{0.01x^2}{x^4 + 0.0256}$

111. $f(x) = \dfrac{4x}{x^2 + 1}$

112. Decide graphically which of the following seems to be the correct derivative of the function of Exercise 111.

$$y_1 = \dfrac{2}{x},$$

$$y_2 = \dfrac{4 - 4x}{x^2 + 1},$$

$$y_3 = \dfrac{4 - 4x^2}{(x^2 + 1)^2},$$

$$y_4 = \dfrac{4x^2 - 4}{(x^2 + 1)^2}$$

2.8 The Chain Rule

OBJECTIVE

■ Differentiate using the Extended Power Rule and the Chain Rule.

The Extended Power Rule

How can we differentiate more complicated functions such as

$$y = (1 + x^2)^3,$$
$$y = (1 + x^2)^{89},$$

or

$$y = (1 + x^2)^{1/3}?$$

For $(1 + x^2)^3$, we can expand and then differentiate. Although this could be done for $(1 + x^2)^{89}$, it would certainly be time-consuming, and such an expansion would not work for $(1 + x^2)^{1/3}$. Not knowing a rule, we might conjecture that the derivative of the function $y = (1 + x^2)^3$ is

$$3(1 + x^2)^2. \tag{1}$$

To check this, we expand $(1 + x^2)^3$ and then differentiate. From algebra, we recall that $(a + h)^3 = a^3 + 3a^2h + 3ah^2 + h^3$, so

$$(1 + x^2)^3 = 1^3 + 3 \cdot 1^2 \cdot (x^2)^1 + 3 \cdot 1 \cdot (x^2)^2 + (x^2)^3$$
$$= 1 + 3x^2 + 3x^4 + x^6.$$

(We could also have done this by finding $(1 + x^2)^2$ and then multiplying again by $1 + x^2$.) It follows that

$$\frac{dy}{dx} = 6x + 12x^3 + 6x^5$$

$$= (1 + 2x^2 + x^4)6x$$

$$= 3(1 + x^2)^2 \cdot 2x. \tag{2}$$

Comparing this with equation (1), we see that the Power Rule is not sufficient for such a differentiation. Note that the factor $2x$ in the actual derivative, equation (2), is the derivative of the "inside" function, $1 + x^2$. This is consistent with the following new rule.

THEOREM 8

The Extended Power Rule

Suppose that $g(x)$ is a differentiable function of x. Then for any real number k,

$$\frac{d}{dx}[g(x)]^k = k[g(x)]^{k-1} \cdot \frac{d}{dx}g(x).$$

Let's differentiate $(1 + x^3)^5$. There are three steps to carry out.

$(1 + x^3)^5$ **1.** Mentally block out the "inside" function, $1 + x^3$.

$5(1 + x^3)^4$ **2.** Differentiate the "outside" function, $(1 + x^3)^5$.

$5(1 + x^3)^4 \cdot 3x^2$ **3.** Multiply by the derivative of the "inside" function.

$= 15x^2(1 + x^3)^4$

CAUTION   Step (3) is most commonly overlooked. Do not forget it!

EXAMPLE 1   Differentiate: $f(x) = (1 + x^3)^{1/2}$.

Solution

$$\frac{d}{dx}(1 + x^3)^{1/2} = \frac{1}{2}(1 + x^3)^{1/2-1} \cdot 3x^2$$

$$= \frac{1}{2}(1 + x^3)^{-1/2} \cdot 3x^2$$

$$= \frac{3x^2}{2\sqrt{1 + x^3}}.$$

Technology Connection

For Example 1, graph f and f'. Check the results of Example 1 graphically.

EXAMPLE 2   Differentiate: $y = (1 - x^2)^3 - (1 - x^2)^2$.

Solution   Here we combine the Difference Rule and the Extended Power Rule:

$$\frac{dy}{dx} = 3(1 - x^2)^2(-2x) - 2(1 - x^2)(-2x).$$ We differentiate each term using the Extended Power Rule.

Thus,

$$\frac{dy}{dx} = -6x(1 - x^2)^2 + 4x(1 - x^2)$$

$$= 2x(1 - x^2)[-3(1 - x^2) + 2] \qquad \text{Here we factor out } 2x(1 - x^2).$$
$$= 2x(1 - x^2)[-3 + 3x^2 + 2]$$
$$= 2x(1 - x^2)(3x^2 - 1).$$

EXAMPLE 3   Differentiate: $f(x) = \sin^3 x \cos^4 x$.

Solution   Since $f(x)$ is the product of the functions $\sin^3 x$ and $\cos^4 x$, we first apply the Product Rule.

$$f'(x) = \sin^3 x \frac{d}{dx} \cos^4 x + \left(\frac{d}{dx} \sin^3 x\right) \cos^4 x.$$

Since $\sin^3 x = (\sin x)^3$ and $\cos^4 x = (\cos x)^4$, we can use the Extended Power Rule to compute the two derivatives indicated above.

$$f'(x) = (\sin^3 x)(4 \cos^3 x)(-\sin x) + 3 \sin^2 x(\cos x) \cos^4 x$$
$$= -4 \sin^4 x \cos^3 x + 3 \sin^2 x \cos^5 x.$$

EXAMPLE 4   Differentiate: $f(x) = \sqrt[4]{\dfrac{x + 3}{x - 1}}$.

Solution   We can use the Quotient Rule to differentiate the inside function, $(x + 3)/(x - 1)$:

$$\frac{d}{dx} \sqrt[4]{\frac{x + 3}{x - 1}} = \frac{d}{dx} \left(\frac{x + 3}{x - 1}\right)^{1/4} = \frac{1}{4} \left(\frac{x + 3}{x - 1}\right)^{1/4 - 1} \left[\frac{(x - 1)1 - 1(x + 3)}{(x - 1)^2}\right]$$

$$= \frac{1}{4} \left(\frac{x + 3}{x - 1}\right)^{-3/4} \left[\frac{x - 1 - x - 3}{(x - 1)^2}\right]$$

$$= \frac{1}{4} \left(\frac{x + 3}{x - 1}\right)^{-3/4} \cdot \frac{-4}{(x - 1)^2}$$

$$= \left(\frac{x - 1}{x + 3}\right)^{3/4} \cdot \frac{-1}{(x - 1)^2}$$

$$= \frac{-1}{(x + 3)^{3/4}(x - 1)^{5/4}}.$$

The Chain Rule

The Extended Power Rule is a special case of a more general rule called the **Chain Rule.**

THEOREM 9

The Chain Rule

If $f(x)$ and $g(x)$ are differentiable, then the derivative of the composition $f \circ g$ is given by

$$\frac{d}{dx}[f \circ g(x)] = \frac{d}{dx}[f(g(x))] = f'(g(x)) \cdot \frac{d}{dx} g(x).$$

To see why the Extended Power Rule is a special case of the Chain Rule, consider the function $f(x) = x^k$. Then for any other function $g(x)$, $(f \circ g)(x) = [g(x)]^k$, and the derivative of the composition is

$$\frac{d}{dx}[g(x)]^k = k[g(x)]^{k-1} \cdot \frac{d}{dx}g(x).$$

The Chain Rule often appears in another form. Suppose that $y = f(u)$ and $u = g(x)$. Then

$$\frac{dy}{dx} = \frac{dy}{du} \cdot \frac{du}{dx}.$$

This form is very helpful when y is expressed in terms of an intermediate variable u.

EXAMPLE 5   For $y = 2 + \sqrt{u}$ and $u = x^3 + 1$, find dy/du, du/dx, and dy/dx.

Solution   First we find dy/du and du/dx:

$$\frac{dy}{du} = \frac{1}{2}u^{-1/2} \quad \text{and} \quad \frac{du}{dx} = 3x^2.$$

Then

$$\frac{dy}{dx} = \frac{dy}{du} \cdot \frac{du}{dx}$$

$$= \frac{1}{2\sqrt{u}} \cdot 3x^2$$

$$= \frac{3x^2}{2\sqrt{x^3 + 1}}. \qquad \text{Substituting } x^3 + 1 \text{ for } u$$

We saw that the Extended Power Rule is a special case of the Chain Rule. In the same way, we can use the Chain Rule to determine the derivative of extended trigonometric functions.

EXAMPLE 6   Compute $\dfrac{d}{dx}\sin(1 + x^2)$.

Solution   Notice that $\sin(1 + x^2)$ is the composition $(f \circ g)(x)$, where $f(x) = \sin x$ and $g(x) = 1 + x^2$. Therefore, $f'(x) = \cos x$. We use the Chain Rule to compute the derivative.

$$\frac{d}{dx}f \circ g(x) = f'(g(x))\frac{d}{dx}g(x)$$

$$= f'(1 + x^2)\frac{d}{dx}(1 + x^2) \qquad \text{Substituting } g(x) = 1 + x^2$$

$$= \cos(1 + x^2)\frac{d}{dx}(1 + x^2)$$

$$= \cos(1 + x^2) \cdot 2x$$

$$= 2x\cos(1 + x^2).$$

Example 6 illustrates one special case of the Chain Rule. If $f(x) = \sin u$, where u is a function of x, then $f'(x) = (\cos u)u'$. The following theorem gives special cases of the Chain Rule involving all the trigonometric functions.

THEOREM 10

Let $u = g(x)$ be a differentiable function. Then

$$\frac{d}{dx} \sin u = (\cos u)u', \quad \frac{d}{dx} \tan u = (\sec^2 u)u', \quad \frac{d}{dx} \sec u = (\sec u \tan u)u',$$

$$\frac{d}{dx} \cos u = -(\sin u)u', \quad \frac{d}{dx} \cot u = -(\csc^2 u)u', \quad \frac{d}{dx} \csc u = -(\csc u \cot u)u'.$$

EXAMPLE 7   Compute the derivatives.

a) $\dfrac{d}{dx} \tan(3x + 2)$

b) $\dfrac{d}{dx} [x \csc(3x + 2)]$

c) $\dfrac{d}{dx} \sec^4(\sqrt{x} + 1)$

Solution

a) We can use the Chain Rule.

$$\frac{d}{dx} \tan(3x + 2) = (\sec^2(3x + 2)) \cdot 3 = 3 \sec^2(3x + 2).$$

b) We first use the Product Rule.

$$\frac{d}{dx} [x \csc(3x + 2)] = x \left[\frac{d}{dx} \csc(3x + 2) \right] + \left(\frac{d}{dx} x \right) \csc(3x + 2)$$

$$= x(-\csc(3x + 2) \cot(3x + 2))(3) + 1 \cdot \csc(3x + 2)$$

$$= \csc(3x + 2) - 3x \csc(3x + 2) \cot(3x + 2).$$

c) Recall that $\sec^4(\sqrt{x} + 1) = (\sec(\sqrt{x} + 1))^4$. We therefore apply the Extended Power Rule.

$$\frac{d}{dx} \sec^4(\sqrt{x} + 1) = \frac{d}{dx} (\sec(\sqrt{x} + 1))^4$$

$$= 4(\sec(\sqrt{x} + 1))^3 \cdot \frac{d}{dx} \sec(\sqrt{x} + 1)$$

$$= 4 \sec^3(\sqrt{x} + 1) \sec(\sqrt{x} + 1) \tan(\sqrt{x} + 1) \frac{d}{dx} (\sqrt{x} + 1)$$

$$= 4 \sec^4(\sqrt{x} + 1) \tan(\sqrt{x} + 1) \frac{1}{2} x^{-1/2}$$

$$= \frac{2 \sec^4(\sqrt{x} + 1) \tan(\sqrt{x} + 1)}{\sqrt{x}}.$$

EXAMPLE 8   Compute $f'(x)$, where $f(x) = \tan(\sin 2x)$.

Solution   It is important to note that $f(x)$ is a composition of functions, not a product of functions. We therefore use the Chain Rule.

$$f'(x) = \frac{d}{dx} \tan(\sin 2x)$$

$$= (\sec^2(\sin 2x)) \frac{d}{dx} \sin 2x$$

$$= (\sec^2(\sin 2x))(\cos 2x) \frac{d}{dx}(2x)$$

$$= 2 \sec^2(\sin 2x) \cos 2x.$$

Exercise Set 2.8

Differentiate.

1. $y = (2x + 1)^2$, three ways

2. $y = (3 - 2x)^2$, three ways

3. $y = (1 - 2x)^{55}$ 4. $y = \tan^2 x$

5. $y = \sec^2 x$ 6. $y = (2x + 3)^{10}$

7. $y = \sqrt{1 - 3x}$ 8. $y = \sqrt[3]{x^2 + 1}$

9. $f(x) = \dfrac{2}{3x^2 + 1}$, two ways

10. $f(x) = \dfrac{3}{\sqrt{2x + 4}}$, two ways

11. $s(t) = t\sqrt{2t + 3}$ 12. $s(t) = t^2 \sqrt[3]{3t + 4}$

13. $s(t) = \sin\left(\dfrac{\pi}{6} t + \dfrac{\pi}{3}\right)$

14. $s(t) = \cos(3t - 4)$

15. $g(x) = (1 + x^3)^3 - (1 + x^3)^4$

16. $y = \sqrt{1 + \sec x}$

17. $y = \sqrt{1 - \csc x}$

18. $g(x) = \sqrt{x} + (x - 3)^3$

19. $g(x) = \sqrt[3]{2x - 1} + (4 - x)^2$

20. $y = x^2 \cos x - 2x \sin x - 2 \cot x$

21. $y = x^3 \sin x + 5x \cos x + 4 \sec x$

22. $f(x) = \sqrt{2x + 3}(x^2 + 3x + 1)$

23. $y = \sqrt{x^2 + x^3}(2x^2 + 3x + 5)$

24. $y = \csc(1/x)$

25. $f(t) = \cos \sqrt{t}$

26. $f(t) = \sin \sqrt{t}$

27. $f(x) = (3x + 2)\sqrt{2x + 5}$

28. $f(x) = (5x - 2)\sqrt{3x + 4}$

29. $y = \sin(\cos x)$

30. $y = \tan(\sin x)$

31. $y = \sqrt{\cos 4t}$

32. $f(x) = \dfrac{(x^2 + 3)^4}{(x^3 - 1)^5}$

33. $f(x) = \dfrac{(x^3 + 2x^2 + 3x - 1)^3}{(2x^4 + 1)^2}$

34. $y = \sqrt[3]{\dfrac{2x + 3}{3x - 5}}$

35. $y = \sqrt{\dfrac{3x - 4}{5x + 3}}$

36. $f(x) = \tan(x\sqrt{x - 1})$

37. $r(x) = x(0.01x^2 + 2.391x - 8.51)^5$

38. $r(x) = (3.21x - 5.87)^3(2.36x - 5.45)^5$

39. $y = \sqrt[5]{\cot 5x - \cos 5x}$

40. $y = \cot x - \sin(\cos^2 x)$

41. $y = \sin(\sec^4(x^2))$

42. $f(x) = \sqrt{\sqrt{\sqrt{2x + 3} + 1}}$

43. $g(x) = \sqrt[3]{\sqrt[4]{\sqrt{x^2 + 2} + 1}}$

44. $y = \dfrac{x}{(x + \sin x)^2}$

45. $y = \dfrac{\sin^3 x}{x^2 + 5}$

46. $y = \tan^2(\sqrt{t + 2})$

47. $f(x) = \cot^3(x \sin(2x + 4))$

48. $y = \sqrt{2 + \cos^2 t}$

49. $y = \sqrt{\sec^4 x + x}$

50. $y = \sqrt{x + \csc x}$

Find $\dfrac{dy}{du}, \dfrac{du}{dx}$, and $\dfrac{dy}{dx}$.

51. $y = \sqrt{u}$ and $u = x^2 - 1$

52. $y = \dfrac{15}{u^3}$ and $u = 2x + 1$

53. $y = u^{50}$ and $u = 4x^3 - 2x^2$

54. $y = \dfrac{u + 1}{u - 1}$ and $u = 1 + \sqrt{x}$

55. $y = u(u + 1)$ and $u = x^3 - 2x$

56. $y = (u + 1)(u - 1)$ and $u = x^3 + 1$

57. Find an equation for the tangent line to the graph of $y = \sqrt{x^2 + 3x}$ at the point $(1, 2)$.

58. Find an equation for the tangent line to the graph of $y = (x^3 - 4x)^{10}$ at the point $(2, 0)$.

59. Find an equation for the tangent line to the graph of $y = x\sqrt{2x + 3}$ at the point $(3, 9)$.

60. Find an equation for the tangent line to the graph of $y = \left(\dfrac{2x + 3}{x - 1}\right)^3$ at the point $(2, 343)$.

61. Find an equation for the tangent line to the graph of $f(x) = \sin^2 x$ at the point $(-\pi/6, 1/4)$.

62. Find an equation for the tangent line to the graph of $f(x) = x \sin 2x$ at the point $(\pi, 0)$.

63. Consider

$$f(x) = \frac{x^2}{(1 + x)^5}.$$

a) Find $f'(x)$ using the Quotient Rule and the Extended Power Rule.
b) Note that $f(x) = x^2(1 + x)^{-5}$. Find $f'(x)$ using the Product Rule and the Extended Power Rule.
c) Compare your answers to parts (a) and (b).

64. Consider

$$g(x) = (x^3 + 5x)^2.$$

a) Find $g'(x)$ using the Extended Power Rule.
b) Note that $g(x) = x^6 + 10x^4 + 25x^2$. Find $g'(x)$.
c) Compare your answers to parts (a) and (b).

65. Let $f(u) = u^3$ and $g(x) = u = 2x^4 + 1$
Find $(f \circ g)'(-1)$.

66. Let $f(u) = \dfrac{u + 1}{u - 1}$ and $g(x) = u = \sqrt{x}$
Find $(f \circ g)'(4)$.

67. Let $f(u) = \sqrt[3]{u}$ and $g(x) = u = 1 - 3x^2$
Find $(f \circ g)'(2)$.

68. Let $f(u) = 2u^5$ and $g(x) = u = \dfrac{3 - x}{4 + x}$
Find $(f \circ g)'(-10)$.

APPLICATIONS

69. *Compound Interest.* If $1000 is invested at interest rate i, compounded annually, it will grow in 3 yr to an amount A given by

$$A = \$1000(1 + i)^3.$$

a) Find the rate of change, dA/di.
tw b) Interpret the meaning of dA/di.

70. *Compound Interest.* If $1000 is invested at interest rate i, compounded quarterly, it will grow in 5 yr to an amount A given by

$$A = \$1000\left(1 + \frac{i}{4}\right)^{20}.$$

a) Find the rate of change, dA/di.
tw b) Interpret the meaning of dA/di.

71. *Chemotherapy Dosage.* The dosage for carboplatin chemotherapy drugs depends on several parameters of the drug as well as the age, weight, and sex of the patient. For a female patient, the formulas giving the dosage for a certain drug are

$$D = 0.85A(c + 25)$$

and

$$c = \frac{(140 - y)w}{72x},$$

where A and x depend on which drug is used, D is the dosage in milligrams (mg), c is called the creatine clearance, y is the patient's age in years, and w is the patient's weight in kg.[11]

a) Suppose a patient is a 45-yr-old woman and the drug has parameters $A = 5$ and $x = 0.6$. Use this information to find formulas for D and c that give D as a function of c and c as a function of w.

b) Use your formulas in part (a) to compute $\dfrac{dD}{dc}$.

c) Use your formulas in part (a) to compute $\dfrac{dc}{dw}$.

d) Compute $\dfrac{dD}{dw}$.

tw e) Interpret the meaning of the derivative $\dfrac{dD}{dw}$.

---

[11]US Oncology. http://www.usoncology.com/Home/

72. *Chemotherapy Dosage.* The dosage for carboplatin chemotherapy drugs depends on several parameters of the drug as well as the age, weight, and sex of the patient. For a male patient, the formulas giving the dosage for a certain drug are

$$D = A(c + 25)$$

and

$$c = \frac{(140 - y)w}{72x},$$

where A and x depend on which drug is used, D is the dosage in milligrams (mg), c is called the creatine clearance, y is the patient's age in years, and w is the patient's weight in kilograms.[12]

a) Suppose a patient is a 45-yr-old man and the drug has parameters $A = 5$ and $x = 0.6$. Use this information to find formulas for D and c that give D as a function of c and c as a function of w.

b) Use your formulas in part (a) to compute $\dfrac{dD}{dc}$.

c) Use your formulas in part (a) to compute $\dfrac{dc}{dw}$.

d) Compute $\dfrac{dD}{dw}$.

tw e) Interpret the meaning of the derivative $\dfrac{dD}{dw}$.

73. *Carbon Dioxide Concentrations.* The concentration (in ppmv, or parts per million by volume) of atmospheric carbon dioxide at Mauna Loa, Hawaii, may be modeled with the function

$$C(t) = 312.7 + 0.74t + 0.01188t^2$$
$$- 0.5407 \sin(2\pi t),$$

where t is the number of years since January 1957.[13]

a) Use this model to predict the rate that the carbon dioxide concentration will be decreasing in January 2009.

b) Use this model to predict the rate that the carbon dioxide concentration will be increasing in July 2009.

---

[12]US Oncology. http://www.usoncology.com/Home/

[13]C. D. Keeling and T. P. Whorf, "Atmospheric CO_2 records from sites in the SIO air sampling network." In *Trends: A Compendium of Data on Global Change. Carbon Dioxide Information Analysis Center,* Oak Ridge National Laboratory, U.S. Department of Energy, Oak Ridge, Tennessee (2002).

74. *Body Temperature.* In a laboratory experiment, the body temperature T of rats was found to follow the function

$$T(t) = 36.3 + 0.82 \sin\left(\frac{\pi}{12}[t + 2]\right),$$

where t is the number of hours after the experiment began. How quickly was the body temperature of rats decreasing 8 hr after the experiment began?[14]

SYNTHESIS

Differentiate.

75. $y = ((x^2 + 4)^8 + 3\sqrt{x})^4$

76. $y = \sqrt{(x^5 + x + 1)^3 + 7 \sec^2 x}$

77. $\sin(\sin(\sin x))$

78. $\cos(\sec(\sin 2x))$

79. $\tan(\cot(\sec 3x))$

80. $\csc(\cos(\sec(\sqrt{x^2 + 1})))$

81. $\sqrt[5]{\sin\left(\frac{3\pi}{2} + 3\right)}$

82. $\sqrt[7]{\csc^3\left(\frac{5\pi}{6} + 2.389\right)}$

83. The Sum Identity for the sine function is

$$\sin(a + x) = \sin a \cos x + \cos a \sin x.$$

Differentiate both sides, thinking of a as a constant and x as the variable. Simplify your answer and show that you get the Sum Identity for the cosine function.

84. The sum formula for the cosine function is

$$\cos(a + x) = \cos a \cos x - \sin a \sin x.$$

Differentiate both sides, thinking of a as a constant and x as the variable. Simplify your answer and show that you get the sum formula for the sine function.

85. The following is the beginning of an alternative proof of the Quotient Rule for finding the derivative of a quotient that uses the Product Rule and the Power Rule. Complete the proof, giving reasons for each step.

---

[14]H. Takeuchi, A. Enzo, and H. Minamitani, "Circadian rhythm changes in heart rate variability during chronic sound stress," *Medical and Biological Engineering and Computing,* Vol. 39, pp. 113–117 (2001).

Proof: Let

$$Q(x) = \frac{N(x)}{D(x)}.$$

Then

$$Q(x) = N(x) \cdot [D(x)]^{-1}.$$

Therefore, ...

tw 86. Describe composition of functions in as many ways as possible.

Technology Connection

For the function in each of Exercises 87 and 88, graph f and f' over the given interval. Then estimate points at which the tangent line is horizontal.

87. $f(x) = 1.68x\sqrt{9.2 - x^2}; [-3, 3]$

88. $f(x) = \sqrt{6x^3 - 3x^2 - 48x + 45}; [-5, 5]$

Find the derivative of each of the following functions analytically. Then use a grapher to check the results.

89. $f(x) = x\sqrt{4 - x^2}$

90. $f(x) = \dfrac{4x}{\sqrt{x - 10}}$

2.9 Higher-Order Derivatives

OBJECTIVE

■ Find derivatives of higher order.

Consider the function given by

$$y = f(x) = x^5 - 3x^4 + x.$$

Its derivative f' is given by

$$y' = f'(x) = 5x^4 - 12x^3 + 1.$$

The derivative function f' can also be differentiated. We can think of its derivative as the rate of change of the slope of the tangent lines of f. We use the notation f'' for the derivative $(f')'$. That is,

$$f''(x) = \frac{d}{dx} f'(x).$$

We call f'' the *second derivative* of f. It is given by

$$y'' = f''(x) = 20x^3 - 36x^2.$$

Continuing in this manner, we have

$$f'''(x) = 60x^2 - 72x, \qquad \text{The third derivative of } f$$

$$f''''(x) = 120x - 72, \qquad \text{The fourth derivative of } f$$

$$f'''''(x) = 120. \qquad \text{The fifth derivative of } f$$

When notation like $f'''(x)$ gets lengthy, we abbreviate it using a symbol in parentheses. Thus, $f^{(n)}(x)$ is the nth derivative. For the function above,

$$f^{(4)}(x) = 120x - 72,$$
$$f^{(5)}(x) = 120,$$
$$f^{(6)}(x) = 0, \quad \text{and}$$
$$f^{(n)}(x) = 0, \quad \text{for any integer } n \geq 6.$$

CAUTION   The notation $y^{(4)}$ means the fourth derivative of y and not $y \cdot y \cdot y \cdot y$.

Leibniz's notation for the second derivative of a function given by $y = f(x)$ is

$$\frac{d^2y}{dx^2}, \quad \text{or} \quad \frac{d}{dx}\left(\frac{dy}{dx}\right),$$

read "the second derivative of y with respect to x." The 2's in this notation are *not* exponents. If $y = x^5 - 3x^4 + x$, then

$$\frac{d^2y}{dx^2} = 20x^3 - 36x^2.$$

Leibniz's notation for the third derivative is d^3y/dx^3; for the fourth derivative, d^4y/dx^4; and so on:

$$\frac{d^3y}{dx^3} = 60x^2 - 72x,$$

$$\frac{d^4y}{dx^4} = 120x - 72,$$

$$\frac{d^5y}{dx^5} = 120.$$

EXAMPLE 1   For $y = 1/x$, find d^2y/dx^2.

Solution   We have $y = x^{-1}$, so

$$\frac{dy}{dx} = -1 \cdot x^{-1-1} = -x^{-2}, \quad \text{or} \quad -\frac{1}{x^2}.$$

Then

$$\frac{d^2y}{dx^2} = (-2)(-1)x^{-2-1} = 2x^{-3}, \quad \text{or} \quad \frac{2}{x^3}. \qquad \blacksquare$$

EXAMPLE 2   For $y = (x^2 + 10x)^{20}$, find y' and y''.

Solution   To find y', we use the Chain Rule:

$$y' = 20(x^2 + 10x)^{19}(2x + 10)$$
$$= 20(x^2 + 10x)^{19} \cdot 2(x + 5)$$
$$= 40(x^2 + 10x)^{19}(x + 5).$$

To find y'', we use the Product Rule and the Chain Rule:

$$y'' = 40(x^2 + 10x)^{19}(1) + 19 \cdot 40(x^2 + 10x)^{18}(2x + 10)(x + 5)$$
$$= 40(x^2 + 10x)^{19} + 760(x^2 + 10x)^{18} \cdot 2(x + 5)(x + 5)$$
$$= 40(x^2 + 10x)^{19} + 1520(x^2 + 10x)^{18}(x + 5)^2$$
$$= 40(x^2 + 10x)^{18}[(x^2 + 10x) + 38(x + 5)^2]$$
$$= 40(x^2 + 10x)^{18}[x^2 + 10x + 38(x^2 + 10x + 25)]$$
$$= 40(x^2 + 10x)^{18}[39x^2 + 390x + 950].$$

We encountered an application of second derivatives when we considered acceleration in Section 2.6. Acceleration can be regarded as a second derivative. As an object moves, its distance from a fixed point after time t is some function of the time, say, $s(t)$. Then

$$v(t) = s'(t) = \text{the velocity at time } t$$

and

$$a(t) = v'(t) = s''(t) = \text{the acceleration at time } t.$$

Whenever a quantity is a function of time, the first derivative gives the rate of change with respect to time and the second derivative gives the acceleration. For example, if $y = P(t)$ gives the number of people in a population at time t, then $P'(t)$ represents how fast the size of the population is changing and $P''(t)$ gives the acceleration in the size of the population, or the rate at which the rate of change is changing.

EXAMPLE 3 *Simple Harmonic Motion.* The vertical position of a weight suspended by a spring is given by

$$y(t) = 10 \cos(3t + 1),$$

where t is time measured in seconds and $y(t)$ is measured in centimeters.

a) Find the velocity function.
b) Find the acceleration function.
c) How are the position and acceleration functions related?

Solution

a) We find the velocity by computing $y'(t)$.

$$v(t) = y'(t) = 10(-\sin(3t + 1))(3) = -30 \sin(3t + 1)$$

b) The acceleration is the second derivative of the position.

$$a(t) = y''(t) = v'(t) = -30(\cos(3t + 1))3 = -90 \cos(3t + 1)$$

c) Rewriting the acceleration function gives

$$a(t) = -90 \cos(3t + 1) = -9(10 \cos(3t + 1)) = -9y(t).$$

The acceleration is -9 times the position. In particular, when the weight is at a positive height, the acceleration is downward and when the position is negative, the acceleration is upward.

Exercise Set 2.9

Find the second derivative.

1. $y = 3x + 5$

2. $y = -4x + 7$

3. $y = \dfrac{-3}{2x + 2}$

4. $y = -\dfrac{1}{3x - 4}$

5. $y = \sqrt[3]{2x + 1}$

6. $f(x) = (3x + 2)^{-3}$

7. $f(x) = (4 - 3x)^{-4}$

8. $y = \sqrt{x - 1}$

9. $y = \sqrt{x + 1}$

10. $f(x) = (3x + 2)^{10}$

11. $f(x) = (2x + 9)^{16}$

12. $g(x) = \tan(2x)$

13. $g(x) = \sec(3x + 1)$

14. $f(x) = 13x^2 + 2x + 7 - \csc x$

15. $f(x) = \sec(2x + 3) + 4x^2 + 3x - 7$

16. $g(x) = mx + b$; m, b are constants

17. $g(x) = ax^2 + bx + c$; a, b, c are constants

18. $y = \sqrt[3]{2x + 4}$

19. $y = \sqrt[4]{(x^2 + 1)^3}$

20. $y = 13x^{3.2} + 12x^{1.2} - 5x^{-0.25}$

21. $f(x) = (4x + 3)\cos x$

22. $s(t) = \sin(at)$; a is a constant

23. $s(t) = \cos(at + b)$; a, b are constants

24. $y = x^{5/2} + x^{3/2} - x^{1/2}$

25. $y = \dfrac{\sqrt{t^2 + 3}}{7} + \sqrt[3]{3t^2 + 1}$

26. $y = \dfrac{1}{\pi^2}\tan \pi t + \dfrac{4}{\pi^2}\sec \pi t$

27. For $y = x^4$, find d^4y/dx^4.

28. For $y = x^5$, find d^4y/dx^4.

29. For $y = x^6 - x^3 + 2x$, find d^5y/dx^5.

30. For $y = x^7 - 8x^2 + 2$, find d^6y/dx^6.

31. For $y = (x^2 - 5)^{10}$, find d^2y/dx^2.

32. For $y = x^k$, find d^5y/dx^5.

33. For $y = \sec(2x + 3)$, find $\dfrac{d^3y}{dx^3}$.

34. For $y = \cot(3x - 1)$, find $\dfrac{d^3y}{dx^3}$.

35. If s is given by $s(t) = 10\cos(3t + 2) - 4\sin(3t + 2)$, find the acceleration.

36. If s is given by $s(t) = 6.8\tan(2.6t - 1)$, find the acceleration.

37. If s is a distance given by $s(t) = t^3 + t^2 + 2t$, find the acceleration.

38. If s is a distance given by $s(t) = t^4 + t^2 + 3t$, find the acceleration.

APPLICATIONS

39. *Weight of Baby.* The median weight of a boy whose age t is between 0 and 36 mo can be approximated by the function

    $$w(t) = 0.000758t^3 - 0.0596t^2 + 1.82t + 8.15,$$

    where t is measured in months and w is measured in pounds. Use this approximation to compute the acceleration in the weight of a boy of median weight.[15]

40. *Weight of Baby.* The median weight of a girl whose age t is between 0 and 36 mo can be approximated by the function

    $$w(t) = 0.0006t^3 - 0.0484t^2 + 1.61t + 7.6,$$

    where t is measured in months and w is measured in pounds. Use this approximation to compute the acceleration in the weight of a girl of median weight.[16]

[15,16]Centers for Disease Control. Developed by the National Center for Health Statistics in collaboration with the National Center for Chronic Disease Prevention and Health Promotion (2000).

41. *Population Growth.* A population grows from an initial size of 100,000 to an amount $P(t)$, given by

$$P(t) = 100{,}000(1 + 0.6t + t^2).$$

What is the acceleration in the size of the population?

42. *Population Growth.* A population grows from an initial size of 100,000 to an amount $P(t)$, given by

$$P(t) = 100{,}000(1 + 0.4t + t^2).$$

What is the acceleration in the size of the population?

SYNTHESIS

Find y', y'', and y'''.

43. $y = \dfrac{x}{\sqrt{x-1}}$

44. $y = \dfrac{\sqrt{x}-1}{\sqrt{x}+1}$

Find $f''(x)$.

45. $f(x) = \dfrac{x}{x-1}$

46. $f(x) = \dfrac{1}{1+x^2}$

47. For $y = \sin x$, find

a) $\dfrac{dy}{dx}$.

b) $\dfrac{d^2y}{dx^2}$.

c) $\dfrac{d^3y}{dx^3}$.

d) $\dfrac{d^4y}{dx^4}$.

e) $\dfrac{d^8y}{dx^8}$.

f) $\dfrac{d^{10}y}{dx^{10}}$.

g) $\dfrac{d^{837}y}{dx^{837}}$.

48. For $y = \cos x$, find

a) $\dfrac{dy}{dx}$.

b) $\dfrac{d^2y}{dx^2}$.

c) $\dfrac{d^3y}{dx^3}$.

d) $\dfrac{d^4y}{dx^4}$.

e) $\dfrac{d^8y}{dx^8}$.

f) $\dfrac{d^{11}y}{dx^{11}}$.

g) $\dfrac{d^{523}y}{dx^{523}}$.

tW 49. The sine function has the property that its second derivative is the negative of the function. That is,

$$\frac{d^2}{dx^2}\sin x = -\sin x.$$ Find as many functions as you

can whose second derivative is the negative of the function.

Find the first through the fifth derivatives. Be sure to simplify at each stage before continuing.

50. $f(x) = \dfrac{x-1}{x+2}$

51. $f(x) = \dfrac{x+3}{x-2}$

Technology Connection

For the function in each of Exercises 52–57, graph f, f', and f'' over the given interval. Analyze and compare the behavior of these functions.

52. $f(x) = 0.1x^4 - x^2 + 0.4;\ [-5,5]$

53. $f(x) = -x^3 + 3x;\ [-3,3]$

54. $f(x) = x^4 + x^3 - 4x^2 - 2x + 4;\ [-3,3]$

55. $f(x) = x^3 - 3x^2 + 2;\ [-3,5]$

56. $f(x) = \tan x;\ [-1.5,\ 1.5]$

57. $f(x) = \sec x;\ [-1.5,\ 1.5]$

Chapter 2   Summary and Review

Terms to Know

Limit, p. 71
Continuity at a point, p. 76
Continuous over an interval, p. 78
Continuous function, p. 78
Average rate of change, p. 93
Secant line, p. 94
Difference quotient, p. 95
Simplified difference quotient, p. 96

Tangent line, p. 102
Derivative, p. 104
Differentiation, p. 104
Leibniz notation, p. 115
Power Rule, p. 116
Sum–Difference Rule, p. 120
Instantaneous rate of change, p. 125
Speed, p. 125

Velocity, p. 126
Acceleration, p. 126
Growth rate, p. 128
Product Rule, p. 133
Quotient Rule, p. 135
Extended Power Rule, p. 143
Chain Rule, p. 144
Higher-order derivative, p. 150

Review Exercises

These review exercises are for test preparation. They can also be used as a lengthened practice test. Answers are at the back of the book. The answers also contain bracketed section references, which tell you where to restudy if your answer is incorrect.

Consider

$$\lim_{x \to -7} f(x), \quad \text{where } f(x) = \frac{x^2 + 4x - 21}{x + 7},$$

for Exercises 1–3.

1. *Limit Numerically.*

   a) Complete the following input–output tables.

$x \to -7^-$	$f(x)$
-8	
-7.5	
-7.1	
-7.01	
-7.001	
-7.0001	

$x \to -7^+$	$f(x)$
-6	
-6.5	
-6.9	
-6.99	
-6.999	
-6.9999	

b) Find $\lim_{x \to -7^-} f(x)$, $\lim_{x \to -7^+} f(x)$, and $\lim_{x \to -7} f(x)$, if each exists.

2. *Limit Graphically.* Graph the function and use the graph to find the limit.

3. *Limit Algebraically.* Find the limit algebraically. Show your work.

Find the limit, if it exists.

4. $\lim_{x \to -2} \dfrac{8}{x}$

5. $\lim_{x \to -1} (2x^4 - 3x^2 + x + 4)$

6. $\lim_{x \to 6} \dfrac{x^2 + 5x - 66}{x - 6}$ **7.** $\lim_{x \to 4} \sqrt{x^2 + 9}$

Determine whether the function is continuous.

8.

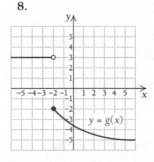

9.

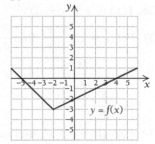

Use the function in Exercise 8 to answer Exercises 10–15.

10. Find $\lim\limits_{x \to 1} g(x)$.　　**11.** Find $g(1)$.

12. Is g continuous at 1?　　**13.** Find $\lim\limits_{x \to -2} g(x)$.

14. Find $g(-2)$.

15. Is g continuous at -2?

16. For $f(x) = x^3 + x^2 - 2x$, find the average rate of change as x changes from -1 to 2.

17. Find a simplified difference quotient for $g(x) = -3x + 5$.

18. Find a simplified difference quotient for $f(x) = 2x^2 - 3$.

19. Find an equation of the tangent line to the graph of $y = x^2 + 3x$ at the point $(-1, -2)$.

20. Find the points on the graph of $y = -x^2 + 8x - 11$ at which the tangent line is horizontal.

21. Find the points on the graph of $y = 5x^2 - 49x + 12$ at which the tangent line has slope 1.

Find dy/dx.

22. $y = 4x^5$　　　　　**23.** $y = 3\sqrt[3]{x}$

24. $y = \dfrac{-8}{x^8}$　　　　**25.** $y = 21x^{4/3}$

26. $y = \sec 5x$　　　　**27.** $y = x \cot(x^2)$

28. $y = 2.3\sqrt{0.4x + 5.3} + 0.01 \sin(0.17x - 0.31)$

Differentiate.

29. $f(x) = \dfrac{1}{6}x^6 + 8x^4 - 5x$

30. $y = \dfrac{x^3 + x}{x}$　　　　**31.** $y = \dfrac{x^2 + 8}{8 - x}$

32. $f(x) = (\tan x)/x$

33. $f(x) = \sqrt{\sin^2 x + x}$

34. $f(x) = \tan^2(x \cos x)$

35. $f(x) = \cot(x - \cos x)$

36. $f(x) = x^2(4x + 3)^{3/4}$

37. For $y = x^3 - \dfrac{2}{x}$, find $\dfrac{d^5 y}{dx^5}$.

38. For $y = \dfrac{1}{42}x^7 - 10x^3 + 13x^2 + 28x - 5 + \cos x$, find $\dfrac{d^4 y}{dx^4}$.

39. For $s(t) = t + t^4$, find each of the following.
a) $v(t)$
b) $a(t)$
c) The velocity and the acceleration when $t = 2$ sec

40. For
$$s(t) = 0.012t^3 - 1.85t + 2.3 + 0.002 \cos\left(\dfrac{\pi}{15}t\right),$$
find each of the following.
a) $v(t)$
b) $a(t)$
c) The velocity and the acceleration when $t = 2.5$ sec

41. *London Eye.* The London Eye is an observation wheel that allows tourists to see downtown London. Tourists enter from a platform 1 m off the ground and are taken 135 m above the ground. The Eye makes one full rotation every half-hour.
a) For a capsule that begins on the top of the Eye, find its height off the ground t hours later. (Ignore the fact that the Eye occasionally stops as passengers get on and off.)
b) How high above the ground will the capsule be 5 min after reaching the bottom?
c) Find the rate of change of the height of the capsule 5 min after reaching the bottom.

42. *Growth Rate.* The population of a city grows from an initial size of 10,000 to an amount P, given by $P = 10,000 + 50t^2$, where t is in years.
a) Find the growth rate.
b) Find the number of people in the city after 20 yr (at $t = 20$).
c) Find the growth rate at $t = 20$.

43. Find $(f \circ g)(x)$ and $(g \circ f)(x)$, given that $f(x) = x^2 + 5$ and $g(x) = 1 - 2x$.

SYNTHESIS

44. Differentiate $y = \dfrac{x\sqrt{1 + 3x}}{1 + x^3}$.

 Technology Connection

Use a grapher that creates input–output tables. Find each of the following limits. Start with Step $= 0.1$ and then go to 0.01, 0.001, and 0.0001. When you think you know the limit, graph, and use the **TRACE** feature to further verify your assertion.

45. $\lim\limits_{x \to 1} \dfrac{2 - \sqrt{x + 3}}{x - 1}$　　**46.** $\lim\limits_{x \to 11} \dfrac{\sqrt{x - 2} - 3}{x - 11}$

47. Graph f and f' over the given interval. Then estimate points at which the tangent line to f is horizontal.
$$f(x) = 3.8x^5 - 18.6x^3; \quad [-3, 3]$$

Chapter 2 Test

Consider

$$\lim_{x \to 6} f(x), \quad \text{where } f(x) = \frac{x^2 - 36}{x - 6}$$

for Exercises 1–3.

1. *Limit Numerically.*

   a) Complete the following input–output tables.

$x \to 6^-$	$f(x)$
5	
5.7	
5.9	
5.99	
5.999	
5.9999	

$x \to 6^+$	$f(x)$
7	
6.5	
6.1	
6.01	
6.001	
6.0001	

   b) Find $\lim\limits_{x \to 6^-} f(x)$, $\lim\limits_{x \to 6^+} f(x)$, and $\lim\limits_{x \to 6} f(x)$, if each exists.

2. *Limit Graphically.* Graph the function and use the graph to find the limit.

3. *Limit Algebraically.* Find the limit algebraically. Show your work.

Limits Graphically. Consider the following graph of function f for Exercises 4–11.

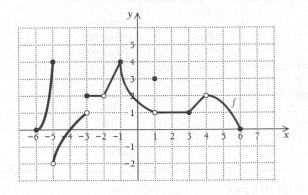

Find the limit, if it exists.

4. $\lim\limits_{x \to -5} f(x)$ **5.** $\lim\limits_{x \to -4} f(x)$

6. $\lim\limits_{x \to -3} f(x)$ **7.** $\lim\limits_{x \to -2} f(x)$

8. $\lim\limits_{x \to -1} f(x)$ **9.** $\lim\limits_{x \to 1} f(x)$

10. $\lim\limits_{x \to 2} f(x)$ **11.** $\lim\limits_{x \to 3} f(x)$

Determine whether the function is continuous.

12. 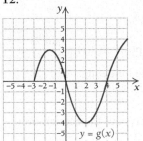   **13.**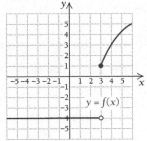

For the function in Exercise 13, answer the following.

14. Find $\lim\limits_{x \to 3} f(x)$.   **15.** Find $f(3)$.

16. Is f continuous at 3?   **17.** Find $\lim\limits_{x \to 4} f(x)$.

18. Find $f(4)$.   **19.** Is f continuous at 4?

Find the limit, if it exists.

20. $\lim\limits_{x \to 1} (3x^4 - 2x^2 + 5)$ **21.** $\lim\limits_{x \to 2^+} \dfrac{x - 2}{x(x^2 - 4)}$

22. $\lim\limits_{x \to 0} \dfrac{7}{x}$

23. Find a simplified difference quotient for
$$f(x) = 2x^2 + 3x - 9.$$

24. Find an equation of the tangent line to the graph of $y = x + (4/x)$ at the point $(4, 5)$.

25. Find the points on the graph of $y = x^3 - 3x^2$ at which the tangent line is horizontal.

Find dy/dx.

26. $y = 10\sqrt{x}$ **27.** $y = \dfrac{-10}{x}$

28. $y = x^{5/4}$

29. $y = -0.5x^2 + 0.61x + 90$

30. $y = \tan 2x$

Differentiate.

31. $f(x) = \dfrac{x}{5 - x}$

32. $y = (x^5 - 4x^3 + x)^{-5}$

33. $f(x) = x\sqrt{x^2 + 5}$

34. $f(x) = \cos(\sqrt{x})$

35. $f(x) = \tan 2x \sec 3x$

36. $f(x) = \dfrac{\tan^2 x}{\cos^2 x - x}$

37. $f(x) = \sin(\cos(x^2))$

38. $f(x) = x^2 + 1 + \sin(2x\sqrt{x + 2})$

39. For $y = x^4 - 3x^2$, find $\dfrac{d^3y}{dx^3}$.

40. Given that an object is at position $s(t) = \dfrac{\sin 2t}{t}$ at any time t, find each of the following.
    a) $v(t)$
    b) $a(t)$
    c) The position, the velocity, and the acceleration when $t = \dfrac{7\pi}{6}$.

41. *Memory.* In a certain memory experiment, a person is able to memorize M words after t minutes, where $M = -0.001t^3 + 0.1t^2$.
    a) Find the rate of change of the number of words memorized with respect to time.

b) How many words are memorized during the first 10 min (at $t = 10$)?

c) What is the memory rate at $t = 10$ min?

42. Find $f \circ g$ and $g \circ f$ for $f(x) = x^2 - x$ and $g(x) = 2x^3$.

SYNTHESIS

43. Differentiate $y = (1 - 3x)^{2/3}(1 + 3x)^{1/3}$.

44. Find $\displaystyle\lim_{x \to 3} \dfrac{x^3 - 27}{x - 3}$.

45. Find $\displaystyle\lim_{x \to 0} \dfrac{\tan t + \sin t}{t}$.

 Technology Connection

46. Graph f and f' over the given interval. Then estimate points at which the tangent line to f is horizontal.
$$f(x) = 5x^3 - 30x^2 + 45x + 5\sqrt{x}; \quad [0, 5]$$

47. Find the following limit using tables on a grapher:
$$\lim_{x \to 0} \dfrac{\sqrt{5x + 25} - 5}{x}.$$

Start with Step $= 0.1$ and then go to 0.01, 0.001, and 0.0001. When you think you know the limit, graph
$$y = \dfrac{\sqrt{5x + 25} - 5}{x},$$

and use the **TRACE** feature to verify your assertion.

Extended Life Science Connection

RATE OF EPIDEMIC SPREAD: SARS

Severe Acute Respiratory Syndrome (SARS) was first detected in late 2002. Starting in March 2003, the World Health Organization periodically compiled the cumulative number of reported SARS cases. Here we will develop a method to approximate a derivative based on the data, which has no formula to differentiate. We will use this method to estimate the rate SARS was spreading, measured in cases per week.

The table at right gives some of this data for March 2003.

Date in 2003	Time (weeks after March 19)	Cumulative number of reported cases $= S(t)$
March 17	$t = -2/7$	167
March 18	$t = 1/7$	219
March 19	$t = 0$	264
March 20	$t = 1/7$	306
March 21	$t = 2/7$	350

Source: WHO Web page
http://www.who.int/csr/sars/country/en/

We interpret the data in the table as approximating a continuous function $S(t)$, where t is time and $S(t)$ is the number of reported cases up to time t. The rate at which SARS spreads is the derivative $S'(t)$.

Recall from page 104 that *difference quotients* are used to define a derivative:

$$f'(0) = \lim_{h \to 0} \frac{f(0 + h) - f(0)}{h}$$
$$= \lim_{h \to 0} \frac{f(h) - f(0)}{h}.$$

The derivative at 0 is the slope of the tangent line, shown as the dashed green line in Figure 1.

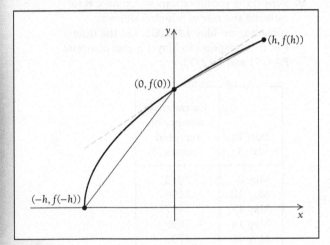

Figure 1

Also recall from page 72 that for the limit $\lim_{h \to 0} \dfrac{f(h) - f(0)}{h}$ to exist, both the left- and right-hand limits must exist, and they must be equal. Let h be positive. The function

$$F_+(h) = \frac{f(h) - f(0)}{h} \qquad (1)$$

(without taking the limit) is called the *forward difference quotient* at $x = 0$. This is the slope of the blue line in Figure 1. Since the limit of $F_+(h)$ as h approaches 0 from the right is the limit of the difference quotient,

$$f'(0) = \lim_{h \to 0^+} \frac{f(h) - f(0)}{h}$$
$$= \lim_{h \to 0^+} F_+(h),$$

we conclude that $F_+(h)$ should give a good approximation for $f'(0)$ for small values of h.

The function

$$F_-(h) = \frac{f(0 - h) - f(0)}{-h}$$
$$= \frac{f(0) - f(-h)}{h}, h > 0 \qquad (2)$$

is called the *backward difference quotient* at $x = 0$, and it is represented by the slope of the red line in Figure 1. For small values of $h > 0$, $F_-(h)$ is also a good approximation to $f'(0)$.

Even though we are trying to develop a procedure for estimating an instantaneous rate of change without knowing the function, let's assume we have one to develop and test our procedures. For example, if $f(x) = \sqrt{4 + x}$,

$$F_+(0.5) = \frac{\sqrt{4 + 0.5} - \sqrt{4}}{0.5}$$
$$= \frac{\sqrt{4.5} - 2}{0.5} \approx 0.242641 \text{ and}$$

$$F_-(0.5) = \frac{\sqrt{4} - \sqrt{4 - 0.5}}{0.5}$$
$$= \frac{2 - \sqrt{3.5}}{0.5} \approx 0.258342.$$

Both are reasonably close to $f'(0) = 0.25$, the actual value of the derivative of the function $f(x) = \sqrt{4 + x}$ at 0.

EXERCISES

1. Verify that if $f(x) = \sqrt{4 + x}$, then $f'(0) = 0.25$.

2. Complete the tables by computing the forward and backward difference quotients for $f(x)$.

a)

h	$\frac{1}{2}$	$\frac{1}{4}$	$\frac{1}{8}$	$\frac{1}{16}$
$F_+(h)$	0.242641			

b)

h	$\frac{1}{2}$	$\frac{1}{4}$	$\frac{1}{8}$	$\frac{1}{16}$
$F_-(h)$	0.258342			

3. a) Use the WHO data to compute the forward and backward difference quotients for March 19, let $t = 0$ correspond to March 19 and $h = 2/7$.

 b) Repeat part (a) with $h = 1/7$. Compare with your answer in (a). Which do you think is more accurate?

4. Let $F_c(h) = \frac{1}{2}[F_+(h) + F_-(h)]$. Use the definitions of $F_+(h)$ and $F_-(h)$ in equations (1) and (2) to show that

$$F_c(h) = \frac{f(h) - f(-h)}{2h}. \qquad (3)$$

The function $F_c(h)$ is called the *central difference quotient* at $x = 0$. The nDeriv function on a TI-83 calculator (see page 111) uses a central difference quotient to establish derivatives.

5. Explain why $F_c(h)$ represents the slope of the solid green line in Figure 2. Why does it make sense that $F_c(h)$ should be closer to $f'(0)$ than to either $F_+(h)$ or $F_-(h)$?

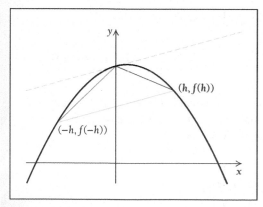

Figure 2

6. Complete the following table for $f(x) = \sqrt{4 + x}$.

h	$\frac{1}{2}$	$\frac{1}{4}$	$\frac{1}{8}$	$\frac{1}{16}$
$F_c(h)$				

7. Compare the numbers you computed in Exercises 2 and 6 to the actual derivative $f'(0) = 0.25$. Which value of h gives the best approximation to the derivative? Which difference quotient $F_+(h)$, $F_-(h)$, or $F_c(h)$ gives the best approximation for this value of h? Does this make sense?

8. Use the WHO data to compute the central difference quotient for March 19. Let $t = 0$ correspond to March 19 and let

 a) $h = 1/7$.

 b) $h = 2/7$.

 c) Compare parts (a) and (b). Which estimate of the rate of change is probably more accurate?

9. Repeat the computations in Exercise 8 to estimate the rate at which SARS was spreading on May 12, 2003. Let the time $t = 0$ correspond to May 12, and compute $F_c(4/7)$ and $F_c(2/7)$.

Date in 2003	Cumulative number of reported cases
May 8	7053
May 10	7296
May 12	7447
May 14	7628
May 16	7739

Source: WHO Web page
http://www.who.int/csr/sars/country/en/

10. The nDeriv function on the TI-83 calculator uses the central difference quotient $F_c(0.001)$ to estimate derivatives. However, $F_c(0.001)$ is only an estimate and may not be exactly correct. For the following functions, compute (if possible) the derivative $f'(0)$ and compare with the answer you get using nDeriv.

 a) $f(x) = x^3 + x$

 b) $f(x) = (10x)^3 + x$

 c) $f(x) = (100x)^3 + x$

 d) $f(x) = (1000x)^3 + x$

 e) $f(x) = |x|$

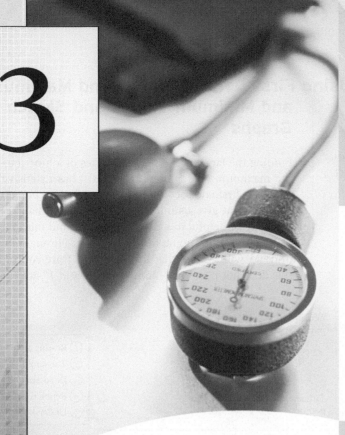

3

Applications of Differentiation

INTRODUCTION *In this chapter, we discover many applications of differentiation. We learn to find maximum and minimum values of functions, and that skill allows us to solve many kinds of problems in which we need to find the largest and/or smallest value of a function. We also apply our differentiation skills to graphing functions.*

AN APPLICATION For a dosage of x cubic centimeters of a certain drug, the resulting blood pressure B is approximated by

$$B(x) - 0.05x^2 - 0.3x^3, \quad 0 \le x \le 0.16.$$

Find the maximum blood pressure and the dosage at which it occurs.

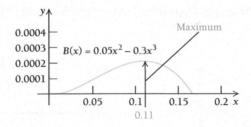

If we find the first derivative, set it equal to 0, and solve

$$B'(x) = 0.1x - 0.9x^2 = 0,$$

the resulting value of x will be the dosage at which the maximum blood pressure occurs.

This problem appears as Exercise 70 in Exercise Set 3.4.

3.1 Using First Derivatives to Find Maximum and Minimum Values and Sketch Graphs

OBJECTIVES

■ Find relative extrema of a continuous function using the First-Derivative Test.
■ Sketch graphs of continuous functions.

Finding the largest and smallest values of a function—that is, the maximum and minimum values—has extensive application. The first and second derivatives of a function are tools of calculus that give us information about the shape of a graph that may be helpful in finding maximum and minimum values of functions and in graphing functions. Throughout this section, we will assume that the functions f are continuous, but this does not necessarily imply that f' and f'' are continuous.

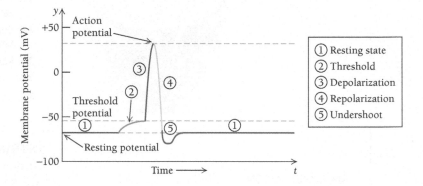

The increasing and decreasing regions of a neuron's action potential correspond to different stages of the sodium and potassium channels in its plasma membrane.[1]

Increasing and Decreasing Functions

If a graph of a function rises from left to right over an interval I, it is said to be **increasing** on I. If the graph drops from left to right, it is said to be **decreasing** on I.

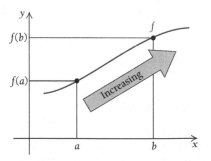

f is an increasing function over I.
If $a < b$, then $f(a) < f(b)$, for all a, b in I.

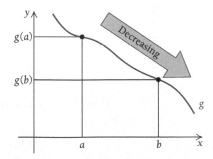

g is a decreasing function over I.
If $a < b$, then $g(a) > g(b)$, for all a, b in I.

---

[1] N. A. Campbell and J. B. Reece, *Biology*, 6th ed. (Benjamin Cummings, New York, 2002).

We can describe this mathematically as follows.

DEFINITION

A function f is **increasing** over the interval I if, for every a and b in I,

$$\text{if } a < b, \quad \text{then } f(a) < f(b).$$

(If the input a is less than the input b, then the output for a is less than the output for b.)

A function f is **decreasing** over the interval I if, for every a and b in I,

$$\text{if } a < b, \quad \text{then } f(a) > f(b).$$

(If the input a is less than the input b, then the output for a is greater than the output for b.)

Note that the directions of the inequalities stay the same for an increasing function, but they differ for a decreasing function.

In Chapter 1, we saw how the slope of a linear function determines whether that function is increasing or decreasing (or neither). For a general function, the derivative yields similar information.

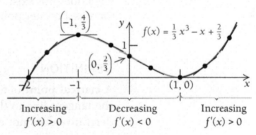

f is increasing over the intervals $(-\infty, -1)$ and $(1, \infty)$; slopes of tangent lines are positive.

f is decreasing over the interval $(-1, 1)$; slopes of tangent lines are negative.

The following theorem shows how we can use derivatives to determine whether a function is increasing or decreasing.

THEOREM 1

If $f'(x) > 0$ for all x in an interval I, then f is increasing over I.

If $f'(x) < 0$ for all x in an interval I, then f is decreasing over I.

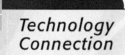

Technology Connection

Exploratory

Graph the function

$$f(x) = -\tfrac{1}{3}x^3 + 6x^2 - 11x - 50$$

and its derivative

$$f'(x) = -x^2 + 12x - 11$$

using a viewing window of $[-10, 25, -100, 150]$, with Xscl $= 5$ and Yscl $= 25$. Then use the TRACE feature, moving from left to right along each graph. As you move the cursor from left to right, note that the x-coordinate always increases. If a function is increasing over an interval, the y-coordinate will be increasing. If a function is decreasing over an interval, the y-coordinate will be decreasing.

Over what intervals is the function increasing?

Over what intervals is the function decreasing?

Over what intervals is the derivative positive?

Over what intervals is the derivative negative?

What can you conjecture?

Critical Points

Consider this graph of a continuous function.

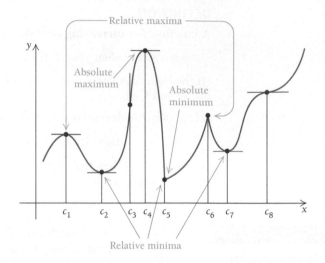

Note the following:

1. $f'(c) = 0$ at points c_1, c_2, c_4, c_7, and c_8. That is, the tangent line to the graph is horizontal at these points.
2. $f'(c)$ does not exist at points c_3, c_5, and c_6. The tangent line is vertical at c_3, and there is a corner point, or sharp point, at both c_5 and c_6.

DEFINITION

A **critical point** of a function is an interior point c of its domain at which the tangent line to the graph at $(c, f(c))$ is horizontal or at which the derivative does not exist. That is, c is a critical point if

$$f'(c) = 0 \quad \text{or} \quad f'(c) \text{ does not exist.}$$

Thus, in the preceding graph:

1. c_1, c_2, c_4, c_7, and c_8 are critical points because $f'(c) = 0$ for each point.
2. c_3, c_5, and c_6 are critical points because $f'(c)$ does not exist at each point.

Note too that a function can change from increasing to decreasing or from decreasing to increasing *only* at a critical point. In the graph above, there are "peaks" and "valleys" at the points c_1, c_2, c_4, c_5, c_6, and c_7. These points separate the intervals over which the function changes from increasing to decreasing or from decreasing to increasing. The points c_3 and c_8 are also critical points, but they do not separate intervals over which the function changes from increasing to decreasing or from decreasing to increasing.

Each of the function values $f(c_1)$, $f(c_4)$, and $f(c_6)$ is called a **relative maximum.** Likewise, each of the function values $f(c_2)$, $f(c_5)$, and $f(c_7)$ is called a **relative minimum.**

Relative minima in a certain energy function have been used to explain the shapes of viruses.[2]

DEFINITION

Suppose that f is a function whose value $f(c)$ exists at input c in the domain of f. Then:

> $f(c)$ is a **relative minimum** if there exists an open interval I_1 containing c in the domain such that $f(c) \leq f(x)$, for all x in I_1;
>
> and
>
> $f(c)$ is a **relative maximum** if there exists an open interval I_2 containing c in the domain such that $f(c) \geq f(x)$, for all x in I_2.

As another example, consider the graph of

$$f(x) = (x - 1)^3 + 2,$$

shown to the right. Note that

$$f'(x) = 3(x - 1)^2,$$

and

$$f'(1) = 3(1 - 1)^2 = 0.$$

The function has a critical point at $c = 1$, but has no relative maximum or minimum at $c = 1$.

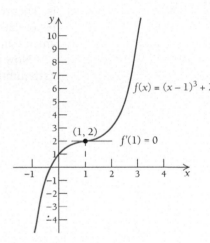

$f(x) = (x-1)^3 + 2$

$(1, 2)$ $f'(1) = 0$

CAUTION   A critical point doesn't have to be either a relative maximum or a relative minimum.

A relative maximum can be thought of as a high point that may or may not be the highest point, or *absolute maximum,* on an interval I. Similarly, a relative minimum can be thought of as a low point that may or may not be the lowest point, or *absolute minimum,* on I. For now we will consider how to find relative maximum or minimum values, stated simply as **relative extrema.**

Look again at the graph on the previous page. The points at which a continuous function has relative extrema are points where the derivative is 0 or where the derivative does not exist—the critical points.

[2]R. F. Bruinsma, W. M. Gelbart, D. Reguera, J. Rudnick, and R. Zandi, "Viral self-assembly as a thermodynamic process," *Physical Review Letters,* Vol. 90, 248101 (2003); C. Day, "Thermodynamics explains the symmetry of spherical viruses," *Physics Today,* pp. 27–29 (December 2004).

THEOREM 2

If a function f has a relative extreme value $f(c)$, then c is a critical point, so

$$f'(c) = 0 \quad \text{or} \quad f'(c) \text{ does not exist.}$$

Theorem 2 is very useful, but it is important to understand it precisely. What it says is that when we are looking for points that are relative extrema, the only points we need to consider are those where the derivative is 0 or where the derivative does not exist. We can think of a critical point as a *candidate* for a relative maximum or minimum, and the candidate might or might not provide a relative extremum. That is, Theorem 2 does not say that if a point is a critical point, its function value must be a relative maximum or minimum. The existence of a critical point does *not* guarantee that a function has a relative maximum or minimum, as discussed above.

Now how can we tell when the existence of a critical point leads us to a relative extremum? The following graph leads us to a test.

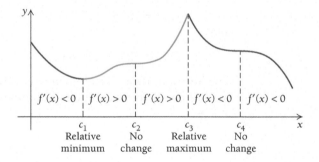

At a critical point for which there is a relative extremum, the function is increasing on one side of the critical point and decreasing on the other side.

Derivatives tell us when a function is increasing or decreasing. This leads us to the **First-Derivative Test.**

THEOREM 3

The First-Derivative Test for Relative Extrema

For any continuous function f that has exactly one critical point c in an open interval (a, b):

F1. f has a relative minimum at c if $f'(x) < 0$ on (a, c) and $f'(x) > 0$ on (c, b). That is, f is decreasing to the left of c and increasing to the right of c.

F2. f has a relative maximum at c if $f'(x) > 0$ on (a, c) and $f'(x) < 0$ on (c, b). That is, f is increasing to the left of c and decreasing to the right of c.

F3. f has neither a relative maximum nor a relative minimum at c if $f'(x)$ has the same sign on (a, c) as on (c, b).

Now let's see how we can use the First-Derivative Test to create and understand graphs and to find relative extrema. As we will see, the First-Derivative Test is easier to use than to memorize by rote.

EXAMPLE 1   Graph the function f given by

$$f(x) = 2x^3 - 3x^2 - 12x + 12$$

and find the relative extrema.

Solution   Suppose that we are trying to graph this function, but don't know any calculus. What can we do? We could plot several points to determine in which direction the graph seems to be turning. Let's guess some x-values and see what happens.

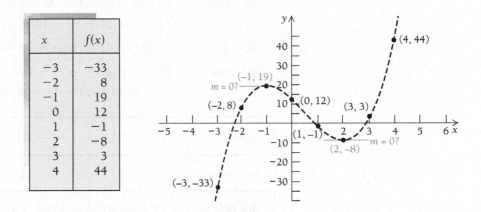

x	$f(x)$
-3	-33
-2	8
-1	19
0	12
1	-1
2	-8
3	3
4	44

We plot the points and use them to sketch a "best guess" of the graph, shown as the dashed line in the figure above. According to this rough sketch, it would seem that the graph has a tangent line with slope 0 somewhere around $x = -1$ and $x = 2$. But how do we know for sure? We can begin by finding a general expression for the derivative:

$$f'(x) = 6x^2 - 6x - 12.$$

We then determine where $f'(x)$ does not exist or where $f'(x) = 0$. We can replace x in $f'(x) = 6x^2 - 6x - 12$ with any real number. Thus, $f'(x)$ exists for all real numbers. So the only possibilities for critical points are where $f'(x) = 0$, at which there are horizontal tangents. To find such points, we solve $f'(x) = 0$:

$$6x^2 - 6x - 12 = 0$$

$$x^2 - x - 2 = 0 \qquad \text{Dividing by 6 on both sides}$$

$$(x + 1)(x - 2) = 0 \qquad \text{Factoring}$$

$$x + 1 = 0 \quad or \quad x - 2 = 0 \qquad \text{Using the Principle of Zero Products}$$

$$x = -1 \quad or \qquad x = 2.$$

The critical points are -1 and 2. Since it is at these points that a relative maximum or minimum will exist, if there is one, we examine the intervals on each side of the critical points. We use the critical points to divide the real-number line into three intervals: $A(-\infty, -1)$, $B(-1, 2)$, and $C(2, \infty)$, as shown below.

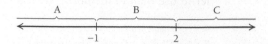

Then we analyze the sign of the derivative on each interval. If $f'(x)$ is positive for one value in the interval, then it will be positive for all numbers in the interval. Similarly, if it is negative for one value, it will be negative for all values in the interval. This is because in order for the derivative to change signs, it must become 0 or be undefined at some point. Such a point will be a critical point. Thus we merely choose a test value in each interval and make a substitution. The test values we choose are -2, 0, and 4.

$$A: \quad \text{Test } -2, \quad f'(-2) = 6(-2)^2 - 6(-2) - 12$$
$$= 24 + 12 - 12 = 24 > 0;$$

$$B: \quad \text{Test } 0, \quad f'(0) = 6(0)^2 - 6(0) - 12 = -12 < 0;$$

$$C: \quad \text{Test } 4, \quad f'(4) = 6(4)^2 - 6(4) - 12$$
$$= 96 - 24 - 12 = 60 > 0.$$

Interval	$(-\infty, -1)$	$(-1, 2)$	$(2, \infty)$
Test Value	$x = -2$	$x = 0$	$x = 4$
Sign of $f'(x)$	$f'(-2) > 0$	$f'(0) < 0$	$f'(4) > 0$
Behavior of f	f is increasing	f is decreasing	f is increasing

Result

Relative maximum given by $f(-1) = 19$

Relative minimum given by $f(2) = -8$

Therefore, by the First-Derivative Test,

f has a relative maximum at $x = -1$ given by

$$f(-1) = 2(-1)^3 - 3(-1)^2 - 12(-1) + 12 \qquad \text{Substituting into the}$$
$$= 19 \qquad\qquad\qquad\qquad\qquad\qquad \text{original function}$$

and f has a relative minimum at $x = 2$ given by

$$f(2) = 2(2)^3 - 3(2)^2 - 12(2) + 12 = -8.$$

Thus there is a relative maximum at $(-1, 19)$ and a relative minimum at $(2, -8)$, as we suspected from the sketch of the graph.

The information we have obtained can be very useful in sketching a graph of the function. We know that this polynomial is continuous, and we know where the function is increasing, where it is decreasing, and where it has relative extrema. We complete the graph by using a calculator to generate some additional function values. Some calculators can actually be programmed to generate function values by first entering a formula and then generating many outputs. The graph of the function, shown below in red, has been scaled to clearly show its curving nature.

Technology Connection

Exploratory

Consider the function f given by

$$f(x) = x^3 - 3x + 2.$$

Graph both f and f' using the same set of axes. Examine the graphs using the **TABLE** and **TRACE** features. Where do you think the relative extrema of $f(x)$ occur? Where is the derivative 0? Where does $f(x)$ have critical points?

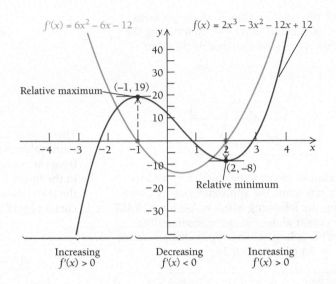

For reference, the graph of the derivative is shown in blue. Note how the critical points reveal themselves. The derivative is 0 where the function has relative extrema.   ■

Keep the following in mind as you find relative extrema.

The *derivative* f' is used to find the critical points of f. The test values, in the intervals defined by the critical points, are substituted into the *derivative* f', and the function values are found using the *original* function f. Use the derivative f' to find information about the shape of the graph of f.

Technology Connection

Finding Relative Extrema

There are several methods for approximating relative extrema on a grapher. As an example, consider finding the relative extrema of

$$f(x) = -0.4x^3 + 6.2x^2 - 11.3x - 54.8.$$

We first graph the function, using a viewing window that reveals the curvature.

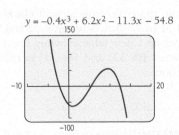

$y = -0.4x^3 + 6.2x^2 - 11.3x - 54.8$

(continued)

Technology Connection (continued)

Method 1: TRACE

Beginning with the window shown on the previous page, we press **TRACE** and move the cursor along the curve, noting where relative extrema might occur.

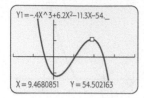

A relative maximum seems to be about Y = 54.5 at X = 9.47. We can refine the approximation by zooming in to obtain the following window. We press **TRACE** and move the cursor along the curve, again noting where the *y*-value is largest. The approximation seems to be about Y = 54.61 at X = 9.34.

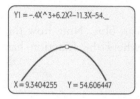 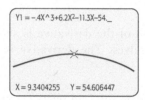

We can continue in this manner until the desired accuracy is achieved.

Method 2: TABLE

We can also use the **TABLE** feature, adjusting starting points and step values to improve accuracy.

X	Y1	
9.3	54.605	
9.31	54.607	
9.32	54.608	
9.33	54.608	
9.34	54.607	
9.35	54.604	
9.36	54.601	
X = 9.32		

Now the approximation seems to be about Y = 54.61 at an *x*-value between 9.32 and 9.33. We could set up a new table showing function values between $f(9.32)$ and $f(9.33)$ to refine the approximation.

Method 3: MAXIMUM, MINIMUM

Using the **MAXIMUM** feature from the **CALC** menu, we find that a relative maximum of about 54.61 occurs at $x \approx 9.32$.

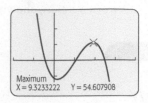

Method 4: fMax or fMin

This feature calculates a relative maximum or minimum value over any specified closed interval. We see from the initial graph that a relative maximum occurs in the interval $[-10, 20]$. Using the fMax feature from the **MATH** menu, we see that a relative maximum occurs on $[-10, 20]$ when $x \approx 9.32$.

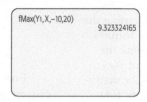

To obtain the maximum value, we evaluate the function at the given *x*-value, obtaining the following.

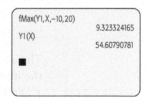

The approximation is about Y = 54.61 at X = 9.32.

Using any of these methods, we find the relative minimum to be about Y = −60.30 at X = 1.01.

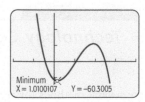

EXERCISE

1. Approximate the relative extrema of the function in Example 1.

EXAMPLE 2   Find the relative extrema of the function f given by
$$f(x) = 2x^3 - x^4.$$
Then sketch the graph.

Solution   First, we must determine the critical points. To do so, we find $f'(x)$:
$$f'(x) = 6x^2 - 4x^3.$$

Next, we find where $f'(x)$ does not exist or where $f'(x) = 0$. We can replace x in $f'(x) = 6x^2 - 4x^3$ with any real number. Thus, $f'(x)$ exists for all real numbers. So the only possibilities for critical points are where $f'(x) = 0$, that is, where there are horizontal tangent lines. To find such points, we solve $f'(x) = 0$:

$$6x^2 - 4x^3 = 0$$
$$2x^2(3 - 2x) = 0 \qquad \text{Factoring}$$
$$2x^2 = 0 \quad or \quad 3 - 2x = 0$$
$$x^2 = 0 \quad or \qquad 3 = 2x$$
$$x = 0 \quad or \qquad x = \tfrac{3}{2}.$$

The critical points are 0 and $\tfrac{3}{2}$. We use these points to divide the real-number line into three intervals: $A(-\infty, 0)$, $B\left(0, \tfrac{3}{2}\right)$, and $C\left(\tfrac{3}{2}, \infty\right)$, as shown below.

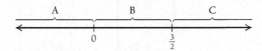

We now analyze the sign of the derivative on each interval. We begin by choosing a test value in each interval and making a substitution. We generally choose as test values numbers for which it is easy to compute outputs of the derivative—in this case, -1, 1, and 2.

$$A: \quad \text{Test } -1, \quad f'(-1) = 6(-1)^2 - 4(-1)^3$$
$$= 6 + 4 = 10 > 0;$$

$$B: \quad \text{Test } 1, \qquad f'(1) = 6(1)^2 - 4(1)^3$$
$$= 6 - 4 = 2 > 0;$$

$$C: \quad \text{Test } 2, \qquad f'(2) = 6(2)^2 - 4(2)^3$$
$$= 24 - 32 = -8 < 0.$$

Technology Connection

EXERCISES
Graph each of the following functions and approximate the relative extrema using a grapher.

1. $f(x) = x^4 - 8x^3 + 18x^2$
2. $f(x) = 2.01x^3 - 0.9x^4$
3. $f(x) = \tfrac{1}{3}x^3 - \tfrac{1}{2}x^2 - 2x + 1$
4. $f(x) = 0.21x^4 - 4.3x^2 + 22$

Interval	$(-\infty, 0)$	$\left(0, \tfrac{3}{2}\right)$	$\left(\tfrac{3}{2}, \infty\right)$
Test Value	$x = -1$	$x = 1$	$x = 2$
Sign of $f'(x)$	$f'(-1) > 0$	$f'(1) > 0$	$f'(2) < 0$
Behavior of f	f is increasing	f is increasing	f is decreasing

Result   └─ No change ─┘└─ Relative ─┘
maximum
given by
$f\left(\tfrac{3}{2}\right) = \tfrac{27}{16}$

Therefore, by the First-Derivative Test, f has neither a relative maximum nor a relative minimum at $x = 0$ since the function is increasing on both sides of 0, and f has a relative maximum at $x = \frac{3}{2}$ given by

$$f\left(\frac{3}{2}\right) = 2\left(\frac{3}{2}\right)^3 - \left(\frac{3}{2}\right)^4 = \frac{27}{16}.$$

Remember to substitute into the original function.

Thus there is a relative maximum at $\left(\frac{3}{2}, \frac{27}{16}\right)$.

We use the information obtained to sketch the graph. Other function values are listed in the table below. (More can be generated by the student.) The graph follows.

x	$f(x)$, approximately
-1	-3
-0.75	-1.16
-0.5	-0.31
-0.25	-0.04
0	0
0.25	0.03
0.5	0.19
0.75	0.53
1	1
1.25	1.46
1.5	1.69
1.75	1.34
2	0

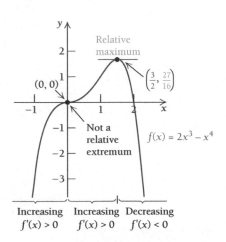

EXAMPLE 3 Find the critical points of the function f given by

$$f(x) = x + \sin 2x.$$

Determine if each critical point is a relative maximum, a relative minimum, or neither. Then sketch the graph on the interval $0 \le x \le 2\pi$.

Solution   First, we must determine the critical points. To do so, we find $f'(x)$:

$$f'(x) = 1 + 2\cos 2x.$$

We then determine where $f'(x)$ does not exist or where $f'(x) = 0$. In this case, $f'(x)$ exists for all real numbers, so the only possibilities for critical points are where $f'(x) = 0$. To find such points, we solve $f'(x) = 0$:

$$1 + 2\cos 2x = 0$$

$$\cos 2x = -\frac{1}{2}$$

$$2x = \frac{2\pi}{3} + 2\pi n \quad or \quad x = \frac{4\pi}{3} + 2\pi n$$

$$x = \frac{\pi}{3} + \pi n \quad or \quad x = \frac{2\pi}{3} + \pi n.$$

The critical points are $x = \frac{\pi}{3} + \pi n$ and $x = \frac{2\pi}{3} + \pi n$, where n is any integer.

On the interval $[0, 2\pi]$, the critical points are $\pi/3$, $2\pi/3$, $4\pi/3$, and $5\pi/3$. We use these points to divide $[0, 2\pi]$ into five intervals: $A(0, \pi/3)$, $B(\pi/3, 2\pi/3)$, $C(2\pi/3, 4\pi/3)$, $D(4\pi/3, 5\pi/3)$, and $E(5\pi/3, 2\pi)$, as shown below.

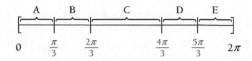

We now analyze the sign of the derivative on each interval. We begin by choosing a test value in each interval and making a substitution. We generally choose as test values numbers for which it is relatively easy to compute outputs of the derivative—in this case, $\pi/6$, $\pi/2$, π, $3\pi/2$, and $11\pi/6$.

A: Test $\pi/6$, $\quad f'(\pi/6) = 1 + 2\cos 2\left(\dfrac{\pi}{6}\right) = 1 + 2\cos\dfrac{\pi}{3} = 2 > 0;$

B: Test $\pi/2$, $\quad f'(\pi/2) = 1 + 2\cos 2\left(\dfrac{\pi}{2}\right) = 1 + 2\cos \pi = -1 < 0;$

C: Test π, $\quad f'(\pi) = 1 + 2\cos 2(\pi) = 1 + 2\cos 2\pi = 3 > 0;$

D: Test $3\pi/2$, $\quad f'(3\pi/2) = 1 + 2\cos 2\left(\dfrac{3\pi}{2}\right) = 1 + 2\cos 3\pi = -1 < 0;$

E: Test $11\pi/6$, $f'(11\pi/6) = 1 + 2\cos 2\left(\dfrac{11\pi}{6}\right) = 1 + 2\cos\dfrac{11\pi}{3} = 2 > 0.$

Interval	$(0, \pi/3)$	$(\pi/3, 2\pi/3)$	$(2\pi/3, 4\pi/3)$	$(4\pi/3, 5\pi/3)$	$(5\pi/3, 2\pi)$
Test Value	$x = \pi/6$	$x = \pi/2$	$x = \pi$	$x = 3\pi/2$	$x = 11\pi/6$
Sign of $f'(x)$	$f'(\pi/6) > 0$	$f'(\pi/2) < 0$	$f'(\pi) > 0$	$f'(3\pi/2) < 0$	$f'(11\pi/6) > 0$
Behavior of f	f is increasing	f is decreasing	f is increasing	f is decreasing	f is increasing

Result   Relative maximum given by $f(\pi/3) \approx 1.91$ Relative minimum given by $f(2\pi/3) \approx 1.23$ Relative maximum given by $f(4\pi/3) \approx 5.05$ Relative minimum given by $f(5\pi/3) \approx 4.37$

Therefore, by the First-Derivative Test, f has a relative maximum at $x = \pi/3$ given by

$$f(\pi/3) = \frac{\pi}{3} + \sin 2\left(\frac{\pi}{3}\right) = \frac{\pi}{3} + \sin\frac{2\pi}{3} \approx 1.91,$$

and another relative maximum at $x = 4\pi/3$ given by

$$f(4\pi/3) = \frac{4\pi}{3} + \sin 2\left(\frac{4\pi}{3}\right) = \frac{4\pi}{3} + \sin\frac{8\pi}{3} \approx 5.05.$$

Furthermore, f has a relative minimum at $x = 2\pi/3$ given by

$$f(2\pi/3) = \frac{2\pi}{3} + \sin 2\left(\frac{2\pi}{3}\right) = \frac{2\pi}{3} + \sin\frac{4\pi}{3} \approx 1.23,$$

and another relative minimum at $x = 5\pi/3$ given by

$$f(5\pi/3) = \frac{5\pi}{3} + \sin 2\left(\frac{5\pi}{3}\right) = \frac{5\pi}{3} + \sin\frac{10\pi}{3} \approx 4.37.$$

Notice that it would have been extremely difficult to ascertain these four points to such precision with only a sketch of the graph.

    We use the information obtained to sketch the graph. Other function values may be obtained by the student. The graph follows for $0 \le x \le 2\pi$.

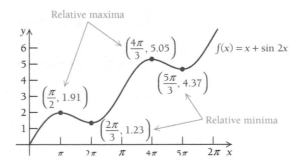

EXAMPLE 4   Find the relative extrema of the function f given by

$$f(x) = (x - 2)^{2/3} + 1.$$

Then sketch the graph.

Solution   First, we determine the critical points. To do so, we find $f'(x)$:

$$f'(x) = \frac{2}{3}(x - 2)^{-1/3}$$

$$= \frac{2}{3\sqrt[3]{x - 2}}.$$

Next, we find where $f'(x)$ does not exist or where $f'(x) = 0$. Note that the derivative $f'(x)$ does not exist at 2, although $f(x)$ does. Thus the number 2 is a critical point. The equation $f'(x) = 0$ has no solution, so the only critical point is 2. We use 2 to divide the real-number line into two intervals: $A(-\infty, 2)$ and $B(2, \infty)$, as shown below.

    We analyze the derivative on each interval. We begin by choosing a test value in each interval and making a substitution. We choose test points 0 and 3. It is not necessary to find an exact value of the derivative; we need only determine the sign. Sometimes we can do this by just examining the formula for the derivative:

Technology Connection

EXERCISES

Consider the function f given by

$$f(x) = 2 - (x - 1)^{2/3}.$$

1. Graph the function using the viewing window $[-4, 6, -2, 4]$.

2. Graph the first derivative. What happens to the graph of the derivative at the critical points?

3. Approximate the relative extrema.

A: Test 0, $f'(0) = \dfrac{2}{3\sqrt[3]{0 - 2}} < 0;$

B: Test 3, $f'(3) = \dfrac{2}{3\sqrt[3]{3 - 2}} > 0.$

Interval	$(-\infty, 2)$	$(2, \infty)$
Test Value	$x = 0$	$x = 3$
Sign of $f'(x)$	$f'(0) < 0$	$f'(3) > 0$
Behavior of f	f is decreasing	f is increasing

Result ⌞——— Relative ———⌝
minimum
given by $f(2) = 1$

Since we have a change from decreasing to increasing, we conclude from the First-Derivative Test that

f has a relative minimum at $x = 2$ given by

$$f(2) = (2 - 2)^{2/3} + 1 = 1.$$

Thus there is a relative minimum at $(2, 1)$. (The graph has *no* tangent line at $(2, 1)$.)

We use the information obtained to sketch the graph. Other function values are listed in the table below. (More can be generated by the student.) The graph follows.

x	$f(x)$, approximately
-1	3.08
-0.5	2.84
0	2.59
0.5	2.31
1	2
1.5	1.63
2	1
2.5	1.63
3	2
3.5	2.31
4	2.59

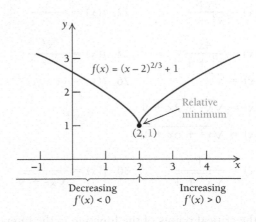

Exercise Set 3.1

Find the relative extrema of the function, if they exist. List your answers in terms of ordered pairs. Then sketch a graph of the function.

1. $f(x) = x^2 - 4x + 5$ 2. $f(x) = x^2 - 6x - 3$

3. $f(x) = 5 + x - x^2$ 4. $f(x) = 2 - 3x - 2x^2$

5. $f(x) = 1 + 6x + 3x^2$

6. $f(x) = 0.5x^2 - 2x - 11$

7. $f(x) = x^3 - x^2 - x + 2$

8. $f(x) = x^3 + \frac{1}{2}x^2 - 2x + 5$

9. $f(x) = x^3 - 3x + 6$ 10. $f(x) = x^3 - 3x^2$

11. $f(x) = 2x^3$ 12. $f(x) = 1 - x^3$

13. $f(x) = 0.02x^2 + 1.3x + 2.31$

14. $f(x) = -0.03x^2 + 1.8x - 3.45$

15. $f(x) = x^4 - 2x^3$ 16. $f(x) = x^4 - 8x^2 + 3$

17. $f(x) = x\sqrt{8 - x^2}$ 18. $f(x) = x\sqrt{16x - x^2}$

19. $f(x) = 1 - x^{2/3}$

20. $f(x) = (x + 3)^{2/3} - 5$

21. $f(x) = \dfrac{-8}{x^2 + 1}$ 22. $f(x) = \dfrac{5}{x^2 + 1}$

23. $f(x) = \dfrac{4x}{x^2 + 1}$ 24. $f(x) = \dfrac{x^2}{x^2 + 1}$

25. $f(x) = \sqrt[3]{x}$ 26. $f(x) = (x + 1)^{1/3}$

27. $f(x) = \sqrt{x^2 + 2x + 5}$

28. $f(x) = \sqrt{x^2 + 6x + 10}$

29. $f(x) = \dfrac{1}{\sqrt{x^2 + 1}}$ 30. $f(x) = \dfrac{1}{\sqrt{x^2 + 9}}$

Find the critical points of the function in the interval $[0, 2\pi]$. Determine if each critical point is a relative maximum, a relative minimum, or neither. Then sketch the graph on the interval $[0, 2\pi]$.

31. $f(x) = \sin x$ 32. $f(x) = \cos x$

33. $f(x) = \sin x - \cos x$ 34. $f(x) = \sin 3x$

35. $f(x) = \cos 2x$ 36. $f(x) = x + 2\sin x$

37. $f(x) = x + \cos 2x$ 38. $f(x) = \dfrac{x}{4} + \sin \dfrac{x}{2}$

39. $f(x) = \dfrac{x}{3} + \cos \dfrac{2x}{3}$ 40. $f(x) = \dfrac{\sin x}{2 + \cos x}$

41. $f(x) = \dfrac{\cos x}{2 - \sin x}$ 42. $f(x) = \sin x \cos x$

43. $f(x) = \sin x - \sin^2 x$

44. $f(x) = \cos x + \cos^2 x$

45. $f(x) = 9 \sin x - 4 \sin^3 x$

46. $f(x) = 3 \cos x - 2 \cos^3 x$

47.–92. Check the results of each of Exercises 1–46 using a grapher.

APPLICATIONS

93. *Path of the Olympic Arrow.* The Olympic flame at the 1992 Summer Olympics was lit by a flaming arrow. As the arrow moved d feet horizontally from the archer, its height h, in feet, was approximated by the function

$$h = -0.002d^2 + 0.8d + 6.6.$$

Find the relative maximum and sketch a graph of the function.

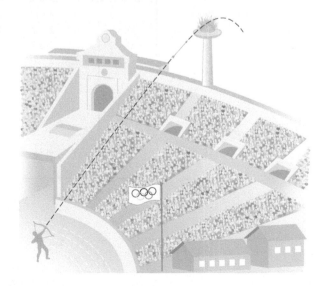

tw 94. Use the graph in Exercise 93 to describe the words "the path of the Olympic arrow." Include the height from which it was launched, its maximum height, and the horizontal distance from the archer at which it hit the ground.

95. *Temperature During an Illness.* The temperature of a person during an intestinal illness is given by

$$T(t) = -0.1t^2 + 1.2t + 98.6, \quad 0 \le t \le 12,$$

where T is the temperature (°F) at time t, in days. Find the relative extrema and sketch a graph of the function.

96. *Solar Eclipse.* On June 21, 2001, there was a total eclipse of the sun in Africa. Between the times 12:00 noon and 1:00 P.M. Universal Time, the center of the eclipse followed a path along the earth described approximately by[3]

$$f(x) = \sqrt{138.1 - 5.025x + 0.2902x^2},$$

where $0 \le x \le 22$ measures the longitude east and $f(x)$ is the latitude south. Find the latitude and longitude of the point farthest north along this path, and plot the function f.

tw 97. Consider this graph.

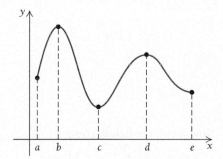

Using the graph and the intervals noted, explain how to relate the concept of the function being increasing or decreasing to the first derivative.

---
[3]NASA.

tw For Exercises 98–100, determine which points are critical points and why. Also, identify the relative maxima and relative minima.

98.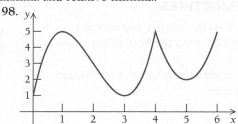

99.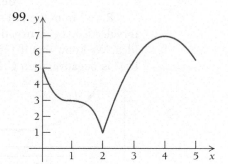

100.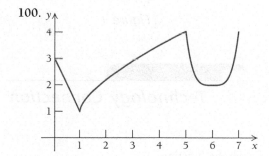

 Technology Connection

Graph the function. Then estimate any relative extrema.

101. $f(x) = x^4 + 4x^3 - 36x^2 - 160x + 400$

102. $f(x) = -x^6 - 4x^5 + 54x^4 + 160x^3 - 641x^2 - 828x + 1200$

3.2 Using Second Derivatives to Find Maximum and Minimum Values and Sketch Graphs

OBJECTIVES

■ Find the relative extrema of a function using the Second-Derivative Test.

■ Sketch the graph of a continuous function.

Concavity: Increasing and Decreasing Derivatives

The graphs of two functions are shown in Fig. 1. The graph in Fig. 1(a) is turning up and the graph in Fig. 1(b) is turning down. Let's see if we can relate this to their derivatives.

In Fig. 1(a), the *slopes* are increasing. That is, f' is increasing over the interval. In Fig. 1(b), the slopes are decreasing.

Recall from the last section that if f' is positive, then f is increasing over an interval. Also, if g' is negative, then g is decreasing over an interval. We can extend this idea. We know that if f'' is positive, then f' is increasing over the interval. Similarly, if g'' is negative, then g' is decreasing over the interval.

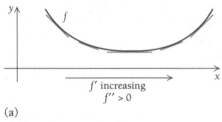

(a)

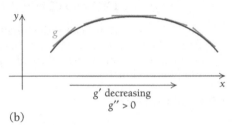

(b)

Figure 1

Technology Connection

Exploratory

Graph the function

$$f(x) = -\tfrac{1}{3}x^3 + 6x^2 - 11x - 50$$

and its second derivative

$$f''(x) = -2x + 12$$

using the viewing window $[-10, 25, -100, 150]$, with Xscl = 5 and Yscl = 25.

Over what intervals is the graph of f concave up?

Over what intervals is the graph of f concave down?

Over what intervals is the graph of f'' positive?

Over what intervals is the graph of f'' negative?

What can you conjecture?

Now graph the first derivative

$$f'(x) = -x^2 + 12x - 11$$

and the second derivative

$$f''(x) = -2x + 12$$

using the viewing window $[-10, 25, -200, 50]$, with Xscl = 5 and Yscl = 25.

Over what intervals is the first derivative f' increasing?

Over what intervals is the first derivative f' decreasing?

Over what intervals is the graph of f'' positive?

Over what intervals is the graph of f'' negative?

What can you conjecture?

DEFINITION

Suppose that f is a function whose derivative f' exists at every point in an open interval I. Then:

1. f is **concave up** on the interval I if f' is increasing over I.
2. f is **concave down** on the interval I if f' is decreasing over I.

For example, the graph in Fig. 1(a) is concave up and the graph in Fig. 1(b) is concave down. We then have the following theorem, which allows us to use second derivatives to determine concavity.

THEOREM 4

A Test for Concavity

1. If $f''(x) > 0$ on an interval I, then the graph of f is concave up on I. (f' is increasing, so the graph is turning up.)
2. If $f''(x) < 0$ on an interval I, then the graph of f is concave down on I. (f' is decreasing, so the graph is turning down.)

A helpful memory device follows.

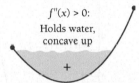

$f''(x) > 0$:
Holds water,
concave up

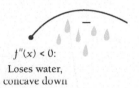

$f''(x) < 0$:
Loses water,
concave down

Finding Relative Extrema Using Second Derivatives

In the following discussion, we see how we can use second derivatives to determine whether a function has a relative extremum on an open interval.

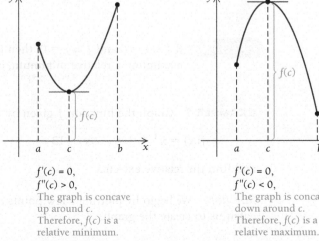

$f'(c) = 0$,
$f''(c) > 0$,
The graph is concave
up around c.
Therefore, $f(c)$ is a
relative minimum.

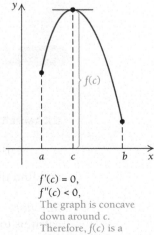

$f'(c) = 0$,
$f''(c) < 0$,
The graph is concave
down around c.
Therefore, $f(c)$ is a
relative maximum.

> **THEOREM 5**
>
> The Second-Derivative Test for Relative Extrema
>
> Suppose that f is a function for which $f'(x)$ exists for every x in an open interval (a, b) contained in its domain, and that there is a critical point c in (a, b) for which $f'(c) = 0$. Then:
>
> 1. $f(c)$ is a relative minimum if $f''(c) > 0$.
> 2. $f(c)$ is a relative maximum if $f''(c) < 0$.
>
> The test fails if $f''(c) = 0$. The First-Derivative Test would then have to be used.

Note that $f''(c) = 0$ does not tell us that there is no relative extremum. It just tells us that we do not know at this point and that we must use some other means, such as the First-Derivative Test, to determine whether there is a relative extremum.

Consider the following graphs. In each one, f' and f'' are both 0 at $c = 2$, but the first function has an extremum and the second function has *no* extremum. Also note that if $f'(c)$ does not exist, then $f''(c)$ does not exist and the Second-Derivative Test cannot be used.

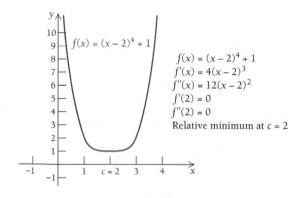

$$f(x) = (x - 2)^4 + 1$$
$$f'(x) = 4(x - 2)^3$$
$$f''(x) = 12(x - 2)^2$$
$$f'(2) = 0$$
$$f''(2) = 0$$
Relative minimum at $c = 2$

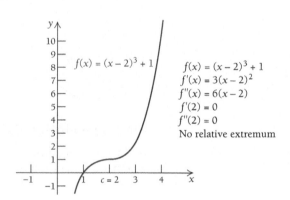

$$f(x) = (x - 2)^3 + 1$$
$$f'(x) = 3(x - 2)^2$$
$$f''(x) = 6(x - 2)$$
$$f'(2) = 0$$
$$f''(2) = 0$$
No relative extremum

CAUTION If $f'(c) = 0$ and $f''(c) = 0$, then it's possible for c to be a relative maximum, a relative minimum, or neither.

EXAMPLE 1 Graph the function f given by

$$f(x) = x^3 + 3x^2 - 9x - 13$$

and find the relative extrema.

Solution We begin by plotting a few points and using them and our knowledge of calculus to create the graph.

x	$f(x)$
-5	-18
-4	7
-3	14
-2	9
-1	-2
0	-13
1	-18
2	-11
3	14

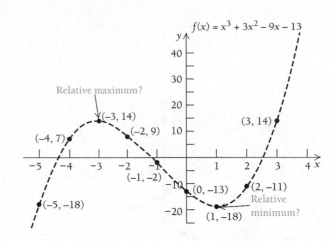

Technology Connection

Exploratory

Consider the function f given by

$$f(x) = x^3 - 3x^2 - 9x - 1.$$

Use a grapher to estimate the relative extrema. Then find the first and second derivatives. Graph both in the same window. Use the **ZERO** feature to determine where the first derivative is zero. Verify that relative extrema occur at those x-values by checking the sign of the second derivative. Then check your work using the analytic method of Example 1.

We plot these points and begin thinking about the shape of the graph. The dashed graph shown above can be considered a first guess. Are the points $(-3, 14)$ and $(1, -18)$ relative extrema? Or are these points merely close to the actual relative extrema? To refine our thoughts, we use calculus. We find both the first and second derivatives, $f'(x)$ and $f''(x)$:

$$f'(x) = 3x^2 + 6x - 9,$$
$$f''(x) = 6x + 6.$$

Then we solve $f'(x) = 0$:

$$3x^2 + 6x - 9 = 0$$
$$x^2 + 2x - 3 = 0 \qquad \text{Dividing by 3 on both sides}$$
$$(x + 3)(x - 1) = 0 \qquad \text{Factoring}$$
$$x + 3 = 0 \quad \text{or} \quad x - 1 = 0 \qquad \text{Using the Principle of Zero Products}$$
$$x = -3 \quad \text{or} \qquad x = 1.$$

We then find second coordinates by substituting in the original function:

$$f(-3) = (-3)^3 + 3(-3)^2 - 9(-3) - 13 = 14;$$
$$f(1) = (1)^3 + 3(1)^2 - 9(1) - 13 = -18.$$

Let's look at the second derivative. We use the Second-Derivative Test with the numbers -3 and 1:

$$f''(-3) = 6(-3) + 6 = -12 < 0; \longrightarrow \quad \text{Relative maximum}$$
$$f''(1) = 6(1) + 6 = 12 > 0. \longrightarrow \quad \text{Relative minimum}$$

Thus there is a relative maximum at $(-3, 14)$ and a relative minimum at $(1, -18)$.

The following figures illustrate the information concerning the function f that can be found from the first and second derivatives of f. The relative extrema are shown in Figs. 2 and 3. In Fig. 3, we see that the x-coordinates of the x-intercepts of f' are the critical points of f. We also see the intervals over which f is increasing and decreasing from the intervals over which f' is positive and negative.

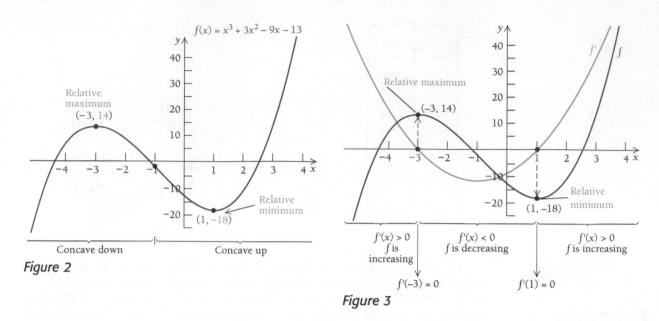

Figure 2

Figure 3

In Fig. 4, the intervals over which f' is increasing and decreasing are seen from the intervals over which f'' is positive and negative. And finally in Fig. 5, we note that when $f''(x) < 0$, f is concave down, and when $f''(x) > 0$, f is concave up.

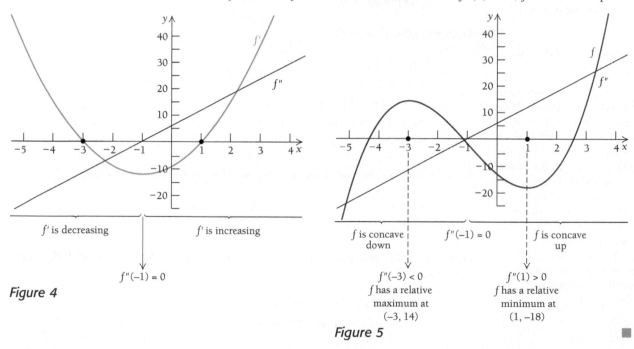

Figure 4

Figure 5

EXAMPLE 2   Find the relative extrema of the function f given by

$$f(x) = 3x^5 - 20x^3$$

and sketch the graph.

Solution   We find both the first and second derivatives, $f'(x)$ and $f''(x)$:

$$f'(x) = 15x^4 - 60x^2,$$
$$f''(x) = 60x^3 - 120x.$$

Then we solve $f'(x) = 0$:

$$15x^4 - 60x^2 = 0$$
$$15x^2(x^2 - 4) = 0$$
$$15x^2(x + 2)(x - 2) = 0 \qquad \text{Factoring}$$
$$15x^2 = 0 \quad or \quad x + 2 = 0 \quad or \quad x - 2 = 0 \qquad \text{Using the Principle of Zero Products}$$

$$x = 0 \quad or \qquad x = -2 \quad or \qquad x = 2.$$

We then find second coordinates by substituting in the original function:

$$f(-2) = 3(-2)^5 - 20(-2)^3 = 64;$$
$$f(2) = 3(2)^5 - 20(2)^3 = -64;$$
$$f(0) = 3(0)^5 - 20(0)^3 = 0.$$

The points $(-2, 64)$, $(2, -64)$, and $(0, 0)$ are candidates for relative extrema. We now use the Second-Derivative Test with the numbers -2, 2, and 0:

$$f''(-2) = 60(-2)^3 - 120(-2) = -240 < 0; \longrightarrow \quad \text{Relative maximum}$$
$$f''(2) = 60(2)^3 - 120(2) = 240 > 0; \longrightarrow \quad \text{Relative minimum}$$
$$f''(0) = 60(0)^3 - 120(0) = 0. \longrightarrow \quad \text{The Second-Derivative Test fails. Use the First-Derivative Test.}$$

Thus there is a relative maximum at $(-2, 64)$ and a relative minimum at $(2, -64)$. Checking the first derivative, we know that the function decreases to the left and to the right of $x = 0$. Thus we know by the First-Derivative Test that it has no relative extremum at the point $(0, 0)$. We complete the graph, plotting other points as needed. The extrema are shown in the graph below.

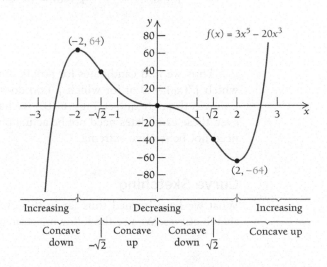

Points of Inflection

A **point of inflection,** or an **inflection point,** is a point across which the direction of concavity changes. For example, in Figs. 6–8, point P is an inflection point.

Technology Connection

Exploratory

Graph the function f given by $f(x) = 3x^5 - 5x^3$ and its second derivative $f''(x) = 60x^3 - 30x$ using $[-3, 3, -10, 10]$ as the viewing window. Estimate any inflection points of f. Where is the second derivative 0?

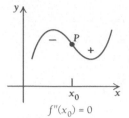

$f''(x_0) = 0$

Figure 6

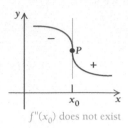

$f''(x_0)$ does not exist

Figure 7

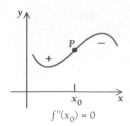

$f''(x_0) = 0$

Figure 8

Although we did not compute them earlier, there are points of inflection in other examples that we have considered in this section. In Example 1, the graph has a point of inflection at $(-1, -2)$, and in Example 2, the graph has points of inflection at $\left(-\sqrt{2}, 28\sqrt{2}\right)$, $(0, 0)$, and $\left(\sqrt{2}, -28\sqrt{2}\right)$.

As we move to the right along the curve in Fig. 6, the concavity changes from concave down, $f''(x) < 0$, on the left of P to concave up, $f''(x) > 0$, on the right of P. Since, as we move through P, $f''(x)$ changes sign from $-$ to $+$, the value of $f''(x_0)$ at P must be 0, as in Fig. 6; or $f''(x_0)$ does not exist, as in Fig. 7. A similar change in concavity occurs at P in Fig. 8.

> **THEOREM 6**
>
> Finding Points of Inflection
>
> If a function f has a point of inflection, it occurs at a point x_0, where
>
> $$f''(x_0) = 0 \quad \text{or} \quad f''(x_0) \text{ does not exist.}$$

Thus we find candidates for points of inflection by looking for numbers x_0 for which $f''(x_0) = 0$ or for which $f''(x_0)$ does not exist. Then if $f''(x)$ changes sign as x moves through x_0 (see Figs. 6–8), we have a point of inflection at $x = x_0$. However, these candidates need not be actual points of inflection, just like critical points need not be relative extrema.

Curve Sketching

What we have learned thus far in this chapter will greatly enhance our ability to sketch curves. We use the following strategy.

Strategy for Sketching Graphs

a) *Derivatives.* Find $f'(x)$ and $f''(x)$.

b) *Critical points of f.* Find the critical points of f by solving $f'(x) = 0$ and finding where $f'(x)$ does not exist. These numbers yield candidates for relative maxima or minima. Find the function values at these points.

c) *Increasing and/or decreasing; relative extrema.* Use the critical points of f from step (b) to define intervals. Determine whether f is increasing or decreasing over the intervals. Do this by selecting test values and substituting into $f'(x)$. Use this information and/or the second derivative to determine the relative maxima and minima.

d) *Inflection points.* Determine candidates for inflection points by finding where $f''(x) = 0$ or where $f''(x)$ does not exist. Find the function values at these points.

e) *Concavity.* Use the candidates for inflection points from step (d) to define intervals. Determine the concavity by checking to see where f' is increasing—that is, where $f''(x) > 0$—and where f' is decreasing—that is, where $f''(x) < 0$. Do this by selecting test values and substituting into $f''(x)$.

f) *Sketch the graph.* Sketch the graph using the information from steps (a) through (e), plotting extra points (computing them with your calculator) if the need arises.

EXAMPLE 3 Find the relative maxima and minima of the function f given by

$$f(x) = x^4 - 2x^2$$

and sketch the graph.

Solution

a) *Derivatives.* Find $f'(x)$ and $f''(x)$:

$$f'(x) = 4x^3 - 4x,$$
$$f''(x) = 12x^2 - 4.$$

b) *Critical points of f.* Find the critical points of f by finding where $f'(x)$ does not exist and by solving $f'(x) = 0$. Since $f'(x) = 4x^3 - 4x$ exists for all values of x, the only critical points of f are where

$$4x^3 - 4x = 0$$
$$4x(x^2 - 1) = 0$$
$$4x = 0 \quad or \quad x^2 - 1 = 0$$
$$x = 0 \quad or \quad x^2 = 1$$
$$x = \pm 1.$$

Now $f(0) = 0$, $f(-1) = -1$, and $f(1) = -1$, which gives the points $(0, 0)$, $(-1, -1)$, and $(1, -1)$ on the graph.

c) *Increasing and/or decreasing; relative extrema.* Find the intervals over which f is increasing and the intervals over which f is decreasing. The critical points are -1, 0, and 1. We therefore use these points to divide the real number line into four intervals: $A(-\infty, -1)$, $B(-1, 0)$, $C(0, 1)$, and $D(1, \infty)$. We choose a test value in each interval and make a substitution. The test values we select are -2, $-\frac{1}{2}$, $\frac{1}{2}$, and 2:

A: Test -2, $\quad f'(-2) = 4(-2)^3 - 4(-2) = -24 < 0$;

B: Test $-\frac{1}{2}$, $\quad f'\left(-\frac{1}{2}\right) = 4\left(-\frac{1}{2}\right)^3 - 4\left(-\frac{1}{2}\right) = \frac{3}{2} > 0$;

C: Test $\frac{1}{2}$, $\quad f'\left(\frac{1}{2}\right) = 4\left(\frac{1}{2}\right)^3 - 4\left(\frac{1}{2}\right) = -\frac{3}{2} < 0$;

D: Test 2, $\quad f'(2) = 4(2)^3 - 4(2) = 24 > 0$.

Interval	$(-\infty, -1)$	$(-1, 0)$	$(0, 1)$	$(1, \infty)$
Test Value	$x = -2$	$x = -\frac{1}{2}$	$x = \frac{1}{2}$	$x = 2$
Sign of $f'(x)$	$f'(-2) < 0$	$f'\left(-\frac{1}{2}\right) > 0$	$f'\left(\frac{1}{2}\right) < 0$	$f'(2) > 0$
Behavior of f	f is decreasing	f is increasing	f is decreasing	f is increasing

Result

Relative minimum given by $f(-1) = -1$

Relative maximum given by $f(0) = 0$

Relative minimum given by $f(1) = -1$

Thus, by the First-Derivative Test, there is a relative maximum at $(0, 0)$ and two relative minima at $(-1, -1)$ and $(1, -1)$. That these are relative extrema can also be verified by the Second-Derivative Test: $f''(0) < 0$, $f''(-1) > 0$, and $f''(1) > 0$.

d) *Inflection points.* Find where $f''(x)$ does not exist and where $f''(x) = 0$. Since $f''(x)$ exists for all real numbers, we just solve $f''(x) = 0$:

$$12x^2 - 4 = 0$$
$$4(3x^2 - 1) = 0$$
$$3x^2 - 1 = 0$$
$$3x^2 = 1$$
$$x^2 = \frac{1}{3}$$
$$x = \pm\sqrt{\frac{1}{3}}$$
$$= \pm\frac{1}{\sqrt{3}}.$$

Now

$$f\left(\frac{1}{\sqrt{3}}\right) = \left(\frac{1}{\sqrt{3}}\right)^4 - 2\left(\frac{1}{\sqrt{3}}\right)^2$$

$$= \frac{1}{9} - \frac{2}{3} = -\frac{5}{9}$$

and

$$f\left(-\frac{1}{\sqrt{3}}\right) = -\frac{5}{9}.$$

This gives the points

$$\left(-\frac{1}{\sqrt{3}}, -\frac{5}{9}\right) \quad \text{and} \quad \left(\frac{1}{\sqrt{3}}, -\frac{5}{9}\right)$$

on the graph—$(-0.6, -0.6)$ and $(0.6, -0.6)$, approximately.

e) *Concavity.* Find the intervals over which f is concave up and concave down. We do this by determining where f' is increasing and decreasing using the numbers found in step (d). Those numbers divide the real-number line into three intervals:

$$A\left(-\infty, -\frac{1}{\sqrt{3}}\right), \quad B\left(-\frac{1}{\sqrt{3}}, \frac{1}{\sqrt{3}}\right), \quad \text{and} \quad C\left(\frac{1}{\sqrt{3}}, \infty\right).$$

Next, we choose a test value in each interval and make a substitution into f''. The test points we select are $-1, 0,$ and 1:

A: Test -1, $f''(-1) - 12(-1)^2 - 4 = 8 > 0$;

B: Test 0, $f''(0) = 12(0)^2 - 4 = -4 < 0$;

C: Test 1, $f''(1) = 12(1)^2 - 4 = 8 > 0$.

Technology Connection

Check the results of Example 3 using a grapher.

Interval	$\left(-\infty, -1/\sqrt{3}\right)$	$\left(-1/\sqrt{3}, 1/\sqrt{3}\right)$	$\left(1/\sqrt{3}, \infty\right)$
Test Value	$x = -1$	$x = 0$	$x = 1$
Sign of $f''(x)$	$f''(-1) > 0$	$f''(0) < 0$	$f''(1) > 0$
Behavior of f	f is concave up	f is concave down	f is concave up

Result   └── Point of ──↑   └── Point of ──↑
                                 inflection   inflection
                                 given by   given by
                              $f(-1/\sqrt{3}) = -5/9$ $f(1/\sqrt{3}) = -5/9$

Therefore, f' is increasing over the interval $\left(-\infty, -1/\sqrt{3}\right)$, decreasing over the interval $\left(-1/\sqrt{3}, 1/\sqrt{3}\right)$, and increasing over the interval $\left(1/\sqrt{3}, \infty\right)$. The function f is concave up over $\left(-\infty, -1/\sqrt{3}\right)$ and $\left(1/\sqrt{3}, \infty\right)$ and concave down over

$(-1/\sqrt{3}, 1/\sqrt{3})$. The graph changes concavity across $(-1/\sqrt{3}, -5/9)$ and $(1/\sqrt{3}, -5/9)$, so these are points of inflection.

f) *Sketch the graph.* Sketch the graph using the information in the following table. By solving $x^4 - 2x^2 = 0$, we can find the x-intercepts easily. They are $(-\sqrt{2}, 0)$, $(0, 0)$, and $(\sqrt{2}, 0)$. This also aids the graphing. Extra function values can be calculated if desired. The graph is shown below.

Technology Connection

EXERCISE

1. Consider $f(x) = x^3(x - 2)^3$. How many relative extrema do you anticipate finding? Where do you think they will be?

    Graph f, f', and f'' using $[-1, 3, -2, 6]$ as a viewing window. Estimate the relative extrema and the inflection points of f. Then check your work using the analytic methods of Examples 3 and 4.

x	$f(x)$, approximately
-2	8
-1.5	0.56
-1	-1
-0.5	-0.44
0	0
0.5	-0.44
1	-1
1.5	0.56
2	8

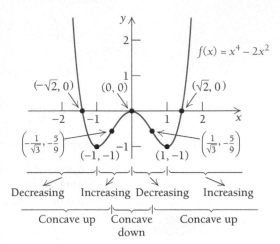

EXAMPLE 4   Find the relative extrema of the function f given by

$$y = \sin^2 x.$$

Then sketch the graph on the interval $0 \le x \le \pi$.

Solution

a) *Derivatives.* Find $f'(x)$ and $f''(x)$:

$$f'(x) = 2 \sin x \cos x,$$
$$f''(x) = 2(\cos x)(\cos x) + 2(\sin x)(-\sin x) = 2 \cos^2 x - 2 \sin^2 x.$$

b) *Critical points of f.* Since $f'(x)$ exists for all values of x, the only critical points of f are where

$$2 \sin x \cos x = 0$$
$$\sin x = 0 \quad or \quad \cos x = 0$$
$$x = n\pi \quad or \quad x = n\pi + \frac{\pi}{2}. \quad \text{n is any integer}$$

In the interval $[0, \pi]$, the critical points are 0, $\pi/2$, and π. Substituting into f, these give the three points $(0, 0)$, $(\pi/2, 1)$, and $(\pi, 0)$.

c) *Increasing and/or decreasing; relative extrema.* The critical points of f divide $[0, \pi]$ into two intervals: A$(0, \pi/2)$ and B$(\pi/2, \pi)$. We choose a test value in each interval and make a substitution. The test values we select are $\pi/4$ and $3\pi/4$.

A: Test $\pi/2$, $f'(\pi/4) = 2 \sin \dfrac{\pi}{4} \cos \dfrac{\pi}{4} = 1 > 0$;

B: Test $3\pi/4$, $f'(3\pi/4) = 2 \sin \dfrac{3\pi}{4} \cos \dfrac{3\pi}{4} = -1 < 0$.

Interval	$(0, \pi/2)$	$(\pi/2, \pi)$
Test Value	$x = \pi/4$	$x = 3\pi/4$
Sign of $f'(x)$	$f'(\pi/4) > 0$	$f'(3\pi/4) < 0$
Behavior of f	f is increasing	f is decreasing

Result   └── Relative ──┘
maximum
given by
$f(\pi/2) = 2$

Thus, by the First-Derivative Test, there is a relative maximum at $(\pi/2, 1)$. Furthermore, since f is increasing on $(0, \pi/2)$, it is clear that $(0, 0)$ is a local minimum. Also, since f is decreasing on $(\pi/2, \pi)$, the point $(\pi, 0)$ is a local minimum as well.

d) *Inflection Points.* Since $f''(x)$ exists for all real numbers, we just solve $f''(x) = 0$:

$$2\cos^2 x - 2\sin^2 x = 0$$
$$\cos^2 x = \sin^2 x$$
$$1 = \tan^2 x$$
$$\tan x = 1 \quad or \quad \tan x = -1$$
$$x = \frac{\pi}{4} + n\pi \quad or \quad x = \frac{3\pi}{4} + n\pi.$$

In the interval $[0, \pi]$, these points are $\pi/4$ and $3\pi/4$. Substituting into f, these give the two points $(\pi/4, 1/2)$ and $(3\pi/4, 1/2)$.

e) *Concavity.* The points found in (d) divide the interval $[0, \pi]$ into three intervals: $A(0, \pi/4)$, $B(\pi/4, 3\pi/4)$, and $C(3\pi/4, \pi)$. We choose a test value in each interval and make a substitution into f''. The test points we select are $\pi/6$, $\pi/2$, and $5\pi/6$:

A: Test $\pi/6$, $f''(\pi/6) = 2\cos^2 \dfrac{\pi}{6} - 2\sin^2 \dfrac{\pi}{6} = 2\left(\dfrac{\sqrt{3}}{2}\right)^2 - 2\left(\dfrac{1}{2}\right)^2 = 1 > 0$

B: Test $\pi/2$, $f''(\pi/2) = 2\cos^2 \dfrac{\pi}{2} - 2\sin^2 \dfrac{\pi}{2} = 2(0)^2 - 2(1)^2 = -2 < 0$

C: Test $5\pi/6$, $f''(5\pi/6) = 2\cos^2 \dfrac{5\pi}{6} - 2\sin^2 \dfrac{5\pi}{6} = 2\left(-\dfrac{\sqrt{3}}{2}\right)^2 - 2\left(\dfrac{1}{2}\right)^2 = 1 > 0$

Interval	$(0, \pi/4)$	$(\pi/4, 3\pi/4)$	$(3\pi/4, \pi)$
Test Value	$x = \pi/6$	$x = \pi/2$	$x = 5\pi/6$
Sign of $f''(x)$	$f''(\pi/6) > 0$	$f''(\pi/2) < 0$	$f''(5\pi/6) > 0$
Behavior of f	f is concave up	f is concave down	f is concave up

Result ⌐— Point of —↑ ⌐— Point of —↑
inflection inflection
given by given by
$f(\pi/4) = 1/2$ $f(3\pi/4) = 1/2$

The function f is concave up over the intervals $(0, \pi/4)$ and $(3\pi/4, \pi)$, but is concave down over the interval $(\pi/4, 3\pi/4)$. The graph changes concavity across $(\pi/4, 1/2)$ and $(3\pi/4, 1/2)$, so these are points of inflection.

f) *Sketch the graph.* A sketch of the graph using steps (a)–(e) is shown below for $0 \le x \le \pi$. This graph turns out to be sinusoidal, as discussed in the exercises. (Recall that sinusoidal functions were discussed in Chapter 1.)

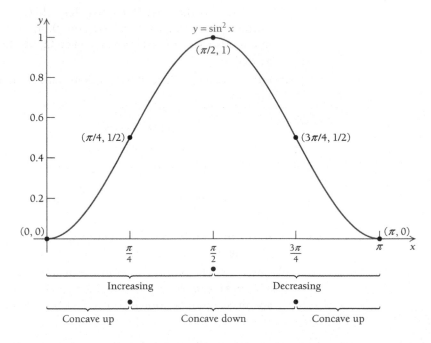

EXAMPLE 5 *Lung Cancer.* The rate of lung and bronchus cancer per 100,000 American males since 1930 is approximated by the function[4]

$$r(x) = -0.000775x^3 + 0.0696x^2 - 0.209x + 4.68,$$

where x is the number of years since 1930. Sketch the graph of $r(x)$.

---

[4]American Cancer Society, Surveillance Research, 2002.

Solution

a) *Derivatives.* Find $r'(x)$ and $r''(x)$:

$$r'(x) = -0.002325x^2 + 0.1392x - 0.209$$
$$r''(x) = -0.00465x + 0.1392.$$

b) *Critical points of r.* Since $r'(x)$ exists for all values of x, the critical points of r are where

$$-0.002325x^2 + 0.1392x - 0.209 = 0.$$

Using the quadratic formula, we find

$$x = \frac{-b \pm \sqrt{b^2 - 4ac}}{2a}$$

$$x = \frac{-0.1392 \pm \sqrt{(0.1392)^2 - 4(-0.002325)(-0.209)}}{2(-0.002325)}$$

$$\approx \frac{-0.1392 \pm 0.132033859}{-0.00465} \quad \text{Using a scientific calculator for approximation}$$

$$x \approx 1.54 \quad \text{or} \quad x \approx 58.33.$$

Now $r(1.54) \approx 4.52$ and $r(58.33) \approx 75.49$, which gives the (approximate) points $(1.54, 4.52)$ and $(58.33, 75.49)$ on the graph.

c) *Increasing and/or decreasing; relative extrema.* Use the critical points of r—namely, 1.54 and 58.33—to divide the real-number line into three intervals: $A(-\infty, 1.54)$, $B(1.54, 58.33)$, and $C(58.33, \infty)$. We choose a test value in each interval and make a substitution. The test values we select are 0, 5, and 100:

A: Test 0, $r'(0) = -0.002325(0)^2 + 0.1392(0) - 0.209 = -0.209 < 0;$

B: Test 5, $r'(5) = -0.002325(5)^2 + 0.1392(5) - 0.209 = 0.428875 > 0;$

C: Test 100, $r'(100) = -0.002325(100)^2 + 0.1392(100) - 0.209 = -9.539 < 0.$

Interval	$(-\infty, 1.54)$	$(1.54, 58.33)$	$(58.33, \infty)$
Test Value	$x = 0$	$x = 5$	$x = 100$
Sign of $r'(x)$	$r'(0) < 0$	$r'(5) > 0$	$r'(100) < 0$
Behavior of r	r is decreasing	r is increasing	r is decreasing

Result ⌞ Relative ⌝ ⌞ Relative ⌝
minimum   maximum
given by   given by
$r(1.54) \approx 4.52$ $r(58.33) \approx 75.49$

Thus, by the First-Derivative Test, there is a relative minimum at approximately $(1.54, 4.52)$ and a relative maximum at approximately $(58.33, 75.49)$. That these are relative extrema can also be verified by the Second-Derivative Test: $r''(1.54) > 0$ and $r''(58.33) < 0.$

d) *Inflection points*. Find where $r''(x)$ does not exist and where $r''(x) = 0$. Since $r''(x)$ exists for all real numbers, we just solve $r''(x) = 0$:

$$-0.00465x + 0.1392 = 0$$
$$x \approx 29.94.$$

Now $r(29.94) \approx 40.01$, and so the point $(29.94, 40.01)$ is (approximately) on the graph.

e) *Concavity*. Find the intervals over which r is concave up and concave down. We do this by determining where r'' is positive and where it is negative using the candidate found in step (d). This number divides the real-number line into two intervals: $A(-\infty, 29.94)$ and $B(29.94, \infty)$. Next, we choose a test value in each interval and make a substitution into r''. The test points we select are 0 and 100:

A: Test 0, $r''(0) = -0.00465(0) + 0.1392 = 0.1392 > 0$;

B: Test 100, $r''(100) = -0.00465(100) + 0.1392 = -0.3258 < 0.$

Interval	$(-\infty, -29.94)$	$(29.94, \infty)$
Test Value	$x = 0$	$x = 100$
Sign of $r''(x)$	$r''(0) > 0$	$r''(100) < 0$
Behavior of r	r is concave up	r is concave down

Result ⌊── Point of ⌐↑
inflection
given by
$r(29.94) = 40.01$

Therefore, r' is increasing over the interval $(-\infty, 29.94)$ and is decreasing over the interval $(29.94, \infty)$. The function r is concave up over $(-\infty, 29.94)$ and concave down over $(29.94, \infty)$. The graph changes concavity at $(29.94, 40.01)$, and so this point is a point of inflection.

f) *Sketch the graph*. A sketch of the graph using steps (a)–(e) is shown below for $0 \le x \le 70$.

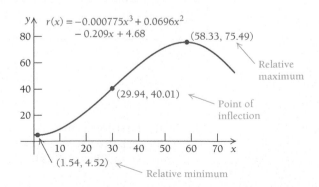

$r(x) = -0.000775x^3 + 0.0696x^2 - 0.209x + 4.68$

(58.33, 75.49) Relative maximum

(29.94, 40.01) Point of inflection

(1.54, 4.52) Relative minimum

Exercise Set 3.2

Find the relative extrema and points of inflection of the function. List your answers in terms of ordered pairs. Use the Second-Derivative Test, where possible. Then sketch the graph.

1. $f(x) = 2 - x^2$ **2.** $f(x) = x^2 + x - 1$

3. $f(x) = 2x^3 - 3x^2 - 36x + 28$

4. $f(x) = 3x^3 - 36x - 3$

5. $f(x) = \frac{8}{3}x^3 - 2x + \frac{1}{3}$

6. $f(x) = 80 - 9x^2 - x^3$

7. $f(x) = 3x^4 - 16x^3 + 18x^2$

8. $f(x) = 3x^4 + 4x^3 - 12x^2 + 5$

9. $f(x) = (x + 1)^{2/3}$ **10.** $f(x) = (x - 1)^{2/3}$

11. $f(x) = x^4 - 6x^2$ **12.** $f(x) = 2x^2 - x^4$

13. $f(x) = 3x^4 + 4x^3$ **14.** $f(x) = x^4 - 2x^3$

15. $f(x) = x^3 - 6x^2 - 135x$

16. $f(x) = x^3 - 3x^2 - 144x - 140$

17. $f(x) = \dfrac{x}{x^2 + 1}$ **18.** $f(x) = \dfrac{3}{x^2 + 1}$

19. $f(x) = (x - 1)^3$ **20.** $f(x) = (x + 2)^3$

21. $f(x) = x^2(1 - x)^2$ **22.** $f(x) = x^2(3 - x)^2$

23. $f(x) = 20x^3 - 3x^5$ **24.** $f(x) = 5x^3 - 3x^5$

25. $f(x) = x\sqrt{4 - x^2}$ **26.** $f(x) = -x\sqrt{1 - x^2}$

27. $f(x) = (x - 1)^{1/3} - 1$ **28.** $f(x) = 2 - x^{1/3}$

Find the critical points of the function in the interval $[0, 2\pi]$. Determine if each critical point is a relative maximum, a relative minimum, or neither. Use the Second-Derivative Test, when possible. Determine the points of inflection in the interval $[0, 2\pi]$. Then sketch the graph on the interval $[0, 2\pi]$.

29. $f(x) = x + \cos 2x$ **30.** $f(x) = x - 2 \sin x$

31. $f(x) = \dfrac{x}{3} - \sin \dfrac{2x}{3}$ **32.** $f(x) = \dfrac{x}{4} + \cos \dfrac{x}{2}$

33. $f(x) = \sin x + \cos x$ **34.** $f(x) = \sin x - \cos x$

35. $f(x) = \sqrt{3} \sin x + \cos x$

36. $f(x) = \sin x - \sqrt{3} \cos x$

37. $f(x) = \dfrac{\sin x}{2 - \cos x}$ **38.** $f(x) = \dfrac{\cos x}{2 + \sin x}$

39. $f(x) = \cos^2 x$ **40.** $f(x) = \sin^4 x$

41. $f(x) = \cos^4 x$

42. $f(x) = 3 \sin x - \sin^3 x$

Find all points of inflection, if they exist.

43. $f(x) = x^3 + 3x + 1$

44. $f(x) = x^3 - 6x^2 + 12x - 6$

45. $f(x) = \frac{4}{3}x^3 - 2x^2 + x$

46. $f(x) = x^4 - 4x^3 + 10$

47. $f(x) = x - \sin x$

48. $f(x) = 2x + 1 + \cos 2x$

49. $f(x) = \tan x$ **50.** $f(x) = \cot x$

51. $f(x) = \tan x + \sec x$ **52.** $f(x) = \cot x + \csc x$

53.–104. Check the results of each of Exercises 1–52 using a grapher.

APPLICATIONS

105. *Coughing Velocity.* A person coughs when a foreign object is in the windpipe. The velocity of the cough depends on the size of the object. Suppose a person has a windpipe with a 20-mm radius. If a foreign object has a radius r, in millimeters, then the velocity V, in millimeters/second, needed to remove the object by a cough is given by

$$V(r) = k(20r^2 - r^3), \quad 0 \le r \le 20,$$

where k is some positive constant. For what size object is the maximum velocity needed to remove the object?

106. *Temperature in January.* Suppose that the temperature T, in degrees Fahrenheit, during a 24-hr day in January is given by

$$T(x) = 0.0027(x^3 - 34x + 240),$$
$$0 \le x \le 24,$$

where x is the number of hours since midnight. Estimate the relative minimum temperature and when it occurs.

107. *New York Temperature.* For any date, the average temperature on that date in New York can be approximated by the function

$$T(x) = 0.0338x^4 - 0.996x^3 + 8.57x^2$$
$$- 18.4x + 43.5,$$

where T represents the temperature in degrees Fahrenheit, $x = 1$ represents the middle of January, $x = 2$ represents the middle of February, and so on.[5]

a) Find the points of inflection of this function.

tw b) What is the significance of these points?

108. *Cesarian Deliveries.* Since 1990, the number N of reported babies delivered by Cesarian section may be modeled by the function

$$N(x) = 748x^3 - 6820x^2 - 5520x + 916,000,$$

where x is the number of years since 1990.

a) Find the relative maximum and minimum of this function.

b) Find the point of inflection of this function.

c) Sketch the graph of $N(x)$.

109. *Hearing Impairments.* The following function[6] approximates the number N, in millions, of hearing-impaired Americans as a function of age x:

$$N(x) = -0.00006x^3 + 0.006x^2$$
$$- 0.1x + 1.9.$$

a) Find the relative maximum and minimum of this function.

b) Find the point of inflection of this function.

c) Sketch the graph of $N(x)$ for $0 \leq x \leq 80$.

SYNTHESIS

tw In each of Exercises 110 and 111, determine which graph is the derivative of the other and explain why.

110.   **111.**

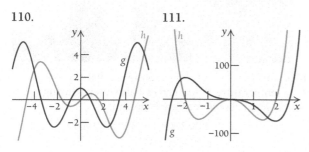

---

tw For Exercises 112–113, identify **a)** relative maxima, **b)** relative minima, **c)** points of inflection, **d)** intervals where the function is increasing, **e)** intervals where the function is decreasing, **f)** intervals where the function is concave up, and **g)** intervals where the function is concave down.

112.

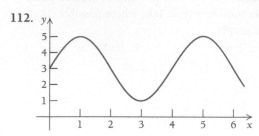

113.

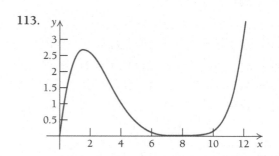

Three Aspects of Love. Researchers at Yale University have suggested that the following graphs may represent three different aspects of love.

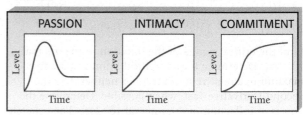

Source: From "A Triangular Theory of Love," by R. J. Sternberg, 1986, *Psychological Review,* 93(2), 119–135. Copyright 1986 by the American Psychological Association, Inc. Reprinted by permission.

tw 114. Analyze each of these graphs in terms of the concepts you have learned in Sections 3.1 and 3.2: relative extrema, concavity, increasing, decreasing, and so on.

tw 115. Do you agree with the researchers regarding the shape of these graphs? Explain your reasons.

---

[5]http://www.worldclimate.com

[6]American Speech-Language Hearing Association.

116. *Trigonometric Identity.* The graph of $f(x) = \sin^2 x$ shown on page 190 appears to be sinusoidal.

    a) Determine a, b, and k so that the graph on page 190 is satisfied by a function of the form $y = a \cos bx + k$.

    tw b) Use Exercise 79 of Section 1.4 to explain the result of part (a).

    tw c) Compute the derivative of $f(x) = \sin^2 x$ and the function $y = a \cos bx + k$ found in (a). Determine a new trigonometric identity by equating the two derivatives.

Technology Connection

Graph the function. Then estimate any relative extrema.

117. $f(x) = 3x^{2/3} - 2x$ 118. $f(x) = 4x - 6x^{2/3}$

119. $f(x) = x^2(x - 2)^3$ 120. $f(x) = x^2(1 - x)^3$

121. $f(x) = x - \sqrt{x}$

122. $f(x) = (x - 1)^{2/3} - (x + 1)^{2/3}$

3.3   Graph Sketching: Asymptotes and Rational Functions

OBJECTIVES

■ Find limits involving infinity.
■ Graph rational functions.

Rational Functions

Thus far we have considered a strategy for graphing a continuous function using the tools of calculus. We now want to consider some discontinuous functions, most of which are rational functions. Our graphing skills will now have to take into account the discontinuities of the graph and certain lines called *asymptotes*.

Let's reconsider the definition of a rational function.

DEFINITION

A **rational function** is a function f that can be described by

$$f(x) = \frac{P(x)}{Q(x)},$$

where $P(x)$ and $Q(x)$ are polynomials and with $Q(x)$ not the zero polynomial. The domain of f consists of all inputs x for which $Q(x) \neq 0$.

Polynomials are themselves a special kind of rational function, since $Q(x)$ can be the polynomial 1. Here we are considering graphs of rational functions in which the denominator is not a constant. Before we do so, however, we need to reconsider limits.

Limits and Infinity

Let's look again at the graph of the rational function $F(x) = 1/x$. We see that

$$\lim_{x \to 0} \frac{1}{x} \quad \text{does not exist.}$$

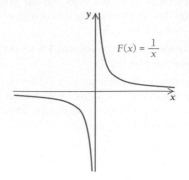

$F(x) = \dfrac{1}{x}$

Looking at the graph above, we note that as x approaches 0 from the right, the outputs increase without bound. These numbers do not approach any real number, though it might be said that the limit from the right is infinity (∞). That is,

$$\lim_{x \to 0^+} \frac{1}{x} = \infty.$$

As x approaches 0 from the left, the outputs become more and more negative without bound. These numbers do not approach any real number, though it might be said that the limit from the left is negative infinity ($-\infty$). That is,

$$\lim_{x \to 0^-} \frac{1}{x} = -\infty.$$

Keep in mind that ∞ and $-\infty$ are not real numbers. We associate ∞ with numbers increasing without bound in a positive direction, as in the interval notation $(2, \infty)$. Thus we associate $-\infty$ with numbers decreasing without bound in a negative direction, as in the interval notation $(-\infty, 8)$.

Occasionally we need to determine limits when the inputs get larger and larger without bound, that is, when they approach infinity. In such cases, we are finding *limits at infinity*. Such a limit is expressed as

$$\lim_{x \to \infty} f(x).$$

EXAMPLE 1 Find

$$\lim_{x \to \infty} \frac{3x - 1}{x}.$$

Solution The function involved is rational. One way to find such a limit is to use an input–output table, as follows, using progressively larger x-values.

Inputs, x	1	10	50	100	2000
Outputs, $\dfrac{3x - 1}{x}$	2.0	2.9	2.98	2.99	2.9995

As the inputs get larger and larger without bound, the outputs get closer and closer to 3. Thus,

$$\lim_{x \to \infty} \frac{3x - 1}{x} = 3.$$

Another way to find this limit is to use some algebra and the fact that

$$\text{as } x \to \infty, \quad \frac{b}{ax^n} \to 0,$$

for any positive number n and any constants a and b, $a \neq 0$. We multiply by 1, using $(1/x) \div (1/x)$. This amounts to dividing both the numerator and the denominator by x:

$$\lim_{x \to \infty} \frac{3x - 1}{x} = \lim_{x \to \infty} \frac{3x - 1}{x} \cdot \frac{(1/x)}{(1/x)}$$

$$= \lim_{x \to \infty} \frac{(3x - 1)\dfrac{1}{x}}{x \cdot \dfrac{1}{x}}$$

$$= \lim_{x \to \infty} \frac{3x \cdot \dfrac{1}{x} - 1 \cdot \dfrac{1}{x}}{1}$$

$$= \lim_{x \to \infty} \left(3 - \frac{1}{x}\right)$$

$$= 3 - 0$$

$$= 3.$$

Technology Connection

1. Verify the limit

$$\lim_{x \to \infty} \frac{3x - 1}{x} = 3$$

by using the **TABLE** feature with larger and larger x-values.

X	Y1	
50	2.98	
150	2.9933	
250	2.996	
350	2.9971	
450	2.9978	
550	2.9982	
650	2.9985	
X = 50		

X	Y1	
500	2.998	
1500	2.9993	
2500	2.9996	
3500	2.9997	
4500	2.9998	
5500	2.9998	
6500	2.9998	
X = 500		

2. Graph the function

$$f(x) = \frac{3x - 1}{x}$$

in **DOT** mode. Then use the **TRACE** feature, moving the cursor along the graph from left to right, and observe the behavior of the y-coordinates.

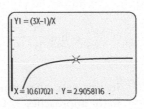

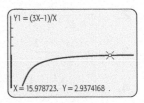

EXERCISES

Consider $\displaystyle\lim_{x \to \infty} \frac{2x + 5}{x}$.

1. Use the **TABLE** feature to find the limit.

2. Graph the function in **DOT** mode and use the **TRACE** feature to find the limit.

EXAMPLE 2 Find

$$\lim_{x \to \infty} \frac{3x^2 - 7x + 2}{7x^2 + 5x + 1}.$$

Solution This function is rational. It involves a quotient of two polynomials. The *degree* of a polynomial is its highest power. Note that the degree of each polynomial above is 2.

The highest power of x in the denominator is x^2. We divide both the numerator and the denominator by x^2:

$$\lim_{x \to \infty} \frac{3x^2 - 7x + 2}{7x^2 + 5x + 1} = \lim_{x \to \infty} \frac{3 - \dfrac{7}{x} + \dfrac{2}{x^2}}{7 + \dfrac{5}{x} + \dfrac{1}{x^2}} = \frac{3 - 0 + 0}{7 + 0 + 0} = \frac{3}{7}.$$

EXAMPLE 3 Find

$$\lim_{x \to \infty} \frac{5x^2 + 7x + 9}{3x^3 + 2x - 4}.$$

Solution The highest power of x in the denominator is x^3. We divide the numerator and the denominator by x^3:

$$\lim_{x \to \infty} \frac{5x^2 + 7x + 9}{3x^3 + 2x - 4} = \lim_{x \to \infty} \frac{\dfrac{5}{x} + \dfrac{7}{x^2} + \dfrac{9}{x^3}}{3 + \dfrac{2}{x^2} - \dfrac{4}{x^3}} = \frac{0 + 0 + 0}{3 + 0 - 0} = \frac{0}{3} = 0.$$

EXAMPLE 4 Find

$$\lim_{x \to \infty} \frac{7x^5 - 8x - 6}{3x^2 + 5x + 2}.$$

Solution The highest power of x in the denominator is x^2. We divide the numerator and the denominator by x^2:

$$\lim_{x \to \infty} \frac{7x^5 - 8x - 6}{3x^2 + 5x + 2} = \lim_{x \to \infty} \frac{7x^3 - \dfrac{8}{x} - \dfrac{6}{x^2}}{3 + \dfrac{5}{x} + \dfrac{2}{x^2}} = \frac{\lim\limits_{x \to \infty} 7x^3 - 0 - 0}{3 + 0 + 0} = \infty.$$

In this case, the numerator increases without bound positively while the denominator approaches 3. This can be checked with an input–output table. Thus the limit is ∞.

Graphs of Rational Functions

Figure 1 shows the graph of the rational function

$$f(x) = \frac{x^2 - 1}{x^2 + x - 6} = \frac{(x - 1)(x + 1)}{(x - 2)(x + 3)}.$$

Technology Connection

EXERCISES

Use the **TABLE** and **TRACE** features to find each of the following limits. Then check your work using the analytic procedure of Examples 1–4.

1. $\lim\limits_{x \to \infty} \dfrac{2x^2 + x - 7}{3x^2 - 4x + 1}$

2. $\lim\limits_{x \to \infty} \dfrac{5x + 4}{2x^3 - 3}$

3. $\lim\limits_{x \to \infty} \dfrac{5x^2 - 2}{4x + 5}$

4. $\lim\limits_{x \to \infty} \dfrac{x^2 - 1}{x^2 + x - 6}$

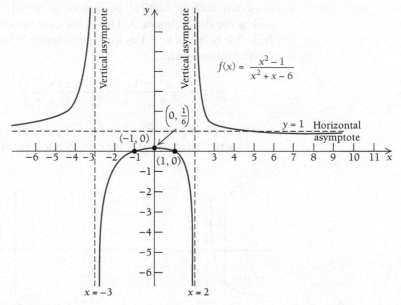

Figure 1

Let's make some observations about this graph.

First, note that as x gets closer to 2 from the left, the function values get smaller and smaller negatively, approaching $-\infty$. As x gets closer to 2 from the right, the function values get larger and larger positively. Thus,

$$\lim_{x \to 2^-} f(x) = -\infty \quad \text{and} \quad \lim_{x \to 2^+} f(x) = \infty.$$

For this graph, we can think of the line $x = 2$ as a "limiting line" called a *vertical asymptote*. Similarly, the line $x = -3$ is a vertical asymptote.

> **DEFINITION**
>
> The line $x = a$ is a **vertical asymptote** if any of the following limit statements is true:
>
> $$\lim_{x \to a^-} f(x) = \infty \quad \text{or} \quad \lim_{x \to a^-} f(x) = -\infty \quad \text{or}$$
> $$\lim_{x \to a^+} f(x) = \infty \quad \text{or} \quad \lim_{x \to a^+} f(x) = -\infty.$$
>
> The graph of a rational function *never* crosses a vertical asymptote. If the expression that defines the rational expression is simplified, meaning the numerator and denominator have no common factors (other than constants), then if a is an input that makes the denominator 0, the line $x = a$ is a vertical asymptote.

For example,

$$f(x) = \frac{x^2 - 1}{x - 1} = \frac{(x - 1)(x + 1)}{x - 1}$$

does not have a vertical asymptote at $x = 1$, even though 1 is an input that makes the denominator 0. This is the case because $(x^2 - 1)/(x - 1)$ is not simplified, that is, it has $x - 1$ as a common factor of the numerator and the denominator. On the other hand,

$$g(x) = \frac{x^2 - 1}{x^2 + x - 6} = \frac{(x + 1)(x - 1)}{(x - 2)(x + 3)}$$

is simplified and has $x = 2$ and $x = -3$ as vertical asymptotes.

Figure 2 shows four ways in which a vertical asymptote can occur.

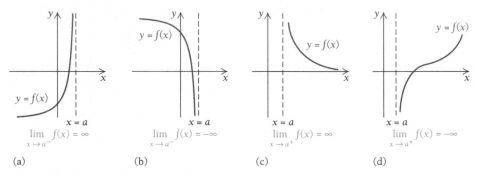

| (a) | (b) | (c) | (d) |

Figure 2

Look again at the graph in Fig. 1. Note that function values get closer and closer to 1 as x approaches $-\infty$, meaning that $f(x) \to 1$ as $x \to -\infty$. Also, function values get closer and closer to 1 as x approaches ∞, meaning that $f(x) \to 1$ as $x \to \infty$. (The graph actually crosses the horizontal asymptote $y = 1$ and then moves back to it.) Thus,

$$\lim_{x \to -\infty} f(x) = 1 \quad \text{and} \quad \lim_{x \to \infty} f(x) = 1.$$

The line $y = 1$ is called a *horizontal asymptote*.

Technology Connection

Use the **GRAPH**, **TABLE**, and **TRACE** features of a grapher to verify that

$$\lim_{x \to \infty} \frac{x^2 - 1}{(x - 2)(x + 3)} = 1.$$

DEFINITION

The line $y = b$ is a **horizontal asymptote** if either or both of the following limit statements is true:

$$\lim_{x \to -\infty} f(x) = b \quad \text{or} \quad \lim_{x \to \infty} f(x) = b.$$

The graph of a rational function may or may not cross a horizontal asymptote. Horizontal asymptotes occur when the degree of the numerator is less than or equal to the degree of the denominator.

In Figs. 3–5, we see three ways in which horizontal asymptotes can occur.

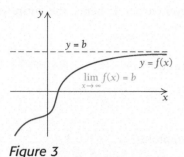

Figure 3

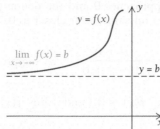

Figure 4

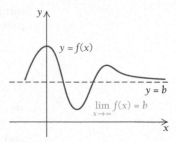

Figure 5

Technology Connection

Asymptotes

Our discussion now allows us to attach the term "vertical asymptote" to those mysterious vertical lines that appear with the graphs of rational functions when the grapher is in **CONNECTED** mode. For example, consider the graph of $f(x) = 8/(x^2 - 4)$, using the viewing window $[-6, 6, -8, 8]$. Vertical asymptotes occur at $x = -2$ and $x = 2$. These lines are not part of the graph.

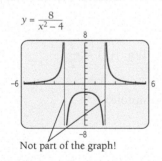

Not part of the graph!

EXERCISES

Graph each of the following in both **DOT** and **CONNECTED** modes. Try to locate the vertical asymptotes visually. Then verify your results using the methods of Examples 1 and 2. You may need to try different viewing windows.

1. $f(x) = \dfrac{x^2 + 7x + 10}{x^2 + 3x - 28}$

2. $f(x) = \dfrac{x^2 + 5}{x^3 - x^2 - 6x}$

Occurrences of Asymptotes

It is important in graphing rational functions to determine where the asymptotes, if any, occur. Vertical asymptotes are easy to locate when the expression is in simplified form and a denominator can be factored. The x-inputs that make a denominator 0 give us the vertical asymptotes.

EXAMPLE 5   Determine the vertical asymptotes:

$$f(x) = \frac{3x - 2}{x(x - 5)(x + 3)}.$$

Solution   The expression is in simplified form. The vertical asymptotes are the lines $x = 0$, $x = 5$, and $x = -3$.   ∎

EXAMPLE 6   Determine the vertical asymptotes:

$$f(x) = \frac{x - 2}{x^3 - x} = \frac{x - 2}{x(x - 1)(x + 1)}.$$

Solution   The expression is in simplified form. The vertical asymptotes are the lines $x = 0$, $x = 1$, and $x = -1$.   ∎

EXAMPLE 7   Find the horizontal asymptotes:

$$f(x) = \frac{2x + 3}{x^3 - 2x^2 + 4}.$$

Solution   Since the degree of the numerator is less than the degree of the denominator, there is a horizontal asymptote. We then divide the numerator and the denominator by x^3 and find the limits as x approaches $-\infty$ and ∞; that is, as $|x|$ gets larger and larger:

$$f(x) = \frac{2x + 3}{x^3 - 2x^2 + 4} = \frac{\dfrac{2}{x^2} + \dfrac{3}{x^3}}{1 - \dfrac{2}{x} + \dfrac{4}{x^3}}.$$

As x approaches either ∞ or $-\infty$, each expression with x or some power of x in it takes on values ever closer to 0. Thus the numerator approaches 0 and the denominator approaches 1; hence the entire expression takes on values ever closer to 0. We have

$$f(x) \approx \frac{0 + 0}{1 - 0 + 0} \quad \text{as } x \text{ approaches either } \infty \text{ or } -\infty,$$

so
$$\lim_{x \to -\infty} f(x) = 0 \quad \text{and} \quad \lim_{x \to \infty} f(x) = 0,$$

and the x-axis, the line $y = 0$, is a horizontal asymptote. ■

EXAMPLE 8 Find the horizontal asymptotes:

$$f(x) = \frac{3x^2 + 2x - 4}{2x^2 - x + 1}.$$

Solution The numerator and the denominator have the same degree, so there is a horizontal asymptote. We divide the numerator and the denominator by x^2:

$$f(x) = \frac{3x^2 + 2x - 4}{2x^2 - x + 1} = \frac{3 + \dfrac{2}{x} - \dfrac{4}{x^2}}{2 - \dfrac{1}{x} + \dfrac{1}{x^2}}.$$

As x approaches either ∞ or $-\infty$, the numerator approaches 3 and the denominator approaches 2. Therefore, the function gets very close to $\frac{3}{2}$. Thus,

$$\lim_{x \to -\infty} f(x) = \frac{3}{2} \quad \text{and} \quad \lim_{x \to \infty} f(x) = \frac{3}{2}.$$

The line $y = \frac{3}{2}$ is a horizontal asymptote. ■

When the degree of the numerator is less than the degree of the denominator, the x-axis, or the line $y = 0$, is a horizontal asymptote.

When the degree of the numerator is the same as the degree of the denominator, the line $y = a/b$ is a horizontal asymptote, where a is the leading coefficient of the numerator and b is the leading coefficient of the denominator.

Technology Connection

EXERCISES

Graph each of the following. Try to locate the horizontal asymptotes using the **TABLE** and **TRACE** features. Verify your results using the methods of Examples 7 and 8.

1. $f(x) = \dfrac{x^2 + 5}{x^3 - x^2 - 6x}$

2. $f(x) = \dfrac{9x^4 - 7x^2 - 9}{3x^4 + 7x^2 + 9}$

3. $f(x) = \dfrac{135x^5 - x^2}{x^7}$

4. $f(x) = \dfrac{3x^2 - 4x + 3}{6x^2 + 2x - 5}$

Oblique Asymptotes

There are asymptotes that are neither vertical nor horizontal. For example, in the graph of

$$f(x) = \frac{x^2 - 4}{x - 1},$$

shown at right, the line $x = 1$ is a vertical asymptote. Note that as x approaches either ∞ or $-\infty$, the curve gets closer and closer to $y = x + 1$. The line $y = x + 1$ is called an *oblique asymptote*.

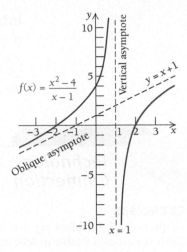

DEFINITION

The line $y = mx + b$ is an **oblique asymptote** of the rational function $f(x) = P(x)/Q(x)$ if $f(x)$ can be expressed as

$$f(x) = (mx + b) + g(x),$$

where both $\lim_{x \to \infty} g(x)$ and $\lim_{x \to -\infty} g(x) = 0$. Oblique asymptotes occur when the degree of the numerator is exactly 1 more than the degree of the denominator. A graph can cross an oblique asymptote.

How can we find an oblique asymptote? One way is by division.

Technology Connection

EXERCISES

Graph each of the following. Try to visually locate the oblique asymptotes. Then use the method in Example 9 to find the oblique asymptote and graph it along with the original function.

1. $f(x) = \dfrac{3x^2 - 7x + 8}{x - 2}$

2. $f(x) = \dfrac{5x^3 + 2x + 1}{x^2 - 4}$

EXAMPLE 9 Find the oblique asymptotes:

$$f(x) = \frac{x^2 - 4}{x - 1}.$$

Solution When we divide the numerator by the denominator, we obtain a quotient of $x + 1$ and a remainder of -3. Thus,

$$f(x) = \frac{x^2 - 4}{x - 1} = (x + 1) + \frac{-3}{x - 1}.$$

$$
\begin{array}{r}
x + 1 \\
x - 1 \overline{) x^2 - 4} \\
\underline{x^2 - x } \\
x - 4 \\
\underline{x - 1} \\
-3
\end{array}
$$

Now we can see that when x gets very large (either positively or negatively), $-3/(x - 1)$ approaches 0. Thus, as x approaches either ∞ or $-\infty$, the expression $x + 1$ is the dominant part of

$$(x + 1) + \frac{-3}{x - 1}.$$

Thus, $y = x + 1$ is an oblique asymptote. ■

Intercepts

If they exist, the x-intercepts of a function occur at those values of x for which $y = f(x) = 0$ and give us points at which the graph crosses the x-axis. If it exists, the y-intercept of a function occurs at the value of y for which $x = 0$ and gives us the point at which the graph crosses the y-axis.

Technology Connection

EXERCISES

Graph each of the following. Use the ZERO feature and a table in ASK mode to find the x- and y-intercepts.

1. $f(x) = \dfrac{x(x - 3)(x + 5)}{(x + 2)(x - 4)}$

2. $f(x) = \dfrac{x^3 + 2x^2 - 3x}{x^2 + 5}$

EXAMPLE 10   Find the intercepts of

$$f(x) = \frac{x^3 - x^2 - 6x}{x^2 - 3x + 2}.$$

Solution   We factor the numerator and the denominator:

$$f(x) = \frac{x(x + 2)(x - 3)}{(x - 1)(x - 2)}.$$

To find the x-intercepts, we solve the equation $f(x) = 0$. Such values occur when the numerator is 0 but the denominator is not. Thus we solve the equation

$$x(x + 2)(x - 3) = 0.$$

The x-values making the numerator 0 are 0, -2, and 3. Since none of these makes the denominator 0, they yield the x-intercepts of the function: $(0, 0)$, $(-2, 0)$, and $(3, 0)$.

To find the y-intercept, we let $x = 0$:

$$f(0) = \frac{0^3 - 0^2 - 6(0)}{0^2 - 3(0) + 2} = 0.$$

In this case, the y-intercept is also an x-intercept, $(0, 0)$.

Sketching Graphs

We can now refine our analytic strategy for graphing.

Strategy for Sketching Graphs

a) *Intercepts.* Find the x-intercept(s) and the y-intercept of the graph.

b) *Asymptotes.* Find the vertical, horizontal, and oblique asymptotes.

c) *Derivatives.* Find $f'(x)$ and $f''(x)$.

d) *Undefined values and critical points of f.* Find the inputs for which the function is not defined, giving denominators of 0. Find also the critical points of f.

e) *Increasing and/or decreasing; relative extrema.* Use the points found in step (d) to determine intervals over which the function f is increasing or decreasing. Use this information and/or the second derivative to determine the relative maxima and minima. A relative extremum can occur only at a point c for which $f(c)$ exists.

f) *Inflection points.* Determine candidates for inflection points by finding points x where $f''(x)$ does not exist or where $f''(x) = 0$. Find the function values at these points. If a function value $f(x)$ does not exist, then the function does not have an inflection point at x.

g) *Concavity.* Use the values c from step (f) as endpoints of intervals. Determine the concavity by checking to see where f' is increasing— that is, $f''(x) > 0$—and where f' is decreasing—that is, $f''(x) < 0$. Do this by selecting test points and substituting into $f''(x)$.

h) *Sketch the graph.* Use the information from steps (a) through (g) to sketch the graph, plotting extra points (computing them with your calculator) as needed.

EXAMPLE 11 Sketch the graph of $f(x) = \dfrac{8}{x^2 - 4}$.

Solution

a) *Intercepts.* The x-intercepts occur at the points where the numerator is 0 but the denominator is not. Since in this case the numerator is the constant 8, there are no x-intercepts. To find the y-intercept, we compute $f(0)$:

$$f(0) = \frac{8}{0^2 - 4} = \frac{8}{-4} = -2.$$

This gives us one point on the graph, $(0, -2)$.

b) *Asymptotes.*

Vertical: The numerator 8 has no common factors with the denominator $x^2 - 4 = (x + 2)(x - 2)$. The denominator is 0 for x-values of -2 and 2. Thus the graph has the lines $x = -2$ and $x = 2$ as vertical asymptotes. We draw them using dashed lines (they are *not* part of the actual graph).

Horizontal: The degree of the numerator is less than the degree of the denominator, so the x-axis, or the line $y = 0$, is a horizontal asymptote. It is already drawn as an axis.

Oblique: There is no oblique asymptote since the degree of the numerator is not 1 more than the degree of the denominator.

c) *Derivatives.* Find $f'(x)$ and $f''(x)$. Using the Quotient Rule, we get

$$f'(x) = \frac{-16x}{(x^2 - 4)^2} \quad \text{and} \quad f''(x) = \frac{16(3x^2 + 4)}{(x^2 - 4)^3}.$$

d) *Undefined values and critical points of f.* The domain of the original function is all real numbers except -2 and 2, where the vertical asymptotes occur. We find the critical points of f by looking for values of x where $f'(x) = 0$ or where $f'(x)$ does not exist. Now $f'(x) = 0$ for values of x for which $-16x = 0$, but the denominator is not 0. The only such number is 0 itself. The derivative $f'(x)$ does not exist at -2 and 2. Thus the undefined values and the critical points are -2, 0, and 2.

e) *Increasing and/or decreasing; relative extrema.* Use the undefined values and the critical points to determine the intervals over which f is increasing and the intervals over which f is decreasing. The points to consider are -2, 0, and 2. These divide the real-number line into four intervals. We choose a test point in each interval and make a substitution into the derivative f':

A: Test -3, $f'(-3) = \dfrac{-16(-3)}{[(-3)^2 - 4]^2} = \dfrac{48}{25} > 0$;

B: Test -1, $f'(-1) = \dfrac{-16(-1)}{[(-1)^2 - 4]^2} = \dfrac{16}{9} > 0$;

C: Test 1, $f'(1) = \dfrac{-16(1)}{[(1)^2 - 4]^2} = \dfrac{-16}{9} < 0$;

D: Test 3, $f'(3) = \dfrac{-16(3)}{[(3)^2 - 4]^2} = \dfrac{-48}{25} < 0$.

Interval	$(-\infty, -2)$	$(-2, 0)$	$(0, 2)$	$(2, \infty)$
Test Value	$x = -3$	$x = -1$	$x = 1$	$x = 3$
Sign of $f'(x)$	$f'(-3) > 0$	$f'(-1) > 0$	$f'(1) < 0$	$f'(3) < 0$
Behavior of f	f is increasing	f is increasing	f is decreasing	f is decreasing

Result   └─ No change ─┘└─ Relative ─┘└─ No change ─┘
                                          maximum
                                          given by
                                          $f(0) = -2$

Now $f(0) = -2$, so there is a relative maximum at $(0, -2)$.

f) *Inflection points.* Determine candidates for inflection points by finding where $f''(x)$ does not exist and where $f''(x) = 0$. Now $f(x)$ does not exist at -2 and 2, so $f''(x)$ does not exist at -2 and 2. We then determine where $f''(x) = 0$, or

$$16(3x^2 + 4) = 0.$$

But $16(3x^2 + 4) > 0$ for all real numbers x, so there are no points of inflection since $f(-2)$ and $f(2)$ do not exist.

g) *Concavity.* Use the values found in step (f) as endpoints of intervals. Determine the concavity by checking to see where f' is increasing and decreasing. The points -2 and 2 divide the real-number line into three intervals. We choose test points in each interval and make a substitution into f'':

A: Test -3, $f''(-3) = \dfrac{16[3(-3)^2 + 4]}{[(-3)^2 - 4]^3} > 0$;

B: Test 0, $f''(0) = \dfrac{16[3(0)^2 + 4]}{[(0)^2 - 4]^3} < 0$;

C: Test 3, $f''(3) = \dfrac{16[3(3)^2 + 4]}{[(3)^2 - 4]^3} > 0$.

Technology Connection

Check the results of Example 11 using a grapher.

Interval	$(-\infty, -2)$	$(-2, 2)$	$(2, \infty)$
Test Value	$x = -3$	$x = 0$	$x = 3$
Sign of $f''(x)$	$f''(-3) > 0$	$f''(0) < 0$	$f''(x) > 0$
Behavior of f	f is concave up	f is concave down	f is concave up

Result ⌐— No point of —↑ ⌐— No point of —↑
inflection since   inflection since
$f(-2)$ does not exist $f(2)$ does not exist

The function is concave up over the intervals $(-\infty, -2)$ and $(2, \infty)$. The function is concave down over the interval $(-2, 2)$.

h) *Sketch the graph.* Sketch the graph using the information in the following table, plotting extra points by computing values from your calculator as needed. The graph follows.

x	$f(x)$, approximately
−5	0.38
−4	0.67
−3	1.6
−1	−2.67
0	−2
1	−2.67
3	1.6
4	0.67
5	0.38

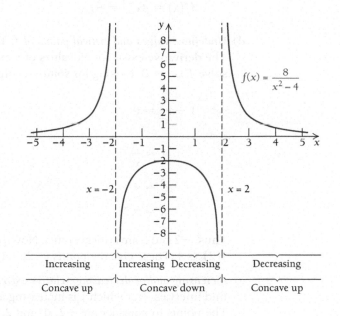

$$f(x) = \frac{8}{x^2 - 4}$$

Increasing   Increasing Decreasing   Decreasing
Concave up   Concave down   Concave up

EXAMPLE 12 Sketch the graph of $f(x) = \dfrac{x^2 + 4}{x}$.

Solution

a) *Intercepts.* The equation $f(x) = 0$ has no real-number solution. Thus there are no x-intercepts. The number 0 is not in the domain of the function. Thus there is no y-intercept.

b) *Asymptotes.*

Vertical: The numerator $x^2 + 4$ has no common factors with the denominator, x. So, the line $x = 0$ is a vertical asymptote.

Horizontal: The degree of the numerator is greater than the degree of the denominator, so there are no horizontal asymptotes.

Oblique: The degree of the numerator is 1 greater than the degree of the denominator, so there is an oblique asymptote. We do the division and express the function in the form

$$f(x) = x + \frac{4}{x}.$$

$$\begin{array}{r} x \\ x{\overline{\smash{\big)}\,x^2 + 4}} \\ \underline{x^2 } \\ 4 \end{array}$$

As $|x|$ gets larger, the term $4/x$ approaches 0, so the line $y = x$ is an oblique asymptote.

c) *Derivatives.* Find $f'(x)$ and $f''(x)$:

$$f'(x) = 1 - 4x^{-2} = 1 - \frac{4}{x^2};$$

$$f''(x) = 8x^{-3} = \frac{8}{x^3}.$$

d) *Undefined values and critical points of f.* The number 0 is not in the domain of f. The derivative exists for all values of x except 0. Thus, to find critical points, we solve $f'(x) = 0$, looking for solutions other than 0:

$$1 - \frac{4}{x^2} = 0$$

$$1 = \frac{4}{x^2}$$

$$x^2 = 4$$

$$x = \pm 2.$$

Thus, -2 and 2 are critical points. Now $f(0)$ does not exist, but $f(-2) = -4$ and $f(2) = 4$. These give the points $(-2, -4)$ and $(2, 4)$ on the graph.

e) *Increasing and/or decreasing; relative extrema.* Use the points found in step (d) to find intervals over which f is increasing and intervals over which f is decreasing. The points to consider are -2, 0, and 2. These divide the real-number line into four intervals. We choose test points in each interval and make a substitution into f':

A: Test -3, $f'(-3) = 1 - \dfrac{4}{(-3)^2} = \dfrac{5}{9} > 0;$

B: Test -1, $f'(-1) = 1 - \dfrac{4}{(-1)^2} = -3 < 0;$

C: Test 1, $f'(1) = 1 - \dfrac{4}{1^2} = -3 < 0;$

D: Test 3, $f'(3) = 1 - \dfrac{4}{3^2} = \dfrac{5}{9} > 0.$

Interval	$(-\infty, -2)$	$(-2, 0)$	$(0, 2)$	$(2, \infty)$
Test Value	$x = -3$	$x = -1$	$x = 1$	$x = 3$
Sign of $f'(x)$	$f'(-3) > 0$	$f'(-1) < 0$	$f'(1) < 0$	$f'(3) > 0$
Behavior of f	f is increasing	f is decreasing	f is decreasing	f is increasing

Result ⌐— Relative —↑ ⌐—No change —↑ ⌐— Relative —↑
maximum   minimum
given by   given by
$f(-2) = -4$ $f(2) = 4$

Now $f(-2) = -4$ and $f(2) = 4$. There is a relative maximum at $(-2, -4)$ and a relative minimum at $(2, 4)$.

f) *Inflection points.* Determine candidates for inflection points by finding where $f''(x)$ does not exist or where $f''(x) = 0$. Now $f''(x)$ does not exist, but because $f(0)$ does not exist, there cannot be an inflection point at 0. Then look for values of x for which $f''(x) = 0$:

$$\frac{8}{x^3} = 0.$$

But this equation has no solution. Thus there are no points of inflection.

g) *Concavity.* Use the values found in step (f) as endpoints of intervals. Determine the concavity by checking to see where f' is increasing and decreasing. The number 0 divides the real-number line into two intervals, $(-\infty, 0)$ and $(0, \infty)$. We could choose a test value in each interval and make a substitution into f''. But note that for any $x < 0$, $x^3 < 0$, so

$$f''(x) = \frac{8}{x^3} < 0. \qquad \text{On the interval } (-\infty, 0).$$

Also, for any $x > 0$, $x^3 > 0$, so

$$f''(x) = \frac{8}{x^3} > 0. \qquad \text{On the interval } (0, \infty).$$

Thus, f is concave down over the interval $(-\infty, 0)$ and concave up over the interval $(0, \infty)$.

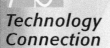

Technology Connection

Check the result of Example 12 using a grapher.

h) *Sketch the graph.* Sketch the graph using the preceding information and any additional computed values of f as needed. The graph follows.

x	$f(x)$, approximately
-6	-6.67
-5	-5.8
-4	-5
-3	-4.3
-2	-4
-1	-5
-0.5	-8.5
0.5	8.5
1	5
2	4
3	4.3
4	5
5	5.8
6	6.67

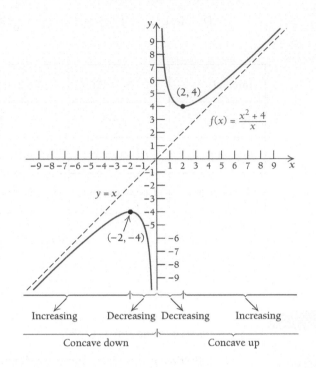

Exercise Set 3.3

Find the limit, if it exists.

1. $\lim\limits_{x\to\infty} \dfrac{2x - 4}{5x}$

2. $\lim\limits_{x\to\infty} \dfrac{3x^2 + 8}{x^3 - 1}$

3. $\lim\limits_{x\to\infty} \left(5 - \dfrac{2}{x}\right)$

4. $\lim\limits_{x\to\infty} \dfrac{3x + 1}{4x}$

5. $\lim\limits_{x\to\infty} \dfrac{2x - 5}{4x + 3}$

6. $\lim\limits_{x\to\infty} \left(7 + \dfrac{3}{x}\right)$

7. $\lim\limits_{x\to-\infty} \dfrac{2x^2 - 5}{3x^2 - x + 7}$

8. $\lim\limits_{x\to-\infty} \dfrac{6x + 1}{5x - 2}$

9. $\lim\limits_{x\to-\infty} \dfrac{4 - 3x}{5 - 2x^2}$

10. $\lim\limits_{x\to-\infty} \dfrac{4 - 3x - 12x^2}{1 + 5x + 3x^2}$

11. $\lim\limits_{x\to\infty} \dfrac{8x^4 - 3x^2}{5x^2 + 6x}$

12. $\lim\limits_{x\to\infty} \dfrac{6x^2 - x}{4x^4 - 3x^3}$

13. $\lim\limits_{x\to\infty} \dfrac{6x^4 - 5x^2 + 7}{8x^6 + 4x^3 - 8x}$

14. $\lim\limits_{x\to\infty} \dfrac{6x^4 - x^3}{4x^2 - 3x^3}$

15. $\lim\limits_{x\to-\infty} \dfrac{11x^5 + 4x^3 - 6x + 2}{6x^3 + 5x^2 + 3x - 1}$

16. $\lim\limits_{x\to-\infty} \dfrac{7x^9 - 6x^3 + 2x^2 - 10}{2x^6 + 4x^2 - x + 23}$

Sketch a graph of the function.

17. $f(x) = \dfrac{4}{x}$

18. $f(x) = -\dfrac{5}{x}$

19. $f(x) = \dfrac{-2}{x-5}$

20. $f(x) = \dfrac{1}{x-5}$

21. $f(x) = \dfrac{1}{x-3}$

22. $f(x) = \dfrac{1}{x+2}$

23. $f(x) = \dfrac{-2}{x+5}$

24. $f(x) = \dfrac{-3}{x-3}$

25. $f(x) = \dfrac{2x+1}{x}$

26. $f(x) = \dfrac{3x-1}{x}$

27. $f(x) = x + \dfrac{9}{x}$

28. $f(x) = x + \dfrac{2}{x}$

29. $f(x) = \dfrac{2}{x^2}$

30. $f(x) = \dfrac{-1}{x^2}$

31. $f(x) = \dfrac{x}{x-3}$

32. $f(x) = \dfrac{x}{x+2}$

33. $f(x) = \dfrac{1}{x^2+3}$

34. $f(x) = \dfrac{-1}{x^2+2}$

35. $f(x) = \dfrac{x-1}{x+2}$

36. $f(x) = \dfrac{x-2}{x+1}$

37. $f(x) = \dfrac{x^2-4}{x+3}$

38. $f(x) = \dfrac{x^2-9}{x+1}$

39. $f(x) = \dfrac{x-1}{x^2-2x-3}$

40. $f(x) = \dfrac{x+2}{x^2+2x-15}$

41. $f(x) = \dfrac{2x^2}{x^2-16}$

42. $f(x) = \dfrac{x^2+x-2}{2x^2+1}$

43. $f(x) = \dfrac{1}{x^2-1}$

44. $f(x) = \dfrac{10}{x^2+4}$

45. $f(x) = \dfrac{x^2+1}{x}$

46. $f(x) = \dfrac{x^3}{x^2-1}$

APPLICATIONS

47. *Cost of Pollution Control.* Cities and companies find the cost of pollution control to increase tremendously with respect to the percentage of pollutants to be removed from a situation. Suppose that the cost C of removing $p\%$ of the pollutants from a chemical dumping site is given by

$$C(p) = \dfrac{\$48{,}000}{100-p}.$$

a) Find $C(0)$, $C(20)$, $C(80)$, and $C(90)$.
b) Find $\lim_{p \to 100^-} C(p)$.
tw c) Explain the meaning of the limit found in part (b).
d) Sketch a graph of C.
tw e) Can the company afford to remove 100% of the pollutants? Explain.

48. *Medication in the Bloodstream.* After an injection, the amount of a medication A in the bloodstream decreases after time t, in hours. Suppose that under certain conditions A is given by

$$A(t) = \dfrac{A_0}{t^2+1},$$

where A_0 is the initial amount of the medication given. Assume that an initial amount of 100 cc is injected.

a) Find $A(0)$, $A(1)$, $A(2)$, $A(7)$, and $A(10)$.
b) Find $\lim_{t \to \infty} A(t)$.
c) Find the maximum value of the injection over the interval $[0, \infty)$.
d) Sketch a graph of the function.
tw e) According to this function, does the medication ever completely leave the bloodstream? Explain your answer.

49. *Temperature During an Illness.* The temperature T of a person during an illness is given by

$$T(t) = \dfrac{6t}{t^2+1} + 98.6,$$

where T is the temperature, in degrees Fahrenheit, at time t, in hours.

a) Find $T(0)$, $T(1)$, $T(2)$, $T(5)$, and $T(10)$.
b) Find $\lim_{t \to \infty} T(t)$.
c) Find the maximum temperature over the interval $[0, \infty)$.

d) Sketch a graph of the function.

tw e) According to this function, does the temperature ever return to 98.6°F? Explain your answer.

50. *Population Growth.* The population P, in thousands, of a small city is given by

$$P(t) = \frac{250t}{t^2 + 4} + 500,$$

where $t \geq 0$ is the time, in years.

a) Find $P(0)$, $P(1)$, $P(2)$, $P(5)$, and $P(10)$.
b) Find $\lim_{t \to \infty} P(t)$.
c) Find the maximum population over the interval $[0, \infty)$.
d) Sketch a graph of the function.

51. *Baseball: Earned-Run Average.* A pitcher's *earned-run average* (the average number of runs given up every 9 innings, or 1 game) is given by

$$E = 9 \cdot \frac{n}{i},$$

where n is the number of earned runs allowed and i is the number of innings pitched. Suppose that we fix the number of earned runs allowed at 4 and let i vary. We get a function given by

$$E(i) = 9 \cdot \frac{4}{i}.$$

a) Complete the following table, rounding to two decimal places.

Innings Pitched (i)	Earned-Run Average (E)
9	
8	
7	
6	
5	
4	
3	
2	
1	
$\frac{2}{3}$ (2 outs)	
$\frac{1}{3}$ (1 out)	

b) Find $\lim_{i \to 0^+} E(i)$.

tw c) On the basis of parts (a) and (b), determine a pitcher's earned-run average if 4 runs were allowed and there were 0 outs.

SYNTHESIS

tw 52. Explain why a vertical asymptote cannot be part of the graph of a function.

tw 53. Using graphs and limits, explain the idea of an asymptote to the graph of a function. Describe three types of asymptotes.

Find the limit, if it exists.

54. $\lim\limits_{x \to -\infty} \dfrac{-3x^2 + 5}{2 - x}$

55. $\lim\limits_{x \to 5} \dfrac{x^2 - 6x + 5}{x^2 - 3x - 10}$

56. $\lim\limits_{x \to -2} \dfrac{x^3 + 8}{x^2 - 4}$

57. $\lim\limits_{x \to \infty} \dfrac{-6x^3 + 7x}{2x^2 - 3x - 10}$

58. $\lim\limits_{x \to -\infty} \dfrac{-6x^3 + 7x}{2x^2 - 3x - 10}$

59. $\lim\limits_{x \to 1} \dfrac{x^3 - 1}{x^2 - 1}$

60. $\lim\limits_{x \to -\infty} \dfrac{7x^5 + x - 9}{6x + x^3}$

61. $\lim\limits_{x \to -\infty} \dfrac{2x^4 + x}{x + 1}$

62. $\lim\limits_{x \to \infty} \sin x$

63. $\lim\limits_{x \to \infty} \cos x$

64. $\lim_{x \to \infty} \dfrac{\sin x}{x}$

65. $\lim_{x \to \infty} \dfrac{\cos x}{x}$

For Exercises 66–71, use the graph of
$f(x) = \dfrac{x^2 - x - 20}{x^2 - 9}$ to compute the limit.

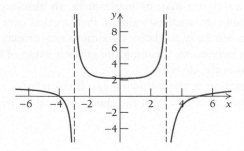

66. $\lim_{x \to 3^+} f(x)$

67. $\lim_{x \to 3^-} f(x)$

68. $\lim_{x \to -3^+} f(x)$

69. $\lim_{x \to -3^-} f(x)$

70. $\lim_{x \to \infty} f(x)$

71. $\lim_{x \to -\infty} f(x)$

For Exercises 72–77, use the graph of
$f(x) = \dfrac{x}{x^2 - x - 2}$ to compute the limit.

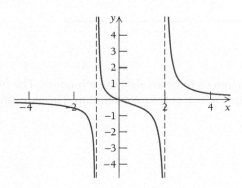

72. $\lim_{x \to 2^+} f(x)$

73. $\lim_{x \to 2^-} f(x)$

74. $\lim_{x \to -1^+} f(x)$

75. $\lim_{x \to -1^-} f(x)$

76. $\lim_{x \to \infty} f(x)$

77. $\lim_{x \to -\infty} f(x)$

 Technology Connection

Graph the function.

78. $f(x) = x^2 + \dfrac{1}{x^2}$

79. $f(x) = \dfrac{x}{\sqrt{x^2 + 1}}$

80. $f(x) = \dfrac{x^3 + 4x^2 + x - 6}{x^2 - x - 2}$

81. $f(x) = \dfrac{x^3 + 2x^2 - 15x}{x^2 - 5x - 14}$

82. $f(x) = \dfrac{x^3 + 2x^2 - 3x}{x^2 - 25}$

83. $f(x) = \left| \dfrac{1}{x} - 2 \right|$

84. Graph the function
$$f(x) = \dfrac{x^2 - 3}{2x - 4}.$$

Using only the **TRACE** and **ZOOM** features:
a) Find all the x-intercepts.
b) Find the y-intercept.
c) Find all the asymptotes.

tw 85. Graph the function
$$f(x) = \dfrac{\sqrt{x^2 + 3x + 2}}{x - 3}.$$

a) Estimate $\lim_{x \to \infty} f(x)$ and $\lim_{x \to -\infty} f(x)$ using the graph and input–output tables as needed to refine your estimates.
b) Describe the outputs of the function over the interval $(-2, -1)$.
c) What appears to be the domain of the function? Explain.
d) Find $\lim_{x \to -2^-} f(x)$ and $\lim_{x \to -1^+} f(x)$.

3.4 Using Derivatives to Find Absolute Maximum and Minimum Values

OBJECTIVES

■ Find absolute extrema using Maximum–Minimum Principle 1.

■ Find absolute extrema using Maximum–Minimum Principle 2.

Absolute Maximum and Minimum Values

A relative minimum may or may not be an absolute minimum, meaning the smallest value of the function over its entire domain. Similarly, a relative maximum may or may not be an absolute maximum, meaning the greatest value of a function over its entire domain.

The function in the following graph has relative minima at interior points c_1 and c_3 of the closed interval $[a, b]$.

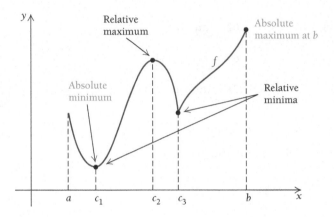

The relative minimum at c_1 also happens to be an absolute minimum. The function has a relative maximum at c_2 but it is not an absolute maximum. The absolute maximum occurs at the endpoint b.

DEFINITION

Suppose that f is a function whose value $f(c)$ exists at input c in an interval I in the domain of f. Then:

$f(c)$ is an **absolute minimum** if $f(c) \leq f(x)$ for all x in I.

$f(c)$ is an **absolute maximum** if $f(x) \leq f(c)$ for all x in I.

Finding Absolute Maximum and Minimum Values Over Closed Intervals

We first consider a continuous function over a closed interval. To do so, look at these graphs and try to determine the points over the closed interval at which the absolute maxima and minima (extrema) occur.

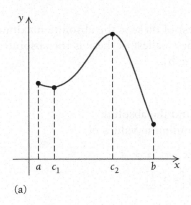

(a)

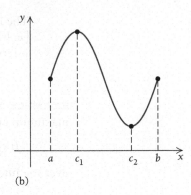

(b)

You may have discovered two theorems. Each of the functions did indeed have an absolute maximum value and an absolute minimum value. This leads us to the following theorem.

> ### THEOREM 7
> #### The Extreme-Value Theorem
> A continuous function f defined over a closed interval $[a, b]$ must have an absolute maximum value and an absolute minimum value at points in $[a, b]$.

Look carefully at the preceding graphs and consider the critical points and the endpoints. In graph (a), the graph starts at $f(a)$ and falls to $f(c_1)$. Then it rises from $f(c_1)$ to $f(c_2)$. From there it falls to $f(b)$. In graph (b), the graph starts at $f(a)$ and rises to $f(c_1)$. Then it falls from $f(c_1)$ to $f(c_2)$. From there it rises to $f(b)$. It seems reasonable that whatever the maximum and minimum values are, they occur among the function values $f(a), f(c_1), f(c_2)$, and $f(b)$. This leads us to a procedure for determining *absolute extrema*.

> ### THEOREM 8
> #### Maximum–Minimum Principle 1
> Suppose that f is a continuous function over a closed interval $[a, b]$. To find the absolute maximum and minimum values of the function over $[a, b]$:
>
> **a)** First find $f'(x)$.
>
> **b)** Then determine the critical points of f in $[a, b]$. That is, find all points c for which
>
> $$f'(c) = 0 \quad \text{or} \quad f'(c) \text{ does not exist.}$$
>
> **c)** List the critical points of f and the endpoints of the interval:
>
> $$a, c_1, c_2, \ldots, c_n, b.$$
>
> **d)** Find the function values at the points in part (c):
>
> $$f(a), f(c_1), f(c_2), \ldots, f(c_n), f(b).$$
>
> *(continued)*

The largest of these is the absolute maximum of f over the interval $[a, b]$. The smallest of these is the absolute minimum of f over the interval $[a, b]$.

EXAMPLE 1 Find the absolute maximum and minimum values of

$$f(x) = x^3 - 3x + 2$$

over the interval $\left[-2, \frac{3}{2}\right]$.

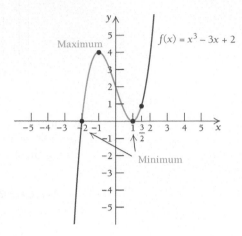

Solution In each of Examples 1–6 of this section, we show the related graph (as here) so that you can see the absolute extrema. The procedures we use do not require the drawing of a graph. Keep in mind that we are considering only the interval $\left[-2, \frac{3}{2}\right]$.

a) Find $f'(x)$:

$$f'(x) = 3x^2 - 3.$$

b) Find the critical points. The derivative exists for all real numbers. Thus we merely solve $f'(x) = 0$:

$$3x^2 - 3 = 0$$
$$3x^2 = 3$$
$$x^2 = 1$$
$$x = \pm 1.$$

c) List the critical points and the endpoints. These points are -2, -1, 1, and $\frac{3}{2}$.

d) Find the function values at the points in step (c):

$$f(-2) = (-2)^3 - 3(-2) + 2 = -8 + 6 + 2 = 0; \longrightarrow \text{Minimum}$$
$$f(-1) = (-1)^3 - 3(-1) + 2 = -1 + 3 + 2 = 4; \longrightarrow \text{Maximum}$$
$$f(1) = (1)^3 - 3(1) + 2 = 1 - 3 + 2 = 0; \longrightarrow \text{Minimum}$$
$$f\left(\frac{3}{2}\right) = \left(\frac{3}{2}\right)^3 - 3\left(\frac{3}{2}\right) + 2 = \frac{27}{8} - \frac{9}{2} + 2 = \frac{7}{8}.$$

The largest of these values, 4, is the maximum. It occurs at $x = -1$. The smallest of these values is 0. It occurs twice: at $x = -2$ and $x = 1$. Thus over the interval $\left[-2, \frac{3}{2}\right]$ the

$$\text{absolute maximum} = 4 \text{ at } x = -1$$

and the

$$\text{absolute minimum} = 0 \text{ at } x = -2 \text{ and } x = 1.$$ ■

Note that an absolute maximum or minimum value can occur at more than one point.

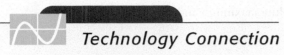

Technology Connection

Finding Absolute Extrema

How can we use a grapher to find absolute extrema? Let's first consider Example 1. We can use any of the methods described in the Technology Connection on pp. 169–170. In this case, we adapt Methods 3 and 4.

Method 3

Method 3 is selected because there are relative extrema in the interval $\left[-2, \frac{3}{2}\right]$. This method gives us approximations for the relative extrema.

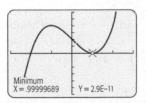

$f(x) = x^3 - 3x + 2$

Maximum
X = −1.000002   Y = 4

Minimum
X = .99999689   Y = 2.9E–11

Then we check function values at these x-values and at the endpoints, using Maximum–Minimum Principle 1 to determine the absolute maximum and minimum values:

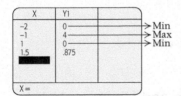

X	Y1	
−2	0	→ Min
−1	4	→ Max
1	0	→ Min
1.5	.875	

X =

Method 4

Example 2 considers the same function as in Example 1, but over a different interval. This time we use fMax and fMin features from the **MATH** menu. The minimum and maximum values occur at the endpoints.

fMin(Y1,X,−3,−1.5)
        −2.999994692
Y1(X)
        −15.99987261

fMax(Y1,X,−3,−1.5)
        −1.500005458
Y1(X)
        3.124979532

EXERCISE

1. Use a grapher to estimate the absolute maximum and minimum values of $f(x) = x^3 - x^2 - x + 2$ first over the interval $\lceil -2, 1 \rceil$ and then over the interval $[-1, 2]$. Then check your work using the methods of Examples 1 and 2.

EXAMPLE 2 Find the absolute maximum and minimum values of

$$f(x) = x^3 - 3x + 2$$

over the interval $\left[-3, -\frac{3}{2}\right]$.

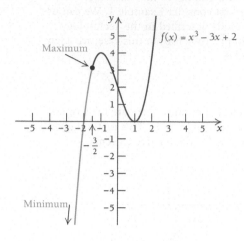

Solution As in Example 1, the critical points are -1 and 1. But neither -1 nor 1 is in the interval $\left[-3, -\frac{3}{2}\right]$, so there are no critical points in this interval. Thus the maximum and minimum values occur at the endpoints:

$$f(-3) = (-3)^3 - 3(-3) + 2$$

$$= -27 + 9 + 2 = -16; \longrightarrow \text{Minimum}$$

$$f\left(-\frac{3}{2}\right) = \left(-\frac{3}{2}\right)^3 - 3\left(-\frac{3}{2}\right) + 2$$

$$= -\frac{27}{8} + \frac{9}{2} + 2 = \frac{25}{8} = 3\frac{1}{8}. \longrightarrow \text{Maximum}$$

Thus, over the interval $\left[-3, -\frac{3}{2}\right]$, the

$$\text{absolute maximum} = 3\frac{1}{8} \text{ at } x = -\frac{3}{2}$$

and the

$$\text{absolute minimum} = -16 \text{ at } x = -3. \quad\blacksquare$$

Finding Absolute Maximum and Minimum Values Over Other Intervals

When there is only one critical point c in I, we may not need to check endpoint values to determine whether the function has an absolute maximum or minimum value at that point.

> **THEOREM 9**
>
> **Maximum–Minimum Principle 2**
>
> Suppose that f is a function such that $f'(x)$ exists for every x in an interval I, and that there is *exactly one* (critical) point c, interior to I, for which $f'(c) = 0$. Then
>
> $f(c)$ is the absolute maximum value over I if $f''(c) < 0$
>
> and
>
> $f(c)$ is the absolute minimum value over I if $f''(c) > 0$.

This theorem holds no matter what the interval I is—whether open, closed, or extending to infinity. If $f''(c) = 0$, either we must use Maximum–Minimum Principle 1 or we must know more about the behavior of the function over the given interval.

Technology Connection

Finding Absolute Extrema

Let's do Example 3 using a grapher. Again we adapt the methods of the Technology Connection on pp. 169–170. Strictly speaking, we cannot use the fMin or fMax features or the **MAXIMUM** or **MINIMUM** features from the **CALC** menu since we do not have a closed interval.

Methods 1 and 2

We create a graph, examine its shape, and use the **TRACE** and/or **TABLE** features. This procedure leads us to see that there is indeed no absolute minimum. We do get an absolute maximum $f(x) = 4$ at $x = 2$.

EXERCISE

1. Use a grapher to estimate the absolute maximum and minimum values of $f(x) = x^2 - 4x$. Then check your work using the method of Example 3.

EXAMPLE 3   Find the absolute maximum and minimum values of

$$f(x) = 4x - x^2.$$

Solution   When no interval is specified, we consider the entire domain of the function. In this case, the domain is the set of all real numbers.

a) Find $f'(x)$:

$$f'(x) = 4 - 2x.$$

b) Find the critical points. The derivative exists for all real numbers. Thus we merely solve $f'(x) = 0$:

$$4 - 2x = 0$$
$$-2x = -4$$
$$x = 2.$$

c) Since there is only one critical point, we can apply Maximum–Minimum Principle 2 using the second derivative:

$$f''(x) = -2.$$

The second derivative is constant. Thus, $f''(2) = -2$, and since this is negative, we have the

$$\text{absolute maximum} = f(2) = 4 \cdot 2 - 2^2$$
$$= 8 - 4 = 4 \text{ at } x = 2.$$

The function has no minimum, as the graph at the top of the following page indicates.

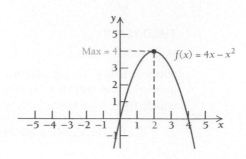

$f(x) = 4x - x^2$

Max = 4

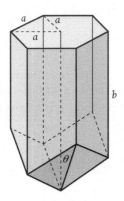

EXAMPLE 4 *Bee Cells.* A honeybee constructs the cells in its comb in such a way that the minimum amount of wax is used.[7] One cell of a honeycomb is shown in the margin. It is a prism whose base (the open part at the top) is a regular hexagon. The bottom comes together at a point. The surface area is given by

$$S(\theta) = 6ab + \frac{3}{2}a^2\frac{\sqrt{3} - \cos\theta}{\sin\theta},$$

where θ is the measure of the so-called angle of inclination and a and b are constants, as shown in the figure. Find the value of θ that minimizes S given that $0° < \theta < 90°$.

Solution

a) Find $S'(\theta)$. Since a and b are constants,

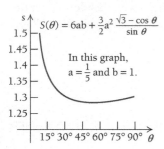

$$S'(\theta) = \frac{3}{2}a^2\left[\frac{(\sin\theta)(\sin\theta) - (\sqrt{3} - \cos\theta)(\cos\theta)}{\sin^2\theta}\right]$$

$$= \frac{3}{2}a^2\left[\frac{\sin^2\theta - \sqrt{3}\cos\theta + \cos^2\theta}{\sin^2\theta}\right]$$

$$= \frac{3}{2}a^2\left[\frac{1 - \sqrt{3}\cos\theta}{\sin^2\theta}\right].$$

b) Find the critical points. The derivative is undefined if $\sin\theta = 0$, and so such values of θ are critical points of S. However, no such values occur for $0° < \theta < 90°$. Thus, we solve $S'(\theta) = 0$ for θ with the assumption that $\sin\theta \neq 0$:

$$S'(\theta) = \frac{3}{2}a^2\left[\frac{1 - \sqrt{3}\cos\theta}{\sin^2\theta}\right] = 0$$

$$\frac{1 - \sqrt{3}\cos\theta}{\sin^2\theta} = 0$$

$$1 - \sqrt{3}\cos\theta = 0$$

$$\cos\theta = \frac{1}{\sqrt{3}} \approx 0.5774.$$

---

[7]This discovery was published in a study by Sir D'Arcy Wentworth Thompson, *On Growth and Form* (Cambridge University Press, 1917).

There is only one angle in the first quadrant whose cosine is $1/\sqrt{3}$. This can be computed on a calculator by using the inverse cosine function. It follows that

$$\theta = \cos^{-1}\left(\frac{1}{\sqrt{3}}\right) \approx 54.74°.$$

c) Since there is only one critical point, we can apply Maximum–Minimum Principle 2:

$$S''(\theta) = \frac{3}{2}a^2\left[\frac{(\sqrt{3}\sin\theta)(\sin^2\theta) - (1 - \sqrt{3}\cos\theta)(2\sin\theta\cos\theta)}{\sin^4\theta}\right]$$

$$= \frac{3}{2}a^2\left[\frac{\sqrt{3}\sin^2\theta - 2(1 - \sqrt{3}\cos\theta)\cos\theta}{\sin^3\theta}\right]$$

Therefore, since a^2 is positive,

$$S''(54.74°) \approx 3.18a^2 > 0.$$

(The exact evaluation of this expression is considered in the exercises). Since this is positive, S has an absolute minimum value at $\theta \approx 54.74°$. Bees indeed tend to use this angle when constructing their honeycombs.   ■

We have thus far restricted the use of Maximum–Minimum Principle 2 to intervals with one critical point. Suppose that a closed interval contains two critical points. Then we could break the interval up into two subintervals, consider maximum and minimum values over those subintervals, and compare. But we would need to consider values at the endpoints, and since we would, in effect, be using Maximum–Minimum Principle 1, we may as well use it at the outset.

A Strategy for Finding Maximum and Minimum Values

The following general strategy can be used when finding maximum and minimum values of continuous functions.

A Strategy for Finding Absolute Maximum and Minimum Values

To find absolute maximum and minimum values of a continuous function over an interval:

a) Find $f'(x)$.

b) Find the critical points.

c) If the interval is closed and there is more than one critical point, use Maximum–Minimum Principle 1.

d) If the interval is closed and there is exactly one critical point, use either Maximum–Minimum Principle 1 or Maximum–Minimum Principle 2. If the function is easy to differentiate, use Maximum–Minimum Principle 2.

e) If the interval is not closed, does not have endpoints, or does not contain its endpoints, such as $(-\infty, \infty)$, $(0, \infty)$, or (a, b), and the function has only one critical point, use Maximum–Minimum Principle 2. In such a case, if the function has a maximum, it will have no minimum; and if it has a minimum, it will have no maximum.

The case of finding absolute maximum and minimum values when more than one critical point occurs in an interval described in step (e) above must be dealt with by a detailed graph or by techniques beyond the scope of this book.

EXAMPLE 5 Find the absolute maximum and minimum values of

$$f(x) = (x - 2)^3 + 1.$$

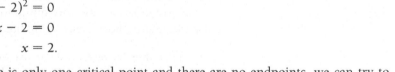

Solution

a) Find $f'(x)$:

$$f'(x) = 3(x - 2)^2.$$

b) Find the critical points. The derivative exists for all real numbers. Thus we solve $f'(x) = 0$:

$$3(x - 2)^2 = 0$$
$$(x - 2)^2 = 0$$
$$x - 2 = 0$$
$$x = 2.$$

c) Since there is only one critical point and there are no endpoints, we can try to apply Maximum–Minimum Principle 2 using the second derivative:

$$f''(x) = 6(x - 2).$$

Now

$$f''(2) = 6(2 - 2) = 0,$$

so Maximum–Minimum Principle 2 fails. We cannot use Maximum–Minimum Principle 1 because there are no endpoints. But note that $f'(x) = 3(x - 2)^2$ is never negative. Thus, $f(x)$ is increasing everywhere except at $x = 2$, so there is no maximum and no minimum. For $x < 2, x - 2 < 0$, so $f''(x) = 6(x - 2) < 0$. Similarly, for $x > 2, x - 2 > 0$, so $f''(x) = 6(x - 2) > 0$. Thus, at $x = 2$, the function has a *point of inflection*. ■

EXAMPLE 6 Find the absolute maximum and minimum values of

$$f(x) = 5x + \frac{35}{x}$$

over the interval $(0, \infty)$.

Solution

a) Find $f'(x)$. We first express $f(x)$ as

$$f(x) = 5x + 35x^{-1}.$$

Then

$$f'(x) = 5 - 35x^{-2}$$
$$= 5 - \frac{35}{x^2}.$$

b) Find the critical points. Now $f'(x)$ exists for all values of x in $(0, \infty)$. Thus the only critical points are those for which $f'(x) = 0$:

$$5 - \frac{35}{x^2} = 0$$

$$5 = \frac{35}{x^2}$$

$$5x^2 = 35 \qquad \text{Multiplying by } x^2, \text{ since } x \neq 0$$

$$x^2 = 7$$

$$x = \pm\sqrt{7} \approx \pm 2.646.$$

c) The interval is not closed and is $(0, \infty)$. Since $-\sqrt{7}$ is outside of this interval, the only critical point in $(0, \infty)$ is $\sqrt{7}$. Therefore, we can apply Maximum–Minimum Principle 2 using the second derivative,

$$f''(x) = 70x^{-3}$$

$$= \frac{70}{x^3},$$

to determine whether we have a maximum or a minimum. Now $f''(x)$ is positive for all values of x in $(0, \infty)$, so $f''(\sqrt{7}) > 0$, and the

$$\text{absolute minimum} = f(\sqrt{7})$$

$$= 5 \cdot \sqrt{7} + \frac{35}{\sqrt{7}}$$

$$= 5\sqrt{7} + \frac{35}{\sqrt{7}} \cdot \frac{\sqrt{7}}{\sqrt{7}}$$

$$= 5\sqrt{7} + \frac{35\sqrt{7}}{7}$$

$$= 5\sqrt{7} + 5\sqrt{7}$$

$$= 10\sqrt{7} \approx 26.458$$

at $x = \sqrt{7}$.

The function has no maximum value.

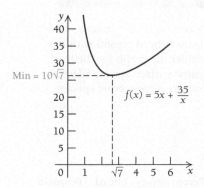

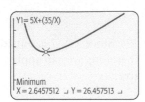

Technology Connection

Finding Absolute Extrema

Let's do Example 6 using the **MAXIMUM** and **MINIMUM** features from the **CALC** menu. The shape of the graph leads us to see that there is no absolute maximum, but there is an absolute minimum.

Y1= 5X+(35/X)

Minimum
X = 2.6457512 ⌐ Y = 26.457513 ⌐

[0, 10, 0, 50]

Note that

$$\sqrt{7} \approx 2.65 \quad \text{and} \quad 10\sqrt{7} \approx 26.458,$$

which confirms the analytic solution.

EXERCISE

1. Use a grapher to estimate the absolute maximum and minimum values of $f(x) = 10x + 1/x$ over the interval $(0, \infty)$. Then check your work using the analytic method of Example 6.

Exercise Set 3.4

1. *Topsoil Nitrogen Levels in Glacier Bay, Alaska.* As glaciers retreat, they leave behind topsoil with low nitrogen content. Consider the graph for the total nitrogen in topsoil against surface age for Glacier Bay, Alaska.[8] Alder trees are a pioneering species

with symbiotic bacteria that increase the nitrogen level in topsoil. In time, spruce trees use this nitrogen, causing the nitrogen level to decrease.

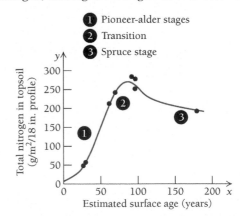

① Pioneer-alder stages
② Transition
③ Spruce stage

a) Estimate the surface age that corresponds to the absolute maximum of the total nitrogen in topsoil.
b) Estimate the surface age that corresponds to the absolute minimum of the total nitrogen in topsoil.
c) What is the total nitrogen in topsoil for land with a surface age of 50 yr?
d) What is the total nitrogen in topsoil for land with a surface age of 150 yr?

[8]N. A. Campbell and J. B. Reece, *Biology*, 6th ed. (Benjamin Cummings, New York, 2002).

Find the absolute maximum and minimum values of
the function, if they exist, over the indicated interval.

2. $f(x) = 4 + x - x^2$; $[0, 2]$

3. $f(x) = x^3 - x^2 - x + 2$; $[0, 2]$

4. $f(x) = x^3 + \frac{1}{2}x^2 - 2x + 5$; $[0, 1]$

5. $f(x) = 3x - 2$; $[-1, 1]$

6. $f(x) = 2x + 4$; $[-1, 1]$

7. $f(x) = 3 - 2x - 5x^2$; $[-3, 3]$

8. $f(x) = 1 + 6x - 3x^2$; $[0, 4]$

9. $f(x) = 1 - x^3$; $[-8, 8]$

10. $f(x) = 2x^3$; $[-10, 10]$

11. $f(x) = 12 + 9x - 3x^2 - x^3$; $[-3, 1]$

12. $f(x) = x^3 - 6x^2 + 10$; $[0, 4]$

13. $f(x) = x^4 - 2x^3$; $[-2, 2]$

14. $f(x) = x^3 - x^4$; $[-1, 1]$

15. $f(x) = x^4 - 2x^2 + 5$; $[-2, 2]$

16. $f(x) = x^4 - 8x^2 + 3$; $[-3, 3]$

17. $f(x) = (x + 3)^{2/3} - 5$; $[-4, 5]$

18. $f(x) = 1 - x^{2/3}$; $[-8, 8]$

19. $f(x) = x + \frac{1}{x}$; $[1, 20]$

20. $f(x) = x + \frac{4}{x}$; $[-8, -1]$

21. $f(x) = \frac{x^2}{x^2 + 1}$; $[-2, 2]$

22. $f(x) = \frac{4x}{x^2 + 1}$; $[-3, 3]$

23. $f(x) = (x + 1)^{1/3}$; $[-2, 26]$

24. $f(x) = \sqrt[3]{x}$; $[8, 64]$

25. $f(x) = \frac{x + 2}{x^2 + 5}$; $[-6, 6]$

26. $f(x) = \frac{x + 2}{x^2 + 3x + 3}$; $[-4, 4]$

27. $f(x) = x\sqrt{x - x^2}$; $[0, 1]$

28. $f(x) = x\sqrt{8x - x^2}$; $[0, 8]$

29. $f(x) = x\sqrt{x + 3}$; $[-3, 6]$

30. $f(x) = x\sqrt{6 - x}$; $[0, 6]$

31. $f(x) = x + 2 \sin x$; $[0, 2\pi]$

32. $f(x) = x - \cos 2x$; $[0, 2\pi]$

33. $f(x) = \frac{\sin x}{2 + \sin x}$; $[0, 2\pi]$

34. $f(x) = \frac{\sin x}{3 - \sin x}$; $[0, 2\pi]$

35. $f(x) = \frac{\sin x}{(1 + \sin x)^2}$; $[0, \pi]$

36. $f(x) = \frac{\cos^2 x}{2 + \cos x}$; $[0, 2\pi]$

37. $f(x) = 2x - \tan x$; $[0, \pi/3]$

38. $f(x) = x + \cot x$; $[\pi/6, 5\pi/6]$

39. $f(x) = 3 \sin x - 2 \sin^3 x$; $[0, 2\pi]$

40. $f(x) = 2 \cos^3 x - 3 \cos x$; $[0, 2\pi]$

Find the absolute maximum and minimum values of the
function, if they exist, over the indicated interval. When
no interval is specified, use the real line $(-\infty, \infty)$.

41. $f(x) = x - \frac{4}{3}x^3$; $(0, \infty)$

42. $f(x) = 16x - \frac{4}{3}x^3$; $(0, \infty)$

43. $f(x) = -0.001x^2 + 4.8x - 60$

44. $f(x) = -0.01x^2 + 1.4x - 30$

45. $f(x) = 2x + \frac{72}{x}$; $(0, \infty)$

46. $f(x) = x + \frac{3600}{x}$; $(0, \infty)$

47. $f(x) = x^2 + \frac{432}{x}$; $(0, \infty)$

48. $f(x) = x^2 + \frac{250}{x}$; $(0, \infty)$

49. $f(x) = (x + 1)^3$

50. $f(x) = (x - 1)^3$

51. $f(x) = 2x - 3$

52. $f(x) = 9 - 5x$

53. $f(x) = x^{2/3};\quad [-1, 1]$

54. $g(x) = x^{2/3}$

55. $f(x) = \frac{1}{3}x^3 - x + \frac{2}{3}$

56. $f(x) = \frac{1}{3}x^3 - \frac{1}{2}x^2 - 2x + 1$

57. $f(x) = x^4 - 2x^2$

58. $f(x) = 2x^4 - 4x^2 + 2$

59. $f(x) = \tan x + \cot x;\quad (0, \pi/2)$

60. $f(x) = \dfrac{\sin x}{1 + \sin x};\quad (-\pi/2, 3\pi/2)$

61. $f(x) = \dfrac{1}{\sin x + \cos x};\quad (-\pi/4, 3\pi/4)$

62. $f(x) = \dfrac{1}{x + 2\cos x};\quad [0, \pi]$

63. $f(x) = \dfrac{1}{x - 2\sin x};\quad (0, \pi/2)$

64. $f(x) = 2\csc x + \cot x;\quad (0, \pi)$

65. $f(x) = 2\csc x + \cot x;\quad (\pi, 2\pi)$

66. $f(x) = \tan x - 2\sec x;\quad (-\pi/2, \pi/2)$

67. $f(x) = \tan x - 2\sec x;\quad (\pi/2, 3\pi/2)$

68. $f(x) = \dfrac{1}{1 + \cos x};\quad (-\pi, \pi)$

69. $f(x) = \dfrac{1}{1 - 2\sin x};\quad (\pi/6, 5\pi/6)$

APPLICATIONS

70. *Blood Pressure.* For a dosage of x cubic centimeters (cc) of a certain drug, the resulting blood pressure B is approximated by

$$B(x) = 0.05x^2 - 0.3x^3, \quad 0 \le x \le 0.16.$$

Find the maximum blood pressure and the dosage at which it occurs.

71. *Minimizing Automobile Accidents.* At travel speed (constant velocity) x, in miles per hour, there are y accidents at nighttime for every 100 million miles of travel, where y is given by

$$y = -6.1x^2 + 752x + 22{,}620.$$

At what travel speed does the greatest number of accidents per 100 million miles traveled occur?

72. *Temperature During an Illness.* The temperature T of a person during an illness is given by

$$T(t) = -0.1t^2 + 1.2t + 98.6, \quad 0 \le t \le 12,$$

where T is the temperature (°F) at time t, in days. Find the maximum value of the temperature and when it occurs.

73. *Death Rate.* The death rate r per 100,000 males may be modeled as a function of the number of hours per day x that the men slept. This equation is given by

$$r(x) = 104.5x^2 - 1501.5x + 6016.$$

At how many hours of sleep per day is the death rate minimized?

Body Surface Area. A cancer patient's surface area is used to determine the dosage level of chemotherapy. Since it is difficult to accurately measure a person's surface area, formulas have been developed to approximate surface area based on a person's height and weight.

Three commonly used formulas[9] for computing the body surface area (in m^2) are stated below for a 70-kg patient with height h (measured in cm):

Mosteller: $\quad S_1(h) = 0.139443\sqrt{h}$

DuBois and DuBois: $S_2(h) = 0.043705h^{0.725}$

Haycock: $\quad S_3(h) = 0.238382h^{0.3964}$

74. We will assume in this exercise that a patient weighs 70 kg.

   a) Let $D(h) = S_1(h) - S_2(h)$. Write the formula for $D(h)$.

   tw b) Find the absolute maximum and minimum of $D(h)$ for $0 \le h \le 215$. What does the maximum and minimum of $D(h)$ say about the two methods of predicting body surface area?

75. We will assume in this exercise that a patient weighs 70 kg.

   a) Let $D(h) = S_1(h) - S_3(h)$. Write the formula for $D(h)$.

   tw b) Find the absolute maximum and minimum of $D(h)$ for $0 \le h \le 215$. What does the maximum and minimum of $D(h)$ say about the two methods of predicting body surface area?

SYNTHESIS

76. Let

$$y = (x - a)^2 + (x - b)^2.$$

For what value of x is y a minimum?

tw 77. Explain the usefulness of the first derivative in finding the absolute extrema of a function.

---

[9]http://www.halls.md/body-surface-area/refs.htm

tw 78. Explain the usefulness of the second derivative in finding the absolute extrema of a function.

79. *Bee Cells.* In Example 4, we used the approximation $\theta_0 = \cos^{-1}(1/\sqrt{3}) \approx 54.74°$ to use Maximum–Minimum Principle 2. In this exercise, we will exactly evaluate $S''(\theta)$ at the critical point.

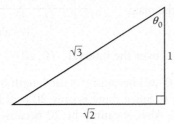

a) Verify that the cosine of the angle θ_0 shown in the figure is $1/\sqrt{3}$.
b) Compute $\sin \theta_0$.
c) Use these values of $\sin \theta_0$ and $\cos \theta_0$ and the expression for $S''(\theta)$ found in Example 4 to show that $S''(\theta_0) = 9a^2\sqrt{2}/4$.

Technology Connection

Use a grapher to graph each function over the given interval. Visually estimate where absolute maximum and minimum values occur. Then use the **TABLE** feature to refine your estimate.

80. $f(x) = x^4 - 4x^3 + 10;\quad [0, 4]$

81. $f(x) = x^{2/3}(x - 5);\quad [1, 4]$

82. $f(x) = \dfrac{3}{4}(x^2 - 1)^{2/3}$

83. $f(x) = x\left(\dfrac{x}{2} - 5\right)^4$

3.5 Maximum-Minimum Problems

OBJECTIVES

■ Solve maximum-minimum problems using calculus.

One very important application of differential calculus is the solving of maximum–minimum problems, that is, finding the absolute maximum or minimum value of some varying quantity Q and the point at which that maximum or minimum occurs.

EXAMPLE 1 *Maximizing Area.* A hobby store has 20 ft of fencing to fence off a rectangular area for an electric train in one corner of its display room. The two sides up against the wall require no fence. What dimensions of the rectangle will maximize the area? What is the maximum area?

Solution At first glance, we might think that it does not matter what dimensions we use: They will all yield the same area. This is not the case. Let's first make a drawing and express the area in terms of one variable. If we let x be the length of one side and y be the length of the other, then since the sum of the lengths must be 20 ft, we have

$$x + y = 20 \quad \text{and} \quad y = 20 - x.$$

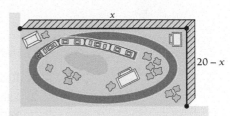

Thus the area is given by

$$A = xy$$
$$= x(20 - x)$$
$$= 20x - x^2.$$

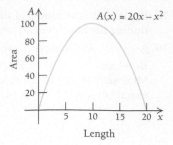

We are trying to find the maximum value of

$$A(x) = 20x - x^2 \quad \text{over the interval} \quad (0, 20).$$

We consider the interval $(0, 20)$ because x is the length of one side and cannot be negative or 0. Since there is only 20 ft of fencing, x cannot be greater than 20. Also, x cannot be 20 because then the length of y would be 0.

a) We first find $A'(x)$:

$$A'(x) = 20 - 2x.$$

b) This derivative exists for all values of x in $(0, 20)$. Thus the only critical points are where

$$A'(x) = 20 - 2x = 0$$
$$-2x = -20$$
$$x = 10.$$

Since there is only one critical point in the interval, we can use the second derivative to determine whether we have a maximum. Note that

$$A''(x) = -2,$$

which is a constant. Thus, $A''(10)$ is negative, so $A(10)$ is a maximum. Now

$$A(10) = 10(20 - 10)$$
$$= 10 \cdot 10$$
$$= 100.$$

Thus the maximum area of 100 ft^2 is obtained using 10 ft for the length of one side and $20 - 10$, or 10 ft for the other. Note that $A(5) = 75$, $A(16) = 64$, and $A(12) = 96$; so length does affect area. ∎

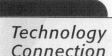

Technology Connection

Exploratory Exercises

1. Complete this table using a grapher.

x	$y = 20 - x$	$A = x(20 - x)$
0		
4		
6.5		
8		
10		
12		
13.2		
20		

2. Graph $A(x) = x(20 - x)$ over the interval $[0, 20]$.

3. Estimate a maximum value and where it would occur.

Here is a general strategy for solving maximum–minimum problems. Although it may not guarantee success, it should certainly improve your chances.

A Strategy for Solving Maximum-Minimum Problems

1. Read the problem carefully. If relevant, make a drawing.

2. Label the picture with appropriate variables and constants, noting what varies and what stays fixed. Be sure to define the variables used.

3. Translate the problem to a mathematical form involving a quantity Q. Find an equation involving Q to be maximized or minimized. Also, find any other equations relating the variables.

4. If Q is a function of more than one variable, try to express Q as a function of *one* variable.

5. Use the procedures developed in Sections 3.1, 3.2, and 3.4 to determine where the maximum and minimum values of Q are and the points at which they occur.

EXAMPLE 2   *Maximizing Volume.* From a thin piece of cardboard 8 in. by 8 in., square corners are cut out so that the sides can be folded up to make a box. What dimensions will yield a box of maximum volume? What is the maximum volume?

Solution   We might again think at first that it does not matter what the dimensions are, but our experience with Example 1 should lead us to think otherwise. We make a drawing.

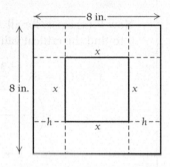 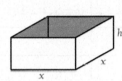

When squares of length h on a side are cut out of the corners, we are left with a square base with sides of length x. The volume of the resulting box is

$$V = x \cdot x \cdot h.$$

We want to express V in terms of one variable. Note that the overall length of a side of the cardboard is 8 in. We see from the figure that

$$h + x + h = 8,$$

or

$$x + 2h = 8.$$

Solving for h, we get

$$2h = 8 - x$$

$$h = \frac{1}{2}(8 - x)$$

$$= \frac{1}{2} \cdot 8 - \frac{1}{2}x$$

$$= 4 - \frac{1}{2}x.$$

Thus,

$$V = x \cdot x \cdot \left(4 - \frac{1}{2}x\right)$$

$$= x^2\left(4 - \frac{1}{2}x\right)$$

$$= 4x^2 - \frac{1}{2}x^3.$$

We are trying to find the maximum value of

$$V(x) = 4x^2 - \frac{1}{2}x^3 \quad \text{over the interval} \quad (0, 8).$$

We first find $V'(x)$:

$$V'(x) = 8x - \frac{3}{2}x^2.$$

Now $V'(x)$ exists for all x in the interval $(0, 8)$, so we set it equal to 0 to find the critical values:

$$V'(x) = 8x - \frac{3}{2}x^2 = 0$$

$$x\left(8 - \frac{3}{2}x\right) = 0$$

$$x = 0 \quad or \quad 8 - \frac{3}{2}x = 0$$

$$x = 0 \quad or \quad -\frac{3}{2}x = -8$$

$$x = 0 \quad or \quad x = -\frac{2}{3}(-8) = \frac{16}{3}.$$

The only critical point in $(0, 8)$ is $\frac{16}{3}$. Thus we can use the second derivative,

$$V''(x) = 8 - 3x,$$

to determine whether we have a maximum. Since

$$V''\left(\frac{16}{3}\right) = 8 - 3 \cdot \frac{16}{3}$$

$$= -8,$$

$V''\left(\frac{16}{3}\right)$ is negative, so $V\left(\frac{16}{3}\right)$ is a maximum, and

$$V\left(\frac{16}{3}\right) = 4 \cdot \left(\frac{16}{3}\right)^2 - \frac{1}{2}\left(\frac{16}{3}\right)^3$$

$$= \frac{1024}{27} = 37\frac{25}{27}.$$

Technology Connection

Exploratory Exercises

1. Complete this table using a grapher.

x	$h = 4 - \frac{1}{2}x$	$V = 4x^2 - \frac{1}{2}x^3$
0		
1		
2		
3		
4		
4.6		
5		
6		
6.8		
7		
8		

2. Graph $V(x) = 4x^2 - \frac{1}{2}x^3$ over the interval $[0, 8]$.

3. Estimate a maximum value and where it would occur.

The maximum volume is $37\frac{25}{27}$ in^3. The dimensions that yield this maximum volume are

$$x = \frac{16}{3} = 5\frac{1}{3}\text{ in.,}\quad\text{by }x = 5\frac{1}{3}\text{ in.,}\quad\text{by }h = 4 - \frac{1}{2}\left(\frac{16}{3}\right) = 1\frac{1}{3}\text{ in.}\quad\blacksquare$$

In the following problem, an open-top container of fixed volume is to be constructed. We want to determine the dimensions that will allow it to be built with the least amount of material. Such a problem could be important from an ecological standpoint.

EXAMPLE 3 *Minimizing Surface Area.* A container firm is designing an open-top rectangular box, with a square base, that will hold 108 cubic centimeters (cc). What dimensions yield the minimum surface area? What is the minimum surface area?

Solution We first make a drawing. The surface area of the box is given by the area of the base plus the area of the four sides, or

$$S = x^2 + 4xy.$$

The volume must be 108 cc, and is given by

$$V = x^2 y = 108.$$

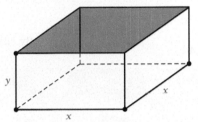

To express S in terms of one variable, we solve $x^2 y = 108$ for y:

$$y = \frac{108}{x^2}.$$

Then

$$S(x) = x^2 + 4x\left(\frac{108}{x^2}\right)$$

$$= x^2 + \frac{432}{x}.$$

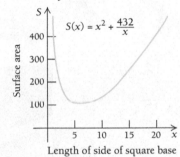

The nature of the application tells us that the domain of S is all positive numbers. Thus we are minimizing S over the interval $(0, \infty)$. We first find dS/dx:

$$\frac{dS}{dx} = 2x - \frac{432}{x^2}.$$

Since dS/dx exists for all x in $(0, \infty)$, the only critical points occur where $dS/dx = 0$. Thus we solve the following equation:

$$2x - \frac{432}{x^2} = 0$$

$$x^2\left(2x - \frac{432}{x^2}\right) = x^2 \cdot 0 \qquad \text{We multiply by } x^2 \text{ to clear the fractions.}$$

$$2x^3 - 432 = 0$$

$$2x^3 = 432$$

$$x^3 = 216$$

$$x = 6.$$

This is the only critical point, so we can use the second derivative to determine whether we have a minimum:

$$\frac{d^2S}{dx^2} = 2 + \frac{864}{x^3}.$$

Note that this is positive for all positive values of x. Thus we have a minimum at $x = 6$. When $x = 6$, it follows that $y = 3$:

$$y = \frac{108}{6^2} = \frac{108}{36} = 3.$$

Thus the surface area is minimized when $x = 6$ cm (centimeters) and $y = 3$ cm. The minimum surface area is

$$S = 6^2 + 4 \cdot 6 \cdot 3$$
$$= 108 \text{ cm}^2.$$

By coincidence, this is the same number as the fixed volume.   ■

EXAMPLE 4   *Determining a Ticket Price.* When a theater owner charges $6 for admission, there is an average attendance of 150 people. For every $1 increase in admission, there is a loss of 20 customers from the average number. Also, every customer spends an average of $1.80 on concessions. What admission price should be charged in order to maximize total revenue?

Solution   Let x be the amount by which the price of $6 should be increased. (If x is negative, the price is decreased.) We first express the total revenue R as a function of x. Note that

$R(x) =$ (Revenue from tickets) + (Revenue from concessions)
$\quad\quad = $ (Number of people) · (Ticket price) + $1.80(Number of people)
$\quad\quad = (150 - 20x)(6 + x) + 1.80(150 - 20x)$
$\quad\quad = 900 + 150x - 120x - 20x^2 + 270 - 36x$
$R(x) = -20x^2 - 6x + 1170.$

We are trying to find the maximum value of R over the set of real numbers. To find x such that $R(x)$ is a maximum, we first find $R'(x)$:

$$R'(x) = -40x - 6.$$

This derivative exists for all real numbers x. Thus the only critical points are where $R'(x) = 0$, so we solve this equation:

$$-40x - 6 = 0$$
$$-40x = 6$$
$$x = -0.15 = -\$0.15.$$

Since this is the only critical point, we can use the second derivative,

$$R''(x) = -40,$$

to determine whether we have a maximum. Since $R''(-0.15)$ is negative, $R(-0.15)$ is a maximum. Therefore, in order to maximize revenue, the theater should charge

$$\$6 + (-\$0.15), \quad \text{or} \quad \$5.85 \quad \text{per ticket.}$$

That is, this reduced ticket price will attract more people to the movie theater, namely,

$$150 - 20(-0.15), \quad \text{or} \quad 153 \text{ people,}$$

and will result in maximum revenue. ∎

EXAMPLE 5   *Flights of Homing Pigeons.* It is known that homing pigeons tend to avoid flying over water in the daytime, perhaps because the downdrafts of air over water make flying difficult. Suppose a homing pigeon is released on an island at point C, which is 3 mi directly out in the water from point B on the shore. Point B is 8 mi downshore from the pigeon's home loft at point A. Assume that a pigeon requires twice the amount of energy per mile to fly over water than flying over land. At what angle θ should the pigeon fly toward the shore in order to minimize the total energy required to get to its home loft?

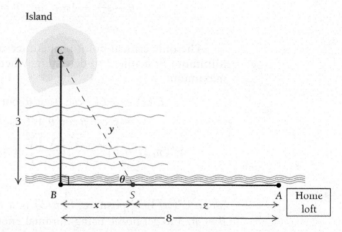

Solution   Let x be the distance between B and the point S toward which the pigeon should fly. Using right triangle $\triangle BCS$, we have

$$\tan \theta = \frac{3}{x}, \quad \text{or} \quad x = 3 \cot \theta.$$

Furthermore, the distance y flown over water is related to θ by

$$\sin \theta = \frac{3}{y}, \quad \text{so} \quad y = 3 \csc \theta.$$

Also, the distance flown over land is

$$z = 8 - x = 8 - 3 \cot \theta.$$

We now express the total energy E as a function of θ.

Suppose that the energy expended to fly over a mile of land is 1, so that the energy expended to fly over a mile of water is 2. Then

$$E(\theta) = (\text{Energy per mile over water}) \cdot (\text{Distance over water})$$
$$+ (\text{Energy per mile over land}) \cdot (\text{Distance over land})$$
$$= 2 \cdot y + 1 \cdot z$$
$$= 6 \csc \theta + 8 - 3 \cot \theta.$$

We are trying to find the minimal value of E over $0 < \theta \leq \pi/2$. To find θ such that $E(\theta)$ is a minimum, we first find $E'(\theta)$:

$$E'(\theta) = -6 \csc \theta \cot \theta + 3 \csc^2 \theta.$$

This derivative exists for all θ in the interval $0 < \theta \leq \pi/2$. Thus the only critical points are where $E'(\theta) = 0$, so we solve this equation:

$$-6 \csc \theta \cot \theta + 3 \csc^2 \theta = 0$$

$$\frac{-6 \cos \theta + 3}{\sin^2 \theta} = 0$$

$$6 \cos \theta = 3$$

$$\cos \theta = \frac{1}{2}$$

$$\theta = \frac{\pi}{3} + 2\pi n \quad or \quad \theta = \frac{5\pi}{3} + 2\pi n.$$

The only critical point in the interval $(0, \pi/2]$ is $\theta = \pi/3$. We use Maximum–Minimum Principle 2 to determine whether it is an absolute minimum or absolute maximum:

$$E''(x) = -6(-\csc \theta \cot \theta) \cot \theta - 6 \csc \theta(-\csc^2 \theta) + 6 \csc \theta(-\csc \theta \cot \theta)$$

$$= 6 \csc \theta \cot^2 \theta + 6 \csc^3 \theta - 6 \csc^2 \theta \cot \theta$$

$$E''(\pi/3) = 6\left(\frac{2\sqrt{3}}{3}\right)\left(\frac{\sqrt{3}}{3}\right)^2 + 6\left(\frac{2\sqrt{3}}{3}\right)^3 - 6\left(\frac{2\sqrt{3}}{3}\right)^2\left(\frac{\sqrt{3}}{3}\right)$$

$$= 4\sqrt{3} > 0.$$

Since $E''(\pi/3)$ is positive, $E(\pi/3)$ is a minimum. The pigeon should fly at angle $\theta = \pi/3$ to get home using minimal energy. ■

Exercise Set 3.5

1. Of all numbers whose sum is 50, find the two that have the maximum product. That is, maximize $Q = xy$, where $x + y = 50$.

2. Of all numbers whose sum is 70, find the two that have the maximum product. That is, maximize $Q = xy$, where $x + y = 70$.

3. In Exercise 1, can there be a minimum product? Explain.

4. In Exercise 2, can there be a minimum product? Explain.

5. Of all numbers whose difference is 4, find the two that have the minimum product.

6. Of all numbers whose difference is 6, find the two that have the minimum product.

7. Maximize $Q = xy^2$, where x and y are positive numbers, such that $x + y^2 = 1$.

8. Maximize $Q = xy^2$, where x and y are positive numbers, such that $x + y^2 = 4$.

9. Minimize $Q = 2x^2 + 3y^2$, where $x + y = 5$.

10. Minimize $Q = x^2 + 2y^2$, where $x + y = 3$.

11. Minimize $Q = x^2 + y^2$, where $x + y = 20$.

12. Minimize $Q = x^2 + y^2$, where $x + y = 10$.

13. Maximize $Q = xy$, where x and y are positive numbers, such that $\frac{4}{3}x^2 + y = 16$.

14. Maximize $Q = xy$, where x and y are positive numbers, such that $x + \frac{4}{3}y^2 = 1$.

15. Maximize $Q = \sqrt{x} + \sqrt{y}$, where $x + y = 1$.

16. Maximize $Q = \sqrt{x} + \sqrt{y}$, where $2x + y = 6$.

APPLICATIONS

17. *Maximizing Area.* A rancher wants to build a rectangular fence next to a river, using 120 yd of fencing. What dimensions of the rectangle will maximize the area? What is the maximum area? (Note that the rancher need not fence in the side next to the river.)

18. *Maximizing Area.* A rancher wants to enclose two rectangular areas near a river, one for sheep and one for cattle. There are 240 yd of fencing available. What is the largest total area that can be enclosed?

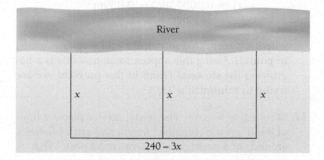

19. *Maximizing Area.* A carpenter is building a rectangular room with a fixed perimeter of 54 ft. What are the dimensions of the largest room that can be built? What is its area?

20. *Maximizing Area.* Of all rectangles that have a perimeter of 34 ft, find the dimensions of the one with the largest area. What is its area?

21. *Maximizing Volume.* From a thin piece of cardboard 30 in. by 30 in., square corners are cut out so that the sides can be folded up to make a box. What dimensions will yield a box of maximum volume? What is the maximum volume?

22. *Maximizing Volume.* From a thin piece of cardboard 20 in. by 20 in., square corners are cut out so that the sides can be folded up to make a box. What dimensions will yield a box of maximum volume? What is the maximum volume?

23. *Minimizing Surface Area.* A container company is designing an open-top, square-based, rectangular box that will have a volume of 62.5 in³. What dimensions yield the minimum surface area? What is the minimum surface area?

24. *Minimizing Surface Area.* A soup company is constructing an open-top, square-based, rectangular metal tank that will have a volume of 32 ft³. What dimensions yield the minimum surface area? What is the minimum surface area?

25. An isosceles triangle has two sides of length 1, and the angle between these sides has measure θ. Determine the angle θ that maximizes the area of the triangle, and find the maximum possible area.

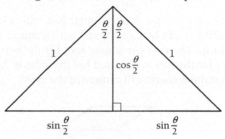

26. A right triangle is formed using the points $(-1, 0)$, a point $(\cos \theta, \sin \theta)$ on the first quadrant of the unit circle, and the point $(\cos \theta, 0)$, as shown in the figure. What is the maximum possible area of the triangle?

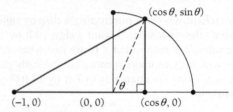

27. *Maximizing Revenue.* A university is trying to determine what price to charge for football tickets. At a price of $6 per ticket, it averages 70,000 people per game. For every increase of $1, it loses 10,000 people from the average number. Every person at the game spends an average of $1.50 on concessions. What price per ticket should be charged in order to maximize revenue? How many people will attend at that price?

28. *Maximizing Profit.* Suppose that you are the owner of a 30-unit motel. All units are occupied when you charge $60 a day per unit. For every increase of x dollars in the daily rate, there are x units vacant. Each occupied room costs $6 per day to service and maintain. What should you charge per unit in order to maximize profit?

29. *Maximizing Yield.* An apple farm yields an average of 30 bushels of apples per tree when 20 trees are planted on an acre of ground. Each time 1 more tree is planted per acre, the yield decreases 1 bu per tree due to the extra congestion. How many trees should be planted in order to get the highest yield?

30. *Maximizing Revenue.* When a theater owner charges $3 for admission, there is an average attendance of

100 people. For every $0.10 increase in admission, there is a loss of 1 customer from the average number. What admission should be charged in order to maximize revenue?

31. *Minimizing Costs.* A rectangular box with a volume of 320 ft³ is to be constructed with a square base and top. The cost per square foot for the bottom is 15¢, for the top is 10¢, and for the sides is 2.5¢. What dimensions will minimize the cost?

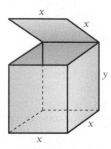

tw 32. A merchant who was purchasing a display sign from a salesclerk said, "I want a sign 10 ft by 10 ft." The salesclerk responded, "That's just what we'll give you; only to make it more aesthetically pleasing, why don't we change it to 7 ft by 13 ft?" Comment.

33. *Maximizing Profit.* The amount of money deposited in a financial institution in savings accounts is directly proportional to the interest rate that the financial institution pays on the money. Suppose that a financial institution can loan *all* the money it takes in on its savings accounts at an interest rate of 18%. What interest rate should it pay on its savings accounts in order to maximize profit?

34. *Maximizing Area.* A page in this book measures 73.125 in². On the average, there is a 0.75-in. margin at the top and at the bottom of each page and a 0.5-in. margin on each of the sides. What should the outside dimensions of each page be so that the printed area is a maximum? Measure the outside dimensions to see whether the actual dimensions maximize the printed area.

35. *Vanity License Plates.* According to a pricing model,[10] increasing the initial fee for vanity license plates by $1 decreases the percentage of a state's population that will request them by 0.04%.

---

[10]E. D. Craft, *The demand for vanity (plates): Elasticities, net revenue maximization, and deadweight loss,* Contemporary Economic Policy, Vol. 20, 133–144 (2002).

a) Recently, the initial fee for vanity license plates in Maryland was $25, and the percentage of the state's population that had vanity plates was 2.13%. Use this information to construct the percentage of Maryland's population that will request vanity license plates as a function of *p*.
b) Find the price *p* of the initial fee that will maximize revenue.

36. *Growth of a Baby.* The median weight of a boy whose age *t* is between 0 and 36 mo can be approximated by the function

$$w(t) = 0.000758t^3 - 0.0596t^2 + 1.82t + 8.15,$$

where *t* is measured in months and *w* is measured in pounds. Using this approximation, when is a boy growing the slowest? (*Hint:* In this problem, we are trying to minimize $w'(t)$.)

37. *Maximizing Volume.* The postal service places a limit of 84 in. on the combined length and girth (distance around) of a package to be sent parcel post. What dimensions of a rectangular box with square cross section will contain the largest volume that can be mailed? (*Hint:* There are two different girths.)

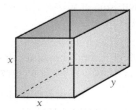

38. *Minimizing Cost.* A rectangular play area is to be fenced off in a person's yard and is to contain 48 yd². The neighbor agrees to pay half the cost of the fence on the side of the play area that lines the lot. What dimensions will minimize the cost of the fence?

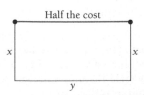

39. *Maximizing Light.* A Norman window is a rectangle with a semicircle on top. Suppose that the perimeter of a particular Norman window is to be 24 ft. What should its dimensions be in order to allow the maximum amount of light to enter through the window?

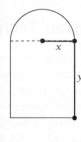

come to the shore in order to minimize cost? Note that S could very well be B or A.

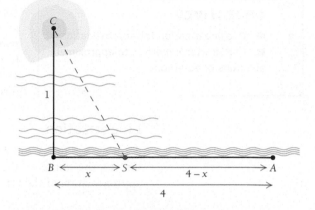

45. *Flights of Homing Pigeons.* Repeat Example 5 by expressing E as a function of x (instead of θ).

46. *Minimizing Distance.* A road is to be built between two cities C_1 and C_2, which are on opposite sides of a river of uniform width r. Because of the river, a bridge must be built. C_1 is a units from the river, and C_2 is b units from the river; $a \leq b$. Where should the bridge be located in order to minimize the travel distance between the cities? Give a general solution using the constants a, b, p, and r in the figure shown here.

40. *Maximizing Light.* Repeat Exercise 39, but assume that the semicircle is to be stained glass, which transmits only half as much light as the semicircle in Exercise 39.

41. For what positive number is the sum of its reciprocal and five times its square a minimum?

42. For what positive number is the sum of its reciprocal and four times its square a minimum?

43. A 24-in. piece of string is cut in two pieces. One piece is used to form a circle and the other to form a square. How should the string be cut so that the sum of the areas is a minimum? a maximum?

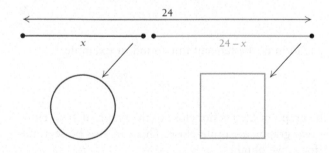

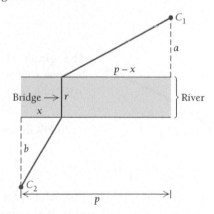

47. Minimize $Q = x^3 + 2y^3$, where x and y are positive numbers, such that $x + y = 1$.

48. Minimize $Q = 3x + y^3$, where $x^2 + y^2 = 2$.

44. *Minimizing Costs.* A power line is to be constructed from a power station at point A to an island at point C, which is 1 mi directly out in the water from a point B on the shore. Point B is 4 mi downshore from the power station at A. It costs $5000 per mile to lay the power line under water and $3000 per mile to lay the line under ground. At what point S downshore from A should the line

3.6 Approximation Techniques

OBJECTIVES

■ Use linearization for approximations.
■ Use Newton's method to approximate solutions of equations.

In this section, we consider ways of using calculus to make approximations. Suppose, for example, that at a certain time in an illness, a patient's temperature is 102° and the instantaneous rate of change is −2° per min. It would seem reasonable to *estimate* the patient's temperature 1 min later to be about 100°, especially if the temperature were following a particular formula.

Linearization

As discussed in Section 2.4, the derivative of a function f is used to find the tangent line to the curve at the point $(a, f(a))$. This tangent line is often an accurate approximation of the function in the vicinity of $x = a$.

EXAMPLE 1 For $f(x) = \sqrt{x}$, find the equation of the tangent line to the curve at the point $(25, 5)$.

Solution To begin, we find $f'(25)$:

$$f'(x) = \frac{1}{2\sqrt{x}}, \quad \text{so that} \quad f'(25) = \frac{1}{10}.$$

Therefore, the equation of the tangent line to the curve at $(25, 5)$ is given by

$$y - 5 = \frac{1}{10}(x - 25)$$

$$y = 5 + \frac{1}{10}(x - 25).$$

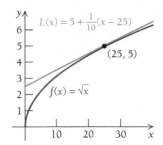

The graph of $f(x) = \sqrt{x}$ and this tangent line are shown in the figure. For reasons that will become apparent shortly, we do not write this equation in slope–intercept form. ■

Let's denote by $L(x)$ the equation of the tangent line found in Example 1:

$$L(x) = 5 + \frac{1}{10}(x - 25).$$

For values of x far from 25, the graph of $L(x)$ is not close to the graph of $f(x)$. However, if x is near 25, then the two graphs are quite close. This can also be seen numerically. If we substitute 27 for x, we obtain

$$L(27) = 5 + \frac{1}{10}(27 - 25)$$

$$= 5 + 0.2$$

$$= 5.2.$$

To five decimal places, $\sqrt{27} \approx 5.19615$. We see that $L(27)$ approximates the true value of $\sqrt{27}$ quite accurately.

Let's investigate further this example of $f(x) = \sqrt{x}$. The number 25 served as an "anchor point," or a point where $f(x)$ could be easily evaluated. We denote this anchor point by a. The number 5 is the value of the function at this anchor point, or $f(a)$. Finally, the number $\frac{1}{10}$ is the value of the derivative at this anchor point, or $f'(a)$. Generalizing these observations leads to the following definition of the **linearization** of a curve at a point.

DEFINITION

The *linearization* of $f(x)$ at $x = a$ is given by

$$L(x) = f(a) + f'(a)(x - a).$$

This is also the equation to the tangent line to the curve $y = f(x)$ at the point $(a, f(a))$. If x is near a, then the value of the linearization $L(x)$ will be a reasonable approximation to the function value $f(x)$.

EXAMPLE 2   Find the linearization of $f(x) = \dfrac{1}{x}$ at the point $a = 2$. Use the linearization to approximate $\dfrac{1}{1.98}$.

Solution   We first compute $f(a) = f(2) = \dfrac{1}{2}$. Also, $f'(x) = -\dfrac{1}{x^2}$, and so $f'(a) = f'(2) = -\dfrac{1}{4}$. Therefore, the linearization is

$$L(x) = \frac{1}{2} + \left(-\frac{1}{4}\right)(x - 2) = \frac{1}{2} - \frac{1}{4}(x - 2).$$

Finally, since 1.98 is close to 2, the value of $\dfrac{1}{1.98}$ may be approximated using this linearization:

$$\frac{1}{1.98} \approx L(1.98)$$

$$= \frac{1}{2} - \frac{1}{4}(1.98 - 2)$$

$$= \frac{1}{2} - \frac{1}{4}(-0.02)$$

$$= 0.505.$$

To five decimal places, $\dfrac{1}{1.98} \approx 0.50505$.   ∎

EXAMPLE 3   Use linearization to approximate $\sqrt[3]{64.6}$.

Solution   Let $f(x) = \sqrt[3]{x}$. While we cannot exactly evaluate $f(64.6)$, we can evaluate $f(64)$:

$$f(64) = \sqrt[3]{64} = 4.$$

Therefore, we take $a = 64$ as the anchor point, so that $f(a) = 4$. Next, we find $f'(a)$:

$$f'(x) = \frac{1}{3}x^{-2/3}, \quad \text{so that} \quad f'(64) = \frac{1}{3}(64)^{-2/3} = \frac{1}{48}.$$

Therefore, the linearization is

$$L(x) = 4 + \frac{1}{48}(x - 64).$$

Since 64.6 is close to 64, the value of $\sqrt[3]{64.6}$ may be approximated using this linearization:

$$\sqrt[3]{64.6} \approx L(64.6) = 4 + \frac{1}{48}(64.6 - 64) = 4.0125.$$

To five decimal places, $\sqrt[3]{64.6} \approx 4.01246$.   ■

EXAMPLE 4   Find the linearization of $f(x) = \sin x$ at the point $a = 0$.

Solution   We first compute $f(a) = f(0) = \sin 0 = 0$. Also, $f'(x) = \cos x$, and so $f'(a) = f'(0) = \cos 0 = 1$. Therefore, the linearization is

$$L(x) = 0 + (1)(x - 0) = x.$$ ■

Linearization uses linear functions to approximate functions. Though we do not do so here, it is possible to use other higher-order polynomials (called *Taylor polynomials*) to approximate functions. These Taylor polynomials may also be used to estimate how accurate linear approximations are.

Newton's Method

So far in this book, we've almost always been able to find *exact* solutions to equations. For example, the equation $x^2 + 5x + 4 = 0$ may be solved either by factoring or by using the quadratic formula. However, the solutions of many equations that arise in practical applications cannot be found exactly using standard algebraic techniques. For such equations, we must resort to finding *approximate* solutions—preferably accurate to several decimal places.

One way of approximating the solution of a given equation is called **Newton's method.** Newton's method starts by making an *initial approximation,* or *initial guess* (denoted x_1). Then we use our initial guess x_1 to construct a second value x_2, which we expect to be closer to the true solution than x_1 is. Continuing, we use x_2 to construct a third value x_3, and so forth. In theory, we construct a *sequence* $x_1, x_2, \cdots, x_n, x_{n+1}, \cdots$, where x_n denotes the nth approximation. We would like this sequence to converge to a solution, say, $\bar{x}$, of the given equation. In practice, we stop the process after a finite number of steps.

CAUTION   Notice that x_{n+1} cannot be replaced by $x_n + 1$ in general. The symbol x_{n+1} means the $(n + 1)$st term of the sequence, while $x_n + 1$ means the nth term of the sequence added to 1.

Newton's method is fairly easy to apply and usually yields very rapid convergence, so that the exact answer can be accurately approximated with only a few steps.

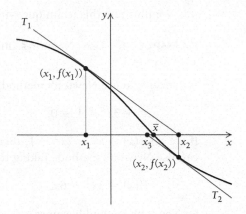

To apply this method, we write the given equation in the form $f(x) = 0$. We will assume that f is differentiable. Let x_1 be our initial approximation, as shown in the figure. Assuming that $f'(x_1) \neq 0$, the tangent line T_1 to the graph of $y = f(x)$ at the point $(x_1, f(x_1))$ will intersect the x-axis at exactly one point. We call this point $(x_2, 0)$. The x-coordinate of this point, x_2, will be our second approximation. Next we locate $(x_2, f(x_2))$ on the graph of $y = f(x)$ and construct the tangent line T_2 to the graph at this new point. Then the tangent line will intersect the x-axis at $(x_3, 0)$ (assuming that $f'(x_2) \neq 0$). The x-coordinate of this point, x_3, is our third approximation to the solution $\bar{x}$. Continuing, we can generate as many approximations as we need.

We now have a geometric method for using any term x_n to construct the next term x_{n+1}. To obtain a formula relating x_n and x_{n+1}, we return to the tangent line T_1. The line is the linearization of the curve $y = f(x)$ at $a = x_1$, and so the equation of T_1 is

$$y = f(x_1) + f'(x_1)(x - x_1).$$

The point $(x_2, 0)$ is on T_1, and so its coordinates must satisfy the equation for T_1. Thus, setting $x = x_2$ and $y = 0$, we have

$$f(x_1) + f'(x_1)(x_2 - x_1) = 0$$
$$f'(x_1)(x_2 - x_1) = -f(x_1)$$
$$x_2 - x_1 = -\frac{f(x_1)}{f'(x_1)} \qquad \text{Since } f'(x_1) \neq 0$$
$$x_2 = x_1 - \frac{f(x_1)}{f'(x_1)}.$$

In general, we have the following.

THEOREM 10

Newton's method for finding an approximate solution to the equation $f(x) = 0$, where f is differentiable, is

$$x_{n+1} = x_n - \frac{f(x_n)}{f'(x_n)},$$

where x_{n+1} is the next approximation after x_n.

We illustrate this technique with a couple of examples.

EXAMPLE 5 Use Newton's method with $x_1 = 1$ to find a solution to $x^3 + 3x^2 = 1$.

Solution   To use Newton's method, we first write the equation in the form

$$x^3 + 3x^2 - 1 = 0.$$

If we set $f(x) = x^3 + 3x^2 - 1$, then the equation takes the form $f(x) = 0$ and we can use Newton's method. Taking the derivative, we get

$$f'(x) = 3x^2 + 6x.$$

So Newton's method becomes

$$x_{n+1} = x_n - \frac{f(x_n)}{f'(x_n)}$$

$$= x_n - \frac{x_n^3 + 3x_n^2 - 1}{3x_n^2 + 6x_n}.$$

Starting with $n = 1$, we find

$$x_{1+1} = x_1 - \frac{x_1^3 + 3x_1^2 - 1}{3x_1^2 + 6x_1}$$

$$x_2 = 1 - \frac{1^3 + 3(1)^2 - 1}{3(1)^2 + 6(1)} \qquad \text{Letting } x_1 = 1$$

$$= \frac{2}{3}.$$

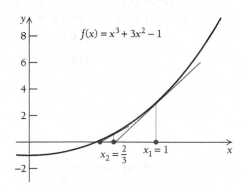

As shown in the figure, this is closer to a solution than $x_1 = 1$. We now let $n = 2$:

$$x_{2+1} = x_2 - \frac{x_2^3 + 3x_2^2 - 1}{3x_2^2 + 6x_2}$$

$$x_3 = \frac{2}{3} - \frac{(2/3)^3 + 3(2/3)^2 - 1}{3(2/3)^2 + 6(2/3)} \qquad \text{Letting } x_2 = \frac{2}{3}$$

$$\approx 0.548611111.$$

As shown in the figure, x_3 is very close to the solution. We continue this procedure for larger values of n:

$$x_4 \approx 0.548611111 - \frac{(0.548611111)^3 + 3(0.548611111)^2 - 1}{3(0.548611111)^2 + 6(0.548611111)}$$

$$\approx 0.532390162$$

$$x_5 \approx 0.532390162 - \frac{(0.532390162)^3 + 3(0.532390162)^2 - 1}{3(0.532390162)^2 + 6(0.532390162)}$$

$$\approx 0.532088989.$$

Notice that x_4 and x_5 agree to three decimal places. We can therefore be confident that a solution of the equation is $\bar{x} = 0.532\dots$. If we continue using Newton's method, we find

$$x_6 \approx 0.532088886$$
$$x_7 \approx 0.532088886.$$

Since x_6 and x_7 agree to nine decimal places, we stop repeating Newton's method at this point. We conclude that an approximate solution of the equation is $x \approx 0.532088886$. This may be verified directly:

$$(0.532088886)^3 + 3(0.532088886)^2 \approx 1. \qquad \blacksquare$$

Newton's method requires a significant amount of computation. The Technology Connection below gives guidance on how Newton's method can be efficiently performed on a calculator.

Technology Connection

Newton's Method

Newton's method can be easily performed on any calculator that has an **Ans** button. For example, if $f(x) = x^3 + 3x^2 - 1$ as in Example 1, first enter 1 and then enter

Ans − (**Ans**^3 + 3
 × **Ans**^2 − 1)/
 (3 × **Ans**^2 + 6 × **Ans**).

This produces the second approximation x_2.

```
1
                              1
Ans − (Ans^3 + 3 * Ans^2 − 1)/(3 *
Ans^2 + 6 * Ans)
                  .6666666667
```

By simply pressing the **Enter** button repeatedly, further approximations can be obtained.

```
                              1
Ans − (Ans^3 + 3 * Ans^2 − 1)/(3 *
Ans^2 + 6 * Ans)
                  .6666666667
                  .5486111111
                  .5323901619
```

Once you see an answer repeated, you can be confident that you have obtained a very accurate approximation to the correct answer.

In the previous example, we found one solution of $x^3 + 3x^2 - 1 = 0$ using the initial approximation $x_1 = 1$. However, as shown in the figure below, there are actually three solutions to this equation. In the next example, we discuss how to find the other two solutions.

EXAMPLE 6   Use Newton's method to determine all points where the curve $y = x^3 + 3x^2 - 1$ crosses the x-axis.

Solution   We seek all values of x so that $x^3 + 3x^2 - 1 = 0$. If we let $f(x) = x^3 + 3x^2 - 1$, we have set up the problem so that Newton's method may be applied. As in Example 5, Newton's method for this problem is

$$x_{n+1} = x_n - \frac{f(x_n)}{f'(x_n)} = x_n - \frac{x_n^3 + 3x_n^2 - 1}{3x_n^2 + 6x_n}.$$

Based on the graph, it appears that there are three solutions, and we approximated one solution, namely, $\bar{x} \approx 0.532088886$, in Example 5. The other two solutions appear to be near -3 and $-\frac{2}{3}$. To find the other two solutions, we repeat Newton's method using these estimates as our initial approximations.

First, we take $x_1 = -3$:

$$x_2 = x_1 - \frac{x_1^3 + 3x_1^2 - 1}{3x_1^2 + 6x_1}$$

$$= -3 - \frac{(-3)^3 + 3(-3)^2 - 1}{3(-3)^2 + 6(-3)} \qquad \text{Using } x_1 = -3$$

$$= -\frac{26}{9}$$

$$x_3 = -\frac{26}{9} - \frac{(-26/9)^3 + 3(-26/9)^2 - 1}{3(-26/9)^2 + 6(-26/9)}$$

$$\approx -2.879451567$$

$$x_4 \approx -2.879385245$$

$$x_5 \approx -2.879385242.$$

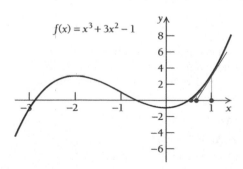

$f(x) = x^3 + 3x^2 - 1$

To find the other solution, we take $x_1 = -\dfrac{2}{3}$:

$$x_2 = -\frac{2}{3} - \frac{(-2/3)^3 + 3(-2/3)^2 - 1}{3(-2/3)^2 + 6(-2/3)} \qquad \text{Using } x_1 = -\tfrac{2}{3}$$

$$\approx -0.652777778$$

$$x_3 \approx -0.652703647$$

$$x_4 \approx -0.652703645.$$

We conclude that the three solutions are approximately given by $\bar{x} = 0.532088886$ (from Example 5), -2.879385242, and -0.652703645. The latter two solutions can be directly checked by substituting into the original function $f(x)$:

$$f(-2.879385242) = (-2.879385242)^3$$
$$+ 3(-2.879385242)^2 - 1 \approx 0,$$

$$f(-0.652703645) = (-0.652703645)^3$$
$$+ 3(-0.652703645)^2 - 1 \approx 0.$$

(The first solution was checked in Example 5.)   ■

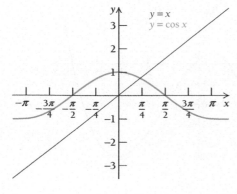

EXAMPLE 7   Use Newton's method to approximate the solution of the equation $\cos x = x$.

Solution   Based on the graphs of $y = \cos x$ and $y = x$, it appears that the two graphs cross near $\pi/4$. We therefore take $f(x) = \cos x - x$ and $x_1 = \pi/4$.

Newton's method for this problem is

$$x_{n+1} = x_n - \frac{f(x_n)}{f'(x_n)} = x_n - \frac{\cos x_n - x_n}{-\sin x_n - 1} = x_n + \frac{\cos x_n - x_n}{\sin x_n + 1}.$$

Therefore,

$$x_2 = x_1 + \frac{\cos x_1 - x_1}{\sin x_1 + 1} = \frac{\pi}{4} + \frac{\cos(\pi/4) - (\pi/4)}{\sin(\pi/4) + 1} \approx 0.739536134$$

$$x_3 \approx 0.739085178$$

$$x_4 \approx 0.739085133$$

$$x_5 \approx 0.739085133.$$

We conclude that the solution is approximately $\bar{x} = 0.739085133$.   ■

In the above examples, we have shown how Newton's method can be used to find approximate solutions of equations. When this technique works, it is very efficient, converging quickly toward a solution. However, there are cases when Newton's method may not work. It is possible for Newton's method to converge to a somewhat unexpected answer or to not converge at all.

There are conditions on the function f that imply that Newton's method does converge as expected; however, these conditions are beyond the scope of this book. Instead, we warn the reader that there can be occasional difficulty with Newton's method; examples of such difficulties are given in the exercises.

Exercise Set 3.6

Find the linearization of $f(x)$ at $x = a$.

1. $f(x) = x^2$, $a = 3$

2. $f(x) = x^3$, $a = 2$

3. $f(x) = \dfrac{1}{x}$, $a = 4$

4. $f(x) = \dfrac{1}{x^2}$, $a = 0.5$

5. $f(x) = x^{3/2}$, $a = 4$

6. $f(x) = \sqrt{x}$, $a = 100$

7. $f(x) = \cos x$, $a = 0$

8. $f(x) = \tan x$, $a = 0$

9. $f(x) = x \cos x$, $a = 0$

10. $f(x) = x \sec x$, $a = 0$

Approximate using linearization.[11]

11. $\sqrt{19}$ 12. $\sqrt{24}$ 13. $\sqrt{99.1}$

14. $\sqrt{103.4}$ 15. $\sqrt[3]{10}$ 16. $\sqrt[3]{0.91}$

17. $\sqrt{97}$ 18. $\sqrt[3]{1729.03}$ 19. $\sin 0.1$

20. $\sin(-0.05)$ 21. $\tan(-0.04)$ 22. $\tan 0.02$

Use the indicated choice of x_1 and Newton's method to solve the given equation. Check your answers.

23. $x = \dfrac{1}{3}x^2 + \dfrac{1}{3}$; $x_1 = 0$

24. $x = \dfrac{1}{3}x^2 + \dfrac{1}{3}$; $x_1 = 3$

25. $x^3 - 3x + 3 = 0$; $x_1 = -3$

26. $x^4 - 4x + 1 = 0$; $x_1 = 0$

27. $x\sqrt{x + 1} = 4$; $x_1 = 2$

28. $x\sqrt{x + 2} = 8$; $x_1 = 2$

29. $\cos 2x = x$; $x_1 = 0$

30. $\sin x = 1 - x$; $x_1 = 0$

31. $\sin x + x = \cos x$; $x_1 = 0$

32. $2x - \sin x = \cos(x^2)$; $x_1 = \pi/4$

Find all real solutions to the given equation. Use the given graph to choose the first approximations in Newton's method.

33. $x^3 - 4x^2 - 5x + 1 = 0$

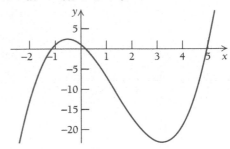

34. $2x^3 + 4x^2 - 6x - 1 = 0$

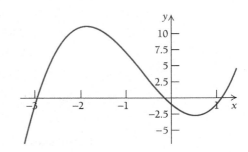

35. $x^4 - 3x^3 - 18x^2 - 2x + 4 = 0$

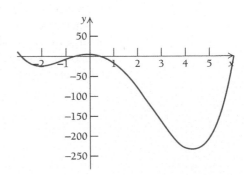

---

[11]An amusing anecdote related to Exercise 18 may be found in *Surely You're Joking, Mr. Feynman!*, written by Nobel Prize-winning physicist Richard P. Feynman (Bantam Books, New York, 1985).

36. $x^7 - 7x + 4 = 0$

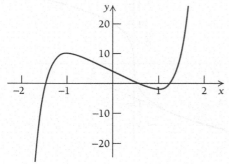

APPLICATIONS

37. *Blood Pressure.* For a dosage of x cubic centimeters of a certain drug, the resulting blood pressure B is approximated by

$$B(x) = 0.05x^2 - 0.3x^3, \quad 0 \le x \le 0.16.$$

Determine the two dosages that result in a blood pressure of 0.0001.

38. *Hearing Impairments.* The number of hearing-impaired Americans (in millions) is approximately

$$N(x) = -0.00006x^3 + 0.006x^2 - 0.1x + 1.9,$$

where x is the age (assumed to be between 0 and 80) and $N(x)$ is measured in millions.[12] Determine the ages x for which there are 3 million hearing-impaired Americans.

39. *Growth of a Baby.* The median weight of a boy whose age t is between 0 and 36 mo can be approximated by the equation

$$w(t) = 8.15 + 1.82t - 0.0596t^2 + 0.000758t^3,$$

where t is measured in months and w is measured in pounds. Determine the age at which the median weight of boys is 15 lb.

40. *Growth of a Baby.* Using the formula in the previous problem, determine the age at which the median weight of boys is 25 lb.

41. *Lung Cancer.* The rate of lung and bronchus cancer per 100,000 males since 1930 is approximated by the function

$$r(x) = -0.000775x^3 + 0.0696x^2 - 0.209x + 4.68,$$

where $0 \le x \le 70$ is the number of years since 1930. Determine the year when the rate was 40 per 100,000 males.

42. *New York Temperature.* For any date, the average temperature on that date in New York can be approximated by the function

$$T(x) = 0.0338x^4 - 0.996x^3 + 8.57x^2 - 18.4x + 43.5,$$

where T represents the temperature in degrees Fahrenheit. Also, $0 \le x \le 12$, where $x = 1$ represents the middle of January, $x = 2$ represents the middle of February, and so on. Determine when the average temperature in New York is 60°F.

43. *Incidence of Breast Cancer.* The incidence of breast cancer per 100,000 women may be approximated by

$$I(x) = -0.000054x^4 + 0.0067x^3 - 0.0997x^2 - 0.84x - 0.25,$$

where x is a woman's age. Use this function to estimate the ages where 300 out of 100,000 women have breast cancer.

Lotka–Leslie Model of Population Analysis. Let r denote the annual growth rate of the size of a species population. For example, if $r = 1.05$, then the population increases by 5% per year. Then r is a solution of the equation

$$r^a - sr^{a-1} - pm\left[1 - \left(\frac{s}{r}\right)^{w-a+1}\right] = 0.$$

In this equation, a is the age of first reproduction, p is the probability that a newborn survives to age a, s is the annual adult survival rate, w is the maximum age of the species, and m is the reproductive rate.[13]

44. For spotted owls (*Strix occidentalis caurina*), $a = 2$, $p = 0.08$, $s = 0.94$, $w = 25$, and $m = 0.24$. Use this information to predict the annual growth rate for spotted owls. Use $r_1 = 1$.

45. For killer whales (*Orcinus orca*), $a = 15$, $p = 0.78$, $s = 0.99$, $w = 50$, and $m = 0.11$. Use this information to predict the annual growth rate for killer whales. Use $r_1 = 1.1$.

46. For Yellowstone elk (*Cervus elaphus*), $a = 3$, $p = 0.66$, $s = 0.99$, $w = 18$, and $m = 0.48$. Use this information to predict the annual growth rate for Yellowstone elk. Use $r_1 = 1.1$.

47. For giant pandas (*Ailuropoda melanoleuca*), $a = 4$, $p = 0.42$, $s = 0.98$, $w = 18$, and $m = 0.42$. Use this information to predict the annual growth rate for giant pandas. Use $r_1 = 1.1$.

---

[12]American Speech-Language Hearing Association.

[13]L. L. Eberhardt, "A paradigm for population analysis of long-lived vertebrates," *Ecology*, Vol. 83, pp. 2841–2854 (2002).

SYNTHESIS

Blood Velocity. Ultrasound measures the velocity of blood (in cm/s) through a blood vessel using

$$v = \frac{77{,}000d \sec t}{f}.$$

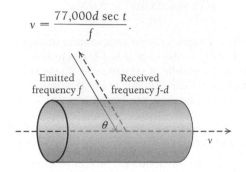

Emitted frequency f Received frequency f-d

In this formula, f is the emitted ultrasound beam frequency, d is the Doppler shift (or the difference between the emitted and received beam frequencies), and t is the angle between the ultrasound beam and the blood vessel. Practically, the ultrasound machine must be positioned so that t is at least 60° (or $\pi/3$ radians). Also, since the precise orientation of a blood vessel within the body is unknown, the angle t can only be approximately known.[14]

48. a) Suppose $f = 5{,}000{,}000$ Hz, $d = 150$ Hz, and $t = \pi/3$ (or 60°). Determine the blood velocity through the vessel.

 b) Suppose the measurement of the angle t is inaccurate by 0.01 radian (about 0.6°). Use linearization to estimate how much the measurement of v will differ from the answer found in part (a).

49. a) Suppose $f = 4{,}000{,}000$ Hz, $d = 100$ Hz, and $t = 4\pi/9$ (or 80°). Determine the blood velocity through the vessel.

 b) Suppose the measurement of the angle t is inaccurate by 0.01 radian (about 0.6°). Use linearization to estimate how much the measurement of v will differ from the answer found in part (a).

 tw c) Compare your answers to part (b) for the previous two exercises. What is the significance of these answers for an ultrasound technician?

50. *Failure of Convergence.* There are examples where Newton's method doesn't converge to an answer.

 a) Attempt to use Newton's method to solve $x^{1/3} = 0$ with $x_1 = 0$.

 b) Based on the graph below, explain why Newton's method does not converge for this function.

---

[14]Triton Technology, Inc.

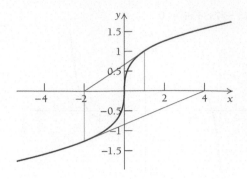

51. *Convergence to a Distant Solution.* Sometimes Newton's method converges to a solution, but it doesn't converge to the closest solution.

 a) Use Newton's method to find a solution to $x^3 - 6x^2 + 8x = 0$. Use $x_1 = 3$.

 tw b) Based on the graph below, explain why Newton's method doesn't converge to either 2 or 4, even though these solutions are closer to 3 than your answer to part (a).

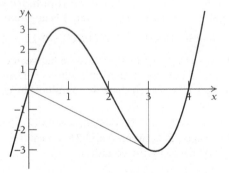

![technology connection icon] *Technology Connection*

Phonetics. When we speak, the position of the tongue changes the shapes of the mouth and throat cavity, thereby changing the sound that is generated. For example, to pronounce the first vowel of the word *father,* the mouth is much wider than the throat cavity. For many vowel sounds, the vocal tract may be modeled by two adjoining tubes, with one end closed (the glottis) and the other end open (the lips). We denote by A_m and L_m the cross-sectional area and length of the mouth, and we denote by A_t and L_t the cross-sectional area and length of the throat cavity.

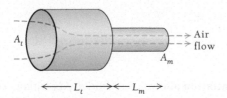

The shape of the vocal tract tends to promote certain sound frequencies.[15] Let c be the speed of sound. If f is a frequency promoted by the vocal tract (in Hertz) and we let $x = 2\pi f/c$, then x is a solution of

$$\frac{1}{A_m}\tan(L_m x) - \frac{1}{A_t}\cot(L_t x) = 0.$$

52. a) For a certain speaker of the first vowel of the word *father,* $A_m = 10A_t$ (that is, the mouth opening is 10 times larger than the throat opening), $L_m = 8$ cm, and $L_t = 9.7$ cm.[16] Use this information to show that

$$\tan(8x) - 10\cot(9.7x) = 0.$$

---

[15]K. N. Stevens, *Acoustic Phonetics* (MIT Press, Cambridge, Massachusetts, 1998).

[16]To verify that these numbers are plausible, you may want to pronounce this vowel. Notice the relative sizes of your mouth and the throat opening formed by your tongue.

b) Use a grapher to sketch the graph of $y = \tan(8x) - 10\cot(9.7x)$ using the graphing window $[0, 0.5, -2, 2]$. From the graph, estimate the first three x-intercepts.

c) Use Newton's method to find the first three solutions. Use $x_1 = 0.1$, $x_1 = 0.2$, and $x_1 = 0.45$ as your three starting points.

d) The speed of sound is approximately 35,400 cm/s. Use this value for c and the relationship $x = 2\pi f/c$ to find the first three natural frequencies of the speaker's vocal tract.

53. For a certain speaker of the vowel in the word *heed,* $A_t = 8A_m$, $L_m = 6$ cm, and $L_t = 9$ cm. Using the two-tube model, find the first three natural frequencies of the vocal tract. (You may wish to use a grapher to get reasonable starting points for Newton's method.)

54. For a certain speaker of the vowel in the word *hat,* $A_m = 8A_t$, $L_m = 13$ cm, and $L_t = 4$ cm. Using the two-tube model, find the first three natural frequencies of the vocal tract.

55. For a certain speaker of the vowel in the word *hot,* $A_m = 7A_t$, $L_m = 8$ cm, and $L_t = 9$ cm. Using the two-tube model, find the first three natural frequencies of the vocal tract.

3.7 Implicit Differentiation and Related Rates*

OBJECTIVES

- Differentiate implicitly.
- Solve related-rates problems.

Implicit Differentiation

Consider the equation

$$y^3 = x.$$

This equation *implies* that y is a function of x, for if we solve for y, we get

$$y = \sqrt[3]{x}$$
$$= x^{1/3}.$$

---

*This section can be omitted without loss of continuity.

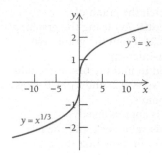

We know from our work in this chapter that

$$\frac{dy}{dx} = \frac{1}{3}x^{-2/3}. \tag{1}$$

A method known as **implicit differentiation** allows us to find dy/dx *without* solving for y. We use the Chain Rule, treating y as a function of x. We use the Extended Power Rule and differentiate both sides of

$$y^3 = x$$

with respect to x:

$$\frac{d}{dx}y^3 = \frac{d}{dx}x.$$

The derivative on the left side is found using the Extended Power Rule:

$$3y^2\frac{dy}{dx} = 1.$$

Then

$$\frac{dy}{dx} = \frac{1}{3y^2}, \quad \text{or} \quad \frac{1}{3}y^{-2}.$$

We can show that this indeed gives us the same answer as equation (1) by replacing y with $x^{1/3}$:

$$\frac{dy}{dx} = \frac{1}{3}y^{-2} = \frac{1}{3}(x^{1/3})^{-2} = \frac{1}{3}x^{-2/3}.$$

CAUTION $\frac{d}{dx}y^3 \neq 3y^2$. Remember that we must use the Chain Rule and treat y as a function of x.

Often, it is difficult or impossible to solve for y, obtaining an explicit expression in terms of x. For example, the equation

$$y^3 + x^2y^5 - x^4 = 27$$

determines y as a function of x, but it would be difficult to solve for y. We can nevertheless find a formula for the derivative of y *without* solving for y. This involves computing $\frac{d}{dx}y^n$ for various integers n, and hence involves the Chain Rule in the form

$$\frac{d}{dx}y^n = ny^{n-1} \cdot \frac{dy}{dx}.$$

EXAMPLE 1 For $y^3 + x^2y^5 - x^4 = 27$:

a) Find dy/dx using implicit differentiation.

b) Find the slope of the tangent line to the curve at the point $(0, 3)$.

Solution

a) We differentiate the term x^2y^5 using the Product Rule. Note that whenever an expression involving y is differentiated, dy/dx must be a factor of the answer. When an expression involving just x is differentiated, there is no factor dy/dx.

$$\frac{d}{dx}(y^3 + x^2y^5 - x^4) = \frac{d}{dx}(27)$$

$$\frac{d}{dx}y^3 + \frac{d}{dx}x^2y^5 - \frac{d}{dx}x^4 = 0$$

$$3y^2 \cdot \frac{dy}{dx} + x^2 \cdot 5y^4 \cdot \frac{dy}{dx} + 2x \cdot y^5 - 4x^3 = 0.$$

Then

$$3y^2 \cdot \frac{dy}{dx} + 5x^2y^4 \cdot \frac{dy}{dx} = 4x^3 - 2xy^5 \qquad \text{Only those terms involving } dy/dx \text{ should appear on one side.}$$

$$(3y^2 + 5x^2y^4)\frac{dy}{dx} = 4x^3 - 2xy^5$$

$$\frac{dy}{dx} = \frac{4x^3 - 2xy^5}{3y^2 + 5x^2y^4} \qquad \text{Solving for } dy/dx. \text{ Leave the answer in terms of } x \text{ and } y.$$

b) To find the slope of the tangent line to the curve at $(0, 3)$, we replace x with 0 and y with 3:

$$\frac{dy}{dx} = \frac{4 \cdot 0^3 - 2 \cdot 0 \cdot 3^5}{3 \cdot 3^2 + 5 \cdot 0^2 \cdot 3^4} = 0.$$

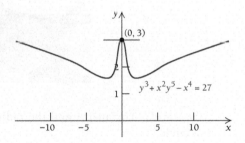

Implicit differentiation may also be used to find dy/dx for equations that include trigonometric functions.

EXAMPLE 2   Find dy/dx using implicit differentiation if

$$x^2 + y \sin y = \tan^2 y.$$

Solution   We differentiate both sides with respect to x. Once again, when an expression involving just x is differentiated, there is no factor dy/dx.

$$2x + \frac{dy}{dx} \cdot \sin y + y \cos y \cdot \frac{dy}{dx} = 2 \tan y \sec^2 y \cdot \frac{dy}{dx}$$

$$\sin y \cdot \frac{dy}{dx} + y \cos y \cdot \frac{dy}{dx} - 2 \tan y \sec^2 y \cdot \frac{dy}{dx} = -2x$$

$$\frac{dy}{dx} = \frac{-2x}{\sin y + y \cos y - 2 \tan y \sec^2 y}. \quad \blacksquare$$

Related Rates

Suppose that y is a function of x, say,

$$y = f(x),$$

and x varies with time t (as a function of time t). Since y depends on x and x depends on t, y also depends on t. That is, y is also a function of time t. The Chain Rule gives the following:

$$\frac{dy}{dt} = \frac{dy}{dx} \cdot \frac{dx}{dt}.$$

Thus the rate of change of y is *related* to the rate of change of x. Let's see how this comes up in problems. It helps to keep in mind that any variable can be thought of as a function of time t, even though a specific expression in terms of t may not be given.

EXAMPLE 3 *Plant Growth.* In one season, the Victoria Water Lily grows circular lily pads with diameters up to 8 ft. Suppose the radius r of a Victoria Lily pad is growing at the rate of 0.4 ft per month at the moment that the radius is 2 ft. At that moment, how fast is the area of the lily pad increasing?

Solution   The area A and the radius r are related by the equation for the area of a circle:

$$A = \pi r^2.$$

We take the derivative of both sides with respect to t:

$$\frac{dA}{dt} = 2\pi r \cdot \frac{dr}{dt}.$$

At the moment in question, $\dfrac{dr}{dt} = 0.4$ ft/mo (feet per month) and $r = 2$ ft, so

$$\frac{dA}{dt} = 2\pi \, (2 \text{ ft}) \left(0.4 \, \frac{\text{ft}}{\text{mo}} \right)$$

$$= 1.6\pi \, \frac{\text{ft}^2}{\text{mo}}$$

$$\approx 5.03 \text{ ft}^2/\text{mo}.$$

EXAMPLE 4 *Photosynthesis.* Through photosynthesis, a plant converts carbon dioxide into organic matter. In a tropical rain forest, trees annually produce approximately 2200 g of organic matter for every square meter. This number is called the *productivity* of the forest. Due to environmental changes, such as increased availability of carbon dioxide, the productivity may change over time.

Suppose that the productivity P of a forest is currently 2200 g/m² and that the forest currently covers an area A of 200,000,000 m². Suppose also that the forest is losing 1,000,000 m² each year, but the productivity is increasing by 1.5 g/(m²-yr). What is the rate of change of the quantity of organic matter produced by the trees of the forest?

Solution   Let Q be the total amount of organic matter produced by the forest. Then Q is the product of the number of square meters A in the forest and the productivity P, or

$$Q = AP.$$

We take the derivative of both sides with respect to t using the Product Rule:

$$\frac{dQ}{dt} = A\frac{dP}{dt} + \frac{dA}{dt}P.$$

At the moment in question, $A = 200,000,000$ m², $dA/dt = -1,000,000$ m²/yr, $P = 2200$ g/m², and $dP/dt = 1.5$ g/(m²·yr). Substituting, we conclude

$$\frac{dQ}{dt} = (200,000,000 \ m^2)\left(1.5\ \frac{g}{m^2 \cdot yr}\right) + \left(-1,000,000\ \frac{m^2}{yr}\right)\left(2200\ \frac{g}{m^2}\right)$$

$$= -1,900,000,000\ \frac{g}{yr}.$$

In other words, the forest will produce approximately 1,900,000,000 fewer grams of organic matter next year than it produced this year.   ■

Exercise Set 3.7

Differentiate implicily to find dy/dx. Then find the slope of the curve at the given point.

1. $xy - x + 2y = 3;\quad \left(-5, \dfrac{2}{3}\right)$

2. $xy + y^2 - 2x = 0;\quad (1, -2)$

3. $x^2 + y^2 = 1;\quad \left(\dfrac{1}{2}, \dfrac{\sqrt{3}}{2}\right)$

4. $x^2 - y^2 = 1;\quad \left(\sqrt{3}, \sqrt{2}\right)$

5. $x^2y - 2x^3 - y^3 + 1 = 0;\quad (2, -3)$

6. $4x^3 - y^4 - 3y + 5x + 1 = 0;\quad (1, -2)$

7. $\sin y + x^2 = \cos y;\quad (1, 2\pi)$

8. $\tan^2 x = \dfrac{2}{3}\cos y;\quad \left(\dfrac{\pi}{6}, \dfrac{\pi}{3}\right)$

9. $x \sin x = y(1 + \cos y)$; $\left(\dfrac{\pi}{2}, \dfrac{\pi}{3}\right)$

10. $\sin^2 x + 3 \cos^2 y = 1$; $\left(\dfrac{\pi}{6}, \dfrac{2\pi}{3}\right)$

Differentiate implicitly to find dy/dx.

11. $2xy + 3 = 0$

12. $x^2 + 2xy = 3y^2$

13. $x^2 - y^2 = 16$

14. $x^2 + y^2 = 25$

15. $y^5 = x^3$

16. $y^3 = x^5$

17. $x^2y^3 + x^3y^4 = 11$

18. $x^3y^2 - x^5y^3 = -19$

19. $\sqrt{x} + \sqrt{y} = 1$

20. $\dfrac{1}{x^2} + \dfrac{1}{y^2} = 5$

21. $y^3 = \dfrac{x - 1}{x + 1}$

22. $y^2 = \dfrac{x^2 - 1}{x^2 + 1}$

23. $x^{3/2} + y^{2/3} = 1$

24. $(x - y)^3 + (x + y)^3 = x^5 + y^5$

25. $\dfrac{x^2y + xy + 1}{2x + y} = 1$ (*Hint:* Simplify first.)

26. $\dfrac{xy}{x + y} = 2$

27. $4 \sin x \cos y = 3$

28. $x \tan^2 y = y^3$

29. $x + y = \sin(\sqrt{y - x})$

30. $\sin(xy) = \cos y$

31. Two variable quantities A and B are found to be related by the equation

$$A^3 + B^3 = 9.$$

What is the rate of change dA/dt at the moment when $A = 2$ and $dB/dt = 3$?

32. Two variable quantities G and H, nonnegative, are found to be related by the equation

$$G^2 + H^2 = 25.$$

What is the rate of change dH/dt when $dG/dt = 3$ and $G = 0$? $G = 1$? $G = 3$?

APPLICATIONS

33. *Rate of Change of a Tumor.* The volume of a tumor is given by

$$V = \frac{4}{3}\pi r^3.$$

The radius is increasing at the rate of 0.03 centimeter per day (cm/day) at the moment when $r = 1.2$ cm. How fast is the volume changing at that moment?

34. *Rate of Change of a Healing Wound.* The area of a healing wound is given by

$$A = \pi r^2.$$

The radius is decreasing at the rate of 1 millimeter per day (-1 mm/day) at the moment when $r = 25$ mm. How fast is the area decreasing at that moment?

Poiseuille's Law. The flow of blood in a blood vessel is faster toward the center of the vessel and slower toward the outside. The speed of the blood V is given by

$$V = \frac{p}{4Lv}(R^2 - r^2),$$

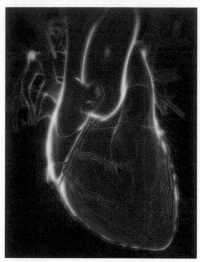

The flow of blood in a blood vessel can be modeled by Poiseuille's Law.

where R is the radius of the blood vessel, r is the distance of the blood from the center of the vessel, and p, L, and v are physical constants related to pressure, length, and viscosity of the blood vessels, respectively. Use this formula for Exercises 35 and 36.

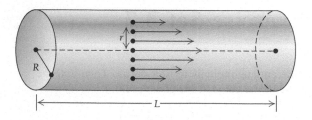

35. Assume that r is a constant as well as p, L, and v.

a) Find the rate of change dV/dt in terms of R and dR/dt when $L = 1$ mm, $p = 100$, and $v = 0.05$.

b) A person goes out into the cold to shovel snow. Cold air has the effect of contracting blood vessels far from the heart. Suppose that a blood vessel contracts at a rate of

$$\frac{dR}{dt} = -0.0015 \text{ mm/min}$$

at a place in the blood vessel where the radius $R = 0.0075$ mm. Find the rate of change dV/dt at that location.

36. Assume that r is a constant as well as p, L, and v.

a) Find the rate of change dV/dt in terms of R and dR/dt when $L = 1$ mm, $p = 100$, and $v = 0.05$.

b) When shoveling snow in cold air, a person with a history of heart trouble can develop angina (chest pains) due to contracting blood vessels. To counteract this, he or she may take a nitroglycerin tablet, which dilates the blood vessels. Suppose that after a nitroglycerin tablet is taken, a blood vessel dilates at a rate of

$$\frac{dR}{dt} = 0.0025 \text{ mm/min}$$

at a place in the blood vessel where the radius $R = 0.02$ mm. Find the rate of change dV/dt.

Body Surface Area. Certain chemotherapy dosages depend on a patient's surface area. According to the Mosteller model,[17]

$$S = \frac{\sqrt{hw}}{60},$$

where h is the person's height in centimeters, w is the person's weight in kilograms, and S is the approximation to the person's surface area in m^2. Use this formula in Exercises 37 and 38.

37. Assume that a male's height is a constant 180 cm, but he is on a diet. If he loses 4 kg per month, how fast is his surface area decreasing at the instant he weighs 85 kg?

38. Assume that a female's height is a constant 160 cm, but she is on a diet. If she loses 3 kg per month, how fast is her surface area decreasing at the instant she weighs 60 kg?

39. Two cars start from the same point at the same time. One travels north at 25 mph, and the other travels east at 60 mph. How fast is the distance between them increasing at the end of 1 hr?

[17]http://www.halls.md/body-surface-area/refs.htm

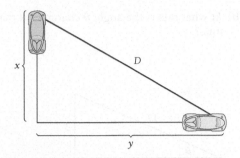

40. A ladder 26 ft long leans against a vertical wall. If the lower end is being moved away from the wall at the rate of 5 ft/sec, how fast is the height of the top decreasing (this will be a negative rate) when the lower end is 10 ft from the wall?

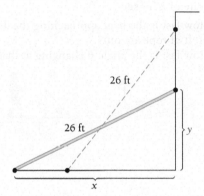

41. An observer watches a hot-air balloon rise 100 m from its liftoff point. At the moment that the angle is $\pi/6$, the angle is increasing at the rate of 0.1 rad/min. How fast is the balloon rising at that moment?

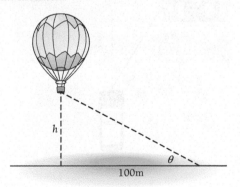

42. A 17-ft ladder is leaning against a house when its base starts to slide away. By the time the base is 8 ft from the house, the base is moving at the rate of 2 ft/sec.

a) At what rate is the top of the ladder falling at that time?

b) At what rate is the angle θ changing at that time?

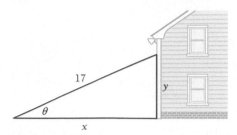

43. A boat is being pulled into a dock by a rope. The dock is 3 ft above the point where the rope is attached to the boat. The rope is being pulled at the rate of 2 ft / sec.

a) How fast is the boat approaching the dock when 10 ft of rope are out?
b) How fast is the angle θ changing at that time?

44. A police officer sits on the side of the road. A car passes 20 ft from her, traveling at the rate of 75 mph (110 ft / sec). How fast is the angle θ changing at that time?

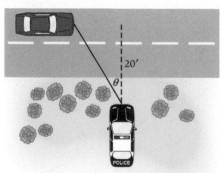

SYNTHESIS

Differentiate implicitly to find dy/dx and d^2y/dx^2.

45. $xy + x - 2y = 4$ 46. $y^2 - xy + x^2 = 5$

47. $x^2 - y^2 = 5$ 48. $x^3 - y^3 = 8$

tw 49. Explain the usefulness of implicit differentiation.

tw 50. Look up the word *implicit* in a dictionary. Explain how that definition can be related to the concept of a function that is defined "implicitly."

 Technology Connection

Use a grapher to graph each of the following equations. On most graphers, equations must be solved for y before they can be entered.

51. $x^2 + y^2 = 4$ *Note:* You will probably need to sketch the graph in two parts:

$$y = \sqrt{4 - x^2} \quad \text{and} \quad y = -\sqrt{4 - x^2}.$$

Then graph the tangent line to the graph at the point $(-1, \sqrt{3})$.

52. $x^4 = y^2 + x^6$

Then graph the tangent line to the graph at the point $(-0.8, 0.384)$.

53. $y^4 = y^2 - x^2$

54. $x^3 = y^2(2 - x)$

55. $y^2 = x^3$

Chapter 3   Summary and Review

Terms to Know

Review Exercises

These exercises are for test preparation. They can also be used as a lengthened practice test. Answers are at the back of the book. The answers also contain bracketed section references, which tell you where to restudy if your answer is incorrect.

Find the relative extrema and points of inflection of the function. List your answers in terms of ordered pairs. Then sketch a graph of the function.

1. $f(x) = 3 - 2x - x^2$

2. $f(x) = x^4 - 2x^2 + 3$

3. $f(x) = \dfrac{-8x}{x^2 + 1}$

4. $f(x) = 4 + (x - 1)^3$

5. $f(x) = \dfrac{1}{2 \sin x + 7}$ on $[0, 2\pi]$

6. $f(x) = 3x^{2/3}$

7. $f(x) = \sin^2 x + 2 \sin x$ on $[0, 2\pi]$

8. $f(x) = 2x + \cos 2x$ on $[0, 2\pi]$

Sketch a graph of the function.

9. $f(x) = \dfrac{-4}{x - 3}$

10. $f(x) = \dfrac{1}{2}x + \dfrac{1}{x}$

11. $f(x) = \dfrac{x^2 - 2x + 2}{x - 1}$

12. $f(x) = \dfrac{x}{x - 2}$

Find the absolute maximum and minimum values of the function, if they exist, over the indicated interval. Where no interval is specified, use the real line.

13. $f(x) = \dfrac{1}{3}x^3 + 3x^2 + 9x + 2$

14. $f(x) = x^2 - 10x + 8;$ $[-2, 6]$

15. $f(x) = 4x^3 - 6x^2 - 24x + 5;$ $[-2, 3]$

16. $f(x) = \dfrac{\sin x}{2 + \sin x + \cos x};$ $[0, 2\pi]$

17. $f(x) = x^2 - \dfrac{2}{x};$ $(-\infty, 0)$

18. $f(x) = 4 \sin^3 x + 3 \sin^2 x;$ $[0, 2\pi]$

19. $f(x) = 9 \tan^2 x + \cot^2 x;$ $(0, \pi/2)$

20. $f(x) = 5x^2 + \dfrac{5}{x^2};$ $(0, \infty)$

21. $f(x) = \cot x - 2 \csc x;$ $(0, \pi)$

22. $f(x) = -x^2 + 5x + 7$

23. Of all numbers whose sum is 60, find the two that have the maximum product.

24. Find the minimum value of $Q = x^2 - 2y^2$, where $x - 2y = 1$.

25. A rectangular box with a square base and a cover is to contain 2500 ft^3. If the cost per square foot for the bottom is \$2, for the top is \$3, and for the sides is \$1, what should the dimensions be in order to minimize the cost?

26. Find the linearization to $f(x) = \dfrac{x}{x-1}$ at $a = 2$.

27. Find the linearization to $f(x) = \dfrac{\sin x}{1 + \cos x}$ at $a = 0$.

28. Use linearization to approximate $\sqrt{63}$.

29. Use Newton's method and $x_1 = 1$ to find an approximate solution of $x^3 - 5x + 3 = 0$.

30. Use Newton's method and $x_1 = 3$ to find an approximate solution of $\sin x = x^2/10$.

31. Find dy/dx if $\sin(y^2) + x = \sqrt{y}$.

32. Differentiate the following implicitly to find dy/dx. Then find the slope of the curve at the given point.

$$2x^3 + 2y^3 = -9xy; \quad (-1, -2)$$

33. A ladder 25 ft long leans against a vertical wall. If the lower end is being moved away from the wall at the rate of 6 ft/sec, how fast is the height of the top decreasing when the lower end is 7 ft from the wall?

SYNTHESIS

34. Find the absolute maximum and minimum values, if they exist, over the indicated interval.

$$f(x) = (x - 3)^{2/5}; \quad (-\infty, \infty)$$

35. Differentiate implicitly to find dy/dx:

$$(x - y)^4 + (x + y)^4 = x^6 + y^6.$$

36. Find the relative maxima and minima of

$$y = x^4 - 8x^3 - 270x^2.$$

 Technology Connection

Use a grapher to estimate the relative extrema of the function.

37. $f(x) = 3.8x^5 - 18.6x^3$

38. $f(x) = \sqrt[3]{|9 - x^2|} - 1$

39. *Incidence of Breast Cancer.* The following table provides data relating the incidence of breast cancer per 100,000 women of various ages.

Age	Incidence per 100,000
0	0
27	10
32	25
37	60
42	125
47	187
52	224
57	270
62	340
67	408
72	437
77	475
82	460
87	420

Source: National Cancer Institute.

a) Use a grapher with a **REGRESSION** feature to fit linear, quadratic, cubic, and quartic functions to the data.

tw b) Decide which function best fits the data and explain your reasons.

tw c) Determine the domain of the function on the basis of the function and the problem situation and explain.

d) Determine the maximum value of the quartic function on the domain. At what age is the incidence of breast cancer the greatest?

Chapter 3 Test

Find the relative extrema and points of inflection of the function. List your answers in terms of ordered pairs. Then sketch a graph of the function.

1. $f(x) = x^2 - 4x - 5$ **2.** $f(x) = 2x^4 - 4x^2 + 1$

3. $f(x) = (x - 2)^{2/3} - 4$ **4.** $f(x) = \dfrac{16}{x^2 + 4}$

5. $f(x) = \dfrac{\sin x}{2 \sin x - 7}$ on $[0, 2\pi]$

6. $f(x) = \cos^2 x - 2 \sin x$ on $[0, 2\pi]$

7. $f(x) = (x + 2)^3$ **8.** $f(x) = x\sqrt{9 - x^2}$

Sketch a graph of the function.

9. $f(x) = \dfrac{2}{x - 1}$ **10.** $f(x) = \dfrac{-8}{x^2 - 4}$

11. $f(x) = \dfrac{x^2 - 1}{x}$ **12.** $f(x) = \dfrac{x + 2}{x - 3}$

Find the absolute maximum and minimum values of the function, if they exist, over the indicated interval. Where no interval is specified, use the real line.

13. $f(x) = x(6 - x)$

14. $f(x) = x^3 + x^2 - x + 1;$ $\left[-2, \tfrac{1}{2}\right]$

15. $f(x) = \cos^2 x - \sin x;$ $[0, 2\pi]$

16. $f(x) = \sin x(1 + \cos x);$ $[0, 2\pi]$

17. $f(x) = 9 \sec^2 x + \csc^2 x;$ $(0, \pi/2)$

18. $f(x) = \sin^2 x - \tan^2 x;$ $(-\pi/2, \pi/2)$

19. $f(x) = x^2 + \dfrac{128}{x};$ $(0, \infty)$

20. Of all numbers whose difference is 8, find the two that have the minimum product.

21. Minimize $Q = x^2 + y^2$, where $x - y = 10$.

22. From a thin piece of cardboard 60 in. by 60 in., square corners are cut out so the sides can be folded up to make a box. What dimensions will yield a box of maximum volume? What is the maximum volume?

23. Find the linearization to $f(x) = x\sqrt{x + 1}$ at $a = 8$.

24. Find the linearization to $f(x) = (\sin x + 1)^2$ at $a = 0$.

25. Use linearization to approximate $\sqrt{104}$.

26. Use Newton's method and $x_1 = 0$ to find an approximate solution of $x^3 + 3x = -5$.

27. Use Newton's method and $x_1 = 1$ to find an approximate solution of $\cos x = x + \sin x$.

28. Differentiate the following implicitly to find dy/dx. Then find the slope of the curve at the given point.

$$x^3 + y^3 = 9; (1, 2)$$

29. A board 13 ft long leans against a vertical wall. If the lower end is being moved away from the wall at the rate of 0.4 ft/sec, how fast is the upper end coming down when the lower end is 12 ft from the wall?

SYNTHESIS

30. Find the absolute maximum and minimum values of the function, if they exist, over the indicated interval.

$$f(x) = \dfrac{x^2}{1 + x^3}; [0, \infty)$$

 Technology Connection

31. Use a grapher to estimate the relative extrema of the function.

$$f(x) = 5x^3 - 30x^2 + 45x + 5\sqrt{x}$$

Extended Life Science Connection

POLYMORPHISM

Polymorphism is "the coexistence of two or more distinct forms of individuals in the same population."[18] We will use linearization to help understand why natural selection sometimes favors polymorphism.

Suppose that an organism has two possible versions of a gene at some location on a chromosome. We refer to the two versions of the gene as *alleles* A and B. Since an organism inherits two genes at each location on the chromosome, in the population there are three possible *genotypes*, or combinations of alleles, AA, AB, and BB. For example, in snapdragons, the three genotypes are seen as red, pink, and white flowers. We let p denote the proportion of A alleles in the population, and we let q be the proportion of B alleles. Since every allele is either an A or a B, $p + q = 1$, or equivalently, $q = 1 - p$ and $p = 1 - q$.

NATURAL SELECTION AND ALLELES

The Hardy–Weinberg Theorem (see Chapter 10) states that with random mating, the next generation will still have a proportion of p A alleles and q B alleles. Furthermore, the proportion of the individuals of each genotype are p^2 for AA, $2pq$ for AB, and q^2 for BB.

The three genotypes may have different traits that affect the probability that an individual survives to reproduce. For example, if a moth of one genotype has a color that better camouflages it than a moth of a different genotype, then one is more likely to survive adolescence to reproduce. In other words, natural selection dictates if each genotype increases or decreases in the next generation.

To model this observation, we assign weights to each genotype giving the relative likelihood of individuals of each genotype to

reproduce. It turns out that the calculations are simpler if we assign weights of 1 for AB, $1 - a$ for AA, and $1 - b$ for BB, where a and b are both less than 1. (The numbers a and b could either be positive or negative.) If we let $w = p^2(1 - a) + 2pq(1) + q^2(1 - b)$, then the actual proportion of individuals of each genotype that survive to reproduce are

$$\text{AA} \quad \frac{p^2(1 - a)}{w},$$

$$\text{AB} \quad \frac{2pq}{w}, \quad \text{and}$$

$$\text{BB} \quad \frac{q^2(1 - b)}{w}.$$

(We divide by w because the weights $1 - a$, 1, and $1 - b$ are relative weights.)

Since each individual of genotype AA has two A alleles and each genotype AB has one A allele, the second-generation adult population would have a proportion

$$f(p) = \frac{2p^2(1 - a) + 2pq}{2w} \qquad (1)$$

of A alleles. For example, if $f(p) > p$, then the proportion of A alleles will be higher in the second generation than in the first generation.

[18]Campbell and Reece, *Biology*, 6th ed., San Francisco, 2002.

EXERCISES

1. Show that
$$w = 1 - p^2 a - q^2 b.$$
(*Hint:* Look for $(p + q)^2$ and remember that $p + q = 1$.)

2. Show that
$$f(p) = p \frac{p(1 - a) + q}{1 - p^2 a - q^2 b}.$$

EQUILIBRIUM

At *equilibrium,* the proportion of A alleles would remain constant from one adult generation to the next adult generation. To find the values of p where the population is at equilibrium, we solve the equation

$$p = f(p), \quad \text{or equivalently,}$$
$$p = p \frac{p(1 - a) + q}{1 - p^2 a - q^2 b}. \quad (2)$$

EXERCISES

3. Show that both $p = 0$ and $p = 1$ are solutions to equation (2). What would a population look like if $p = 0$? Would the population be polymorphic? What if $p = 1$?

4. Show that the only other possible solution to equation (2) is $p = \dfrac{b}{a + b}$. Would this solution represent a polymorphic population?

5. a) Show that if a and b have the same sign, then p and q are positive; and if a and b have opposite signs, then either p or q is negative.

b) Explain why polymorphism cannot occur if a and b have opposite signs.

6. Compute all the values of p that are at equilibrium given that

a) $a = 0.5$ and $b = 0.5$.

b) $a = -0.5$ and $b = 0.5$.

c) $a = -0.5$ and $b = -0.5$.

STABILITY OF EQUILIBRIUM

An equilibrium is *stable* if, whenever the value of p is close to an equilibrium value, then $f(p)$ (the next adult generation value of p) becomes closer to the equilibrium value. In the next two exercises, we investigate the stability of the equilibrium at $p = 0$. That is, we assume the proportion of A alleles in the population is very small, but positive, and determine if the proportion of A alleles grows or decays.

EXERCISES

7. Replace q with $1 - p$ in equation (1) and show that the linearization of $f(p)$ at $p = 0$ is
$$y = \frac{1}{1 - b} p.$$

Keep in mind that p is the variable, while a and b are constants.

The linearization approximates the function $f(p)$. So, the proportion of A alleles in the next generation will be approximately $\dfrac{1}{1 - b} p$. If b is positive, then $1 - b < 1$, so $\dfrac{1}{1 - b} > 1$. This means that for the next generation, the value of p is multiplied by a number bigger than 1. Consequently, $f(p) > p$ if p is positive and close to 0. In other words, when b is positive, then $p = 0$ does not give a stable equilibrium.

EXERCISES

8. Show that if b is negative, then the equilibrium at $p = 0$ is stable.

9. We now investigate the stability of the equilibrium at $p = 1$. Show that the linearization of $f(p)$ at $p = 1$ is
$$y = 1 + \frac{1}{1 - a} (p - 1).$$

10. Show that the equilibrium at $p = 1$ is unstable if $a > 0$ and is stable if $a < 0$.

The following chart summarizes our findings.

	$a > 0$	$a < 0$
$b > 0$	$p = 0$ and $p = 1$ unstable	Stable at $p = 1$ only
	Stable polymorphism equilibrium exists	No polymorphism; population becomes all AA
$b < 0$	Stable at $p = 0$ only	Stable at $p = 0$ and $p = 1$
	No polymorphism; population becomes all BB	Unstable polymorphism equilibrium exists

Notice that when both $a > 0$ and $b > 0$, both $p = 0$ and $p = 1$ are unstable. In this case, the population tends toward the stable equilibrium of $p = \dfrac{b}{a + b}$. With $a > 0$ and $b > 0$, natural selection favors having two alleles and three genotypes in the population.

EXERCISES

11. Recall that the relative survival likelihoods for genotypes AA, AB, and BB are $1 - a$, 1, and $1 - b$, respectively. What does $a > 0$ and $b > 0$ say about the relative likelihood of each genotype surviving to adulthood? Give an intuitive explanation of why natural selection favors polymorphism if $a > 0$ and $b > 0$.

12. (Challenging.) Show that the linearization of $f(p)$ at $p = \dfrac{b}{a + b}$ is

$$y = \frac{b}{a + b} + \left(1 - \frac{ab}{a + b - ab}\right)$$
$$\times \left(p - \frac{b}{a + b}\right).$$

13. (Challenging.) Show that if a and b are both positive, then the equilibrium at $p = \dfrac{b}{a + b}$ is stable; but if both a and b are negative, then this equilibrium is unstable.

4

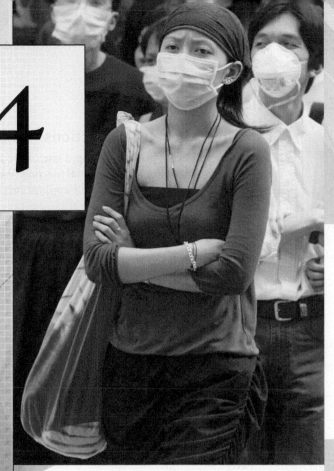

Exponential and Logarithmic Functions

INTRODUCTION *In this chapter, we consider two types of functions that are closely related: exponential functions and logarithmic functions. We will learn how to find derivatives of such functions. Both are rich in applications such as population growth, decay, interest compounded continuously, spread of rumors, and carbon dating.*

AN APPLICATION SARS, a respiratory disease, was first detected in late 2002. Starting in March 2003, the World Health Organization kept track of the number of reported cases throughout the world. The following data shows the total number of cases reported to the WHO through the given day.[1] What is the doubling time?

Date	Days since March 17	Cumulative Number of Reported Cases
March 27	10	1408
March 31	14	1622
April 15	29	3235
April 21	35	3861
April 25	39	4649
May 1	45	5865
May 6	50	6727
May 10	54	7296

This exercise appears as Exercise 33 in the Chapter Summary and Review.

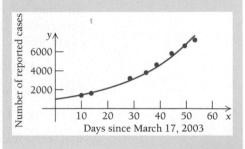

[1]WHO Web page http://www.who.int/csr/sars/country/en/

4.1 Exponential Functions

OBJECTIVES
- Graph exponential functions.
- Differentiate exponential functions.

Graphs of Exponential Functions

Consider the following graph. The rapid rise of the graph indicates that it approximates an *exponential function*. We now consider such functions and many of their applications.

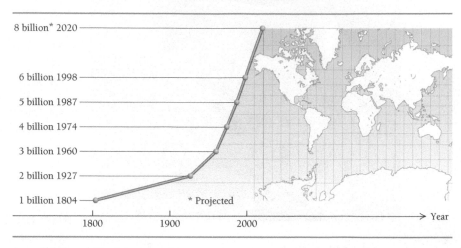

World Population Growth

8 billion* 2020

6 billion 1998

5 billion 1987

4 billion 1974

3 billion 1960

2 billion 1927

1 billion 1804

* Projected

1800   1900   2000   Year

Source: U.S. Bureau of the Census.

Let's review definitions of such expressions as a^x, where x is a rational number. For example,

$$a^{2.34} \quad \text{or} \quad a^{234/100}$$

means "raise a to the 234th power and then take the 100th root $\left(\sqrt[100]{a^{234}}\right)$."

What about expressions with irrational exponents, such as $2^{\sqrt{2}}$, 2^{π}, or $2^{-\sqrt{3}}$? An irrational number is a number named by an infinite, nonrepeating decimal. Let's consider 2^{π}. We know that π is irrational with infinite, nonrepeating decimal expansion:

$$3.141592653\ldots.$$

This means that π is approached as a limit by the rational numbers

$$3, 3.1, 3.14, 3.141, 3.1415, \ldots,$$

so it seems reasonable that 2^{π} should be approached as a limit by the rational powers

$$2^3, 2^{3.1}, 2^{3.14}, 2^{3.141}, 2^{3.1415}, \ldots.$$

Estimating each power with a calculator, we get the following:

$$8, 8.574188, 8.815241, 8.821353, 8.824411, \ldots.$$

In general, a^x is approximated by the values of a^r for rational numbers r near x; a^x is the limit of a^r as r approaches x through rational values. In summary, for $a > 0$, the definition of a^x for rational numbers x can be extended to arbitrary real numbers x in such a way that the usual laws of exponents, such as

$$a^x \cdot a^y = a^{x+y}, \qquad a^x \div a^y = a^{x-y}, \qquad (a^x)^y = a^{xy}, \qquad \text{and} \quad a^{-x} = \frac{1}{a^x},$$

still hold. Moreover, the function so obtained, $f(x) = a^x$, is continuous.

Technology Connection

We can approximate 2^π using the exponential keys on a grapher:

$$2^\pi \approx 8.824977827.$$

EXERCISES

Approximate.

1. 5^π **2.** $5^{\sqrt{3}}$

3. $7^{-\sqrt{2}}$ **4.** $18^{-\pi}$

DEFINITION

An **exponential function** f is given by

$$f(x) = a^x,$$

where x is any real number, $a > 0$, and $a \neq 1$. The number a is called the **base**.

The following are examples of exponential functions:

$$f(x) = 2^x, \qquad f(x) = \left(\tfrac{1}{2}\right)^x, \qquad f(x) = (0.4)^x.$$

Note that in contrast to power functions like $y = x^2$ or $y = x^3$, the variable in an exponential function is in the exponent, not the base. Exponential functions have extensive application. Let's consider their graphs.

EXAMPLE 1   Graph: $y = g(x) = 2^x$.

Solution

a) First, we find some function values.

Note: For:

	$y = g(x)$
x	(or 2^x)
0	1
$\frac{1}{2}$	1.4
1	2
2	4
3	8
-1	$\frac{1}{2}$
-2	$\frac{1}{4}$

$$x = 0, \qquad y = 2^0 = 1;$$
$$x = \tfrac{1}{2}, \qquad y = 2^{1/2} = \sqrt{2} \approx 1.4;$$
$$x = 1, \qquad y = 2^1 = 2;$$
$$x = 2, \qquad y = 2^2 = 4;$$
$$x = 3, \qquad y = 2^3 = 8;$$
$$x = -1, \qquad y = 2^{-1} = \tfrac{1}{2};$$
$$x = -2, \qquad y = 2^{-2} = \frac{1}{2^2} = \tfrac{1}{4}.$$

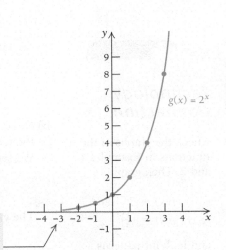

The curve comes very close to the x-axis, but does not touch or cross it. The x-axis is a horizontal asymptote.

b) Next, we plot the points and connect them with a smooth curve. The graph is continuous, increasing, and concave up. We see too that the x-axis is a horizontal asymptote (see Section 3.3); that is,

$$\lim_{x \to -\infty} g(x) = 0 \quad \text{and} \quad \lim_{x \to \infty} g(x) = \infty.$$

EXAMPLE 2 Graph: $y = f(x) = \left(\frac{1}{2}\right)^x$.

Solution

a) First, we find some function values. Before we do so, note that

$$y = f(x) = \left(\frac{1}{2}\right)^x$$
$$= (2^{-1})^x$$
$$= 2^{-x}.$$

This will ease our work.

x	0	$\frac{1}{2}$	1	2	-1	-2	-3
y	1	0.7	$\frac{1}{2}$	$\frac{1}{4}$	2	4	8

Note: For:

$$x = 0, \quad y = 2^{-0} = 1;$$
$$x = \frac{1}{2}, \quad y = 2^{-1/2} = \frac{1}{2^{1/2}}$$
$$= \frac{1}{\sqrt{2}} \approx 0.7;$$
$$x = 1, \quad y = 2^{-1} = \frac{1}{2};$$
$$x = 2, \quad y = 2^{-2} = \frac{1}{4};$$
$$x = -1, \quad y = 2^{-(-1)} = 2;$$
$$x = -2, \quad y = 2^{-(-2)} = 4;$$
$$x = -3, \quad y = 2^{-(-3)} = 8.$$

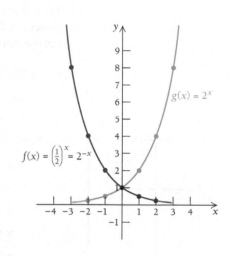

Technology Connection

Check the graphs of the functions in Examples 1 and 2. Then graph

$$f(x) = 3^x \quad \text{and}$$
$$g(x) = \left(\frac{1}{3}\right)^x$$

and look for patterns.

b) Next, we plot these points and connect them with a smooth curve, as shown by the red curve in the figure. The graph is continuous, decreasing, and concave up. We see too that the x-axis is a horizontal asymptote; that is,

$$\lim_{x \to \infty} f(x) = 0 \quad \text{and} \quad \lim_{x \to -\infty} f(x) = \infty.$$

The graph of $g(x) = 2^x$, the blue curve, is shown for comparison. ∎

The following are some properties of the exponential function.

1. The function $f(x) = a^x$, where $a > 1$, is a positive, increasing, continuous function. As x gets smaller, a^x approaches 0. The graph

is concave up, as shown below. The x-axis is a horizontal asymptote.

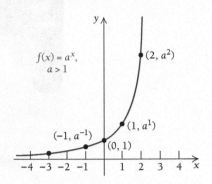

2. The function $f(x) = a^x$, where $0 < a < 1$, is a positive, decreasing, continuous function. As x gets larger, a^x approaches 0. The graph is concave up, as shown below. The x-axis is a horizontal asymptote.

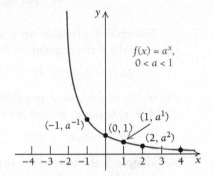

When $a = 1$, $f(x) = a^x = 1^x = 1$, and is a constant function. This is why we do not allow 1 to be the base of an exponential function.

Exponential Growth and the Number e

EXAMPLE 3   Suppose that the number of bacteria in a sample grows exponentially at a rate of $r = 10\%$ every hour. If the initial population is 1000 bacteria, find the number of bacteria in the sample after 1 hr, after 2 hr, and after 10 hr.

Solution   After each hour the population increases by 10% over the previous hour. After 1 hr, the population is

$$1000 + 1000(0.1) = 1000(1 + 0.1)$$
$$= 1000(1.1)$$
$$= 1100 \text{ bacteria.}$$

We see that each hour the population is multiplied by 1.1. The population after 2 hr is

$$1000(1.1)^2 = 1210 \text{ bacteria.}$$

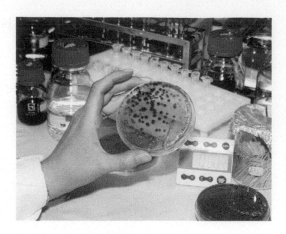

The population after 10 hr is

$$1000(1.1)^{10} \approx 1000(2.594)$$
$$= 2594 \text{ bacteria.}$$

    Example 3 provides an example of *exponential growth*. A population *grows exponentially* if the population $N(t)$ at time t is

$$N(t) = N_0(1 + r)^t,$$

where N_0 is the initial population size. The *base* of the exponential function is $a = 1 + r$. Using a, we can rewrite the exponential growth function as

$$N(t) = N_0 a^t.$$

In Example 3, the base of the exponential function is 1.1.

    Recall from Chapter 2 that there is a difference between average growth rate and instantaneous growth rate. In Example 3, the growth rate of 10% is not the instantaneous growth rate, but rather the average rate over an hour. We now investigate the *instantaneous* growth rate for an exponential function.

    If a population's relative growth is divided equally n times per hour, then there will be nt stages of population growth, and the growth rate for each stage will be $\dfrac{r}{n}$. Therefore, the population growth follows the function

$$N(t) = N_0\left(1 + \frac{r}{n}\right)^{nt},$$

where, as above, N_0 is the population size at time $t = 0$, r is the relative growth rate, and t is time. We look closely at the case $r = 1$:

$$N(t) = N_0\left(1 + \frac{1}{n}\right)^{nt}.$$

    If we take $n = 2$, then

$$N(1) = N_0\left(1 + \frac{1}{2}\right)^{2 \cdot 1}$$
$$= N_0(1.5)^2$$
$$= N_0(2.25).$$

Instead of growing 100% over the hour, the population grew at a relative rate of $2.25 - 1 = 1.25$, or 125%.

Let's try dividing the growth rate equally among the 60 minutes of an hour; that is, we let $n = 60$.

$$N(1) = N_0\left(1 + \frac{1}{60}\right)^{60\cdot1}$$

$$\approx N_0(1.06666667)^{60}$$

$$\approx N_0(2.69597) \qquad \text{Using a scientific calculator for approximation}$$

We approximate a population that has an *instantaneous* growth rate of 100% per hour by dividing the growth evenly into shorter and shorter time intervals. We seek

$$\lim_{n\to\infty} N_0\left(1 + \frac{1}{n}\right)^n.$$

The table in the margin shows the value of $\left(1 + \frac{1}{n}\right)^n$ for very large values of n.

n	$\left(1 + \dfrac{1}{n}\right)^n$
10	2.593742460
100	2.704813829
1000	2.716923932
10,000	2.718145927
100,000	2.718268237
1,000,000	2.718280469
10,000,000	2.718281693
100,000,000	2.718181815
1,000,000,000	2.718281827
10,000,000,000	2.718281828
100,000,000,000	2.718281828

To visualize this data, we plot the function $y = \left(1 + \frac{1}{x}\right)^x$. Note that there is a horizontal asymptote at approximately $y = 2.72$.

We note in the table that when n is very large, the value of $\left(1 + \frac{1}{n}\right)^n$ becomes close to 2.718281828. Based on this observation, we make the following definition.

DEFINITION

$$e = \lim_{n\to\infty}\left(1 + \frac{1}{n}\right)^n$$

$$\approx 2.718281828.$$

We have just seen that if a population has an instantaneous growth rate of 100% per hour, then the population is given by $N(t) = N_0e^t$, where N_0 is the initial population and t is time.

CAUTION   The number e is irrational. This means that the decimal expansion of e neither stops nor continues to repeat the same pattern. In fact, a closer approximation for e is 2.7182818284590450908.

Even though e is irrational, it is usually the preferred base for exponential functions because of the simple formula for its derivative, which we will now investigate.

The Derivative of the Exponential Function

Let's find the formula for the derivative of the function $f(x) = e^x$. We use the definition of the derivative:

$$f'(x) = \lim_{h \to 0} \frac{f(x + h) - f(x)}{h} \qquad \text{Definition of derivative}$$

$$= \lim_{h \to 0} \frac{e^{x+h} - e^x}{h} \qquad \text{Substituting for } f(x + h) \text{ and } f(x)$$

$$= \lim_{h \to 0} \frac{e^x(e^h - 1)}{h}$$

$$= \lim_{h \to 0} e^x \frac{e^h - 1}{h}$$

$$= e^x \left[\lim_{h \to 0} \frac{e^h - 1}{h} \right]. \qquad \text{Since } e^x \text{ does not depend on } h$$

In order to complete the calculation of $\dfrac{d}{dx} e^x$, we need to compute

$$\lim_{h \to 0} \frac{e^h - 1}{h}.$$

We compute $(e^h - 1)/h$ for various values of h in the table, and we plot the graph of $y = \dfrac{e^h - 1}{h}$.

h	$\dfrac{e^h - 1}{h}$
0.1	1.0517092
0.01	1.0050167
0.001	1.0005002
0.0001	1.0000500
0.00001	1.0000050
0.000001	1.0000005

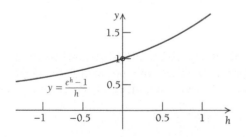

We can see from both the table and the graph that as h gets very close to 0, $(e^h - 1)/h$ becomes very close to 1. That is,

$$\lim_{h \to 0} \frac{e^h - 1}{h} = 1.$$

Thus,

$$f'(x) = e^x \cdot \left[\lim_{h \to 0} \frac{e^h - 1}{h} \right]$$

$$= e^x \cdot 1$$

$$= e^x,$$

and we get the following theorem.

THEOREM 1

$$\frac{d}{dx} e^x = e^x$$

In retrospect, the formula for the derivative of e^x is expected. If $N(t) = e^t$, then the instantaneous growth rate is 100%. Since the instantaneous growth rate is measured by the derivative, the derivative should be 100% of $N(t)$.

Let's find some other derivatives.

EXAMPLE 4

$$\frac{d}{dx}3e^x = 3\frac{d}{dx}e^x$$
$$= 3e^x$$

■

EXAMPLE 5

$$\frac{d}{dx}(x^2 e^x) = x^2 \cdot e^x + 2x \cdot e^x \qquad \text{By the Product Rule}$$
$$= xe^x(x + 2) \qquad \text{Factoring}$$

■

EXAMPLE 6

$$\frac{d}{dx}\left(\frac{e^x}{x^3}\right) = \frac{x^3 \cdot e^x - 3x^2 \cdot e^x}{x^6} \qquad \text{By the Quotient Rule}$$
$$= \frac{x^2 e^x(x - 3)}{x^6} \qquad \text{Factoring}$$
$$= \frac{e^x(x - 3)}{x^4} \qquad \text{Simplifying}$$

■

EXAMPLE 7

$$\frac{d}{dx}\sin e^x = (\cos e^x)\frac{d}{dx}e^x \qquad \text{Chain Rule}$$
$$= (\cos e^x) \cdot e^x$$

■

Technology Connection

Check the results of Examples 4–7 using a grapher. This assumes that you have a grapher that graphs f and f'.

Then differentiate $f(x) = e^x/x^2$ and check your answer using the grapher.

Suppose that we have a more complicated function in the exponent, such as

$$h(x) = e^{x^2 - 5x}.$$

This is a composition of functions. In general, we have

$$h(x) = e^{f(x)} = g[f(x)], \quad \text{where} \quad g(x) = e^x.$$

Now $g'(x) = e^x$. Then by the Chain Rule (Section 2.8), we have

$$h'(x) = g'[f(x)] \cdot f'(x)$$
$$= e^{f(x)} \cdot f'(x).$$

For the case above, $f(x) = x^2 - 5x$, so $f'(x) = 2x - 5$. Then

$$h'(x) = g'[f(x)] \cdot f'(x)$$
$$= e^{f(x)} \cdot f'(x)$$
$$= e^{x^2 - 5x}(2x - 5).$$

The next rule, which we have proven using the Chain Rule, allows us to find derivatives of functions like the one above.

THEOREM 2

$$\frac{d}{dx} e^{f(x)} = e^{f(x)} f'(x),$$

or

$$\frac{d}{dx} e^u = e^u \cdot \frac{du}{dx}$$

The derivative of e to some power is the derivative of the power times e to the power.

EXAMPLE 8

$$\frac{d}{dx} e^{5x} = 5e^{5x}$$

EXAMPLE 9

$$\frac{d}{dx} e^{-x^2+4x-7} = (-2x + 4)e^{-x^2+4x-7}$$

EXAMPLE 10

$$\frac{d}{dx} e^{\sqrt{x^2-3}} = \frac{d}{dx} e^{(x^2-3)^{1/2}}$$

$$= \frac{1}{2}(x^2 - 3)^{-1/2} \cdot 2x \cdot e^{(x^2-3)^{1/2}}$$

$$= x(x^2 - 3)^{-1/2} \cdot e^{\sqrt{x^2-3}}$$

$$= \frac{xe^{\sqrt{x^2-3}}}{\sqrt{x^2 - 3}}$$

EXAMPLE 11   Let $f(x) = (1 + \cos x)e^{\sin 2x}$. Find $f'(x)$.

Solution

$$f'(x) = \frac{d}{dx}[(1 + \cos x)e^{\sin 2x}]$$

$$= -(\sin x)e^{\sin 2x} + (1 + \cos x)e^{\sin 2x}\left(\frac{d}{dx}\sin 2x\right) \quad \text{Product Rule}$$

$$= -\sin x\, e^{\sin 2x} + (1 + \cos x)e^{\sin 2x}(\cos 2x) \cdot 2$$

$$= -\sin x\, e^{\sin 2x} + (2\cos 2x + 2\cos x \cos 2x)e^{\sin 2x}$$

$$= (2\cos 2x + 2\cos x \cos 2x - \sin x)e^{\sin 2x}$$

Graphs of e^x, e^{-x}, and $1 - e^{-kx}$

Now that we know how to find the derivative of $f(x) = e^x$, let's look at the graph of $f(x) = e^x$ from the standpoint of calculus concepts and the curve-sketching techniques discussed in Section 3.2.

EXAMPLE 12   Graph: $f(x) = e^x$. Analyze the graph using calculus.

Solution   If all we want is a quick graph, we might simply find some function values using a calculator, plot the points, and sketch the graph as shown below.

x	$f(x)$
-2	0.135
-1	0.368
0	1
1	2.718
2	7.389

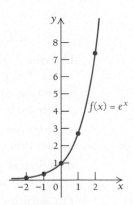

However, we can analyze the graph using calculus as follows.

a) *Derivatives.* Since $f(x) = e^x$, it follows that $f'(x) = e^x$, so $f''(x) = e^x$.

b) *Critical points of f.* Since $f'(x) = e^x > 0$ for all real numbers x, we know that the derivative exists for all real numbers and there is no solution of the equation $f'(x) = 0$. There are no critical points and therefore no maximum or minimum values.

c) *Increasing.* We have $f'(x) = e^x > 0$ for all real numbers x, so the function f is increasing over the entire real line, $(-\infty, \infty)$.

d) *Inflection points.* We have $f''(x) = e^x > 0$ for all real numbers x, so the equation $f''(x) = 0$ has no solution and there are no points of inflection.

e) *Concavity.* Since $f''(x) = e^x > 0$ for all real numbers x, the function f' is increasing and the graph is concave up over the entire real line.   ■

EXAMPLE 13   Graph: $g(x) = e^{-x}$. Analyze the graph using calculus.

Solution   First, we find some function values using a calculator, plot the points, and sketch the graph as shown below.

x	$g(x)$
-2	7.389
-1	2.718
0	1
1	0.368
2	0.135

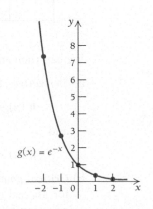

We can then analyze the graph using calculus as follows.

a) *Derivatives.* Since $g(x) = e^{-x}$, we have

$$g'(x) = (-1)e^{-x} = -e^{-x},$$

so

$$g''(x) = (-1)(-1)e^{-x} = e^{-x}.$$

b) *Critical points of g.* Since the expression $e^{-x} > 0$, the derivative $g'(x) = -e^{-x} < 0$ for all real numbers x. Thus the derivative exists for all real numbers, and the equation $g'(x) = 0$ has no solution. There are no critical points and therefore no maximum or minimum values.

c) *Decreasing.* Since the derivative $g'(x) = -e^{-x} < 0$ for all real numbers x, the function g is decreasing over the entire real line, $(-\infty, \infty)$.

d) *Inflection points.* We have $g''(x) = e^{-x} > 0$, so the equation $g''(x) = 0$ has no solution and there are no points of inflection.

e) *Concavity.* We also know that since $g''(x) = e^{-x} > 0$ for all real numbers x, the function g' is increasing and the graph is concave up over the entire real line.

◼

Functions of the type $f(x) = 1 - e^{kx}$, $x \geq 0$, also have important applications.

EXAMPLE 14   Graph: $h(x) = 1 - e^{-2x}$, $x \geq 0$. Analyze the graph using calculus.

Solution   First, we find some function values using a calculator, plot the points, and sketch the graph as shown below.

x	$h(x)$
0	0
1	0.865
2	0.982
3	0.998
4	0.9997
5	0.99995

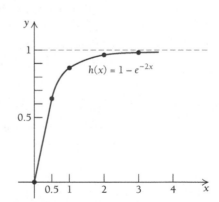

$h(x) = 1 - e^{-2x}$

We can then analyze the graph using calculus as follows.

a) *Derivatives.* Since $h(x) = 1 - e^{-2x}$,

$$h'(x) = (-2)(-1)e^{-2x} = 2e^{-2x}$$

and

$$h''(x) = (-2)(2e^{-2x}) = -4e^{-2x}.$$

b) *Critical points.* Since the expression $e^{-2x} > 0$, the derivative $h'(x) = 2e^{-2x} > 0$ for all real numbers x. Thus the derivative exists for all real numbers, and the equation $h'(x) = 0$ has no solution. There are no critical points.

c) *Increasing.* Since the derivative $h'(x) = 2e^{-2x} > 0$ for all real numbers x, h is increasing over the entire real line.

d) *Inflection points.* Since $h''(x) = -4e^{-2x} < 0$, we know that the equation $h''(x) = 0$ has no solution, so there are no points of inflection.

e) *Concavity.* Since $h''(x) = -4e^{-2x} < 0$ for all real numbers x, the function h' is decreasing and the graph is concave down over the entire real line.   ■

In general, the graph of $h(x) = 1 - e^{-kx}$, for $k > 0$ and $x \geq 0$, is increasing, which we expect since $h'(x) = ke^{-kx}$ is always positive. We also see that $h(x)$ approaches 1 as x approaches ∞; that is, $\lim\limits_{x \to \infty} (1 - e^{-kx}) = 1$.

CAUTION   Functions of the type a^x (for example, 2^x, 3^x, and e^x) are different from functions of the type x^a (for example, x^2, x^3, $x^{1/2}$). For a^x, the variable is in the exponent. For x^a, the variable is in the base. The derivative of a^x is not xa^{x-1}. In particular, we have the following:

$$\frac{d}{dx}e^x \neq xe^{x-1}, \quad \text{but} \quad \frac{d}{dx}e^x = e^x.$$

Exercise Set 4.1

Graph.

1. $y = 4^x$

2. $y = 5^x$

3. $y = (0.4)^x$

4. $y = (0.2)^x$

5. $x = 4^y$

6. $x = 5^y$

A population grows exponentially at an average growth rate of r percent each year and an initial population of N_0. Compute the population at times $t = 1$, $t = 2$, and $t = 5$ given

7. $N_0 = 1000, r = 20\%$.

8. $N_0 = 50, r = 5\%$.

9. $N_0 = 286{,}000{,}000, r = 2.1\%$.

10. $N_0 = 50{,}000{,}000, r = 1.9\%$.

Differentiate.

11. $f(x) = e^{3x}$

12. $f(x) = e^{2x}$

13. $f(x) = 5e^{-2x}$

14. $f(x) = 4e^{-3x}$

15. $f(x) = 3 - e^{-x}$

16. $f(x) = 2 - e^{-x}$

17. $f(x) = -7e^x$

18. $f(x) = -4e^x$

19. $f(x) = \frac{1}{2}e^{2x}$

20. $f(x) = \frac{1}{4}e^{4x}$

21. $f(x) = x^4 e^x$

22. $f(x) = x^5 e^x$

23. $f(x) = (x^2 + 3x - 9)e^x$

24. $f(x) = (x^2 - 2x + 2)e^x$

25. $f(x) = (\sin x)e^x$

26. $f(x) = (\cos x)e^x$

27. $f(x) = \dfrac{e^x}{x^4}$

28. $f(x) = \dfrac{e^x}{x^5}$

29. $f(x) = e^{-x^2 + 7x}$

30. $f(x) = e^{-x^2 + 8x}$

31. $f(x) = e^{-x^2/2}$

32. $f(x) = e^{x^2/2}$

33. $y = e^{\sqrt{x-7}}$

34. $y = e^{\sqrt{x-4}}$

35. $y = \sqrt{e^x - 1}$

36. $y = \sqrt{e^x + 1}$

37. $y = \tan(e^x + 1)$

38. $y = \sin e^{2x}$

39. $y = e^{\tan x}$

40. $y = e^{\cos x}$

41. $y = (2x + \cos x)e^{3x+1}$

42. $y = (\sec x + \tan x)e^x$

43. $y = xe^{-2x} + e^{-x} + x^3$

44. $y = e^x + x^3 - xe^x$

45. $y = 1 - e^{-x}$

46. $y = 1 - e^{-3x}$

47. $y = 1 - e^{-kx}$

48. $y = 1 - e^{-mx}$

49. $y = (e^{3x} + 1)^5$

50. $y = (e^{x^2} - 2)^4$

51. $y = \dfrac{e^{3t} - e^{7t}}{e^{4t}}$

52. $y = \sqrt[3]{e^{3t} + t}$

53. $y = \dfrac{e^x}{x^2 + 1}$

54. $y = \dfrac{e^x}{1 - e^x}$

55. $f(x) = e^{\sqrt{x}} + \sqrt{e^x}$

56. $f(x) = \dfrac{1}{e^x} + e^{1/x}$

57. $f(x) = e^{x/2} \cdot \sqrt{x - 1}$

58. $f(x) = \dfrac{xe^{-x}}{1 + x^2}$

59. $f(x) = \dfrac{e^x - e^{-x}}{e^x + e^{-x}}$

60. $f(x) = e^{e^x}$

Graph the function. Then analyze the graph using calculus.

61. $f(x) = e^{2x}$

62. $f(x) = e^{(1/2)x}$

63. $f(x) = e^{-2x}$

64. $f(x) = e^{-(1/2)x}$

65. $f(x) = 3 - e^{-x}$, for nonnegative values of x

66. $f(x) = 2(1 - e^{-x})$, for nonnegative values of x

 67.–72. For each of Exercises 61–66, graph the function and its first and second derivatives using a grapher.

73. Find the tangent line to the graph of $f(x) = e^x$ at the point $(0, 1)$.

74. Find the tangent line to the graph of $f(x) = 2e^{-3x}$ at the point $(0, 2)$.

75. and 76. For each of Exercises 73 and 74, graph the function and the tangent line using a grapher.

APPLICATIONS

77. *Medication Concentration.* The concentration C, in parts per million, of a medication in the body t hours after ingestion is given by the function

$$C(t) = 10t^2 e^{-t}.$$

a) Find the concentration after 0 hr; 1 hr; 2 hr; 3 hr; 10 hr.

b) Sketch a graph of the function for $0 \leq t \leq 10$.

c) Find the rate of change of the concentration $C'(t)$.

d) Find the maximum value of the concentration and where it occurs.

tw e) Interpret the meaning of the derivative.

78. *Ebbinghaus Learning Model.* Suppose that you are given the task of learning 100% of a block of knowledge. Human nature tells us that we would retain only a percentage P of the knowledge t weeks after we have learned it. The *Ebbinghaus learning model* asserts that P is given by

$$P(t) = Q + (100\% - Q)\, e^{-kt},$$

where Q is the percentage that we would never forget and k is a constant that depends on the knowledge learned. Suppose that $Q = 40\%$ and $k = 0.7$.

a) Find the percentage retained after 0 wk; 1 wk; 2 wk; 6 wk; 10 wk.

b) Find $\lim_{t \to \infty} P(t)$.

c) Sketch a graph of P.

d) Find the rate of change of P with respect to time t, $P'(t)$.

tw e) Interpret the meaning of the derivative.

Insect Growth. The Logan model for the relationship between the temperature and an organism's growth rate is

$$r(T) = A(e^{b(T-T_L)} - e^{b(T_U - T_L) - c(T_U - T)}),$$

where T is the temperature in degrees Celsius; A, b, and c are constants that depend on the organism; T_L is the lowest temperature that allows the organism to grow; and T_U is the highest temperature that allows the organism to grow. Values for the numbers A, b, c, T_L, and T_U were experimentally determined for the parasitoid *Diglyphus isaea*. (*D. isaea* is a wasp used as a natural insecticide to control leaf miners.) This function will be used in Exercises 79 and 80.[2]

79. During metamorphosis, an insect is in its pupa stage. For male *D. isaea* pupa, $A = 0.153$, $b = 0.141$, $c = 0.153$, $T_L = 9.5°C$, and $T_U = 39.4°C$.

a) Write the function for the Logan model with these parameters.

b) Compute the derivative of the function you obtained in part (a).

80. During metamorphosis, an insect is in its pupa stage. For female *D. isaea* pupa, $A = 0.124$, $b = 0.129$, $c = 0.144$, $T_L = 9.5°C$, and $T_U = 41.5°C$.

a) Write the function for the Logan model with these parameters.

b) Compute the derivative of the function you obtained in part (a).

SYNTHESIS

Differentiate.

81. $y = x\sqrt{x}e^{3x^3 + 2x - 1}$

82. $y = x\sqrt[3]{x^2}e^{x + 1/x}$

83. $y = \sqrt{x} + \sqrt{e^x} + \sqrt{xe^x}$

84. $y = 3x^{-2} - \dfrac{1}{e^{2x}} + 5\sqrt{x} - 2$

85. $y = \sin(\cos e^x)$

86. $y = e^{\sin e^x}$

87. $y = 1 + e^{1 + e^{1 + e^x}}$

88. $y = 1 + e^{1 + e^{1 + e}}$

Each of the following is an expression for e. Find the function values that are approximations for e. Round to five decimal places.

89. $e = \lim_{t \to 0} f(t);\ f(t) = (1 + t)^{1/t}$
Find $f(1), f(0.5), f(0.2), f(0.1)$, and $f(0.001)$.

90. $e = \lim_{t \to 1} g(t);\ g(t) = t^{1/(t-1)}$
Find $g(0.5), g(0.9), g(0.99), g(0.999)$, and $g(0.9998)$.

91. Find the maximum value of $f(x) = x^2 e^{-x}$ over $[0, 4]$.

92. Find the minimum value of $f(x) = xe^x$ over $[-2, 0]$.

---

[2]Bazzocchi et al., "Effects of temperature and host on the pre-imaginal development of the parasitoid *Diglyphus isaea*," *Biological Control*, 26 (2003) 74–82.

93. *Maximum Growth Rate.* The rate of growth of the fungus *Fusarium verticilloides* is
$$F(t) = e^{-(9/(t-15)+0.56/(35-t))},$$
where t is the temperature in degrees Celsius and $15 < t < 35$.[3]

a) Compute $\lim\limits_{t\to 15^+} F(t)$.

b) Compute $\lim\limits_{t\to 35^-} F(t)$.

c) Maximize the function $F(t)$ for $15 < t < 35$. Use parts (a) and (b) to verify that you found a maximum.

94. *Maximum Growth Rate.* The rate of growth of the fungus *Fusarium graminearum* is proportional to
$$F(t) = e^{-(9/(t-15)+0.69/(31-t))},$$
where t is the temperature in degrees Celsius and $15 < t < 31$.[4]

a) Compute $\lim\limits_{t\to 15^+} F(t)$.

b) Compute $\lim\limits_{t\to 31^-} F(t)$.

c) Maximize the function $F(t)$ for $15 < t < 31$. Use parts (a) and (b) to verify that you found a maximum.

Death Rate. The annual death rate (per thousand) in Mexico may be modeled by the function
$$D(t) = 34.4 - \frac{30.48}{1 + 29.44e^{-0.072t}},$$
where t is the number of years after 1900.[5]

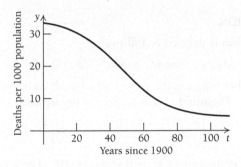

95. a) According to this model, what will Mexico's death rate be in 2010? in 2050?

b) Find and interpret $\lim\limits_{t\to\infty} D(t)$.

96. a) Find $D'(t)$ and interpret its meaning.

b) Between 2007 and 2008, by how much will Mexico's death rate decline?

97. Use Newton's method to solve the equation $5 = 3.2e^{1.07x}$.

98. Use Newton's method to solve the equation $2e^{0.04t} = 6.3t$.

99. *Capture Probability.* The probability that a tadpole with mass 0.1 g is captured by notonectids (aquatic insects) in an experiment is
$$P = 1 - \frac{1}{1 + e^{-(0.055+0.083T)}},$$
where T is the water temperature in degrees Celsius. Suppose that when the water temperature is 20°C, it is rising at the rate of 2 degrees per hour. How fast is the probability of capture decreasing?[6]

100. *Capture Probability.* In an experiment where the water temperature is 25°C, the probability that a tadpole is captured by notonectids (aquatic insects) is
$$P = 1 - \frac{1}{1 + e^{-5.297+31.669m}},$$
where m is the tadpole's mass in grams. Suppose that when the tadpole has a mass of 0.2 g, it is growing at a rate of 0.1 g/wk. How fast is the probability of capture increasing?[7]

[3,4]Stewart, Reid, Nicol, and Schaafsma, "A mathematical simulation of growth of *Fusarium* in Maize ears after artificial inoculation," *Analytical and Theoretical Plant Pathology*, Vol. 92, No. 5, pp. 534–541 (2002).

[5]B. J. L. Berry, L. S. Hall, R. Hernandez-Guerrero, and P. H. Martin, "México's Demographic Transition: Public Policy and Spatial Process," *Population and Environment*, Vol. 21, 363–383 (2000).

[6,7]Anderson, Kiesecker, Chivers, and Blaustein, "The direct and indirect effects of temperature on a predator-prey relationship," *Can. J. Zool*, Vol 79, pp. 1834–1841 (2001).

tw 101. A student made the following error on a test:

$$\frac{d}{dx}e^x = xe^{x-1}.$$

Explain the error and how to correct it.

tw 102. Describe the differences in the graphs of $f(x) = 3^x$ and $g(x) = x^3$.

103. Use the linear approximation for the function $f(x) = (1 + x)^n$ at $x = 0$ and the definition of e to conclude without using a calculator that $e > 2$.

 Technology Connection

104. Graph

$$f(x) = \left(1 + \frac{1}{x}\right)^x.$$

Use the **TABLE** feature and very large values of x to confirm that e is approached as a limit.

Graph each of the following and find the relative extrema.

105. $f(x) = x^2 e^{-x}$ 106. $f(x) = e^{-x^2}$

For each of the following functions, graph f, f', and f''.

107. $f(x) = e^x$ 108. $f(x) = e^{-x}$

109. $f(x) = 2e^{0.3x}$ 110. $f(x) = 1000e^{-0.08x}$

4.2 Logarithmic Functions

OBJECTIVES

■ Convert between exponential and logarithmic equations.
■ Solve exponential equations.
■ Solve problems involving exponential and logarithmic functions.
■ Differentiate functions involving natural logarithms.

Graphs of Logarithmic Functions

Suppose that we want to solve the equation

$$10^y = 1000.$$

We are trying to find that power of 10 that will give 1000. We can see that the answer is 3. The number 3 is called the "logarithm, base 10, of 1000."

DEFINITION

A **logarithm** is defined as follows:

$$y = \log_a x \quad \text{means} \quad x = a^y, \quad a > 0, a \neq 1.$$

The number $\log_a x$ is the power y to which we raise a to get x. The number a is called the **logarithmic base.**

For logarithms base 10, $\log_{10} x$ is the power y such that $x = 10^y$. Therefore, a logarithm can be thought of as an exponent. We can convert from a logarithmic equation to an exponential equation, and conversely, as follows.

Logarithmic Equation	Exponential Equation
$\log_a M = N$	$a^N = M$
$\log_{10} 100 = 2$	$10^2 = 100$
$\log_{10} 0.01 = -2$	$10^{-2} = 0.01$
$\log_{49} 7 = \frac{1}{2}$	$49^{1/2} = 7$

In order to graph a logarithmic equation, we can graph its equivalent exponential equation.

EXAMPLE 1   Graph: $y = \log_2 x$.

Solution   We first write the equivalent exponential equation:

$$x = 2^y.$$

We select values for y and find the corresponding values of 2^y. Then we plot points, remembering that x is still the first coordinate, and connect the points with a smooth curve.

x, or 2^y	y
1	0
2	1
4	2
8	3
$\frac{1}{2}$	-1
$\frac{1}{4}$	-2

① Select y.
② Compute x.

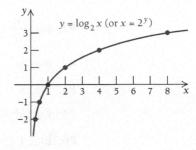

$y = \log_2 x$ (or $x = 2^y$)

Technology Connection

Exploratory

Graph $f(x) = 2^x$. Then use the **TABLE** feature to find the coordinates of points on the graph. How can each ordered pair help you make a hand drawing of a graph of $g(x) = \log_2 x$?

The graphs of $f(x) = 2^x$ and $g(x) = \log_2 x$ are shown below using the same set of axes. Note that we can obtain the graph of g by reflecting the graph of f across the line $y = x$. Graphs obtained in this manner are known as *inverses* of each other.

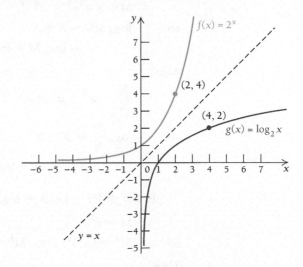

Technology Connection

Graphing Logarithmic Functions

To graph $y = \log_2 x$, we first graph $y = 2^x$. We then use the **DRAWINV** feature. Both graphs are drawn together.

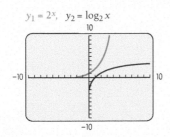

$y_1 = 2^x$, $y_2 = \log_2 x$

EXERCISES

Graph.

1. $y = \log_3 x$ **2.** $y = \log_5 x$
3. $f(x) = \log_e x$ **4.** $f(x) = \log_{10} x$

Although we cannot develop inverses in detail here, it is of interest to note that they "undo" each other. For example,

$$f(3) = 2^3 = 8$$ The input 3 gives the output 8.

and $g(8) = \log_2 8 = 3.$ The input 8 gets us back to 3.

Basic Properties of Logarithms

The following are some basic properties of logarithms. The proofs follow from properties of exponents.

THEOREM 3

Properties of Logarithms

For any positive numbers M, N, and a, $a \neq 1$, and any real number k:

P1. $\log_a MN = \log_a M + \log_a N$

P2. $\log_a \dfrac{M}{N} = \log_a M - \log_a N$

P3. $\log_a M^k = k \cdot \log_a M$

P4. $\log_a a = 1$

P5. $\log_a a^k = k$

P6. $\log_a 1 = 0$

P7. $a^{\log_a M} = M$

Proof of P1 and P2: Let $X = \log_a M$ and $Y = \log_a N$. Writing the equivalent exponential equations, we then have

$$M = a^X \quad \text{and} \quad N = a^Y.$$

Then by the properties of exponents, we have

$$MN = a^X a^Y = a^{X+Y},$$

so $\quad \log_a MN = X + Y$

$$= \log_a M + \log_a N.$$

Also,

$$\frac{M}{N} = a^X \div a^Y = a^{X-Y},$$

so $\quad \log_a \dfrac{M}{N} = X - Y$

$$= \log_a M - \log_a N.$$

Proof of P3: Let $X = \log_a M$. Then

$$M = a^X,$$

so $\quad M^k = (a^X)^k, \quad \text{or} \quad M^k = a^{Xk}.$

Thus,

$$\log_a M^k = Xk = kX = k \cdot \log_a M.$$

Proof of P4: $\log_a a = 1$ because $a^1 = a$.

Proof of P5: $\log_a (a^k) = k$ because $(a^k) = a^k$.

Proof of P6: $\log_a 1 = 0$ because $a^0 = 1$.

Proof of P7:   Let $y = \log_a M$. Then

$$M = a^y \qquad \text{By definition of } \log_a M$$
$$= a^{\log_a M}. \qquad \text{Substituting for } y$$

Let's illustrate these properties.

EXAMPLE 2   Given

$$\log_a 2 = 0.301 \quad \text{and} \quad \log_a 3 = 0.477,$$

find each of the following.

a) $\log_a 6$

$$\begin{aligned}
\log_a 6 &= \log_a (2 \cdot 3) \\
&= \log_a 2 + \log_a 3 \qquad \text{By P1} \\
&= 0.301 + 0.477 \\
&= 0.778
\end{aligned}$$

b) $\log_a \frac{2}{3}$

$$\begin{aligned}
\log_a \frac{2}{3} &= \log_a 2 - \log_a 3 \qquad \text{By P2} \\
&= 0.301 - 0.477 \\
&= -0.176
\end{aligned}$$

c) $\log_a 81$

$$\begin{aligned}
\log_a 81 &= \log_a 3^4 \\
&= 4 \log_a 3 \qquad \text{By P3} \\
&= 4(0.477) \\
&= 1.908
\end{aligned}$$

d) $\log_a \frac{1}{3}$

$$\begin{aligned}
\log_a \frac{1}{3} &= \log_a 1 - \log_a 3 \qquad \text{By P2} \\
&= 0 - 0.477 \qquad \text{By P6} \\
&= -0.477
\end{aligned}$$

e) $\log_a \sqrt{a}$

$$\log_a \sqrt{a} = \log_a a^{1/2} = \tfrac{1}{2} \qquad \text{By P5}$$

f) $\log_a 2a$

$$\begin{aligned}
\log_a 2a &= \log_a 2 + \log_a a \qquad \text{By P1} \\
&= 0.301 + 1 \qquad \text{By P4} \\
&= 1.301
\end{aligned}$$

g) $\log_a 5$

No way to find using just these properties.
$(\log_a 5 \ne \log_a 2 + \log_a 3)$

CAUTION

$$\log_a (M + N) \ne \log_a (M) + \log_a (N)$$
$$\log_a (M + N) \ne \log_a (M) \log_a (N)$$

The logarithm of a sum is neither the sum nor the product of the logarithms.

h) $\dfrac{\log_a 3}{\log_a 2}$

$$\frac{\log_a 3}{\log_a 2} = \frac{0.477}{0.301} \approx 1.58$$

We simply divided and used none of the properties.

Common Logarithms

The number $\log_{10} x$ is called the **common logarithm** of x and is abbreviated $\log x$; that is:

Technology Connection

EXERCISE

1. Graph $f(x) = 10^x$, $y = x$, and $g(x) = \log_{10} x$ using the same set of axes. Then find $f(3)$, $f(0.699)$, $g(5)$, and $g(1000)$.

DEFINITION

For any positive number x,

$$\log x = \log_{10} x.$$

Thus, when we write $\log x$ with no base indicated, base 10 is understood. Note the following comparison of common logarithms and powers of 10.

$1000 = 10^3$	The common	$\log 1000 \;= 3$
$100 = 10^2$	logarithms at	$\log 100 \;\;\;= 2$
$10 = 10^1$	the right follow	$\log 10 \;\;\;\;= 1$
$1 = 10^0$	from the powers at the left.	$\log 1 \;\;\;\;\;= 0$
$0.1 = 10^{-1}$		$\log 0.1 \;\;\;= -1$
$0.01 = 10^{-2}$		$\log 0.01 \;\;= -2$
$0.001 = 10^{-3}$		$\log 0.001 = -3$

Since $\log 100 = 2$ and $\log 1000 = 3$, it seems reasonable that $\log 500$ is somewhere between 2 and 3. Tables were originally used for such approximations, but with the advent of the calculator, that method of finding logarithms is used infrequently. Using a calculator with a $\boxed{\text{log}}$ key, we find that $\log 500 \approx 2.6990$, rounded to four decimal places.

Before calculators and computers became so readily available, common logarithms were used extensively to do certain kinds of computations. In fact, computation is the reason logarithms were developed. Since the standard notation we use for numbers is based on 10, it is logical that base-10, or common, logarithms were used for computations. Today, computations with common logarithms are mainly of historical interest; the logarithmic functions, which use base e, are of modern importance.

Natural Logarithms

The number e, which is approximately 2.718282, was developed in Section 4.1 and has extensive application in many fields. The number $\log_e x$ is called the **natural logarithm** of x and is abbreviated $\ln x$; that is:

DEFINITION

For any positive number x,

$$\ln x = \log_e x.$$

The following is a restatement of the basic properties of logarithms in terms of natural logarithms.

THEOREM 4

P1. $\ln MN = \ln M + \ln N$

P2. $\ln \dfrac{M}{N} = \ln M - \ln N$

P3. $\ln a^k = k \cdot \ln a$

P4. $\ln e = 1$

P5. $\ln e^k = k$

P6. $\ln 1 = 0$

P7. $e^{\ln M} = M$

Let's illustrate these properties.

EXAMPLE 3   Given

$$\ln 2 = 0.6931 \quad \text{and} \quad \ln 3 = 1.0986,$$

find each of the following.

a) $\ln 6$

$$\begin{aligned}
\ln 6 = \ln (2 \cdot 3) &= \ln 2 + \ln 3 \quad \text{By P1}\\
&= 0.6931 + 1.0986\\
&= 1.7917
\end{aligned}$$

b) $\ln 81$

$$\begin{aligned}
\ln 81 = \ln (3^4) &\\
&= 4 \ln 3 \quad \text{By P3}\\
&= 4(1.0986)\\
&= 4.3944
\end{aligned}$$

c) $\ln \frac{2}{3}$

$$\begin{aligned}
\ln \tfrac{2}{3} &= \ln 2 - \ln 3 \quad \text{By P2}\\
&= 0.6931 - 1.0986\\
&= -0.4055
\end{aligned}$$

d) $\ln \frac{1}{3}$

$$\begin{aligned}
\ln \tfrac{1}{3} &= \ln 1 - \ln 3 \quad \text{By P2}\\
&= 0 - 1.0986 \quad \text{By P6}\\
&= -1.0986
\end{aligned}$$

e) $\ln 2e$

$$\begin{aligned}
\ln 2e &= \ln 2 + \ln e \quad \text{By P1}\\
&= 0.6931 + 1 \quad \text{By P4}\\
&= 1.6931
\end{aligned}$$

f) $\ln \sqrt{e^3}$

$$\begin{aligned}
\ln \sqrt{e^3} &= \ln e^{3/2}\\
&= \tfrac{3}{2} \quad \text{By P5}
\end{aligned}$$

Finding Natural Logarithms Using a Calculator

You should have a calculator with a $\boxed{\text{ln}}$ key. You can find natural logarithms directly using this key.

EXAMPLE 4 Find each logarithm on your calculator. Round to six decimal places.

a) $\ln 5.24 = 1.656321$

b) $\ln 0.00001277 = -11.268412$

Exponential Equations

If an equation contains a variable in an exponent, we call the equation *exponential*. We can use logarithms to manipulate or solve **exponential equations.**

EXAMPLE 5 Solve $e^{-0.04t} = 0.05$ for t.

Solution We have

$$\ln e^{-0.04t} = \ln 0.05 \qquad \text{Taking the natural logarithm on both sides}$$

$$-0.04t = \ln 0.05 \qquad \text{By P5}$$

$$t = \frac{\ln 0.05}{-0.04}$$

$$t = \frac{-2.995732}{-0.04}$$

$$t \approx 75.$$

For purposes of space and explanation, we have rounded the value of $\ln 0.05$ to -2.995732 in an intermediate step. When using your calculator, you should find

$$\frac{\ln 0.05}{-0.04}$$

directly, without rounding. Answers at the back of the book have been found in this manner. Remember, the number of places in a table or on a calculator may affect the accuracy of the answer. Usually, your answer should agree to at least three digits.

EXAMPLE 6 Solve $2e^{3x} - 5e^{0.5x} = 0$ for x.

Solution We have

$$2e^{3x} = 5e^{0.5x}$$

$$e^{3x} = \frac{5}{2} e^{0.5x}.$$

We now take the natural logarithm to obtain

$$\ln e^{3x} = \ln\left(\frac{5}{2}e^{0.5x}\right)$$

$$3x = \ln\left(\frac{5}{2}\right) + \ln e^{0.5x} \quad \text{By P1}$$

$$3x \approx 0.91629 + 0.5x \quad \text{By P5 and using a scientific calculator for approximation}$$

$$2.5x \approx 0.91629$$

$$x \approx \frac{0.91629}{2.5}$$

$$x \approx 0.366516.$$

We can check our answer by substituting 0.366516 into the original equation:

$$2e^{3(0.366516)} - 5e^{0.5(0.366516)} \approx -0.0000044 \approx 0.$$

This answer is very close to zero; however, it is not *exactly* zero since we used the approximation $\ln(5/2) \approx 0.91629$. ∎

Technology Connection

Solving Exponential Equations

Let's solve the equation $e^t = 40$ using a grapher.

Method 1: The INTERSECT Feature

We change the variable to x and consider the system of equations $y_1 = e^x$ and $y_2 = 40$. We graph the equations in the viewing window $[-1, 8, -10, 70]$ in order to see the curvature and possible points of intersection.

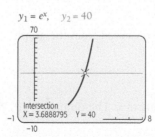

$$y_1 = e^x, \quad y_2 = 40$$

Then we use the INTERSECT feature to find the point of intersection, about $(3.7, 40)$. The x-coordinate, 3.7, is the solution of the original equation $e^t = 40$.

Method 2: The ZERO Feature

We change the variable to x and get a 0 on one side of the equation: $e^x - 40 = 0$. Then we graph $y = e^x - 40$ in the window $[-1, 8, -10, 10]$.

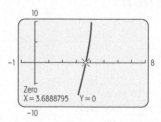

Using the ZERO feature, we see that the x-intercept is about $(3.7, 0)$, so 3.7 is the solution of the original equation $e^t = 40$.

EXERCISES

Solve using a grapher.

1. $e^t = 1000$
2. $e^{-x} = 60$
3. $e^{-0.04t} = 0.05$
4. $e^{0.23x} = 41{,}378$
5. $15e^{0.2x} = 34{,}785.13$

Graphs of Natural Logarithmic Functions

There are two ways in which we might obtain the graph of $y = f(x) = \ln x$. One is by writing its equivalent equation $x = e^y$. Then we select values for y and use a calculator to find the corresponding values of e^y. We then plot points, remembering that x is still the first coordinate. The graph follows.

x, or e^y	y
0.1	-2
0.4	-1
1.0	0
2.7	1
7.4	2
20.1	3

① Select y.

② Compute x.

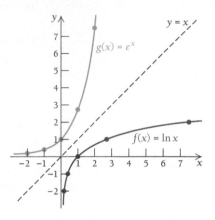

The figure above shows the graph of $g(x) = e^x$ for comparison. Note again that the functions are inverses of each other. That is, the graph of $y = \ln x$, or $x = e^y$, is a reflection, or mirror image, across the line $y = x$ of the graph of $y = e^x$. Any ordered pair (a, b) on the graph of g yields an ordered pair (b, a) on f. Note too that $\lim\limits_{x \to 0^+} \ln x = -\infty$ and the y-axis is a vertical asymptote.

The second method of graphing $y = \ln x$ is to use a calculator to find function values. For example, when $x = 2$, then $y = \ln 2 \approx 0.6931 \approx 0.7$. This gives the pair $(2, 0.7)$ on the graph, which follows.

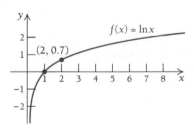

The following properties can be observed from the graph.

Technology Connection

Graphing Logarithmic Functions

To graph $y = \ln x$, we can use the $\boxed{\text{LN}}$ key and enter the function as $Y_1 = \ln(X)$.

THEOREM 5

$\ln x$ exists only for positive numbers x. The domain is $(0, \infty)$.

$\quad \ln x < 0$ for $0 < x < 1$.

$\quad \ln x = 0$ when $x = 1$.

$\quad \ln x > 0$ for $x > 1$.

The range is the entire real line, $(-\infty, \infty)$.

Derivatives of Natural Logarithmic Functions

Technology Connection

Exploratory

Consider $f(x) = \ln x$. Graph f and f' using the viewing window $[0, 10, -3, 3]$. Make an input–output table for the functions. Compare a given x-value with its corresponding y-value for f and f'. What do you observe?

Let's find the derivative of

$$f(x) = \ln x. \tag{1}$$

We first write its equivalent exponential equation:

$$e^{f(x)} = x. \quad \text{By P7} \tag{2}$$

Now we differentiate on both sides of this equation:

$$\frac{d}{dx} e^{f(x)} = \frac{d}{dx} x$$

$$f'(x) \cdot e^{f(x)} = 1 \qquad \text{By the Chain Rule}$$

$$f'(x) \cdot x = 1 \qquad \text{Substituting } x \text{ for } e^{f(x)} \text{ from equation (2)}$$

$$f'(x) = \frac{1}{x}.$$

Thus we have the following.

THEOREM 6

For any positive number x,

$$\frac{d}{dx} \ln x = \frac{1}{x}.$$

Theorem 6 asserts that for the function $f(x) = \ln x$, to find the slope of the tangent line at x, we need only take the reciprocal of x. This is true only for positive values of x, since $\ln x$ is defined only for positive numbers. For negative numbers x, this derivative formula becomes

$$\frac{d}{dx} \ln |x| = \frac{1}{x}.$$

Let's find some derivatives.

EXAMPLE 7

$$\frac{d}{dx} 3 \ln x = 3 \frac{d}{dx} \ln x$$

$$= \frac{3}{x}$$

EXAMPLE 8

$$\frac{d}{dx} (x^2 \ln x + 5x) = x^2 \cdot \frac{1}{x} + 2x \cdot \ln x + 5 \qquad \text{Using the Product Rule on } x^2 \ln x$$

$$= x + 2x \cdot \ln x + 5 \qquad \text{Simplifying}$$

$$= x(1 + 2 \ln x) + 5$$

EXAMPLE 9

$$\frac{d}{dx} \frac{\ln x}{x^3} = \frac{x^3 \cdot (1/x) - (3x^2)(\ln x)}{x^6} \qquad \text{By the Quotient Rule}$$

$$= \frac{x^2 - 3x^2 \ln x}{x^6}$$

$$= \frac{x^2(1 - 3 \ln x)}{x^6} \qquad \text{Factoring}$$

$$= \frac{1 - 3 \ln x}{x^4} \qquad \text{Simplifying} \qquad ■$$

Suppose that we want to differentiate a more complicated function, such as

$$h(x) = \ln (x^2 - 8x).$$

This is a composition of functions. In general, we have

$$h(x) = \ln f(x) = g[f(x)], \quad \text{where} \quad g(x) = \ln x.$$

Now $g'(x) = 1/x$. Then by the Chain Rule (Section 2.8), we have

$$h'(x) = g'[f(x)] \cdot f'(x) = \frac{1}{f(x)} \cdot f'(x).$$

For the above case, $f(x) = x^2 - 8x$, so $f'(x) = 2x - 8$. Then

$$h'(x) = \frac{1}{x^2 - 8x} \cdot (2x - 8) = \frac{2x - 8}{x^2 - 8x}.$$

The following rule, which we have proven using the Chain Rule, allows us to find derivatives of functions like the one above.

THEOREM 7

$$\frac{d}{dx} \ln f(x) = \frac{1}{f(x)} f'(x) = \frac{f'(x)}{f(x)},$$

or

$$\frac{d}{dx} \ln u = \frac{1}{u} \cdot \frac{du}{dx}.$$

The derivative of the natural logarithm of a function is the derivative of the function divided by the function.

EXAMPLE 10

$$\frac{d}{dx} \ln 3x = \frac{3}{3x} = \frac{1}{x}$$

Note that we could have done this another way, using property 1:

$$\ln 3x = \ln 3 + \ln x;$$

then

$$\frac{d}{dx}\ln 3x = \frac{d}{dx}\ln 3 + \frac{d}{dx}\ln x = 0 + \frac{1}{x} = \frac{1}{x}.$$

EXAMPLE 11

$$\frac{d}{dx}\ln(\sec x + \tan x) = \frac{\sec x \tan x + \sec^2 x}{\sec x + \tan x}$$

$$= \frac{\sec x(\tan x + \sec x)}{\sec x + \tan x}$$

$$= \sec x$$

EXAMPLE 12

$$\frac{d}{dx}\ln(\ln x) = \frac{1}{x}\cdot\frac{1}{\ln x} = \frac{1}{x\ln x}$$

EXAMPLE 13

$$\frac{d}{dx}\ln\left(\frac{x^3 + 4}{x}\right) = \frac{d}{dx}\left[\ln(x^3 + 4) - \ln x\right]$$

By P2. This avoids use of the Quotient Rule.

$$= \frac{3x^2}{x^3 + 4} - \frac{1}{x}$$

$$= \frac{3x^2}{x^3 + 4}\cdot\frac{x}{x} - \frac{1}{x}\cdot\frac{x^3 + 4}{x^3 + 4}$$

$$= \frac{(3x^2)x - (x^3 + 4)}{x(x^3 + 4)}$$

$$= \frac{3x^3 - x^3 - 4}{x(x^3 + 4)} = \frac{2x^3 - 4}{x(x^3 + 4)}$$

Technology Connection

Check the results of Examples 10–13 using a grapher. Then differentiate

$$y = \ln\left(\frac{x^5 - 2}{x}\right)$$

and check your answer using the grapher.

Applications

EXAMPLE 14 *Forgetting.* In a psychological experiment, students were shown a set of nonsense syllables, such as POK, RTZ, PDQ, and so on, and asked to recall them every second thereafter. The percentage $R(t)$ who retained the syllables after t seconds was found to be given by the logarithmic learning model

$$R(t) = 80 - 27 \ln t, \quad \text{for } t \geq 1.$$

Strictly speaking, the function is not continuous, but in order to use calculus, we "fill in" the graph with a smooth curve, considering $R(t)$ to be defined for any number $t \geq 1$.

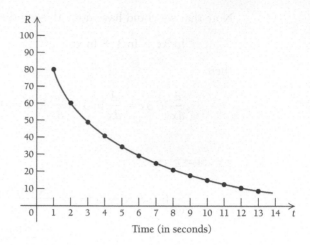

Time (in seconds)

Technology Connection

Graph $y = 80 - 27 \ln x$ using the viewing window $[1, 14, -1, 100]$. Trace along the graph. Describe the meaning of each coordinate in an ordered pair.

a) What percentage retained the syllables after 1 sec?

b) Find $R'(t)$, the rate of change of R with respect to t.

c) Find the maximum and minimum values, if they exist.

Solution

a) $R(1) = 80 - 27 \cdot \ln 1 = 80 - 27 \cdot 0 = 80\%$

b) $R'(t) = -\dfrac{27}{t}$

c) $R'(t)$ exists for all values of t in the interval $[1, \infty)$. Note that for $t \geq 1$, $-27/t < 0$. Thus there are no critical points and R is decreasing. Then R has a maximum value at the endpoint 1. This maximum value is $R(1)$, or 80%. There is no minimum value. ■

Log-Log Plots

Often in science, it is easier to understand the relationship between two quantities by plotting a graph of the logarithm of their values. This allows one to see relationships that are not obvious from the plots of the quantities themselves. The plot of the logarithms of both variables is called the **log-log plot.**

EXAMPLE 15 $\frac{3}{2}$ *Thinning Rule.* In a field with uniform growing conditions, one kind of plant is allowed to grow in each of five identical regions within the field. The data shows the number of plants in each region and the average mass of each plant in kilograms.

Use the data to find a relationship between W, the average mass of a plant, and N, the number of plants growing in a fixed region of land. Then use this relationship to estimate the average mass of a plant if the number of plants were 700.

Solution   We plot a graph using $x = \log N$ and $y = \log W$. The table shows the logarithms of the values for N and the values for W.

Number of Plants (N)	Average Mass of Plant (in kg) (W)
52	0.263
97	0.097
200	0.039
304	0.020
392	0.012
523	0.009
599	0.007
1023	0.003

$x = \log N$	$y = \log W$
1.716	−0.58004
1.9868	−1.01323
2.30103	−1.40894
2.48287	−1.69897
2.59329	−1.92082
2.7185	−2.04576
2.77743	−2.1549
3.00988	−2.52288

We now plot the points (x, y) from the table. Note that the red line appears to be a good fit for the data. Using the points $(3.00988, -2.52288)$ and $(1.716, -0.58004)$ on the line, the slope is

$$m = \frac{-2.52288 - (-0.58004)}{3.00988 - 1.716} \approx -1.5.$$

Using the point–slope form of the equation of a line, we have

$$y - (-2.52288) = -1.5(x - 3.00988)$$
$$y + 2.52288 = -1.5x + 4.51482$$
$$y = -1.5x + 1.9919.$$

Since $x = \log N$ and $y = \log W$, we see that (approximately) $\log W = -1.5 \log N + 2$.

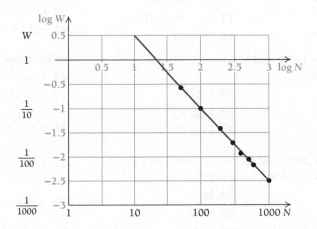

We now convert this relationship between $\log N$ and $\log W$ into one between N and W. To do this, we use property P7:

$$
\begin{aligned}
W &= 10^{\log W} && \text{P7} \\
&= 10^{-1.5 \log N + 2} \\
&= 10^{(\log N)(-1.5)} \times 10^2 \\
&= (10^{\log N})^{-1.5} \times 100 \\
&= 100 \times N^{-1.5}. && \text{P7}
\end{aligned}
$$

We have $W = 100N^{-1.5}$. If $N = 700$, then we would expect

$$W = 100(700)^{-1.5} \approx 0.0054 \text{ kg.}$$

Note that the average weight of a plant is proportional to the number of plants to the $-1.5 = -\frac{3}{2}$ power. In ecology, this observation is called the $\frac{3}{2}$ *thinning rule*. ∎

If positive variables are related by $y = Ax^c$ for some numbers $A > 0$ and c, then $\log y = \log A + c \log x$. The slope of the log-log plot gives the value of c and the y-intercept is $\log A$.

Exercise Set 4.2

Write an equivalent exponential equation.

1. $\log_2 8 = 3$
2. $\log_3 81 = 4$
3. $\log_8 2 = \frac{1}{3}$
4. $\log_{27} 3 = \frac{1}{3}$
5. $\log_a K = J$
6. $\log_a J = K$
7. $-\log_{10} h = p$
8. $-\log_b V = w$

Write an equivalent logarithmic equation.

9. $e^M = b$
10. $e^t = p$
11. $10^2 = 100$
12. $10^3 = 1000$
13. $10^{-1} = 0.1$
14. $10^{-2} = 0.01$
15. $M^p = V$
16. $Q^n = T$

Given $\log_b 3 = 1.099$ and $\log_b 5 = 1.609$, find each of the following.

17. $\log_b 15$
18. $\log_b \frac{3}{5}$
19. $\log_b \frac{1}{5}$
20. $\log_b \sqrt{b^3}$
21. $\log_b 5b$
22. $\log_b 75$

Given $\ln 4 = 1.3863$ and $\ln 5 = 1.6094$, find each of the following. Do not use a calculator.

23. $\ln 20$
24. $\ln \frac{5}{4}$
25. $\ln \frac{1}{4}$
26. $\ln 4e$
27. $\ln \sqrt{e^8}$
28. $\ln 100$

Find the logarithm. Round to six decimal places.

29. $\ln 3927$
30. $\ln 12,000$
31. $\ln 0.0182$
32. $\ln 0.00087$
33. $\ln 8100$
34. $\ln 0.011$

Solve for t.

35. $e^t = 100$
36. $e^t = 1000$
37. $e^t = 60$
38. $e^t = 90$
39. $e^{-t} = 0.1$
40. $e^{-t} = 0.01$
41. $e^{-0.02t} = 0.06$
42. $e^{0.07t} = 2$

Differentiate.

43. $y = -6 \ln x$
44. $y = -4 \ln x$
45. $y = x^4 \ln x - \frac{1}{2}x^2$
46. $y = x^5 \ln x - \frac{1}{4}x^4$
47. $y = \dfrac{\ln x}{x^4}$
48. $y = \dfrac{\ln x}{x^5}$
49. $y = \ln \dfrac{x}{4}$ $\left(Hint: \ln \dfrac{x}{4} = \ln x - \ln 4. \right)$
50. $y = \ln \dfrac{x}{2}$
51. $g(x) = \ln \cos x$
52. $g(x) = \ln \tan x$
53. $f(x) = \ln (\ln 4x)$

54. $f(x) = \ln (\ln 3x)$

55. $f(x) = \ln \left(\dfrac{x^2 - 7}{x} \right)$

56. $f(x) = \ln \left(\dfrac{x^2 + 5}{x} \right)$

57. $f(x) = e^x \ln x$

58. $f(x) = e^{2x} \ln x$

59. $f(x) = e^x \sec x$

60. $f(x) = (\sin 2x)e^x$

61. $f(x) = \ln (e^x + 1)$

62. $f(x) = \ln (e^x - 2)$

63. $f(x) = (\ln x)^2$

64. $f(x) = (\ln x)^3$

65. $y = (\ln x)^{-4}$

66. $y = (\ln x)^n$

67. $f(t) = \ln (t^3 + 1)^5$

68. $f(t) = \ln (t^2 + t)^3$

69. $f(x) = [\ln (x + 5)]^4$

70. $f(x) = \ln [\ln (\ln 3x)]$

71. $f(t) = \ln [(t^3 + 3)(t^2 - 1)]$

72. $f(t) = \ln \dfrac{1 - t}{1 + t}$

73. $y = \ln \dfrac{x^5}{(8x + 5)^2}$

74. $y = \ln \sqrt{5 + x^2}$

75. $f(t) = \dfrac{\ln \sin x}{\sin x}$

76. $f(x) = \dfrac{1}{5} x^5 \left(\ln x - \dfrac{1}{5} \right)$

77. $y = \dfrac{x^{n+1}}{n + 1} \left(\ln x - \dfrac{1}{n + 1} \right)$

78. $y = \dfrac{x \ln x - x}{x^2 + 1}$

79. $y = \ln \left(t + \sqrt{1 + t^2} \right)$

80. $f(x) = \ln \dfrac{1 + \sqrt{x}}{1 - \sqrt{x}}$

81. $y = \sin(\ln x)$

82. $y = \cot(2 \ln x)$

83. $y = (\sin x) \ln (\tan x)$

84. $y = (\cos x) \ln (\cot x)$

85. $y = \ln (\sec 2x + \tan 2x)$

86. $y = \ln (\csc 3x + \cot 3x)$

87. $y = \ln \dfrac{e^x - e^{-x}}{e^x + e^{-x}}$

88. $y = \ln \dfrac{e^x + e^{-x}}{e^x - e^{-x}}$

In Exercises 89–90 construct a log-log plot of the given data. Then approximate a relationship of the form $y = Ax^c$.

89.

x	y
10	164
20	465
30	854
40	1316
50	1839
60	2417
70	3045

90.

x	y
100	0.300
200	0.212
500	0.134
700	0.113
1100	0.0904
1500	0.0775

APPLICATIONS

In Exercises 91–94 construct a log-log plot of the given data. Then approximate a relationship of the form $y = Ax^c$.

91. *Species Richness.* The table below shows the number of bird species found in some North American land areas.[8]

x = Land area (acres)	y = Bird species count
30	25
200	30
20,000	80
25,000,000	170
1,000,000,000	250

92. *Species Richness.* The table below shows the number of plant species found on some islands near the Galápogos Islands.[9]

x = Area (sq. mi)	y = Plant species count
2	15
2	30
9	50
17	110
180	175
2500	270

[8]N. A. Campbell and J. B. Reece, *Biology,* 6th ed. (San Francisco: Benjamin Cummings, 2002), p. 1194.
[9]N. A. Campbell and J. B. Reece, *Biology,* 6th ed. (San Francisco: Benjamin Cummings, 2002), p. 1195.

93. *Zipf's Law: United States.* The table below shows the number of U.S. cities with a 2000 population larger than x. For example, the last line means that, in 2000, there were 29 American cities with a population of at least 500,000.[10]

Population, x	Number of Cities y
25,000	1079
50,000	588
100,000	243
200,000	90
500,000	29

94. *Zipf's Law: Texas.* The table below shows the number of Texas cities with a 2000 population larger than x. For example, the last line means that, in 2000, there were seven Texas cities with a population of at least 300,000.[11]

Population, x	Number of Cities y
10,000	205
25,000	68
50,000	43
100,000	24
300,000	7

95. *Acceptance of a New Medicine.* The percentage P of doctors who accept a new medicine is given by

$$P(t) = 100(1 - e^{-0.2t}),$$

where t is the time, in months.

a) Find P(1) and P(6).
b) Find P'(t).
c) How many months will it take for 90% of the doctors to accept the new medicine?
tw d) Find $\lim_{t \to \infty} P(t)$ and discuss its meaning.

96. *The Reynolds Number.* For many kinds of animals, the Reynolds number R is given by

$$R = A \ln r - Br,$$

where A and B are positive constants and r is the radius of the aorta. Find the maximum value of R.

97. *Forgetting.* Students in college botany took a final exam. They took equivalent forms of the exam in

monthly intervals thereafter. The average score S(t), in percent, after t months was found to be given by

$$S(t) = 68 - 20 \ln (t + 1), \quad t \ge 0.$$

a) What was the average score when they initially took the test, t = 0?
b) What was the average score after 4 mos?
c) What was the average score after 24 mos?
d) What percentage of the initial score did they retain after 2 yr (24 mos)?
e) Find S'(t).
f) Find the maximum and minimum values, if they exist.
tw g) Find $\lim_{t \to \infty} S(t)$ and discuss its meaning.

98. *Forgetting.* Students in college zoology took a final exam. They took equivalent forms of the exam in monthly intervals thereafter. The average score S(t), in percent, after t months was found to be given by

$$S(t) = 78 - 15 \ln (t + 1), \quad t \ge 0.$$

a) What was the average score when they initially took the test, t = 0?
b) What was the average score after 4 mos?
c) What was the average score after 24 mos?
d) What percentage of the initial score did they retain after 2 yr (24 mos)?
e) Find S'(t).
f) Find the maximum and minimum values, if they exist.
tw g) Find $\lim_{t \to \infty} S(t)$ and discuss its meaning.

99. *Walking Speed.* Bornstein and Bornstein found in a study that the average walking speed v of a person living in a city of population p, in thousands, is given by

$$v(p) = 0.37 \ln p + 0.05,$$

where v is in feet per second.[12]

a) The population of Seattle is 531,000. What is the average walking speed of a person living in Seattle?
b) The population of New York is 7,900,000. What is the average walking speed of a person living in New York?
c) Find v'(p).
tw d) Interpret v'(p) found in part (c).

[10,11]U.S. Bureau of the Census.

[12]M. H. Bornstein and H. G. Bornstein, "The Pace of Life," *Nature,* Vol. 259, pp. 557–559 (1976).

100. *The Hullian Learning Model.* A typist learns to type W words per minute after t weeks of practice, where W is given by

$$W(t) = 100(1 - e^{-0.3t}).$$

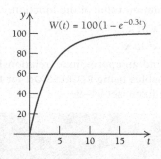

a) Find $W(1)$ and $W(8)$.

b) Find $W'(t)$.

c) After how many weeks will the typist's speed be 95 words per min?

tw d) Find $\lim_{t \to \infty} W(t)$ and discuss its meaning.

SYNTHESIS

Differentiate.

101. $f(x) = \ln [\ln x]^3$ **102.** $f(x) = \dfrac{\ln x}{1 + (\ln x)^2}$

103. Find $\lim_{h \to 0} \dfrac{\ln (1 + h)}{h}$.

Solve for t.

104. $P = P_0 e^{-kt}$ **105.** $P = P_0 e^{kt}$

Verify each of the following.

106. $\log x = \dfrac{\ln x}{\ln 10} \approx 0.4343 \ln x$

107. $\ln x = \dfrac{\log x}{\log e} \approx 2.3026 \log x$

Insect Growth. The Logan model for the relationship between the temperature and an organism's growth rate is

$$r(T) = A\left(e^{b(T-T_L)} - e^{b(T_U-T_L)-c(T_U-T)}\right),$$

where T is the temperature in degrees Celsius, A, b, and c are constants that depend on the organism, T_L is the lowest temperature that allows the organism to grow, and T_U is the highest temperature that allows the organism to grow. Values for the numbers A, b, c, T_L, and T_U were experimentally determined for parasitoid

Diglyphus isaea. This function will be used in Exercises 108–111.[13]

108. For male pupa, $A = 0.153$, $b = 0.141$, $c = 0.153$, $T_L = 9.5°C$, and $T_U = 39.4°C$.

a) Write the formula for the Logan model with these parameters.

b) Compute the derivative with respect to T of the function you obtained in part (a).

c) Use part (b) to find the temperature that maximizes the rate of growth.

109. For female pupa, $A = 0.124$, $b = 0.129$, $c = 0.144$, $T_L = 9.5°C$, and $T_U = 41.5°C$.

a) Write the formula for the Logan model with these parameters.

b) Compute the derivative with respect to T of the function you obtained in part (a).

c) Use part (b) to find the temperature that maximizes the rate of growth.

110. Compute the derivative $r'(T)$, keeping in mind that the numbers A, b, c, T_L, and T_U are all constants. Check your answer by substituting in the values given in Exercise 108 or 109 to see if the answers agree.

111. Use the formula you obtained in Exercise 110 to find a simple formula for the critical point for the function $r(T)$. Check your answer by substituting the values in Exercise 108 or 109 to see if your answers agree.

Semi-logarithm plots. For a semi-logarithm plot, the points $(x, \log_{10} f(x))$ are plotted. For a function of the form $f(x) = 10^{mx+b}$, a semi-logarithm plot will be the usual graph of $y = mx + b$.

112. Let $f(x) = 10^{2x-1}$.

a) Plot the points $(0.5, \log f(0.5))$, $(1, \log f(1))$, $(1.5, \log f(1.5))$, and $(2, \log f(2))$.

b) Use part (a) to sketch a semi-logarithm plot for $f(x)$. Your answer should be a straight line.

tw c) Check that the line in part (b) is the graph of $y = 2x - 1$. Explain why this is expected.

113. Let $f(x) = 10^{-3x+1}$.

a) Plot the points $\left(\frac{1}{3}, \log f\left(\frac{1}{3}\right)\right)$, $(1, \log f(1))$, $(2, \log f(2))$, and $(3, \log f(3))$.

b) Use part (a) to sketch a semi-logarithm plot for $f(x)$. Your answer should be a straight line.

tw c) Check that the line in part (b) is the graph of $y = -3x + 1$. Explain why this is expected.

---

[13]Bazzocchi et al., "Effects of temperature and host on the preimaginal development of the parasitoid *Diglyphus isaea,*" *Biological Control*, 26 (2003) 74–82.

114. *Bacteria Growth.* The table below gives the number of bacteria in a culture.

Time in Days t	Number of Bacteria N
1	201
2	246
3	360
4	446
5	540

a) Make a semi-logarithm plot of the data.
b) Use the first and last rows to estimate the slope and the y-intercept of a line approximating the graph in part (a).
c) Use part (b) to write an approximate relationship between t and N.

115. *World Population.* The table below gives the approximate population of the world.

Year t	World Population (in billions) N
1927	2
1960	3
1974	4
1987	5
1998	6
2003	6.3

a) Make a semi-logarithm plot of the data.
b) Use the first and last rows to estimate the slope and the y-intercept of a line approximating the graph in part (a).
c) Use part (b) to write an approximate relationship between t and N.

tw 116. Explain how the graph of $y = \ln x$ could be used to find the graph of $y = e^x$.

tw 117. Consider the true statement

$$\frac{d}{dx} \ln 4x = \frac{1}{x}.$$

Explain what this means graphically.

 Technology Connection

118. Which is larger, e^π or π^e?

119. Find $\sqrt[e]{e}$. Compare it to other expressions of the type $\sqrt[x]{x}$, $x > 0$. What can you conclude?

Use input–output tables to find the limit.

120. $\lim\limits_{x \to 1} \ln x$ **121.** $\lim\limits_{x \to \infty} \ln x$

Graph each function f and its derivative f'.

122. $f(x) = \ln x$ **123.** $f(x) = x \ln x$

124. $f(x) = x^2 \ln x$ **125.** $f(x) = \dfrac{\ln x}{x^2}$

Find the minimum value of the function.

126. $f(x) = x \ln x$

127. $f(x) = x^2 \ln x$

128.–131. Find an approximate relationship between the variables using **REGRESSION** for the log-log plots in Exercises 91–94.

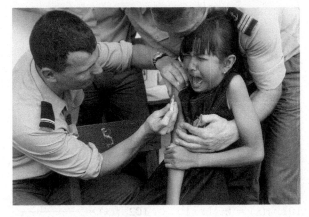

Modeling an Epidemic. The solution x of the equation

$$k \ln x - k \ln P - x + P = 0$$

represents the number of individuals unaffected by an epidemic that affects a population of size P. The constant k is determined by the rate at which the epidemic spreads. (This equation will be derived in the Extended Life Science Connection of Chapter 9.)

   Use Newton's method to solve for x given the values of k and P.

132. $k = 10{,}000, P = 20{,}000$

133. $k = 500, P = 1000$

134. $k = 300, P = 1000$

135. $k = 1500, P = 5000$

Lotka–Volterra Model. Let x be the number of predators (in thousands) in a closed habitat. Then x satisfies the equation

$$r \ln x - sx + C = 0,$$

where r, s, and C are constants. (The Lotka–Volterra model is described in Chapter 9.) Use Newton's method to solve for x given the values of r, s, and C.

136. $r = 1, s = 0.4, C = 2$

137. $r = 2, s = 0.5, C = 5$

138. $r = 0.5, s = 0.2, C = 1$

139. $r = 0.4, s = 0.2, C = 1$

4.3 Applications: The Uninhibited Growth Model, $dP/dt = kP$

OBJECTIVES

■ Find functions that satisfy $dP/dt = kP$.
■ Convert between growth rate and doubling time.
■ Solve application problems involving exponential growth.

Exponential Growth

Consider the function

$$f(x) = 2e^{3x}.$$

Differentiating, we get

$$f'(x) = 3 \cdot 2e^{3x}$$
$$= 3 \cdot f(x).$$

Graphically, this says that the derivative, or slope of the tangent line, is simply the constant 3 times the function value.

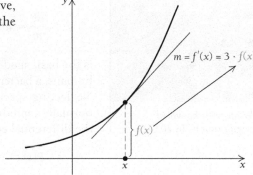

The exponential function $f(x) = ce^{kx}$ is the only function whose derivative is a constant times the function itself.

THEOREM 8

A function $y = f(x)$ satisfies the equation

$$\frac{dy}{dx} = ky \qquad [f'(x) = k \cdot f(x)]$$

if and only if

$$y = ce^{kx} \qquad [f(x) = ce^{kx}]$$

for some constant c.

When an equation involves a derivative, then that equation is called a **differential equation.** A function that satisfies a differential equation is called a *solution* to the differential equation. Theorem 8 states that if the differential equation has the form $\dfrac{dy}{dx} = ky$, then we know all of its solutions.

EXAMPLE 1 Find all solutions to the differential equation

$$\frac{dA}{dt} = 3A.$$

Solution In this equation we are using A instead of y, t instead of x, and 3 instead of k. Otherwise, the differential equation is identical to the differential equation in Theorem 8. The solutions are of the form $A = ce^{3t}$, or equivalently, $A(t) = ce^{3t}$. To check this, notice

$$\frac{d}{dt}A(t) = \frac{d}{dt}(ce^{3t}) = c \cdot (3e^{3t}) = 3ce^{3t} = 3A(t).$$

Note that for every value of c we have a solution.　■

What will the world population be in 2020?

The equation

$$\frac{dP}{dt} = kP, \quad k > 0 \qquad [P'(t) = k \cdot P(t), \quad k > 0]$$

is the basic model of uninhibited **exponential** growth, whether it be a population of humans, a bacteria culture, or money invested at interest compounded continuously. Neglecting special inhibiting and stimulating factors, we know that a population normally reproduces itself at a rate proportional to its size, and this is exactly what the differential equation $dP/dt = kP$ says. The solution of the equation is

$$P(t) = ce^{kt}, \tag{1}$$

where t is the time. At $t = 0$, we have some "initial" population $P(0)$ that we will represent by P_0. We can rewrite equation (1) in terms of P_0 as

$$P_0 = P(0) = ce^{k \cdot 0} = ce^0 = c \cdot 1 = c.$$

Thus, $P_0 = c$, so we can express $P(t)$ as

$$P(t) = P_0 e^{kt}.$$

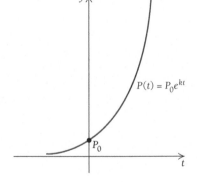

Its graph shows how uninhibited growth produces a "population explosion."

A problem where we are given a differential equation and enough information to determine a unique solution is called an **initial-value problem.**

EXAMPLE 2 Find the function that solves the initial-value problem

$$\frac{dP}{dt} = 0.1P$$

with $P_0 = 25$.

Solution The solution has the form

$$P(t) = P_0 e^{kt}.$$

Since $P_0 = 25$ and $k = 0.1$, we can write the solution as

$$P(t) = 25e^{0.1t}.$$

This solution may be checked by finding $\dfrac{dP}{dt}$ and $P(0)$; this exercise is left to the reader.   ■

EXAMPLE 3   Find the solution to the initial-value problem

$$f'(Q) = 0.01f(Q)$$

with $f(0) = 1.03$.

Solution   The solution has the form $f(Q) = ce^{0.01Q}$. Since $1.03 = f(0) = c$, we see that $c = 1.03$. The solution is $f(Q) = 1.03e^{0.01Q}$.   ■

The constant k is called the **exponential growth rate,** or simply the **growth rate.** This is not the rate of change of the population size, which is

$$\frac{dP}{dt} = kP,$$

but the constant by which P must be multiplied in order to get its rate of change. It is thus a different use of the word *rate*. It is like the *interest rate* paid by a bank. If the interest rate is 7%, or 0.07, we do not mean that your bank balance P is growing at the rate of 0.07 dollars per year, but at the rate of $0.07P$ dollars per year. We therefore express the rate as 7% per year, rather than 0.07 dollars per year. We could say that the rate is 0.07 dollars *per dollar* per year. When interest is **compounded continuously,** the interest rate is a true exponential growth rate.

EXAMPLE 4   *Interest Compounded Continuously.* Suppose that an amount P_0 is invested in a savings account where interest is compounded continuously at 7% per year. That is, the balance P grows at the rate given by

$$\frac{dP}{dt} = 0.07P.$$

a) Find the function that satisfies the equation. List it in terms of P_0 and 0.07.
b) Suppose that $100 is invested. What is the balance after 1 yr?
c) After what period of time will an investment of $100 double itself?

Solution

a) $P(t) = P_0e^{0.07t}$

b) $P(1) = 100e^{0.07(1)} = 100e^{0.07}$

$$\approx \$107.25$$

Under ideal conditions, the growth rate of this rabbit population might be 11.7% per day. When will this population of rabbits double?

c) We are looking for that time T for which $P(T) = \$200$. The number T is called the **doubling time.** To find T, we solve the equation

$$200 = 100e^{0.07 \cdot T}$$

$$2 = e^{0.07T}.$$

We use natural logarithms to solve this equation:

$$\ln 2 = \ln e^{0.07T}$$

$$\ln 2 = 0.7T \qquad \text{By P5}$$

$$\frac{\ln 2}{0.07} = T$$

$$\frac{0.693147}{0.07} = T$$

$$9.9 \approx T.$$

Thus, $\$100$ will double itself in 9.9 yr.

We can find a general expression relating the growth rate k and the doubling time T by solving the following equation:

$$2P_0 = P_0 e^{kT}$$

$$2 = e^{kT} \qquad \text{Dividing by } P_0$$

$$\ln 2 = \ln e^{kT}$$

$$\ln 2 = kT.$$

Biologists often use the term **generation time** instead of *doubling time,* but they both mean the time required for the population to double.

THEOREM 9

The *growth rate k* and the *generation time* (or *doubling time*) are related by the following equivalent equations:

$$kT = \ln 2 \approx 0.693147,$$

$$k = \frac{\ln 2}{T} \approx \frac{0.693147}{T}, \quad \text{and}$$

$$T = \frac{\ln 2}{k} \approx \frac{0.693147}{k}.$$

Note that the relationship between k and T does not depend on P_0.

How do biologists measure the growth rate of small organisms growing in water? One method is to grow a culture in a *chemostat*. Water is slowly added and allowed to drain off the top. Organisms are lost in the drained water, but at the same time, the organisms in the chemostat are growing. The flow of water is adjusted using a photocell so that the loss rate just balances the growth rate, producing an equilibrium.

Let V be the volume of the chemostat, N the number of organisms in the chemostat, and X the volume of water added to (and simultaneously drained from) the chemostat. The density of organisms in the chemostat is $\frac{N}{V}$, so that the number of organisms drained is $\frac{N}{V}X$. The rate the organisms are drained is thus

$$\frac{d}{dt}\left(\frac{N}{V}X\right) = \frac{N}{V}X'$$

At equilibrium, the growth rate $N' = kN$ matches the rate organisms are drained off. So,

$$\frac{N}{V}X' = kN \text{ or } k = \frac{X'}{V}.$$

That is, when an equilibrium is reached, the rate that water is added divided by the volume of the chemostat gives the growth rate k.

While measuring growth rates of *Pseudomonas aeruginosa*, a bacteria that causes respiratory disease, the volume was 2 l and the rate that water was added at equilibrium was 0.4 l per hour. The growth rate is $k = \frac{0.4}{2} = 0.2/\text{hr}$.[14]

EXAMPLE 5   *Growth of E. Coli.* A sample of *E. Coli* is growing exponentially. The generation time is 40 min. What is the exponential growth rate?

Solution   We have

$$k = \frac{\ln 2}{T}$$

$$= \frac{\ln 2}{40}$$

$$\approx 0.01732 \text{ per min.} \quad \text{Using a scientific calculator for approximation}$$

The exponential growth rate is approximately 1.7% per min.   ∎

The Rule of 69

The relationship between doubling time T and interest rate k is the basis of a rule often used in the investment world, called the *Rule of 69*. To estimate how long it will take to double your money at varying rates of return, divide 69 by the rate of return. To see how this works, let the interest rate $k = r\%$. Then

$$T = \frac{\ln 2}{k} = \frac{0.693147}{r\%}$$

$$= \frac{0.693147}{r \times 0.01} \cdot \frac{100}{100}$$

$$= \frac{69.3147}{r} \approx \frac{69}{r}.$$

EXAMPLE 6   Recall that in Example 4 the interest rate was 7%. Use the Rule of 69 to estimate the doubling time.

Solution   The Rule of 69 gives the doubling time as approximately $\frac{69}{7} \approx 9.9$. In Example 4 we found the doubling time to be 9.9 yr.   ∎

EXAMPLE 7   *World Population Growth.* According to the U.S. Bureau of the Census, the world population in 2004 was 6.393 billion people and the world population in 2010 is projected to be 6.973 billion people. Assume that the world population is growing exponentially. (How are these projections computed? The answer is in the model we develop in the following Technology Connection.) Thus,

$$\frac{dP}{dt} = kP,$$

[14]Beyenal, Chen, and Lewandowski, "The double substrate growth kinetics of *Pseudomonas aeruginosa*," *Enzyme and Microbal Technology,* 32 (2003) pp. 92–98.

where t is time in years after 2004, P is the world population in billions, and k is the exponential growth constant. We assume for simplicity that the population data is for the beginning of the year.

a) Find the function that satisfies the initial-value problem.

b) Estimate the world population in 2024.

c) After what period of time will the world population double?

Solution

a) The solution has the form $P(t) = 6.393e^{kt}$. We are given that $P(6) = 6.973$. So we have

$$6.973 = 6.393e^{6k}$$

$$\frac{6.973}{6.393} = e^{6k}$$

$$\ln\left(\frac{6.973}{6.393}\right) = \ln e^{6k}$$

$$\ln(1.0907) \approx 6k$$

$$\frac{\ln(1.0907)}{6} \approx k$$

$$0.01447 \approx k.$$

So $P(t) = 6.393e^{0.01447t}$.

b) $P(20) = 6.393e^{0.01447(20)} \approx 6.393(1.336) \approx 8.54$ billion.

c) $T = \dfrac{\ln 2}{k} = \dfrac{\ln 2}{0.01447} \approx 47.9$ yr.

According to this model, the world population will double by the end of 2051. There is a world population calculator at http://www.ibiblio.org/lunarbin/worldpop. Try comparing the above population estimates to the Web calculator. ■

Technology Connection

Exponential Regression: Modeling World Population Growth

The table below shows data regarding world population growth. (A graph illustrating these data, along with the projected population in 2020, appeared in Section 4.1.)

Year	World Population (in billions)
1927	2
1960	3
1974	4
1987	5
1998	6
2003	6.3

How were the data projected in 2020? Let's use the data from 1927 to 2003 to find an exponential model. The graph shows a rapidly growing population that can be modeled with an exponential function. We carry out the regression procedure very much as we did in Section 1.2, but here we choose ExpReg rather than LinReg.

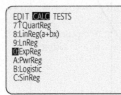
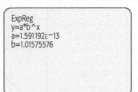

   Note that this gives us an exponential model of the type $y = a \cdot b^x$, but the base is not the number e. We can make a conversion to an exponential function, base e, using the formula $b^x = e^{x(\ln b)}$.
   The exponential regression gives us the model

$$y = (1.591192 \cdot 10^{-13})(1.01575576)^x. \quad (1)$$

Using $b^x = e^{x(\ln b)}$, we obtain

$$b^x = (1.01575576)^x = e^{x(\ln b)}$$
$$= e^{x \ln(1.01575576)} \approx e^{x(0.0156329266)}$$
$$= e^{0.0156329266x}.$$

We have converted equation (1) to

$$y = (1.591192 \cdot 10^{-13})e^{0.0156329266x}. \quad (2)$$

The advantage of this form is that we see the growth rate. Here the world population growth rate is approximately 0.016, or 1.6%. To estimate the world population in 2010, we can substitute 2010 for x into either equation (1) or (2). We choose equation (2):

$$y = (1.591192 \cdot 10^{-13})e^{0.0156329266(2010)}$$
$$\approx 7.0502 \text{ billion.}$$

EXERCISES

Use equation (1) or (2) to predict world population in each year.

1. 2020   **2.** 2030   **3.** 2040   **4.** 2050

Projected Income. For Exercises 5 and 6, use the table below, which gives average income for persons holding a bachelor's degree.[15]

Years since 1990	Average Salary
0	31,112
1	31,323
2	32,629
3	35,121
4	37,224
5	36,980
6	38,112
7	40,478
8	43,782
9	45,678
10	49,674

5. Use the REGRESSION feature to fit an exponential function $y = a \cdot b^x$ to the data. Then convert that formula to an exponential function, base e, and determine the exponential growth rate.

6. Use either of the exponential functions found in Exercise 5 to predict the average income of a person with a bachelor's degree in 2010, 2025, and 2040.

---

[15]U.S. Census Bureau Web site (www.census.gov/hhes /income/histinc/p28)

Some myths about alcohol. *It's a fact — the blood alcohol concentration (BAC) in the human body is measurable. And there's no cure for its effect on the central nervous system except time. It takes time for the body's metabolism to recover. That means a cup of coffee, a cold shower, and fresh air can't erase the effect of several drinks.*

There are variables, of course: a person's body weight, how many drinks have been consumed in a given time, how much has been eaten, and so on. These account for different BAC levels. But the myth that some people can "handle their liquor" better than others is a gross rationalization — especially when it comes to driving. Some people can act more sober than others. But an automobile doesn't act; it reacts.

Modeling Other Phenomena

EXAMPLE 8 *Alcohol Absorption and the Risk of Having an Accident.* Extensive research has provided data relating the risk R (in percent) of having an automobile accident to the blood alcohol level b (in percent). Using two representative points $(0, 1\%)$ and $(0.14, 20\%)$, we can approximate the data with an exponential function. The modeling assumption is that the rate of change of the risk R with respect to the blood alcohol level b is given by

$$\frac{dR}{db} = kR.$$

a) Find the function that satisfies the equation. Assume that $R_0 = 1\%$.

b) Find k using the data point $R(0.14) = 20\%$.

c) Rewrite $R(b)$ in terms of k.

d) At what blood alcohol level will the risk of having an accident be 100%? Round to the nearest hundredth.

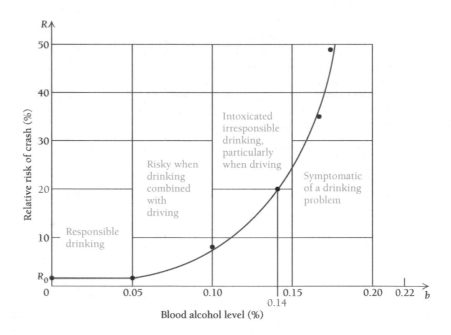

Solution

a) Since both R and b are percents, we omit the % symbol to simplify computation. The solution is

$$R(b) = e^{kb}, \quad \text{since } R_0 = 1.$$

We have made use of the data point $(0, 1)$.

b) We now use the second data point, $(0.14, 20)$, to determine k. We solve the equation

$$R(b) = e^{kb}$$

for k, using natural logarithms:

$$20 = e^{k(0.14)} = e^{0.14k}$$

$$\ln 20 = \ln e^{0.14k}$$

$$\ln 20 = 0.14k$$

$$\frac{\ln 20}{0.14} = k$$

$$21.4 \approx k. \qquad \text{Rounded to the nearest tenth}$$

c) $R(b) = e^{21.4b}$.

d) We substitute 100 for $R(b)$ and solve the equation for b:

$$100 = e^{21.4b}$$

$$\ln 100 = \ln e^{21.4b}$$

$$\ln 100 = 21.4b$$

$$\frac{\ln 100}{21.4} = b$$

$$0.22 \approx b. \qquad \text{Rounded to the nearest hundredth}$$

The calculations done in this problem can be performed more conveniently on your calculator if you do not stop to round. For example, in part (b), we find $\ln 20$ and divide by 0.14, obtaining 21.39808767.... We then use that value for k in part (d). Answers will be found that way in the exercises. You may note some variance in the last one or two decimal places if you round as you go.

Thus when the blood alcohol level is 0.22%, according to this model, the risk of an accident is 100%. From the above graph, we see that this would occur for a 160-lb man after 12 1-oz drinks of 86-proof whiskey. "Theoretically," the model tells us that after 12 drinks of whiskey, one is "sure" to have an accident. This might be questioned in reality, since a person who has had 12 drinks might not be able to drive at all. ∎

Models of Limited Growth

The growth model $P(t) = P_0 e^{kt}$ has many applications to unlimited population growth as we have seen thus far in this section. However, it seems reasonable that there can be factors that prevent a population from exceeding some limiting value L—perhaps a limitation on food, living space, or other natural resources. One model of such growth is

$$P(t) = \frac{L}{1 + be^{-kt}},$$

which is called the **logistic equation.**

EXAMPLE 9 *Ginseng Roots.* Ginseng is a genus of plants believed to have medicinal properties. As the plant ages, the length of the root approaches a limiting value. The annual growth of American ginseng roots can be modeled using the logistic equation. For a first-year growth, the values of L and b were determined to give the logistic growth formula

$$C(t) = \frac{71.4}{1 + 149e^{-0.085t}},$$

where $C(t)$ is the length of the root as a percentage of its limiting value, and t is the number of days after a fixed starting date.[16]

a) Find the percentage of the root length at times $t = 10$, $t = 20$, $t = 50$, and $t = 100$.

b) At what time is the root's length 25% of its limiting value?

c) Find the rate of change $C'(t)$.

d) Graph the function $C(t)$.

Solution

a) We use a calculator to find the function values, listing them in a table as shown in the margin.

b) The root's length will be 25% its limiting value when $C(t) = 25\%$. We solve for t.

$$25 = \frac{71.4}{1 + 149e^{-0.085t}}$$

$$25(1 + 149e^{-0.085t}) = 71.4$$

$$1 + 149e^{-0.085t} = \frac{71.4}{25} = 2.856$$

$$149e^{-0.085t} = 1.856$$

$$e^{-0.085t} = \frac{1.856}{149} \approx 0.012456$$

$$\ln(e^{-0.085t}) \approx \ln(0.012456)$$

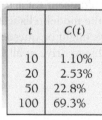

t	$C(t)$
10	1.10%
20	2.53%
50	22.8%
100	69.3%

---

[16] Terence P. McGonigle, "On the use of non-linear regression with the logistic equation for changes with time of percentage root length colonized by arbuscular mycorrhizal fungi," *Mycorrhiza* (2001) 10:249–254.

$$-0.085t \approx -4.3855 \qquad \text{Using a scientific calculator for approximation}$$

$$t \approx \frac{-4.3855}{-0.085} \approx 51.6 \text{ days}$$

We can check our answer by substituting $t = 51.6$ into the original equation: $C(51.6) \approx 25.008\%$. So the root's length is 25% of its limiting value after growing approximately 52 days.

c) We find the rate of change $C'(t)$ using the Quotient Rule.

$$C'(t) = \frac{(1 + 149e^{-0.085t}) \cdot 0 - (149)(-0.085)e^{-0.085t} \cdot (71.4)}{(1 + 149e^{-0.085t})^2}$$

$$= \frac{904.281e^{-0.085t}}{(1 + 149e^{-0.085t})^2}$$

d) We consider the curve-sketching procedures discussed in Section 3.2. The derivative $C'(t)$ exists for all real numbers t; thus, there are no critical points for which the derivative does not exist. The equation $C'(t) = 0$ holds only when the numerator is 0. But the numerator is positive for all real numbers t, since $e^{-0.085t} > 0$ for all t. Thus, the function has no critical points and hence no relative extrema.

Since both the numerator and the denominator are positive in the formula for $C'(t)$, $C(t)$ is an increasing function over the entire real line. In the exercises, you will be asked to show that the point of inflection for the logistic equation occurs when $t = \frac{\ln b}{k}$. In our case, $b = 149$ and $k = 0.085$, so the point of inflection occurs at $t = \frac{\ln 149}{0.085} \approx 58.9$. The point of inflection is approximately $(58.9, 35.7)$. The graph is concave up (C' is increasing) over the interval $(0, 58.9)$ and concave down (C' is decreasing) over the interval $(58.9, \infty)$. Also note that

$$\lim_{t \to \infty} \frac{71.4}{1 + 149e^{-0.085t}} = \frac{71.4}{1 + 0} = 71.4,$$

and so there is a horizontal asymptote at $y = 71.4$. The graph is the S-shaped curve shown below.

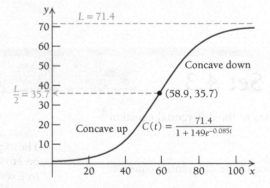

The logistic equation's graph has the general S shape of the graph of $C(t)$ above. A function whose graph has this shape is often referred to as a *sigmoidal function*. A sigmoidal logistic function has a point of inflection at the point $\left(\frac{\ln b}{k}, \frac{L}{2}\right)$. Note that there are two horizontal asymptotes; one is the x-axis and the other is the line $y = L$. Also, the point of inflection occurs when the value of the function is half its limiting value.

Another model of limited growth is provided by the function

$$P(t) = L(1 - e^{-kt}),$$

which is graphed below. This function also increases over the entire interval $[0, \infty)$.

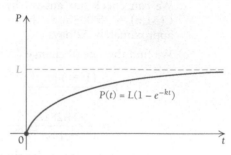

As we can see in the graph, the limiting value of the population is L.

EXAMPLE 10 The population of deer in a certain woods is

$$P(t) = 1000(1 - e^{-1.1t}),$$

where t is time measured in years.

a) Find the population at time $t = 1$.
b) Find the population at time $t = 5$.
c) Find the population at time $t = 10$.

Solution

a) $P(1) = 1000(1 - e^{-1.1(1)}) = 1000(1 - e^{-1.1}) \approx 667$
b) $P(5) = 1000(1 - e^{-1.1(5)}) = 1000(1 - e^{-5.5}) \approx 996$
c) $P(10) = 1000(1 - e^{-1.1(10)}) = 1000(1 - e^{-11}) \approx 1000$. Note that by year 10, the population is at or very close to its limiting value. ■

Exercise Set 4.3

1. Find all solutions to the differential equation
$$\frac{dQ}{dt} = kQ.$$

2. Find all solutions to the differential equation
$$\frac{dR}{dt} = kR.$$

3. Find the solution to the initial-value problem
$$\frac{dy}{dt} = 2y \text{ given that } y = 5 \text{ when } t = 0.$$

4. Find the solution to the initial-value problem
$$\frac{dH}{dt} = 0.5H \text{ given that } H = 10 \text{ when } t = 0.$$

APPLICATIONS

The exponential growth rate of an organism depends on environmental conditions, such as the temperature. In Exercises 5 and 6 you are given the exponential growth rate k and the size of the initial population P_0. The time units are minutes. Assume the temperature remains constant and determine:

a) The equation that gives the population at time t.
b) The population at time $t = 30$.
c) The population after 1 hr.
d) The population after 1 day.
e) The generation time.

5. For *E. Coli,* a bacteria present in a healthy human intestine, at 40°C $k = 0.033$. Assume that $P_0 = 1000$.[17]

6. For *Giardia lambia,* the protozoa most dreaded by hikers, $k = 0.000641$ at 37°C. Assume that $P_0 = 1000$.[18]

In Exercises 7 and 8 you are given the generation time of an organism and the initial population. Assuming that the temperature remains the same and that the growth is uninhibited, determine

a) The exponential growth rate k.
b) The population after 3 hr.
c) The population after 1 day.
d) The time it takes the population to triple.

7. For the bacteria *Staphylococcus aureus* at 37°C, the generation time is 0.47 hr. Assume the initial population is 200.

8. For the fungi *Saccharomyces cerevisiae* at 30°C, the generation time is 2 hr. Assume the initial population is 500.

9. *Planning an Experiment.* For an experiment, a biologist wishes to have a sample of 20,000 *E. Coli* O157:H7, a dangerous strain of *E. Coli* that can occur in mishandled beef. The experiment is to be conducted at 3:00 P.M. How many *E. Coli* O157:H7 organisms should be placed in a sample at 1:00 P.M. in order for the sample to contain the right number for the experiment? Assume the conditions in Exercise 5.

10. *Planning an Experiment.* A biologist wishes to have a sample of 10,000 *Staphylococcus aureus* for an experiment at 10:30 A.M. How many *Staphylococcus aureus* bacteria should be placed into a sample at 8:00 A.M. in order for the sample to contain the right number for the experiment? Assume the conditions in Exercise 7.

11. *Population Growth.* The population of the United States was 281 million in 2000. At that point, it was estimated that the population P was growing exponentially with growth rate 0.9% per year—that is,

$$\frac{dP}{dt} = 0.009P,$$

where t is the number of years after 2000.[19]

a) Find the function that satisfies the initial-value problem.

b) Estimate the population of the United States in 2015.
c) After what period of time will the population be double that in 2000?

12. *Population Growth.* The population of the United Kingdom in July 2002 was estimated to be 59.8 million. At that time, it was estimated that the population was growing according to the differential equation

$$\frac{dP}{dt} = 0.0021P,$$

where t is the number of years, after July 2002.[20]

a) Find the function that satisfies the initial-value problem.
b) Estimate the population of the United Kingdom in 2015.
c) After what period of time will the population be double that in 2002?

13. *Compound Interest.* Suppose that P_0 is invested in a savings account in which interest is compounded continuously at 6.5% per year. That is, the balance P grows at the rate given by

$$\frac{dP}{dt} = 0.065P.$$

a) Find the function that satisfies the equation. List it in terms of P_0 and 0.065.
b) Suppose that $1000 is invested. What is the balance after 1 yr? after 2 yr?
c) When will an investment of $1000 double itself?

14. *Compound Interest.* Suppose that P_0 is invested in a savings account in which interest is compounded continuously at 8% per year. That is, the balance P grows at the rate given by

$$\frac{dP}{dt} = 0.08P.$$

a) Find the function that satisfies the equation. List it in terms of P_0 and 0.08.
b) Suppose that $20,000 is invested. What is the balance after 1 yr? after 2 yr?
c) When will an investment of $20,000 double itself?

15. *Annual Interest Rate.* A bank advertises that it compounds interest continuously and that it will double your money in 10 yr. What is its exponential growth rate?

[17,18]Prescott, Harley, Klein, *Microbiology* (New York: McGraw-Hill, 2002).
[19]U.S. Bureau of the Census.

[20]CIA—The World Fact Book.

16. *Annual Interest Rate.* A bank advertises that it compounds interest continuously and that it will double your money in 12 yr. What is its annual interest rate?

Population Growth. For Exercises 17–22, complete the following.

Population	Growth Rate, k	Doubling Time, T
17. Mexico	3.5% per year	
18. Europe		69.31 yr
19. Oil reserves		6.931 yr
20. Coal reserves		17.3 yr
21. Alaska	2.794% per year	
22. Central America		19.8 yr

23. *Blood Alcohol Level.* Refer to Example 8 (on alcohol absorption). At what blood alcohol level will the risk of an accident be 80%?

24. *Blood Alcohol Level.* Refer to Example 8. At what blood alcohol level will the risk of an accident be 90%?

25. *Bicentennial Growth of the United States.* The population of the United States in 1776 was about 2,508,000. In its bicentennial year, the population was about 216,000,000.

   a) Assuming the exponential model, what was the growth rate of the United States through its bicentennial year?

tw b) Is this a reasonable assumption? Explain.

26. *Population Growth.* The table below shows the population of Denton County, Texas, as reported in the 10-year U.S. Census reports.

Year	Years since 1960	Denton Co. Population
1960	0	47,432
1970	10	75,633
1980	20	143,126
1990	30	273,525
2000	40	432,976

   a) Use the **REGRESSION** feature to fit an exponential function $y = a \cdot b^t$ to the data. Give an interpretation of the number b. (See the Technology Connection on p. 303.)

   b) Convert your answer in part (a) to the form $y = a \cdot e^{kt}$. (Recall from p. 283 that $b = e^{\ln b}$ is needed to make this conversion.)

   c) Project the population of Denton County in the year 2010.

   d) When do you project the population of Denton County to reach 1 million?

   e) What is the doubling time for Denton County?

27. *Population Growth.* Refer to the data in Exercise 26.

   a) Find an exponential function to fit the data by using the points (0, 47,432) and (40, 432,976).

   b) Use part (a) to project the population in Denton County in the year 2010.

   c) Use part (a) to project when the population of Denton County will reach 1 million.

   d) Use part (a) to approximate the doubling time for Denton County.

   e) Compare your answers to the answers you obtained in Exercise 26.

28. *Algae Growth.* Algae growth is a problem common to swimming pools. The relative number of algae in a solution is measured by shining a light through the solution and measuring the amount of light absorbed. The table below gives the relative number of algae in the solution.

Time in Hours	Relative Number of Algae
0	85
1	92
2	98
3	102
4	109

   a) Use the **REGRESSION** feature to fit an exponential function $y = a \cdot b^t$ to the data. Give an interpretation of the number b. (See the Technology Connection on p. 303.)

   b) Convert your answer in part (a) to the form $y = a \cdot e^{kt}$. (Recall from p. 283 that $b = e^{\ln b}$ is needed to make this conversion.)

   c) Give a projection for the relative number of algae at 5 hr.

   d) When do you project the relative number of algae to reach 1000?

   e) What is the generation time?

29. *Algae Growth.* Refer to the data in Exercise 28.

   a) Find an exponential function to fit the data using the points (0, 85) and (4, 109).

   b) Use part (a) to project the relative number of algae at 5 hr.

   c) Use part (a) to project when the relative number of algae will reach 1000.

   d) Use part (a) to approximate the generation time.

   e) Compare your answers to the answers you obtained in Exercise 28.

30. *Atmospheric Pressure.* The table below gives the atmospheric pressure at various heights above sea level. Let x be the pressure in pounds per square inch (psi) and $A(x)$ the altitude in feet.

Atmospheric Pressure (pounds per square inch)	Altitude Above Sea Level (feet)
15.2	−1000
14.7	0
13.7	2000
12.7	4000
11.8	6000
10.9	8000
10.1	10,000
8.29	15,000
6.76	20,000
4.37	30,000

a) Use the **REGRESSION** feature on a grapher to fit the data to the function

$$A(x) = b + c \ln x.$$

This is called *logarithmic regression* and it may be computed using **LnReg** on the TI-83.

b) Use the grapher to plot the function $A(x)$ from part (a).

c) Compute $A'(x)$.

tw d) Notice in the table that when the pressure is approximately 14 psi, every decrease of 1 psi increases the altitude by 2000 ft. Based on part (c), give an explanation of this observation.

31. *Atmospheric Pressure.* Reverse the columns in the table used for Exercise 30, so we view the pressure as a function of altitude. Let x be the altitude in feet and $P(x)$ the pressure in psi.

a) Use the exponential regression feature on a grapher to fit the data to the function

$$P(x) = a \cdot b^x.$$

b) Convert your answer in part (a) to the form $P(x) = a \cdot e^{kx}$. (Recall from p. 283 that $b = e^{\ln b}$ is needed to make this conversion.)

c) Use the grapher to plot the function $P(x)$.

d) Compute $P'(x)$.

tw e) Notice in the table that when the altitude is approximately at sea level, an increase of 2000 ft of altitude decreases the pressure by approximately 1 psi. Based on part (c), give an explanation of this observation.

tw f) Using the function $P(x)$ from this exercise and $A(x)$ from Exercise 30, compute and simplify the compositions $P(A(x))$ and $A(P(x))$. Explain your answers.

32. *Limited Population Growth.* A lake is stocked with 400 fish of a new variety. The size of the lake, the availability of food, and the number of other fish restrict growth in the lake to a *limiting value* of 2500. The population of fish in the lake after time t, in months, is given by

$$P(t) = \frac{2500}{1 + 5.25e^{-0.32t}}.$$

a) Find the population after 0 mo; 1 mo; 5 mo; 10 mo; 15 mo; 20 mo.

b) Find the rate of change $P'(t)$.

c) Sketch a graph of the function.

33. *Oil Absorbsion of French Fries.* The cumulated oil distribution of a french fry fried at 185° for 1 min is approximated by the function

$$G(x) = \frac{100}{1 + 43.3e^{-0.0425x}},$$

where x is the distance from the surface of the potato, in μm ($1\ \mu$m $= 10^{-6}$ m), and $G(x)$ is the percent of oil the potato absorbed within a distance of x from the surface. For example, $G(100) \approx 21.1$ indicates that 21.1% of the absorbed oil is in the first 100 μm of the potato.[21]

a) Use the formula to compute the percent of oil within 100 μm from the surface, 150 μm from the surface, and 300 μm from the surface.

b) Compute the derivative $G'(x)$.

c) Find the point of inflection for the function $G(x)$.

d) Sketch the graph of the function.

34. *American Ginseng Growth.* The logistic curve for the root length of American ginseng growing under certain conditions was given in Example 9. Under different conditions, the parameters in the model change. In this exercise, use the logistic growth equation

$$C(t) = \frac{40.2}{1 + 335e^{-0.092t}},$$

where $C(t)$ is the length of the root measure as a percent of its limiting value and t is time in days.[22]

a) Find the root length at time $t = 20$, $t = 50$, and $t = 100$. Your answers will be a percent of the maximum root length.

---

[21]Bouchon et al., "Oil Distribution in Fried Potatoes Monitored by Infrared Microscopy," *Journal of Food Science*, Vol. 66, No. 7, 2001, pp. 918–923.

[22]Terence P. McGonigle, "On the use of non-linear regression with the logistic equation for changes with time of percentage root length colonized by arbuscular mycorrhizal fungi," *Mycorrhiza* (2001) 10:249–254.

b) At what time is the root 30% of its maximum length?

c) Find the rate of change $C'(t)$.

d) Find the point of inflection for the function $C(t)$.

e) Sketch the graph of the function $C(t)$.

35. *Maize Leaf Growth.* The growth of maize (corn) leaves is modeled by the logistic equation. For the sixth leaf on a maize plant, it was found that the function

$$A(t) = \frac{105}{1 + 32900e^{-0.04t}}$$

is a good approximation of the leaf's area $A(t)$ (in cm^2) from the time the leaf forms until it stops growing.[23] The number t is the number of degree days since the plant sprouted. (Degree days measure time weighted by the temperature. The number of degree days for maize is computed by multiplying the time by the number of degrees above 8°C.)

a) Use the logistic equation to approximate the leaf area after 100 degree days, after 150 degree days, and after 200 degree days.

b) Find the rate of change of the area $A'(t)$.

c) Find the point of inflection for the function $A(t)$.

d) Sketch the graph of $A(t)$.

e) According to this model, what is the maximum area of the sixth leaf on a maize plant?

36. *Hullian Learning Model.* The Hullian learning model asserts that the probability p of mastering a task after t learning trials is given by

$$p(t) = 1 - e^{-kt},$$

where k is a constant that depends on the task to be learned. Suppose that a new task is introduced in a factory on an assembly line. For that particular task, the constant $k = 0.28$.

a) What is the probability of learning the task after 1 trial? 2 trials? 5 trials? 11 trials? 16 trials? 20 trials?

b) Find the rate of change $p'(t)$.

c) Sketch a graph of the function.

37. *Diffusion of Information.* Pharmaceutical firms spend an immense amount of money in order to test a new medication. After the drug is approved by the Federal Drug Administration, it still takes time for physicians to fully accept and make use of the medication. The use approaches a *limiting value* of 100%, or 1, after time t, in months. Suppose that for a new cancer medication the percentage P of physicians using the product after t months is given by

$$P(t) = 100\%(1 - e^{-0.4t}).$$

a) What percentage of doctors have accepted the medication after 0 mo? 1 mo? 2 mo? 3 mo? 5 mo? 12 mo? 16 mo?

b) Find the rate of change $P'(t)$.

c) Sketch a graph of the function.

38. *Spread of a Rumor.* The rumor "People who study math are people you can count on" spreads across a college campus. Data in the table below shows the number of students N who have heard the rumor after time t, in days.

Time, t (in days)	Number Who Have Heard the Rumor
1	1
2	2
3	4
4	7
5	12
6	18
7	24
8	26
9	28
10	28
11	29
12	30

a) Use the **REGRESSION** feature on a grapher to fit a logistic equation

$$N(t) = \frac{c}{1 + ae^{-bt}}$$

to the data.

b) Graph the function and estimate the limiting value of the function. At most, how many students will hear the rumor?

c) Find the rate of change $N'(t)$.

tw d) Find $\lim\limits_{t \to \infty} N'(t)$ and explain its meaning.

---

[23]J. I. Lizaso et al., *Field Crops Research*, 80 (2003) 1–17.

SYNTHESIS

Effective Annual Yield. Suppose that $100 is invested at
7%, compounded continuously, for 1 yr. We know from
Example 4 that the balance will be $107.25. This is the
same as if $100 were invested at 7.25% and com-
pounded once a year (simple interest). The 7.25% is
called the *effective annual yield*. In general, if P_0 is invested at $k\%$ compounded continuously, then the effective annual yield is that number i satisfying
$P_0(1 + i) = P_0 e^k$. Then $1 + i = e^k$, or

$$\text{Effective annual yield} = i = e^k - 1.$$

39. An amount is invested at 7.3% per year com-
    pounded continuously. What is the effective annual
    yield?

40. An amount is invested at 8% per year compounded
    continuously. What is the effective annual yield?

41. The effective annual yield on an investment com-
    pounded continuously is 9.24%. At what rate was it
    invested?

42. The effective annual yield on an investment com-
    pounded continuously is 6.61%. At what rate was it
    invested?

43. Find an expression relating the growth rate k and
    the *tripling time T_3*.

44. Find an expression relating the growth rate k and
    the *quadrupling time T_4*.

45. Gather data concerning population growth in your
    city. Estimate the population in 2010; in 2020.

46. A quantity Q_1 grows exponentially with a doubling
    time of 1 yr. A quantity Q_2 grows exponentially
    with a doubling time of 2 yr. If the initial amounts
    of Q_1 and Q_2 are the same, when will Q_1 be twice
    the size of Q_2?

47. A growth rate of 100% per day corresponds to what
    exponential growth rate per hour?

48. Show that any two measurements of an exponen-
    tially growing population will determine k. That is,
    show that if y has the values y_1 at t_1 and y_2 at t_2,
    then

    $$k = \frac{\ln (y_2/y_1)}{t_2 - t_1}.$$

tw 49. Complete the table below, which relates growth rate
    k and doubling time T.

Growth Rate, k (in percent per year)	1%	2%			14%
Doubling Time, T (in years)				15	10

Graph $T = \ln 2/k$. Is this a linear relationship?
Explain.

tw 50. Explain the differences in the graphs of an expo-
    nential function and a logistic function.

tw 51. Explain the Rule of 69 to a fellow student.

4.4 Applications: Decay

OBJECTIVES

■ Find a function that satisfies *dP/dt = −kP*.
■ Convert between decay rate and half-life.
■ Solve applied problems involving exponential decay.

In the equation of population growth $dP/dt = kP$, the constant k is actually given by

$$k = (\text{Birthrate}) - (\text{Death rate}).$$

Thus a population "grows" only when the *birthrate* is greater than the *death rate*. When the birthrate is less than the death rate, k will be negative so the population will be decreasing, or "decaying," at a rate proportional to its size. For convenience in our computations, we will express such a negative value as $-k$, where $k > 0$. The equation

$$\frac{dP}{dt} = -kP, \quad \text{where } k > 0,$$

shows P to be *decreasing* as a function of time, and the solution

$$P(t) = P_0 e^{-kt}$$

shows it to be decreasing exponentially. This is called **exponential decay.** The amount present initially at $t = 0$ is again P_0. The constant k is called the **decay rate.**

Technology Connection

Exploratory

Using the same set of axes, graph $y_1 = e^{2x}$ and $y_2 = e^{-2x}$. Compare the graphs.

Then, using the same set of axes, graph $y_1 = 100e^{-0.06x}$ and $y_2 = 100e^{0.06x}$. Compare these graphs.

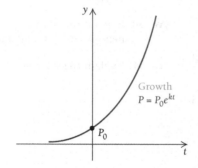

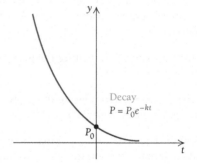

Radioactive Decay

Radioactive elements decay exponentially; that is, they disintegrate at a rate that is proportional to the amount present.

EXAMPLE 1 Strontium-90 is a cancer-causing isotope that is formed in fission reactors and nuclear bomb blasts. It has a decay rate of 2.41% per yr. The rate of change of an amount N is given by

$$\frac{dN}{dt} = -0.0241N.$$

a) Find the function that satisfies the equation in terms of N_0 (the amount present at $t = 0$).

b) Suppose that 1000 grams (g) of strontium-90 is present at $t = 0$. How much will remain after 100 yr?

c) After how long will half of the 1000 g remain?

Solution

a) $N(t) = N_0 e^{-0.0241t}$

b) $N(100) = 1000e^{-0.0241(100)}$

$\qquad = 1000e^{-2.41}$

$\qquad \approx 89.8 \text{ g}$

c) We are asking, "At what time T will $N(T) = 500$?" The number T is called the **half-life.** To find T, we solve the equation

$$500 = 1000e^{-0.0241T}$$

$$\frac{1}{2} = e^{-0.0241T}$$

$$\ln \frac{1}{2} = \ln e^{-0.0241T}$$

$$\ln 1 - \ln 2 = -0.0241T$$

$$0 - \ln 2 = -0.0241T$$

$$\frac{-\ln 2}{-0.0241} = T$$

$$28.8 \approx T.$$

Thus the half-life of strontium-90 is 28.8 yr. ∎

We can find a general expression relating the decay rate k and the half-life T by solving the equation

$$\tfrac{1}{2}P_0 = P_0 e^{-kT}$$

$$\tfrac{1}{2} = e^{-kT}$$

$$\ln \tfrac{1}{2} = \ln e^{-kT}$$

$$\ln 1 - \ln 2 = -kT$$

$$0 - \ln 2 = -kT$$

$$-\ln 2 = -kT$$

$$\ln 2 = kT.$$

Again, we have the following.

THEOREM 10

The *decay rate* k and the *half-life* T are related by the following equivalent equations:

$$kT = \ln 2 \approx 0.693147,$$

$$k = \frac{\ln 2}{T}, \quad \text{and}$$

$$T = \frac{\ln 2}{k}.$$

Thus the half-life T depends only on the decay rate k. In particular, it is independent of the initial population size.

The effect of half-life is shown in the radioactive decay curve that follows.

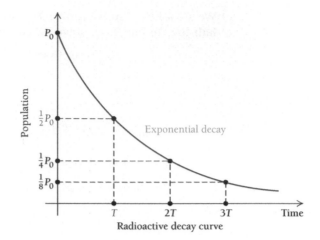

Radioactive decay curve

Note that the exponential function gets close to, but never reaches, 0 as t gets larger. Thus, in theory, a radioactive substance never completely decays.

EXAMPLE 2 *Half-life.* Plutonium-239, a common product and ingredient of nuclear reactors, is of great concern to those who are against the building of nuclear reactors. Its decay rate is 0.00286% per year. What is its half-life?

Solution We have

$$T = \frac{\ln 2}{k} = \frac{\ln 2}{0.0000286} \approx 24{,}200 \text{ yr.}$$

Thus the half-life of plutonium-239 is 24,200 yr. ■

EXAMPLE 3 *Carbon Dating.* The radioactive element carbon-14 has a half-life of 5750 yr. The percentage of carbon-14 present in the remains of plants and animals can be used to determine age. Archaeologists found that the linen wrapping from one of the Dead Sea Scrolls had lost 22.3% of its carbon-14. How old was the linen wrapping?

Solution

a) Find the decay rate k:

$$k = \frac{\ln 2}{T} = \frac{\ln 2}{5750} \approx 0.0001205, \quad \text{or} \quad 0.01205\% \text{ per year.}$$

b) Find the exponential equation for the amount $N(t)$ that remains from an initial amount N_0 after t years:

$$N(t) = N_0 e^{-0.0001205t}.$$

Note: This equation can be used for any subsequent carbon-dating problem.

How can scientists determine that an animal bone has lost 30% of its carbon-14? In order to determine this, we need to assume that the percentage of carbon-14 in the atmosphere is a constant. Since plants get their carbon from the atmosphere and animals get their carbon from plants, the percentage of carbon-14 in living animal bones is very close to the percentage of carbon-14 in the atmosphere. To determine the percentage of carbon-14, a small bone sample is exposed to a low-energy accelerator beam. Carbon ions are emitted, and heavier ions are deflected less than lighter ions. Counters are used to count how many ions of carbon-14 are emitted compared to the ions of the more common isotope of carbon-12.

Before accelerators were used, scientists had to destroy the sample and rely on counting the number of disintegrations of carbon-14 per minute. Using the old technique, it was very difficult to use carbon dating for samples older than 40,000 years. Furthermore, the sample was destroyed. With accelerators, it is possible to accurately date samples as old as 100,000 years, and only a tiny sample is required.

In 1947, a Bedouin youth looking for a stray goat climbed into a cave at Kirbet Qumran on the shores of the Dead Sea near Jericho and came upon earthenware jars containing an incalculable treasure of ancient manuscripts. Shown here are fragments of those so-called Dead Sea Scrolls, a portion of some 600 or so texts found so far that concern the Jewish books of the Bible. Officials date them before A.D. 70, making them the oldest Biblical manuscripts by 1000 years.

c) If the Dead Sea Scrolls lost 22.3% of their carbon-14 from an initial amount P_0, then 77.7% (P_0) is the amount present. To find the age t of the scrolls, we solve the following equation for t:

$$77.7\% \, N_0 = N_0 e^{-0.0001205t}$$

$$0.777 = e^{-0.0001205t}$$

$$\ln 0.777 = \ln e^{-0.0001205t}$$

$$\ln 0.777 = -0.0001205t$$

$$\frac{\ln 0.777}{0.0001205} = t$$

$$2094 \approx t.$$

Thus the linen wrapping of the Dead Sea Scrolls is about 2094 yr old.   ∎

Newton's Law of Cooling

A hot cup of soup, at a temperature of 200°, is placed in a room whose temperature is 70°. The temperature of the soup cools over time t, in minutes, according to the mathematical model, or equation, called **Newton's Law of Cooling.**

The temperature T of a cooling object drops at a rate that is proportional to the difference $T - C$, where C is the constant temperature of the surrounding medium. Thus,

$$\frac{dT}{dt} = -k(T - C). \tag{1}$$

Suppose we let $P = T - C$. Since C is a constant, $\dfrac{dP}{dt} = \dfrac{dT}{dt}$. Consequently, we can rewrite equation (1):

$$\frac{dP}{dt} = \frac{dT}{dt}$$
$$= -k(T - C)$$
$$= -kP.$$

This is simply the decay equation that we have already studied; its solution is $P(t) = P_0 e^{-kt}$. Since $T = P + C$, we have the following theorem.

THEOREM 11

Newton's Law of Cooling. The temperature of an object cooling in a surrounding medium with constant temperature C is

$$T = P_0 e^{-kt} + C,$$

where $P_0 = T(0) - C$. The constant P_0 is the initial temperature of the object minus the temperature of the surrounding medium.

EXAMPLE 4   *Scalding Coffee.* McDivett's Pie Shoppes, a national restaurant firm, finds that the temperature of its freshly brewed coffee is 130°. They are naturally concerned that if a customer spills hot coffee on themselves, a lawsuit might result. Room temperature in the restaurants is generally 72°. The temperature of the coffee cools to 120° after 4.3 min. The company determines that it is safer to serve the coffee at a temperature of 105°. How long does it take a cup of coffee to cool to 105°?

Solution

a) We first find the value of a P_0 in Newton's Law of Cooling. At $t = 0$, $T = 130°$. We see that

$$P_0 = 130 - 72 = 58.$$

Next, we find k using the fact that at $t = 4.3$, $T = 120°$. We solve the following equation for k:

$$120 = 58e^{-k \cdot (4.3)} + 72$$

$$48 = 58e^{-4.3k}$$

$$\frac{48}{58} = e^{-4.3k}$$

$$\ln \frac{48}{58} = \ln e^{-4.3k}$$

$$\ln \frac{48}{58} = -4.3k$$

$$0.044 \approx k$$

b) To see how long it will take the coffee to cool to 105°, we solve for t:

$$105 = 58e^{-0.044t} + 72$$

$$33 = 58e^{-0.044t}$$

$$\frac{33}{58} = e^{-0.044t}$$

$$\ln \frac{33}{58} = \ln e^{-0.044t}$$

$$\ln \frac{33}{58} = -0.044t$$

$$t \approx 12.8 \text{ min.}$$

Thus, if the coffee is allowed to cool for about 13 min, then it will be "safe" to serve. ◼

The graph of $T(t) = P_0e^{-kt} + C$ shows that $\lim_{t \to \infty} T(t) = C$. The temperature of the object decreases toward the temperature of the surrounding medium.

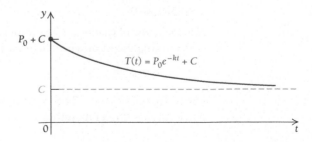

Mathematically, this model tells us that the temperature never reaches C, but in practice this happens eventually. At least, the temperature of the cooling object gets so close to that of the surrounding medium that no device could detect a difference. Let's now see how Newton's Law of Cooling can be used in solving a crime.

EXAMPLE 5   *When Was the Murder Committed?* The police discover the body of a calculus professor. Critical to solving the crime is determining when the murder was committed. The police call the coroner, who arrives at 12 noon. She immediately takes the temperature of the body and finds it to be 94.6°. She waits 1 hr, takes the temperature again, and finds it to be 93.4°. She also notes that the temperature of the room is 70°. When was the murder committed?

Solution   We first find P_0 in the equation $T(t) = P_0 e^{-kt} + C$. Assuming that the temperature of the body was normal when the murder occurred, we have $T = 98.6°$ at $t = 0$. Thus,

$$P_0 = 98.6 - 70 = 28.6.$$

Thus, T is given by $T(t) = 28.6e^{-kt} + 70$.

We want to find the number of hours N since the murder was committed. To do so, we must first determine k. From the two temperature readings the coroner made, we have

$$94.6 = 28.6e^{-kN} + 70, \quad \text{or} \quad 24.6 = 28.6e^{-kN}; \qquad (2)$$
$$93.4 = 28.6e^{-k(N+1)} + 70, \quad \text{or} \quad 23.4 = 28.6e^{-k(N+1)}. \qquad (3)$$

Dividing equation (2) by equation (3), we get

$$\frac{24.6}{23.4} = \frac{28.6e^{-kN}}{28.6e^{-k(N+1)}}$$
$$= e^{-kN+k(N+1)}$$
$$= e^{-kN+kN+k} = e^{k}.$$

We solve this equation for k:

$$\ln \frac{24.6}{23.4} = \ln e^{k} \qquad \text{Taking the natural logarithm on both sides}$$
$$0.05 \approx k.$$

Next, we substitute back into equation (2) and solve for N:

$$24.6 = 28.6e^{-0.05N}$$
$$\frac{24.6}{28.6} = e^{-0.05N}$$
$$\ln \frac{24.6}{28.6} = \ln e^{-0.05N}$$
$$\ln \frac{24.6}{28.6} \approx -0.05N$$
$$3 \approx N.$$

Since the coroner arrived at 12 noon, the murder was committed at about 9:00 A.M.

In summary, we have added several functions to our candidates for curve fitting. Let's review them.

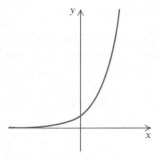

Exponential:
$f(x) = ab^x$, or ae^{kx}
$a, b > 0, k > 0$

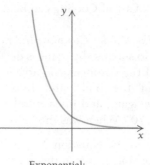

Exponential:
$f(x) = ab^{-x}$, or ae^{-kx}
$a, b > 0, k > 0$

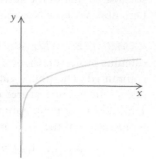

Logarithmic:
$f(x) = a + b \ln x$
$b > 0$

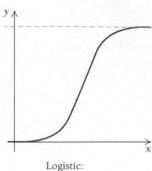

Logistic:
$f(x) = \dfrac{a}{1 + be^{-kx}}$

Now, when we analyze a set of data, we can consider these models, as well as linear, quadratic, polynomial, rational and trigonometric functions, for curve fitting.

Exercise Set 4.4

APPLICATIONS

Population Growth and Decay. For each of the following scatterplots, determine which, if any, of these functions might be used as a model for the data.

a) Quadratic: $f(x) = ax^2 + bx + c$
b) Polynomial, not quadratic
c) Exponential: $f(x) = ab^x$, or ae^{kx}, $k > 0$
d) Exponential: $f(x) = ab^x$, or ae^{-kx}, $k > 0$
e) Logarithmic: $f(x) = a + b \ln x$

f) Logistic: $f(x) = \dfrac{a}{1 + be^{-kx}}$

1.

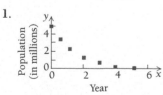

2.

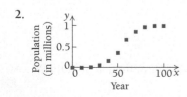

3.

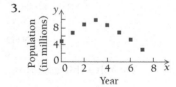

4.

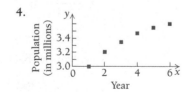

5.

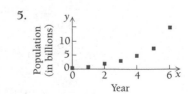

Life and Physical Sciences

Radioactive Decay. For Exercises 6–11, complete the following.

Radioactive Substance	Decay Rate, k	Half-life, T
6. Polonium-218		3.1 min
7. Lead-210		22 yr
8. Iodine-131	8.6% per day	
9. Strontium-92		2.7 hr
10. Uranium-238		4.47×10^9 yr
11. Plutonium-239		24,200 yr

12. *Half-Life.* Of an initial amount of 1000 g of lead, how much will remain after 100 yr? See Exercise 7 for the value of k.

13. *Half-Life.* Of an initial amount of 1000 g of polonium, how much will remain after 20 min? See Exercise 6 for the value of k.

14. *Carbon Dating.* How old is an ivory tusk that has lost 40% of its carbon-14?

15. *Carbon Dating.* How old is a piece of wood that has lost 90% of its carbon-14?

16. *Carbon Dating.* How old is a skeleton that has lost 50% of its carbon-14?

17. *Carbon Dating.* How old is a Chinese artifact that has lost 60% of its carbon-14?

18. *Ötzi the Iceman.* In 1992, two hikers in the Alps found a dead body. Only well after the police had arrived did anyone realize that the body was not from a recent death. Using carbon dating, it was found that the body was 5300 yr old. In 2001, a flint arrowhead was found in Ötzi's shoulder, indicating the cause of death.[24]

    a) What percentage of Ötzi's carbon-14 had he lost from the time he died until his body was discovered?

    b) What percentage of Ötzi's carbon-14 had he lost from the time he died until his cause of death was discovered?

19. *Siberian Yuribei Mammoth.* A sample of hair from a frozen Siberian Yuribei mammoth was carbon-dated and found to be approximately 10,000 yr old. In order for scientists to arrive at this conclusion, what percent of the carbon-14 in the mammoth's hair had decayed?[25]

---

[24]http://www.bris.ac.uk/Depts/Archaeology/events/iceman.html
[25]http://www.museum.state.il.us/exhibits/larson/mammuthus.html

20. *Cancer Treatment.* Iodine-125 is often used to treat cancer and has a half-life of 60.1 days.

    a) Suppose in a sample the amount of iodine-125 decreased by 25% while in storage. How long was that sample sitting on the shelf?

    b) Suppose that a sample of iodine-125 was on the shelf for 48 days. What percent of the iodine-125 remains in the sample?

A method to estimate the age of a rock is based on the fact that potassium K^{40} decomposes into argon Ar^{40} with a half-life of 1.3×10^9 yr. Since argon is a gas, it will escape from a molten rock. After the rock solidifies, the argon gas produced from K^{40} is trapped in the rock and can therefore be measured. By looking at the amount of K^{40} compared to the amount of Ar^{40} in the rock, one can estimate the percentage of K^{40} that has decayed since the rock formed, thus establishing the age of the rock.

21. *Radioactive Dating.* A meteorite is found to have only 8.84% of the K^{40} it contained when it was formed. Estimate its age based on the potassium-argon method. *Note:* Your answer gives one estimate of the earth's age since it is assumed that the earth formed at approximately the same time as the meteorites.

22. *Radioactive Dating.* A rock was found to have 30.2% of its original K^{40}. Estimate the age of the rock using the potassium-argon method.

23. *Radioactive Dating.* A rock was found to have 65.3% of its original K^{40}. Estimate the age of the rock using the potassium-argon method.

24. In a *chemical reaction*, substance A decomposes at a rate proportional to the amount of A present.

    a) Write an equation relating A to the amount left of an initial amount A_0 after time t.

    b) It is found that 10 lb of A will reduce to 5 lb in 3.3 hr. After how long will there be only 1 lb left?

25. In a *chemical reaction*, substance A decomposes at a rate proportional to the amount of A present.

    a) Write an equation relating A to the amount left of an initial amount A_0 after time t.

    b) It is found that 8 g of A will reduce to 4 g in 3 hr. After how long will there be only 1 g left?

26. *Weight Loss.* The initial weight of a starving animal is W_0. Its weight W after t days is given by

$$W = W_0 e^{-0.009t}.$$

    a) What percentage of its weight does it lose each day?

    b) What percentage of its initial weight remains after 30 days?

27. *Weight Loss.* The initial weight of a starving animal is W_0. Its weight W after t days is given by

$$W = W_0e^{-0.008t}.$$

a) What percentage of its weight does it lose each day?
b) What percentage of its initial weight remains after 30 days?

The Beer-Lambert Law

A beam of light enters a medium such as water or smoky air with initial intensity I_0. Its intensity is decreased depending on the thickness (or concentration) of the medium. The intensity I at a depth (or concentration) of x units is given by

$$I = I_0e^{-\mu x}.$$

The constant μ ("mu"), called the *coefficient of absorption,* varies with the medium.

---

28. *Light Through Smog.* Particulate concentrations of pollution reduce sunlight. In a smoggy area, $\mu = 0.01$ and x is the concentration of particulates measured in micrograms per cubic meter. What percentage of an initial amount I_0 of sunlight passes through smog that has a concentration of 100 micrograms per cubic meter?

29. *Sea water.* Light through sea water has $\mu = 1.4$ when x is measured in meters. What percentage of I_0 remains at a depth of sea water that is 1 m? 2 m? 3 m?

30. *Cooling.* The temperature of a hot liquid is 100°. The liquid is placed in a refrigerator where the temperature is 40°, and it cools to 90° in 5 min.

a) Find the value of the constant P_0 in Newton's Law of Cooling.
b) Find the value of the constant k.
c) What is the temperature after 10 min?
d) How long does it take the liquid to cool to 41°?
tw e) Find the rate of change of the temperature and interpret its meaning.

31. *Cooling.* The temperature of a hot liquid is 100° and the room temperature is 75°. The liquid cools to 90° in 10 min.

a) Find the value of the constant P_0 in Newton's Law of Cooling.
b) Find the value of the constant k. Round to the nearest hundredth.
c) What is the temperature after 20 min?
d) How long does it take the liquid to cool to 80°?
tw e) Find the rate of change of the temperature and interpret its meaning.

32. *Cooling Body.* The coroner arrives at the scene of a murder at 2 A.M. He takes the temperature of the body and finds it to be 61.6°. He waits 1 hr, takes the temperature again, and finds it to be 57.2°. The body is in a meat freezer, where the temperature is 10°. When was the murder committed?

33. *Cooling Body.* The coroner arrives at the scene of a murder at 11 P.M. She takes the temperature of the body and finds it to be 85.9°. She waits 1 hr, takes the temperature again, and finds it to be 83.4°. She notes that the room temperature is 60°. When was the murder committed?

34. *Population Decrease of Russia.* The population of Russia dropped from 150 million in 1995 to 144 million in 2002. Assume the population is decreasing according to the exponential-decay model.[26]

a) Find the value of k, and write the equation.
b) Estimate the population of Russia in 2010.
c) When will the population of Russia be 100 million?

35. *Population Decrease of the Ukraine.* The population of the Ukraine dropped from 51.9 million in 1995 to 48.8 million in 2001. Assume the population is decreasing according to the exponential-decay model.[27]

a) Find the value of k, and write the equation.
b) Estimate the population of the Ukraine in 2015.

---

[26]CIA—The World Factbook.
[27]CIA—The World Factbook.

c) After how many years will the population of the Ukraine be 1, according to this model?

36. *Atmospheric Pressure.* Atmospheric pressure P at altitude a is given by

$$P = P_0 e^{-0.00005a},$$

where P_0 is the pressure at sea level. Assume that $P_0 = 14.7 \ lb/in^2$ (pounds per square inch).

a) Find the pressure at an altitude of 1000 ft.
b) Find the pressure at an altitude of 20,000 ft.
c) At what altitude is the pressure $1.47 \ lb/in^2$?
tw d) Find the rate of change of the pressure and interpret its meaning.

37. *Satellite Power.* The power supply of a satellite is a radioisotope. The power output P, in watts, decreases at a rate proportional to the amount present; P is given by

$$P = 50 e^{-0.004t},$$

where t is the time, in days.

a) How much power will be available after 375 days?
b) What is the half-life of the power supply?
c) The satellite's equipment cannot operate on fewer than 10 watts of power. How long can the satellite stay in operation?
d) How much power did the satellite have to begin with?
tw e) Find the rate of change of the power and interpret its meaning.

38. *Carbon Dating.* Recently, while digging in Chaco Canyon, New Mexico, archeologists found corn pollen that was 4000 yr old. This was evidence that Indians had begun cultivating crops in the Southwest centuries earlier than scientists had thought. What percentage of the carbon-14 had been lost from the pollen?

39. *Forgetting.* In an art class, students were tested at the end of the course on a final exam. Then they were retested with an equivalent test at subsequent time intervals. Their scores after time t, in months, are given in the following table.

Time, t (in months)	Score, y
1	84.9%
2	84.6%
3	84.4%
4	84.2%
5	84.1%
6	83.9%

a) Use the **REGRESSION** feature on a grapher to fit a logarithmic function $y = a + b \ln x$ to the data.
b) Use the function to predict test scores after 8 mo; 10 mo; 24 mo; 36 mo.
c) After how long will the test scores fall below 82%?
tw d) Find the rate of change of the scores and interpret its meaning.

40. *Decline in Beef Consumption.* The annual consumption of beef per person has declined from 72 lb in 1980 to 64 lb in 2000. Assume consumption is decreasing according to the exponential-decay model.[28]

a) Find the value of k and write the equation.
b) Estimate the consumption of beef in 2010.
c) In what year (theoretically) will the consumption of beef be 20 lb per person?

41. *Decline in Pork Consumption.* The annual consumption of pork per person has declined from 52 lb in 1980 to 48 lb in 2000. Assume consumption is decreasing according to the exponential-decay model.[29]

a) Find the value of k and write the equation.
b) Estimate the consumption of pork in 2010.
c) In what year (theoretically) will the consumption of pork be 10 lb per person?

SYNTHESIS

42. *Newton's Law of Cooling.* Consider the following exploratory situation. Draw a glass of hot tap water. Place a thermometer in the glass and check the temperature. Check the temperature every 30 min thereafter. Plot your data on the graph on the following page, and connect the points with a smooth curve.

[28]http://www.armedia.org/vegstats.htm
[29]http://www.armedia.org/vegstats.htm

(Start) Time (in minutes)

a) What was the temperature when you began?
b) At what temperature does there seem to be a leveling off of the graph?
c) What is the difference between your answers to parts (a) and (b)?
d) How does the temperature in part (b) compare with the room temperature?
e) Find an equation that "fits" the data. Use this equation to check values of other data points. How do they compare?
f) Is it ever "theoretically" possible for the temperature of the water to be the same as the room temperature? Explain.

tw g) Find the rate of change of the temperature and interpret its meaning.

4.5 The Derivatives of a^x and $\log_a x$

The Derivative of a^x

OBJECTIVES

■ Convert exponential expressions to powers of e.
■ Differentiate functions involving a^x and $\log_a x$.

To find the derivative of a^x, for any base $a > 0$, we first express a^x as a power of e. Recall that property P7 for logarithms states that

$$a = e^{\ln a}$$

Raising both sides to the power x, we obtain,

$$a^x = (e^{\ln a})^x = e^{x \cdot \ln a}.$$

Thus we have the following.

THEOREM 12

$$a^x = e^{x \cdot \ln a}$$

EXAMPLE 1 Express as a power of e.

a) 3^2

$$3^2 = e^{2 \cdot \ln 3}$$
$$\approx e^{2.1972}$$

b) 10^x $10^x = e^{x \cdot \ln 10}$

$\qquad\qquad \approx e^{2.3026x}$

Now we can differentiate.

EXAMPLE 2

$$\frac{d}{dx} 2^x = \frac{d}{dx} e^{x \cdot \ln 2} \qquad\qquad \text{Theorem 12}$$

$$= \left[\frac{d}{dx}(x \cdot \ln 2)\right] \cdot e^{x \cdot \ln 2}$$

$$= (\ln 2)(e^{\ln 2})^x$$

$$= (\ln 2)2^x$$

We completed this by taking the derivative of $x \ln 2$ and replacing $e^{x \ln 2}$ with 2^x.

In general,

$$\frac{d}{dx} a^x = \frac{d}{dx} e^{x \cdot \ln a} \qquad \text{Theorem 12}$$

$$= \left[\frac{d}{dx}(x \cdot \ln a)\right] \cdot e^{x \cdot \ln a}$$

$$= (\ln a)\, a^x.$$

Thus we have the following.

THEOREM 13

$$\frac{d}{dx} a^x = (\ln a)\, a^x$$

EXAMPLE 3

$$\frac{d}{dx} 3^x = (\ln 3)\, 3^x$$

EXAMPLE 4

$$\frac{d}{dx}(1.4)^x = (\ln 1.4)(1.4)^x$$

Compare these formulas:

$$\frac{d}{dx} a^x = (\ln a)\, a^x \quad \text{and} \quad \frac{d}{dx} e^x = e^x.$$

It is the simplicity of the latter formula that is a reason for the use of base e in calculus. The many applications of e in natural phenomena provide other reasons.

The Derivative of $\log_a x$

Just as the derivative of a^x is expressed in terms of $\ln a$, so too is the derivative of $\log_a x$. To find this derivative, we first express $\log_a x$ in terms of $\ln a$ using property 7:

$$a^{\log_a x} = x.$$

Then

$$\ln (a^{\log_a x}) = \ln x$$

$$(\log_a x) \cdot \ln a = \ln x \qquad \text{By P3, treating } \log_a x \text{ as an exponent}$$

and

$$\log_a x = \boxed{\frac{1}{\ln a}} \cdot \ln x.$$

$$\underset{\text{constant}}{\underline{\qquad}}$$

The derivative of $\log_a x$ follows.

THEOREM 14

$$\frac{d}{dx} \log_a x = \frac{1}{\ln a} \cdot \frac{1}{x}$$

Comparing this with

$$\frac{d}{dx} \ln x = \frac{1}{x},$$

we again see a reason for the use of base e in calculus.

EXAMPLE 5

$$\frac{d}{dx} \log_3 x = \frac{1}{\ln 3} \cdot \frac{1}{x}$$

■

EXAMPLE 6   *Acidity.* The pH or acidity of a solution is defined to be $-\log [H^+]$, the negative logarithm of the hydronium ion concentration (measured in moles per liter). Recall that $\log x = \log_{10} x$.

a) Given that the pH of a solution is 5.1, what is the hydronium ion concentration?

b) Find the rate of change of the pH with respect to the hydronium ion concentration.

Technology Connection

Check the results of Examples 3–5 using a grapher. Then differentiate $y = \log_2 x$ and check the result on the grapher.

Azaleas prefer slightly acidic soil.

Solution

a) Since the solution has a pH of 5.1, we have:

$$5.1 = -\log[H^+]$$
$$-5.1 = \log[H^+]$$
$$10^{-5.1} = [H^+]$$
$$7.94 \times 10^{-6} \approx [H^+] \qquad \text{Using a scientific calculator for approximation}$$

b) Let $x = [H^+]$. Then we are asked to compute $\dfrac{d}{dx}(-\log_{10} x)$.

$$\frac{d}{dx}(-\log x) = -\frac{1}{\ln(10)}\frac{1}{x}.$$

EXAMPLE 7

$$\frac{d}{dx}x^2 \log x = x^2 \frac{1}{\ln 10} \cdot \frac{1}{x} + 2x \log x \qquad \text{By the Product Rule}$$

$$= \frac{x}{\ln 10} + 2x \log x, \quad \text{or} \quad x\left(\frac{1}{\ln 10} + 2\log x\right)$$

The formula

$$\log_a x = \frac{\ln x}{\ln a}$$

is a special case of the *change of base formula* given by

$$\log_a x = \frac{\log_b x}{\log_b a}.$$

EXAMPLE 8   Write $\log_2 x$ in terms of $\log_{10} x$.

Solution   The change of base formula states that

$$\log_2 x = \frac{\log_{10} x}{\log_{10} 2}.$$

EXAMPLE 9   Compute $\dfrac{d}{dx}\log_x 5$.

Solution   We use the change of base formula to write

$$\frac{d}{dx}\log_x 5 = \frac{d}{dx}\frac{\ln 5}{\ln x}$$

$$= \ln 5 \cdot \frac{d}{dx}(\ln x)^{-1}$$

$$= \ln 5(-1)(\ln x)^{-2} \cdot \frac{1}{x}$$

$$= -\frac{\ln 5}{x(\ln x)^2}.$$

Exercise Set 4.5

Express as a power of e.

1. 5^4
2. 2^3
3. $(3.4)^{10}$
4. $(5.3)^{20}$
5. 4^k
6. 5^R
7. 8^{kT}
8. 10^{kR}

Differentiate.

9. $y = 6^x$
10. $y = 7^x$
11. $f(x) = 10^x$
12. $f(x) = 100^x$
13. $f(x) = x(6.2)^x$
14. $f(x) = x(5.4)^x$
15. $y = x^3 10^x$
16. $y = x^4 5^x$
17. $y = \log_4 x$
18. $y = \log_5 x$
19. $f(x) = 2 \log x$
20. $f(x) = 5 \log x$
21. $f(x) = \log \dfrac{x}{3}$
22. $f(x) = \log \dfrac{x}{5}$
23. $y = x^3 \log_8 x$
24. $y = x \log_6 x$
25. $y = \csc x \log_2 x$
26. $y = \cot x \log_5 (2x)$
27. $y = \log_{10} \sin x$
28. $y = \log_2 \sec x$
29. $y = \log_x 3$
30. $y = x \log_x 10$
31. $g(x) = (\log_x 10)(\log_{10} x)$
32. $g(x) = 2^x \log_2 x$

APPLICATIONS

Business and Economics

33. *Recycling Aluminum Cans.* It is known that one-fourth of all aluminum cans distributed will be recycled each year. A beverage company distributes 250,000 cans. The number still in use after time t, in years, is given by
$$N(t) = 250{,}000\left(\tfrac{1}{4}\right)^t.$$
    a) Find $N'(t)$.
    tw b) Interpret the meaning of $N'(t)$.

34. *Double Declining-Balance Depreciation.* An office machine is purchased for $5200. Under certain assumptions, its salvage value V depreciates according to a method called double declining balance, basically 80% each year, and is given by
$$V(t) = \$5200(0.80)^t,$$
    where t is the time, in years.
    a) Find $V'(t)$.
    tw b) Interpret the meaning of $V'(t)$.

Earthquake Magnitude. The magnitude R (measured on the Richter scale) of an earthquake of intensity I is defined as
$$R = \log \frac{I}{I_0},$$

where I_0 is a minimum intensity used for comparison. When one earthquake is 10 times as intense as another, its magnitude on the Richter scale is 1 higher. If one earthquake is 100 times as intense as another, its magnitude on the Richter scale is 2 higher, and so on. Thus an earthquake whose magnitude is 7 on the Richter scale is 10 times as intense as an earthquake whose magnitude is 6. Earthquakes can be interpreted as multiples of the minimum intensity I_0.

35. In 1986, there was an earthquake near Cleveland, Ohio. It had an intensity of $10^5 \cdot I_0$. What was its magnitude on the Richter scale?

36. On October 17, 1989, there was an earthquake in San Francisco, California, during the World Series. It had an intensity of $10^{6.9} \cdot I_0$. What was its magnitude on the Richter scale?

This photograph shows part of the damage in the San Francisco, California, area earthquake in 1989.

37. *Earthquake Intensity.* The intensity I of an earthquake is given by
$$I = I_0 10^R,$$
    where R is the magnitude on the Richter scale and I_0 is the minimum intensity, where $R = 0$, used for comparison.
    a) Find I, in terms of I_0, for an earthquake of magnitude 7 on the Richter scale.
    b) Find I, in terms of I_0, for an earthquake of magnitude 8 on the Richter scale.
    c) Compare your answers to parts (a) and (b).
    d) Find the rate of change dI/dR.
    tw e) Interpret the meaning of dI/dR.

38. *Intensity of Sound.* The intensity of a sound is given by
$$I = I_0 10^{0.1L},$$

where L is the loudness of the sound as measured in decibels and I_0 is the minimum intensity detectable by the human ear.

a) Find I, in terms of I_0, for the loudness of a power mower, which is 100 decibels.
b) Find I, in terms of I_0, for the loudness of just audible sound, which is 10 decibels.
c) Compare your answers to parts (a) and (b).
d) Find the rate of change dI/dL.
tw e) Interpret the meaning of dI/dL.

39. *Earthquake Magnitude.* The magnitude R (measured on the Richter scale) of an earthquake of intensity I is defined as

$$R = \log \frac{I}{I_0},$$

where I_0 is the minimum intensity (used for comparison).

a) Find the rate of change dR/dI.
tw b) Interpret the meaning of dR/dI.

40. *Loudness of Sound.* The loudness L of a sound of intensity I is defined as

$$L = 10 \log \frac{I}{I_0},$$

where I_0 is the minimum intensity detectable by the human ear and L is the loudness measured in decibels. (The exponential form of this definition is given in Exercise 38.)

a) Find the rate of change dL/dI.
tw b) Interpret the meaning of dL/dI.

41. *Response to Drug Dosage.* The response y to a dosage x of a drug is given by

$$y = m \log x + b.$$

The response may be hard to measure with a number. The patient might perspire more, have an increase in temperature, or faint.

a) Find the rate of change dy/dx.
tw b) Interpret the meaning of dy/dx.

42. *Acidity.* Pure water is neutral with a pH of 7.

a) What is the concentration of the hydronium ions in pure water?
b) Suppose that during an experiment the concentration of the hydronium ions is given by $x = 0.001t + 10^{-7}$, where $0 \le t \le 100$ is the time measured in seconds. Find a formula relating the pH of the solution to the time t.
c) Use your answer to part (b) to compute the rate of change of the pH of the solution.
d) What is the pH at time 0?

e) At what time does the pH of the solution change most rapidly? What is the pH of the solution at that time?

43. *Hydroxide and Hydronium Concentrations.* For a solution in water, $[\text{H}^+][\text{OH}^-] = 10^{-14}$ where $[\text{H}^+]$ is the concentration of the hydronium ions and $[\text{OH}^-]$ is the concentration of the hydroxide ions, both measured in moles per liter. Suppose that a base is added to a solution in such a way that the concentration of the hydroxide ions is given by

$$[\text{OH}^-] = x = 0.002t + 10^{-7},$$

where $0 \le t \le 60$ is time measured in seconds. Use the definition of pH given above to answer the questions.

a) What is the concentration of the hydroxide ions at time 0?
b) What is the concentration of the hydronium ions at time 0?
c) What is the pH at time 0?
d) Find a formula that gives the pH at time $0 \le t \le 60$.
e) Find the rate of change of the pH in this solution.
f) At what time does the pH of the solution change most rapidly? What is the pH of the solution at that time?

SYNTHESIS

Differentiate.

44. $y = 2^{x^4}$

45. $y = x^x, x > 0$

46. $y = \log_3 (x^2 + 1)$

47. $f(x) = x^{e^x}, x > 0$

48. $y = a^{f(x)}$

49. $y = \log_a f(x), f(x)$ positive

50. $y = [f(x)]^{g(x)}, f(x)$ positive

tw 51. Explain in your own words how to justify the formula for finding the derivative of $f(x) = a^x$.

tw 52. Explain in your own words how to justify the formula for finding the derivative of $f(x) = \log_a x$.

Chapter 4 Summary and Review

Terms to Know

Exponential function, p. 265
Base, p. 265
The number e, p. 269
Logarithm, p. 278
Logarithmic base, p. 278
Common logarithm, p. 282
Natural logarithm, p. 282
Exponential equation, p. 284

Log-log plot, p. 290
Differential equation, p. 297
Exponential growth, p. 297
Initial-value problem, p. 298
Exponential growth rate, p. 299
Interest compounded continuously, p. 299
Doubling time, p. 300

Generation time, 300
Logistic equation, p. 305
Exponential decay, p. 314
Decay rate, p. 314
Half-life, p. 315
Carbon dating, p. 316
Newton's Law of Cooling, p. 317

Review Exercises

These review exercises are for test preparation. They can also be used as a lengthened practice test. Answers are at the back of the book. The answers also contain bracketed section references, which tell you where to restudy if your answer is incorrect.

Differentiate.

1. $y = \ln x$

2. $y = e^x$

3. $y = \ln (x^4 + 5)$

4. $y = e^{2\sqrt{x}}$

5. $f(x) = \ln (\sin x + x)$

6. $f(x) = e^{4x} + x^4$

7. $f(x) = \dfrac{\ln x}{\tan x}$

8. $f(x) = e^{x^2} \cdot \ln 4x$

9. $f(x) = e^{4x} - \ln \dfrac{x}{4}$

10. $g(x) = x^8 - 8 \ln x$

11. $y = \dfrac{\ln e^x}{e^x}$

Given $\log_a 2 = 1.8301$ and $\log_a 7 = 5.0999$, find each of the following.

12. $\log_a 14$

13. $\log_a \frac{2}{7}$

14. $\log_a 28$

15. $\log_a 3.5$

16. $\log_a \sqrt{7}$

17. $\log_a \frac{1}{4}$

18. Find the function that satisfies $dQ/dt = kQ$. List the answer in terms of Q_0.

19. *Population Growth.* The population of Boomtown doubled in 16 yr. What was the growth rate of the city? Round to the nearest tenth of a percent.

20. *Interest Compounded Continuously.* Suppose that $8300 is deposited in a savings and loan associa-

tion in which the interest rate is 6.8%, compounded continuously. How long will it take for the $8300 to double itself? Round to the nearest tenth of a year.

21. *Cost of a Prime-Rib Dinner.* The average cost C of a prime-rib dinner was $4.65 in 1962. In 2005, it was $25.38. Assuming that the exponential-growth model applies:

  a) Find the exponential-growth rate, and write the equation.
  b) What will the cost of such a dinner be in 2015? in 2020?

22. *Franchise Growth.* A clothing firm is selling franchises throughout the United States and Canada. It is estimated that the number of franchises N will increase at the rate of 12% per year, that is,

$$\frac{dN}{dt} = 0.12N,$$

where t is the time, in years.

  a) Find the function that satisfies the equation, assuming that the number of franchises in 2000 ($t = 0$) is 60.
  b) How many franchises will there be in 2010?
  c) After how long will the number of franchises be 120? Round to the nearest tenth of a year.

23. *Decay Rate.* The decay rate of a certain radioactive isotope is 13% per year. What is its half-life? Round to the nearest tenth of a year.

24. *Half-Life.* The half-life of radon-222 is 3.8 days. What is its decay rate? Round to the nearest tenth of a percent.

25. *Decay Rate.* A certain radioactive element has a decay rate of 7% per day, that is,

$$\frac{dA}{dt} = -0.07A,$$

where A is the amount of the element present at time t, in days.

a) Find a function that satisfies the equation if the amount of the element present at $t = 0$ is 800 g.
b) After 20 days, how much of the 800 g will remain? Round to the nearest gram.
c) After how long does half the original amount remain?

26. *The Hullian Learning Model.* The probability p of mastering a certain assembly line task after t learning trials is given by

$$p(t) = 1 - e^{-0.7t}.$$

a) What is the probability of learning the task after 1 trial? 2 trials? 5 trials? 10 trials? 14 trials?
b) Find the rate of change $p'(t)$.
tw c) Interpret the meaning of $p'(t)$.
d) Sketch a graph of the function.

Differentiate.
27. $y = 3^x$ **28.** $f(x) = \log_{15} x$

SYNTHESIS

29. Differentiate: $y = \dfrac{e^{2x} + e^{-2x}}{e^{2x} - e^{-2x}}$.

30. Find the minimum value of $f(x) = x^4 \ln 4x$.

 Technology Connection

31. Graph: $f(x) = \dfrac{e^{1/x}}{(1 + e^{1/x})^2}$.

32. Find $\displaystyle\lim_{x \to 0} \dfrac{e^{1/x}}{(1 + e^{1/x})^2}$.

33. *Spread of SARS.* SARS, a respiratory disease, was first detected in late 2002. Starting in March 2003, the World Health Organization kept track of the number of reported cases throughout the world. The following data shows the total number of cases reported to the WHO through the given day.[30]

Date	Days since March 17	Cumulative Number of Reported Cases
March 27	10	1408
March 31	14	1622
April 15	29	3235
April 21	35	3861
April 25	39	4649
May 1	45	5865
May 6	50	6727
May 10	54	7296

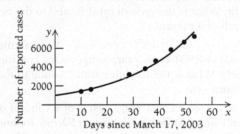

a) Use the **REGRESSION** feature on a grapher to fit an exponential function $y = a \cdot b^t$ to the data. Then convert that formula to an exponential function, base e, where t is the number of days since March 14. What is the exponential growth rate?
b) Assuming the spread of SARS continued according to the model of part (a), estimate the number of cases reported after 100 days; after 200 days.
c) What is the doubling time?
d) Based on your answer in part (a), determine how long until there are 6.2 billion reported cases. (Note that 6.2 billion is the approximate population of the world in 2003.)
tw e) Is the model you developed realistic? Why or why not?

---

[30]WHO Web page http://www.who.int/csr/sars/country/en/

Chapter 4 Test

Differentiate.

1. $y = e^x$

2. $y = (\tan x) \ln x$

3. $f(x) = e^{-x^2}$

4. $f(x) = \ln \dfrac{x}{7}$

5. $f(x) = e^x - 5x^3$

6. $f(x) = 3e^x \ln x$

7. $y = \ln (e^x - \sin x)$

8. $y = \dfrac{\ln x}{e^x}$

Given $\log_b 2 = 0.2560$ and $\log_b 9 = 0.8114$, find each of the following.

9. $\log_b 18$ **10.** $\log_b 4.5$ **11.** $\log_b 3$

12. Find the function that satisfies $dM/dt = kM$. List the answer in terms of M_0.

13. The doubling time of a certain bacteria culture is 4 hr. What is the growth rate? Round to the nearest tenth of a percent.

14. *Interest Compounded Continuously.* An investment is made at 6.931% per year, compounded continuously. What is the doubling time? Round to the nearest year.

15. *Cost of Bread.* The average cost C of a 1-lb loaf of white bread in January 1993 was $0.748. In January 2003, it was $1.042. Assume the exponential-growth model applies.[31]

  a) Find the exponential-growth rate, and write the equation.
  b) Find the cost of a loaf of bread in 2010 and 2020.
  c) What was the cost of a loaf of bread the year you were born?

16. *Drug Dosage.* A dose of a drug is injected into the body of a patient. The drug amount in the body decreases at the rate of 10% per hour, that is,

$$\frac{dA}{dt} = -0.1A,$$

where A is the amount in the body and t is the time, in hours.

  a) A dose of 3 cubic centimeters (cc) is administered. Find the function that satisfies the equation.

  b) How much of the initial dose of 3 cc will remain after 10 hr?
  c) After how long does half the original dose remain?

17. *Decay Rate.* The decay rate of zirconium-88 is 0.83% per day. What is its half-life?

18. *Half-Life.* The half-life of cesium-135 is 2,300,000 yr. What is its decay rate?

19. *Effect of Advertising.* A company introduces a new product on a trial run in a city. They advertise the product on television and find that the percentage P of people who buy the product after t ads have been run satisfies the function

$$P(t) = \frac{100\%}{1 + 24e^{-0.28t}}.$$

  a) What percentage buys the product without having seen the ad ($t = 0$)?
  b) What percentage buys the product after the ad has been run 1 time? 5 times? 10 times? 15 times? 20 times? 30 times? 35 times?
  c) Find the rate of change $P'(t)$.
tw **d)** Interpret the meaning of $P'(t)$.
  e) Sketch a graph of the function.

Differentiate.

20. $f(x) = 20^x$

21. $y = \log_{20} x$

SYNTHESIS

22. Differentiate: $y = x (\ln x)^2 - 2x \ln x + 2x$.

23. Find the maximum and minimum values of $f(x) = x^4 e^{-x}$ over $[0, 10]$.

 Technology Connection

24. Graph: $f(x) = \dfrac{e^x - e^{-x}}{e^x + e^{-x}}$.

25. Find $\lim\limits_{x \to 0} \dfrac{e^x - e^{-x}}{e^x + e^{-x}}$.

---

[31]U.S. Department of Labor, Bureau of Labor Statistics (http://www.bls.gov/).

26. *Growth Rate.* The bacteria *Clostridium perfringens* is a common bacterial agent in foodborn-disease outbreaks. The following data measures the growth of *Clostridium perfringens* at 40.6°C.[32]

Hours	Number of Bacteria
2	7
3	330
4	3640
5	60,000

[32]M. Juneja, "Predictive model for growth of *Clostridium perfringens* during cooling of cooked chicken," *Food Microbiology*, Vol 19, 313–327 (2002).

a) Use the REGRESSION feature on a grapher to fit an exponential function $y = a \cdot b^t$ to the data. Then convert that formula to an exponential function, base e, where t is the number of hours.

b) Estimate the number of bacteria after 6 hr; after 7 hr.

c) After how many hours will the number of bacteria be approximately the population of people on Earth (6.3 billion)?

d) What is the generation time for the bacteria?

Extended Life Science Connection

MAXIMUM SUSTAINABLE HARVEST

In certain situations, biologists are able to determine what is called a **reproduction curve.** This is a function

$$y = f(P)$$

such that if P is the population at a certain time t, then the population 1 yr later, at time $t + 1$, is $f(P)$. Such a curve is shown below.

description of f, then we know that the population stays the same from year to year. But the graph of f above lies mostly above the line, indicating that, in this case, the population is increasing.

Too many deer in a forest can deplete the food supply and eventually cause the

The line $y = P$ is significant for two reasons. First, if $y = P$ is a

population to decrease for lack of food. Often in such cases and with some controversy, hunters are allowed to "harvest" some of the deer. Then with a greater food supply, the remaining deer population might actually prosper and increase.

We know that a population P will grow to a population $f(P)$ in 1 yr. If this were a population of fur-bearing animals and the population were increasing, then one could "harvest" the amount

$$f(P) - P$$

each year without depleting the initial population P. If the population were remaining the same or decreasing, then such a harvest would deplete the population.

Suppose that we want to know the value of P_0 that would allow the harvest to be the largest. If we could determine that P_0, then we could let the population grow until it reached that level, and then begin harvesting year after year the amount $f(P_0) - P_0$.

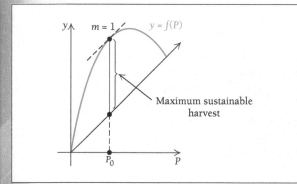

Let the harvest function H be given by

$$H(P) = f(P) - P.$$

Then $\qquad H'(P) = f'(P) - 1.$

Now, if we assume that $H'(P)$ exists for all values of P and that there is only one critical point, it follows that the **maximum sustainable harvest** occurs at that value P_0 such that

$$H'(P_0) = f'(P_0) - 1 = 0$$

and $\qquad H''(P_0) = f''(P_0) < 0.$

Or equivalently, we have the following.

THEOREM

The **maximum sustainable harvest** occurs at P_0 such that

$$f'(P_0) = 1 \quad \text{and} \quad f''(P_0) < 0,$$

and is given by

$$H(P_0) = f(P_0) - P_0.$$

EXERCISES

For each reproduction curve in Exercises 1–3, do the following.

a) Graph the reproduction function, the function $y = P$, and the harvest function using the same set of axes or viewing window.
b) Find the population at which the maximum sustainable harvest occurs. Use both a graphical solution and a calculus solution.
c) Find the maximum sustainable harvest.

1. $f(P) = \ln(20P)$, where P is measured in thousands.
2. $f(P) = -0.025P^2 + 4P$, where P is measured in thousands. This is the reproduction curve in the Hudson bay area for the snowshoe hare, a fur-bearing animal.

3. $f(P) = \ln 8x(x + 1)$, where P is measured in thousands.

For each reproduction curve in Exercises 4 and 5, do the following.

a) Graph the reproduction function, the function $y = P$, and the harvest function using the same set of axes or viewing window.
b) Find the population at which the maximum sustainable harvest occurs. Use just a graphical solution.
c) Find the maximum sustainable harvest.

4. $f(P) = 40\sqrt{P}$, where P is measured in thousands. Assume that this is the reproduction curve for the brown trout population in a large lake.

5. $f(P) = 5.27(e^{1.03x(1-x)} - 1)$, where P is measured in thousands.

6. The following table lists data regarding the reproduction of a certain animal.

a) Use the REGRESSION feature on a grapher to fit a cubic polynomial to these data.
b) Graph the reproduction function, the function $y = P$, and the harvest function using the same set of axes or viewing window.
c) Find the population at which the maximum sustainable harvest occurs. Use just a graphical solution.

Population, P (in thousands)	Population, $f(P)$, One Year Later
10	9.7
20	23.1
30	37.4
40	46.2
50	42.6

5

Integration

5.1 Integration
5.2 Areas and Accumulations
5.3 The Fundamental Theorem of Calculus
5.4 Properties of Definite Integrals
5.5 Integration Techniques: Substitution
5.6 Integration Techniques: Integration by Parts
5.7 Integration Techniques: Tables and Technology
5.8 Volume
5.9 Improper Integrals

AN APPLICATION The forces (in newtons) on the human gastrocnemius tendon during traction and during recoil can be approximated by the functions

$$f(x) = 71.3x - 4.15x^2 + 0.434x^3 \quad \text{and}$$
$$g(x) = 71.0x - 10.3x^2 + 0.986x^3,$$

respectively, where $0 \leq x \leq 11$ is the tendon elongation in millimeters. Compute the *elastic strain energy*, given by the area bounded by the graphs of the two functions.[1]

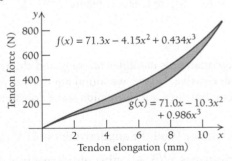

This problem appears as Exercise 33 in Section 5.4.

[1]N. Maganaris and J. P. Paul, "Tensile properties of the *in vivo* human gastrocnemius tendon," *Journal of Biomechanics*, 35 (2002), pp. 1639–1646.

INTRODUCTION *Suppose that we do the reverse of differentiating; that is, suppose we try to find a function whose derivative is a given function. This process is called* antidifferentiation, *or* integration; *it is the main topic of this chapter and the second main branch of calculus, the first being differentiation. We will see that we can use integration to find area under a curve over a closed interval, to find the accumulation of a certain quantity over an interval, to calculate the average value of a function, and to compute volume.*

We first consider the meaning of integration and then learn several techniques for integrating.

5.1 Integration

OBJECTIVES

■ Find indefinite integrals and antiderivatives.
■ Solve applied problems involving antiderivatives.

In Chapters 2, 3, and 4, we have considered several interpretations of the derivative, some of which are listed below.

Function	Derivative
Distance	Velocity
Velocity	Acceleration
Population	Population growth rate

Many applications of mathematics involve the reverse of differentiation, called *antidifferentiation*. For example, if we know a velocity function in a particular situation, we can use antidifferentiation to find a distance function.

The Antiderivative

Suppose that y is a function of x and that the derivative is the constant 8. Can we find y? It is easy to see that one such function is $8x$. That is, $8x$ is a function whose derivative is 8. Is there another function whose derivative is 8? In fact, there are many. Here are some examples:

$$8x + 3, \qquad 8x - 10, \qquad 8x + 7.4, \qquad 8x + \sqrt{2}.$$

All these functions are $8x$ plus some constant. There are no other functions having a derivative of 8 other than those of the form $8x + C$. Another way of saying this is that any two functions having a derivative of 8 must differ by a constant. In general, any two functions having the same derivative differ by a constant.

THEOREM 1

If two functions F and G have the same derivative over an interval, then

$$F(x) = G(x) + C, \quad \text{where } C \text{ is a constant.}$$

The reverse of differentiating is *antidifferentiating,* and the result of antidifferentiating is called an **antiderivative.** Above, we found antiderivatives of the function 8. There are several of them, but they are all $8x$ plus some constant.

EXAMPLE 1 Antidifferentiate (find the antiderivatives of) x^2.

Solution One antiderivative is $x^3/3$. All other antiderivatives differ from this by a constant, so we can denote them as follows:

$$\frac{x^3}{3} + C.$$

To check this, differentiate $x^3/3 + C$:

$$\frac{d}{dx}\left(\frac{x^3}{3} + C\right) = 3\left(\frac{x^2}{3}\right) = x^2.$$

So $x^3/3 + C$ is an antiderivative of x^2. ∎

The solution to Example 1 is the *general form* of the antiderivative of x^2.

Integrals and Integration

The process of antidifferentiating is often called **integration.** To indicate that the antiderivative of x^2 is $x^3/3 + C$, we write

$$\int x^2 \, dx = \frac{x^3}{3} + C, \tag{1}$$

and we note that

$$\int f(x) \, dx$$

is the symbolism we will use from now on to call for the antiderivative of a function $f(x)$. More generally, we can write

$$\int f(x) \, dx = g(x) + C, \tag{2}$$

where $g(x) + C$ is the general form of the antiderivative of $f(x)$. Equation (2) is read "the *indefinite integral* of $f(x)$ is $g(x) + C$." The constant C is called the *constant of integration* and can have any fixed value; hence the use of the word "indefinite."

The symbolism $\int f(x) \, dx$, from Leibniz, is called an *integral,* or more precisely, an **indefinite integral.** The symbol $\int$ is called an *integral sign.* The left side of equation (2) is often read "the integral of $f(x)$, dx" or more briefly, "the integral of $f(x)$." In this context, $f(x)$ is called the *integrand.* Think of "dx" as indicating that the variable involved in the integration is x.

EXAMPLE 2   Evaluate: $\int x^9 \, dx$.

Solution   We have

$$\int x^9 \, dx = \frac{x^{10}}{10} + C.$$

We can check by differentiating the antiderivative. The derivative of $\dfrac{x^{10}}{10} + C$ is x^9, which is the integrand. ∎

EXAMPLE 3   Evaluate: $\int 5e^{4x} \, dx$.

Solution   We have

$$\int 5e^{4x} \, dx = \frac{5}{4}e^{4x} + C.$$

We can check by differentiating the antiderivative. The derivative of $\frac{5}{4}e^{4x} + C$ is $5e^{4x}$, which is the integrand. ∎

CAUTION   Look closely at the difference between Example 2 and Example 3. In Example 2, the base of the exponent is the variable and the power is a constant. In Example 3, the situation is reversed, and so a different rule applies.

$$\int e^x \, dx \neq \frac{e^{x+1}}{x+1} + C$$

To integrate (or antidifferentiate), we make use of differentiation formulas, in effect, reading them in reverse. Below are some of these stated as integration formulas. Each can be checked by differentiating the right-hand side and noting that the result is the integrand.

THEOREM 2

Basic Integration Formulas

1. $\displaystyle \int k \, dx = kx + C$ (k is a constant)

2. $\displaystyle \int x^r \, dx = \frac{x^{r+1}}{r+1} + C,$ provided $r \neq -1$

   (To integrate a power of x other than -1, increase the power by 1 and divide by the increased power.)

3. $\displaystyle \int x^{-1} \, dx = \int \frac{1}{x} \, dx = \int \frac{dx}{x} = \ln |x| + C$

   (When $x < 0$, $\ln x$ is not defined. In order to handle the cases when $x > 0$ and when $x < 0$ in one formula, we use $|x|$.)

4. $\displaystyle \int e^{ax} \, dx = \frac{1}{a} e^{ax} + C$

5. $\displaystyle \int \sin ax \, dx = -\frac{1}{a} \cos ax + C$

6. $\displaystyle \int \cos ax \, dx = \frac{1}{a} \sin ax + C$

7. $\displaystyle \int \sec^2 ax \, dx = \frac{1}{a} \tan ax + C$

8. $\displaystyle \int \csc^2 ax \, dx = -\frac{1}{a} \cot ax + C$

9. $\displaystyle \int \sec ax \tan ax \, dx = \frac{1}{a} \sec ax + C$

10. $\displaystyle \int \csc ax \cot ax \, dx = -\frac{1}{a} \csc ax + C$

The following rules combined with the preceding formulas allow us to find many integrals. They can be derived by reversing familiar differentiation rules.

THEOREM 3

Rule A. $\displaystyle\int kf(x)\,dx = k\int f(x)\,dx$

(The integral of a constant times a function is the constant times the integral of the function.)

Rule B. $\displaystyle\int [f(x) + g(x)]\,dx = \int f(x)\,dx + \int g(x)\,dx$

(The integral of a sum is the sum of the integrals.)

Rule C. $\displaystyle\int [f(x) - g(x)]\,dx = \int f(x)\,dx - \int g(x)\,dx$

(The integral of a difference is the difference of the integrals.)

CAUTION There is no Product or Quotient Rule for integration.

EXAMPLE 4

$$\int (5x + 4x^3)\,dx = \int 5x\,dx + \int 4x^3\,dx \qquad \text{Rule B}$$

$$= 5\int x\,dx + 4\int x^3\,dx \qquad \text{Rule A}$$

Then

$$\int (5x + 4x^3)\,dx = 5\cdot\frac{x^2}{2} + 4\cdot\frac{x^4}{4} + C = \frac{5}{2}x^2 + x^4 + C.$$

Don't forget the constant of integration. It is not necessary to write two constants of integration. If we did, we could add them and consider C as the sum.

We can check this as follows:

$$\frac{d}{dx}\left(\frac{5}{2}x^2 + x^4 + C\right) = 2\cdot\frac{5}{2}\cdot x + 4x^3 = 5x + 4x^3.$$

∎

We can always check an integration by differentiating.

EXAMPLE 5

$$\int \left(7e^{6x} - \sqrt{x}\right)dx = \int 7e^{6x}\,dx - \int \sqrt{x}\,dx$$

$$= \int 7e^{6x}\,dx - \int x^{1/2}\,dx$$

$$= \frac{7}{6}e^{6x} - \frac{x^{(1/2)+1}}{\frac{1}{2}+1} + C$$

$$= \frac{7}{6}e^{6x} - \frac{x^{3/2}}{\frac{3}{2}} + C$$

$$= \frac{7}{6}e^{6x} - \frac{2}{3}x^{3/2} + C$$

∎

EXAMPLE 6

$$\int \left(1 - \frac{3}{x} + \frac{1}{x^4}\right) dx = \int 1 \, dx - 3 \int \frac{dx}{x} + \int x^{-4} \, dx$$

$$= x - 3 \ln |x| + \frac{x^{-4+1}}{-4+1} + C$$

$$= x - 3 \ln |x| - \frac{1}{3x^3} + C \qquad \blacksquare$$

EXAMPLE 7

$$\int (2 \sin 5x + 3 \cos 4x) \, dx = 2 \int \sin 5x \, dx + 3 \int \cos 4x \, dx$$

$$= -\frac{2}{5} \cos 5x + \frac{3}{4} \sin 4x + C \qquad \blacksquare$$

Another Look at Antiderivatives

The graphs of the antiderivatives of x^2 are the graphs of the functions

$$y = \int x^2 \, dx = \frac{x^3}{3} + C$$

for the various values of the constant C.

As we can see in the figures below, x^2 is the derivative of each function; that is, the tangent line at the point

$$\left(a, \frac{a^3}{3} + C\right)$$

has slope a^2. The curves $(x^3/3) + C$ fill up the plane, with exactly one curve going through any given point (x_0, y_0).

Technology Connection

Exploratory

Using the same set of axes, graph

$$y = x^4 - 4x^3$$

and

$$y = x^4 - 4x^3 + 7.$$

Compare the graphs. Draw a tangent line to each graph at $x = -1$, $x = 0$, and $x = 1$ and find the slope of each line. Compare the slopes. Then find the derivative of each function. Compare the derivatives. Is either function an antiderivative of $y = 4x^3 - 12x^2$?

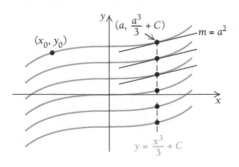

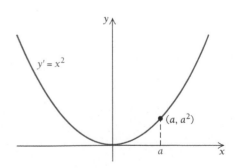

Suppose that we look for an antiderivative of x^2 with a specified value at a certain point—say, $f(-1) = 2$. We find that there is only one such function.

EXAMPLE 8   Find the function f such that

$$f'(x) = x^2$$

and

$$f(-1) = 2.$$

Solution

a) We find $f(x)$ by integrating:

$$f(x) = \int x^2 \, dx = \frac{x^3}{3} + C.$$

b) The condition $f(-1) = 2$ allows us to find C:

$$f(-1) = \frac{(-1)^3}{3} + C = 2.$$

Solving for C, we get

$$-\tfrac{1}{3} + C = 2$$
$$C = 2 + \tfrac{1}{3} = \tfrac{7}{3}.$$

Thus, $f(x) = \dfrac{x^3}{3} + \dfrac{7}{3}.$ ■

The equation $f'(x) = x^2$ is a *differential equation*, because it involves the derivative of a function. Furthermore, by specifying that $f(-1) = 2$, we have specified an *initial condition* or *initial value* for the solution to the differential equation. Sometimes the term *boundary value* is used instead of *initial value*. A differential equation together with an initial value constitute an **initial-value problem.**

Finding Velocity and Distance from Acceleration

Recall that the position coordinate at time t of an object moving along a number line is $s(t)$. Then

$$s'(t) = v(t) = \text{the } \textit{velocity} \text{ at time } t,$$
$$v'(t) = a(t) = \text{the } \textit{acceleration} \text{ at time } t.$$

EXAMPLE 9   *Distance.* Suppose that $v(t) = 5t^4$ and $s(0) = 9$. Find $s(t)$.

Solution

a) We find $s(t)$ by integrating:

$$s(t) = \int v(t) \, dt = \int 5t^4 \, dt = t^5 + C.$$

b) We determine C by using the initial condition $s(0) = 9$, which is the position of s at time $t = 0$:

$$s(0) = 0^5 + C = 9$$
$$C = 9.$$

Thus, $s(t) = t^5 + 9.$ ■

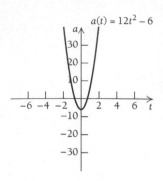

EXAMPLE 10 *Distance.* Suppose that $a(t) = 12t^2 - 6$, the initial velocity is $v(0) = 5$ and the initial position is $s(0) = 10$. Find $s(t)$.

Solution

a) We find $v(t)$ by integrating $a(t)$:

$$v(t) = \int a(t) \, dt$$

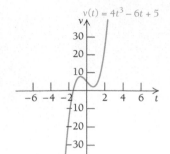

$$= \int (12t^2 - 6) \, dt$$

$$= 4t^3 - 6t + C_1.$$

b) The condition $v(0) = 5$ allows us to find C_1:

$$v(0) = 4 \cdot 0^3 - 6 \cdot 0 + C_1 = 5$$

$$C_1 = 5.$$

Thus, $v(t) = 4t^3 - 6t + 5$.

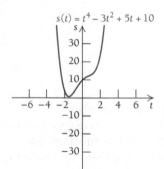

c) We find $s(t)$ by integrating $v(t)$:

$$s(t) = \int v(t) \, dt$$

$$= \int (4t^3 - 6t + 5) \, dt$$

$$= t^4 - 3t^2 + 5t + C_2.$$

d) The condition $s(0) = 10$ allows us to find C_2:

$$s(0) = 0^4 - 3 \cdot 0^2 + 5 \cdot 0 + C_2 = 10$$

$$C_2 = 10.$$

Thus, $s(t) = t^4 - 3t^2 + 5t + 10$. ∎

Finding Populations from Growth Rates

EXAMPLE 11 A population of bacteria grows at the rate of $P'(t) = 150e^{0.03t}$, where t is the time measured in hours. Also, $P(0) = 7000$.

a) Find $P(t)$, the number of bacteria in the population at time t.

b) Find $P(4)$, the number of bacteria at time 4 hr.

Solution

a) We find $P(t)$ by integrating $P'(t)$.

$$P(t) = \int P'(t) \, dt$$

$$= \int 150e^{0.03t} \, dt$$

$$= \frac{150}{0.03} e^{0.03t} + C$$

$$= 5000e^{0.03t} + C.$$

We now determine the constant C.

$$7000 = P(0)$$
$$7000 = 5000 + C$$
$$2000 = C.$$

So the number of bacteria is $P(t) = 5000e^{0.03t} + 2000$.

b) $P(4) = 5000e^{0.03(4)} + 2000$

$$\approx 7637. \qquad \text{Using a scientific calculator for approximation}$$

Exercise Set 5.1

Evaluate.

1. $\displaystyle\int x^6\, dx$

2. $\displaystyle\int x^7\, dx$

3. $\displaystyle\int 2\, dx$

4. $\displaystyle\int 4\, dx$

5. $\displaystyle\int x^{1/4}\, dx$

6. $\displaystyle\int x^{1/3}\, dx$

7. $\displaystyle\int (x^2 + x - 1)\, dx$

8. $\displaystyle\int (x^2 - x + 2)\, dx$

9. $\displaystyle\int (t^2 - 2t + 3)\, dt$

10. $\displaystyle\int (3t^2 - 4t + 7)\, dt$

11. $\displaystyle\int 5e^{8x}\, dx$

12. $\displaystyle\int 3e^{5x}\, dx$

13. $\displaystyle\int (w^3 - w^{8/7})\, dw$

14. $\displaystyle\int (t^4 - t^{6/5})\, dt$

15. $\displaystyle\int \frac{1000}{r}\, dr$

16. $\displaystyle\int \frac{500}{x}\, dx$

17. $\displaystyle\int \frac{dx}{x^2}$

18. $\displaystyle\int \frac{dx}{x^3}$

19. $\displaystyle\int \sqrt{s}\, ds$

20. $\displaystyle\int \sqrt[3]{x^2}\, dx$

21. $\displaystyle\int \frac{-6}{\sqrt[3]{x^2}}\, dx$

22. $\displaystyle\int \frac{20}{\sqrt[5]{x^4}}\, dx$

23. $\displaystyle\int 8e^{-2x}\, dx$

24. $\displaystyle\int 7e^{-0.25x}\, dx$

25. $\displaystyle\int \left(x^2 - \frac{3}{2}\sqrt{x} + x^{-4/3}\right) dx$

26. $\displaystyle\int \left(x^4 + \frac{1}{8\sqrt{x}} - \frac{4}{5}x^{-2/5}\right) dx$

27. $\displaystyle\int 5 \sin 2\pi\theta\, d\theta$

28. $\displaystyle\int \frac{1}{4}\cos 2\theta\, d\theta$

29. $\displaystyle\int (5 \sin 5x - 4 \cos 2x)\, dx$

30. $\displaystyle\int (3 \cos 2\pi x) - 8 \sin \pi x)\, dx$

31. $\displaystyle\int 3 \sec^2 3x\, dx$

32. $\displaystyle\int 5 \csc^2 2x\, dx$

33. $\displaystyle\int \frac{1}{3}\sec\frac{x}{9}\tan\frac{x}{9}\, dx$

34. $\displaystyle\int 4 \sec 2x \tan 2x\, dx$

35. $\displaystyle\int (\sec x + \tan x)\sec x\, dx$

36. $\displaystyle\int (\csc x + \cot x)\csc x\, dx$

37. $\displaystyle\int \left[\frac{1}{t} + \frac{1}{t^2} - \frac{1}{e^t}\right] dt$

38. $\displaystyle\int [4e^{7w} - w^{-2/5} + 7w^{100}]\, dw$

Find f such that:

39. $f'(x) = x - 3,\quad f(2) = 9$

40. $f'(x) = x - 5,\quad f(1) = 6$

41. $f'(x) = x^2 - 4,\quad f(0) = 7$

42. $f'(x) = x^2 + 1,\quad f(0) = 8$

43. $f'(x) = 2 \cos 3x,\quad f(0) = 1$

44. $f'(x) = 3 \sin 8x,\quad f(0) = 3$

45. $f'(x) = 5e^{2x},\quad f(0) = -10$

46. $f'(x) = 3e^{-x},\quad f(0) = 6$

APPLICATIONS

47. *Population Growth.* A city grows at a rate of $157t + 1000$ people per year, where t is time in years after the beginning of 2005. Given that the population at the beginning of 2005 is 156,239, find the population at the beginning of 2007.

48. *Population Growth.* A county grows at a rate of $500 + 10t + 0.3t^3$ people per year, where t is the number of months after the beginning of 2006. On January 1, 2006, the population of the county is 452,937. Find the population at the end of 2010.

Find $s(t)$.

49. $v(t) = 3t^2$, $s(0) = 4$ **50.** $v(t) = 2t$, $s(0) = 10$

Find $v(t)$.

51. $a(t) = 4t$, $v(0) = 20$ **52.** $a(t) = 6t$, $v(0) = 30$

Find $s(t)$.

53. $a(t) = -2t + 6$, $v(0) = 6$, and $s(0) = 10$

54. $a(t) = -6t + 7$, $v(0) = 10$, and $s(0) = 20$

55. *Distance.* For a freely falling object, $a(t) = -32$ ft/sec², $v(0) =$ initial velocity $= v_0$, and $s(0) =$ initial height $= s_0$. Find a general expression for $s(t)$ in terms of v_0 and s_0.

56. *Time.* A ball is thrown from a height of 10 ft, where $s(0) = 10$, at an initial velocity of 80 ft/sec, where $v(0) = 80$. How long will it take before the ball hits the ground? (See Exercise 55.)

57. *Distance.* A car with constant acceleration goes from 0 to 60 mph in $\frac{1}{2}$ min. How far does the car travel during that time?

58. *Braking Distance.* A car can brake with an acceleration of -68.5 ft/sec². It travels on the highway at 70 mi/hr (or 102.7 ft/sec). How far does the car travel from the time the brakes are applied until the car stops?

59. *Braking Distance.* A car can brake with an acceleration of -68.5 ft/sec². It travels on a highway at 90 mi/hr (or 132 ft/sec). How far does the car travel from the time the brakes are applied until the car stops?

60. *Area of a Healing Wound.* The area A of a healing wound is decreasing at a rate given by

$$A'(t) = -43.4t^{-2}, \quad 1 \le t \le 7,$$

where t is the time, in days, and A is in square centimeters.

a) Find $A(t)$ if $A(1) = 39.7$.
b) Find the area of the wound after 7 days.

61. *Memory.* In a certain memory experiment, the rate of memorizing is given by

$$M'(t) = 0.2t - 0.003t^2,$$

where $M(t)$ is the number of Spanish words memorized in t minutes.

a) Find $M(t)$ if it is known that $M(0) = 0$.
b) How many words are memorized in 8 min?

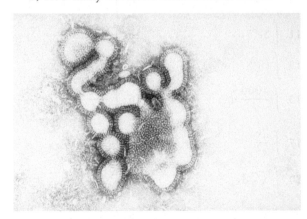

62. *Spread of Influenza.* During 34 weeks in the 2000–2001 flu season, the rate of reported influenza per 100,000 population in Ireland could be approximated by

$$I'(t) = 3.389e^{0.1049t},$$

where I is the total number of people per 100,000 who have contracted influenza and t is time measured in weeks.[2]

---

[2]Department of Public Health.

a) Estimate $I(t)$, the total number who have contracted influenza by time t per 100,000 population of Ireland given that $I(0) = 0$.

b) Approximately how many people per 100,000 contracted influenza during the first 27 wk?

c) Approximately how many people per 100,000 contracted influenza during the whole 34 wk?

d) Approximately how many people per 100,000 contracted influenza during the last 7 wk of the 34 wk?

63. *Spread of SARS.* For a few months, the rate that SARS spread could be approximated by

$$N'(t) = 38.2e^{0.0376t},$$

where $N(t)$ is the total number of cases reported to the World Health Organization up to t days after March 17, 2003.[3]

a) Given that $N(0) = 1622$, find the function $N(t)$.

b) Compute $N(30)$.

c) Compute $N(60)$.

SYNTHESIS

Find f.

64. $f'(t) = \sqrt{t} + \dfrac{1}{\sqrt{t}}, \quad f(4) = 0$

65. $f'(t) = t^{\sqrt{3}}, \quad f(0) = 8$

---

[3]WHO Web page http://www.who.int/csr/sars/country/en/

Evaluate. Each of the following can be integrated using the rules developed in this section, but some algebra may be required beforehand.

66. $\displaystyle\int (5t + 4)^2\, dt$

67. $\displaystyle\int (x - 1)^2 x^3\, dx$

68. $\displaystyle\int (1 - t)\sqrt{t}\, dt$

69. $\displaystyle\int \dfrac{(t + 3)^2}{\sqrt{t}}\, dt$

70. $\displaystyle\int \dfrac{x^4 - 6x^2 - 7}{x^3}\, dx$

71. $\displaystyle\int (t + 1)^3\, dt$

72. $\displaystyle\int \dfrac{1}{\ln 10}\,\dfrac{dx}{x}$

73. $\displaystyle\int be^{ax}\, dx$

74. $\displaystyle\int (3x - 5)(2x + 1)\, dx$

75. $\displaystyle\int \sqrt[3]{64x^4}\, dx$

76. $\displaystyle\int \dfrac{x^2 - 1}{x + 1}\, dx$

77. $\displaystyle\int \dfrac{t^3 + 8}{t + 2}\, dt$

78. $\displaystyle\int \cos x \tan x\, dx$

79. $\displaystyle\int (\cos^3 x + \cos x \sin^2 x)\, dx$

80. $\displaystyle\int \tan^2 3x\, dx$ (*Hint:* Use a trigonometric identity.)

81. $\displaystyle\int \cot^2 2x\, dx$

tW 82. On a test, a student makes the statement, "The function $f(x) = x^2$ has a unique integral." Discuss.

tW 83. Describe the graphical interpretation of an antiderivative.

5.2 Areas and Accumulations

OBJECTIVES

- Approximate areas with rectangles.
- Define Riemann sums and definite integrals.
- Find accumulations using integrals.

We now pursue a completely different approach to integration. This approach is based on the concept of area, but it is also applicable to a wide range of computations. These include volume and applications that involve accumulations.

Limits of Sums

Let's consider the area under a curve over an interval $[a, b]$. We divide $[a, b]$ into subintervals of equal length and construct rectangles, the sum of whose areas approximates the area under the curve.

In the following figure, $[a, b]$ has been divided into 4 subintervals, each having width $\Delta x = (b - a)/4$.

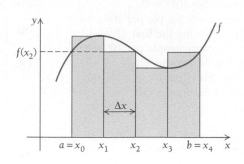

The area under a curve can be approximated by a sum of rectangular areas.

The heights of the rectangles shown are

$$f(x_1), \quad f(x_2), \quad f(x_3), \quad \text{and} \quad f(x_4).$$

The area of the region under the curve is approximately the sum of the areas of the four rectangles:

$$f(x_1)\,\Delta x + f(x_2)\,\Delta x + f(x_3)\,\Delta x + f(x_4)\,\Delta x.$$

We can denote this sum with **summation notation,** which uses the Greek capital letter sigma, Σ:

$$\sum_{i=1}^{4} f(x_i)\,\Delta x, \quad \text{or} \quad \sum_{i=1}^{4} f(x_i)\,\Delta x.$$

This is read "the sum of the numbers $f(x_i)\,\Delta x$ from $i = 1$ to $i = 4$." To recover the original expression, we substitute the numbers 1 through 4 successively into $f(x_i)\,\Delta x$ and write plus signs between the results.

Before we continue, let's consider some examples involving summation notation.

EXAMPLE 1 Write summation notation for $2 + 4 + 6 + 8 + 10$.

Solution

$$2 + 4 + 6 + 8 + 10 = \sum_{i=1}^{5} 2i.$$

EXAMPLE 2 Write summation notation for

$$g(x_1)\,\Delta x + g(x_2)\,\Delta x + \cdots + g(x_{19})\,\Delta x.$$

Solution

$$g(x_1)\,\Delta x + g(x_2)\,\Delta x + \cdots + g(x_{19})\,\Delta x = \sum_{i=1}^{19} g(x_i)\,\Delta x.$$

EXAMPLE 3 Express

$$\sum_{i=1}^{4} 3^i$$

without using summation notation.

Solution

$$\sum_{i=1}^{4} 3^i = 3^1 + 3^2 + 3^3 + 3^4 = 120.$$

EXAMPLE 4 Express

$$\sum_{i=1}^{30} h(x_i)\,\Delta x$$

without using summation notation.

Solution

$$\sum_{i=1}^{30} h(x_i)\,\Delta x = h(x_1)\,\Delta x + h(x_2)\,\Delta x + \cdots + h(x_{30})\,\Delta x.$$

Approximation of area by rectangles becomes more accurate as we use smaller subintervals and hence more rectangles, as shown in the following figures.

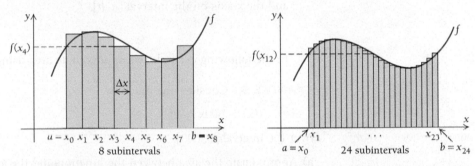

In general, suppose that the interval $[a, b]$ is divided into n equal subintervals, each of width $\Delta x = (b - a)/n$. We construct rectangles with heights

$$f(x_1), f(x_2), \ldots, f(x_n).$$

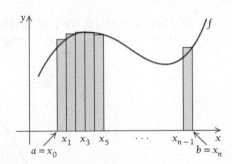

The width of each rectangle is Δx, so the first rectangle has area

$$f(x_1)\,\Delta x,$$

the second rectangle has area

$$f(x_2)\,\Delta x,$$

and so on. The area of the region under the curve is approximated by the sum of the areas of the rectangles:

$$\sum_{i=1}^{n} f(x_i)\,\Delta x.$$

This sum is called a **Riemann sum.** We obtain the actual area by finding the limit of the Riemann sums[4] as n grows without bound.

Historical Note: Riemann sums are named after the German mathematician G. F. Bernhard Riemann, 1826–1866. Riemann received his doctorate at the University of Göttingen and did much of his research in geometry.

DEFINITION

Let f be a continuous function and $a < b$. The limit as $n \to \infty$ of the Riemann sum is the **definite integral** of f from a to b. We write the definite integral as

$$\int_{a}^{b} f(x)\,dx = \lim_{n\to\infty} \sum_{i=1}^{n} f(x_i)\Delta x.$$

The numbers a and b are the **limits of integration.** If $f(x) \geq 0$ on $[a, b]$, then this definite integral represents the area between the curve $y = f(x)$ and the x-axis on the interval $[a, b]$.

In the following example, we approximate area using Riemann sums.

Historically, integral notation comes from the notation used for Riemann sums. The integral sign $\int$, which looks like an elongated "S," in the integral replaces the sigma used in the Riemann sum (Σ). Furthermore, the Δx in the Riemann sum is replaced by the notation dx.

EXAMPLE 5   Consider the function

$$f(x) = 600x - x^2$$

over the interval $[0, 600]$.

a) Approximate the area between the function and the x-axis by dividing the interval into 6 subintervals.

b) Approximate the area between the function and the x-axis by dividing the interval into 12 subintervals.

Solution

a) The interval $[0, 600]$ is divided into 6 subintervals, each of length $\Delta x = (600 - 0)/6 = 100$, and x_i ranges from $x_0 = 0$ to $x_6 = 600$.

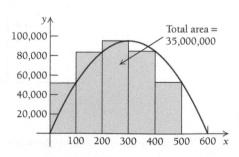

---

[4]The definition we give for Riemann sums assumes a continuous function. The definition is also valid for bounded functions that are continuous except at a finite number of points. To extend the definition to more general functions, we would have to allow the lengths of subintervals to be different and allow any point in a subinterval to determine the rectangle height.

i	x_i	$f(x_i)$
1	100	50,000
2	200	80,000
3	300	90,000
4	400	80,000
5	500	50,000
6	600	0

i	x_i	$f(x_i)$
1	50	27,500
2	100	50,000
3	150	67,500
4	200	80,000
5	250	87,500
6	300	90,000
7	350	87,500
8	400	80,000
9	450	67,500
10	500	50,000
11	550	27,500
12	600	0

Thus we have

$$\sum_{i=1}^{6} f(x_i)\,\Delta x = f(100) \cdot 100 + f(200) \cdot 100 + f(300) \cdot 100 + f(400) \cdot 100$$
$$+ f(500) \cdot 100 + f(600) \cdot 100$$
$$= 50{,}000 \cdot 100 + 80{,}000 \cdot 100 + 90{,}000 \cdot 100 + 80{,}000 \cdot 100$$
$$+ 50{,}000 \cdot 100 + 0 \cdot 100$$
$$= 35{,}000{,}000.$$

b) The interval $[0, 600]$ is divided into 12 subintervals, each of length $\Delta x = (600 - 0)/12 = 50$, and x_i ranges from $x_0 = 0$ to $x_{12} = 600$.

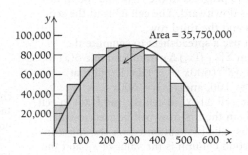

Thus we have

$$\sum_{i=1}^{12} f(x_i)\,\Delta x = f(50) \cdot 50 + f(100) \cdot 50 + f(150) \cdot 50$$
$$+ f(200) \cdot 50 + f(250) \cdot 50 + f(300) \cdot 50 + f(350) \cdot 50$$
$$+ f(400) \cdot 50 + f(450) \cdot 50 + f(500) \cdot 50$$
$$+ f(550) \cdot 50 + f(600) \cdot 50$$

$$= 27{,}500 \cdot 50 + 50{,}000 \cdot 50 + 67{,}500 \cdot 50 + 80{,}000 \cdot 50$$
$$+ 87{,}500 \cdot 50 + 90{,}000 \cdot 50 + 87{,}500 \cdot 50 + 80{,}000 \cdot 50$$
$$+ 67{,}500 \cdot 50 + 50{,}000 \cdot 50 + 27{,}500 \cdot 50 + 0 \cdot 50$$

$$= 35{,}750{,}000.$$

In Example 5, we approximated the integral $\int_0^{600} (600x - x^2)\, dx$. We will see in Section 5.3 that the actual integral is 36,000,000. The approximation with 12 rectangles is more accurate than the approximation with 6 rectangles. In general, as the number of rectangles increases, the accuracy of the approximation given by the Riemann sum becomes closer to the definite integral.

Technology Connection

Riemann Sums

To make the computations easier, a Riemann sum can be organized in a spreadsheet. We will use Microsoft *Excel*. With minor modifications, the directions below apply to any spreadsheet, including the spreadsheet on the TI-83+ and certain other graphing calculators.

Each *cell* on the spreadsheet is labeled with a letter and a number. The upper-left cell is labeled A1. The letters progress to the right and the numbers progress downward. The cell B3 is in the second column and the third row.

We first use a spreadsheet to compute the Riemann sum $\sum_{i=1}^{6} f(x_i)\,\Delta x$ from Example 5(a) to approximate $\int_0^{600}(600x - x^2)\,dx$. Recall that $x_0 = 0$, $\Delta x = 100$, and $f(x) = 600x - x^2$.

Type **i** in cell A1, **xi** in cell B1, and **f(xi)** in cell C1. The cells in the first row will be used to label their columns. The first column gives the subscript on x, the second column is the value of x_i, and the third column is $f(x_i)$.

	A	B	C
1	*i*	*xi*	*f(xi)*

In column A, type **0, 1, 2, 3, 4, 5, 6** in rows 2, 3, 4, 5, 6, 7, and 8 as shown below to indicate the subscript values.

	A
1	*i*
2	0
3	1
4	2
5	3
6	4
7	5
8	6

In cell B2, type 0 to indicate that $x_0 = 0$. Cell B3 should contain x_1. Since $\Delta x = x_1 - x_0$, we know that $x_1 = x_0 + \Delta x = x_0 + 100$. We enter this formula by typing **=B2+100** in cell B3. The equal sign tells the spreadsheet to compute the value.

SUM		▼ × ✓ *f*
	A	B
1	*i*	*xi*
2	0	0
3	1	=B2+100

We do not need the value for $f(x_0)$, so we leave cell C2 blank. To enter the value for $f(x_1)$, in cell C3, type **=600*B3-B3^2** as shown below. It is important to put the equal sign at the beginning to let the spreadsheet know to evaluate the cell and * is needed to indicate multiplication. Cell C3 computes 600 times the value of B3 minus the value of B3 squared. That is, C3 shows the value of $600x_1 - x_1^2$.

SUM		▼ × ✓ *fx* =600*B3–B3^2
	A	B	C
1	*i*	*xi*	*f(xi)*
2	0	0	
3	1	100	=600*B3–B3^2

We could type all the formulas we need, but this would be tedious. Spreadsheets have the capability of copying a cell and changing the cell references to reflect the position change. Let's try using this feature to complete the column for x_i. Use the mouse to highlight cells B3 through B8 in column B. In the **EDIT** menu select **FILL**, then select **DOWN**. The resulting column is shown below.

	B
	xi
	0
	100
	200
	300
	400
	500
	600

Click on cell B4 to check that the formula is **=B3+100**. That is, the value in cell B4 is 100 more than the value in the cell just above it, in the same way that the value in cell B3 was set to be 100 more than the value of the cell just above it. The **FILL DOWN** operation copies the top cell into all the highlighted cells, but it changes cell references to be relative to the target cell.

Highlight cells C3 through C8 in column C and use the **FILL DOWN** feature to compute the value of the function $f(x_i)$ for each $1 \leq i \leq 6$. Click on cell C4 to check that the formula is **=600*B4–B4^2**. Notice how the formula in cell C3 was copied to cells C4 through C8, but the cell referenced in each case is the cell just to the left.

(continued)

We compute the sum of the values in cells C3 through C8 by typing =**SUM(C3:C8)** in cell C9. In cell A9, type **TOTAL** to label the sum.

SUM		▼ ✕ ✓ *fx* =SUM(C3:C8)	
	A	B	C
1	*i*	*xi*	*f(xi)*
2	0	0	
3	1	100	50000
4	2	200	80000
5	3	300	90000
6	4	400	80000
7	5	500	50000
8	6	600	0
9	Total		=SUM(C3:C8)

We now multiply the total by Δx to complete the calculation.

$$\sum_{i=1}^{6} f(x_i)\,\Delta x = \Delta x \sum_{i=1}^{6} f(x_i)$$
$$= 100(350{,}000)$$
$$= 35{,}000{,}000$$

We can do the last step in the spreadsheet by typing =**100*C9** in cell C11. Label this as the integral by typing **INTEGRAL** in cell A11.

	A	B	C
1	*i*	*xi*	*f(xi)*
2	0	0	
3	1	100	50000
4	2	200	80000
5	3	300	90000
6	4	400	80000
7	5	500	50000
8	6	600	0
9	Total		350000
10			
11	Integral		35000000

EXERCISES

1. Do Example 5(b) using a spreadsheet.
2. With a spreadsheet, estimate $\int_0^{600} (600x - x^2)\,dx$ using the Riemann sum $\sum_{i=1}^{24} f(x_i)\,\Delta x$.
3. Use a spreadsheet to estimate $\int_1^4 (2x + 3x^2 + x^3)\,dx$ using the Riemann sum $\sum_{i=1}^{5} f(x_i)\,\Delta x$.
4. Repeat Exercise 3 using the Riemann sum $\sum_{i=1}^{10} f(x_i)\,\Delta x$.

In Example 5, we computed two different Riemann sums. We now compute the limit of Riemann sums to determine a definite integral.

EXAMPLE 6   Use limits of Riemann sums to compute $\displaystyle\int_0^4 x^2\,dx$. Interpret the answer as an area.

Solution   Instead of computing the Riemann sum for a fixed value of n, we seek a formula for the Riemann sum for a general value n. This will allow us to compute the limit as n approaches infinity. Since we integrate over the interval $[0, 4]$, each subinterval has length

$$\Delta x = \frac{4 - 0}{n} = \frac{4}{n}.$$

As seen in the graph,

$$x_0 = 0,$$
$$x_1 = x_0 + \Delta x = \frac{4}{n},$$
$$x_2 = x_1 + \Delta x = \frac{4}{n} + \frac{4}{n} = 2 \cdot \frac{4}{n},$$
$$x_3 = x_2 + \Delta x = 2 \cdot \frac{4}{n} + \frac{4}{n} = 3 \cdot \frac{4}{n}, \quad \text{and so on.}$$

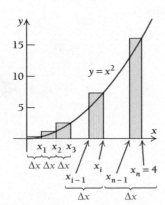

In general,

$$x_i = i \cdot \frac{4}{n}.$$

Using $f(x) = x^2$, we write the Riemann sum.

$$\sum_{i=1}^{n} f(x_i)\,\Delta x = \sum_{i=1}^{n} f\left(i \cdot \frac{4}{n}\right)\frac{4}{n}$$

$$= \sum_{i=1}^{n}\left(i \cdot \frac{4}{n}\right)^2 \frac{4}{n}$$

$$= \sum_{i=1}^{n} i^2 \cdot \left(\frac{4}{n}\right)^3$$

$$= 1^2 \cdot \left(\frac{4}{n}\right)^3 + 2^2 \cdot \left(\frac{4}{n}\right)^3 + 3^2 \cdot \left(\frac{4}{n}\right)^3 + \cdots + n^2 \cdot \left(\frac{4}{n}\right)^3$$

$$= (1^2 + 2^2 + 3^2 + \cdots + n^2)\left(\frac{4}{n}\right)^3.$$

The sum $1^2 + 2^2 + 3^2 + \cdots + n^2$ can be rewritten using the formula

$$1^2 + 2^2 + 3^2 + \cdots + n^2 = \frac{n(n + 1)(2n + 1)}{6}.$$

This simplifies the Riemann sum.

$$\sum_{i=1}^{n} f(x_i)\,\Delta x = (1^2 + 2^2 + 3^2 + \cdots n^2)\left(\frac{4}{n}\right)^3$$

$$= \frac{n(n + 1)(2n + 1)}{6}\frac{4^3}{n^3}$$

$$= \frac{32(n + 1)(2n + 1)}{3n^2}.$$

To compute the definite integral, we let n approach infinity.

$$\int_0^4 x^2\,dx = \lim_{n\to\infty} \sum_{i=1}^{n} f(x_i)\,\Delta x$$

$$= \lim_{n\to\infty} \frac{32(n + 1)(2n + 1)}{3n^2}$$

$$= \frac{32}{3} \lim_{n\to\infty} \frac{(n + 1)(2n + 1)}{n^2}$$

$$= \frac{32}{3} \lim_{n\to\infty} (2n^2 + 3n + 1) \cdot \frac{1}{n^2}$$

$$= \frac{32}{3} \lim_{n \to \infty} \left(2 + \frac{3}{n} + \frac{1}{n^2} \right)$$

$$= \frac{32}{3} \cdot 2$$

$$= \frac{64}{3}$$

The area under the graph of the function $f(x) = x^2$ between the x-values 0 and 4 is $\frac{64}{3}$.   ■

Signed Area

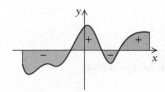

The definite integral represents area as long as the function is positive. However, when the function is negative, then the integral is the negative of the area between the x-axis and the function. In general, the definite integral gives the **signed area**, which is positive when the function is positive and negative when the function is negative. Signed area is illustrated with the figure in the margin.

EXAMPLE 7   Compute $\int_{-1}^{1} x^3 \, dx$.

Solution   The integral assigns a positive value to the area right of the y-axis and a negative value to the area left of the y-axis. Since the two areas have the same magnitude, they add to zero.

$$\int_{-1}^{1} x^3 \, dx = 0$$ ■

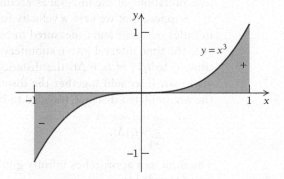

EXAMPLE 8   In each of the following figures, decide visually whether

$$\int_{a}^{b} f(x) \, dx$$

is positive, negative, or zero.

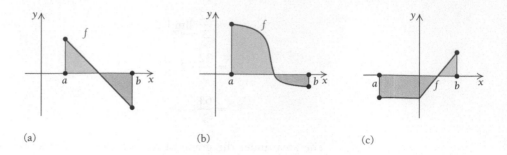

(a)   (b)   (c)

Solution

a) In this figure, it appears that there is the same area above the x-axis as below. Thus,

$$\int_a^b f(x)\,dx = 0.$$

b) In this figure, there is more area above the x-axis than below. Thus,

$$\int_a^b f(x)\,dx > 0.$$

c) In this figure, there is more area below the x-axis than above. Thus,

$$\int_a^b f(x)\,dx < 0.$$

Accumulations

We defined the definite integral by interpreting the Riemann sum as an accumulation of area that eventually gives the total signed area. There are many other useful interpretations of the integral as accumulations of quantities.

Suppose that we have a velocity function $v(t)$, where $a \le t \le b$ with v measured in miles per hour and t measured in hours. As we did in the case of area, we can divide the time interval into n subintervals, all of equal length. For one subinterval of time, t_i to $t_{i+1} = t_i + \Delta t$, the distance traveled during that time is approximately $v(t_i) \cdot \Delta t$. If we add together the distances traveled over the n subintervals, we get the accumulated distance traveled to be approximately

$$\sum_{i=1}^{n} v(t_i)\Delta t.$$

The limit as n approaches infinity gives the exact distance traveled. That is, the distance traveled is

$$\lim_{n\to\infty} \sum_{i=1}^{n} v(t_i)\,\Delta t = \int_a^b v(t)\,dt.$$

EXAMPLE 9   A particle begins at the origin and travels with velocity function $v(t) = t^2$ mi/hr. Find the distance traveled from $t = 0$ to $t = 4$ using **a)** definite integration and **b)** indefinite integration.

Solution

a) The distance traveled from $t = 0$ to $t = 4$ may be represented by the definite integral $\int_0^4 t^2 \, dt$. In Example 6, we determined this integral (using the variable) x instead of t to be $\frac{64}{3}$. So the distance traveled is $\frac{64}{3}$ mi.

b) The antiderivative of $v(t)$ is the position function $s(t)$:

$$s(t) = \int v(t) \, dt = \int t^2 \, dt = \frac{t^3}{3} + C.$$

We determine C by using the initial condition $s(0) = 0$:

$$s(0) = \frac{(0)^3}{3} + C, \quad \text{so that} \quad C = 0.$$

Thus, $s(t) = \frac{1}{3}t^3$. The distance traveled at $t = 4$ is

$$s(4) = \frac{(4)^3}{3} = \frac{64}{3} \text{ mi.}$$

Notice that the same answer was obtained using either technique. This suggests that antiderivatives may be used to compute definite integrals, thus avoiding cumbersome calculations such as the computations of Example 6. We will further develop the connection between these two types of integration in Section 5.3. ∎

In general, the definite integral of a rate of change gives the accumulated change. In the next example, we use a Riemann sum to estimate an accumulated change from a rate of change.

EXAMPLE 10 *Candy Sales.* A candy store finds that its sales rate of Valentine's Day candy is given by

$$S'(t) = 200 - (t - 13)^2,$$

where $S(t)$ is the accumulated pounds of Valentine's Day candy sold until time t and $0 \le t \le 14$ is the number of days since the beginning of February. Approximate the total amount of Valentine's Day candy sold the first 14 days of February using the Riemann sum

$$\sum_{i=1}^{4} S'(t_i) \, \Delta t.$$

Solution

Since we divide the time interval $[0,14]$ into four subintervals, each subinterval has length $\Delta t = \frac{14}{4} = 3.5$ days. We use a table to organize our calculations of

$$\sum_{i=1}^{4} S'(t_i) \, \Delta t.$$

We multiply the total by $\Delta t = 3.5$ to get an accumulated sales of

$$697.5 \cdot 3.5 = 2441.25 \text{ lb.} \quad ∎$$

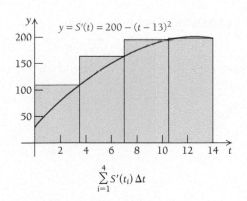

$$\sum_{i=1}^{4} S'(t_i) \, \Delta t$$

i	t_i	$S'(t_i)$
1	3.5	109.75
2	7	164
3	10.5	193.75
4	14	199
Total		697.5

Technology Connection

Approximating Definite Integrals

There are two methods for evaluating definite integrals using a grapher. Let's consider the function of Example 10,

$$s'(t) = 200 - (t - 13)^2.$$

Method 1: fnInt

We first enter $y_1 = 200 - (x - 13)^2$. Then we select the fnInt feature from the **MATH** menu. Next, we enter the function, the variable, and the endpoints of the interval over which we are integrating. The grapher returns the value of the definite integral found in Example 10.

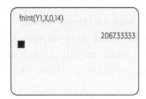

Method 2: ∫ f(x) dx

We first graph $y_1 = 200 - (x - 13)^2$. Then we select the ∫ $f(x)\, dx$ feature from the **CALC** menu and enter the lower and upper limits of integration. The grapher shades the area and returns the value of the definite integral.

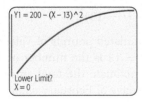

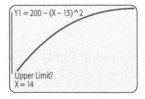

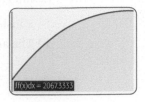

EXERCISES
Evaluate the definite integral.

1. $\displaystyle\int_{-1}^{2} (x^2 - 1)\, dx$

2. $\displaystyle\int_{-2}^{3} (x^3 - 3x + 1)\, dx$

3. $\displaystyle\int_{1}^{6} \frac{\ln x}{x^2}\, dx$

4. $\displaystyle\int_{-8}^{2} \frac{4}{(1 + e^x)^2}\, dx$

5. $\displaystyle\int_{-10}^{10} (0.002x^4 - 0.3x^2 + 4x - 7)\, dx$

Exercise Set 5.2

1. a) Approximate

$$\int_1^7 \frac{dx}{x^2}$$

by computing the area of each rectangle to four decimal places and then adding.

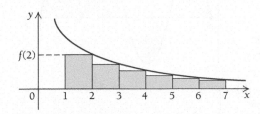

b) Approximate

$$\int_1^7 \frac{dx}{x^2}$$

by computing the area of each rectangle to four decimal places and then adding.

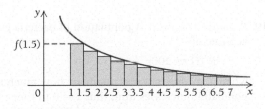

2. a) Approximate

$$\int_0^5 (x^2 + 1)\, dx$$

by computing the area of each rectangle and then adding.

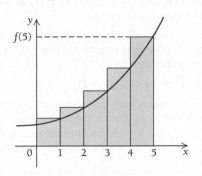

b) Approximate

$$\int_0^5 (x^2 + 1)\, dx$$

by computing the area of each rectangle and then adding.

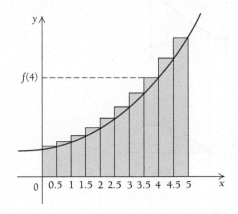

Compare your answers to parts (a) and (b).

In each case, give two interpretations of the shaded region other than as area.

3.   **4.**

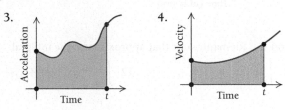

5.

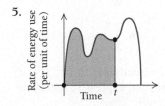

6.

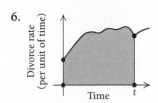

7.

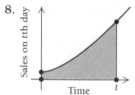

Concentration of a drug (in milligrams per cubic centimeter)

Volume of blood (in cubic centimeters)

v

8.

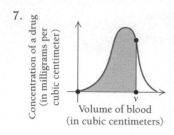

Sales on *t*th day

Time
t

9.

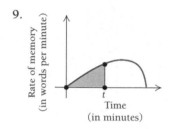

Rate of memory (in words per minute)

t

Time (in minutes)

10.

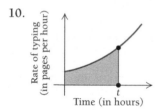

Rate of typing (in pages per hour)

t

Time (in hours)

Find the Riemann sum that approximates the integral.

11. $\int_0^2 x^2\,dx;\ n = 4$ **12.** $\int_{-1}^1 x^2\,dx;\ n = 4$

13. $\int_4^5 x\,dx;\ n = 6$ **14.** $\int_0^1 x\,dx;\ n = 6$

15. $\int_0^\pi \sin x\,dx;\ n = 4$ **16.** $\int_0^\pi \cos x\,dx;\ n = 4$

In each exercise, determine visually whether $\int_a^b f(x)\,dx$ is positive, negative, or zero, and express $\int_a^b f(x)\,dx$ in terms of A. Explain your result.

tw 17. a)

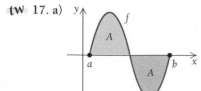

b)

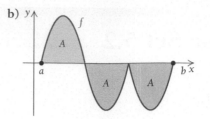

tw 18. a)

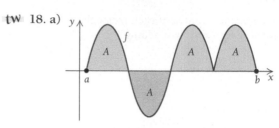

b)

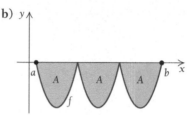

APPLICATIONS

19. *Population Growth.* A population of bacteria grows at a rate of
$$P'(t) = 200e^{-t},$$
where t is time in hours. Estimate how much the population increases from time $t = 0$ to time $t = 2$ by approximating the integral $\int_0^2 P'(t)\,dt$ with a Riemann sum using $n = 6$.

20. *Population Growth.* A population of algae grows at a rate of
$$P'(t) = 10e^t,$$
where t is time in hours. Estimate how much the population increases from time $t = 0$ to time $t = 3$ by approximating the integral $\int_0^3 P'(t)\,dt$ with a Riemann sum using $n = 6$.

21. *Population Decay.* The population of a city decreases (its growth is negative). The rate of increase of the population is
$$P'(t) = -500(20 - t),$$
where t is time in years and $0 \le t \le 20$. Estimate how much the population decreased from time $t = 0$ until time $t = 20$ by approximating the integral $\int_0^{20} P'(t)\,dt$ with a Riemann sum using $n = 5$.

22. *Population Decay.* The population of a city decreases (its growth is negative). The rate of increase of the population is

$$P'(t) = -50t^2,$$

where t is time in years and $0 \le t \le 10$. Estimate how much the population decreased from time $t = 0$ until time $t = 10$ by approximating the integral $\int_0^{10} P'(t)\, dt$ with a Riemann sum using $n = 5$.

23. *Velocity.* A particle starts out from the origin. Its velocity at time t is given by

$$v(t) = 3t^2 + 2t.$$

Estimate how far it travels between the times $t = 1$ and $t = 5$ using a Riemann sum with $n = 4$.

24. *Velocity.* A particle starts out from the origin. Its velocity at time t is given by

$$v(t) = 4t^3 + 2t.$$

Estimate how far it travels between the times $t = 0$ and $t = 3$ using a Riemann sum with $n = 6$.

SYNTHESIS

When computing limits of Riemann sums, the formulas below are sometimes very helpful.

a) $a + a + a + \cdots + a = na$, where there are n summands of a

b) $1 + 2 + 3 + \cdots + n = \dfrac{n(n + 1)}{2}$

c) $1^2 + 2^2 + 3^2 + \cdots + n^2 = \dfrac{n(n + 1)(2n + 1)}{6}$

d) $1^3 + 2^3 + 3^3 + \cdots + n^3 = \dfrac{n^2(n + 1)^2}{4}$

Compute the integrals by finding the limit of the Riemann sums.

25. $\displaystyle\int_0^2 x\, dx$

26. $\displaystyle\int_0^1 2x\, dx$

27. $\displaystyle\int_0^1 3x^2\, dx$

28. $\displaystyle\int_0^3 x^2\, dx$

29. $\displaystyle\int_0^4 x^3\, dx$

30. $\displaystyle\int_0^1 4x^3\, dx$

31. $\displaystyle\int_1^3 x^2\, dx$

32. $\displaystyle\int_1^4 x^3\, dx$

33. $\displaystyle\int_0^2 (x + x^2)\, dx$

Technology Connection

34.–42. Estimate each integral in Exercises 25–33 using a grapher.

Graph the function and estimate the integral using a grapher.

43. $\displaystyle\int_0^\pi \sin x\, dx$

44. $\displaystyle\int_0^\pi \cos x\, dx$

45. $\displaystyle\int_0^4 \sqrt{x}\, dx$

46. $\displaystyle\int_0^3 \frac{2}{\sqrt{x + 1}}\, dx$

47. $\displaystyle\int_2^4 \ln x\, dx$

48. $\displaystyle\int_1^e \frac{1}{x}\, dx$

5.3 The Fundamental Theorem of Calculus

OBJECTIVES

■ Compute definite integrals using the Fundamental Theorem of Calculus.

■ Find the average value of a function.

In Section 5.1, we defined the indefinite integral of a function to be its antiderivative. In Section 5.2, we defined the definite integral of a function as a limit of Riemann sums, which is often realized as an area. In this section, we find a connection between the two concepts of integral, which facilitates the computation of definite integrals.

The Derivative of Area Functions

For $a < b$, we can interpret the integral $\int_a^b f(x)\, dx$ as the signed area bounded by the function and the x-axis that is between $x = a$ and $x = b$.

We extend the definition of the definite integral to include cases where $a \geq b$.

a) $\int_a^a f(x)\, dx = 0$. (The area under a point is 0.)

b) $\int_a^b f(x)\, dx = -\int_b^a f(x)\, dx$. (Reversing the limits of integration changes the sign of the integral.)

Let's define the *area function* to be the integral

$$A(x) = \int_a^x f(t)\, dt,$$

where a is a fixed number. We can interpret the area function as representing the signed area between a and x that lies between the horizontal axis and the graph of the function $y = f(x)$. Notice that $A(a) = \int_a^a f(t)\, dt = 0$.

EXAMPLE 1   Find the area function $A(x)$ and compute $A'(x)$.

a) $f(x) = 3; a = 0$

b) $f(x) = x; a = 2$

Solution

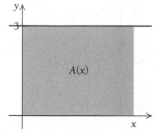

a) The area function is given by the integral

$$A(x) = \int_a^x f(t)\, dt = \int_0^x 3\, dt.$$

For $x > 0$, this integral is simply the area of a rectangle with height 3 and width x. The area is given by

$$A(x) = 3x.$$

Note that this formula is correct if $x = 0$ since $A(0) = 0$. It can also be checked that the formula is correct when $x < 0$.

The derivative of the area function is

$$A'(x) = \frac{d}{dx}(3x) = 3.$$

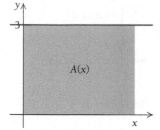

b) The area function is given by the integral

$$A(x) = \int_a^x f(t)\, dt = \int_2^x t\, dt.$$

For $x > 2$, this integral is the area of a trapezoid. That is,

$$A(x) = \frac{1}{2}(x + 2)(x - 2) = \frac{1}{2}(x^2 - 4) = \frac{1}{2}x^2 - 2.$$

This formula is valid even if $x \leq a$. The derivative of the area function is

$$A'(x) = \frac{d}{dx}\left(\frac{1}{2}x^2 - 2\right) = x.$$

Note that in both cases, $A'(x) = f(x)$. ∎

In Example 1, we computed the derivative of the area function for two functions and in each case we saw that $A'(x) = f(x)$. The **Fundamental Theorem of Calculus** says that this is always the case for continuous functions.

THEOREM 4

First Form of the Fundamental Theorem of Calculus

Let f be a continuous function and let

$$A(x) = \int_a^x f(t)\, dt.$$

Then

$$A'(x) = f(x).$$

Proof: The situation described in the theorem is shown in the following figure.

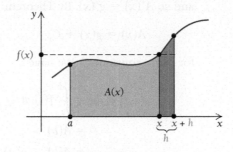

The derivative of $A(x)$ is, by the definition of a derivative,

$$A'(x) = \lim_{h \to 0} \frac{A(x+h) - A(x)}{h}.$$

Note from the figure that $A(x + h) - A(x)$ is the area of the small, dark orange, vertical strip. Thus we have

$$A(x + h) - A(x) \approx f(x) \cdot h.$$

Now

$$
\begin{aligned}
A'(x) &= \lim_{h \to 0} \frac{A(x+h) - A(x)}{h} \\
&\approx \lim_{h \to 0} \frac{f(x) \cdot h}{h} \\
&= \lim_{h \to 0} f(x) = f(x).
\end{aligned}
$$

EXAMPLE 2   Find $\dfrac{d}{dx} \displaystyle\int_0^x t \sin^2 t\, dt.$

Solution   The function to be integrated is $f(t) = t \sin^2 t$. The Fundamental Theorem of Calculus states that

$$\frac{d}{dx} \int_0^x t \sin^2 t\, dt = A'(x) = f(x) = x \sin^2 x.$$

■

It is possible to use the First Form of the Fundamental Theorem of Calculus to compute definite integrals. However, there is a second form that makes the computation easier.

THEOREM 5

Second Form of the Fundamental Theorem of Calculus

Let $f(x)$ be a continuous function and suppose that $g'(x) = f(x)$. In other words, suppose that $g(x)$ is an antiderivative of $f(x)$. Then

$$\int_a^b f(x)\, dx = g(b) - g(a).$$

Proof: We let $A(x) = \int_a^x f(t)\, dt$. The First Form of the Fundamental Theorem of Calculus states that $A'(x) = f(x)$. By assumption, we also know that $g'(x) = f(x)$ and so $A'(x) = g'(x)$. By Theorem 1,

$$A(x) = g(x) + C,$$

for some constant C. We have

$$
\begin{aligned}
\int_a^b f(x)\, dx &= \int_a^b f(t)\, dt && \text{Area doesn't depend on variable name} \\
&= A(b) \\
&= A(b) - A(a) && \text{Since } A(a) = 0 \\
&= (g(b) + C) - (g(a) + C) \\
&= g(b) - g(a).
\end{aligned}
$$
∎

The Second Form of the Fundamental Theorem of Calculus gives us an effective method of computing the definite integral $\int_a^b f(x)\, dx$ for many functions *without explicitly using Riemann sums*. The procedure is:

1. Find an antiderivative $g(x)$ for the function $f(x)$.

2. Substitute b and a to find the difference $g(b) - g(a)$. The difference is the definite integral $\int_a^b f(x)\, dx$. For notational convenience, we write

$$[g(x)]_a^b = g(x)\big|_a^b = g(b) - g(a).$$

EXAMPLE 3   Compute the definite integrals.

a) $\displaystyle\int_0^2 (x + x^2)\, dx$

b) $\displaystyle\int_0^1 e^x\, dx$

c) $\displaystyle\int_{\pi/2}^{\pi} \sin x\, dx$

Solution   We compute the antiderivative and use the Second Form of the Fundamental Theorem of Calculus.

a) $\displaystyle\int_0^2 (x + x^2)\, dx = \left[\frac{1}{2}x^2 + \frac{1}{3}x^3\right]_0^2$ Compute antiderivative.

$$= \left(\frac{1}{2}(2)^2 + \frac{1}{3}(2)^3\right) - \left(\frac{1}{2}(0)^2 + \frac{1}{3}(0)^3\right) \qquad \text{Substitute 2 and 0.}$$

$$= \left(2 + \frac{8}{3}\right) - (0 + 0)$$

$$= \frac{14}{3}$$

b) $\displaystyle\int_0^1 e^x\, dx = e^x \Big|_0^1$ Compute antiderivative.

$$= e^1 - e^0 \qquad \text{Substitute 0 and 1.}$$

$$= e - 1$$

c) $\displaystyle\int_{\pi/2}^{\pi} \sin x\, dx = -\cos x \Big|_{\pi/2}^{\pi}$ Compute antiderivative.

$$= (-\cos \pi) - (-\cos \pi/2) \qquad \text{Substitute } \pi \text{ and } \pi/2.$$

$$= -(-1) - (0)$$

$$= 1$$

EXAMPLE 4   Find the area under the graph $f(x) = 3 \cos x$ over the interval $[-\pi/4, \pi/4]$.

Solution   Since $\cos x \geq 0$ over the interval $[-\pi/4, \pi/4]$, we can compute the area as the integral

$$\int_{-\pi/4}^{\pi/4} 3 \cos x\, dx = 3 \int_{-\pi/4}^{\pi/4} \cos x\, dx$$

$$= 3 \sin x \Big|_{-\pi/4}^{\pi/4}$$

$$= 3(\sin(\pi/4) - \sin(-\pi/4))$$

$$= 3\left(\frac{\sqrt{2}}{2} - \left(-\frac{\sqrt{2}}{2}\right)\right)$$

$$= 3\sqrt{2}.$$

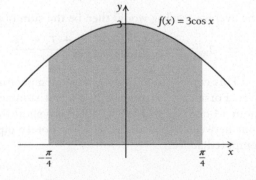

$$\text{Area} = \int_{-\pi/4}^{\pi/4} 3 \cos x\, dx$$

$$= 3\sqrt{2}$$

EXAMPLE 5 Evaluate: $\int_{-1}^{2}(x^3 - 3x + 1)\,dx$. Interpret the results in terms of area.

Solution We have

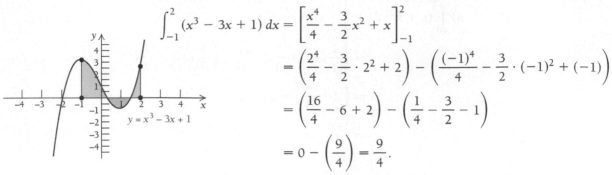

$$\int_{-1}^{2}(x^3 - 3x + 1)\,dx = \left[\frac{x^4}{4} - \frac{3}{2}x^2 + x\right]_{-1}^{2}$$

$$= \left(\frac{2^4}{4} - \frac{3}{2}\cdot 2^2 + 2\right) - \left(\frac{(-1)^4}{4} - \frac{3}{2}\cdot(-1)^2 + (-1)\right)$$

$$= \left(\frac{16}{4} - 6 + 2\right) - \left(\frac{1}{4} - \frac{3}{2} - 1\right)$$

$$= 0 - \left(\frac{9}{4}\right) = \frac{9}{4}.$$

We can graph the function $f(x) = x^3 - 3x + 1$ over the interval and shade the area between the curve and the x-axis. The sum of the areas above the axis minus the area below is $\frac{9}{4}$. ∎

The Average Value of a Function

Suppose that

$$T = f(t)$$

is the temperature at time t recorded at a weather station on a certain day. The station uses a 24-hr clock, so the domain of the temperature function is the interval $[0, 24]$. The function is continuous, as shown below.

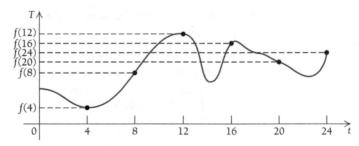

To find the average temperature for the given day, we might take six temperature readings at 4-hr intervals, starting at midnight:

$$T_1 = f(4), \qquad T_2 = f(8),$$
$$T_3 = f(12), \qquad T_4 = f(6),$$
$$T_5 = f(20), \qquad T_6 = f(4).$$

The average reading would then be the sum of these six readings divided by 6:

$$T_{av} = \frac{T_1 + T_2 + T_3 + T_4 + T_5 + T_6}{6}.$$

This computation of the average temperature may not give the most useful answer. For example, suppose it is a hot summer day, and at 2:00 in the afternoon (hour 14 on the 24-hr clock), there is a short thunderstorm that cools the air for an hour between our readings. This temporary dip would not show up in the average computed above.

What can we do? We could take 48 readings at half-hour intervals. This should give us a better result. In fact, the shorter the time between readings, the better the result should be. It seems reasonable that we might define the **average value** of T over the interval $[0, 24]$ to be the limit, as n approaches ∞, of the average of n values:

$$\text{Average value of } T = \lim_{n \to \infty} \frac{1}{n} \sum_{i=1}^{n} T_i$$

$$= \lim_{n \to \infty} \frac{1}{n} \sum_{i=1}^{n} f(t_i).$$

Note that this is not too far from our definition of an integral. All we would need is to get Δt, which is $(24 - 0)/n$, or $24/n$, into the summation:

$$\text{Average value of } T = \lim_{n \to \infty} \frac{1}{n} \sum_{i=1}^{n} f(t_i)$$

$$= \lim_{n \to \infty} \frac{1}{\Delta t} \cdot \frac{1}{n} \sum_{i=1}^{n} f(t_i)\, \Delta t$$

$$= \lim_{n \to \infty} \frac{n}{24} \cdot \frac{1}{n} \sum_{i=1}^{n} f(t_i)\, \Delta t \qquad \Delta t = \frac{24}{n}, \text{ or } \frac{1}{\Delta t} = \frac{n}{24}$$

$$= \frac{1}{24} \lim_{n \to \infty} \sum_{i=1}^{n} f(t_i)\, \Delta t$$

$$= \frac{1}{24} \int_{0}^{24} f(t)\, dt.$$

DEFINITION

Let f be a continuous function over a closed interval $[a, b]$. Its **average value, y_{av},** over $[a, b]$ is given by

$$y_{av} = \frac{1}{b - a} \int_{a}^{b} f(x)\, dx.$$

Let's consider average value in another way. If we multiply both sides of this definition by $b - a$, we get

$$(b - a)y_{av} = \int_{a}^{b} f(x)\, dx.$$

Now the expression on the left will give the area of a rectangle of length $b - a$ and height y_{av}. The area of such a rectangle is the same as the area under the graph of $y = f(x)$ over the interval $[a, b]$, as shown in the figure.

EXAMPLE 6   Find the average value of $f(x) = x^2$ over the interval $[0, 2]$.

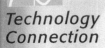

Technology Connection

Graph $f(x) = x^4$. Compute the average value of the function over the interval $[0, 2]$, using the method of Example 6. Then use that value y_{av} and draw a graph of it as a horizontal line using the same set of axes. What does this line represent in comparison to the graph of $f(x) = x^4$?

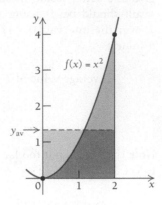

Solution   The average value is

$$\frac{1}{2-0} \int_0^2 x^2 \, dx = \frac{1}{2}\left[\frac{x^3}{3}\right]_0^2$$

$$= \frac{1}{2}\left(\frac{2^3}{3} - \frac{0^3}{3}\right)$$

$$= \frac{1}{2} \cdot \frac{8}{3} = \frac{4}{3}.$$

Note that although the values of $f(x)$ increase from 0 to 4 over $[0, 2]$, we would not expect the average value to be $(0 + 4)/2 = 2$. We see from the graph that $f(x)$ is less than 2 over more than half the interval.   ■

EXAMPLE 7   *Engine Emissions.* The emissions of an engine are given by

$$E(t) = 2t^2,$$

where $E(t)$ is the engine's rate of emission, in billions of pollution particulates per year, at time t, in years. Find the average emissions from $t = 1$ to $t = 5$.

Solution   The average emissions are

$$\frac{1}{5-1} \int_1^5 2t^2 \, dt = \frac{1}{4}\left[\frac{2}{3}t^3\right]_1^5$$

$$= \frac{1}{4} \cdot \frac{2}{3}(5^3 - 1^3)$$

$$= \frac{1}{6}(125 - 1)$$

$$= 20\tfrac{2}{3} \text{ billion pollution particulates per year.}$$ ■

Exercise Set 5.3

Find the area under the given curve over the indicated interval.

Evaluate. Then interpret the results.

1. $y = 4$; $[1, 3]$

2. $y = 5$; $[1, 3]$

15. $\int_0^{1.5} (x - x^2)\, dx$

16. $\int_0^2 (x^2 - x)\, dx$

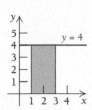

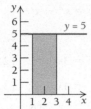

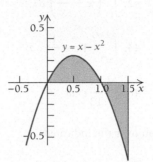

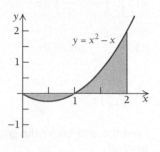

3. $y = 2x$; $[1, 3]$

4. $y = x^2$; $[0, 3]$

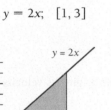

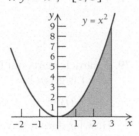

17. $\int_0^{3\pi/2} \cos x\, dx$

5. $y = x^2$; $[0, 5]$

6. $y = x^3$; $[0, 2]$

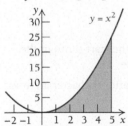

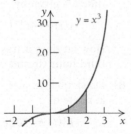

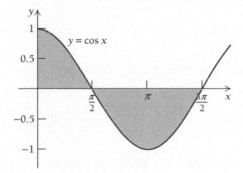

7. $y = x^3$; $[0, 1]$

8. $y = 1 - x^2$; $[-1, 1]$

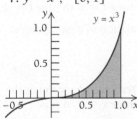

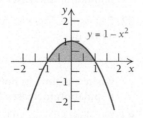

18. $\int_0^b -2e^{3x}\, dx$

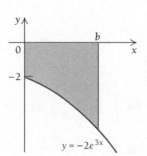

9. $y = 4 - x^2$; $[-2, 2]$

10. $y = e^x$; $[0, 2]$

11. $y = e^x$; $[0, 3]$

12. $y = \dfrac{2}{x}$; $[1, 4]$

13. $y = \dfrac{3}{x}$; $[1, 6]$

14. $y = 3 \sin x$; $\left[0, \dfrac{\pi}{4}\right]$

19.–36. Check the results of each of Exercises 1–18 using a grapher.

Evaluate.

37. $\int_a^b e^t \, dt$

38. $\int_0^a (ax - x^2) \, dx$

39. $\int_a^b 3t^2 \, dt$

40. $\int_{-5}^2 4t^3 \, dt$

41. $\int_1^e \left(x + \frac{1}{x} \right) dx$

42. $\int_1^e \left(x - \frac{1}{x} \right) dx$

43. $\int_0^{\pi/6} \frac{5}{2} \sin 2x \, dx$

44. $\int_{\pi/3}^{2\pi/3} 4 \cos \frac{1}{2} x \, dx$

45. $\int_{-4}^1 \frac{10}{17} t^3 \, dt$

46. $\int_0^1 \frac{12}{13} t^2 \, dt$

Find the area under the graph over the indicated interval.

47. $y = x^3$; $[0, 2]$

48. $y = x^4$; $[0, 1]$

49. $y = x^2 + x + 1$; $[2, 3]$

50. $y = 2 - x - x^2$; $[-2, 1]$

51. $y = 5 - x^2$; $[-1, 2]$ 52. $y = e^x$; $[-2, 3]$

53. $y = e^x$; $[-1, 5]$ 54. $y = 2x + \frac{1}{x^2}$; $[1, 4]$

Evaluate. Some algebra may be required before finding the integral.

55. $\int_2^3 \frac{x^2 - 1}{x - 1} \, dx$

56. $\int_1^5 \frac{x^5 - x^{-1}}{x^2} \, dx$

57. $\int_4^{16} (x - 1) \sqrt{x} \, dx$

58. $\int_0^1 (x + 2)^3 \, dx$

59. $\int_1^8 \frac{\sqrt[3]{x^2} - 1}{\sqrt[3]{x}} \, dx$

60. $\int_0^1 \frac{x^3 + 8}{x + 2} \, dx$

61. $\int_1^2 (4x + 3)(5x - 2) \, dx$

62. $\int_2^5 \left(t + \sqrt{3} \right)\left(t - \sqrt{3} \right) dt$

63. $\int_0^1 (t + 1)^3 \, dt$

64. $\int_1^3 \left(x - \frac{1}{x} \right)^2 dx$

65. $\int_1^3 \frac{t^5 - t}{t^3} \, dt$

66. $\int_4^9 \frac{t + 1}{\sqrt{t}} \, dt$

67. $\int_3^5 \frac{x^2 - 4}{x - 2} \, dx$

68. $\int_0^1 \frac{t^3 + 1}{t + 1} \, dt$

Find the average value over the given interval.

69. $y = 2x^3$; $[-1, 1]$

70. $y = 4 - x^2$; $[-2, 2]$

71. $y = e^x$; $[0, 1]$

72. $y = e^{-x}$; $[0, 1]$

73. $y = x^2 - x + 1$; $[0, 2]$

74. $y = x^2 + x - 2$; $[0, 4]$

75. $y = 3x + 1$; $[2, 6]$

76. $y = 4x + 1$; $[3, 7]$

77. $y = x^n$; $[0, 1]$

78. $y = x^n$; $[1, 2]$

APPLICATIONS

79. A particle starts out from the origin. Its velocity at time t is given by
$$v(t) = 3t^2 + 2t.$$
How far does it travel from $t = 1$ to $t = 5$?

80. A particle starts out from the origin. Its velocity at time t is given by
$$v(t) = 4t^3 + 2t.$$
How far does it travel from the start through the third hour (from $t = 0$ to $t = 3$)?

81. *Population Growth.* A population of bacteria grows at a rate of
$$P'(t) = 200e^{-t},$$
where t is time in hours. Determine how much the population increases from time $t = 0$ to time $t = 2$.

82. *Population Growth.* A population of algae grows at a rate of
$$P'(t) = 10e^t,$$
where t is time in hours. Determine how much the population increases from time $t = 0$ to time $t = 3$.

83. *Population Decay.* The population of a city decreases (its growth is negative). The rate of decrease of the population is
$$P'(t) = -500(20 - t),$$
where t is time in years and $0 \le t \le 20$. How much did the population decrease from time $t = 0$ until time $t = 10$?

84. *Population Decay.* The population of a city decreases (its growth is negative). The rate of decrease of the population is

$$P'(t) = -50t^2,$$

where t is time in years and $0 \le t \le 10$. How much did the population decrease from time $t = 0$ until time $t = 6$?

85. *Tendon Strain.* The human gastrocnemius tendon is located in the calf. The force on the gastrocnemius tendon as it is stretched can be approximated by the function

$$f(x) = 71.3x - 4.15x^2 + 0.434x^3,$$

where x is the tendon elongation in millimeters and $f(x)$ is the force exerted by the tendon measured in newtons (N). The area under the curve gives the amount of work done on the tendon. Compute the work done on the tendon for $2 \le x \le 11$. The units are N · mm, which is equivalent to millijoules.[5]

86. *Tendon Strain.* The force on the human gastrocnemius tendon while it contracts can be approximated by the function

$$f(x) = 71.0x - 10.3x^2 + 0.986x^3,$$

where x is the tendon elongation in millimeters and $f(x)$ is the force exerted by the tendon measured in newtons (N). The area under the curve gives the amount of work done. Compute the work for $2 \le x \le 11$. The units are N · mm, which is equivalent to millijoules.[6]

87. *Average Drug Dose.* The concentration of phenylbutazone in the plasma of a calf injected with this anti-inflammatory agent is given approximately by

$$C(t) = 42.03e^{-0.01050t},$$

where C is the concentration in μg/ml and $0 \le t \le 120$ is the number of hours after the injection.[7]

a) What is the initial dosage?

b) What is the average amount of phenylbutazone in the calf's body for the time between 10 and 120 hr?

88. *New York Temperature.* Historically, the average temperature in New York can be approximated by the function $T(x) = 43.5 - 18.4x + 8.57x^2 - 0.996x^3 + 0.0338x^4$, where T represents the temperature in degrees Fahrenheit and x is the number of months since the beginning of the year. Compute the average temperature in New York over the whole year to the nearest degree.[8]

SYNTHESIS

89. *Circumference and Area of a Circle.* The area of a circle of radius r is $A = \pi r^2$ and its circumference is $C = 2\pi r$.

a) Show that the integral of the circumference between $r = 0$ and $r = b$ gives the area of a circle of radius b.

tw b) Draw a picture and explain why this is expected.

90. *Surface Area and Volume of a Sphere.* The volume of a sphere of radius r is $V = \frac{4}{3}\pi r^3$ and its surface area is $S = 4\pi r^2$.

a) Show that the integral of the surface area between $r = 0$ and $r = b$ gives the volume of a sphere of radius b.

tw b) Explain why this is expected.

Find the error in each of the following. Explain.

tw **91.** $\displaystyle\int_1^2 (x^2 + x + 1)\, dx = \left[\frac{1}{3}x^3 + \frac{1}{2}x^2 + x \right]_1^2$

$$= \left(\frac{1}{3} \cdot 2^3 + \frac{1}{2} \cdot 2^2 + 2 \right)$$

$$= \frac{10}{3}$$

tw **92.** $\displaystyle\int_1^2 (\ln x - e^x)\, dx = \left[\frac{1}{x} - e^x \right]_1^2$

$$= \left(\frac{1}{2} - e^2 \right) - (1 - e^1)$$

$$= e - e^2 - \frac{1}{2}$$

[5]N. Maganaris and J. P. Paul, "Tensile properties of the *in vivo* human gastrocnemius tendon," *Journal of Biomechanics, 35* (2002), pp. 1639–1646.

[6]N. Maganaris and J. P. Paul, "Tensile properties of the *in vivo* human gastrocnemius tendon," *Journal of Biomechanics, 35* (2002), pp. 1639–1646.

[7]A. Arifah, P. Lees, "Pharmacodynamics and pharmacokinetics of phenylbutazone in calves," *Journal of Veterinary Pharmacology and Therapeutics, 25,* 299–309, 2002.

[8]http://www.worldclimate.com/

Technology Connection

Evaluate.

93. $\int_{-2}^{3} (x^3 - 4x)\, dx$

94. $\int_{-1.2}^{6.3} (x^3 - 9x^2 + 27x + 50)\, dx$

95. $\int_{-8}^{1.4} (x^4 + 4x^3 - 36x^2 - 160x + 300)\, dx$

96. $\int_{-1}^{20} \frac{4x}{x^2 + 1}\, dx$

97. $\int_{-1}^{1} \left(1 - \sqrt{1 - x^2}\right) dx$

98. $\int_{-2}^{2} \sqrt{4 - x^2}\, dx$

99. $\int_{0}^{10} (x^3 - 6x^2 + 15x)\, dx$

100. $\int_{0}^{8} x(x - 5)^4\, dx$

101. $\int_{-2}^{2} x^{2/3}\left(\frac{5}{2} - x\right) dx$

102. $\int_{2}^{4} \frac{x^2 - 4}{x^2 - 3}\, dx$

103. $\int_{-10}^{10} \frac{8}{x^2 + 4}\, dx$

5.4 Properties of Definite Integrals

OBJECTIVES

- Use properties of definite integrals.
- Find the area between curves.
- Solve applied problems involving definite integrals.

Properties of Definite Integrals

The following properties of definite integrals can be derived from the definition of a definite integral.

PROPERTY 1

$$\int_{a}^{b} k \cdot f(x)\, dx = k \cdot \int_{a}^{b} f(x)\, dx$$

The integral of a constant times a function is the constant times the integral of the function. That is, a constant can be "factored out" of the integrand.

PROPERTY 2

$$\int_a^b [f(x) + g(x)]\, dx = \int_a^b f(x)\, dx + \int_a^b g(x)\, dx$$

The integral of a sum is the sum of the integrals.

EXAMPLE 1

$$\int_0^5 (100e^x + x)\, dx = 100 \int_0^5 e^x\, dx + \int_0^5 x\, dx$$

$$= 100[e^x]_0^5 + \left[\frac{x^2}{2}\right]_0^5$$

$$= 100(e^5 - e^0) + \left(\frac{25}{2} - 1\right)$$

$$= 100(e^5 - 1) + \frac{25}{2}$$

$$\approx 14{,}753.82$$

PROPERTY 3

For $a < c < b$,

$$\int_a^b f(x)\, dx = \int_a^c f(x)\, dx + \int_c^b f(x)\, dx.$$

For any number c between a and b, the integral from a to b is the integral from a to c plus the integral from c to b.

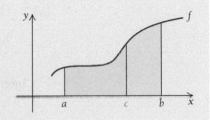

Property 3 can be used to integrate a function defined piecewise.

EXAMPLE 2   Find the area under the graph of $y = f(x)$ from -4 to 5, where

$$f(x) = \begin{cases} 9, & \text{if } x < 3, \\ x^2, & \text{if } x \geq 3. \end{cases}$$

Solution

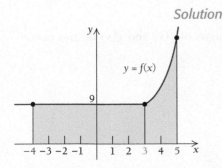

$$\int_{-4}^5 f(x)\, dx = \int_{-4}^3 f(x)\, dx + \int_3^5 f(x)\, dx = \int_{-4}^3 9\, dx + \int_3^5 x^2\, dx$$

$$= 9 \int_{-4}^3 dx + \int_3^5 x^2\, dx = 9[x]_{-4}^3 + \left[\frac{x^3}{3}\right]_3^5$$

$$= 9[3 - (-4)] + \left(\frac{5^3}{3} - \frac{3^3}{3}\right)$$

$$= 95\tfrac{2}{3}.$$

The Area of a Region Bounded by Two Graphs

Suppose that we want to find the area of a region bounded by the graphs of two functions, $y = f(x)$ and $y = g(x)$, as shown in Fig. 1(a).

Note that the area of the desired region A is that of A_2 in Fig. 1(b) minus that of A_1 in Fig. 1(c).

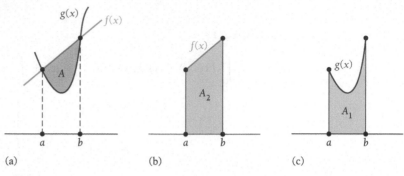

(a)   (b)   (c)

Figure 1

Thus,

$$A = \int_a^b f(x)\, dx - \int_a^b g(x)\, dx,$$

or

$$A = \int_a^b [f(x) - g(x)]\, dx.$$

In general, we have the following.

THEOREM 6

Let f and g be continuous functions and suppose that $f(x) \geq g(x)$ over the interval $[a, b]$. Then the area of the region between the two curves, from $x = a$ to $x = b$, is

$$\int_a^b [f(x) - g(x)]\, dx.$$

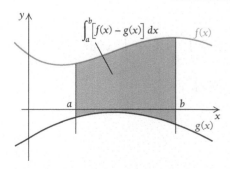

As long as $f(x) \geq g(x)$, the signs of $f(x)$ and $g(x)$ do not matter, and the integral

$$\int_a^b [f(x) - g(x)]\, dx$$

is the area between the functions.

EXAMPLE 3   Find the area of the region bounded by the graphs of $f(x) = 2x - 2$ and $g(x) = x^2 - 2$.

Solution

a) We first find the points where the graphs intersect by solving the equation $f(x) = g(x)$.

$$2x - 2 = x^2 - 2$$
$$2x = x^2$$
$$2x - x^2 = 0$$
$$x(2 - x) = 0$$
$$x = 0 \quad or \quad x = 2.$$

We see that there are two points of intersection. At $x = 0$ we have $f(0) = g(0) = -2$, and at $x = 2$ we have $f(2) = g(2) = 2$. The two points of intersection are $(0, -2)$ and $(2, 2)$.

b) We next make a reasonably accurate sketch to ensure that we have the right configuration. For our calculation, we need to identify which is the *upper* graph. In this case, $2x - 2 \geq x^2 - 2$ over the interval $[0, 2]$.

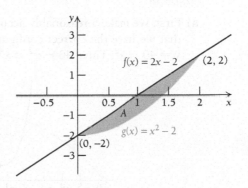

c) We may now compute the area. The graphs of the two functions cut the plane into five regions, only one of which is bounded. This region, shaded orange in the figure, is between $x = 0$ and $x = 2$. Its area is

$$\int_0^2 [(2x - 2) - (x^2 - 2)]\, dx = \int_0^2 (2x - x^2)\, dx$$

$$= \left[x^2 - \frac{x^3}{3} \right]_0^2$$

$$= \left(2^2 - \frac{2^3}{3} \right) - \left(0^2 - \frac{0^3}{3} \right)$$

$$= 4 - \frac{8}{3}$$

$$= \frac{4}{3}.$$

Technology Connection

To find the area bounded by two graphs, such as $y_1 = -2x - 7$ and $y_2 = -x^2 - 4$, we graph each function and use the **INTERSECT** feature from the **CALC** menu to determine the points of intersection.

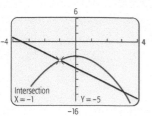

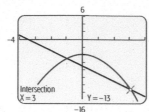

Next, we use the fnInt feature from the **MATH** menu.

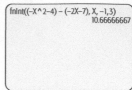

The area bounded by the two curves is $10\frac{2}{3}$.

Find the area of the region in Example 3 using a grapher.

EXAMPLE 4 Find the area of the region bounded by
$$y = x^4 - 3x^3 - 4x^2 + 10, \qquad y = 40 - x^2, \qquad x = 1, \quad \text{and} \quad x = 3.$$

Solution

a) First, we make a reasonably accurate sketch, as in the following figure, to ensure that we have the correct configuration. Note that over $[1, 3]$, the upper graph is $y = 40 - x^2$. Thus, $40 - x^2 \geq x^4 - 3x^3 - 4x^2 + 10$ over $[1, 3]$.

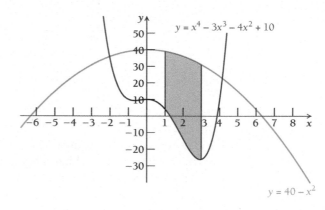

b) The limits of integration are stated, so we can compute the area as follows:

$$\int_1^3 [(40 - x^2) - (x^4 - 3x^3 - 4x^2 + 10)]\, dx = \int_1^3 (-x^4 + 3x^3 + 3x^2 + 30)\, dx$$

$$= \left[-\frac{x^5}{5} + \frac{3}{4}x^4 + x^3 + 30x \right]_1^3$$

$$= \left(-\frac{3^5}{5} + \frac{3}{4} \cdot 3^4 + 3^3 + 30 \cdot 3 \right)$$

$$\quad - \left(-\frac{1^5}{5} + \frac{3}{4} \cdot 1^4 + 1^3 + 30 \cdot 1 \right)$$

$$= 97.6.$$

An Application

EXAMPLE 5   *Emission Control.* A clever college student develops an engine that is believed to meet federal standards for emission control. The engine's rate of emission is given by

$$E(t) = 2t^2,$$

where $E(t)$ is the emissions, in billions of pollution particulates per year, at time t, in years. The emission rate of a conventional engine is given by

$$C(t) = 9 + t^2.$$

The graphs of both curves are shown below.

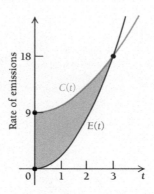

a) At what point in time will the emission rates be the same?

b) What is the reduction in emissions resulting from using the student's engine between time 0 and when the emission rates are the same?

Solution

a) The rate of emission will be the same when $E(t) = C(t)$, or

$$2t^2 = 9 + t^2$$
$$t^2 - 9 = 0$$
$$(t - 3)(t + 3) = 0$$
$$t = 3 \quad or \quad t = -3.$$

Since negative time has no meaning in this problem, the emission rates will be the same when $t = 3$ yr.

b) The reduction in emissions is represented by the area of the shaded region in the figure above. It is the area between $C(t) = 9 + t^2$ and $E(t) = 2t^2$, from $t = 0$ to $t = 3$, and is computed as follows:

$$\int_0^3 [(9 + t^2) - 2t^2] \, dt = \int_0^3 (9 - t^2) \, dt = \left[9t - \frac{t^3}{3} \right]_0^3$$

$$= \left(9 \cdot 3 - \frac{3^3}{3} \right) - \left(9 \cdot 0 - \frac{0^3}{3} \right)$$

$$= 27 - 9$$

$$= 18 \text{ billion pollution particulates.}$$

Exercise Set 5.4

Find the area of the region bounded by the given graphs.

1. $y = x$, $y = x^3$, $x = 0$, $x = 1$

2. $y = x$, $y = x^4$

3. $y = x + 2$, $y = x^2$

4. $y = x^2 - 2x$, $y = x$

5. $y = 6x - x^2$, $y = x$

6. $y = x^2 - 6x$, $y = -x$

7. $y = 2x - x^2$, $y = -x$

8. $y = x^2$, $y = \sqrt{x}$

9. $y = x$, $y = \sqrt{x}$

10. $y = 3$, $y = x$, $x = 0$

11. $y = 5$, $y = \sqrt{x}$, $x = 0$

12. $y = x^2$, $y = x^3$

13. $y = 4 - x^2$, $y = 4 - 4x$

14. $y = x^2 + 1$, $y = x^2$, $x = 1$, $x = 3$

15. $y = x^2 + 3$, $y = x^2$, $x = 1$, $x = 2$

16. $y = \cos x$, $y = \sin x$, $x = 0$, $x = \dfrac{\pi}{4}$

17. $y = x + 3$, $y = \sin x + \cos x$, $x = 0$, $x = \dfrac{\pi}{2}$

18. $y = \sec^2 x$, $y = 1$, $x = 0$, $x = \dfrac{\pi}{4}$

19. $y = \csc^2 x$, $y = \sin x$, $x = \dfrac{\pi}{4}$, $x = \dfrac{\pi}{2}$

20. $y = 2x^2 - x - 3$, $y = x^2 + x$

21. $y = 2x^2 - 6x + 5$, $y = x^2 + 6x - 15$

Find the area of the shaded region.

22. $f(x) = 2x + x^2 - x^3$, $g(x) = 0$
    Hint: Write two separate integrals and use Property 3.

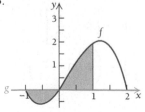

23. $f(x) = x^3 + 3x^2 - 9x - 12$, $g(x) = 4x + 3$

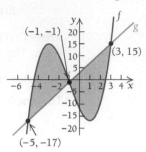

24. $f(x) = x^4 - 8x^3 + 18x^2$, $g(x) = x + 28$

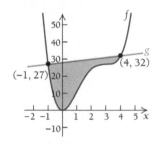

25. $f(x) = 4x - x^2$, $g(x) = x^2 - 6x + 8$

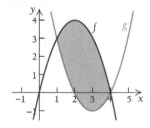

Find the area under the graph over the interval $[-2, 3]$, where:

26. $f(x) = \begin{cases} x^2, & \text{if } x < 1, \\ 1, & \text{if } x \geq 1. \end{cases}$

27. $f(x) = \begin{cases} 4 - x^2, & \text{if } x < 0, \\ 4, & \text{if } x \geq 0. \end{cases}$

Find the area of the region bounded by the given graphs.

28. $y = x^2$, $y = x^{-2}$, $x = 1$, $x = 5$

29. $y = e^x$, $y = e^{-x}$, $x = 0$, $x = 1$

30. $y = x^2$, $y = \sqrt[3]{x^2}$, $x = 1$, $x = 8$

31. $y = x^2, y = x^3, x = -1, x = 1$
32. $x + 2y = 2, y - x = 1, 2x + y = 7$

APPLICATIONS

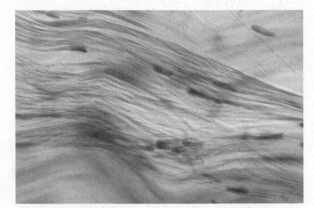

A human tendon.

33. *Elastic Strain Energy.* The forces on the human gastrocnemius tendon during traction and during recoil can be approximated by the functions

$$f(x) = 71.3x - 4.15x^2 + 0.434x^3 \quad \text{and}$$
$$g(x) = 71.0x - 10.3x^2 + 0.986x^3,$$

respectively, where $0 \le x \le 11$ is the tendon elongation in millimeters. The forces are measured in newtons (N). Integrate from $x = 0$ to $x = 11$ to compute the *elastic strain energy,* given by the area bounded by the graph of the two functions. Your units will be N · mm, which is equivalent to millijoules.[9]

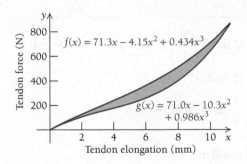

---

[9]N. Maganaris and J. P. Paul, "Tensile properties of the *in vivo* human gastrocnemius tendon," *Journal of Biomechanics,* 35 (2002), pp. 1639–1646.

34. *Degree Days.* For maize, the number of *degree days* is given by the area above the line $f(t) = 8°C$ and below the graph of temperature as a function of time, where time is measured in days and temperature is measured in degrees Celsius. During a day ($0 \le t \le 1$), the temperature is given by

$$T(t) = 5.76 + 24t - 16t^2.$$

Integrate to determine the number of degree days.

35. *Degree Days.* Refer to Exercise 34 for the definition of degree days. For 1 day ($0 \le t \le 1$), the temperature function is given by

$$T(t) = 25 + 3e^{-t} - 20(t - 0.5)^2,$$

where t is time in days since midnight and $T(t)$ is the temperature in degrees Celsius at time t. Integrate to determine the number of degree days.

36. *Memorizing.* In a certain memory experiment, subject A is able to memorize words at the rate given by

$$m'(t) = -0.009t^2 + 0.2t \quad \text{(words per minute)}.$$

In the same memory experiment, subject B is able to memorize at the rate given by

$$M'(t) = -0.003t^2 + 0.2t \quad \text{(words per minute)}.$$

a) Which subject has the higher rate of memorization?
b) How many more words does that subject memorize from $t = 0$ to $t = 10$ (during the first 10 min)?

SYNTHESIS

37. Find the area of the region bounded by $y = 3x^5 - 20x^3$, the x-axis, and the first coordinates of the relative maximum and minimum values of the function.

38. Find the area of the region bounded by $y = x^3 - 3x + 2$, the x-axis, and the first coordinates of the relative maximum and minimum values of the function.

39. *Poiseuille's Law.* The flow of blood in a blood vessel is faster toward the center of the vessel and slower toward the outside. The speed of the blood is given by

$$V = \frac{p}{4Lv}(R^2 - r^2),$$

where R is the radius of the blood vessel, r is the distance of the blood from the center of the vessel, and p, v, and L are physical constants related to the pressure and viscosity of the blood and the length of the blood vessel. If R is constant, we can think of V as a function of r:

$$V(r) = \frac{p}{4Lv}(R^2 - r^2).$$

The *total blood flow* Q is given by

$$Q = \int_0^R 2\pi \cdot V(r) \cdot r \cdot dr.$$

Find Q.

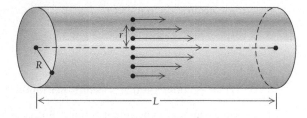

 Technology Connection

Find the area of the region bounded by the given graphs.

40. $y = x + 6$, $y = -2x$, $y = x^3$

41. $y = x^2 + 4x$, $y = \sqrt{16 - x^2}$

42. $y = x\sqrt{4 - x^2}$, $y = \dfrac{-4x}{x^2 + 1}$, $x = 0$, $x = 2$

43. $y = 2x^2 + x - 4$, $y = 1 - x + 8x^2 - 4x^4$

44. $y = \sqrt{1 - x^2}$, $y = 1 - x^2$, $x = -1$, $x = 1$

45. Consider the following functions:
$$f(x) = 3.8x^5 - 18.6x^3,$$
$$g(x) = 19x^4 - 55.8x^2.$$

   a) Graph these functions in the window $[-4, 4, -70, 70]$, with Yscl $= 10$.
   b) Estimate the first coordinates a, b, and c of the three points of intersection of the two graphs.
   c) Find the area between the curves on the interval $[a, b]$.
   d) Find the area between the curves on the interval $[b, c]$.

5.5 Integration Techniques: Substitution

OBJECTIVES

■ Evaluate integrals using substitution.
■ Solve applied problems involving integration by substitution.

The following formulas provide examples of an integration technique called **substitution**.

A. $\displaystyle\int u^r\, du = \frac{u^{r+1}}{r+1} + C$, provided $r \neq -1$

B. $\displaystyle\int e^u\, du = e^u + C$

C. $\displaystyle\int \frac{1}{u}\, du = \ln|u| + C$

Recall that the Chain Rule states that

$$\frac{d}{dx}f(u) = f'(u) \cdot u',$$

where u is a function of x. Since the right side is the derivative of $f(u)$, the anti-derivative of the right side is simply $f(u) + C$. That is,

$$\int f'(u) \cdot u'\, dx = f(u) + C. \tag{1}$$

Equation (1) is the basis for a technique of integration called *substitution* or *change of variable*.

EXAMPLE 1 Compute $\int 2xe^{x^2}\, dx$.

Solution   We need to guess what to use for u and what to use for $f(u)$. If we let $u = x^2$, then $u' = 2x$.

$$\int 2xe^{x^2}\, dx = \int u'e^u\, dx$$

$$= \int e^u \cdot u'\, dx$$

If we let $f(u) = e^u$, then $f'(u) = e^u$. Using equation (1), we see that

$$\int 2xe^{x^2}\, dx = e^u + C$$

$$= e^{x^2} + C.$$ ■

Integration by substitution can always be done using equation (1). However, it is often easier to use the dx notation. Rewriting equation (1) we have

$$\int f'(u) \cdot u'\, dx = \int f'(u) \cdot \frac{du}{dx}\, dx.$$

Even though $\dfrac{du}{dx}$ is not a fraction, we may "cancel" the dx's to obtain

$$\int f'(u) \cdot \frac{du}{dx}\, dx = \int f'(u)\, du$$

$$= f(u) + C.$$

Although we have not given a definition of dx or du, we will formally use them when computing integrals.

NOTATION

$$du = u'\, dx = \frac{du}{dx}\, dx$$

EXAMPLE 2   For $u = g(x) = \ln x$, find du.

Solution   We have

$$\frac{du}{dx} = g'(x) = \frac{1}{x},$$

so

$$du = g'(x)\, dx = \frac{1}{x}\, dx, \quad \text{or} \quad \frac{dx}{x}.$$ ■

EXAMPLE 3 For $y = f(x) = e^{x^2}$, find dy.

Solution Using the Chain Rule, we have

$$\frac{dy}{dx} = f'(x) = 2xe^{x^2},$$

so

$$dy = f'(x)\, dx = 2xe^{x^2}\, dx.$$ ■

To illustrate how the dx notation can be used to integrate, we redo Example 1 using this notation.

EXAMPLE 4 Use substitution and the dx notation to integrate $\displaystyle\int 2xe^{x^2}\, dx$.

Solution We use the same substitution as before, $u = x^2$. Then $du = 2x\, dx$. Using this substitution, we have

$$\int 2xe^{x^2}\, dx = \int e^{x^2}\, (2x)\, dx$$

$$= \int e^u\, du \qquad \underline{\text{Substitution}} \quad \begin{aligned} u &= x^2, \\ du &= 2x\, dx \end{aligned}$$

$$= e^u + C$$

$$= e^{x^2} + C.$$

Be sure to switch back to a function of x at the end of the solution. Since the problem asks for the antiderivative of $2xe^{x^2}$, the answer should be a function of x. ■

Integration is a skill acquired only after much practice. If you try a substitution that doesn't result in an integral that can be easily computed, try another substitution. There are many integrals that cannot be carried out using substitution. We do know that any integration that fits rule A, B, or C on page 380 can be done with substitution.

Let's consider some additional examples.

EXAMPLE 5 Evaluate: $\displaystyle\int \frac{2x\, dx}{1 + x^2}$.

Solution

$$\int \frac{2x\, dx}{1 + x^2} = \int \frac{du}{u} \qquad \underline{\text{Substitution}} \quad \begin{aligned} u &= 1 + x^2, \\ du &= 2x\, dx. \end{aligned}$$

$$= \ln |u| + C$$

$$= \ln |1 + x^2| + C$$

$$= \ln (1 + x^2) + C. \qquad \text{Since } 1 + x^2 > 0$$ ■

EXAMPLE 6 Evaluate: $\displaystyle\int \frac{2x\, dx}{(1 + x^2)^2}$.

Solution

$$\int \frac{2x\,dx}{(1+x^2)^2} = \int \frac{du}{u^2} \qquad \underline{\text{Substitution}} \quad \begin{aligned} u &= 1 + x^2, \\ du &= 2x\,dx \end{aligned}$$

$$= \int u^{-2}\,du$$

$$= -u^{-1} + C$$

$$= -\frac{1}{u} + C$$

$$= \frac{-1}{1+x^2} + C.$$

EXAMPLE 7 Evaluate: $\displaystyle\int \frac{\ln 3x\,dx}{x}$.

Solution

$$\int \frac{\ln 3x\,dx}{x} = \int u\,du \qquad \underline{\text{Substitution}} \quad \begin{aligned} u &= \ln 3x, \\ du &= \frac{1}{x}\,dx \end{aligned}$$

$$= \frac{u^2}{2} + C$$

$$= \frac{(\ln 3x)^2}{2} + C.$$

EXAMPLE 8 Evaluate: $\displaystyle\int x\sqrt{x^2 + 1}\,dx$.

Solution Suppose we try

$$u = x^2 + 1;$$

then we have

$$du = 2x\,dx.$$

We don't have $2x$ in $\int x\sqrt{x^2 + 1}\,dx$. We need to supply a 2. We do so by multiplying by 1, using $\frac{1}{2} \cdot 2$, to obtain

$$\int x\sqrt{x^2 + 1}\,dx = \frac{1}{2} \cdot 2 \cdot \int x\sqrt{x^2 + 1}\,dx$$

$$= \frac{1}{2} \int 2x\sqrt{x^2 + 1}\,dx$$

$$= \frac{1}{2} \int \sqrt{x^2 + 1}\,(2x)\,dx \qquad \underline{\text{Substitution}} \quad \begin{aligned} u &= x^2 + 1, \\ du &= 2x\,dx \end{aligned}$$

$$= \frac{1}{2} \int u^{1/2}\,du$$

$$= \frac{1}{2} \cdot \frac{2}{3} \cdot u^{3/2} + C$$

$$= \frac{1}{3}(x^2 + 1)^{3/2} + C.$$

EXAMPLE 9 Evaluate: $\displaystyle\int \frac{1}{\sqrt{x}} \sin \sqrt{x}\, dx$.

Solution

$$\int \frac{1}{\sqrt{x}} \sin \sqrt{x}\, dx = 2\int (\sin \sqrt{x})\left(\frac{1}{2\sqrt{x}}\right) dx \qquad \underline{\text{Substitution}} \quad \begin{array}{l} u = \sqrt{x}, \\ du = 1/(2\sqrt{x})\, dx \end{array}$$

$$= 2\int \sin u\, du$$

$$= -2\cos u + C$$

$$= -2\cos \sqrt{x} + C.$$

EXAMPLE 10 Evaluate: $\displaystyle\int x^2(x^3 + 1)^{10}\, dx$.

Solution

$$\int x^2(x^3 + 1)^{10}\, dx = \frac{1}{3}\int (x^3 + 1)^{10}\,(3x^2)\, dx \qquad \underline{\text{Substitution}} \quad \begin{array}{l} u = x^3 + 1, \\ du = 3x^2\, dx \end{array}$$

$$= \frac{1}{3}\int u^{10}\, du$$

$$= \frac{1}{3}\cdot\frac{u^{11}}{11} + C$$

$$= \frac{1}{33}(x^3 + 1)^{11} + C.$$

Technology Connection

EXERCISE

1. Use a grapher to evaluate
$$\int_0^1 x^2(x^3 + 1)^{10}\, dx.$$

EXAMPLE 11 Use substitution to evaluate $\displaystyle\int \tan x\, dx$.

Solution We first rewrite $\tan x$ in terms of the sine and cosine functions.

$$\int \tan x\, dx = \int \frac{\sin x}{\cos x}\, dx$$

$$= -\int \frac{-\sin x}{\cos x}\, dx \qquad \underline{\text{Substitution}} \quad \begin{array}{l} u = \cos x, \\ du = -\sin x\, dx \end{array}$$

$$= -\int \frac{du}{u}$$

$$= -\ln |u| + C$$

$$= -\ln |\cos x| + C.$$

Exercise Set 5.5

Evaluate using a substitution. (Be sure to check by differentiating!)

1. $\int \dfrac{3x^2\,dx}{7 + x^3}$

2. $\int \dfrac{3x^2\,dx}{1 + x^3}$

3. $\int e^{4x}\,dx$

4. $\int e^{3x}\,dx$

5. $\int e^{x/2}\,dx$

6. $\int e^{x/3}\,dx$

7. $\int x^3 e^{x^4}\,dx$

8. $\int x^4 e^{x^5}\,dx$

9. $\int t^2 e^{-t^3}\,dt$

10. $\int t e^{-t^2}\,dt$

11. $\int \dfrac{\ln 4x\,dx}{x}$

12. $\int \dfrac{\ln 5x\,dx}{x}$

13. $\int \dfrac{dx}{1 + x}$

14. $\int \cos(2x + 3)\,dx$

15. $\int 3 \sin(4x + 2)\,dx$

16. $\int 2 \sec^2(2x + 3)\,dx$

17. $\int \csc(2x + 3) \cot(2x + 3)\,dx$

18. $\int \cot^2(2 - x)\,dx$

19. $\int \dfrac{dx}{4 - x}$

20. $\int \dfrac{dx}{1 - x}$

21. $\int t^2(t^3 - 1)^7\,dt$

22. $\int t(t^2 - 1)^5\,dt$

23. $\int x \sin x^2\,dx$

24. $\int x^2 \cos x^3\,dx$

25. $\int (x + 1) \sec^2(x^2 + 2x + 3)\,dx$

26. $\int (x^2 + 2x + 2) \csc(x^3 + 3x^2 + 6x - 7)$
    $\times \cot(x^3 + 3x^2 + 6x - 7)\,dx$

27. $\int \dfrac{e^x\,dx}{4 + e^x}$

28. $\int \dfrac{e^t\,dt}{3 + e^t}$

29. $\int \dfrac{\ln x^2}{x}\,dx$

30. $\int \dfrac{(\ln x)^2}{x}\,dx$

31. $\int \dfrac{dx}{x \ln x}$

32. $\int \dfrac{dx}{x \ln x^2}$

33. $\int \sqrt{ax + b}\,dx$

34. $\int x\sqrt{ax^2 + b}\,dx$

35. $\int b e^{ax}\,dx$

36. $\int P_0 e^{kt}\,dt$

37. $\int a \sin(bx + c)\,dx$

38. $\int a \cos(bx + c)\,dx$

39. $\int \dfrac{3x^2\,dx}{(1 + x^3)^5}$

40. $\int \dfrac{x^3\,dx}{(2 - x^4)^7}$

41. $\int \cot x\,dx$

42. $\int 12x\sqrt[5]{1 + 6x^2}\,dx$

43. $\int 5t\sqrt{1 - 4t^2}\,dt$

44. $\int e^t \sin e^t\,dt$

45. $\int \dfrac{e^{\sqrt{w}}}{\sqrt{w}}\,dw$

46. $\int \dfrac{x^2}{e^{x^3}}\,dx$

47. $\int \sin^2 x \cos x\,dx$

48. $\int \cos^2\theta \sin\theta\,d\theta$

49. $\int e^{-\sin t} \cos t\,dt$

50. $\int \dfrac{\sin t}{e^{\cos t}}\,dt$

51. $\int r^2 \sin(3r^3 + 7)\,dr$

52. $\int \dfrac{1}{x} \sec^2 \ln x\,dx$

Evaluate.

53. $\int_0^1 2x e^{x^2}\,dx$

54. $\int_0^1 3x^2 e^{x^3}\,dx$

55. $\int_0^1 x(x^2 + 1)^5\,dx$

56. $\int_1^2 x(x^2 - 1)^7\,dx$

57. $\int_1^3 \dfrac{dt}{1 + t}$

58. $\int_1^3 e^{2x}\,dx$

59. $\int_1^4 \dfrac{2x + 1}{x^2 + x - 1}\,dx$

60. $\int_1^3 \dfrac{2x + 3}{x^2 + 3x}\,dx$

61. $\int_0^b e^{-x}\,dx$

62. $\int_0^b 2e^{-2x}\,dx$

63. $\int_0^b m e^{-mx}\,dx$

64. $\int_0^b k e^{-kx}\,dx$

65. $\int_0^4 (x - 6)^2\,dx$

66. $\int_0^3 (x - 5)^2\,dx$

67. $\int_{-1/3}^0 \cos(\pi x + \pi/3)\,dx$

68. $\int_0^{\sqrt{\pi/6}} 2x \sin 3x^2\,dx$

69. $\int_0^2 \dfrac{3x^2\,dx}{(1 + x^3)^5}$

70. $\int_{-1}^0 \dfrac{x^3\,dx}{(2 - x^4)^7}$

71. $\int_0^{\sqrt{7}} 7x\sqrt[3]{1 + x^2}\,dx$

72. $\int_0^1 12x\sqrt[5]{1 - x^2}\,dx$

APPLICATIONS

73. *Volume of a Tree.* The volume V between heights $x = h_0$ and $x = h_1$ of a tree with total height H can be approximated by the integral

$$V = K \int_{h_0}^{h_1} (H - x)^{3/2}\, dx, \qquad (1)$$

where K is a constant.[10]

a) Compute $K \int_0^H (H - x)^{3/2}\, dx$, the total volume of the tree. Your answer will include the constant H and the constant K.

b) Compute $K \int_0^{H/2} (H - x)^{3/2}\, dx$, the volume of the lower half of the tree.

c) What proportion of the total volume is in the lower half of the tree? Your simplified answer should include neither H nor K.

d) What proportion of the total volume is in the upper half of the tree?

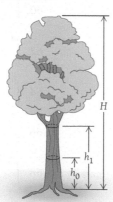

---

[10]Y. Zhang, B. Borders, and R. Bailey, "Derivation, fitting, and implication of a compatible stem taper-volume-weight system for intensively managed, fast growing loblolly pine," *Forest Science*, 48(3), 2002, pp. 595–607.

74. *Divorce.* The annual divorce rate in the United States is approximated by

$$D(t) = 100{,}000e^{0.025t},$$

where $D(t)$ is the divorce rate at time t and t is the number of years measured from 1900. That is, $t = 0$ corresponds to 1900, $t = 98\frac{9}{365}$ corresponds to January 9, 1998, and so on.

a) Find the total number of divorces from 1900 to 1999. Note that this is

$$\int_0^{99} D(t)\, dt.$$

b) Find the total number of divorces from 1980 to 1999. Note that this is

$$\int_{80}^{99} D(t)\, dt.$$

SYNTHESIS

Find the area of the shaded region.

75.

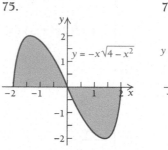

$y = -x\sqrt{4 - x^2}$

76.

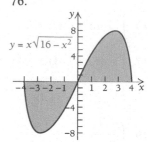

$y = x\sqrt{16 - x^2}$

77. $\displaystyle\int \frac{t^2 + 2t}{(t + 1)^2}\, dt$

$$\left(Hint: \frac{t^2 + 2t}{(t + 1)^2} = \frac{t^2 + 2t + 1 - 1}{t^2 + 2t + 1}\right.$$
$$\left. = 1 - \frac{1}{(t + 1)^2}.\right)$$

78. $\displaystyle\int \frac{x^2 + 6x}{(x + 3)^2}\, dx$

79. $\displaystyle\int \frac{x + 3}{x + 1}\, dx$

$$\left(Hint: \frac{x + 3}{x + 1} = 1 + \frac{2}{x + 1}.\right)$$

80. $\displaystyle\int \frac{t - 5}{t - 4}\, dt$

81. $\displaystyle\int \frac{dx}{x(\ln x)^n}, \; n \neq 1$

82. $\int \dfrac{dx}{e^x + 1}$

$\left(\text{Hint: } \dfrac{1}{e^x + 1} = \dfrac{e^{-x}}{1 + e^{-x}}.\right)$

83. $\int \dfrac{dx}{x \ln x \, [\ln (\ln x)]}$

84. $\int \dfrac{e^{-mx}}{1 + ae^{-mx}} \, dx$

85. $\int \sec x \, dx$ $\left(\text{Hint: Multiply by } \dfrac{\tan x + \sec x}{\tan x + \sec x}.\right)$

86. $\int \csc x \, dx$

tw 87. Determine whether the following is a theorem:

$$\int [f(x)]^2 \, dx = 2 \int f(x) \, dx.$$

5.6 Integration Techniques: Integration by Parts

OBJECTIVES

■ Evaluate integrals using the formula for integration by parts.

■ Solve applied problems involving integration by parts.

Recall the Product Rule for derivatives:

$$\frac{d}{dx} uv = u\frac{dv}{dx} + v\frac{du}{dx}.$$

Integrating both sides with respect to x, we get

$$uv = \int u\frac{dv}{dx} \, dx + \int v\frac{du}{dx} \, dx$$

$$= \int u \, dv + \int v \, du.$$

Solving for $\int u \, dv$, we get the following.

THEOREM 7

The Integration-by-Parts Formula

$$\int u \, dv = uv - \int v \, du$$

This equation can be used as a formula for integrating in certain situations—that is, situations in which an integrand is a product of two functions, and one of the functions can be integrated using the techniques we have already developed. For example,

$$\int xe^x \, dx$$

can be considered as

$$\int x(e^x \, dx) = \int u \, dv,$$

where we let

$$u = x \quad \text{and} \quad dv = e^x \, dx.$$

If so, we have

$$du = dx, \quad \text{by differentiating}$$

and

$$v = e^x, \quad \text{by integrating and using the simplest antiderivative.}$$

Then the Integration-by-Parts Formula gives us

$$\int \overset{u}{(x)} \overset{dv}{(e^x \, dx)} = \overset{u}{(x)} \overset{v}{(e^x)} - \int \overset{v}{(e^x)} \overset{du}{(dx)}$$

$$= xe^x - e^x + C.$$

This method of integrating is called **integration by parts.**

Note that integration by parts, like substitution, is a trial-and-error process. In the preceding example, suppose that we had reversed the roles of x and e^x. We would have obtained

$$u = e^x, \qquad dv = x \, dx,$$

$$du = e^x \, dx, \qquad v = \frac{x^2}{2},$$

and

$$\int (e^x)(x \, dx) = (e^x)\left(\frac{x^2}{2}\right) - \int \left(\frac{x^2}{2}\right)(e^x \, dx).$$

Now the integrand on the right is more difficult to integrate than the one with which we began. When we can integrate *both* factors of an integrand, and thus have a choice as to how to apply the Integration-by-Parts Formula, it can happen that only one (and maybe none) of the possibilities will work.

Tips on Using Integration by Parts

1. If you have tried substitution and have had no success, then try integration by parts.

2. Use integration by parts when an integral is of the form

$$\int f(x) \, g(x) \, dx.$$

Then match it with an integral of the form

$$\int u \, dv$$

by choosing a function to be $u = f(x)$, where $f(x)$ can be differentiated, and the remaining factor to be $dv = g(x) \, dx$, where $g(x)$ can be integrated.

3. Find du by differentiating and v by integrating.

4. If the resulting integral is more complicated than the original, make some other choice for u and dv.

Let's consider some additional examples.

EXAMPLE 1   Evaluate: $\int \ln x \, dx$.

Solution   Note that $\int (dx/x) = \ln |x| + C$, but we do not yet know how to find $\int \ln x \, dx$ since we do not know how to find an antiderivative of $\ln x$. Since we can differentiate $\ln x$, we let

$$u = \ln x \quad \text{and} \quad dv = 1 \cdot dx.$$

Then

$$du = \frac{1}{x} \, dx \quad \text{and} \quad v = x.$$

Using the Integration-by-Parts Formula gives

$$\overset{u}{\int} \overset{dv}{(\ln x)(dx)} = \overset{u}{(\ln x)}\overset{v}{x} - \int \overset{v}{x} \left(\overset{du}{\frac{1}{x} \, dx} \right)$$

$$= x \ln x - \int dx$$

$$= x \ln x - x + C.$$

EXAMPLE 2   Evaluate: $\int x \ln x \, dx$.

Solution   Let's examine several choices, as follows.

Choice 1:   We let

$$u = 1 \quad \text{and} \quad dv = x \ln x \, dx.$$

This will not work because we are back to our original integral, in which we do not know how to integrate $dv = x \ln x \, dx$.

Choice 2:   We let

$$u = x \ln x \qquad\qquad \text{and} \quad dv = 1 \cdot dx.$$

Then

$$du = \left[x\left(\frac{1}{x}\right) + 1(\ln x) \right] dx \quad \text{and} \quad v = x.$$

$$= (1 + \ln x) \, dx$$

Using the Integration-by-Parts Formula, we have

$$\int u \, dv = uv - \int v \, du$$

$$= (x \ln x)x - \int x(1 + \ln x) \, dx$$

$$= x^2 \ln x - \int (x + x \ln x) \, dx.$$

This integral seems more complicated than the original.

Choice 3: We let

$$u = \ln x \quad \text{and} \quad dv = x \, dx.$$

Then

$$du = \frac{1}{x} \, dx \quad \text{and} \quad v = \frac{x^2}{2}.$$

Using the Integration-by-Parts Formula, we have

$$\int u \, dv = uv - \int v \, du = (\ln x) \frac{x^2}{2} - \int \frac{x^2}{2} \cdot \frac{1}{x} \, dx$$

$$= \frac{x^2}{2} \ln x - \int \frac{x}{2} \, dx$$

$$= \frac{x^2}{2} \ln x - \frac{x^2}{4} + C.$$

This choice of u and dv allows us to evaluate the integral. ■

EXAMPLE 3 Evaluate: $\int x\sqrt{x+1} \, dx$.

Solution We let

$$u = x \quad \text{and} \quad dv = (x+1)^{1/2} \, dx.$$

Then

$$du = dx \quad \text{and} \quad v = \tfrac{2}{3}(x+1)^{3/2}.$$

Note that we had to use substitution in order to integrate dv. We see this as follows:

$$\int (x+1)^{1/2} \, dx = \int w^{1/2} \, dw \qquad \underline{\text{Substitution}} \quad \begin{array}{l} w = x+1, \\ dw = dx \end{array}$$

$$= \frac{w^{1/2+1}}{\frac{1}{2}+1} = \tfrac{2}{3}w^{3/2} = \tfrac{2}{3}(x+1)^{3/2}.$$

(We did not include the constant of integration since the Integration-by-Parts Formula requires only one antiderivative v.)

Using the Integration-by-Parts Formula gives us

$$\int x\sqrt{x+1} \, dx = x \cdot \frac{2}{3}(x+1)^{3/2} - \int \frac{2}{3}(x+1)^{3/2} \, dx$$

$$= \frac{2}{3}x(x+1)^{3/2} - \frac{2}{3}\int w^{3/2} \, dw \qquad \underline{\text{Substitution}} \quad \begin{array}{l} w = x+1, \\ dw = dx \end{array}$$

$$= \frac{2}{3}x(x+1)^{3/2} - \left(\frac{2}{3}\right)\left(\frac{2}{5}w^{5/2}\right) + C$$

$$= \frac{2}{3}x(x+1)^{3/2} - \frac{4}{15}(x+1)^{5/2} + C.$$

Integration by parts gives us the correct answer, but we could also have solved the problem by applying the substitution $u = x + 1$.

$$\int x\sqrt{x+1}\,dx = \int (u-1)\sqrt{u}\,du \qquad \underline{\text{Substitution}} \quad \begin{array}{l} u = x+1, \\ du = dx \end{array}$$

$$= \int (u^{3/2} - u^{1/2})\,du$$

$$= \frac{2}{5}u^{5/2} - \frac{2}{3}u^{3/2} + C$$

$$= \frac{2}{5}(x+1)^{5/2} - \frac{2}{3}(x+1)^{3/2} + C$$

In the exercises, we will show that these two apparently different answers are actually equal.   ◼

EXAMPLE 4   Evaluate: $\int_0^\pi x \sin x\,dx$.

Solution

a) First we find the indefinite integral $\int x \sin x\,dx$. We let

$$u = x \quad \text{and} \quad dv = \sin x\,dx.$$

Then

$$du = dx \quad \text{and} \quad v = -\cos x.$$

The Integration-by-Parts Formula gives us

$$\int x \sin x\,dx = -x \cos x - \int -\cos x\,dx$$

$$= -x \cos x + \sin x + C.$$

b) Then we compute the definite integral:

$$\int_0^\pi x \sin x\,dx = [-x \cos x + \sin x]_0^\pi$$

$$= (-\pi \cos \pi + \sin \pi) - (-0 \cos 0 + \sin 0)$$

$$= \pi. \qquad ◼$$

Repeated Integration by Parts

In some cases, we may need to apply the Integration-by-Parts Formula more than once.

EXAMPLE 5   Evaluate $\int_0^7 x^2 e^{-x}\,dx$ to find the area of the shaded region below.

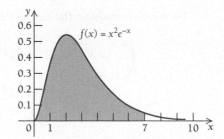

Technology Connection

EXERCISE

1. Use a grapher to evaluate

$$\int_0^7 x^2 e^{-x}\,dx.$$

Solution

a) We let

$$u = x^2 \qquad \text{and} \quad dv = e^{-x}\, dx.$$

Then

$$du = 2x\, dx \quad \text{and} \quad v = -e^{-x}.$$

Using the Integration-by-Parts Formula gives

$$\int u\, dv = uv - \int v\, du$$

$$= x^2(-e^{-x}) - \int -e^{-x}(2x\, dx)$$

$$= -x^2 e^{-x} + \int 2x e^{-x}\, dx. \qquad (1)$$

To evaluate the integral on the right, we can apply integration by parts again, as follows. We let

$$u = 2x \qquad \text{and} \quad dv = e^{-x}\, dx.$$

Then

$$du = 2\, dx \quad \text{and} \quad v = -e^{-x}.$$

Using the Integration-by-Parts Formula once again, we get

$$\int u\, dv = uv - \int v\, du$$

$$= 2x(-e^{-x}) - \int -e^{-x}(2\, dx)$$

$$= -2x e^{-x} - 2e^{-x} + C.$$

Thus the original integral becomes

$$\int x^2 e^{-x}\, dx = -x^2 e^{-x} + \int 2x e^{-x}\, dx$$

$$= -x^2 e^{-x} - 2x e^{-x} - 2e^{-x} + C \qquad \text{Equation (1)}$$

$$= -e^{-x}(x^2 + 2x + 2) + C.$$

b) We now evaluate the definite integral:

$$\int_0^7 x^2 e^{-x}\, dx = \left[-e^{-x}(x^2 + 2x + 2) \right]_0^7$$

$$= \left[-e^{-7}(7^2 + 2(7) + 2) \right] - \left[-e^{-0}(0^2 + 2(0) + 2) \right]$$

$$= -65e^{-7} + 2 \approx 1.94.$$

Tabular Integration by Parts

In situations like that in Example 5, we have an integral

$$\int f(x)\, g(x)\, dx,$$

where $f(x)$ is a polynomial and $g(x)$ can be repeatedly integrated easily. In such cases, we can use integration by parts more than once to evaluate the integral. As we saw in Example 5, the procedures can get complicated. When this happens, we can use **tabular integration,** which is illustrated in Example 6.

EXAMPLE 6   Evaluate: $\int x^3 e^x \, dx$.

Solution   We let $f(x) = x^3$ and $g(x) = e^x$. Then we make a tabulation as follows.

$f(x)$ and Repeated Derivatives		$g(x)$ and Repeated Integrals
x^3	(+)	e^x
$3x^2$	(−)	e^x
$6x$	(+)	e^x
6	(−)	e^x
0		e^x

We end the table when we have 0 in the derivative column. This is because the Integration-by-Parts Formula multiplies the antiderivative v times the derivative du, so if either is 0, their product is zero.

We add products along the arrows, making the alternating sign changes to obtain

$$\int x^3 e^x \, dx = x^3 e^x - 3x^2 e^x + 6xe^x - 6e^x + C.$$

Exercise Set 5.6

Evaluate using integration by parts. Check by differentiating.

1. $\displaystyle\int 5xe^{5x}\, dx$

2. $\displaystyle\int 2xe^{2x}\, dx$

3. $\displaystyle\int x \sin x \, dx$

4. $\displaystyle\int x \cos x \, dx$

5. $\displaystyle\int xe^{2x}\, dx$

6. $\displaystyle\int xe^{3x}\, dx$

7. $\displaystyle\int xe^{-2x}\, dx$

8. $\displaystyle\int xe^{-x}\, dx$

9. $\displaystyle\int x^2 \ln x \, dx$

10. $\displaystyle\int x^3 \ln x \, dx$

11. $\int x \ln x^2 \, dx$

12. $\int x^2 \ln x^3 \, dx$

13. $\int \ln (x + 3) \, dx$

14. $\int \ln (x + 1) \, dx$

15. $\int (x + 2) \ln x \, dx$

16. $\int (x + 1) \ln x \, dx$

17. $\int (x - 1) \sin x \, dx$

18. $\int (x - 2) \cos x \, dx$

19. $\int x\sqrt{x + 2} \, dx$

20. $\int x\sqrt{x + 4} \, dx$

21. $\int x^3 \ln 2x \, dx$

22. $\int x^2 \ln 5x \, dx$

23. $\int x^2 e^x \, dx$

24. $\int (\ln x)^2 \, dx$

25. $\int x^2 \sin 2x \, dx$

26. $\int x^{-5} \ln x \, dx$

27. $\int x^3 e^{-2x} \, dx$

28. $\int x^5 \cos 4x \, dx$

29. $\int x \sec^2 x \, dx$

30. $\int x \csc^2 x \, dx$

Evaluate using integration by parts.

31. $\int_1^2 x^2 \ln x \, dx$

32. $\int_1^2 x^3 \ln x \, dx$

33. $\int_2^6 \ln (x + 3) \, dx$

34. $\int_0^5 \ln (x + 1) \, dx$

35. $\int_0^1 x e^x \, dx$

36. $\int_0^1 x e^{-x} \, dx$

37. $\int_0^{5\pi/6} 3x \cos x \, dx$

38. $\int_0^{4\pi/3} 5x \sin x \, dx$

APPLICATIONS

39. *Growth of a Tree.* The rate of change of the mass M of a tree is approximated by

$$M'(t) = 10t\sqrt{t + 15},$$

where the mass is measured in grams and t is time in days. At time $t = 0$, the mass of the tree is 150,000 g. Find the mass of the tree after 10 days.

40. *Bacteria Growth.* The rate of change of the number of bacteria in a culture is

$$N'(t) = 1000t^2 e^{-0.2t},$$

where t in hours.

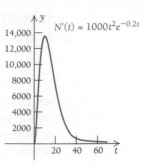

Find the number of bacteria in the culture at time $t = 60$, given that $N(0) = 2000$.

41. *Electrical Energy Use.* The rate of electrical energy used by a family, in kilowatts, is given by

$$K(t) = 10te^{-t},$$

where t is the time, in hours. That is, t is in the interval $[0, 24]$.

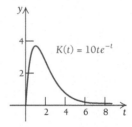

a) How many kilowatt hours does the family use in the first T hours of a day ($t = 0$ to $t = T$)?

b) How many kilowatt hours does the family use in the first 4 hr of the day?

42. *Drug Dosage.* Suppose that an oral dose of a drug is taken. From that time, the drug is assimilated in the body and excreted through the urine. The total amount of the drug that has passed through the body in time T is given by

$$\int_0^T E(t) \, dt,$$

where E is the rate of excretion of the drug. A typical rate-of-excretion function is

$$E(t) = te^{-kt},$$

where $k > 0$ and t is the time, in hours.

a) Use integration by parts to find a formula for

$$\int_0^T E(t) \, dt.$$

b) Find

$$\int_0^{10} E(t)\, dt, \quad \text{when } k = 0.2 \text{ mg/hr.}$$

43. In Example 3, we computed $\int x\sqrt{x+1}\, dx$ two different ways. Use algebra to show that the two answers are equal. (*Hint*: $(x+1)^{5/2} =$ $(x+1)(x+1)^{3/2} = x(x+1)^{3/2} + (x+1)^{3/2}$.)

44. a) Integrate $\int x\sqrt[3]{x+2}\, dx$ using a substitution.
   b) Use integration by parts to compute $\int x\sqrt[3]{x+2}\, dx$.
   c) Use algebra to show that your answers in parts (a) and (b) are equal.

Evaluate using integration by parts.

45. $\displaystyle\int \sqrt{x}\, \ln x\, dx$

46. $\displaystyle\int x^n \ln x\, dx$

47. $\displaystyle\int \frac{te^t}{(t+1)^2}\, dt$

48. $\displaystyle\int x^2(\ln x)^2\, dx$

49. $\displaystyle\int \frac{\ln x}{\sqrt{x}}\, dx$

50. $\displaystyle\int x^n\,(\ln x)^2\, dx$

51. $\displaystyle\int \frac{13t^2 - 18}{\sqrt[5]{4t+7}}\, dt$

52. $\displaystyle\int (27x^3 + 83x - 2)\sqrt[6]{3x+8}\, dx$

53. Verify that for any positive integer n,

$$\int x^n e^x\, dx = x^n e^x - n \int x^{n-1} e^x\, dx.$$

54. Verify that for any positive integer n,

$$\int (\ln x)^n\, dx = x(\ln x)^n - n \int (\ln x)^{n-1}\, dx.$$

55. Verify that for any positive integer n,

$$\int x^n \sin x\, dx = -x^n \cos x + n \int x^{n-1} \cos x\, ds.$$

56. Verify that for any positive integer n,

$$\int x^n \cos x\, dx = x^n \sin x - n \int x^{n-1} \sin x\, dx.$$

tw 57. Determine whether the following is a theorem:

$$\int f(x)g(x)\, dx = \int f(x)\, dx \cdot \int g(x)\, dx.$$

Explain.

tw 58. Compare the procedures of differentiation and integration. Which seems to be the most complicated or difficult and why?

 Technology Connection

59. Use a grapher to evaluate

$$\int_1^{10} x^5 \ln x\, dx.$$

5.7 Integration Techniques: Tables and Technology

OBJECTIVES

■ Evaluate integrals using a table of integration formulas.
■ Evaluate integrals using technology.

Tables of Integration Formulas

You have probably noticed that, generally speaking, integration is more difficult and "tricky" than differentiation. Because of this, integral formulas that are reasonable or important have been gathered into tables. Table 1, listed at the back of the book, though quite brief, is such an example. Entire books of integration formulas are available in libraries, and lengthy tables are also available in mathematics handbooks. Such tables are usually classified by the form of the integrand. The idea is to properly match the integral in question with a formula in the table. Sometimes some algebra or a technique such as integration by substitution or parts may be needed as well as a table.

EXAMPLE 1 Evaluate:

$$\int \frac{5x}{7x-8}\,dx.$$

Solution If we first factor 5 out of the integral, then the integral fits *Formula 30* in Table 1:

$$\int \frac{x}{ax+b}\,dx = \frac{x}{a} - \frac{b}{a^2}\ln|ax+b| + C.$$

In our integral, $a = 7$ and $b = -8$, so we have, by the formula,

$$\int \frac{5x}{7x-8}\,dx = 5\int \frac{x}{7x-8}\,dx$$

$$= 5\left[\frac{x}{7} - \frac{-8}{7^2}\ln|7x-8|\right] + C$$

$$= 5\left[\frac{x}{7} + \frac{8}{49}\ln|7x-8|\right] + C$$

$$= \frac{5x}{7} + \frac{40}{49}\ln|7x-8| + C.$$

EXAMPLE 2 Evaluate:

$$\int \sqrt{16x^2+3}\,dx.$$

Solution This integral almost fits *Formula 34* in Table 1:

$$\int \sqrt{x^2 \pm a^2}\,dx = \tfrac{1}{2}\left[x\sqrt{x^2 \pm a^2} \pm a^2\ln|x + \sqrt{x^2 \pm a^2}|\right] + C.$$

But the x^2-coefficient needs to be 1. To achieve this, we first factor out 16. Then we apply *Formula 34*:

$$\int \sqrt{16x^2+3}\,dx = \int \sqrt{16\left(x^2 + \tfrac{3}{16}\right)}\,dx$$

$$= \int 4\sqrt{x^2 + \tfrac{3}{16}}\,dx$$

$$= 4\int \sqrt{x^2 + \tfrac{3}{16}}\,dx$$

$$= 4 \cdot \tfrac{1}{2}\left[x\sqrt{x^2 + \tfrac{3}{16}} + \tfrac{3}{16}\ln|x + \sqrt{x^2 + \tfrac{3}{16}}|\right] + C$$

$$= 2\left[x\sqrt{x^2 + \tfrac{3}{16}} + \tfrac{3}{16}\ln|x + \sqrt{x^2 + \tfrac{3}{16}}|\right] + C.$$

In our integral, $a^2 = 3/16$ and $a = \sqrt{3}/4$, though we did not need to use a in this form when applying the formula.

EXAMPLE 3   Evaluate: $\int (\ln x)^3 \, dx$.

Solution   This integral fits *Formula 9* in Table 1:

$$\int (\ln x)^n \, dx = x(\ln x)^n - n \int (\ln x)^{n-1} \, dx + C, \quad n \neq -1.$$

We must apply the formula three times:

$$\int (\ln x)^3 \, dx = x(\ln x)^3 - 3 \int (\ln x)^2 \, dx + C \qquad \text{Formula 9}$$

$$= x(\ln x)^3 - 3\left[x(\ln x)^2 - 2 \int \ln x \, dx \right] + C \qquad \begin{array}{l}\text{Applying Formula 9}\\ \text{again}\end{array}$$

$$= x(\ln x)^3 - 3\left[x(\ln x)^2 - 2\left(x \ln x - \int dx \right) \right] + C \qquad \begin{array}{l}\text{Applying}\\ \text{Formula 9 for}\\ \text{the third time}\end{array}$$

$$= x(\ln x)^3 - 3x(\ln x)^2 + 6x \ln x - 6x + C. \qquad \blacksquare$$

Technology Connection

Integration Using Technology

Mathematical computer programs, such as *Mathematica* and *Maple,* and even some calculators use sophisticated algorithms to compute indefinite integrals. These programs are called *Computer Algebra Systems* (CAS). Using CAS to compute integrals is simply a matter of typing in the integral. Wolfram Research provides a Web-based integrator at its Web site http://integrals.wolfram.com/.

We illustrate its use by computing the integral

$$\int x e^x \, dx.$$

Before entering the function, we need to be aware of some conventions. The number e is entered as a capital E. Also, the natural logarithm, $\ln x$, is denoted by Log [x]. In general, the first letter of a function is uppercase. For logarithm base a, the notation is Log [a, x]. Before computing many integrals, take a few minutes to briefly read the function input directions given at the Web site.

To have the integrator compute $\int x e^x dx$, we enter X*E^X. The integrator gives the answer

$e^x(-1 + x)$. The integrator never includes the constant of integration. The user must keep in mind that the integrator just gives one antiderivative, and any other antiderivative can be obtained by adding a constant.

Next try computing

$$\int \frac{1}{x} \, dx$$

using the integrator. The integrator gives an answer of Log[x]. It assumes that x is positive and leaves out the absolute value symbols and the constant.

Compute the following using the integrator.

1. $\int x^3 e^x \, dx$ **2.** $\int x^5 e^x \, dx$

3. $\int x^{10} e^x \, dx$ **4.** $\int e^x \cos x \, dx$

5. $\int \dfrac{1}{\sqrt{x} + \sqrt[3]{x}} \, dx$ (Enter x^(1/2) for $\sqrt{x}$ and x^(1/3) for $\sqrt[3]{x}$.)

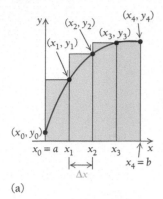

(a)

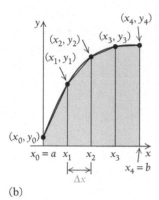

(b)

Figure 1

Trapezoid Rule

There are times when we wish to compute a definite integral, but we are unable to compute the antiderivative. This happens for some functions that arise in applications. For example, $\int e^{-x^2} dx$ cannot be computed in *closed form*. This means that although the integral exists, it cannot be expressed using the usual functions from calculus. Even so, in Chapter 10 we will need to approximate $\int_0^a e^{-x^2} dx$ for various values of a.

For such integrals, we resort to *estimating* the definite integral, as opposed to using the Fundamental Theorem of Calculus. Recall from Section 5.2 that a definite integral is the limit of the sum of n approximating rectangular areas, as illustrated in Fig. 1(a). Instead of using rectangles, here we use n trapezoids to improve the accuracy of the approximation, as illustrated in Fig. 1(b). We let $x_0 = a$ and $x_n = b$. The edges of the trapezoids occur along the x-axis at positions $x_0, x_1, \ldots, x_n$, equally spaced between a and b. The height of each trapezoid (measured left to right) is $\Delta x = (b - a)/n$. We let $y_k = f(x_k)$. The kth trapezoid has left-side base length y_{k-1} and right-side base length y_k.

Comparing Figs. 1(a) and 1(b), it seems that the area is better approximated using trapezoids instead of rectangles. The area A_k of the kth trapezoid is the average of the two base lengths multiplied by the height, or

$$A_k = \left(\frac{y_k + y_{k-1}}{2} \right) \Delta x.$$

Adding the areas of all the trapezoids together gives us the **Trapezoid rule**.

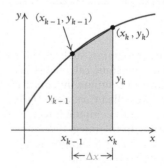

DEFINITION

Trapezoid Rule

The Trapezoid rule approximation for $\int_a^b f(x)\, dx$ using n trapezoids is

$$T_n = \frac{\Delta x}{2} [y_0 + 2y_1 + 2y_2 + \cdots + 2y_{n-1} + y_n],$$

where $\Delta x = \dfrac{b - a}{n}$, $x_k = a + k\Delta x$ and $y_k = f(x_k)$.

The *weights* are the coefficients $1, 2, 2, 2, \ldots, 2, 1$ and the *terms* are the weights multiplied by y_k. For example, the weight of y_0 is 1 and the first term is $1 \cdot y_0 = y_0$.

EXAMPLE 4   Estimate the integral $\displaystyle\int_{-1}^{1} \sqrt{1 - x^2}\, dx$ using

a) the Trapezoid rule with $n = 4$.

b) the Trapezoid rule with $n = 8$.

Solution

a) To use the Trapezoid rule with $n = 4$, we compute

$$\Delta x = \frac{b - a}{n} = \frac{1 - (-1)}{4} = \frac{1}{2} = 0.5,$$

since $a = -1$ and $b = 1$.

We compile the needed data in the table below.

k	$x_k = a + k\Delta x$	$y_k = \sqrt{1 - x_k^2}$	Trapezoid Weight	Trapezoid Term
0	-1	$\sqrt{1 - (-1)^2} = 0$	1	$1 \cdot 0 = 0.000000$
1	$-1/2$	$\sqrt{1 - (-1/2)^2} \approx 0.866025$	2	$2 \cdot 0.866025 = 1.732050$
2	0	$\sqrt{1 - (0)^2} = 1$	2	$2 \cdot 1 = 2.000000$
3	$1/2$	$\sqrt{1 - (-1/2)^2} \approx 0.866025$	2	$2 \cdot 0.866025 = 1.732050$
4	1	$\sqrt{1 - (1)^2} = 0$	1	$1 \cdot 0 = 0.000000$
			Total	5.464100

We can now estimate the integral using the Trapezoid rule.

$$T_4 = \frac{\Delta x}{2}(y_0 + 2y_1 + 2y_2 + 2y_3 + y_4)$$

$$\approx \frac{0.5}{2}(5.464100)$$

$$= 1.366025$$

b) For $n = 8$,

$$\Delta x = \frac{1 - (-1)}{8} = 0.25.$$

The table below summarizes the data needed to compute T_8.

k	$x_k = a + k\Delta x$	$y_k = \sqrt{1 - x_k^2}$	Trapezoid Weight	Trapezoid Term
0	-1	$\sqrt{1 - (-1)^2} = 0$	1	$1 \cdot 0 = 0.000000$
1	$-3/4$	$\sqrt{1 - (-3/4)^2} \approx 0.661438$	2	$2 \cdot 0.661438 = 1.322876$
2	$-1/2$	$\sqrt{1 - (1/2)^2} \approx 0.866025$	2	$2 \cdot 0.866025 = 1.732050$
3	$-1/4$	$\sqrt{1 - (1/4)^2} \approx 0.968246$	2	$2 \cdot 0.968246 = 1.936492$
4	0	$\sqrt{1 - (0)^2} = 1$	2	$2 \cdot 1 = 2.000000$
5	$1/4$	$\sqrt{1 - (1/4)^2} \approx 0.968246$	2	$2 \cdot 0.968246 = 1.936492$
6	$1/2$	$\sqrt{1 - (1/2)^2} \approx 0.866025$	2	$2 \cdot 0.866025 = 1.732050$
7	$3/4$	$\sqrt{1 - (3/4)^2} \approx 0.661438$	2	$2 \cdot 0.661438 = 1.322876$
8	1	$\sqrt{1 - (1)^2} = 0$	1	$1 \cdot 0 = 0.000000$
		Total		11.982836

The Trapezoid rule gives

$$T_8 = \frac{\Delta x}{2}(y_0 + 2y_1 + 2y_2 + 2y_3 + 2y_4 + 2y_5 + 2y_6 + 2y_7 + y_8)$$

$$\approx \frac{0.25}{2} \cdot 11.982836$$

$$= 1.497855.$$

The graph of the function $f(x) = \sqrt{1 - x^2}$ for $-1 \le x \le 1$ is shown below. It is half of a circle of radius 1, so the integral is half of the area of the circle, or $\frac{1}{2}\pi \cdot 1^2 \approx 1.570796$. We see that the Trapezoid rule gives an estimate T_8 that differs by approximately 0.07 from the actual answer.

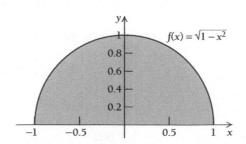

Technology Connection

Trapezoid Rule

Organizing the data in a table as in Example 4 suggests using a spreadsheet such as Microsoft *Excel*. This gives an efficient way to estimate an integral using the Trapezoid rule. You may first wish to review some of the basic features of a spreadsheet in the Technology Connection on pages 352–353.

1. Do Example 4(a) using a spreadsheet by following the steps below.

    a) In the first row, label the columns k, x_k, y_k, Trapezoid Weight, and Trapezoid Term.

    b) Enter the values in the columns labeled k and Trapezoid Weight one cell at a time, as in the table in Example 4(a).

    c) Enter -1 for the value of x_0.

    d) Enter the formula for x_1 that adds Δx to x_0. Recall that $\Delta x = 0.5$ in Example 4(a). Use the **FILL DOWN** option to complete the column.

    e) Enter the formula for $y_0 = f(x_0)$ and use the **FILL DOWN** option to complete the column. Use ^(1/2) to indicate a square root.

    f) Enter the Trapezoid Weights (1 for the first and last rows, 2 otherwise) in the next column.

    g) Enter the formula for the Trapezoid Term y_0. (Multiply the contents of the two cells to the left.) Then use **FILL DOWN** to complete the column.

    h) Below the Trapezoid Terms, enter a formula that gives the sum of the weights.

    i) Use the sum of the weights to compute the Trapezoid rule in another cell.

2. Repeat Exercise 1 for Example 4(b).

Simpson's Rule

Riemann sums approximate the function using horizontal line segments, while the Trapezoid rule approximates the function with lines between points on the graph. To improve the accuracy, we could use parabolas. In the figure, the red dots are the points $(x_{k-1}, f(x_{k-1}))$, $(x_k, f(x_k))$, $(x_{k+1}, f(x_{k+1}))$. A parabola is drawn to pass through the three red points. The light area indicates the approximation using the Trapezoid rule. The light and dark areas together give the approximation to the integral using a parabola to approximate the function. We see that using a parabola should be superior to the Trapezoid rule if the function is curved. This method is called **Simpson's rule.**

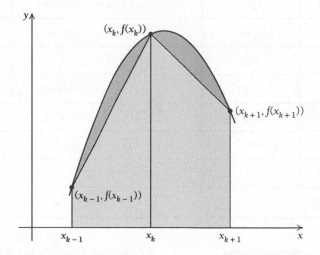

DEFINITION

Simpson's Rule

Simpson's approximation for $\int_a^b f(x)\,dx$ using n subintervals is

$$S_n = \frac{\Delta x}{3}[y_0 + 4y_1 + 2y_2 + 4y_3 + 2y_4$$
$$+ \cdots + 2y_{n-2} + 4y_{n-1} + y_n],$$

where n is even, $\Delta x = \dfrac{b - a}{n}$, $x_k = a + k\Delta x$, and $y_k = f(x_k)$. To use Simpson's rule, n must be even.

EXAMPLE 5 Estimate the integral $\int_{-1}^{1} \sqrt{1 - x^2}\, dx$ using Simpson's rule for $n = 8$.

Solution We can use Simpson's rule since 8 is even. As in Example 4, $\Delta x = \dfrac{1 - (-1)}{8} = 0.25$. We modify the table from Example 4 to fit Simpson's rule.

k	$x_k = a + k\Delta x$	$y_k = \sqrt{1 - x_k^2}$	Simpson Weight	Simpson Term
0	-1	$\sqrt{1 - (-1)^2} = 0$	1	$1 \cdot 0 = 0.000000$
1	$-3/4$	$\sqrt{1 - (-3/4)^2} \approx 0.661438$	4	$4 \cdot 0.661438 = 2.645752$
2	$-1/2$	$\sqrt{1 - (1/2)^2} \approx 0.866025$	2	$2 \cdot 0.866025 = 1.732050$
3	$-1/4$	$\sqrt{1 - (1/4)^2} \approx 0.968246$	4	$4 \cdot 0.968246 = 3.872984$
4	0	$\sqrt{1 - (0)^2} = 1$	2	$2 \cdot 1 = 2.000000$
5	$1/4$	$\sqrt{1 - (1/4)^2} \approx 0.968246$	4	$4 \cdot 0.968246 = 3.872984$
6	$1/2$	$\sqrt{1 - (1/2)^2} \approx 0.866025$	2	$2 \cdot 0.866025 = 1.732050$
7	$3/4$	$\sqrt{1 - (3/4)^2} \approx 0.661438$	4	$4 \cdot 0.661438 = 2.645752$
8	1	$\sqrt{1 - (1)^2} = 0$	1	$1 \cdot 0 = 0.000000$
			Total	18.501572

Technology Connection

Simpson's Rule

Use a spreadsheet to verify the calculations used in Example 5. You may wish to use the Technology Connection on page 401 as a guide.

We can now estimate the integral using Simpson's rule.

$$S_8 = \frac{\Delta x}{3}(y_0 + 4y_1 + 2y_2 + 4y_3 + 2y_4 + 4y_5 + 2y_6 + 4y_7 + y_8)$$

$$\approx \frac{0.25}{3}(18.501572)$$

$$\approx 1.541798.$$

Since this definite integral represents half the area of a circle with radius 1, the actual answer is $\pi/2 \approx 1.570796$. Simpson's rule gives an error of approximately 0.03. Recall that in Example 4 we estimated the same integral using the Trapezoid rule and the error was approximately 0.07, more than twice the error found from the Simpson's rule. Generally, Simpson's rule gives a better approximation than the Trapezoid rule if the same number of subintervals is used. ∎

EXAMPLE 6 *Degree Days.* The number of *degree days* is one factor in determining the growth of a crop. For maize, it is computed as the area above the line $y = 8$ and below the graph of temperature (in degrees Celsius) as a function of time (in days). The following table[11] shows the temperature every 2 hr in Denton, Texas, during the

---

[11]http://weather.com

Technology Connection

Simpson's Rule and Cubic Polynomials

1. Let $f(x) = x^3$.
   a) Compute $\int_0^4 f(x)\,dx$ exactly.
   b) Using a calculator, apply Simpson's rule with $n = 4$ to approximate $\int_0^4 f(x)\,dx$. How does your answer compare to part (a)?
   c) Using a calculator, apply Simpson's rule with $n = 2$ to approximate $\int_0^4 f(x)\,dx$. How does your answer compare to part (a)?

2. Repeat Exercise 1 using $f(x) = 2x^3 + 3x^2 + x - 1$ and integrating $\int_{-2}^3 f(x)\,dx$.

3. Based on your observations in Exercises 1 and 2, how accurate is Simpson's rule when applied to a cubic polynomial? (See Exercise 61 in Exercise Set 5.7).

day June 10, 2003. Use the Trapezoid rule and Simpson's rule to estimate the number of degree days.

Hours After Midnight	Temperature (°C)
0	26
2	25
4	24
6	23
8	24
10	27
12	29
14	31
16	32
18	32
20	31
22	27
24	26

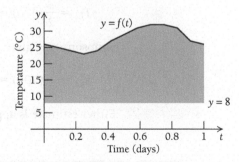

Solution   Let $f(t)$ be the temperature at time t measured in days. Since the temperature on June 10 did not dip below 8°C, we wish to compute

$$\int_0^1 [f(t) - 8]\,dt.$$

We convert time to days, subtract 8 from the temperature, and fill in the table on the next page. Since 2 hr pass between each recorded temperature, $\Delta t = 2/24 = 1/12$ days.

k	Time (days) $t_k = a + k\Delta t$	Temperature -8 (°C) y_k	Trapezoid Weight	Trapezoid Term	Simpson Weight	Simpson Term
0	0	$26 - 8 = 18$	1	$1(18) = 18$	1	$1(18) = 18$
1	$2/24 \approx 0.083$	$25 - 8 = 17$	2	$2(17) = 34$	4	$4(17) = 68$
2	$4/24 \approx 0.167$	$24 - 8 = 16$	2	$2(16) = 32$	2	$2(16) = 32$
3	$6/24 = 0.25$	$23 - 8 = 15$	2	$2(15) = 30$	4	$4(15) = 60$
4	$8/24 \approx 0.333$	$24 - 8 = 16$	2	$2(16) = 32$	2	$2(16) = 32$
5	$10/24 \approx 0.417$	$27 - 8 = 19$	2	$2(19) = 38$	4	$4(19) = 76$
6	$12/24 = 0.5$	$29 - 8 = 21$	2	$2(21) = 42$	2	$2(21) = 42$
7	$14/24 \approx 0.583$	$31 - 8 = 23$	2	$2(23) = 46$	4	$4(23) = 92$
8	$16/24 \approx 0.667$	$32 - 8 = 24$	2	$2(24) = 48$	2	$2(24) = 48$
9	$18/24 = 0.75$	$32 - 8 = 24$	2	$2(24) = 48$	4	$4(24) = 96$
10	$20/24 \approx 0.833$	$31 - 8 = 23$	2	$2(23) = 46$	2	$2(23) = 46$
11	$22/24 \approx 0.917$	$27 - 8 = 19$	2	$2(19) = 38$	4	$4(19) = 76$
12	$24/24 = 1.0$	$26 - 8 = 18$	1	$1(18) = 18$	1	$1(18) = 18$
	Total			470		704

The Trapezoid rule gives

$$T_{12} = \frac{\frac{1}{12}}{2}(470) \approx 19.583 \text{ degree days.}$$

Simpson's rule gives

$$S_{12} \approx \frac{\frac{1}{12}}{3}(704) \approx 19.556 \text{ degree days.}$$

Either estimate is approximately 19.6 degree days.

Exercise Set 5.7

Evaluate using Table 1.

1. $\displaystyle\int xe^{-3x}\, dx$

2. $\displaystyle\int xe^{4x}\, dx$

3. $\displaystyle\int 5^x\, dx$

4. $\displaystyle\int \frac{1}{\sqrt{x^2 - 9}}\, dx$

5. $\displaystyle\int \frac{1}{16 - x^2}\, dx$

6. $\displaystyle\int \frac{1}{x\sqrt{4 + x^2}}\, dx$

7. $\displaystyle\int \frac{x}{5 - x}\, dx$

8. $\displaystyle\int \frac{x}{(1 - x)^2}\, dx$

9. $\displaystyle\int \frac{1}{x(5-x)^2}\,dx$

10. $\displaystyle\int \sqrt{x^2+9}\,dx$

11. $\displaystyle\int \ln 3x\,dx$

12. $\displaystyle\int \ln \frac{4}{5}x\,dx$

13. $\displaystyle\int x^4 e^{5x}\,dx$

14. $\displaystyle\int x^3 e^{-2x}\,dx$

15. $\displaystyle\int x^3 \sin x\,dx$

16. $\displaystyle\int 5x^4 \ln x\,dx$

17. $\displaystyle\int \sec 2x\,dx$

18. $\displaystyle\int \csc 3x\,dx$

19. $\displaystyle\int 2\tan(2x+1)\,dx$

20. $\displaystyle\int 4\cot(3x-2)\,dx$

21. $\displaystyle\int \frac{dx}{\sqrt{x^2+7}}$

22. $\displaystyle\int \frac{3\,dx}{x\sqrt{1-x^2}}$

23. $\displaystyle\int \frac{10\,dx}{x(5-7x)^2}$

24. $\displaystyle\int \frac{2}{5x(7x+2)}\,dx$

25. $\displaystyle\int \frac{-5}{4x^2-1}\,dx$

26. $\displaystyle\int \sqrt{9t^2-1}\,dt$

27. $\displaystyle\int \sqrt{4m^2+16}\,dm$

28. $\displaystyle\int \frac{3\ln x}{x^2}\,dx$

29. $\displaystyle\int \frac{-5\ln x}{x^3}\,dx$

30. $\displaystyle\int (\ln x)^4\,dx$

31. $\displaystyle\int \frac{e^x}{x^{-3}}\,dx$

32. $\displaystyle\int \frac{3}{\sqrt{4x^2+100}}\,dx$

33. $\displaystyle\int x\sqrt{1+2x}\,dx$

34. $\displaystyle\int x\sqrt{2+3x}\,dx$

Evaluate using CAS.

35. $\displaystyle\int (\ln x)^4\,dx$

36. $\displaystyle\int x^3(\ln x)^2\,dx$

37. $\displaystyle\int \frac{1}{x(x^2-1)}\,dx$

38. $\displaystyle\int \frac{1}{\sqrt[3]{x}+\sqrt[4]{x}}\,dx$

39. $\displaystyle\int x^4 \sin 3x\,dx$

40. $\displaystyle\int x^6 \cos 2x\,dx$

41. $\displaystyle\int e^{2x} \sin 3x\,dx$

42. $\displaystyle\int e^{-x} \cos 2x\,dx$

Estimate using

a) the Trapezoid rule.
b) Simpson's rule.

43. $\displaystyle\int_0^1 e^{-x^2}\,dx$, use $n=4$.

44. $\displaystyle\int_1^4 \frac{e^x}{x}\,dx$, use $n=4$.

45. $\displaystyle\int_{-1}^1 \frac{e^x}{x+5}\,dx$, use $n=6$.

46. $\displaystyle\int_0^1 \frac{1}{1+x^2}\,dx$, use $n=6$.

47. $\displaystyle\int_{0.1}^2 \frac{\sin x}{x}\,dx$, use $n=8$.

48. $\displaystyle\int_{0.1}^{0.72} \frac{\cos 2x}{\sqrt{x}}\,dx$, use $n=8$.

49. $\displaystyle\int_{0.25}^{0.5} \frac{1}{\sqrt{1-x^2}}\,dx$, use $n=8$.

50. $\displaystyle\int_{0.25}^{0.75} \frac{x}{(x+1)(x^2+1)}\,dx$, use $n=8$.

APPLICATIONS

51. *Learning Rate.* The rate of change of the probability that an employee learns a task on a new assembly line is given by

$$p'(t) = \frac{1}{t(2+t)^2},$$

where p is the probability of learning the task after time t, in months. Find the function $p(t)$ given that $p = 0.8267$ when $t = 2$.

In Exercises 52–53, estimate the number of degree days, as defined in Example 6 for maize.

 52. *Degree Days.* Use the Trapezoid rule to estimate the number of degree days between 9:00 P.M. July 5 and 9:00 P.M. July 7, 2003, in Fort Wayne, Indiana. Temperatures every 4 hr are given in the table below.[12]

Temperatures in Fort Wayne	
Hours After 9:00 P.M. July 5	**Temperature (°C)**
0	26.7
4	22.2
8	21.2
12	21.1
16	25
20	29
24	22
28	19
32	19
36	23
40	21
44	26.7
48	27.2

 53. *Degree Days.* Use Simpson's rule to estimate the number of degree days between 9:00 P.M. July 5 and 9:00 P.M. July 7, 2003, in Ames, Iowa. Temperatures every 4 hr are given in the table below.[13]

Temperatures in Ames	
Hours After 9:00 P.M. July 5	**Temperature (°C)**
0	27.8
4	22
8	21.2
12	19
16	26.1
20	28.3
24	28.3
28	25.6
32	25
36	25
40	30
44	30.6
48	23.3

SYNTHESIS

Evaluate using Table 1. You may also need to use substitution.

54. $\displaystyle\int \frac{8}{3x^2 - 2x}\, dx$

55. $\displaystyle\int \frac{x\, dx}{4x^2 - 12x + 9}$

56. $\displaystyle\int \frac{dx}{x^3 - 4x^2 + 4x}$

57. $\displaystyle\int e^x \sqrt{e^{2x} + 1}\, dx$

58. $\displaystyle\int \frac{-e^{-2x}\, dx}{9 - 6e^{-x} + e^{-2x}}$

59. $\displaystyle\int \frac{\sqrt{(\ln x)^2 + 49}}{2x}\, dx$

60. *Simpson's Rule and Quadratic Polynomials.* Let

$$f(x) = ax^2 + bx + c, \tag{2}$$

where a, b, and c are constants.

a) Compute $\int_{-h}^{h} f(x)\, dx$. Your answer will contain the constants a, b, c, and h.

---

[12]http://weather.noaa.gov
[13]http://weather.noaa.gov

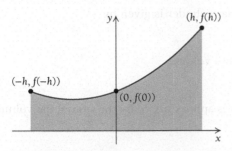

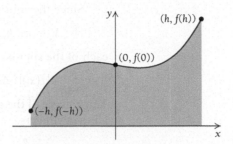

b) Substitute $x = h$ into equation (2) to find an expression for $f(h)$ in terms of $a, b, c,$ and h. Do the same to find expressions for $f(0)$ and $f(-h)$.

c) Use parts (a) and (b) to show that

$$\int_{-h}^{h} f(x)\, dx = \frac{h}{3}[f(-h) + 4f(0) + f(h)].$$

tw d) Explain why part (c) gives Simpson's rule.

61. *Simpson's Rule and Cubic Polynomials.* Let

$$f(x) = ax^3 + bx^2 + cx + d, \qquad (3)$$

where $a, b, c,$ and d are constants.

a) Compute $\int_{-h}^{h} f(x)\, dx$. Your answer will contain the constants $a, b, c, d,$ and h.

b) Substitute $x = h$ into equation (3) to find an expression for $f(h)$ in terms of $a, b, c, d,$ and h. Do the same to find an expression for $f(0)$ and $f(-h)$.

c) Use parts (a) and (b) to show that

$$\int_{-h}^{h} f(x)\, dx = \frac{h}{3}[f(-h) + 4f(0) + f(h)].$$

tw d) Explain why part (c) implies that Simpson's rule gives the exact answer for cubic polynomials.

5.8 Volume

OBJECTIVES

■ Find the volume of a solid of revolution.

■ Find the volume of a solid using cross-sectional area.

Consider the graph of $y = f(x)$. If the upper half-plane is rotated about the x-axis, then each point on the graph has a circular path, and the whole graph sweeps out a certain surface, called a *surface of revolution*.

The plane region bounded by the graph, the x-axis, and the lines $x = a$ and $x = b$ sweeps out a *solid of revolution*. To calculate the **volume** of this solid, we first approximate it by a finite sum of thin right circular cylinders (Fig. 1). We divide the interval $[a, b]$ into equal subintervals, each of length Δx. Thus the height of each cylinder is Δx (Fig. 2). The radius of each cylinder is $f(x_i)$ if f is nonnegative, or $|f(x_i)|$ in general, where x_i is the right-hand endpoint of the subinterval that determines that cylinder.

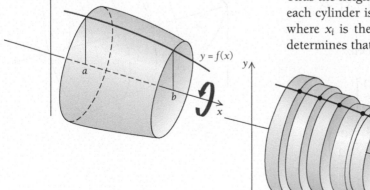

Figure 1

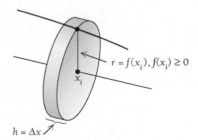

Figure 2

Since the volume of a right circular cylinder is given by

$$V = \pi r^2 h,$$

each of the approximating cylinders has volume

$$\pi |f(x_i)|^2 \, \Delta x = \pi [f(x_i)]^2 \, \Delta x.$$

The volume of the solid of revolution is approximated by the sum of the volumes of all the cylinders:

$$V \approx \sum_{i=1}^{n} \pi [f(x_i)]^2 \, \Delta x.$$

The actual volume is the limit as the thickness of the cylinders approaches zero, or equivalently, the number of them approaches infinity:

$$V = \lim_{n \to \infty} \sum_{i=1}^{n} \pi [f(x_i)]^2 \, \Delta x = \int_a^b \pi [f(x)]^2 \, dx.$$

(See Section 5.2.) That is, the volume is the value of the definite integral of the function $y = \pi [f(x)]^2$ from a to b.

> **THEOREM 8**
>
> For a continuous function f defined on an interval $[a, b]$, the **volume V of the solid of revolution** obtained by rotating the area under the graph of f over $[a, b]$ is given by
>
> $$V = \int_a^b \pi [f(x)]^2 \, dx.$$

EXAMPLE 1   Find the volume of the solid of revolution generated by rotating the region under the graph of $y = \sqrt{x}$ from $x = 0$ to $x = 1$ about the x-axis.

Solution

$$V = \int_0^1 \pi [f(x)]^2 \, dx$$

$$= \int_0^1 \pi [\sqrt{x}]^2 \, dx$$

$$= \int_0^1 \pi x \, dx$$

$$= \pi \left[\frac{x^2}{2} \right]_0^1$$

$$= \pi \left(\frac{1^2}{2} - \frac{0^2}{2} \right)$$

$$= \frac{\pi}{2}.$$

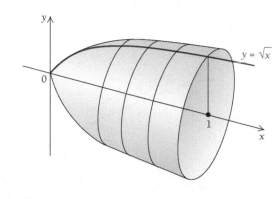

Explain how this could be interpreted as a solid of revolution.

EXAMPLE 2   Find the volume of the solid of revolution generated by rotating the region under the graph of

$$y = e^x$$

from $x = -1$ to $x = 2$ about the x-axis.

Solution

$$V = \int_{-1}^{2} \pi[f(x)]^2 \, dx$$

$$= \int_{-1}^{2} \pi[e^x]^2 \, dx$$

$$= \int_{-1}^{2} \pi e^{2x} \, dx$$

$$= \left[\frac{\pi}{2} e^{2x} \right]_{-1}^{2}$$

$$= \frac{\pi}{2} (e^{2 \cdot 2} - e^{2(-1)})$$

$$= \frac{\pi}{2} (e^4 - e^{-2}).$$

EXAMPLE 3   Find the volume of the solid of revolution generated by rotating the region under the graph of

$$y = \sin x$$

from $x = 0$ to $x = \pi$ about the x-axis.

Solution

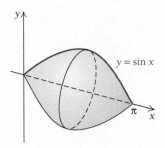

$$V = \int_{0}^{\pi} \pi[\sin x]^2 \, dx$$

$$= \pi \int_{0}^{\pi} \sin^2 x \, dx$$

$$= \pi \int_{0}^{\pi} \frac{\cos 2x + 1}{2} \, dx \qquad \text{Double-angle formula}$$

$$= \pi \left[\frac{1}{4} \sin 2x + \frac{1}{2} x \right]_{0}^{\pi}$$

$$= \pi \left[\left(\frac{1}{4} \cdot 0 + \frac{\pi}{2} \right) - \left(\frac{1}{4} \cdot 0 + \frac{1}{2} \cdot 0 \right) \right]$$

$$= \frac{\pi^2}{2}.$$

Cross Section

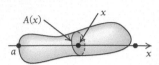

Often we wish to compute the volume of a solid that is not a solid of revolution. We derived Theorem 8 by approximating the volume of each thin cylinder as $\pi[(f(x))]^2 \cdot \Delta x$. The function $\pi[f(x)]^2$ is simply the area of the surface obtained from slicing the solid with a plane perpendicular to the x-axis. Even if a solid is not a solid of revolution, we can still approximate the volume of a thin slice as $A(x)\,\Delta x$, where $A(x)$ is the **cross-sectional area.** This leads to the following theorem.

> **THEOREM 9**
>
> If the cross-sectional area of a solid sliced perpendicular to the x-axis is $A(x)$ for $a \le x \le b$, then the volume of the solid is
>
> $$\int_a^b A(x)\,dx.$$

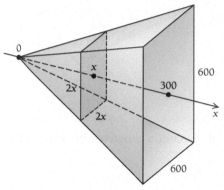

Figure 3

EXAMPLE 4   *The Pyramid of Memphis (Tennessee).* A pyramid in Memphis, Tennessee, used for sporting events, is 300 ft tall with a 600-ft-by-600-ft square base. Find the volume of the pyramid.

Solution   We orient the pyramid as in Fig. 3. The pyramid starts at $x = 0$ and ends at $x = 300$. Each cross section for x between 0 and 300 is a square. Furthermore, the side length L of the square is proportional to x, so that $L = Kx$. Also, we know that $L = 600$ when $x = 300$. Solving for K,

$$600 = K \cdot 300$$
$$2 = K.$$

The cross section at x is a square with side length $2x$, so the area is

$$A(x) = (2x)^2$$
$$= 4x^2.$$

We can now compute the volume using Theorem 9.

$$V = \int_a^b A(x)\,dx$$
$$= \int_0^{300} 4x^2\,dx$$
$$= \left[\frac{4}{3}x^3\right]_0^{300}$$
$$= \frac{4}{3}(300^3 - 0^3)$$
$$= 36{,}000{,}000 \text{ ft}^3.$$ ■

Exercise Set 5.8

Find the volume generated by revolving about the x-axis the regions bounded by the graphs of the following equations.

1. $y = x, x = 0, x = 1$

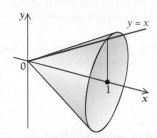

2. $y = \sqrt{x}, x = 0, x = 2$
3. $y = \sqrt{\sin x}, x = 0, x = \pi/2$
4. $y = \sqrt{\cos x}, x = 0, x = \pi/4$
5. $y = e^x, x = -2, x = 5$
6. $y = e^x, x = -3, x = 2$
7. $y = \dfrac{1}{x}, x = 1, x = 3$
8. $y = \dfrac{1}{x}, x = 1, x = 4$
9. $y = \dfrac{1}{\sqrt{x}}, x = 1, x = 3$
10. $y = \dfrac{1}{\sqrt{x}}, x = 1, x = 4$
11. $y = 4, x = 1, x = 3$
12. $y = 5, x = 1, x = 3$
13. $y = x^2, x = 0, x = 2$
14. $y = x + 1, x = -1, x = 2$
15. $y = \cos x, x = 0, x = \dfrac{\pi}{2}$
16. $y = \sec 2x, x = -\dfrac{\pi}{8}, x = \dfrac{\pi}{8}$
17. $y = \tan x, x = 0, x = \dfrac{\pi}{4}$
18. $y = \csc x, x = \dfrac{\pi}{4}, x = \dfrac{\pi}{2}$
19. $y = \sqrt{1 + x}, x = 2, x = 10$
20. $y = 2\sqrt{x}, x = 1, x = 2$
21. $y = \sqrt{4 - x^2}, x = -2, x = 2$
22. $y = \sqrt{r^2 - x^2}, x = -r, x = r$

Find the volume of the solids with given cross-sectional areas.

23. $A(x) = \dfrac{1}{2}x^2, 0 \le x \le 6$

24. $A(x) = 1 - x^2, -1 \le x \le 1$

25. $A(x) = \dfrac{\sqrt{3}}{2}x^2, 0 \le x \le 9$

26. $A(x) = \dfrac{1}{x^2}, 1 \le x \le 4$

Find the volume of the solids.

27. At each point $0 \le x \le 4$, the cross section is a square with side length x^2.

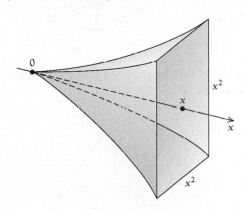

28. At each point $0 \le x \le 3$, the cross section is a square with side length $2x^2$.

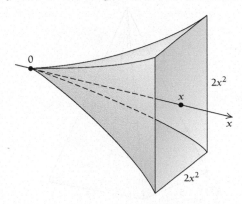

29. At each point $0 \leq x \leq 5$, the cross section is a rectangle with base x and height $2x$.

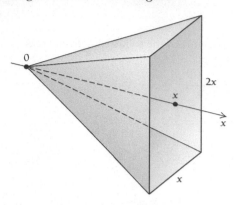

30. At each point $0 \leq x \leq 8$, the cross section is a rectangle with base $\frac{1}{2}x$ and height $\frac{1}{4}x$.

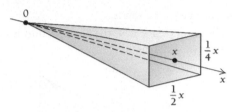

31. A pyramid with height 8 and base a square with side length 10

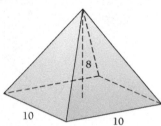

32. A pyramid with height 12 and base a square with side length 5

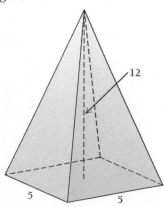

APPLICATIONS

Volume of a Tree. **The radius of a tree trunk at height x can be approximated by**

$$r(x) = K(H - x)^{3/4},$$

where K is a constant and H is the height of the tree. Use this model to do Exercises 33 and 34.[14]

33. The height of a tree is 75 ft and the radius at the base of the tree is 1.5 ft. Compute the volume.

34. The height of the tree is 50 ft and the radius at the base of the tree is 1 ft. Compute the volume.

SYNTHESIS

Find the volume generated by revolving about the x-axis the regions bounded by the following graphs.

35. $y = \sqrt{\ln x}, x = e, x = e^3$

36. $y = \sqrt{xe^{-x}}, x = 1, x = 2$

37. In this exercise you will find the volume formula for a common solid. Let r and h be fixed numbers. The line $y = \frac{r}{h}x$ is graphed below for $0 \leq x \leq h$.

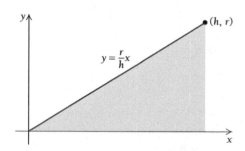

a) What is the name of the solid of revolution obtained when the shaded region is rotated about the x-axis? Identify the base radius and the height of the solid.

b) Compute the volume of the solid of revolution.

[14]Y. Zhang, B. Borders, and R. Bailey, "Derivation, fitting, and implication of a compatible stem taper-volume-weight system for intensively managed, fast growing loblolly pine," *Forest Science*, Vol. 48, pp. 595–607 (2002).

Technology Connection

38. The cross-sectional area of a human leg is given in the table below.[15]

Height (cm)	Cross-Sectional Area (cm^2)
0	130
10	67
20	71
30	80
40	92
50	104
60	102

[15]J. Norton, N. Donaldson, L. Dekker, "3D whole body scanning to determine mass properties of legs," *Journal of Biometrics*, Vol. 35, pp. 81–86 (2002).

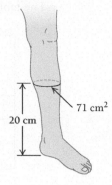

71 cm^2

20 cm

a) Use the Trapezoid rule to approximate the volume of the leg.

b) Use Simpson's rule to approximate the volume of a leg.

5.9 Improper Integrals

OBJECTIVES

■ Determine whether an improper integral is convergent or divergent.

■ Solve applied problems involving improper integrals.

Let's try to find the area of the region under the graph of $y = 1/x^2$ over the interval $[1, \infty)$.

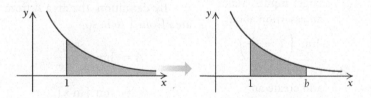

Note that this region is of infinite extent. We have not yet considered how to find the area of such a region. Let's find the area under the curve over the interval from 1 to b, and then see what happens as b gets very large. The area under the graph over $[1, b]$ is

$$\int_1^b \frac{dx}{x^2} = \left[-\frac{1}{x} \right]_1^b$$

$$= \left(-\frac{1}{b} \right) - \left(-\frac{1}{1} \right)$$

$$= -\frac{1}{b} + 1$$

$$= 1 - \frac{1}{b}.$$

The table shows the area for various values of b.

b	$1 - \frac{1}{b}$
10	$1 - \frac{1}{10} = 0.9$
100	$1 - \frac{1}{100} = 0.99$
1000	$1 - \frac{1}{1000} = 0.999$
10^6	$1 - \frac{1}{10^6} = 0.999999$

It appears from the table that the area is approaching 1. Indeed, this limit can also be found algebraically:

$$\lim_{b \to \infty} \int_1^b \frac{dx}{x^2} = \lim_{b \to \infty} \left(1 - \frac{1}{b} \right) = 1.$$

We *define* the area from 1 to infinity to be this limit. Here we have an example of an infinitely long region with a finite area.

Such areas may not always be finite. Let's try to find the area of the region under the graph of $y = 1/x$ over the interval $[1, \infty)$.

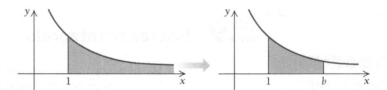

By definition, the area A from 1 to infinity is the limit as b approaches ∞ of the area from 1 to b, so

$$A = \lim_{b \to \infty} \int_1^b \frac{dx}{x}$$

$$= \lim_{b \to \infty} \left[\ln x \right]_1^b$$

$$= \lim_{b \to \infty} (\ln b - \ln 1)$$

$$= \lim_{b \to \infty} \ln b.$$

In Section 4.2, we learned that $\ln b$ increases indefinitely as b increases. Therefore, the limit does not exist.

Thus we have an infinitely long region with an infinite area. Note that the graphs of $y = 1/x^2$ and $y = 1/x$ have similar shapes, but the region under one of them has a finite area and the other does not.

An integral such as

$$\int_a^\infty f(x)\, dx,$$

with an upper limit of infinity, is called an **improper integral**. Its value is defined to be the following limit.

Technology Connection

EXERCISE

1. Graph $y = 1 - 1/x$ and create an input–output table. Consider larger and larger inputs. Make an assertion about
$$\lim_{x \to \infty} \left(1 - \frac{1}{x} \right).$$

2. Graph $y = \ln x$ and create an input–output table. Consider larger and larger inputs. Make an assertion about
$$\lim_{x \to \infty} (\ln x).$$

DEFINITION

$$\int_a^\infty f(x)\, dx = \lim_{b \to \infty} \int_a^b f(x)\, dx$$

If the limit exists, then we say that the improper integral **converges,** or is **convergent.** If the limit does not exist, then we say that the improper integral **diverges,** or is **divergent.** Thus,

$$\int_1^\infty \frac{dx}{x^2} = 1 \quad converges \quad \text{and} \quad \int_1^\infty \frac{dx}{x} \quad diverges.$$

EXAMPLE 1 Determine whether the following integral is convergent or divergent, and calculate its value if it is convergent:

$$\int_0^\infty 2e^{-2x}\, dx.$$

Solution   We have

$$
\begin{aligned}
\int_0^\infty 2e^{-2x}\, dx &= \lim_{b \to \infty} \int_0^b 2e^{-2x}\, dx \\
&= \lim_{b \to \infty} \left[\frac{2}{-2} e^{-2x} \right]_0^b \\
&= \lim_{b \to \infty} \left[-e^{-2x} \right]_0^b \\
&= \lim_{b \to \infty} \left[-e^{-2b} - (-e^{-2 \cdot 0}) \right] \\
&= \lim_{b \to \infty} (-e^{-2b} + 1) \\
&= \lim_{b \to \infty} \left(1 - \frac{1}{e^{2b}} \right).
\end{aligned}
$$

Now as b approaches ∞, we know that e^{2b} approaches ∞ (from Chapter 4), so

$$\frac{1}{e^{2b}} \to 0.$$

Thus,

$$\int_0^\infty 2e^{-2x}\, dx = \lim_{b \to \infty} \left(1 - \frac{1}{e^{2b}} \right) = 1 - 0 = 1.$$

The integral is convergent. ∎

Following are definitions of two other types of improper integrals.

DEFINITIONS

1. $\displaystyle\int_{-\infty}^{b} f(x)\, dx = \lim_{a \to -\infty} \int_{a}^{b} f(x)\, dx$

2. $\displaystyle\int_{-\infty}^{\infty} f(x)\, dx = \int_{-\infty}^{c} f(x)\, dx + \int_{c}^{\infty} f(x)\, dx$, where c is any real number.

In order for $\int_{-\infty}^{\infty} f(x)\, dx$ to converge, both integrals on the right in Definition 2 above must converge. Typically, c is chosen to be 0 in Definition 2.

EXAMPLE 2   Compute the improper integrals.

a) $\displaystyle\int_{-\infty}^{0} \frac{x}{(x^2+1)^2}\, dx$ b) $\displaystyle\int_{-\infty}^{\infty} \frac{x}{(x^2+1)^2}\, dx$

c) $\displaystyle\int_{-\infty}^{0} \frac{x}{x^2+1}\, dx$ d) $\displaystyle\int_{-\infty}^{\infty} \frac{x}{x^2+1}\, dx$

Solution

a) We first compute the antiderivative:

$$\int \frac{x}{(x^2+1)^2}\, dx = \int \frac{1}{2} \frac{1}{u^2}\, du \qquad \underline{\text{Substitution}} \qquad \begin{aligned} u &= x^2 + 1, \\ du &= 2x\, dx \end{aligned}$$

$$= \frac{1}{2}\left(\frac{-1}{u}\right) + C$$

$$= \frac{-1}{2(x^2+1)} + C.$$

Using the definition,

$$\int_{-\infty}^{0} \frac{x}{(x^2+1)^2}\, dx = \lim_{a \to -\infty} \int_{a}^{0} \frac{x}{(x^2+1)^2}\, dx$$

$$= \lim_{a \to -\infty} \left[\frac{-1}{2(x^2+1)}\right]_{a}^{0}$$

$$= \lim_{a \to -\infty} \left[-\frac{1}{2(0^2+1)} - \frac{-1}{2(a^2+1)}\right]$$

$$= -\frac{1}{2}.$$

b) The definition says

$$\int_{-\infty}^{\infty} \frac{x}{(x^2+1)^2}\, dx = \int_{-\infty}^{0} \frac{x}{(x^2+1)^2}\, dx + \int_{0}^{\infty} \frac{x}{(x^2+1)^2}\, dx.$$

We have already computed one integral. We now compute the other.

$$\int_0^\infty \frac{x}{(x^2 + 1)^2} \, dx = \lim_{b \to \infty} \int_0^b \frac{x}{(x^2 + 1)^2} \, dx$$

$$= \lim_{b \to \infty} \left[\frac{-1}{2(x^2 + 1)} \right]_0^b$$

$$= \lim_{b \to \infty} \left[\frac{-1}{2(b^2 + 1)} - \frac{-1}{2(0^2 + 1)} \right]$$

$$= \frac{1}{2}.$$

We have

$$\int_{-\infty}^\infty \frac{x}{(x^2 + 1)^2} \, dx = \int_{-\infty}^0 \frac{x}{(x^2 + 1)^2} \, dx + \int_0^\infty \frac{x}{(x^2 + 1)^2} \, dx$$

$$= -\frac{1}{2} + \frac{1}{2} = 0.$$

c) We first compute the antiderivative:

$$\int \frac{x}{x^2 + 1} \, dx = \int \frac{1}{2u} \, du \qquad \underline{\text{Substitution}} \quad u = x^2, \\ du = 2x \, dx$$

$$= \frac{1}{2} \ln |u| + C$$

$$= \frac{1}{2} \ln |x^2 + 1| + C$$

$$= \frac{1}{2} \ln (x^2 + 1) + C. \qquad \text{Since } x^2 + 1 > 0$$

We have

$$\int_{-\infty}^0 \frac{x}{x^2 + 1} \, dx = \lim_{a \to -\infty} \int_a^0 \frac{x}{x^2 + 1} \, dx$$

$$= \lim_{a \to -\infty} \left[\frac{1}{2} \ln (x^2 + 1) \right]_a^0$$

$$= \lim_{a \to -\infty} \left[\frac{1}{2} \ln(1) - \frac{1}{2} \ln (a^2 + 1) \right]$$

$$= -\frac{1}{2} \lim_{a \to -\infty} \ln (a^2 + 1)$$

$$= -\infty.$$

This improper integral does not exist.

d) Since $\int_{-\infty}^0 \frac{x}{x^2 + 1} \, dx$ does not exist, $\int_{-\infty}^\infty \frac{x}{x^2 + 1} \, dx$ also does not exist. ∎

Application to Radioctive Decay

EXAMPLE 3 *Radioactive Decay Energy.* Plutonium has a decay rate of 0.00286% per year. The amount of energy released from a radioactive sample is measured in rem units and is given by $\int_0^a P_0 e^{-kt}\, dt$, where k is the decay rate, a is the number of years, and P_0 is the initial rate of released energy. For a given sample of plutonium, $P_0 = 150$ rem/yr.

a) Find the number of rems released the first 100 yr.

b) Compute the total number of rems released from the present time on.

c) Explain why it makes sense that the improper integral in part (b) is convergent.

Solution

a) $\displaystyle \int_0^{100} P_0 e^{-kt}\, dt = \int_0^{100} 150 e^{-0.0000286t}\, dt$

$\displaystyle = 150 \int_0^{100} e^{-0.0000286t}\, dt$

$\displaystyle = -\frac{150}{0.0000286} [e^{-0.0000286t}]_0^{100}$

$\displaystyle = -\frac{150}{0.0000286} (e^{-0.003} - e^0)$

$\displaystyle \approx 14979$ Using a scientific calculator for approximation

b) To compute the energy released from the present time on, we integrate from 0 to infinity.

$\displaystyle \int_0^{\infty} P_0 e^{-kt}\, dt = \lim_{b \to \infty} \int_0^b P_0 e^{-kt}\, dt$

$\displaystyle = \lim_{b \to \infty} \int_0^b 150 e^{-0.0000286t}\, dt$

$\displaystyle = \lim_{b \to \infty} -\frac{150}{0.0000286} [e^{-0.0000286t}]_0^b$

$\displaystyle \approx -5{,}240{,}000 \lim_{b \to \infty} (e^{-0.0000286b} - e^0)$

$\displaystyle = -5{,}240{,}000 \lim_{b \to \infty} (e^{-0.0000286b} - 1)$

$\displaystyle = 5{,}240{,}000$ Since $\lim_{x \to \infty} e^{-x} = 0$

Approximately, 5,240,000 rem will be released from this sample.

c) It makes sense that the improper integral in part (b) is convergent, because we started with a finite sample of plutonium. Each atom decays once and there are only a finite (but large) number of atoms in the sample. ∎

In Example 3, a fixed sample of plutonium was considered. We next consider the situation of an amount P of radioactive material being released into the atmosphere annually. The amount present at time T is given by

$$\int_0^T Pe^{-kt}\,dt = \frac{P}{k}(1 - e^{-kT}).$$

As T approaches ∞ (the radioactive material is to be released forever), the buildup of radioactive material approaches a limiting value P/k.

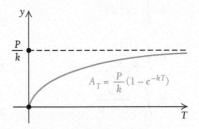

Improper Integrals and L'Hôpital's Rule

Theorem 10 is often very useful when an improper integral involves a polynomial and an exponential. Theorem 10 is a special case of a more general theorem called L'Hôpital's Rule; its proof can be found in advanced texts.

THEOREM 10

If r is a positive and n is any number, then

$$\lim_{x \to \infty} x^n e^{-rx} = \lim_{x \to \infty} \frac{x^n}{e^{rx}} = 0.$$

EXAMPLE 4   Compute the improper integral $\displaystyle\int_0^\infty xe^{-x}\,dx$.

Solution   We first compute the definite integral. We integrate by parts using

$$u = x \quad \text{and} \quad dv = e^{-x}\,dx$$
$$du = dx \quad \text{and} \quad v = -e^{-x}.$$

$$\int xe^{-x}\,dx = -xe^{-x} - \int -e^{-x}\,dx$$
$$= -xe^{-x} - e^{-x} + C.$$

So,

$$\int_0^{-\infty} xe^{-x}\,dx = \lim_{b \to \infty} \left[-xe^{-x} - e^{-x}\right]_0^b$$
$$= \lim_{b \to \infty} \left[-be^{-b} - e^{-b} - (0 \cdot e^0 - e^0)\right]$$
$$= (0 - 0) - (0 - 1) \qquad\qquad \text{Theorem 10}$$
$$= 1.$$

Exercise Set 5.9

Determine whether the improper integral is convergent or divergent, and calculate its value if it is convergent.

1. $\int_2^\infty \dfrac{dx}{x^2}$

2. $\int_2^\infty \dfrac{dx}{x}$

3. $\int_4^\infty \dfrac{dx}{x}$

4. $\int_0^\infty x^2\, dx$

5. $\int_{-\infty}^{-1} \dfrac{dt}{t^2}$

6. $\int_{-\infty}^{-4} \dfrac{dx}{x}$

7. $\int_0^\infty 4e^{-4x}\, dx$

8. $\int_0^\infty \dfrac{du}{1 + u}$

9. $\int_0^\infty e^x dx$

10. $\int_{-\infty}^0 e^{2x}\, dx$

11. $\int_{-\infty}^\infty e^{2x}\, dx$

12. $\int_{-\infty}^\infty \dfrac{x\, dx}{(1 + x^2)^2}$

13. $\int_{-\infty}^\infty \dfrac{t\, dt}{(1 + t^2)^3}$

14. $\int_{-1}^\infty \dfrac{2x + 3}{(x^2 + 3x + 6)^4}\, dx$

15. $\int_{-\infty}^\infty 2xe^{-3x^2}\, dx$

16. $\int_{-\infty}^\infty x^2 e^{-x^3}\, dx$

17. $\int_0^\infty 2t^2 e^{-2t}\, dt$

18. $\int_{-\infty}^0 xe^x\, dx$

19. $\int_{-\infty}^1 2xe^{3x}\, dx$

20. $\int_0^\infty \dfrac{dt}{t^{2/3}}$

21. $\int_1^\infty \dfrac{dx}{\sqrt{x}}$

22. $\int_0^\infty \dfrac{e^t}{(5 + e^t)^2}\, dt$

23. $\int_0^\infty \dfrac{1 - \sin x}{(x + \cos x)^2}\, dx$

24. $\int_0^\infty \sin \theta\, d\theta$

25. $\int_1^\infty \dfrac{1}{x(x + 1)^2}\, dx$

26. $\int_5^\infty \dfrac{dt}{t^2 - 16}$

27. $\int_1^\infty \dfrac{x}{\sqrt{x + 2}}\, dx$

28. $\int_e^\infty \dfrac{\ln x}{x^2}\, dx$

29. $\int_e^\infty \dfrac{\ln x}{x}\, dx$

30. $\int_2^\infty \dfrac{dt}{t\sqrt{t^2 + 1}}$

31. $\int_{-\infty}^\infty 3xe^{-x^2/2}\, dx$

32. $\int_{-\infty}^\infty 2xe^{-x^2}\, dx$

33. $\int_0^\infty me^{-mx}\, dx,\ m > 0$

34. $\int_0^\infty Ae^{-kt}\, dt,\ k > 0$

35. Find the area, if it exists, of the region under the graph of $y = 1/x^2$ over the interval $[2, \infty)$.

36. Find the area, if it exists, of the region under the graph of $y = 1/x$ over the interval $[2, \infty)$.

37. Find the area, if it exists, of the region bounded by $y = 2xe^{-x^2}$ and the lines $x = 0$ and $y = 0$.

38. Find the area, if it exists, of the region bounded by $y = 1/\sqrt{(3x - 2)^3}$ and the lines $x = 6$ and $y = 0$.

APPLICATIONS

39. *Cancer Treatment.* Implants of iodine-125 with a half-life of 60.1 days are used to treat prostate

cancer. The implants are left in the patient and never removed. The amount of energy from the implant that is transmitted to the body is measured in rem units and is given by $\int_0^a P_0 e^{-kt}\, dt$, where k is the decay constant for the radioactive material, a is the number of years since the implant, and P_0 is the initial rate that energy is being transmitted.

a) Find the decay rate k of iodine-125.

b) How much energy (measured in rems) is given off the first month if the initial energy rate is 10 rem per yr?

c) What is the total amount of energy that the implant will give off over all time?

40. *Cancer Treatment.* Implants of palladium-103 with a half-life of 16.99 days are used to treat prostate cancer. The implants are left in the patient and never removed. The amount of energy from the implant that is transmitted to the body is measured in rem units and is given by $\int_0^a P_0 e^{-kt}\, dt$, where k is the decay constant for radioactive material, a is the number of years since the implant, and P_0 is the initial rate that energy is being transmitted.

a) Find the decay rate k of palladium-103.

b) How much energy (measured in rems) is given off the first month if the initial energy rate is 15 rem per yr?

c) What is the total amount of energy that the implant will give off?

41. *Radioactive Buildup.* Plutonium has a decay rate of 0.00286% per year. Suppose that a nuclear accident causes plutonium to be released into the atmosphere each year perpetually at the rate of 1 lb per year. What is the limiting value of the radioactive buildup?

42. *Radioactive Buildup.* Cesium-137 has a decay rate of 2.3% per year. Suppose that a nuclear accident causes cesium-137 to be released into the atmo-

sphere each year perpetually at the rate of 1 lb per year. What is the limiting value of the radioactive buildup?

SYNTHESIS

43. Determine for which values of r the integral $\int_1^\infty x^r\, dx$ is convergent and for which values of r it is divergent.

44. Determine for which values of r the integral $\int_e^\infty \frac{(\ln x)^r}{x}\, dx$ is convergent and for which values of r it is divergent.

Drug Dosage. Suppose that an oral dose of a drug is taken. From that time, the drug is assimilated in the body and excreted through the urine. The total amount of the drug that has passed through the body in time T is given by

$$\int_0^T E(t)\, dt,$$

where E is the rate of excretion of the drug. A typical rate-of-excretion function is $E(t) = te^{-kt}$, where $k > 0$ and t is the time, in hours.

45. Find $\int_0^\infty E(t)\, dt$ and interpret the answer. That is, what does the integral represent?

46. A physician prescribes a dosage of 100 mg. Find k.

 47. Consider the functions

$$y = \frac{1}{x^2} \quad \text{and} \quad y = \frac{1}{x}.$$

Suppose that you go to a paint store to buy paint to cover the region under each graph over the interval $[1, \infty)$. Discuss whether you could be successful and why or why not.

Technology Connection

48. Graph the function E and shade the area under the curve for the situation in Exercises 45 and 46.

Use a grapher to approximate the integral. Choose 1E99 to represent the upper limit.

49. $\int_1^\infty \frac{4}{1 + x^2}\, dx$

50. $\int_1^\infty \frac{6}{5 + e^x}\, dx$

Chapter 5 Summary and Review

Terms to Know

Review Exercises

These review exercises are for test preparation. They can also be used as a lengthened practice test. Answers are at the back of the book. The answers also contain bracketed section references, which tell you where to restudy if your answer is incorrect.

Evaluate.

1. $\displaystyle\int 8x^4\,dx$

2. $\displaystyle\int (3x^2 + 2\sin x + 3)\,dx$

3. $\displaystyle\int \left(3t^2 + 7t + \frac{1}{t}\right)\,dt$

Find the area under the curve over the indicated interval.

4. $y = 4 - x^2$; $[-2, 1]$

5. $y = \cos x$; $[0, \pi/4]$

In each case, give two interpretations of the shaded region.

6.
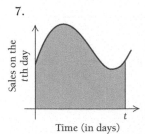
Shorthand speed (in words per minute) vs. Time (in minutes)

7.
Sales on the tth day vs. Time (in days)

Evaluate.

8. $\displaystyle\int_a^b x^5\,dx$

9. $\displaystyle\int_{-1}^{1} (x^3 - x^4)\,dx$

10. $\displaystyle\int_0^1 (e^x + x)\,dx$

11. $\displaystyle\int_1^3 \frac{3}{x}\,dx$

12. $\displaystyle\int_0^{\pi/6} \sec^2\theta\,d\theta$

Decide whether $\int_a^b f(x)\,dx$ is positive, negative, or zero.

13.
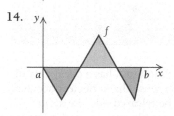

14.
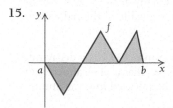

15.

16. Find the area of the region bounded by $y = 3x^2$ and $y = 9x$.

Evaluate using substitution. Do not use Table 1.

17. $\int x^3 e^{x^4}\, dx$

18. $\int \dfrac{24t^5}{4t^6 + 3}\, dt$

19. $\int \dfrac{\ln 4x}{2x}\, dx$

20. $\int 2e^{-3x}\, dx$

21. $\int x^2 \csc^2 x^3\, dx$

Evaluate using integration by parts. Do not use Table 1.

22. $\int 3x\, e^{3x}\, dx$

23. $\int \ln x^7\, dx$

24. $\int 3x^2 \ln x\, dx$

25. $\int 2x^2 \sin x\, dx$

Evaluate using Table 1 or CAS.

26. $\int \dfrac{1}{49 - x^2}\, dx$

27. $\int x^2 e^{5x}\, dx$

28. $\int \dfrac{x}{7x + 1}\, dx$

29. $\int \dfrac{dx}{\sqrt{x^2 - 36}}$

30. $\int x^6 \ln x\, dx$

31. $\int xe^{8x}\, dx$

32. $\int x^5 \cos x\, dx$

33. Approximate $\int_1^4 (2/x)\, dx$ by computing the Riemann sum with $n = 3$.

34. Find the average value of $y = e^{-x} + 5$ over $[0, 2]$.

35. A particle starts out from the origin. Its velocity (in km/hr) at any time t, $t \geq 0$, is given by $v(t) = 3t^2 + 2t$. Find the distance that the particle travels during the first 4 hr (from $t = 0$ to $t = 4$).

36. *Plant Growth.* During June, the growth rate of a stalk of corn is given by

$$L'(t) = 0.1e^{0.03t},$$

where $L(t)$ is the length of the stalk in inches at the end of the tth day of June. Find the accumulated change in the length of the stalk during June.

Integrate using any method.

37. $\int x^3 e^{0.1x}\, dx$

38. $\int \dfrac{12t^2}{4t^3 + 7}\, dt$

39. $\int \dfrac{x\, dx}{\sqrt{4 + 5x}}$

40. $\int 5x^4 e^{x^5}\, dx$

41. $\int \dfrac{dx}{x + 9}$

42. $\int t^7 (t^8 + 3)^{11}\, dt$

43. $\int \ln 7x\, dx$

44. $\int x \ln 8x\, dx$

45. $\int t^4 \tan(t^5 + 1) \sec(t^5 + 1)\, dt$

46. $\int \tan(3x + 7)\, dx$

Determine whether the improper integral is convergent or divergent and calculate its value if it is convergent.

47. $\int_1^{\infty} \dfrac{1}{x^2}\, dx$

48. $\int_1^{\infty} e^{4x}\, dx$

49. $\int_0^{\infty} xe^{-2x}\, dx$

50. $\int_1^{\infty} \dfrac{3}{x^{1.01}}\, dx$

Find the volume generated by revolving about the x-axis the region bounded by each of the following graphs.

51. $y = x^3, x = 1, x = 2$

52. $y = \dfrac{1}{x + 2}, x = 0, x = 1$

53. Find the volume of a pyramid of height 50 ft whose base is a square with side length 40 ft.

54. Find the volume of a pyramid whose base is an equilateral triangle with side length 3 m and whose height is 4 m. Recall that for an equilateral triangle, the height is $\sqrt{3}/2$ times the side length.

SYNTHESIS

Evaluate.

55. $\int \dfrac{t^4 \ln (t^5 + 3)}{t^5 + 3}\, dt$

56. $\int \dfrac{dx}{e^x + 2}$

57. $\int \dfrac{\ln \sqrt{x}}{x}\, dx$

58. $\int x^{91} \ln x\, dx$

59. $\int \ln \left(\dfrac{x - 3}{x - 4} \right) dx$

60. $\int \dfrac{dx}{x(\ln x)^4}$

Technology Connection

61. Use a grapher to approximate the area between the following curves:

$$y = 2x^2 - 2x,$$
$$y = 12x^2 - 12x^3,$$
$$x = 0, \quad x = 1.$$

62. Use the Trapezoid rule with $n = 6$ to approximate $\int_2^5 x(\ln x)e^{x/2}\, dx$.

63. Use Simpson's rule with $n = 6$ to approximate $\int_2^5 x(\ln x)e^{x/2}\, dx$.

Chapter 5 Test

Evaluate.

1. $\int dx$

2. $\int 1000x^4 \, dx$

3. $\int \left(e^x + \dfrac{1}{x} + x^{3/8} \right) dx$

Find the area under the curve over the indicated interval.

4. $y = x - x^2$; $[0, 1]$

5. $y = \dfrac{4}{x}$; $[1, 3]$

6. Give two interpretations of the shaded area.

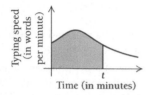

Evaluate.

7. $\displaystyle\int_{-1}^{2} (2x + 3x^2) \, dx$

8. $\displaystyle\int_{0}^{1} e^{-2x} \, dx$

9. $\displaystyle\int_{a}^{b} \dfrac{dx}{x}$

10. Decide whether $\int_a^b f(x)\, dx$ is positive, negative, or zero.

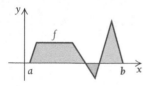

Evaluate using substitution. Do not use Table 1 or CAS.

11. $\int \dfrac{dx}{x + 8}$

12. $\int e^{-0.5x} \, dx$

13. $\int e^x \sin e^x \, dx$

Evaluate using integration by parts. Do not use Table 1 or CAS.

14. $\int x \sin 5x \, dx$

15. $\int x^3 \ln x^4 \, dx$

Evaluate using Table 1 or CAS.

16. $\int 2^x \, dx$

17. $\int \dfrac{dx}{x(7 - x)}$

18. Find the average value of $y = 4t^3 + 2t$ over $[-1, 2]$.

19. Find the area of the region bounded by $y = x$, $y = x^5$, $x = 0$, and $x = 1$.

20. Approximate

$$\int_{0}^{5} (25 - x^2) \, dx$$

using a Riemann sum with $n = 5$.

21. *Tree Growth.* The diameter of a tree grows at the rate of

$$D'(t) = 0.35 + 0.001t,$$

where $D(t)$ is the diameter of the trunk of the tree in inches t years after 1910. In 1910, the tree had a diameter of 2 in. What will the diameter of the tree be in 2010?

22. *Learning Curve.* A typist's speed over a 4-min interval is given by

$$W(t) = -6t^2 + 12t + 90, \quad t \text{ in } [0, 4],$$

where $W(t)$ is the speed, in words per minute, at time t. How many words are typed during the second minute (from $t = 1$ to $t = 2$)?

Integrate using any method.

23. $\int \dfrac{dx}{x(10 - x)}$

24. $\int x^5 e^x \, dx$

25. $\int x^5 e^{x^6} \, dx$

26. $\int \sqrt{x} \ln x \, dx$

27. $\int x^3 \sqrt{x^2 + 4} \, dx$

28. $\int \dfrac{dx}{64 - x^2}$

29. $\int x^4 e^{0.1x} \, dx$

30. $\int x^2 \cos x \, dx$

Find the volume generated by revolving about the x-axis the regions bounded by the following graphs.

31. $y = \dfrac{1}{\sqrt{x}}, x = 1, x = 5$

32. $y = \sqrt{2 + x}, x = 0, x = 1$

33. Find the volume of a pyramid 7 ft high with a 4-ft-by-4-ft square base.

Determine whether the improper integral is convergent or divergent, and calculate its value if it is convergent.

34. $\displaystyle\int_{1}^{\infty} \frac{1}{x^3}\, dx$

35. $\displaystyle\int_{0}^{\infty} \frac{3}{1+x}\, dx$

SYNTHESIS

Evaluate using any method.

36. $\displaystyle\int \frac{[(\ln x)^3 - 4(\ln x)^2 + 5]}{x}\, dx$

37. $\displaystyle\int \ln\left(\frac{x+3}{x+5}\right) dx$

38. $\displaystyle\int \frac{8x^3 + 10}{\sqrt[3]{5x - 4}}\, dx$

Technology Connection

39. Use a grapher to approximate the area between the following curves.

$$y = 3x - x^2,$$
$$y = 2x^3 - x^2 - 5x,$$
$$x = -2, \quad x = 0.$$

40. Approximate the integral $\displaystyle\int_{0}^{2} e^{-x^2}\, dx$ using

a) the Trapezoid rule with $n = 6$.
b) Simpson's rule with $n = 6$.

Extended Life Science Connection

ESTIMATING DINOSAUR MASS

How would we estimate the mass of an animal that has been extinct for over 150 million years? We will follow a method recently developed by Frank Seebacher that uses integration.[16]

To compute the mass, first the volume is estimated. Then the volume is multiplied by the density to compute the mass. For animals living today, the density is very close to 1000 kg/m^3, so we will assume the same density for dinosaurs.

FIRST ESTIMATE: SOLID OF REVOLUTION

To estimate the body volume, skeletal measurements were taken of the body depth along the length of the animal. Half the depth is then plotted as a function of the distance along the animal. If the animal were a solid of revolution, its body volume could be computed by integrating.

$$V = \int_{0}^{L} \pi[f(x)]^2\, dx, \qquad (1)$$

where L is the total body length, and $f(x)$ is the radius (half the depth) at position x along the animal. As a first approximation, we

[16]F. Seebacher, "A new method to calculate allometric length-mass relationships of dinosaurs," *Journal of Vertebrate Paleontology* 21(1), 2001, pp. 51–60.

compute the volume assuming the animal's shape is a solid of revolution.

To simplify our calculations, we will ignore the volume and mass of legs. To get a better approximation of the total mass, similar techniques could be applied to the animal's legs.

The tables below give measurements from dinosaur fossils.[17]

Hypsilophodon	
Distance Along Body (m)	Depth (m)
0	0
0.25	0.078
0.5	0.184
0.75	0.290
1	0.251
1.25	0.101
1.5	0.100
1.75	0.047
2	0.025
2.25	0.011
2.5	0

Diplodocus	
Distance Along Body (m)	Depth (m)
0	0
1	0.308
2	0.410
3	0.470
4	0.556
5	0.641
6	0.701
7	1.111
8	2.179
9	2.479
10	2.009
11	1.496
12	1.026
13	0.786
14	0.641
15	0.556
16	0.427
17	0.299
18	0.171
19	0.128
20	0.085
21	0.064
22	0

EXERCISES

1. We will use the Trapezoid rule and Simpson's rule to approximate equation (1) for the Hypsilophodon's data. This will give an estimate of the volume of the Hypsilophodon's body volume.

   a) Find the values of n and Δx.
   b) Identify the values $y_0, y_1, y_2, \ldots, y_n$. Recall that the values of $f(x)$ are half of the depth.
   c) Apply the Trapezoid rule to estimate the volume of the body of the Hypsilophodon.
   d) Repeat part (c) using Simpson's rule.

2. Estimate the volume of a Diplodocus by repeating Exercise 1 using the second table.

---

[17]Personal correspondence from Frank Seebacher.

SECOND ESTIMATE: PARABOLIC CROSS-SECTIONS

We next refine our estimate of the volume by taking into account that a dinosaur's body is *not* a volume of revolution. The body width and body depth are in fact different. We make the assumption that the ratio r of the body width to the body depth is a constant throughout the length of the animal. For a Hypsilophodon, $r = 0.65$, and for a Diplodocus, $r = 0.9$. That is, the width of a Hypsilophodon is 0.65 times its depth, and the width of a Diplodocus is 0.9 times its depth. We make the further assumption that a cross section of the animal's body looks like the area shown below between the graphs of $y = 1 - ax^2$ (red) and $y = -1 + ax^2$ (green).

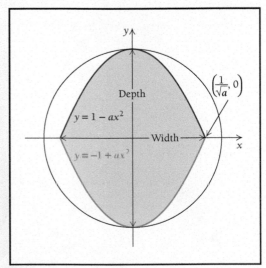

Figure 1

EXERCISES

3. Show that the points where the graph of $y = 1 - ax^2$ intersects $y = -1 + ax^2$ are $\left(\dfrac{1}{\sqrt{a}}, 0\right)$ and $\left(-\dfrac{1}{\sqrt{a}}, 0\right)$.

4. By Exercise 3, the ratio of the body width to the body depth is $r = \dfrac{1}{\sqrt{a}}$. Find the value of a for

a) the Hypsilophodon.
b) the Diplodocus.

5. Compute the area between the graphs of $y = 1 - ax^2$ and $y = -1 + ax^2$ using the value of a for

a) the Hypsilophodon.
b) the Diplodocus.

6. Let p be the ratio of the area computed in Exercise 5 to the area of a circle of radius 1. Compute p for

a) the Hypsilophodon.
b) the Diplodocus.

tw 7. The volume of the dinosaur is approximated by p times the volumes computed in Exercises 1 and 2. Explain why this makes sense.

8. Use the values computed in Exercises 1 and 2 with Simpson's rule and the values of p computed in Exercise 6 to improve the volume approximation for

a) the Hypsilophodon.
b) the Diplodocus.

9. The mass of a dinosaur is its volume times its density. Assume that the density of a dinosaur is 1000 kg/m³, approximately the same as animals alive today. Estimate the body mass of

a) the Hypsilophodon.
b) the Diplodocus.

tw 10. List the various assumptions we made throughout to finally arrive at the answer in Exercise 9. Do the assumptions seem reasonable? Can you think of anything we could do to get a better estimate?

6

Matrices

INTRODUCTION *In this chapter, we learn essential properties of matrices. We discover how matrices can be used to model population growth and other biological phenomena. These ideas will be used again when we extend the concepts of calculus to functions with more than one variable.*

AN APPLICATION A population of cerulean warblers is grouped into hatchlings (H), second year (S), and after second year (A). Their Leslie matrix for one location is given below, showing estimates of the survivability and fecundity.[1]

$$G = \begin{bmatrix} 0 & 0.75 & 0.95 \\ 0.3 & 0 & 0 \\ 0 & 0.7 & 0.95 \end{bmatrix}$$

Determine the long-term growth rate.

This problem appears as Exercise 46 in Exercise Set 6.4.

---

[1]Jones, J. Barg, T. Sillett, M. Veit, and J. Robertson, "Minimum estimates of survival and population growth for cerulean warblers (*dendroica cerulea*) breeding in Ontario, Canada," *The Auk* Vol. 121, No. 1, Research Library, pp. 15–22 (Jan. 2004).

6.1 Matrix Operations

OBJECTIVES

■ Model populations using Leslie matrices.
■ Multiply matrices.
■ Add and subtract matrices.

Matrices provide a compact way of recording information needed for calculations. As we will see, they are useful in a variety of settings, including solving systems of linear equations, population growth, and solving difference and differential equations.

Matrices

A **matrix** is a rectangular array of numbers. We will normally denote a matrix using a boldface capital letter. If the matrix $\mathbf{A}$ has n rows and k columns, then we will say that the **size of the matrix** is $n \times k$, read "n by k." If the number of rows is the same as the number of columns, then we say the matrix is *square*. To refer to a number in the matrix, we will generally use a lowercase letter with two subscripts indicating the position in the matrix. For example, the entry a_{23} represents the number in the second row and third column of the matrix $\mathbf{A}$. We will also write $\mathbf{A} = [a_{ij}]$.

EXAMPLE 1 Let $\mathbf{A} = [a_{ij}] = \begin{bmatrix} 0 & 4 \\ 3 & -5 \\ -2 & 1 \end{bmatrix}$.

a) Find the size of $\mathbf{A}$.
b) Find the entry a_{12}.
c) Find the entry a_{21}.
d) Find the entry a_{32}.
e) Find the entry a_{23}.

Solution

a) The matrix $\mathbf{A}$ has three rows and two columns, and so its size is 3×2.

b) We are to find the entry in the first row and second column: $a_{12} = 4$.

c) The entry in the second row and the first column is $a_{21} = 3$.

d) The entry in the third row and second column is $a_{32} = 1$.

e) We are to find the entry in the second row and the third column. Since $\mathbf{A}$ has only two columns, there is no entry that we would label a_{23}. ■

Leslie Matrices

In Chapter 4, we studied properties of the exponential model of population growth. This model assumes that the growth rate is proportional to the population. This assumption may not be valid since the very young and the very old of a species generally do not reproduce. A more refined model, incorporating the proportion of individuals who survive and their reproduction rate, might be a better predictor of population growth.

We can use matrices to store information about a population's growth. The proportion of the population that survives from one season to the next is the

Technology Connection

Storing Matrices

Many graphers have the capability to store and manipulate matrices. We can accomplish this on a TI-83 by touching the [MATRX] key, moving the cursor to EDIT, and then moving the cursor to the matrix we wish to edit. In this case, we move the cursor to [A]. When we touch the [ENTER] key, we can edit the matrix. We first enter the size of the matrix and then the entries. The screen below shows how to enter the Leslie matrix in Example 2.

```
MATRIX [A] 2×2
[.8         1.2
[.6         .5        ]
```

We can view the matrix by touching the [MATRX] key, highlighting [A], and then touching the [ENTER] key.

```
[A]
            [[.8 1.2]
             [.6 .5 ]]
```

EXERCISES

1. Enter the matrix $\begin{bmatrix} -1 & 2 \\ -3 & 4 \end{bmatrix}$ as matrix B.

2. Enter the matrix $\begin{bmatrix} 0.1 \\ -1.2 \\ -2.1 \end{bmatrix}$ as matrix C.

3. Enter the matrix $[2 \quad -6 \quad 0]$ as matrix D.

4. Enter the matrix $\begin{bmatrix} 0 & 1 & 2 \\ 1 & 9 & 0 \\ -2 & 0 & 1 \end{bmatrix}$ as matrix F.

survivability. The number of offspring is the **fecundity** of an organism. A **Leslie matrix** or **population projection matrix** stores the survivability and fecundity information about a species. To construct a Leslie matrix, we first divide the population into groups, usually based on age. For example, we could divide the population of a bird species into two groups. One group consists of the hatchlings and the other consists of the adults. The Leslie matrix is formed by filling in the survivability and fecundity in the different age groups, as follows:

$$\begin{bmatrix} \text{Average fecundity of hatchlings next year} & \text{Average fecundity of adults next year} \\ \text{Survivability of hatchlings} & \text{Survivability of older birds} \end{bmatrix}.$$

EXAMPLE 2 *Bird Population.* The Leslie matrix for a bird population consisting of 100 hatchlings and 200 adults is
$$G = \begin{bmatrix} 0.8 & 1.2 \\ 0.6 & 0.5 \end{bmatrix}.$$

a) Interpret the four numbers in the matrix **G**.

b) Estimate the number of hatchlings and the number of adults next year.

Solution

a) First, the entry 0.8 indicates that the average number of hatchlings next year from hatchlings this year is 0.8. Since there are 100 hatchlings this year, we would expect a total of 0.8 × 100 = 80 hatchlings next year from this year's hatchlings.

Second, the average fecundity of this year's adult birds measured next year is 1.2. Since there are 200 adult birds in this year's population, then we would expect a total of 1.2 × 200 = 240 hatchlings next year from this year's adults.

Third, hatchling survivability is 0.6. This means that of all the hatchlings, 60% survive until the next year. Since there are 100 hatchlings this year, we expect 0.6 × 100 = 60 of this year's hatchlings to survive to be adults next year.

Fourth, the adult survivability is 0.5. This means that 50% of this year's adults will survive until next year. Since there are 200 adults this year, we would expect a total of 0.5 × 200 = 100 adults to survive until next year.

b) Next year's hatchlings are of two types: those whose parents are hatchlings this year and those whose parents are adults this year. In (a), we calculated these two numbers to be 80 and 240, respectively, giving a total of 80 + 240 = 320 hatchlings.

Also, next year's adults are of two types. Some are hatchlings this year, and some are adults this year. In (a), we calculated these two numbers to be 60 and 100, respectively, giving a total of 60 + 100 = 160 adults. ■

Matrix Products

A **row matrix** is a matrix with just one row, and a **column matrix** is a matrix with just one column. We will also refer to column matrices as **vectors.** We will often use boldface lowercase letters, such as **v**, to denote vectors.

An example of a vector is the **population vector.** Each entry in the population vector is the number of individuals in the group. In Example 2, the population vector this year is $\begin{bmatrix} 100 \\ 200 \end{bmatrix}$, while the population vector for next year is predicted to be $\begin{bmatrix} 320 \\ 160 \end{bmatrix}$. The calculation to find next year's population vector is an example of a **product** of matrices. We can define the product of two matrices **A** and **B**, as long as the number of columns in **A** is the same as the number of rows in **B**.

DEFINITION

Let **A** be an $n \times k$ matrix and **B** a $k \times m$ matrix. Then the product $\mathbf{C} = \mathbf{AB}$ is an $n \times m$ matrix so that

$$c_{ij} = a_{i1}b_{1j} + a_{i2}b_{2j} + a_{i3}b_{3j} + \cdots + a_{ik}b_{kj}.$$

In other words, the entry in row i and column j is computed by adding the products of corresponding entries in the ith row of **A** with the jth column of **B**. If the number of columns of **A** is not equal to the number of rows in **B**, then the product **AB** is not defined.

The definition of matrix product seems rather strange and unnatural at first. However, in many applications, this definition is exactly what is needed. Example 3 demonstrates one application of a matrix product.

EXAMPLE 3   *Bird Population.* Let **G** be the Leslie matrix in Example 2 and let **v** be this year's population vector.

a) Compute the matrix product **Gv**.

b) Interpret the product **Gv**.

Solution

a) The Leslie matrix is $\mathbf{G} = \begin{bmatrix} 0.8 & 1.2 \\ 0.6 & 0.5 \end{bmatrix}$ and the population vector is $\mathbf{v} = \begin{bmatrix} 100 \\ 200 \end{bmatrix}$. Since the size of **G** is 2×2 and the size of **v** is 2×1 the size of the product is 2×1. Therefore, **Gv** is a vector. The first entry in **Gv** is obtained by using the first row of **G** and the first (and only) column of **v**, while the second entry in **Gv** is obtained using the second row of **G** and the column of **v**.

$$\mathbf{Gv} = \begin{bmatrix} 0.8 & 1.2 \\ 0.6 & 0.5 \end{bmatrix} \begin{bmatrix} 100 \\ 200 \end{bmatrix}$$

$$= \begin{bmatrix} (0.8)(100) + (1.2)(200) \\ (0.6)(100) + (0.5)(200) \end{bmatrix}$$

$$= \begin{bmatrix} 320 \\ 160 \end{bmatrix}$$

Technology Connection

Matrix Multiplication

Many graphers are able to multiply matrices. Let's do Example 4a) on the TI-83.

First, we enter $A = \begin{bmatrix} 1 & -3 & 0 \\ -2 & -1 & 4 \end{bmatrix}$ and

$B = \begin{bmatrix} 2 & -1 \\ 5 & 3 \\ -4 & -5 \end{bmatrix}$ as described in the Tech-

nology Connection on page 431. We use the $\boxed{\text{MATRX}}$ key to enter the product as shown on the screen below. The second screen shows how to store the product as matrix **C**.

```
[A] * [B]
              [[-13  -10]
               [-25  -21]]
```

```
[A] * [B]→[C]
              [[-13  -10]
               [-25  -21]]
```

EXERCISES

Compute the product **AB**.

1. $A = \begin{bmatrix} 1 & 8 \\ 0 & 12 \end{bmatrix}, B = \begin{bmatrix} -3 & 4 \\ 7 & -10 \end{bmatrix}$

2. $A = \begin{bmatrix} 8 & 2 & -7 \\ 9 & -3 & 1 \end{bmatrix},$

$B = \begin{bmatrix} 6 & -3 & -7 \\ 4 & 0 & 12 \\ 7 & 9 & -5 \end{bmatrix}$

3. $A = \begin{bmatrix} -2 & 5 & 6 \end{bmatrix}, B = \begin{bmatrix} 8 \\ -1 \\ 6 \end{bmatrix}$

b) We see from the calculation that the matrix product calculations are identical to the calculations we used to predict next year's population in Example 2. The vector $\mathbf{w} = \mathbf{Gv} = \begin{bmatrix} 320 \\ 160 \end{bmatrix}$ is next year's population vector. ∎

As seen in Example 3, the product of a Leslie matrix and a population vector gives next year's population vector.

EXAMPLE 4   Compute the product $\mathbf{C} = \mathbf{AB}$.

a) $A = \begin{bmatrix} 1 & -3 & 0 \\ -2 & -1 & 4 \end{bmatrix}, B = \begin{bmatrix} 2 & -1 \\ 5 & 3 \\ -4 & -5 \end{bmatrix}$

b) $A = \begin{bmatrix} 1 & -3 & 0 \\ -2 & -1 & 4 \end{bmatrix}, B = \begin{bmatrix} 2 & -1 & 5 \\ 3 & -4 & -5 \end{bmatrix}$

Solution

a) Since **A** has size 2×3 and **B** has size 3×2, the product **C** has size 2×2. To find c_{21}, we use the second row of **A** and the first column of **B**:

$$c_{21} = (-2)(2) + (-1)(5) + (4)(-4).$$

We find the other entries of **C** similarly:

$$C = \begin{bmatrix} 1 & -3 & 0 \\ -2 & -1 & 4 \end{bmatrix}\begin{bmatrix} 2 & -1 \\ 5 & 3 \\ -4 & -5 \end{bmatrix}$$

$$= \begin{bmatrix} (1)(2) + (-3)(5) + (0)(-4) & (1)(-1) + (-3)(3) + (0)(-5) \\ (-2)(2) + (-1)(5) + (4)(-4) & (-2)(-1) + (-1)(3) + (4)(-5) \end{bmatrix}$$

$$= \begin{bmatrix} -13 & -10 \\ -25 & -21 \end{bmatrix}.$$

b) We cannot multiply these two matrices since the number of columns in **A** is different from the number of rows in **B**. ∎

> **CAUTION**   In order to multiply two matrices, the number of columns in the first matrix must equal the number of rows in the second.

Multiple-Year Projections

Sometimes biologists use a diagram, called a **Leslie diagram,** to display the information in a Leslie matrix. Each group within the population is represented by a circle. The entry g_{ij} in the Leslie matrix is used to label an arrow from group j to group i. We do not draw an arrow if the corresponding entry in the Leslie matrix is 0.

> **CAUTION** The number g_{ij} labels an arrow from j to i, not from i to j.

EXAMPLE 5 *Ovenbirds.* The Leslie diagram for ovenbirds in central Missouri can be approximated by the diagram below, where (H) represents hatchlings and (A) represents adults.[2]

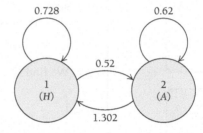

a) Write the Leslie matrix.

b) There are 36 adult ovenbirds with a total of 60 hatchlings in a wood. What do we predict for the number of adults and hatchlings next year?

c) How many adults and hatchlings do we expect in year 3?

Solution

a) The survivability values are given as $g_{21} = 0.52$ and $g_{22} = 0.62$. The fecundity values are $g_{11} = 0.728$ and $g_{12} = 1.302$. The Leslie matrix is

$$G = \begin{bmatrix} 0.728 & 1.302 \\ 0.52 & 0.62 \end{bmatrix}.$$

[2]Jones, J., Barg, J. J., Sillett, T. S., Veit, M. L., and Robertson, R. J., "Season-Long Fecundity, Survival, and Viability of Ovenbirds in Fragmented and Unfragmented Landscapes," *Conservation Biology,* Vol. 13, No. 2, pp. 1151–1161 (1999).

b) The population vector the first year is $P = \begin{bmatrix} 60 \\ 36 \end{bmatrix}$. To find the population vector the second year, we compute the product Gp.

$$Gp = \begin{bmatrix} 0.728 & 1.302 \\ 0.52 & 0.62 \end{bmatrix} \begin{bmatrix} 60 \\ 36 \end{bmatrix} = \begin{bmatrix} (0.728)(60) + (1.302)(36) \\ (0.52)(60) + (0.62)(36) \end{bmatrix} = \begin{bmatrix} 90.552 \\ 53.52 \end{bmatrix}.$$

We expect approximately 91 hatchlings and 54 adults the second year.

c) The second-year population vector is estimated to be $\begin{bmatrix} 90.552 \\ 53.52 \end{bmatrix}$. We multiply the Leslie matrix by the second-year population vector to get the third-year population vector.

$$G(Gp) = \begin{bmatrix} 0.728 & 1.302 \\ 0.52 & 0.62 \end{bmatrix} \begin{bmatrix} 90.552 \\ 53.52 \end{bmatrix} = \begin{bmatrix} (0.728)(90.552) + (1.302)(53.52) \\ (0.52)(90.552) + (0.62)(53.52) \end{bmatrix} \approx \begin{bmatrix} 135.6 \\ 80.3 \end{bmatrix}.$$

We expect approximately 136 hatchlings and 80 adults the third year. ∎

In Example 5, we computed $G(Gp)$ to estimate the population in year 3. When multiplying three numbers, we know that the associative law $a(bc) = (ab)c$ holds. The associative law also holds for matrix multiplication. That is, $A(BC) = (AB)C$, as long as the matrices $A, B,$ and C have sizes that allow the products.

EXAMPLE 6   *Ovenbirds.*   Use the Leslie matrix and initial population of Example 5 to compute the following.

a) GG

b) (GG)p

c) Interpret the matrix (GG)p.

Solution

a) We compute GG.

$$GG = \begin{bmatrix} 0.728 & 1.302 \\ 0.52 & 0.62 \end{bmatrix} \begin{bmatrix} 0.728 & 1.302 \\ 0.52 & 0.62 \end{bmatrix}$$

$$= \begin{bmatrix} 0.728(0.728) + 1.302(0.52) & 0.728(1.302) + 1.302(0.62) \\ 0.52(0.728) + 0.62(0.52) & 0.52(1.302) + 0.62(0.62) \end{bmatrix}$$

$$= \begin{bmatrix} 1.207 & 1.755 \\ 0.701 & 1.061 \end{bmatrix}.$$

b) We computed GG in part (a). To compute $(GG)p$, we multiply the matrix computed in part (a) by p.

$$(GG)p = \begin{bmatrix} 1.207 & 1.755 \\ 0.701 & 1.061 \end{bmatrix} \begin{bmatrix} 60 \\ 36 \end{bmatrix} = \begin{bmatrix} 1.207(60) + 1.755(36) \\ 0.701(60) + 1.061(36) \end{bmatrix} = \begin{bmatrix} 135.6 \\ 80.3 \end{bmatrix}.$$

As expected, this is the same as the product $G(Gp)$, which we found in part c) of Example 5.

c) The product of the matrix GG and the population vector gives the population estimate for two years later. ∎

If A is a square matrix, we let $A^2 = AA$, $A^3 = AAA$, and so on. As we see from Example 6, the powers of the Leslie matrix give us a way to estimate the population after more than one year. For example, $G^{10}p$ would estimate the population after 10 years.

There is one other common type of multiplication involving matrices. If c is a number and A is a matrix, we define the **scalar product** $cA = Ac$ as the matrix obtained by multiplying each entry in A by c. Note that a scalar product multiplies a number times a matrix to give a matrix. For example,

$$5\begin{bmatrix} 2 & 3 \\ -1 & 7 \end{bmatrix} = \begin{bmatrix} 5(2) & 5(3) \\ 5(-1) & 5(7) \end{bmatrix} = \begin{bmatrix} 10 & 15 \\ -5 & 35 \end{bmatrix}.$$

The scalar product $(-1)A$ multiplies each entry of A by -1. Multiplying a number by -1 is equivalent to changing its sign. We sometimes write $-A$ instead of $(-1)A$ if we wish to change the sign of each entry of A.

The scalar product $(0)A$ is a matrix with all entries 0. We call this a **zero matrix.**

Matrix Addition and Subtraction

We can define the sum and difference of two matrices with the same size. To compute the sum, we add corresponding entries, and to compute the difference, we subtract corresponding entries.

Technology Connection

Matrix Operations

The TI-83, and many other graphers, are capable of adding and subtracting matrices. Furthermore, many graphers are able to compute powers of matrices and scalar products. The operations are entered just as we would enter the operations for numbers. Refer to your calculators owner's manual for details.

EXERCISES

Let $A = \begin{bmatrix} 2 & -3 \\ 4 & 6 \end{bmatrix}$ and

$B = \begin{bmatrix} -5 & 6 \\ 2 & 1 \end{bmatrix}$.

1. Compute $A + B$.
2. Compute $A - B$.
3. Compute $A + AB$.
4. Compute $3A$.
5. Compute A^5.

> **DEFINITION**
>
> Let A and B be $n \times m$ matrices. Then the *sum* $S = A + B$ is also an $n \times m$ matrix defined by
>
> $$s_{ij} = a_{ij} + b_{ij}.$$
>
> Similarly, the *difference* $D = A - B$ is an $n \times m$ matrix defined by
>
> $$d_{ij} = a_{ij} - b_{ij}.$$

EXAMPLE 7   Let $A = \begin{bmatrix} 3 & 0 & -6 \\ -1 & 7 & -3 \end{bmatrix}$, $B = \begin{bmatrix} 4 & -2 & 0 \\ 5 & -2 & 10 \end{bmatrix}$, and $C = \begin{bmatrix} 3 & 9 \\ 2 & 1 \end{bmatrix}$. Compute the following.

a) $A + B$

b) $A - B$

c) $-A + A$

d) $A + C$

Solution

a) We compute the sum by adding corresponding entries.

$$A + B = \begin{bmatrix} 3 & 0 & -6 \\ -1 & 7 & -3 \end{bmatrix} + \begin{bmatrix} 4 & -2 & 0 \\ 5 & -2 & 10 \end{bmatrix}$$

$$= \begin{bmatrix} 3+4 & 0+(-2) & -6+0 \\ -1+5 & 7+(-2) & -3+10 \end{bmatrix} = \begin{bmatrix} 7 & -2 & -6 \\ 4 & 5 & 7 \end{bmatrix}.$$

b) We compute the difference by subtracting corresponding entries.

$$\mathbf{A} - \mathbf{B} = \begin{bmatrix} 3 & 0 & -6 \\ -1 & 7 & -3 \end{bmatrix} - \begin{bmatrix} 4 & -2 & 0 \\ 5 & -2 & 10 \end{bmatrix}$$

$$= \begin{bmatrix} 3-4 & 0-(-2) & -6-0 \\ -1-5 & 7-(-2) & -3-10 \end{bmatrix}$$

$$= \begin{bmatrix} -1 & 2 & -6 \\ -6 & 9 & -13 \end{bmatrix}.$$

c) We compute $-\mathbf{A}$ and add this matrix to $\mathbf{A}$.

$$-\mathbf{A} + \mathbf{A} = -\begin{bmatrix} 3 & 0 & -6 \\ -1 & 7 & -3 \end{bmatrix} + \begin{bmatrix} 3 & 0 & -6 \\ -1 & 7 & -3 \end{bmatrix}$$

$$= \begin{bmatrix} -3 & -0 & 6 \\ 1 & -7 & 3 \end{bmatrix} + \begin{bmatrix} 3 & 0 & -6 \\ -1 & 7 & -3 \end{bmatrix}$$

$$= \begin{bmatrix} 0 & 0 & 0 \\ 0 & 0 & 0 \end{bmatrix}.$$

We see that $-\mathbf{A} + \mathbf{A}$ is the zero matrix.

d) The matrices $\mathbf{A}$ and $\mathbf{C}$ do not have the same size, so we cannot add them. ◼

Identity Matrix

For each n, there is a special $n \times n$ matrix is called the **identity matrix I**. This matrix is defined so that $\mathbf{AI} = \mathbf{A}$ and $\mathbf{IA} = \mathbf{A}$.

Technology Connection

Identity Matrix

The identity matrix can be entered on the TI-83 by touching the MATRX key and selecting **EDIT**. Then select **IDENTITY(** and enter the number of rows in the identity matrix. The screen below shows how to store the 5×5 identity matrix as matrix **D**.

```
identity(5)→[D]
[[1 0 0 0 0]
 [0 1 0 0 0]
 [0 0 1 0 0]
 [0 0 0 1 0]
 [0 0 0 0 1]]
```

DEFINITION

An *identity matrix* **I** is a square matrix of the form

$$\mathbf{I} = \begin{bmatrix} 1 & 0 & 0 & \cdots & 0 \\ 0 & 1 & 0 & \cdots & 0 \\ 0 & 0 & 1 & \cdots & 0 \\ \vdots & \vdots & \vdots & \ddots & \vdots \\ 0 & 0 & 0 & \cdots & 1 \end{bmatrix}$$

That is, the identity matrix has all entries zero except for the entries on the **main diagonal,** which are all 1.

CAUTION   Identity matrices of different sizes are not equal.

EXAMPLE 8   Let $\mathbf{A} = \begin{bmatrix} -3 & 2 & -1 \\ -2 & 4 & 5 \end{bmatrix}$.

a) Compute $\mathbf{AI}$.

b) Compute $\mathbf{IA}$.

Solution

a) Since **A** is 2×3, it is understood that **I** is a 3×3 matrix for this product.

$$\mathbf{AI} = \begin{bmatrix} -3 & 2 & -1 \\ -2 & 4 & 5 \end{bmatrix} \begin{bmatrix} 1 & 0 & 0 \\ 0 & 1 & 0 \\ 0 & 0 & 1 \end{bmatrix}$$

$$= \begin{bmatrix} (-3)(1) + (2)(0) + (-1)(0) & (-3)(0) + (2)(1) + (-1)(0) & (-3)(0) + (2)(0) + (-1)(1) \\ (-2)(1) + (4)(0) + (5)(0) & (-2)(0) + (4)(1) + (5)(0) & (-2)(0) + (4)(0) + (5)(1) \end{bmatrix}$$

$$= \begin{bmatrix} -3 & 2 & -1 \\ -2 & 4 & 5 \end{bmatrix}$$

We see that $\mathbf{AI} = \mathbf{A}$.

b) Since **I** is on the right of **A**, here it is understood that **I** is 2×2.

$$\mathbf{IA} = \begin{bmatrix} 1 & 0 \\ 0 & 1 \end{bmatrix} \begin{bmatrix} -3 & 2 & -1 \\ -2 & 4 & 5 \end{bmatrix}$$

$$= \begin{bmatrix} (1)(-3) + (0)(-2) & (1)(2) + (0)(4) & (1)(-1) + (0)(5) \\ (0)(-3) + (1)(-2) & (0)(2) + (1)(4) & (0)(-1) + (1)(5) \end{bmatrix}$$

$$= \begin{bmatrix} -3 & 2 & -1 \\ -2 & 4 & 5 \end{bmatrix}$$

We see that $\mathbf{IA} = \mathbf{A}$.

Matrix Algebra

We noted in Examples 5 and 6 that the associative law holds for matrices as long as their sizes allow the products. That is, $\mathbf{A}(\mathbf{BC}) = (\mathbf{AB})\mathbf{C}$.

Another important property is the distributive law. For numbers a, b, and c, the distributive law says that $a(b + c) = ab + ac$. For matrices, there are two forms of the distributive law: $\mathbf{A}(\mathbf{B} + \mathbf{C}) = \mathbf{AB} + \mathbf{AC}$ and $(\mathbf{B} + \mathbf{C})\mathbf{A} = \mathbf{BA} + \mathbf{CA}$. We illustrate the first version with Example 9. We will verify in the exercises that the second version of the distributive law holds.

EXAMPLE 9 Let $\mathbf{A} = \begin{bmatrix} 4 & 2 \\ 1 & -2 \end{bmatrix}$, $\mathbf{B} = \begin{bmatrix} -2 & -3 \\ 4 & 1 \end{bmatrix}$, and $\mathbf{C} = \begin{bmatrix} 6 & 1 \\ -3 & 2 \end{bmatrix}$.

a) Compute $\mathbf{A}(\mathbf{B} + \mathbf{C})$.

b) Compute $\mathbf{AB} + \mathbf{AC}$.

Solution

a) We first compute $\mathbf{B} + \mathbf{C}$ and then multiply by **A**.

$$\mathbf{B} + \mathbf{C} = \begin{bmatrix} -2 & -3 \\ 4 & 1 \end{bmatrix} + \begin{bmatrix} 6 & 1 \\ -3 & 2 \end{bmatrix} = \begin{bmatrix} 4 & -2 \\ 1 & 3 \end{bmatrix}$$

$$\mathbf{A}(\mathbf{B} + \mathbf{C}) = \begin{bmatrix} 4 & 2 \\ 1 & -2 \end{bmatrix} \begin{bmatrix} 4 & -2 \\ 1 & 3 \end{bmatrix}$$

$$= \begin{bmatrix} (4)(4) + & (2)(1) & (4)(-2) + & (2)(3) \\ (1)(4) + & (-2)(1) & (1)(-2) + & (-2)(3) \end{bmatrix}$$

$$= \begin{bmatrix} 18 & -2 \\ 2 & -8 \end{bmatrix}.$$

b) Here we multiply A by B and C and then add.

$$\mathbf{AB} = \begin{bmatrix} 4 & 2 \\ 1 & -2 \end{bmatrix} \begin{bmatrix} -2 & -3 \\ 4 & 1 \end{bmatrix} = \begin{bmatrix} (4)(-2) + & (2)(4) & (4)(-3) + & (2)(1) \\ (1)(-2) + & (-2)(4) & (1)(-3) + & (-2)(1) \end{bmatrix}$$

$$= \begin{bmatrix} 0 & -10 \\ -10 & -5 \end{bmatrix},$$

$$\mathbf{AC} = \begin{bmatrix} 4 & 2 \\ 1 & -2 \end{bmatrix} \begin{bmatrix} 6 & 1 \\ -3 & 2 \end{bmatrix} = \begin{bmatrix} (4)(6) + & (2)(-3) & (4)(1) + & (2)(2) \\ (1)(6) + & (-2)(-3) & (1)(1) + & (-2)(2) \end{bmatrix}$$

$$= \begin{bmatrix} 18 & 8 \\ 12 & -3 \end{bmatrix},$$

$$\mathbf{AB} + \mathbf{AC} = \begin{bmatrix} 0 & -10 \\ -10 & -5 \end{bmatrix} + \begin{bmatrix} 18 & 8 \\ 12 & -3 \end{bmatrix} = \begin{bmatrix} 18 & -2 \\ 2 & -8 \end{bmatrix}. \quad \blacksquare$$

Unlike the associative and distributive laws, the commutative law of multiplication fails for matrices. In general, if $\mathbf{A}$ and $\mathbf{B}$ are two matrices, then $\mathbf{AB} \neq \mathbf{BA}$. If $\mathbf{A}$ and $\mathbf{B}$ are not square, then this is obvious. For example, if $\mathbf{A}$ is 2×3 and $\mathbf{B}$ is 3×2, then $\mathbf{AB}$ is 2×2, but $\mathbf{BA}$ is 3×3. Even if $\mathbf{A}$ and $\mathbf{B}$ are square matrices of the same size, generally speaking $\mathbf{AB} \neq \mathbf{BA}$. (There are some pairs of matrices where $\mathbf{AB} = \mathbf{BA}$, but these are not typical cases.)

EXAMPLE 10   Let $\mathbf{A} = \begin{bmatrix} 6 & 4 \\ 2 & 5 \end{bmatrix}$ and $\mathbf{B} = \begin{bmatrix} 3 & -5 \\ -4 & 1 \end{bmatrix}$. Compute the following.

a) $\mathbf{AB}$

b) $\mathbf{BA}$

Solution

a) $\mathbf{AB} = \begin{bmatrix} 6 & 4 \\ 2 & 5 \end{bmatrix} \begin{bmatrix} 3 & -5 \\ -4 & 1 \end{bmatrix} = \begin{bmatrix} (6)(3) + (4)(-4) & (6)(-5) + (4)(1) \\ (2)(3) + (5)(-4) & (2)(-5) + (5)(1) \end{bmatrix}$

$= \begin{bmatrix} 2 & -26 \\ -14 & -5 \end{bmatrix}.$

b) $\mathbf{BA} = \begin{bmatrix} 3 & -5 \\ -4 & 1 \end{bmatrix} \begin{bmatrix} 6 & 4 \\ 2 & 5 \end{bmatrix} = \begin{bmatrix} (3)(6) + (-5)(2) & (3)(4) + (-5)(5) \\ (-4)(6) + (1)(2) & (-4)(4) + (1)(5) \end{bmatrix}$

$= \begin{bmatrix} 8 & -13 \\ -22 & -11 \end{bmatrix}.$

We see that $\mathbf{AB} \neq \mathbf{BA}$. $\quad \blacksquare$

CAUTION   Although matrix algebra has many of the basic properties of ordinary algebra, we must be careful not to reverse the order when multiplying matrices.

Exercise Set 6.1

Let $A = \begin{bmatrix} 4 & -1 \\ 7 & -9 \end{bmatrix}$, $B = \begin{bmatrix} 3 & 9 \\ 2 & -2 \end{bmatrix}$, $C = \begin{bmatrix} 8 & 3 \\ 0 & -3 \end{bmatrix}$,

$D = \begin{bmatrix} 5 & -6 & 1 \\ 10 & 3 & -1 \end{bmatrix}$, $E = \begin{bmatrix} -7 & 4 \\ -3 & 2 \\ 2 & -1 \end{bmatrix}$,

$F = \begin{bmatrix} -4 & 2 & 3 \\ 0 & -1 & 2 \\ -7 & -2 & 5 \end{bmatrix}$, and $v = \begin{bmatrix} 2 \\ -3 \end{bmatrix}$.

1. Compute $A + B$ and $B + A$. (This is an example of the commutative property of addition.)

2. Compute $A + C$.   3. Compute $B + C$.

4. Compute $A + D$.   5. Compute $A + E$.

6. Compute $3A$.   7. Compute $B2$.

8. Compute $(-1)B$ and $B(-1)$. Are they equal?

9. Compute $A - B$ and $A + (-1)B$. Compare your answers.

10. Compute $B - C$ and $B + (-1)C$. Compare your answers.

11. Compute AD.   12. Compute A^2v.

13. Compute A^3v.   14. Compute B^2v.

15. Compute B^3v.   16. Compute BF.

17. Compute $A + DE$.

18. Compute $(DE)A$ and $D(EA)$ to verify the associative property for these matrices.

19. Compute $(BD)F$ and $B(DF)$ to verify the associative property for these matrices.

20. Compute AI and IA to verify the multiplicative identity property for this matrix.

21. Compute $A(B + C)$ and $AB + AC$ to verify the distributive property for these matrices.

22. Compute $(B + C)D$ and $BD + CD$ to verify the distributive property for these matrices.

Let $A = \begin{bmatrix} 1 & 6 & -2 \\ 4 & -2 & -1 \\ 0 & 3 & -5 \end{bmatrix}$, $B = \begin{bmatrix} 0 & 4 & -2 \\ 5 & 0 & -3 \\ 6 & 2 & 1 \end{bmatrix}$, and

$C = \begin{bmatrix} 4 & 4 & 0 \\ -2 & 3 & 8 \\ 1 & -3 & 6 \end{bmatrix}$.

23. Compute AB and BA. Are these products equal?

24. Compute $A(B + C)$ and $AB + AC$ to verify that these matrices satisfy the distributive property.

25. Compute $A(B - C)$ and $AB - AC$ to verify that these matrices satisfy the distributive property.

26. Compute:

   a) $A \begin{bmatrix} 0 \\ 5 \\ 6 \end{bmatrix}$.

   b) $A \begin{bmatrix} 4 \\ 0 \\ 2 \end{bmatrix}$.

   c) $A \begin{bmatrix} -2 \\ -3 \\ 1 \end{bmatrix}$.

 tw d) Compare parts (a), (b), and (c) with Exercise 23. Explain.

27. Compute:

   a) $B \begin{bmatrix} 1 \\ 4 \\ 0 \end{bmatrix}$.

   b) $B \begin{bmatrix} 6 \\ -2 \\ 3 \end{bmatrix}$.

   c) $B \begin{bmatrix} -2 \\ -1 \\ -5 \end{bmatrix}$.

 tw d) Compare parts (a), (b), and (c) with Exercise 23. Explain.

28. Compute $A - 3I$.   29. Compute $B - 4I$.

APPLICATIONS

30. *Bird Population.* The Leslie matrix for a bird population is $G = \begin{bmatrix} 1.2 & 2.5 \\ 0.8 & 0.8 \end{bmatrix}$. In year 1, the hatchling population is 240 and the adult population is 124.

   a) Draw and label the Leslie diagram.

   b) Estimate the population of hatchlings and adults in year 2.

   c) Estimate the population of hatchlings and adults in year 3.

31. *Bird Population.* The Leslie matrix for a bird population is $\mathbf{G} = \begin{bmatrix} 1.4 & 1.8 \\ 0.5 & 0.4 \end{bmatrix}$. In year 1, the hatchling population is 56 and the adult population is 20.

a) Draw and label the Leslie diagram.
b) Estimate the population of hatchlings and adults in year 2.
c) Estimate the population of hatchlings and adults in year 3.

32. *Bird Population.* A population of birds with the Leslie diagram given below has 100 hatchlings and 48 adults in year 1.

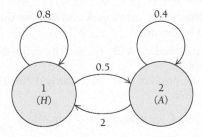

a) Write the Leslie matrix.
b) Estimate the population of hatchlings and adults in year 2.
c) Estimate the population of hatchlings and adults in year 3.

33. *Bird Population.* A population of birds with the Leslie diagram given below has 1200 hatchlings and 1520 adults in year 1.

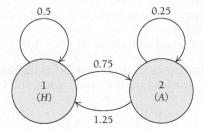

a) Write the Leslie matrix.
b) Estimate the population of hatchlings and adults in year 2.
c) Estimate the population of hatchlings and adults in year 3.

34. *Cerulean Warblers.* The population of cerulean warblers is grouped into hatchlings (H), second year (S), and after second year (A). The Leslie diagram below shows estimates of the survivability

and fecundity.[3] The population in year 1 is $H = 579$, $S = 80$, and $A = 609$.

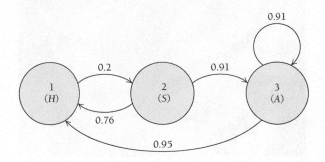

a) Write the Leslie matrix.
b) Estimate the population of each group in year 2.
c) Estimate the population of each group in year 3.

35. *Dinophilus Gyrociliatus.* *Dinophilus gyrociliatus* is a small species that lives in the fouling community of harbor environments. On average, a female has approximately 30 eggs during her first 6 wk of life. If she survives her first 6 wk, she has on average 15 eggs her second 6 wk of life. Furthermore, approximately 80% of the females survive their first 6 wk and none survive beyond the second 6 wk.[4] Assume half the eggs are female and for simplicity, assume that all the eggs are hatched at once at the beginning of each 6 wk period. Ignore the male population and make the two groups females under 6 wk old and females over 6 wk old.

a) Draw and label the Leslie diagram.
b) Find the Leslie matrix.
c) Twenty hatchlings are introduced into an area. Estimate the population of the two groups after 6 wk.
d) Estimate the population of the two groups after 12 wk.

[3]Jones, J., Barg, J.J., Sillett, T.S., Veit, M.L., and Robertson, R.J., "Season-Long Fecundity, Survival, and Viability of Ovenbirds in Fragmented and Unfragmented Landscapes," *Conservation Biology,* Vol. 13, No. 2, pp. 1151–1161 (1999).
[4]D. Prevedelli and R. Zunarelli Vandini, "Survival, fecundity and sex ratio of *Dinophilus gyrociliatus* (Polychaeta: Dinophilidae) under different dietary conditions," *Marine Biology* Vol. 133, pp. 231–236 (1999).

36. *Fairy Shrimp.* San Diego fairy shrimp live in ponds that fill and dry many times during a year. While the ponds are dry, the fairy shrimp survive as cysts. The population is divided into three groups. Cysts that survive one dry period (1DP) are in group 1, cysts that survive two dry periods (2DP) are in group 2, and cysts that survive three or more dry periods (3DP) are in group 3. A Leslie diagram estimating the survivability and fecundity between dry periods is given below.[5]

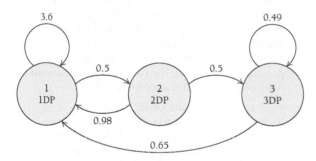

a) Find the Leslie matrix.
b) Twenty cysts in group 1 are introduced into a pond. How many cysts are expected in each of the three groups after a dry period?
c) Estimate the population of the three groups after two dry periods.
d) Estimate the population of the three groups after three dry periods.

SYNTHESIS

37. Let $A = \begin{bmatrix} 3 & -2 \\ 4 & 5 \end{bmatrix}$ and $B = \begin{bmatrix} -2 & 7 \\ 1 & -3 \end{bmatrix}$.

   a) Compute A^2.
   b) Compute B^2.
   c) Compute $(A + B)^2$.
   d) Compute $A^2 + 2AB + B^2$.
   tw e) Explain why the answers to parts (c) and (d) are different.
   tw f) Find a correct formula for $(A + B)^2$. Explain.

38. Let A and B be as in Exercise 37.

   a) Compute $(A - B)^2$.
   b) Compute $A^2 - 2AB + B^2$.
   tw c) Explain why the answers to parts (a) and (b) are different.
   tw d) Find a correct formula for $(A - B)^2$. Explain.

If A is a matrix, its transpose $B = A^T$ is the matrix obtained by switching the rows and columns of A. For example, $\begin{bmatrix} 1 & 2 & 3 \\ 4 & 5 & 6 \end{bmatrix}^T = \begin{bmatrix} 1 & 4 \\ 2 & 5 \\ 3 & 6 \end{bmatrix}$.

39. Let $A = \begin{bmatrix} a_{11} & a_{12} & a_{13} & \cdots & a_{1n} \end{bmatrix}$ and

$$B = \begin{bmatrix} b_{11} \\ b_{21} \\ b_{31} \\ \vdots \\ b_{n1} \end{bmatrix}.$$

   a) What are the sizes of A^T and B^T?
   tw b) Show that $A \cdot B = B^T \cdot A^T$.

40. The sizes of the matrices A and B allow the product AB to be defined.

   tw a) Is the product $A^T B^T$ necessarily defined? Explain.
   tw b) Is the product $B^T A^T$ necessarily defined? Explain.

Let A, B, and C be arbitrary $n \times n$ matrices.

tw 41. Explain why $(A + B) + C = A + (B + C)$.

tw 42. Explain why $A + B = B + A$.

tw 43. Explain why $A + (-A) = 0$.

tw 44. Explain why $(A + B)^T = A^T + B^T$.

 Technology Connection

45.–49. Repeat Exercises 23–27 using a grapher.

50. *Bird Population.* Compute the population vector for the birds in Exercise 30 after 10 yr.

51. *Bird Population.* Compute the population vector for the birds in Exercise 31 after 12 yr.

52. *Bird Population.* Compute the population vector for the birds in Exercise 32 after 8 yr.

53. *Bird Population.* Compute the population vector for the birds in Exercise 33 after 20 yr.

54. *Warbler Population.* Compute the population vector for the warblers in Exercise 34 after 10 yr.

55. *Fairy Shrimp Population.* Compute the population vector for the fairy shrimp in Exercise 36 after 15 dry periods.

6.2 **Solving Systems of Linear Equations**

OBJECTIVES

■ Solve systems of linear equations using substitution.
■ Solve systems of linear equations using elimination.
■ Solve systems of linear equations using row operations.

Matrices have many important uses, including solving systems of linear equations. Although there are many ways to solve a system of linear equations, using matrix row operations is one of the most efficient. In this section, we learn how to solve these systems using matrices.

Linear Systems

In Example 5 on page 434, we predicted the population vector next year based on this year's population vector $\mathbf{p} = \begin{bmatrix} 60 \\ 36 \end{bmatrix}$ and the Leslie matrix $\begin{bmatrix} 0.728 & 1.302 \\ 0.52 & 0.62 \end{bmatrix}$. To estimate what the population vector was *last* year, we need to find values x and y so that

$$G\begin{bmatrix} x \\ y \end{bmatrix} = \begin{bmatrix} 60 \\ 36 \end{bmatrix}$$

$$\begin{bmatrix} 0.728 & 1.302 \\ 0.52 & 0.62 \end{bmatrix}\begin{bmatrix} x \\ y \end{bmatrix} = \begin{bmatrix} 60 \\ 36 \end{bmatrix} \tag{1}$$

$$\begin{bmatrix} 0.728x + 1.302y \\ 0.52x + 0.62y \end{bmatrix} = \begin{bmatrix} 60 \\ 36 \end{bmatrix}.$$

Equating entries of the left-hand vector and the right-hand vector gives us two *linear equations:*

$$0.728x + 1.302y = 60$$
$$0.52x + 0.62y = 36.$$

In other words, we converted the matrix equation (1) into a **system of linear equations.** A *solution* of the system is an assignment of values to the variables so that each equation is true. (We will solve this system in Example 3.)

EXAMPLE 1 Verify that $x = 1$, $y = 2$, and $z = -2$ is a solution of the system

$$x - 3y + 2z = -9$$
$$3x + 2y + 5z = -3$$
$$y + z = 0.$$

Solution We substitute $x = 1$, $y = 2$, and $z = -2$ into the left-hand side of each equation and simplify to see if we get the right-hand side. For the first equation,

$$x - 3y + 2z = 1 - 3(2) + 2(-2) = -9.$$

For the second equation,

$$3x + 2y + 5z = 3(1) + 2(2) + 5(-2) = -3.$$

For the third equation,

$$y + z = 2 + (-2) = 0.$$

Since each equation is satisfied, $x = 1$, $y = 2$, and $z = -2$ is a solution. ■

Example 1 illustrates a system of linear equations with more than two variables. We can write the solution of this system as $(1, 2, -2)$, where it is understood that the first number is the value for x, the second is the value for y, and the third is the value for z. If other variables are involved, then we need to specify the order that we list the variables.

Geometry of Linear Systems

How many solutions can a system of linear equations have? To begin, let's look at the case where we have two linear equations and two unknowns:

$$a_1 x + b_1 y = d_1$$
$$a_2 x + b_2 y = d_2.$$

The graph of each equation is a line. A solution to the system is a point that lies on each line. Generally, two lines intersect in a point, but it is also possible that the lines are parallel and do not intersect. It is further possible that the lines are the same. If there is no solution, then we say the system is **inconsistent.** If the lines are the same, then we say that the system is **dependent,** otherwise the system is **independent.**

Consistent and independent

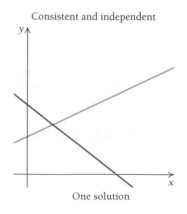

One solution

Inconsistent and independent

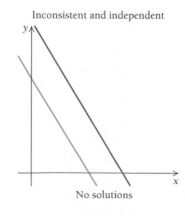

No solutions

Consistent and dependent

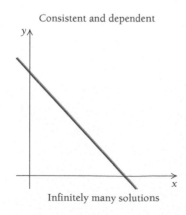

Infinitely many solutions

It is difficult to visualize the graph of a linear system in three variables. If there are more than three variables, visualization is impossible. Even so, it is sometimes helpful to think about the geometry of the solutions of systems with two variables to help our understanding of systems with more variables.

For example, it is impossible for a system of linear equations to have exactly two solutions. Looking at the preceding graph, we see that either there is no solution, there is exactly one solution, or there are an infinite number of solutions. This is true in general, regardless of the number of linear equations and the number of variables. Therefore, if a system of linear equations has at least two solutions, then it has infinitely many solutions.

Substitution Method

One technique used for solving a system of linear equations is called the **substitution method.** The idea behind the substitution method is to solve for a variable in one equation and then substitute that variable into all of the other equations. We illustrate how the substitution method works with an example.

EXAMPLE 2   Solve the system

$$3x - 2y = 1 \tag{2}$$

$$-x + y = 1. \tag{3}$$

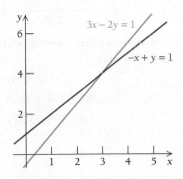

Solution   These linear equations are sketched in the figure and intersect at one point. Let's use algebra to determine this point. Solving for y in equation (3), we find

$$-x + y = 1$$

$$y = x + 1. \tag{4}$$

We now substitute $x + 1$ for y into equation (2) and solve for x.

$$3x - 2y = 1$$

$$3x - 2(x + 1) = 1$$

$$3x - 2x - 2 = 1$$

$$x = 3.$$

To find y, we substitute $x = 3$ into any of the equations involving y. We use equation (4).

$$y = x + 1 = 3 + 1 = 4.$$

The solution is $(3, 4)$, in accordance with the previous graph. ∎

In the previous example, we could have begun by solving for x instead of y. We also could have used the first equation instead of the second. The student is invited to verify that solving Example 2 in these different ways results in the same solution.

EXAMPLE 3   *Ovenbirds.* The Leslie matrix **G** for ovenbirds[6] and this year's population vector **p** are given by

$$G = \begin{bmatrix} 0.728 & 1.302 \\ 0.520 & 0.620 \end{bmatrix} \quad \text{and} \quad p = \begin{bmatrix} 60 \\ 36 \end{bmatrix}.$$

Find last year's population vector.

Solution   Let x be the number of hatchlings and y the number of adults last year. As we saw at the beginning of this section, to find last year's population vector, we solve the system of linear equations:

$$0.728x + 1.302y = 60 \tag{5}$$
$$0.520x + 0.620y = 36. \tag{6}$$

We solve equation (5) for x.

$$x = \frac{60 - 1.302y}{0.728} \approx 82.4 - 1.79y$$

Substituting this expression for x into equation (6), we have an equation we can solve to find y.

$$0.520x + 0.620y = 36$$
$$0.520(82.4 - 1.79y) + 0.620y \approx 36$$
$$42.8 - 0.931y + 0.620y \approx 36$$
$$-0.311y = -6.8$$
$$y = 21.9$$

We substitute this value of y into the equation $x = 82.4 - 1.79y$.

$$x \approx 82.4 - 1.79y \approx 82.4 - 1.79(21.9) \approx 43.2.$$

We estimate that last year's population was approximately 43 hatchlings and 22 adult ovenbirds. ∎

Elimination Method

Substitution can also be used for systems with more than two equations. The key is to solve for a variable and substitute into all the other equations. However, other methods are generally more efficient than substitution when there are more than two variables.

---

[6]P. A. Porneluzi and J. Faaborg, "Season-Long Fecundity, Survival, and Viability of Ovenbirds in Fragmented and Unfragmented Landscapes," *Conservation Biology*, Vol. 13, No. 2, pp. 1151–1161 (1999).

Another technique for solving systems of linear equations is called the **elimination method.** The elimination method is based on the idea of **equivalent systems.** Two systems of equations are *equivalent* if they have exactly the same solutions. For example, the system

$$x + y = 2 \tag{7}$$
$$x - y = 0 \tag{8}$$

is equivalent to the system

$$x + y = 2 \tag{7}$$
$$2x \quad\;\; = 2. \tag{9}$$

Using the substitution method, we can see that $(1, 1)$ is the solution of both systems. Of course, the second system is easier to solve since x may be found directly from equation (9).

In order to simplify a system of linear equations into a system that is easier to solve, we can use three basic operations. By using a combination of these operations, we can find an equivalent system that is much easier to solve.

THEOREM 1

The following operations on a system of equations yield equivalent systems.

1. Change the order in which the equations are written, but do not change the equations.
2. Multiply an equation by a constant other than 0.
3. Add a multiple of one equation to another equation, but do not change any other equation.

Of course, changing the order in which we write the equations does not change the solutions, so operation 1 yields an equivalent system. Multiplying an equation by a constant other than 0 in no way changes its solutions, so operation 2 also yields an equivalent system. We will investigate operation 3 in the exercises.

EXAMPLE 4   Solve the system in Example 2 using Theorem 1.

$$3x - 2y = 1 \tag{10}$$
$$-x + \;\; y = 1 \tag{11}$$

Solution   We will attempt to find an equivalent system that is easier to solve. We first multiply equation (11) by -1 to get an equivalent system. The purpose of this is to make the x-coefficient $+1$ instead of -1.

$$3x - 2y = 1 \tag{10}$$
$$x - \;\; y = -1 \tag{12}$$

Next we interchange the equations.

$$x - y = -1 \tag{12}$$
$$3x - 2y = 1 \tag{10}$$

If we multiply equation (12) by -3 and add it to equation (10), then the coefficient of x in equation (10) becomes 0. In other words, we eliminate x from equation (10).

$$-3(x - y) + (3x - 2y) = (-3)(-1) + 1$$
$$3x + (-3)x - 2y + (-3)(-y) = 4$$

We simplify and replace (10) with this new equation:

$$x - y = -1 \tag{12}$$
$$y = 4. \tag{13}$$

Equation (13) states that the value of y is 4. To find x, we **back-substitute** 4 for y into equation (12).

$$x - y = -1$$
$$x - 4 = -1$$
$$x = 3$$

The solution is $(3, 4)$. As expected, this agrees with the result of Example 2. ∎

Row Operations

Looking back at Example 4, we could save some time and some writing by using matrix notation. Instead of writing a linear equation, we write the coefficients in a matrix, being careful to keep the coefficients of x in the first column and the coefficients of y in the second column. If there are more variables, we reserve one column for each variable. We then *augment* the coefficient matrix by drawing a vertical line and placing the constants to the right of the line.

The system in Example 4 can be written as an **augmented matrix:**

$$\begin{bmatrix} 3 & -2 & | & 1 \\ -1 & 1 & | & 1 \end{bmatrix}.$$

Operations on equations are interpreted as row operations on matrices with the understanding that a row corresponds to an equation and a column corresponds to a variable.

EXAMPLE 5 Solve the system in Example 4 using an augmented matrix.

Solution We perform the same operations as in Example 4, but we simply leave out the variables and use matrix notation. The matching equation numbers are also given.

$$\begin{bmatrix} 3 & -2 & | & 1 \\ -1 & 1 & | & 1 \end{bmatrix} \tag{10, 11}$$

$$\begin{bmatrix} 3 & -2 & | & 1 \\ 1 & -1 & | & -1 \end{bmatrix} \quad \text{Multiply row 2 by } -1. \tag{10, 12}$$

$$\begin{bmatrix} 1 & -1 & | & -1 \\ 3 & -2 & | & 1 \end{bmatrix} \quad \text{Exchange rows.} \tag{12, 10}$$

$$\begin{bmatrix} 1 & -1 & | & -1 \\ 0 & 1 & | & 4 \end{bmatrix} \quad \text{Add } -3 \text{ times row 1 to row 2.} \tag{12, 13}$$

In the last augmented matrix, row 2 says $0x + 1y = 4$, or $y = 4$. Similarly, row 1 says that $1x - 1y = -1$. Back-substituting $y = 4$ into this equation yields $x - 4 = -1$, so $x = 3$.   ■

We list the three **elementary row operations** that correspond to the three operations given in Theorem 1.

Elementary Row Operations

1. Any row may be switched with any other row.
2. Any row may be multiplied by any number other than 0.
3. A multiple of any row may be added to a different row.

If we can convert one matrix to another using elementary row operations, then we say the two matrices are **row-equivalent.**

The last matrix we obtained in Example 5 is said to be in **row-echelon** form.

DEFINITION

A matrix is in *row-echelon form* if the conditions below are satisfied.

1. If a row does not have all zeros, then the first nonzero entry is a 1. This is called the *leading one.*
2. The leading one in a row is to the right of the leading one in the row above.
3. All rows with only zero entries are at the bottom of the matrix.

The process of finding a row-equivalent matrix in row-echelon form is called **Gaussian elimination.** Generally, the strategy employed in Gaussian elimination is to work on one column at a time, starting with the leftmost and working to the right. We illustrate with an example.

EXAMPLE 6   Solve the system of linear equations.

$$3x + 18y + 8z = 31$$
$$x + 5y + 2z = 7$$
$$2x + 4y + 5z = 4$$

Solution   We first rewrite the system as an augmented matrix.

$$\begin{bmatrix} 3 & 18 & 8 & | & 31 \\ 1 & 5 & 2 & | & 7 \\ 2 & 4 & 5 & | & 4 \end{bmatrix}$$

We now perform row operations to simplify the augmented matrix. Our first goal is to put the first column in the proper form. We wish to have every entry in the first column be 0 except in the first row, where we want a 1.

$$\begin{bmatrix} 1 & 5 & 2 & | & 7 \\ 3 & 18 & 8 & | & 31 \\ 2 & 4 & 5 & | & 4 \end{bmatrix}$$ Switch rows 1 and 2.

$$\begin{bmatrix} 1 & 5 & 2 & | & 7 \\ 3+(-3)(1) & 18+(-3)(5) & 8+(-3)(2) & | & 31+(-3)(7) \\ 2 & 4 & 5 & | & 4 \end{bmatrix}$$ Add −3 times row 1 to row 2.

$$\begin{bmatrix} 1 & 5 & 2 & | & 7 \\ 0 & 3 & 2 & | & 10 \\ 2 & 4 & 5 & | & 4 \end{bmatrix}$$ Simplify

$$\begin{bmatrix} 1 & 5 & 2 & | & 7 \\ 0 & 3 & 2 & | & 10 \\ 2+(-2)(1) & 4+(-2)(5) & 5+(-2)(2) & | & 4+(-2)(7) \end{bmatrix}$$ Add −2 times row 1 to row 3.

$$\begin{bmatrix} 1 & 5 & 2 & | & 7 \\ 0 & 3 & 2 & | & 10 \\ 0 & -6 & 1 & | & -10 \end{bmatrix}$$ Simplify

We now have the first row and the first column in the proper form. We do not use the first row in any more row operations. Instead, we focus on the second and third rows. We next make the second entry on the main diagonal 1.

$$\begin{bmatrix} 1 & 5 & 2 & | & 7 \\ 0 & 1 & \frac{2}{3} & | & \frac{10}{3} \\ 0 & -6 & 1 & | & -10 \end{bmatrix}$$ Multiply row 2 by 1/3.

If we can make the first two entries in the last row both 0, then the last row will give the value of z.

$$\begin{bmatrix} 1 & 5 & 2 & | & 7 \\ 0 & 1 & \frac{2}{3} & | & \frac{10}{3} \\ 0+6(0) & -6+6(1) & 1+6\left(\frac{2}{3}\right) & | & -10+6\left(\frac{10}{3}\right) \end{bmatrix}$$ Add 6 times row 2 to row 3.

$$\begin{bmatrix} 1 & 5 & 2 & | & 7 \\ 0 & 1 & \frac{2}{3} & | & \frac{10}{3} \\ 0 & 0 & 5 & | & 10 \end{bmatrix}$$ Simplify

We now have the second column and the second row in the proper form, so we concentrate on the third column.

$$\begin{bmatrix} 1 & 5 & 2 & | & 7 \\ 0 & 1 & \frac{2}{3} & | & \frac{10}{3} \\ 0 & 0 & 1 & | & 2 \end{bmatrix}$$ Multiply row 3 by 1/5.

We now read off the value of $z = 2$ from the last row. To find y, we back-substitute $z = 2$ into the equation determined by the second row:

$$y + \frac{2}{3}z = \frac{10}{3}$$

$$y + \frac{2}{3}(2) = \frac{10}{3}$$

$$y = 2.$$

We now back-substitute $z = 2$ and $y = 2$ into the equation determined by the first row.

$$x + 5y + 2z = 7$$
$$x + 5(2) + 2(2) = 7$$
$$x = -7$$

The solution is $(-7, 2, 2)$. ∎

EXAMPLE 7   Solve the system

$$
\begin{aligned}
x \quad\quad + z &= 2 \\
y + z &= 3 \\
-2x + 2y \quad\quad &= 2
\end{aligned}
$$

using Gaussian elimination.

Solution   We convert the system of equations into an augmented matrix and perform row operations.

$$\left[\begin{array}{ccc|c} 1 & 0 & 1 & 2 \\ 0 & 1 & 1 & 3 \\ -2 & 2 & 0 & 2 \end{array}\right]$$

$$\left[\begin{array}{ccc|c} 1 & 0 & 1 & 2 \\ 0 & 1 & 1 & 3 \\ 0 & 2 & 2 & 6 \end{array}\right] \quad \text{Add 2 times row 1 to row 3.}$$

$$\left[\begin{array}{ccc|c} 1 & 0 & 1 & 2 \\ 0 & 1 & 1 & 3 \\ 0 & 0 & 0 & 0 \end{array}\right] \quad \text{Add } -2 \text{ times row 2 to row 3.}$$

We now have the matrix in row-echelon form. Since the last row gives the equation $0 = 0$, it gives us no information. We therefore consider only the first two rows. The equation from row 2 is

$$y + z = 3, \quad \text{or} \quad y = 3 - z.$$

The equation from row 1 gives the value of x.

$$x + z = 2, \quad \text{or} \quad x = 2 - z$$

We conclude that the solutions are of the form $(2 - z, 3 - z, z)$ for any value of z. There are an infinite number of solutions. For example, setting z equal to $0, 2, 3$, and 5, we get the solutions $(2, 3, 0), (0, 1, 2), (-1, 0, 3)$, and $(-3, -2, 5)$, respectively. ∎

As seen in Example 7, if a row of all zeros appears during Gaussian elimination, then the system is dependent and the row is ignored when finding the solutions.

EXAMPLE 8 Solve the system.

$$
\begin{aligned}
x \qquad + z &= 2 \\
y + z &= 3 \\
-2x + 2y \qquad &= 3
\end{aligned}
$$

Solution This system is almost the same as the system in Example 7. We write it as an augmented matrix and apply Gaussian elimination.

$$
\left[\begin{array}{ccc|c}
1 & 0 & 1 & 2 \\
0 & 1 & 1 & 3 \\
-2 & 2 & 0 & 3
\end{array}\right]
$$
  Converting the system to an augmented matrix

$$
\left[\begin{array}{ccc|c}
1 & 0 & 1 & 2 \\
0 & 1 & 1 & 3 \\
0 & 2 & 2 & 7
\end{array}\right]
$$
  Add 2 times row 1 to row 3.

$$
\left[\begin{array}{ccc|c}
1 & 0 & 1 & 2 \\
0 & 1 & 1 & 3 \\
0 & 0 & 0 & 1
\end{array}\right]
$$
  Add -2 times row 2 to row 3.

The matrix is in row-echelon form. The last row gives the equation

$$0 = 1.$$

This is a false statement, so there are no values of x, y, and z that can satisfy this system. We conclude that there are no solutions. The system is inconsistent. ■

> If while performing Gaussian elimination we obtain a row of all zeros (including the augmentation), then we can ignore that row. On the other hand, if we obtain a row of all zeros except that the augmentation is not zero, then the system is inconsistent and there are no solutions.

Reduced Row-Echelon Form

When an augmented matrix is in row-echelon form, we can use back-substitution to find the solution. However, if we perform a few more row operations, then we can put the matrix into a form where we can read off the solution without back-substitution. We illustrate this method with an example.

EXAMPLE 9 Solve the system:

$$
\begin{aligned}
x + 2y + 3z &= 1 \\
x + 3y + 2z &= 6 \\
2x + 5y + 6z &= 5
\end{aligned}
$$

Technology Connection

Row Reduction

Graphers that handle matrices can perform row operations and both Gaussian elimination and Gauss–Jordan elimination. Let's do Example 9 using the TI-83. First we enter the 3×4 augmented matrix

$$\left[\begin{array}{ccc|c} 1 & 2 & 3 & 1 \\ 1 & 3 & 2 & 6 \\ 2 & 5 & 6 & 5 \end{array}\right]$$

as matrix **A**. On the calculator, we do not enter the vertical line. From the $\boxed{\text{MATRX}}$ menu we select MATH and then RREF(. We then fill in [A]. The result is shown below.

```
rref( [A] )
   [[1  0  0  1 ]
    [0  1  0  3 ]
    [0  0  1  -2]]
```

We see that the solution is $(1, 3, -2)$.

If we wished to perform Gaussian elimination, we would use REF(instead of RREF(.

EXERCISES

Use Gauss–Jordan elimination on the grapher to solve the following systems.

1.
$$\begin{aligned} 4x + y - 3z &= 3 \\ x - 5y + 7z &= 3 \\ 8x + 2y - 10z &= -6 \end{aligned}$$

2.
$$\begin{aligned} 7x + 6y - 5z &= 0 \\ 4x - 10y + 4z &= 8 \\ 3x + 16y - 9z &= -8 \end{aligned}$$

3.
$$\begin{aligned} 0.18x - 1.3y - 4.7z &= 1.1 \\ 1.3x + 2.1y - 3.5z &= 10.1 \\ -2.6x + 3.6y - 1.4z &= -2.6 \end{aligned}$$

Solution We write the augmented matrix and perform row operations.

$$\left[\begin{array}{ccc|c} 1 & 2 & 3 & 1 \\ 1 & 3 & 2 & 6 \\ 2 & 5 & 6 & 5 \end{array}\right]$$

$$\left[\begin{array}{ccc|c} 1 & 2 & 3 & 1 \\ 0 & 1 & -1 & 5 \\ 2 & 5 & 6 & 5 \end{array}\right] \quad \text{Add } -1 \text{ times row 1 to row 2.}$$

$$\left[\begin{array}{ccc|c} 1 & 2 & 3 & 1 \\ 0 & 1 & -1 & 5 \\ 0 & 1 & 0 & 3 \end{array}\right] \quad \text{Add } -2 \text{ times row 1 to row 3.}$$

$$\left[\begin{array}{ccc|c} 1 & 2 & 3 & 1 \\ 0 & 1 & -1 & 5 \\ 0 & 0 & 1 & -2 \end{array}\right] \quad \text{Add } -1 \text{ times row 2 to row 3.}$$

We have reduced the matrix to row-echelon form. Let's continue to do row operations with the goal of making the matrix to the left of the augmentation the identity matrix.

$$\left[\begin{array}{ccc|c} 1 & 0 & 5 & -9 \\ 0 & 1 & -1 & 5 \\ 0 & 0 & 1 & -2 \end{array}\right] \quad \text{Add } -2 \text{ times row 2 to row 1.}$$

$$\left[\begin{array}{ccc|c} 1 & 0 & 0 & 1 \\ 0 & 1 & -1 & 5 \\ 0 & 0 & 1 & -2 \end{array}\right] \quad \text{Add } 5 \text{ times row 3 to row 1.}$$

$$\left[\begin{array}{ccc|c} 1 & 0 & 0 & 1 \\ 0 & 1 & 0 & 3 \\ 0 & 0 & 1 & -2 \end{array}\right] \quad \text{Add 1 times row 3 to row 2.}$$

We can now read off the solution simply by writing the equations corresponding to the rows. The solution is $x = 1, y = 3$, and $z = -2$, or $(1, 3, -2)$. ■

In Example 9, we stopped with the identity matrix left of the augmentation. This is not always possible. For example, if the matrix is not square, then it cannot be row reduced to the identity, since the identity matrix is square. It may also turn out that a row contains only zeros, again making it impossible to row reduce to the identity.

Instead of requiring the matrix to the left of the augmentation to be the identity, we make it as close as we can. We say that a matrix is in **reduced row-echelon form** if in addition to being in row-echelon form, each column that contains a leading 1 has all other entries 0. The process of row reducing a matrix into reduced row-echelon form is called **Gauss–Jordan elimination**.

EXAMPLE 10 *Equilibrium.* Three mg of glactosyl human serum albumin (Tc-GSA) is injected into the bloodstream. Let B and L be the amount (in mg) of Tc-GSA in the blood and liver, respectively. The quantities B and L are related by[7]

$$3 = B + L$$

$$\frac{dB}{dt} = -0.06B + 0.03L$$

$$\frac{dL}{dt} = 0.06B - 0.03L.$$

At equilibrium, the quantities B and L are constant. Find the equilibrium amount of Tc-GSA in the blood and the liver.

Solution Since the derivative of a constant is 0, we set $\dfrac{dB}{dt} = \dfrac{dL}{dt} = 0$. We then have a system of three equations in two unknowns.

$$
\begin{aligned}
B + \quad\; L &= 3 \\
-0.06B + 0.03L &= 0 \\
0.06B - 0.03L &= 0
\end{aligned}
$$

We solve by writing the associated augmented matrix and row reducing.

$$
\left[\begin{array}{cc|c}
1 & 1 & 3 \\
-0.06 & 0.03 & 0 \\
0.06 & -0.03 & 0
\end{array}\right]
$$

$$
\left[\begin{array}{cc|c}
1 & 1 & 3 \\
-0.06 & 0.03 & 0 \\
0 & 0 & 0
\end{array}\right]
\qquad \text{Add row 2 to row 3.}
$$

$$
\left[\begin{array}{cc|c}
1 & 1 & 3 \\
0 & 0.09 & 0.18 \\
0 & 0 & 0
\end{array}\right]
\qquad \text{Add 0.06 times row 1 to row 2.}
$$

$$
\left[\begin{array}{cc|c}
1 & 1 & 3 \\
0 & 1 & 2 \\
0 & 0 & 0
\end{array}\right]
\qquad \text{Multiply row 2 by 1/0.09.}
$$

$$
\left[\begin{array}{cc|c}
1 & 0 & 1 \\
0 & 1 & 2 \\
0 & 0 & 0
\end{array}\right]
\qquad \text{Add } -1 \text{ times row 2 to row 1.}
$$

The solution is $B = 1$ mg and $L = 2$ mg. ∎

---

[7]K. Kinoshita, M. Ukikusa, K. Iwaisako, A. Arimoto, N. Fujisawa, T. Ozaki, H. Tanaka, S. Seo, M. Naitoh, A. Nomura, T. Inomoto, T. Kitai, K. Ino, H. Higashiyama, T. Hanafusa, and Y. Nakajima, "Preoperative assessment of heptic function: Utility of a new convenient two-compartment model analysis using galactosyl human serum albumin scintigraphy," *Journal of Gastroenterology and Hepatology,* Vol. 18, pp. 99–104 (2003).

We have investigated three ways to solve a system of equations. First, when there are only two variables, it is often simplest to use the substitution method. However, when there are more than two variables, it is generally easier to convert the system of equations into an augmented matrix and apply elementary row operations. Second, we could perform Gaussian elimination and stop when the matrix is in row-echelon form. The solution is then found using back-substitution. Third, we could use Gauss–Jordan elimination. In this case, we continue performing row operations until the matrix is in reduced row-echelon form. We can then read off the solution as the augmentation.

Exercise Set 6.2

Solve the following system of equations using the substitution method.

1. $x + 2y = 5$
   $2x - y = 0$

2. $3x - 2y = 20$
   $4x - y = 35$

3. $5w - z = 14$
   $2w + 3z = 26$

4. $-2r - 5s = -8$
   $3r + 2s = 1$

5. $s - t = 7$
   $-2s + 2t = -5$

6. $4a - b = 2$
   $12a - 3b = 6$

7. $x - y = 7$
   $-2x + 2y = -14$

8. $4x - y = 2$
   $12x - 3y = 5$

9.–16. Solve Exercises 1–8 using Gaussian elimination.

Solve using Gaussian elimination.

17. $y + 3z = -1$
   $x \qquad + 6z = 37$
   $2y + z = -2$

18. $-1.5y - z = 5$
   $-x - 1.5y - 2.5z = 7$
   $- y - 0.5z = -2$

19. $7x - y - 9z = 1$
   $2x \qquad - 4z = -4$
   $-4x \qquad + 6z = -3$

20. $y + 3z = 2$
   $x + y + 6z = 0$
   $x \qquad + 2z = 4$

21. $2x - 2y + 3z = 3$
   $4x - 3y + 3z = 2$
   $-x + y - z = 4$

22. $x + 2y - z = 6$
   $-2x \qquad + z = 3$
   $x - y \qquad = -4$

23.–28. Solve the systems in Exercises 17–22 using Gauss–Jordan elimination.

Solve using Gaussian elimination.

29. $x + y + 2z = 5$
   $x + y + z = -10$
   $2x + 3y + 4z = 2$

30. $3x + y - 2z = -10$
   $-2x - y + 3z = 14$
   $5x - 3y - 3z = -29$

31. $x + y - 2z = 4$
   $4x + 7y + 3z = 3$
   $14x + 23y + 5z = 17$

32. $x + y - 2z = 4$
   $4x + 7y + 3z = 3$
   $14x + 23y + 5z = 10$

33. $x - y + 3z = 2$
   $2x + 3y - z = 5$
   $-x - 9y + 11z = 1$

34. $x - y + 3z = 2$
   $2x + 3y - z = 5$
   $-x - 9y + 11z = -4$

35. $x - 2y - 5z = 0$
   $2x + 3y + 15z = 0$
   $-2x - y - 8z = 1$

36. $9x + 11y + 15z = 30$
$14x + 18y + 25z = 20$
$4x + 5y + 7z = 10$

37. $x + y + z + w = 5$
$x \quad + z + w = 6$
$y + z + w = 4$
$x \quad + z \quad = 3$

38. $x - 2y \quad + 3w = \quad 5$
$2x + 3y - z - 2w = -7$
$y - 3z + 4w = \quad 21$
$x - 2y + 5z - w = -16$

39. $-2x - y + 6z - w = \quad 2$
$-3x - 5y + 6z + w = -3$
$x + y - 2z \quad = 4$
$y - z \quad = 1$

40. $x + y + 6z \quad = -2$
$x + y + 5z + 2w = 0$
$x + y + 5z + w = 3$
$2x + 3y + 13z + 4w = 7$

APPLICATIONS

41. *Home Air Quality.* Chemicals enter a house's basement air. Let $F(t)$ and $B(t)$ be the amount of the chemical (in mg/m^3) in the first-floor air and the basement air after t minutes, respectively. The rate of change of F and B are given by the equations[8]

$B' = -0.01B + 0.1$ and $F' = 0.01B - 0.02F$.

Find the equilibrium values for B and F.

42. *Bird Population.* The Leslie matrix for a bird population is $G = \begin{bmatrix} 1.2 & 2.5 \\ 0.8 & 0.8 \end{bmatrix}$. In year 2, the hatchling population is 240 and the adult population is 124. Find the population of both groups in year 1.

43. *Cerulean Warblers.* The population of cerulean warblers is grouped into hatchlings (H), second year (S), and after second year (A). Their Leslie diagram is given below, showing estimates of the survivability and fecundity.[9] The population in year 2

is $H = 579$, $S = 80$, and $A = 609$. Estimate the population of each group in year 1.

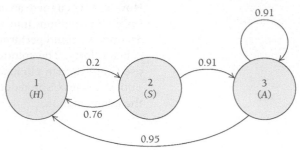

44. *Fairy Shrimp.* San Diego fairy shrimp live in ponds that fill and dry many times during a year. While the ponds are dry, the fairy shrimp survive as cysts. The population is divided into three groups. Cysts that survive one dry period (1DP) are in group 1, cysts that survive two dry periods (2DP) are in group 2, and cysts that survive three or more dry periods (3DP) are in group 3. The Leslie matrix representing the survivability and fecundity is given below.[10] There are 1000, 2317, and 1493 individuals in groups 1, 2, and 3, respectively. How many were in each group in the previous dry period?

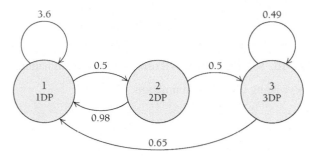

SYNTHESIS

tw 45. In this exercise we will explore part (c) of Theorem 1.

a) Explain why a solution of a linear system also is a solution to the system obtained using part (c) of Theorem 1.

b) Explain why a solution of the system obtained using part (c) of Theorem 1 is also a solution to the original system.

c) Explain how parts (a) and (b) of this exercise imply part (c) of Theorem 1.

[8]D. A. Olson and R. L. Corsi, "Characterizing exposure to chemicals from soil vapor intrusion using a two-compartment model," *Atmospheric Environment*, Vol. 35, pp. 4201–4209 (2001).

[9]Jones, J., Barg, J. J., Sillett, T. S., Veit, M. L., and Robertson, R. J., "Season-Long Fecundity, Survival, and Viability of Oven-birds in Fragmented and Unfragmented Landscapes," *Conservation Biology*, Vol. 13, No. 2, pp. 1151–1161 (1999).

[10]B. J. Ripley, J. Holtz, and M. Simovich, "Cyst bank life-history model for a fairy shrimp from ephemeral ponds," *Freshwater Biology* Vol. 49, pp. 221–231 (2004).

46. Let $A = \begin{bmatrix} 0 & 1 & 0 \\ 1 & 0 & 0 \\ 0 & 0 & 1 \end{bmatrix}$. The matrix A is the identity

matrix with the first two rows exchanged.

a) Let $B = \begin{bmatrix} 1 & 2 & 3 \\ 4 & 5 & 6 \\ 7 & 8 & 9 \end{bmatrix}$. Compute AB.

b) Let $C = \begin{bmatrix} 4 & -2 & 7 \\ 1 & 1 & -3 \\ -4 & 3 & 6 \end{bmatrix}$. Compute AC.

tw c) Based on parts (a) and (b), what is the effect of multiplying A on the left with another 3×3 matrix? Explain why.

47. Let $A = \begin{bmatrix} 1 & 0 & 0 \\ 0 & 4 & 0 \\ 0 & 0 & 1 \end{bmatrix}$.

a) Let $B = \begin{bmatrix} 1 & 2 & 3 \\ 4 & 5 & 6 \\ 7 & 8 & 9 \end{bmatrix}$. Compute AB.

b) Let $C = \begin{bmatrix} 4 & -2 & 7 \\ 1 & 1 & -3 \\ -4 & 3 & 6 \end{bmatrix}$. Compute AC.

tw c) Based on parts (a) and (b), what is the effect of multiplying A on the left with another 3×3 matrix? Explain why.

48. Let $A = \begin{bmatrix} 1 & 0 & 0 \\ 3 & 1 & 0 \\ 0 & 0 & 1 \end{bmatrix}$.

a) Let $B = \begin{bmatrix} 1 & 2 & 3 \\ 4 & 5 & 6 \\ 7 & 8 & 9 \end{bmatrix}$. Compute AB.

b) Let $C = \begin{bmatrix} 4 & -2 & 7 \\ 1 & 1 & -3 \\ -4 & 3 & 6 \end{bmatrix}$. Compute AC.

tw c) Based on parts (a) and (b), what is the effect of multiplying A on the left with another 3×3 matrix? Explain why.

49. Assume that (x_0, y_0, z_0) and (x_1, y_1, z_1) are both solutions to $ax + by + cz = d$.

a) Show that $(tx_0 + (1 - t)x_1, ty_0 + (1 - t)y_1, tz_0 + (1 - t)z_1)$ is also a solution to $ax + by + cz = d$ for any value of t.

tw b) If a system of linear equations in three variables has two solutions, explain why it has an infinite number of solutions.

tw c) Explain why part (b) generalizes to any number of variables.

 Technology Connection

50.–55. Solve the systems of linear equations in Exercises 35–40 using a grapher.

6.3 Finding a Matrix Inverse and Determinant

OBJECTIVES

■ Compute the inverse of a matrix.
■ Compute the determinant of a matrix.

In Section 6.2, we used augmented matrices to solve systems of linear equations. In this section, we investigate another way to use a matrix to solve a system of linear equations.

The Inverse Matrix

Let $A = \begin{bmatrix} 1 & 0 \\ 2 & 1 \end{bmatrix}$ and $B = \begin{bmatrix} 1 & 0 \\ -2 & 1 \end{bmatrix}$. We may compute the product AB:

$$AB = \begin{bmatrix} 1 & 0 \\ 2 & 1 \end{bmatrix}\begin{bmatrix} 1 & 0 \\ -2 & 1 \end{bmatrix} = \begin{bmatrix} (1)(1) + (0)(-2) & (1)(0) + (0)(1) \\ (2)(1) + (1)(-2) & (2)(0) + (1)(1) \end{bmatrix} = \begin{bmatrix} 1 & 0 \\ 0 & 1 \end{bmatrix} = I.$$

Furthermore,

$$BA = \begin{bmatrix} 1 & 0 \\ -2 & 1 \end{bmatrix}\begin{bmatrix} 1 & 0 \\ 2 & 1 \end{bmatrix} = \begin{bmatrix} (1)(1) + (0)(2) & (1)(0) + (0)(1) \\ (-2)(1) + (1)(2) & (-2)(0) + (1)(1) \end{bmatrix} = \begin{bmatrix} 1 & 0 \\ 0 & 1 \end{bmatrix} = I.$$

Because $\mathbf{AB} = \mathbf{I}$ and $\mathbf{BA} = \mathbf{I}$, we say the matrix $\mathbf{B}$ is the **inverse** of $\mathbf{A}$.

> **DEFINITION**
>
> Let $\mathbf{A}$ be a square matrix. If it exists the *multiplicative inverse* of $\mathbf{A}$, or simply its *inverse*, is a matrix $\mathbf{A}^{-1}$ with the property that
>
> $$\mathbf{AA}^{-1} = \mathbf{I} \quad \text{and} \quad \mathbf{A}^{-1}\mathbf{A} = \mathbf{I}.$$
>
> If $\mathbf{A}$ has an inverse, we say that $\mathbf{A}$ is invertible, or **nonsingular.** If $\mathbf{A}$ does not have an inverse, we say that $\mathbf{A}$ is **singular.**

As shown above, the inverse of the matrix $\mathbf{A} = \begin{bmatrix} 1 & 0 \\ 2 & 1 \end{bmatrix}$ is $\mathbf{A}^{-1} = \begin{bmatrix} 1 & 0 \\ -2 & 1 \end{bmatrix}$.

EXAMPLE 1   Find the inverse of the matrix $\mathbf{A} = \begin{bmatrix} 1 & 3 \\ 2 & -1 \end{bmatrix}$.

Solution   Suppose the inverse matrix is

$$\mathbf{A}^{-1} = \begin{bmatrix} x & z \\ y & w \end{bmatrix}.$$

We need to solve for x, y, z, and w in the matrix equation $\mathbf{AA}^{-1} = \mathbf{I}$. Multiplying,

$$\begin{bmatrix} 1 & 3 \\ 2 & -1 \end{bmatrix}\begin{bmatrix} x & z \\ y & w \end{bmatrix} = \begin{bmatrix} 1 & 0 \\ 0 & 1 \end{bmatrix}$$

$$\begin{bmatrix} x + 3y & z + 3w \\ 2x - y & 2z - w \end{bmatrix} = \begin{bmatrix} 1 & 0 \\ 0 & 1 \end{bmatrix}.$$

Equating entries, we see that we have two systems of linear equations to solve:

$$\begin{aligned} x + 3y &= 1 \\ 2x - y &= 0 \end{aligned} \quad \text{and} \quad \begin{aligned} z + 3w &= 0 \\ 2z - w &= 1. \end{aligned}$$

In order to solve these two systems, we write the augmented matrices:

$$\begin{bmatrix} 1 & 3 & | & 1 \\ 2 & -1 & | & 0 \end{bmatrix} \quad \text{and} \quad \begin{bmatrix} 1 & 3 & | & 0 \\ 2 & -1 & | & 1 \end{bmatrix}.$$

Since the only difference in the two matrices is the augmentation column, we combine the two matrices by keeping the common coefficient matrix and including both augmentation columns. We then use Gauss–Jordan elimination.

$$\begin{bmatrix} 1 & 3 & | & 1 & 0 \\ 2 & -1 & | & 0 & 1 \end{bmatrix} \qquad \text{Include both augmentations}$$

$$\begin{bmatrix} 1 & 3 & | & 1 & 0 \\ 0 & -7 & | & -2 & 1 \end{bmatrix} \qquad \text{Add } -2 \text{ times row 1 to row 2.}$$

$$\begin{bmatrix} 1 & 3 & | & 1 & 0 \\ 0 & 1 & | & \frac{2}{7} & -\frac{1}{7} \end{bmatrix} \qquad \text{Multiply row 2 by } -1/7.$$

$$\begin{bmatrix} 1 & 0 & | & \frac{1}{7} & \frac{3}{7} \\ 0 & 1 & | & \frac{2}{7} & -\frac{1}{7} \end{bmatrix} \qquad \text{Add } -3 \text{ times row 2 to row 1.}$$

We can now read off the solution: $x = 1/7$, $y = 2/7$, $z = 3/7$, and $w = -1/7$. The inverse matrix is

$$\mathbf{A}^{-1} = \begin{bmatrix} \frac{1}{7} & \frac{3}{7} \\ \frac{2}{7} & -\frac{1}{7} \end{bmatrix}.$$

Notice that the inverse is simply the augmentation of the reduced matrix. We leave it to the student to check that both $\mathbf{AA}^{-1} = \mathbf{I}$ and $\mathbf{A}^{-1}\mathbf{A} = \mathbf{I}$.   ∎

Example 1 illustrates the following theorem.

THEOREM 2

Let $\mathbf{A}$ be a square matrix. If we can row reduce the augmented matrix $[\mathbf{A}|\mathbf{I}]$ to the form $[\mathbf{I}|\mathbf{B}]$, then $\mathbf{A}^{-1} = \mathbf{B}$. Otherwise, the matrix $\mathbf{A}$ does not have an inverse.

EXAMPLE 2   Compute $\mathbf{A}^{-1}$, where $\mathbf{A} = \begin{bmatrix} 4 & -1 & 3 \\ 3 & -1 & 0 \\ -1 & \frac{1}{2} & \frac{1}{2} \end{bmatrix}$.

Solution   We begin with the augmented matrix $[\mathbf{A}|\mathbf{I}]$

$$\left[\begin{array}{ccc|ccc} 4 & -1 & 3 & 1 & 0 & 0 \\ 3 & -1 & 0 & 0 & 1 & 0 \\ -1 & \frac{1}{2} & \frac{1}{2} & 0 & 0 & 1 \end{array}\right]$$

$$\left[\begin{array}{ccc|ccc} -1 & \frac{1}{2} & \frac{1}{2} & 0 & 0 & 1 \\ 3 & -1 & 0 & 0 & 1 & 0 \\ 4 & -1 & 3 & 1 & 0 & 0 \end{array}\right] \quad \text{Switch rows 1 and 3.}$$

$$\left[\begin{array}{ccc|ccc} 1 & -\frac{1}{2} & -\frac{1}{2} & 0 & 0 & -1 \\ 3 & -1 & 0 & 0 & 1 & 0 \\ 4 & -1 & 3 & 1 & 0 & 0 \end{array}\right] \quad \text{Multiply row 1 by } -1.$$

$$\left[\begin{array}{ccc|ccc} 1 & -\frac{1}{2} & -\frac{1}{2} & 0 & 0 & -1 \\ 0 & \frac{1}{2} & \frac{3}{2} & 0 & 1 & 3 \\ 4 & -1 & 3 & 1 & 0 & 0 \end{array}\right] \quad \text{Add } -3 \text{ times row 1 to row 2.}$$

$$\left[\begin{array}{ccc|ccc} 1 & -\frac{1}{2} & -\frac{1}{2} & 0 & 0 & -1 \\ 0 & \frac{1}{2} & \frac{3}{2} & 0 & 1 & 3 \\ 0 & 1 & 5 & 1 & 0 & 4 \end{array}\right] \quad \text{Add } -4 \text{ times row 1 to row 3.}$$

$$\left[\begin{array}{ccc|ccc} 1 & -\frac{1}{2} & -\frac{1}{2} & 0 & 0 & -1 \\ 0 & 1 & 3 & 0 & 2 & 6 \\ 0 & 1 & 5 & 1 & 0 & 4 \end{array}\right] \quad \text{Multiply row 2 by 2.}$$

Technology Connection

Matrix Inverse

Let's find the inverse of the matrix in Example 2 using the TI-83. We first enter

the matrix $\mathbf{A} = \begin{bmatrix} 4 & -1 & 3 \\ 3 & -1 & 0 \\ -1 & \frac{1}{2} & \frac{1}{2} \end{bmatrix}$. We then

use the $\boxed{\mathbf{x}^{-1}}$ key on the calculator as indicated below.

```
[A]⁻¹
[[ -.5   2    3 ]
 [-1.5   5    9 ]
 [  .5  -1   -1 ]]
```

EXERCISES

Compute $\mathbf{A}^{-1}$.

1. $\mathbf{A} = \begin{bmatrix} 2 & 4 & 9 \\ -1 & 6 & 2 \\ -8 & 4 & -1 \end{bmatrix}$

2. $\mathbf{A} = \begin{bmatrix} -5 & 9 & -2 \\ -8 & 0 & 10 \\ -12 & 7 & 5 \end{bmatrix}$

3. $\mathbf{A} = \begin{bmatrix} 3 & 5 & -2 & 7 \\ -10 & 6 & -3 & 4 \\ -2 & 8 & 9 & 11 \\ 3 & 1 & -9 & 6 \end{bmatrix}$

$$\left[\begin{array}{ccc|ccc} 1 & -\frac{1}{2} & -\frac{1}{2} & 0 & 0 & -1 \\ 0 & 1 & 3 & 0 & 2 & 6 \\ 0 & 0 & 2 & 1 & -2 & -2 \end{array}\right]$$

Add −1 times row 2 to row 3.

$$\left[\begin{array}{ccc|ccc} 1 & -\frac{1}{2} & -\frac{1}{2} & 0 & 0 & -1 \\ 0 & 1 & 3 & 0 & 2 & 6 \\ 0 & 0 & 1 & \frac{1}{2} & -1 & -1 \end{array}\right]$$

Multiply row 3 by 1/2.

$$\left[\begin{array}{ccc|ccc} 1 & 0 & 1 & 0 & 1 & 2 \\ 0 & 1 & 3 & 0 & 2 & 6 \\ 0 & 0 & 1 & \frac{1}{2} & -1 & -1 \end{array}\right]$$

Add 1/2 times row 2 to row 1.

$$\left[\begin{array}{ccc|ccc} 1 & 0 & 0 & -\frac{1}{2} & 2 & 3 \\ 0 & 1 & 3 & 0 & 2 & 6 \\ 0 & 0 & 1 & \frac{1}{2} & -1 & -1 \end{array}\right]$$

Add −1 times row 3 to row 1.

$$\left[\begin{array}{ccc|ccc} 1 & 0 & 0 & -\frac{1}{2} & 2 & 3 \\ 0 & 1 & 0 & -\frac{3}{2} & 5 & 9 \\ 0 & 0 & 1 & \frac{1}{2} & -1 & -1 \end{array}\right]$$

Add −3 times row 3 to row 2.

We found that $\mathbf{A}^{-1} = \left[\begin{array}{ccc} -\frac{1}{2} & 2 & 3 \\ -\frac{3}{2} & 5 & 9 \\ \frac{1}{2} & -1 & -1 \end{array}\right]$. The student is encouraged to verify that

$\mathbf{A}\mathbf{A}^{-1} = \mathbf{I}$ and $\mathbf{A}^{-1}\mathbf{A} = \mathbf{I}$. ■

CAUTION   Only a square matrix can have an inverse.

Some matrices, even though square, do not have inverses. If during Gauss–Jordan elimination we get a row with all zeros to the left of the augmentation, then we know the inverse does not exist.

EXAMPLE 3   Find the inverse of $\mathbf{A} = \left[\begin{array}{cc} 1 & 4 \\ 3 & 12 \end{array}\right]$.

Solution   We perform Gauss–Jordan elimination on the augmented matrix [A|I].

$$\left[\begin{array}{cc|cc} 1 & 4 & 1 & 0 \\ 3 & 12 & 0 & 1 \end{array}\right]$$

$$\left[\begin{array}{cc|cc} 1 & 4 & 1 & 0 \\ 0 & 0 & -3 & 1 \end{array}\right]$$

Add −3 times row 1 to row 2.

Since we have a row of zeros to the left of the augmentation, we conclude that **A** does not have an inverse. ■

Inverse Matrices and Linear Systems

In Section 6.2, we developed several techniques for solving a system of linear equations. Another technique involves the inverse of a matrix, as illustrated in the next example.

EXAMPLE 4   Solve the system

$$x + 3y = 14$$
$$2x - y = 49.$$

Solution   The system of linear equations is equivalent to the matrix equation

$$A\begin{bmatrix} x \\ y \end{bmatrix} = \begin{bmatrix} 14 \\ 49 \end{bmatrix}, \quad \text{where } A = \begin{bmatrix} 1 & 3 \\ 2 & -1 \end{bmatrix}.$$

In Example 1, we found that $A^{-1} = \begin{bmatrix} \frac{1}{7} & \frac{3}{7} \\ \frac{2}{7} & -\frac{1}{7} \end{bmatrix}$. Multiplying on the left of both sides of the above equation by A^{-1}, we find

$$A^{-1} A \begin{bmatrix} x \\ y \end{bmatrix} = A^{-1} \begin{bmatrix} 14 \\ 49 \end{bmatrix}$$

$$I \begin{bmatrix} x \\ y \end{bmatrix} = \begin{bmatrix} \frac{1}{7} & \frac{3}{7} \\ \frac{2}{7} & -\frac{1}{7} \end{bmatrix} \begin{bmatrix} 14 \\ 49 \end{bmatrix}$$

$$\begin{bmatrix} x \\ y \end{bmatrix} = \begin{bmatrix} 23 \\ -3 \end{bmatrix}.$$

The student is encouraged to check this answer by either substituting into the original equations or by solving the system using a different method.   ■

Example 4 illustrates how the inverse can be used to solve linear equations.

THEOREM 3

Let A be an $n \times n$ matrix and $\mathbf{v}$ a vector. If A is invertible and $A\mathbf{v} = \mathbf{b}$, then $\mathbf{v} = A^{-1}\mathbf{b}$.

Theorem 3 can be used to solve a matrix equation or a system of linear equations.

EXAMPLE 5   *Cerulean Warblers.* The population of cerulean warblers is grouped into hatchlings (H), second year (S), and after second year (A). The Leslie matrix is[11]

$$G = \begin{bmatrix} 0 & 0.76 & 0.95 \\ 0.2 & 0 & 0 \\ 0 & 0.91 & 0.91 \end{bmatrix}.$$

---

[11]Jones, J., Barg, J. J., Sillett, T. S., Veit, M. L., and Robertson, R. J., "Season-Long Fecundity, Survival, and Viability of Ovenbirds in Fragmented and Unfragmented Landscapes," *Conservation Biology,* Vol. 13, No. 2, pp. 1151–1161 (1999).

Technology Connection

Solving Systems

Theorem 3 provides a convenient way to solve a system of n linear equations in n unknowns on a grapher. Let's do Example 5 with the TI-83.

We first enter the matrix

$$A = \begin{bmatrix} 0 & 0.76 & 0.95 \\ 0.2 & 0 & 0 \\ 0 & 0.91 & 0.91 \end{bmatrix} \text{ and the}$$

population vector $B = \begin{bmatrix} 408 \\ 20 \\ 428 \end{bmatrix}$. We

then multiply $A^{-1}B$.

Due to round-off error, this answer is slightly different from the answer we computed in Example 5. However, they round to the same estimate of $H = 100$, $S = 204$, and $A = 266$.

EXERCISES

Use inverse matrices to solve the systems.

1. $\begin{aligned} 4x + y - 3z &= 3 \\ x - 5y + 7z &= 3 \\ 8x + 2y - 10z &= -6 \end{aligned}$

2. $\begin{aligned} 7x + 6y - 5z &= 0 \\ 4x - 10y + 4z &= 8 \\ 3x + 22y - 15z &= 13 \end{aligned}$

3. $\begin{aligned} 0.18x - 1.3y - 4.7z &= 1.1 \\ 1.3x + 2.1y - 3.5z &= 10.1 \\ -2.6x + 3.6y - 1.4z &= -2.6 \end{aligned}$

The population in year 2 is $H = 408$, $S = 20$, and $A = 428$. Estimate the population vector in year 1.

Solution We wish to solve the matrix equation

$$G \begin{bmatrix} H \\ S \\ A \end{bmatrix} = \begin{bmatrix} 408 \\ 20 \\ 428 \end{bmatrix},$$

so the solution is

$$\begin{bmatrix} H \\ S \\ A \end{bmatrix} = G^{-1} \begin{bmatrix} 408 \\ 20 \\ 428 \end{bmatrix},$$

provided G^{-1} exists. We compute G^{-1}.

$$\left[\begin{array}{ccc|ccc} 0 & 0.76 & 0.95 & 1 & 0 & 0 \\ 0.2 & 0 & 0 & 0 & 1 & 0 \\ 0 & 0.91 & 0.91 & 0 & 0 & 1 \end{array} \right]$$

$$\left[\begin{array}{ccc|ccc} 0.2 & 0 & 0 & 0 & 1 & 0 \\ 0 & 0.76 & 0.95 & 1 & 0 & 0 \\ 0 & 0.91 & 0.91 & 0 & 0 & 1 \end{array} \right] \quad \text{Exchange rows 1 and 2.}$$

$$\left[\begin{array}{ccc|ccc} 1 & 0 & 0 & 0 & 5 & 0 \\ 0 & 0.76 & 0.95 & 1 & 0 & 0 \\ 0 & 0.91 & 0.91 & 0 & 0 & 1 \end{array} \right] \quad \text{Multiply row 1 by 5.}$$

$$\left[\begin{array}{ccc|ccc} 1 & 0 & 0 & 0 & 5 & 0 \\ 0 & 1 & 1.25 & 1.32 & 0 & 0 \\ 0 & 0.91 & 0.91 & 0 & 0 & 1 \end{array} \right] \quad \text{Multiply row 2 by } 1/0.76.$$

$$\left[\begin{array}{ccc|ccc} 1 & 0 & 0 & 0 & 5 & 0 \\ 0 & 1 & 1.25 & 1.32 & 0 & 0 \\ 0 & 0 & -0.228 & -1.20 & 0 & 1 \end{array} \right] \quad \begin{array}{l} \text{Add } -0.91 \text{ times row 2 to} \\ \text{row 3.} \end{array}$$

$$\left[\begin{array}{ccc|ccc} 1 & 0 & 0 & 0 & 5 & 0 \\ 0 & 1 & 1.25 & 1.32 & 0 & 0 \\ 0 & 0 & 1 & 5.26 & 0 & -4.40 \end{array} \right] \quad \begin{array}{l} \text{Multiply row 3 by} \\ 1/(-0.228). \end{array}$$

$$\left[\begin{array}{ccc|ccc} 1 & 0 & 0 & 0 & 5 & 0 \\ 0 & 1 & 0 & -5.26 & 0 & 5.50 \\ 0 & 0 & 1 & 5.26 & 0 & -4.40 \end{array} \right] \quad \begin{array}{l} \text{Add } -1.25 \text{ times row 3 to} \\ \text{row 2.} \end{array}$$

The inverse matrix is $G^{-1} \approx \begin{bmatrix} 0 & 5 & 0 \\ -5.26 & 0 & 5.50 \\ 5.26 & 0 & -4.40 \end{bmatrix}$. The

population vector in year 1 is

$$\begin{bmatrix} H \\ S \\ A \end{bmatrix} = G^{-1} \begin{bmatrix} 408 \\ 20 \\ 428 \end{bmatrix} \approx \begin{bmatrix} 0 & 5 & 0 \\ -5.26 & 0 & 5.50 \\ 5.26 & 0 & -4.340 \end{bmatrix} \begin{bmatrix} 480 \\ 20 \\ 428 \end{bmatrix} \approx \begin{bmatrix} 100 \\ 204 \\ 266 \end{bmatrix}.$$

We estimate that there were 100 hatchlings, 204 second-year warblers, and 266 beyond-second-year warblers in the population in year 1. ■

Determinants

How can we tell if a square matrix is singular or nonsingular? One way is to attempt to compute its inverse using Theorem 2. If we can row reduce the matrix to the identity, then we know that the matrix is nonsingular. If we cannot, then the matrix is not invertible, so it is singular.

The **determinant** of a matrix provides another, and often quicker, way of assessing if a matrix is singular or nonsingular. We denote the determinant of a matrix A as $|A|$ or $\det(A)$.

DEFINITION

The *determinant* of the 1×1 matrix $[a]$ is a. That is, the determinant of a matrix with just one entry is the entry itself.

For a 2×2 matrix, the determinant is

$$\begin{vmatrix} a & b \\ c & d \end{vmatrix} = ad - bc.$$

For a 3×3 matrix $A = \begin{bmatrix} a_{11} & a_{12} & a_{13} \\ a_{21} & a_{22} & a_{23} \\ a_{31} & a_{32} & a_{33} \end{bmatrix}$, the determinant is

$$\det(A) = a_{11}a_{22}a_{33} + a_{12}a_{23}a_{31} + a_{21}a_{32}a_{13}$$
$$-a_{13}a_{22}a_{31} - a_{12}a_{21}a_{33} - a_{11}a_{23}a_{32}.$$

The formula for a 2×2 matrix is not difficult to remember, but the figure below helps us to remember it.

Positive   Negative
term   term

The figures below help us to remember the formula for 3×3 determinants. Notice that the positive terms are all oriented parallel to the main diagonal and the negative terms are not.

Positive terms   Negative terms   or   Positive terms   Negative terms

There are formulas for determinants for larger square matrices, but the formulas are somewhat complicated. We will investigate determinants of larger matrices later.

EXAMPLE 6   Compute the determinants.

a) $\det[-2]$

b) $\begin{vmatrix} 3 & 8 \\ 2 & -3 \end{vmatrix}$

c) $\det A$, where $A = \begin{bmatrix} 4 & 2 & 3 \\ -2 & 7 & 1 \\ 0 & 5 & -3 \end{bmatrix}$

Solution

a) Because the determinant of a 1×1 matrix is its entry, $\det[-2] = -2$.

b) We use the formula for a 2×2 determinant.

$$\begin{vmatrix} 3 & 8 \\ 2 & -3 \end{vmatrix} = (3)(-3) - (8)(2) = -25.$$

c) We use the figure on the previous page to remind us of the formula for a 3×3 determinant.

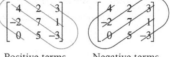

Positive terms   Negative terms

$$\det(A) = \begin{vmatrix} 4 & 2 & 3 \\ -2 & 7 & 1 \\ 0 & 5 & -3 \end{vmatrix}$$

$$= a_{11}a_{22}a_{33} + a_{12}a_{23}a_{31} + a_{21}a_{32}a_{13} - a_{13}a_{22}a_{31} - a_{12}a_{21}a_{33} - a_{11}a_{23}a_{32}$$

$$= (4)(7)(-3) + (2)(1)(0) + (-2)(5)(3) - (3)(7)(0) - (2)(-2)(-3) - (4)(1)(5)$$

$$= -146. \qquad \blacksquare$$

CAUTION   Although the notation $|A|$ looks like absolute value, a determinant can be negative.

The next theorem describes how the determinant can be used to assess whether a matrix is invertible.

THEOREM 4

A square matrix is invertible if and only if its determinant is not zero.

EXAMPLE 7   Determine if the matrix is invertible.

a) $A = \begin{bmatrix} 3 & 7 \\ 9 & 2 \end{bmatrix}$

b) $B = \begin{bmatrix} 2 & -1 & 3 \\ 4 & 4 & 6 \\ 10 & 7 & 15 \end{bmatrix}$

Solution   To determine if a square matrix is invertible, we compute the determinant.

a) Since $\det A = \begin{vmatrix} 3 & 7 \\ 9 & 2 \end{vmatrix} = (3)(2) - (7)(9) = -57 \neq 0$, the matrix **A** is invertible.

However, to actually determine A^{-1}, we need to use Theorem 2.

b) In this case,

$$\det \begin{bmatrix} 2 & -1 & 3 \\ 4 & 4 & 6 \\ 10 & 7 & 15 \end{bmatrix} = (2)(4)(15) + (-1)(6)(10) + (3)(4)(7)$$
$$- (3)(4)(10) - (-1)(4)(15) - (2)(6)(7) = 0.$$

We conclude that the matrix **B** is not invertible. (Notice that this assessment is easier than attempting to directly compute the inverse!)   ■

In Section 6.4, we will need to find nonzero vectors **x** that solve a matrix equation of the form **Ax** = **0**, where **0** is the zero vector. Of course, **x** = **0** is a solution, but we will need to find other solutions. The next theorem tells us when there are nonzero solutions.

THEOREM 5

If **A** is a square matrix and **Ax** = **0** for some vector **x** $\neq$ **0**, then $\det(A) = 0$.

Proof: If **A** is invertible, then by Theorem 3,

$$\mathbf{x} = A^{-1}\mathbf{0} = \mathbf{0}.$$

However, since we are given that **x** $\neq$ **0**, we conclude that **A** cannot be invertible. Therefore, $\det(A) = 0$.   ■

Determinants of Larger Matrices (Optional)

The definition we are using for determinants only covers the cases of 3 × 3 and smaller square matrices. To define the determinant of larger matrices, we first define a **minor** and a **cofactor.**

DEFINITION

Let **A** be a square $n \times n$ matrix. If $1 \leq i \leq n$ and $1 \leq j \leq n$, then the *ijth minor* of **A** is the determinant M_{ij} of the matrix obtained by deleting the *i*th row and the *j*th column of **A**.

The *ijth cofactor* of **A** is $C_{ij} = (-1)^{i+j}M_{ij}$.

EXAMPLE 8 Let $A = \begin{bmatrix} 4 & 2 & 3 \\ -2 & 7 & 1 \\ 0 & 5 & -3 \end{bmatrix}$. Compute the following minors and cofactors.

a) M_{11} and C_{11}

b) M_{12} and C_{12}

c) M_{13} and C_{13}

Solution

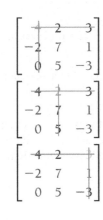

a) We compute the determinant of the 2×2 matrix obtained by deleting row 1 and column 1 from A, as illustrated to the left.

$$M_{11} = \begin{vmatrix} 7 & 1 \\ 5 & -3 \end{vmatrix} = (7)(-3) - (1)(5) = -26.$$

We next compute C_{11}.

$$C_{11} = (-1)^{1+1}M_{11} = (1)(-26) = -26.$$

b) For M_{12} we delete row 1 and column 2 and find the determinant of the resulting 2×2 matrix.

$$M_{12} = \begin{vmatrix} -2 & 1 \\ 0 & -3 \end{vmatrix} = (-2)(-3) - (1)(0) = 6$$

$$C_{12} = (-1)^{1+2}M_{12} = (-1)(6) = -6.$$

c) We delete row 1 and column 3 to compute M_{13}.

$$M_{13} = \begin{vmatrix} -2 & 7 \\ 0 & 5 \end{vmatrix} = (-2)(5) - (7)(0) = -10$$

$$C_{13} = (-1)^{1+3}M_{13} = (1)(-10) = -10.$$ ■

We can define the determinant of square matrices with $n \geq 2$ rows using cofactors.

DEFINITION

Cofactor Definition of Determinant

Let $A = [a_{ij}]$ be a square matrix with $n \geq 2$ rows. Then the determinant of A is given by

$$\det(A) = a_{11}C_{11} + a_{12}C_{12} + a_{13}C_{13} + \cdots + a_{1n}C_{1n}.$$

That is, the determinant is the sum of each entry in the first row multiplied by its cofactor.

The figure below illustrates the cofactor definition of the determinant for a 5×5 matrix. The signs indicate which sign is used in the definition of cofactor. The blue square indicates the entry of the matrix that is multiplied by the red cofactor.

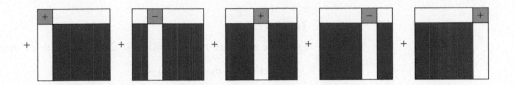

For a 2×2 matrix $A = \begin{bmatrix} a & b \\ c & d \end{bmatrix}$, we have $C_{11} = (-1)^{1+1}d = d$ and $C_{12} = (-1)^{1+2}c = -c$. The cofactor definition of determinant says

$$\det(A) = aC_{11} + bC_{12} = (a)(d) + (b)(-c) = ad - bc.$$

This verifies that the definition involving cofactors is consistent with the earlier formula we used for 2×2 matrices. We will verify in the exercises that the cofactor definition is the same as the first definition we studied for 3×3 matrices.

EXAMPLE 9 Let $A = \begin{bmatrix} 4 & 2 & 3 \\ -2 & 7 & 1 \\ 0 & 5 & -3 \end{bmatrix}$. Compute $\det(A)$ using the cofactor definition and check that the answer agrees with the formula for a 3×3 determinant.

Solution   The cofactor definition of determinant says

$$\det(A) = a_{11}C_{11} + a_{12}C_{12} + a_{13}C_{13}.$$

In Example 8, we found these cofactors to be $C_{11} = -26, C_{12} = -6$, and $C_{13} = -10$. Substituting these values gives

$$\det \begin{bmatrix} 4 & 2 & 3 \\ -2 & 7 & 1 \\ 0 & 5 & -3 \end{bmatrix} = (4)(-26) + (2)(-6) + (3)(-10) = -146.$$

This agrees with the answer found in part (c) of Example 6, confirming that the cofactor definition gives the same answer as the formula for this 3×3 matrix. ∎

It turns out that if we use *any* row or column in the cofactor definition of determinant, we get the same answer. For example, let's use the second column instead of the first row on the matrix $A = \begin{bmatrix} a & b \\ c & d \end{bmatrix}$:

$$\begin{aligned} \det(A) &= bC_{12} + dC_{22} \\ &= (b)(-1)^{1+2}c + (d)(-1)^{2+2}(a) \\ &= -bc + da \\ &= ad - bc. \end{aligned}$$

We get the determinant of A.

THEOREM 6

Let A be a square matrix with $n \geq 2$ rows. Then for any $1 \leq k \leq n$,

$$\det(A) = a_{k1}C_{k1} + a_{k2}C_{k2}$$
$$+ a_{k3}C_{k3} + \cdots + a_{kn}C_{kn}$$

and

$$\det(A) = a_{1k}C_{1k} + a_{2k}C_{2k}$$
$$+ a_{3k}C_{3k} + \cdots + a_{nk}C_{nk}.$$

That is, we can compute the determinant using cofactors along any row or any column.

Technology Connection

Computing Determinants

Let's do Example 10 using a grapher. We first enter the matrix

$$A = \begin{bmatrix} 1 & 0 & 2 & 1 \\ 2 & 0 & 0 & -1 \\ 0 & 0 & 1 & -1 \\ 2 & -3 & 0 & 1 \end{bmatrix}. \text{ We press the}$$

 MATRX key and select DET(under the MATH menu.

EXERCISES

Compute the determinants using a grapher.

1. $\det \begin{bmatrix} 4 & -2 & 0 \\ -10 & 5 & 2 \\ 5 & 9 & 4 \end{bmatrix}$

2. $\det \begin{bmatrix} 5 & 9 & 1 & -7 \\ 4 & 10 & -3 & -1 \\ 4 & 6 & -9 & -8 \\ 2 & -1 & 5 & 3 \end{bmatrix}$

EXAMPLE 10 Let $A = \begin{bmatrix} 1 & 0 & 2 & 1 \\ 2 & 0 & 0 & -1 \\ 0 & 0 & 1 & -1 \\ 2 & -3 & 0 & 1 \end{bmatrix}$. Compute $\det(A)$.

Solution We may expand along any row or any column. We use the second column since it has three zeros.

$$\det(A) = \det \begin{bmatrix} 1 & 0 & 2 & 1 \\ 2 & 0 & 0 & -1 \\ 0 & 0 & 1 & -1 \\ 2 & -3 & 0 & 1 \end{bmatrix}$$

$$= (-1)^{1+2}(0) \cdot \det \begin{bmatrix} 2 & 0 & -1 \\ 0 & 1 & -1 \\ 2 & 0 & 1 \end{bmatrix}$$

$$+ (-1)^{2+2}(0) \cdot \det \begin{bmatrix} 1 & 2 & 1 \\ 0 & 1 & -1 \\ 2 & 0 & 1 \end{bmatrix}$$

$$+ (-1)^{3+2}(0) \cdot \det \begin{bmatrix} 1 & 2 & 1 \\ 2 & 0 & -1 \\ 2 & 0 & 1 \end{bmatrix}$$

$$+ (-1)^{4+2}(-3) \cdot \det \begin{bmatrix} 1 & 2 & 1 \\ 2 & 0 & -1 \\ 0 & 1 & -1 \end{bmatrix}$$

$$= 0 + 0 + 0 + (1)(-3)(7)$$

$$= -21.$$

Exercise Set 6.3

Compute the inverse matrix.

1. $\begin{bmatrix} 1 & 1 \\ -1 & 0 \end{bmatrix}$

2. $\begin{bmatrix} 2 & 3 \\ 3 & 5 \end{bmatrix}$

3. $\begin{bmatrix} 0 & 1 \\ 1 & 0 \end{bmatrix}$

4. $\begin{bmatrix} -1 & 0 \\ 0 & 2 \end{bmatrix}$

5. $\begin{bmatrix} 3 & 4 \\ 5 & 7 \end{bmatrix}$

6. $\begin{bmatrix} 2 & 1 \\ -4 & 7 \end{bmatrix}$

7. $\begin{bmatrix} 3 & 7 \\ 8 & -2 \end{bmatrix}$

8. $\begin{bmatrix} 4 & -8 \\ -1 & 5 \end{bmatrix}$

9. $\begin{bmatrix} -2 & 2 & 1 \\ 1 & 2 & 0 \\ 0 & -1 & 0 \end{bmatrix}$

10. $\begin{bmatrix} 0 & 1 & 2 \\ 0 & 0 & -1 \\ 1 & 2 & 6 \end{bmatrix}$

11. $\begin{bmatrix} 0 & 0 & -1 \\ 1 & 2 & 6 \\ 0 & 1 & 2 \end{bmatrix}$

12. $\begin{bmatrix} 0 & -1 & 0 \\ -2 & 2 & 1 \\ 1 & 2 & 0 \end{bmatrix}$

13. $\begin{bmatrix} 0 & -1 & -3 \\ 1 & 5 & 16 \\ 1 & 2 & 8 \end{bmatrix}$

14. $\begin{bmatrix} 2 & -3 & 6 \\ 7 & 10 & 7 \\ -8 & -7 & -11 \end{bmatrix}$

15. $\begin{bmatrix} -5 & -7 & -5 \\ -1 & 0 & -2 \\ 9 & 10 & 11 \end{bmatrix}$

16. $\begin{bmatrix} 1 & 2 & 1 \\ -5 & -6 & -7 \\ \frac{9}{2} & \frac{9}{2} & \frac{13}{2} \end{bmatrix}$

17. $\begin{bmatrix} 7 & -1 & -9 \\ 2 & 0 & -4 \\ -4 & 0 & 6 \end{bmatrix}$

18. $\begin{bmatrix} 1 & -1 & -1 \\ 5 & -1 & -7 \\ 4 & -2 & -6 \end{bmatrix}$

19. $\begin{bmatrix} -4.5 & -4.5 & -4 \\ 8.5 & 7.5 & 8 \\ -2 & -1 & -2 \end{bmatrix}$

20. $\begin{bmatrix} -14 & -20 & -27 \\ 5 & 7 & 10 \\ 7 & 10 & 13 \end{bmatrix}$

21. $\begin{bmatrix} 11 & -4 & -3 \\ 8 & -1 & -3 \\ 26 & -8 & -8 \end{bmatrix}$

22. $\begin{bmatrix} 17 & -4 & -6 \\ 14 & -1 & -6 \\ 44 & -8 & -17 \end{bmatrix}$

23. $\begin{bmatrix} 1 & 8 & 28 & -1 \\ -2 & 9 & 26 & 1 \\ 0 & -3 & -10 & 0 \\ 1 & -1 & -1 & 0 \end{bmatrix}$

24. $\begin{bmatrix} 1 & 1 & 6 & 0 \\ 1 & 2 & 8 & -1 \\ -2 & 0 & -4 & 1 \\ 1 & -1 & -1 & 0 \end{bmatrix}$

Compute the determinant of each matrix. Determine if the matrix is invertible without computing the inverse.

25. $\begin{bmatrix} 2 & 1 \\ 3 & 6 \end{bmatrix}$

26. $\begin{bmatrix} 4 & -2 \\ -4 & 3 \end{bmatrix}$

27. $\begin{bmatrix} 9 & 2 \\ -5 & 3 \end{bmatrix}$

28. $\begin{bmatrix} 7 & 2 \\ 8 & -9 \end{bmatrix}$

29. $\begin{bmatrix} 2 & 3 \\ 6 & 9 \end{bmatrix}$

30. $\begin{bmatrix} -12 & -8 \\ 3 & 2 \end{bmatrix}$

31. $\begin{bmatrix} 1 & 8 & 3 \\ -2 & 5 & -3 \\ 7 & -4 & 2 \end{bmatrix}$

32. $\begin{bmatrix} 3 & 1 & 6 \\ -5 & 0 & -2 \\ 4 & 6 & -1 \end{bmatrix}$

33. $\begin{bmatrix} 1 & 1 & 1 \\ 2 & 2 & 1 \\ 3 & 3 & 1 \end{bmatrix}$

34. $\begin{bmatrix} -1 & 4 & 0 \\ 0 & 3 & -2 \\ 5 & 8 & 7 \end{bmatrix}$

35. $\begin{bmatrix} 0 & 4 & 2 \\ -1 & 6 & 2 \\ -1 & 14 & 6 \end{bmatrix}$

36. $\begin{bmatrix} 2 & 4 & 2 \\ -9 & 6 & 3 \\ -3 & -5 & 8 \end{bmatrix}$

37. $\begin{bmatrix} 2 & -4 & 8 \\ -3 & -5 & 7 \\ -4 & -1 & 5 \end{bmatrix}$

38. $\begin{bmatrix} 4 & 0 & 4 \\ 5 & 2 & 8 \\ 1 & 2 & 4 \end{bmatrix}$

39. $\begin{bmatrix} 3 & 5 & 2 & 1 \\ 0 & 1 & 0 & 5 \\ 2 & 0 & 1 & -1 \\ 2 & -1 & 3 & 0 \end{bmatrix}$

40. $\begin{bmatrix} 4 & 0 & 2 & 1 \\ 0 & 2 & -2 & 3 \\ 5 & 0 & 6 & 4 \\ 2 & -1 & 1 & 6 \end{bmatrix}$

41. $\begin{bmatrix} 0 & 1 & 0 & -1 & 0 \\ -1 & 0 & 2 & 0 & 2 \\ 0 & 1 & 2 & 0 & 3 \\ 1 & 0 & 1 & 0 & 1 \\ 2 & 3 & 0 & 0 & -1 \end{bmatrix}$

42. $\begin{bmatrix} 3 & 1 & 0 & 0 & 0 \\ 1 & 0 & 2 & -1 & 3 \\ 0 & 0 & -1 & 1 & 1 \\ 4 & -2 & 0 & 0 & 0 \\ 0 & -2 & 0 & 1 & 0 \end{bmatrix}$

APPLICATIONS

43. *Bird Population.* The Leslie matrix for a bird population of hatchlings and adults is

$\begin{bmatrix} 0.5 & 1.25 \\ 0.75 & 0.25 \end{bmatrix}$. In year 2 the population vector is

$\begin{bmatrix} 156 \\ 48 \end{bmatrix}$, meaning there are 156 hatchlings and

48 adults. Use the inverse matrix to estimate the population vector in year 1.

44. *Cerulean Warblers.* The population of cerulean warblers is grouped into hatchlings (H), second year (S), and after second year (A). The Leslie matrix for one location is given below, showing estimates of the survivability and fecundity.[12] The population in year 3 is $H = 206$, $S = 55$, and $A = 211$.

$$G = \begin{bmatrix} 0 & 0.75 & 0.95 \\ 0.3 & 0 & 0 \\ 0 & 0.9 & 0.95 \end{bmatrix}$$

a) Find the inverse of the Leslie matrix.
b) Estimate the population of hatchlings, second years, and third-year adults in year 2.
c) Estimate the population of hatchlings, second years, and third-year adults in year 1.

45. *Dinophilus Gyrociliatus. Dinophilus gyrociliatus* is a small species that lives in the fouling community of harbor environments. The female population is grouped into those 6 wk old or less and those that are more than 6 wk old. The Leslie matrix is given below.[13]

$$G = \begin{bmatrix} 15 & 7.5 \\ 0.8 & 0 \end{bmatrix}$$

During the third 6-wk period, the population of group 1 is 58,815 and the population of group 2 is 3060.

a) Find the inverse of the Leslie matrix.
b) Estimate the population of the two groups during the second 6 wk.
c) Estimate the population of the two groups during the first 6 wk.

46. *Fairy Shrimp.* San Diego fairy shrimp live in ponds that fill and dry many times during a year. While the ponds are dry, the fairy shrimp survive as cysts. The population is divided into three groups. Cysts that survive one dry period are in group 1, cysts that survive two dry periods are in group 2 and cysts that survive three or more dry periods are in group 3. The Leslie matrix representing the survivability and fecundity is given below.[14]

$$G = \begin{bmatrix} 3.6 & 0.98 & 0.65 \\ 0.5 & 0 & 0 \\ 0 & 0.5 & 0.49 \end{bmatrix}$$

In the third dry period there are 8365, 1095, and 310 individuals in groups 1, 2, and 3, respectively.

a) Find the inverse of the Leslie matrix.
b) Estimate the population of each group during the second dry spell.
c) Estimate the population of each group during the first dry spell.

SYNTHESIS

47. Let $A = \begin{bmatrix} a_{11} & a_{12} & a_{13} \\ a_{21} & a_{22} & a_{23} \\ a_{31} & a_{32} & a_{33} \end{bmatrix}$. Show that the cofactor

definition of determinant is the same as the formula for a 3×3 determinant given on page 463.

In Exercises 48 and 49, we investigate inverses and determinants of matrix products.

48. *Inverse of Products.* Show that $(AB)^{-1} = B^{-1}A^{-1}$ by checking that $B^{-1}A^{-1}$ satisfies the definition for an inverse of AB.

49. *Determinant of Products.* In this exercise, we verify that $\det(AB) = \det(A)\det(B)$ by looking at examples.

a) Let $A = \begin{bmatrix} 2 & 5 \\ 4 & 7 \end{bmatrix}$ and $B = \begin{bmatrix} 1 & 5 \\ -2 & 3 \end{bmatrix}$. Verify that

$\det(AB) = \det(A)\det(B)$.

b) Let $A = \begin{bmatrix} 2 & -4 & 6 \\ 1 & 0 & -1 \\ 4 & -2 & 3 \end{bmatrix}$ and

$B = \begin{bmatrix} 2 & -7 & 2 \\ 5 & 3 & -1 \\ 4 & 0 & 2 \end{bmatrix}$. Verify that

$\det(AB) = \det(A)\det(B)$.

[12]Jones, J., Barg, J.J., Sillett, T.S., Veit, M.L., and Robertson, R.J., "Season-Long Fecundity, Survival, and Viability of Ovenbirds in Fragmented and Unfragmented Landscapes," *Conservation Biology*, Vol. 13, No. 2, pp. 1151–1161 (1999).

[13]D. Prevedelli and R. Zunarelli Vandini, "Survival, fecundity and sex ratio of *Dinophilus gyrociliatus* (Polychaeta: Dinophilidae) under different dietary conditions," *Marine Biology*, Vol. 133, pp. 231–236 (1999).

[14]B.J. Ripley, J. Holtz, and M. Simovich, "Cyst bank life-history model for a fairy shrimp from ephemeral ponds," *Freshwater Biology*, Vol. 49, pp. 221–231 (2004).

c) Pick another pair of square matrices **A** and **B**. Verify that det(**AB**) = det(**A**) det(**B**).

50. *Upper Triangular Matrices.* A matrix is *upper triangular* if every entry below the main diagonal is 0. Let $A = |a_{ij}|$ be an upper triangular $n \times n$ matrix. Use Theorem 6 to show det(A) = $a_{11}a_{22}a_{33}\cdots a_{nn}$, the product of the diagonal entries.

Technology Connection

51.–54. Do Exercises 21–24 using a grapher.

55.–58. Do Exercises 39–42 using a grapher.

6.4 Computing Eigenvalues and Eigenvectors

OBJECTIVES

- Multiply matrices and vectors.
- Compute eigenvalues and eigenvectors.
- Compute growth rate.

We have seen that a Leslie matrix can be used to predict future populations. However, these calculations are time-consuming and they often do not shed much light on the population growth pattern. The population growth rate may be quantified by computing the *eigenvalues* and *eigenvectors* of the Leslie matrix. In this section we will learn how to compute eigenvalues and eigenvectors and how to interpret them to predict the long-term population growth pattern.

Vectors

Recall that a vector is a column matrix. In terms of its size, a vector is an $n \times 1$ matrix. In this case, we say the vector is an n-vector. Suppose that **A** is an $n \times n$ matrix and **v** is an n-vector. Then **Av** is another n-vector.

EXAMPLE 1   Let $A = \begin{bmatrix} -9 & 6 & 20 \\ 2 & 2 & -4 \\ -6 & 3 & 13 \end{bmatrix}$, $u = \begin{bmatrix} 2 \\ -6 \\ 5 \end{bmatrix}$, and $v = \begin{bmatrix} 6 \\ 1 \\ 3 \end{bmatrix}$. Compute

a) **Au**.

b) **Av**.

Solution

a) $Au = \begin{bmatrix} -9 & 6 & 20 \\ 2 & 2 & -4 \\ -6 & 3 & 13 \end{bmatrix}\begin{bmatrix} 2 \\ -6 \\ 5 \end{bmatrix} = \begin{bmatrix} (-9)(2) + (6)(-6) + (20)(5) \\ (2)(2) + (2)(-6) + (-4)(5) \\ (-6)(2) + (3)(-6) + (13)(5) \end{bmatrix} = \begin{bmatrix} 46 \\ -28 \\ 35 \end{bmatrix}$.

b) $Av = \begin{bmatrix} -9 & 6 & 20 \\ 2 & 2 & -4 \\ -6 & 3 & 13 \end{bmatrix}\begin{bmatrix} 6 \\ 1 \\ 3 \end{bmatrix} = \begin{bmatrix} (-9)(6) + (6)(1) + (20)(3) \\ (2)(6) + (2)(1) + (-4)(3) \\ (-6)(6) + (3)(1) + (13)(3) \end{bmatrix} = \begin{bmatrix} 12 \\ 2 \\ 6 \end{bmatrix}$.   ■

Eigenvectors and Eigenvalues

Let's take a closer look at Example 1. The vector **Au** doesn't look much like the vector **u**. Typically, it is difficult to see a relationship between a vector **x** and the product **Ax**. However, this is not the case with the vector **v**. We can see that **Av** = 2**v**. If the product of a matrix and a vector is a multiple of the vector, then we call the vector an *eigenvector*.

DEFINITION
Suppose that A is an $n \times n$ matrix and v is an n-vector with at least one entry not 0. If there is a number r such that $Av = rv$, then we say that v is an **eigenvector** with **eigenvalue** r.

In Example 1, we saw that

$$\begin{bmatrix} -9 & 6 & 20 \\ 2 & 2 & -4 \\ -6 & 3 & 13 \end{bmatrix} \begin{bmatrix} 6 \\ 1 \\ 3 \end{bmatrix} = \begin{bmatrix} 12 \\ 2 \\ 6 \end{bmatrix} = 2 \begin{bmatrix} 6 \\ 1 \\ 3 \end{bmatrix}.$$

We say that the vector $\begin{bmatrix} 6 \\ 1 \\ 3 \end{bmatrix}$ is an eigenvector for the matrix $\begin{bmatrix} -9 & 6 & 20 \\ 2 & 2 & -4 \\ -6 & 3 & 13 \end{bmatrix}$ with eigenvalue 2.

Determining Eigenvalues

How do we find eigenvectors and eigenvalues? The following theorem provides a polynomial whose roots are the eigenvalues. By finding the roots of this polynomial, we can find all of the eigenvalues.

THEOREM 7
Let A be an $n \times n$ square matrix. If the number r is an eigenvalue for A, then $\det(A - rI) = 0$.

Proof: Suppose that r is an eigenvalue for A. Then for some nonzero vector v, $Av = rv$. We can rewrite this equation as

$$Av = rv$$
$$Av = rIv \qquad \text{I is the identity matrix.}$$
$$Av - rIv = 0$$
$$(A - rI)v = 0 \qquad \text{Distributive property}$$

Since the eigenvector v is not zero, and $A - rI$ is a square matrix, we conclude that $\det(A - rI) = 0$ by Theorem 5. ∎

Thinking of r as a variable, $\det(A - rI)$ is a polynomial of degree n called the **characteristic polynomial.** The **characteristic equation** is the equation $\det(A - rI) = 0$.

EXAMPLE 2 Find all eigenvalues for the matrix $A = \begin{bmatrix} -7 & 18 \\ -3 & 8 \end{bmatrix}$.

Solution   According to Theorem 7, we need to solve the characteristic equation $\det(A - rI) = 0$.

$$A - rI = \begin{bmatrix} -7 & 18 \\ -3 & 8 \end{bmatrix} - \begin{bmatrix} r & 0 \\ 0 & r \end{bmatrix} = \begin{bmatrix} -7 - r & 18 \\ -3 & 8 - r \end{bmatrix}$$

$$\begin{aligned} \det(A - rI) &= \begin{vmatrix} -7 - r & 18 \\ -3 & 8 - r \end{vmatrix} \\ &= (-7 - r)(8 - r) - (18)(-3) \\ &= -56 - r + r^2 + 54 \\ &= r^2 - r - 2 \\ &= (r - 2)(r + 1) \end{aligned}$$

Solving $(r - 2)(r + 1) = 0$ yields the two eigenvalues $r = 2$ and $r = -1$.   ∎

Determining Eigenvectors

Once we know an eigenvalue r, we can find the corresponding eigenvectors by solving the matrix equation

$$Av = rv$$

for **v**. This equation is equivalent to the equation

$$(A - rI)v = 0.$$

THEOREM 8

The eigenvectors **v** associated with the eigenvalue r are the nonzero solutions to the matrix equation

$$(A - rI)v = 0.$$

EXAMPLE 3   Find all the eigenvectors corresponding to each eigenvalue of the matrix $A = \begin{bmatrix} -7 & 18 \\ -3 & 8 \end{bmatrix}$.

Solution   In Example 2, we found the eigenvalues to be $r = 2$ and $r = -1$.

1. We start with $r = 2$ and seek vectors $v = \begin{bmatrix} x \\ y \end{bmatrix}$ so that $Av = 2v$. By Theorem 8, the eigenvectors are nonzero solutions to the matrix equation $(A - 2I)v = 0$. The matrix equation is

$$\left(\begin{bmatrix} -7 & 18 \\ -3 & 8 \end{bmatrix} - 2 \begin{bmatrix} 1 & 0 \\ 0 & 1 \end{bmatrix} \right) \begin{bmatrix} x \\ y \end{bmatrix} = \begin{bmatrix} 0 \\ 0 \end{bmatrix}$$

$$\begin{bmatrix} -7 - 2 & 18 - 0 \\ -3 - 0 & 8 - 2 \end{bmatrix} \begin{bmatrix} x \\ y \end{bmatrix} = \begin{bmatrix} 0 \\ 0 \end{bmatrix}$$

$$\begin{bmatrix} -9 & 18 \\ -3 & 6 \end{bmatrix} \begin{bmatrix} x \\ y \end{bmatrix} = \begin{bmatrix} 0 \\ 0 \end{bmatrix}.$$

We write this as an augmented matrix and perform Gaussian elimination:

$$\begin{bmatrix} -9 & 18 & | & 0 \\ -3 & 6 & | & 0 \end{bmatrix}$$

$$\begin{bmatrix} 1 & -2 & | & 0 \\ -3 & 6 & | & 0 \end{bmatrix} \qquad \text{Multiply row 1 by } -1/9.$$

$$\begin{bmatrix} 1 & -2 & | & 0 \\ 0 & 0 & | & 0 \end{bmatrix} \qquad \text{Add 3 times row 1 to row 2.}$$

The first row says that $x = 2y$. Therefore, all eigenvectors corresponding to the eigenvalue 2 are of the form $\begin{bmatrix} 2y \\ y \end{bmatrix} = y \begin{bmatrix} 2 \\ 1 \end{bmatrix}$, where $y \neq 0$. We cannot allow $y = 0$ since the zero vector is not considered an eigenvector. The eigenvectors are the nonzero multiples of the vector $\begin{bmatrix} 2 \\ 1 \end{bmatrix}$.

2. We next find an eigenvector with eigenvalue -1. By Theorem 8, we can find the solution by solving the matrix equation

$$\left(\begin{bmatrix} -7 & 18 \\ -3 & 8 \end{bmatrix} - (-1) \begin{bmatrix} 1 & 0 \\ 0 & 1 \end{bmatrix} \right) \begin{bmatrix} x \\ y \end{bmatrix} = \begin{bmatrix} 0 \\ 0 \end{bmatrix}$$

$$\begin{bmatrix} -6 & 18 \\ -3 & 9 \end{bmatrix} \begin{bmatrix} x \\ y \end{bmatrix} = \begin{bmatrix} 0 \\ 0 \end{bmatrix}.$$

We write this as an augmented matrix and perform Gaussian elimination:

$$\begin{bmatrix} -6 & 18 & | & 0 \\ -3 & 9 & | & 0 \end{bmatrix}$$

$$\begin{bmatrix} 1 & -3 & | & 0 \\ -3 & 9 & | & 0 \end{bmatrix} \qquad \text{Multiply row 1 by } -1/6.$$

$$\begin{bmatrix} 1 & -3 & | & 0 \\ 0 & 0 & | & 0 \end{bmatrix} \qquad \text{Add 3 times row 1 to row 2.}$$

The first row says $x = 3y$. Therefore, all eigenvectors for the eigenvalue -1 are of the form $\begin{bmatrix} 3y \\ y \end{bmatrix} = y \begin{bmatrix} 3 \\ 1 \end{bmatrix}$, where $y \neq 0$. The eigenvectors are the nonzero multiples of the vector $\begin{bmatrix} 3 \\ 1 \end{bmatrix}$.   ∎

Often we can express the eigenvectors for a given eigenvalue as nonzero multiples of a fixed vector. In Example 3, we expressed the eigenvectors corresponding to the eigenvalue $r = -1$ as nonzero multiples of the vector $\begin{bmatrix} 3 \\ 1 \end{bmatrix}$. This is the same as the nonzero multiples of the vector $\begin{bmatrix} 1 \\ \frac{1}{3} \end{bmatrix}$, since each is a nonzero multiple of the other. When we compare our answer with answers that others obtain, we must keep this in mind.

| CAUTION | The multiples of the vector **v** are the same as the multiples of the vector x**v** for any $x \neq 0$. |

EXAMPLE 4   Let $A = \begin{bmatrix} 1 & -2 & -2 \\ 4 & -5 & -2 \\ 8 & -4 & 5 \end{bmatrix}$.

a) Find all the eigenvalues for **A**.

b) Find the eigenvectors associated with each eigenvalue.

Solution

a) We find the characteristic polynomial by computing $\det(A - rI)$.

$$\det(A - rI) = \det \begin{bmatrix} 1-r & -2 & -2 \\ 4 & -5-r & -2 \\ 8 & -4 & 5-r \end{bmatrix}$$

$$= (1-r)(-5-r)(5-r) + 4(-4)(-2) + 8(-2)(-2)$$
$$\quad - (-2)(-5-r)(8) - (-2)(4)(5-r) - (1-r)(-2)(-4)$$
$$= -(1-r)(25-r^2) + 32 + 32 + 16(-5-r) + 8(5-r)$$
$$\quad - 8(1-r)$$
$$= -25 + 25r + r^2 - r^3 + 64 - 80 - 16r + 40 - 8r - 8 + 8r$$
$$= -9 + 9r + r^2 - r^3$$

We set the characteristic polynomial equal to zero and solve for r.

$$-9 + 9r + r^2 - r^3 = 0$$
$$r^3 - r^2 - 9r + 9 = 0$$
$$r^2(r - 1) - 9(r - 1) = 0$$
$$(r^2 - 9)(r - 1) = 0$$
$$(r + 3)(r - 3)(r - 1) = 0$$

b) The eigenvalues are -3, 3, and 1.

1. To find the eigenvectors with eigenvalue -3, we solve the matrix equation $(A - (-3)I)V = 0$.

$$\begin{bmatrix} 1-(-3) & -2 & -2 \\ 4 & -5-(-3) & -2 \\ 8 & -4 & 5-(-3) \end{bmatrix} \begin{bmatrix} x \\ y \\ z \end{bmatrix} = \begin{bmatrix} 0 \\ 0 \\ 0 \end{bmatrix} \quad \text{or}$$

$$\begin{bmatrix} 4 & -2 & -2 \\ 4 & -2 & -2 \\ 8 & -4 & 8 \end{bmatrix} \begin{bmatrix} x \\ y \\ z \end{bmatrix} = \begin{bmatrix} 0 \\ 0 \\ 0 \end{bmatrix}.$$

We now solve using row reduction.

$$\begin{bmatrix} 4 & -2 & -2 & | & 0 \\ 4 & -2 & -2 & | & 0 \\ 8 & -4 & 8 & | & 0 \end{bmatrix}$$

$$\begin{bmatrix} 0 & 0 & 0 & | & 0 \\ 4 & -2 & -2 & | & 0 \\ 8 & -4 & 8 & | & 0 \end{bmatrix} \quad \text{Add } -1 \text{ times row 2 to row 1.}$$

$$\begin{bmatrix} 0 & 0 & 0 & | & 0 \\ 2 & -1 & -1 & | & 0 \\ 2 & -1 & 2 & | & 0 \end{bmatrix} \quad \begin{matrix} \text{Multiply row 2 by 1/2.} \\ \text{Multiply row 3 by 1/4.} \end{matrix}$$

$$\begin{bmatrix} 0 & 0 & 0 & | & 0 \\ 2 & -1 & -1 & | & 0 \\ 0 & 0 & 3 & | & 0 \end{bmatrix} \quad \text{Add } -1 \text{ times row 2 to row 3.}$$

$$\begin{bmatrix} 2 & -1 & -1 & | & 0 \\ 0 & 0 & 3 & | & 0 \\ 0 & 0 & 0 & | & 0 \end{bmatrix} \quad \text{Exchange order of rows.}$$

$$\begin{bmatrix} 1 & -\frac{1}{2} & -\frac{1}{2} & | & 0 \\ 0 & 0 & 1 & | & 0 \\ 0 & 0 & 0 & | & 0 \end{bmatrix} \quad \begin{matrix} \text{Multiply row 1 by 1/2.} \\ \text{Multiply row 2 by 1/3.} \end{matrix}$$

Row 2 says that $z = 0$. Back-substituting $z = 0$ into the equation obtained from row 1, we have

$$x - \frac{1}{2}y - \frac{1}{2}(0) = 0$$

$$y = 2x.$$

The eigenvectors with eigenvalue -3 have the form $\begin{bmatrix} x \\ 2x \\ 0 \end{bmatrix} = x\begin{bmatrix} 1 \\ 2 \\ 0 \end{bmatrix}$, where $x \neq 0$.

2. We next find the eigenvectors with eigenvalue 3. We solve the matrix equation $(A - 3I)v = 0$.

$$\begin{bmatrix} 1-3 & -2 & -2 \\ 4 & -5-3 & 2 \\ 8 & -4 & 5-3 \end{bmatrix} \begin{bmatrix} x \\ y \\ z \end{bmatrix} = \begin{bmatrix} 0 \\ 0 \\ 0 \end{bmatrix}$$

$$\begin{bmatrix} -2 & -2 & -2 \\ 4 & -8 & 2 \\ 8 & -4 & 2 \end{bmatrix} \begin{bmatrix} x \\ y \\ z \end{bmatrix} = \begin{bmatrix} 0 \\ 0 \\ 0 \end{bmatrix}.$$

We write the corresponding augmented matrix and row reduce.

$$\left[\begin{array}{ccc|c} -2 & -2 & -2 & 0 \\ 4 & -8 & -2 & 0 \\ 8 & -4 & 2 & 0 \end{array}\right]$$

$$\left[\begin{array}{ccc|c} 1 & 1 & 1 & 0 \\ 2 & -4 & -1 & 0 \\ 4 & -2 & 1 & 0 \end{array}\right]$$
Multiply row 1 by $-1/2$.
Multiply row 2 by $1/2$.
Multiply row 3 by $1/2$.

$$\left[\begin{array}{ccc|c} 1 & 1 & 1 & 0 \\ 0 & -6 & -3 & 0 \\ 0 & -6 & -3 & 0 \end{array}\right]$$
Add -2 times row 1 to row 2.
Add -4 times row 1 to row 3.

$$\left[\begin{array}{ccc|c} 1 & 1 & 1 & 0 \\ 0 & 1 & \frac{1}{2} & 0 \\ 0 & 1 & \frac{1}{2} & 0 \end{array}\right]$$
Multiply row 2 by $-1/6$.
Multiply row 3 by $-1/6$.

$$\left[\begin{array}{ccc|c} 1 & 1 & 1 & 0 \\ 0 & 1 & \frac{1}{2} & 0 \\ 0 & 0 & 0 & 0 \end{array}\right]$$
Add -1 times row 2 to row 3.

Row 2 says that $y + \frac{1}{2}z = 0$, or $z = -2y$. Back-substituting z into the equation for row 1 gives

$$x + y + -2y = 0$$

$$x = y.$$

The eigenvectors with eigenvalue 3 have the form $\begin{bmatrix} y \\ y \\ -2y \end{bmatrix} = y \begin{bmatrix} 1 \\ 1 \\ -2 \end{bmatrix}$, where $y \neq 0$.

3. Finally, we compute the eigenvectors with eigenvalue 1. We solve the matrix equation $(A - 1I)v = 0$.

$$\begin{bmatrix} 1-1 & -2 & -2 \\ 4 & -5-1 & -2 \\ 8 & -4 & 5-1 \end{bmatrix} \begin{bmatrix} x \\ y \\ z \end{bmatrix} = \begin{bmatrix} 0 \\ 0 \\ 0 \end{bmatrix}$$

$$\begin{bmatrix} 0 & -2 & -2 \\ 4 & -6 & 2 \\ 8 & -4 & 4 \end{bmatrix} \begin{bmatrix} x \\ y \\ z \end{bmatrix} = \begin{bmatrix} 0 \\ 0 \\ 0 \end{bmatrix}$$

Technology Connection

Computing Eigenvectors

Solving the linear equations required to find eigenvectors is tedious and time-consuming. Let's use a grapher to find the eigenvectors of Example 4. To find the eigenvectors associated with the eigenvalue -3, we enter the matrix

$$A = \begin{bmatrix} 1 & -2 & -2 \\ 4 & -5 & -2 \\ 8 & -4 & 5 \end{bmatrix}.$$ We then use the

RREF(option to convert $A - (-3)I = A + 3I$ to reduced row-echelon form. We do not need to include the augmentation since all the entries are zero.

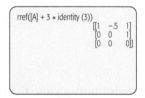

rref([A] + 3 * identity (3))

$$\begin{bmatrix} 1 & -.5 & 1 \\ 0 & 0 & 1 \\ 0 & 0 & 0 \end{bmatrix}$$

This gives the solutions $z = 0$ and $y = 2x$. The eigenvectors are of the form

$$x \begin{bmatrix} 1 \\ 2 \\ 0 \end{bmatrix},$$ where $x \neq 0$. Notice that the

reduced matrix is different from the matrix we found on page 476. Even though the reduced matrices have different forms, they have the same solutions.

EXERCISES

Let $A = \begin{bmatrix} 4 & -2 & -2 \\ 4 & -2 & -2 \\ 8 & -4 & 8 \end{bmatrix}$.

1. Find the eigenvectors associated with the eigenvalue 3.

2. Find the eigenvectors associated with the eigenvalue 1.

We write the corresponding augmented matrix and row reduce.

$$\begin{bmatrix} 0 & -2 & -2 & | & 0 \\ 4 & -6 & -2 & | & 0 \\ 8 & -4 & 4 & | & 0 \end{bmatrix}$$

$$\begin{bmatrix} 0 & 1 & 1 & | & 0 \\ 2 & -3 & -1 & | & 0 \\ 2 & -1 & 1 & | & 0 \end{bmatrix}$$
Multiply row 1 by $-1/2$.
Multiply row 2 by $1/2$.
Multiply row 3 by $1/4$.

$$\begin{bmatrix} 2 & -3 & -1 & | & 0 \\ 2 & -1 & 1 & | & 0 \\ 0 & 1 & 1 & | & 0 \end{bmatrix}$$
Change row order.

$$\begin{bmatrix} 2 & -3 & -1 & | & 0 \\ 0 & 2 & 2 & | & 0 \\ 0 & 1 & 1 & | & 0 \end{bmatrix}$$
Add -1 times row 1 to row 2.

$$\begin{bmatrix} 2 & -3 & -1 & | & 0 \\ 0 & 1 & 1 & | & 0 \\ 0 & 1 & 1 & | & 0 \end{bmatrix}$$
Multiply row 2 by $1/2$.

$$\begin{bmatrix} 2 & -3 & -1 & | & 0 \\ 0 & 1 & 1 & | & 0 \\ 0 & 0 & 0 & | & 0 \end{bmatrix}$$
Add -1 times row 2 to row 3.

Row 2 says that $y + z = 0$, or that $y = -z$. Back-substituting y into the equation for row 1 gives

$$2x - 3(-z) - z = 0$$
$$x = -z.$$

The eigenvectors with eigenvalue 1 have the form $\begin{bmatrix} -z \\ -z \\ z \end{bmatrix} =$

$z \begin{bmatrix} -1 \\ -1 \\ 1 \end{bmatrix}$, where $z \neq 0$. ∎

Eigenvalues and Eigenvectors of 2 X 2 Matrices

As we can see from Example 4, finding eigenvalues and eigenvectors requires a good deal of work. In the case of 2×2 matrices, it is possible to give fairly simple formulas for finding the eigenvectors and eigenvalues. For 3×3 and larger matrices, there is no simple formula.

THEOREM 9

To find the eigenvalues and eigenvectors of the 2×2 matrix

$$A = \begin{bmatrix} a & b \\ c & d \end{bmatrix},$$

we may use the following procedure:

1. The eigenvalue(s) may be found by solving the characteristic equation

$$r^2 - (a + d)r + (ad - bc) = 0.$$

The term $(a + d)$ is called the **trace** of A, while $(ad - bc)$ is the determinant of A.
If either $b = 0$ or $c = 0$ (or both), then the eigenvalues are simply a and d.

2. The eigenvectors may be found as follows:

   a) If A is a multiple of the identity matrix I, then all vectors are eigenvectors of A.

   b) Otherwise, the eigenvectors associated with r are either nonzero multiples of $\begin{bmatrix} b \\ r - a \end{bmatrix}$ or nonzero multiples of $\begin{bmatrix} r - d \\ c \end{bmatrix}$.

Let's verify the formula for a matrix we have already studied.

EXAMPLE 5   Use Theorem 9 to find the eigenvalues and eigenvectors of the matrix from Example 3, $A = \begin{bmatrix} -7 & 18 \\ -3 & 8 \end{bmatrix}$.

Solution   The trace is $-7 + 8 = 1$, and the determinant is $(-7)(8) - (18)(-3) = -2$. The characteristic polynomial is $r^2 - (1)r + (-2) = r^2 - r - 2$. We solve the characteristic equation

$$r^2 - r - 2 = 0, \quad \text{or} \quad (r - 2)(r + 1) = 0.$$

The eigenvalues are $r = 2$ and $r = -1$.

The eigenvectors for $r = 2$ are multiples of $\begin{bmatrix} b \\ r - a \end{bmatrix} = \begin{bmatrix} 18 \\ 2 - (-7) \end{bmatrix} = \begin{bmatrix} 18 \\ 9 \end{bmatrix}$. Factoring out the common factor of 9, we see that the eigenvectors for the eigenvalue $r = 2$ are nonzero multiples of $\begin{bmatrix} 2 \\ 1 \end{bmatrix}$. We also could have written the eigenvectors as nonzero multiples of the vector $\begin{bmatrix} r - d \\ c \end{bmatrix} = \begin{bmatrix} 2 - 8 \\ -3 \end{bmatrix} = \begin{bmatrix} -6 \\ -3 \end{bmatrix}$. This answer is equivalent to nonzero multiples of $\begin{bmatrix} 2 \\ 1 \end{bmatrix}$.

The eigenvalues for $r = -1$ are nonzero multiples of $\begin{bmatrix} b \\ r - a \end{bmatrix} = \begin{bmatrix} 18 \\ -1 - (-7) \end{bmatrix} = \begin{bmatrix} 18 \\ 6 \end{bmatrix}$. Factoring out the common factor of 6, we see that the eigenvectors for the eigenvalue $r = -1$ are nonzero multiples of $\begin{bmatrix} 3 \\ 1 \end{bmatrix}$. We could have

also written the eigenvectors as nonzero multiples of the vector $\begin{bmatrix} r - d \\ c \end{bmatrix} =$

$\begin{bmatrix} -1 - 8 \\ -3 \end{bmatrix} = \begin{bmatrix} -9 \\ -3 \end{bmatrix}$. Again, this answer is equivalent to nonzero multiples of $\begin{bmatrix} 3 \\ 1 \end{bmatrix}$.

This answer agrees with the answer found in Example 3.   ■

EXAMPLE 6   Find the eigenvalues and eigenvectors for the matrix $\begin{bmatrix} 2 & 0 \\ 1 & 3 \end{bmatrix}$.

Solution   Since $b = 0$, part 1 of Theorem 9 says that the eigenvalues are the entries on the main diagonal, so that $r = 2$ and $r = 3$.

We set $r = 2$ to find the eigenvectors for the eigenvalue 2. Since the vector

$\begin{bmatrix} b \\ r - a \end{bmatrix} = \begin{bmatrix} 0 \\ 2 - 2 \end{bmatrix} = \begin{bmatrix} 0 \\ 0 \end{bmatrix}$ gives us the zero vector, we must use the other form

given in Theorem 9. The other form is $\begin{bmatrix} r - d \\ c \end{bmatrix} = \begin{bmatrix} 2 - 3 \\ 1 \end{bmatrix} = \begin{bmatrix} -1 \\ 1 \end{bmatrix}$. The eigen-

vectors corresponding to the eigenvalue $r = 2$ are the nonzero multiples of $\begin{bmatrix} -1 \\ 1 \end{bmatrix}$.

We next set $r = 3$ to find the eigenvectors for the eigenvalue 3. This time, the

first form gives us the nonzero vector $\begin{bmatrix} b \\ r - a \end{bmatrix} = \begin{bmatrix} 0 \\ 3 - 2 \end{bmatrix} = \begin{bmatrix} 0 \\ 1 \end{bmatrix}$. The eigenvectors

corresponding to the eigenvalue $r = 3$ are the nonzero multiples of $\begin{bmatrix} 0 \\ 1 \end{bmatrix}$.   ■

Eigenvalues and Growth

Eigenvalues and eigenvectors have many applications in science. In Chapter 9, we will see that they are very useful in solving systems of differential equations that arise in life science settings. Here we show how eigenvalues can be used to model population growth rates. The key fact that we need is stated in the following theorem.

THEOREM 10

Suppose that the Leslie matrix G for a population has eigenvectors $v_1, v_2, v_3, \ldots, v_m$ with associated eigenvalues $r_1, r_2, r_3, \ldots, r_m$, respectively. If the initial population vector is $\mathbf{p} = a_1 v_1 + a_2 v_2 + a_3 v_3 + \cdots + a_m v_m$, then after n time periods, the population vector is

$$a_1 r_1^n v_1 + a_2 r_2^n v_2 + a_3 r_3^n v_3 + \cdots + a_m r_m^n v_m.$$

Proof: The population vector after n time periods is $\mathbf{G}^n\mathbf{p}$, since we multiply by the Leslie matrix once for each time period. Note that multiplying once by $\mathbf{G}$ gives

$$\begin{aligned} \mathbf{Gp} &= \mathbf{G}(a_1\mathbf{v}_1 + a_2\mathbf{v}_2 + a_3\mathbf{v}_3 + \cdots + a_m\mathbf{v}_m) \\ &= \mathbf{G}a_1\mathbf{v}_1 + \mathbf{G}a_2\mathbf{v}_2 + \mathbf{G}a_3\mathbf{v}_3 + \cdots + \mathbf{G}a_m\mathbf{v}_m \\ &= a_1\mathbf{G}\mathbf{v}_1 + a_2\mathbf{G}\mathbf{v}_2 + a_3\mathbf{G}\mathbf{v}_3 + \cdots + a_m\mathbf{G}\mathbf{v}_m \\ &= a_1r_1\mathbf{v}_1 + a_2r_2\mathbf{v}_2 + a_3r_3\mathbf{v}_3 + \cdots + a_mr_m\mathbf{v}_m. \end{aligned}$$

We see that the effect of applying $\mathbf{G}$ is to multiply the coefficient of $\mathbf{v}_k$ by r_k. If we apply $\mathbf{G}$ n times, then

$$\mathbf{G}^n\mathbf{p} = a_1r_1^n\mathbf{v}_1 + a_2r_2^n\mathbf{v}_2 + a_3r_3^n\mathbf{v}_3 + \cdots + a_mr_m^n\mathbf{v}_m. \qquad \blacksquare$$

An expression of the form $a_1\mathbf{v}_1 + a_2\mathbf{v}_2 + a_3\mathbf{v}_3 + \cdots + a_m\mathbf{v}_m$ is called a **linear combination** of the vectors $\mathbf{v}_1, \mathbf{v}_2, \mathbf{v}_3, \ldots, \mathbf{v}_m$. For example, $\begin{bmatrix} 17 \\ 27 \end{bmatrix} = 3\begin{bmatrix} 1 \\ 2 \end{bmatrix} + 7\begin{bmatrix} 2 \\ 3 \end{bmatrix}$ is a linear combination of the vectors $\begin{bmatrix} 1 \\ 2 \end{bmatrix}$ and $\begin{bmatrix} 2 \\ 3 \end{bmatrix}$. In the next example, we illustrate how to write a vector as a linear combination of eigenvectors.

EXAMPLE 7   *Bird Population Growth.*   The Leslie matrix for a population of hatchling and adult birds has eigenvalues 2 and $\frac{1}{2}$. The corresponding eigenvectors are $\begin{bmatrix} 2 \\ 1 \end{bmatrix}$ and $\begin{bmatrix} 1 \\ 1 \end{bmatrix}$, respectively. The initial population vector is $\mathbf{p} = \begin{bmatrix} \text{Hatchlings} \\ \text{Adults} \end{bmatrix} = \begin{bmatrix} 400 \\ 300 \end{bmatrix}$.

a) Write the initial population vector as a linear combination of eigenvectors.

b) Find the population vector after 10 yr.

Solution

a) We wish to find numbers a and b such that

$$\begin{bmatrix} 400 \\ 300 \end{bmatrix} = a\begin{bmatrix} 2 \\ 1 \end{bmatrix} + b\begin{bmatrix} 1 \\ 1 \end{bmatrix}.$$

Rewriting this matrix equation, we have

$$\begin{bmatrix} 400 \\ 300 \end{bmatrix} = \begin{bmatrix} 2a + b \\ a + b \end{bmatrix}.$$

This can be solved by equating corresponding entries and solving the system of linear equations

$$2a + b = 400 \tag{1}$$

$$a + b = 300. \tag{2}$$

Subtracting equation (2) from equation (1), we have $a = 100$. Substituting $a = 100$ into equation (2) gives $b = 200$.

We can write $\begin{bmatrix} 400 \\ 300 \end{bmatrix} = 100 \begin{bmatrix} 2 \\ 1 \end{bmatrix} + 200 \begin{bmatrix} 1 \\ 1 \end{bmatrix}$. The student can check this answer by simplifying the right-hand side.

b) According to Theorem 10, the population vector after 10 yr is

$$100(2)^{10} \begin{bmatrix} 2 \\ 1 \end{bmatrix} + 200 \left(\frac{1}{2} \right)^{10} \begin{bmatrix} 1 \\ 1 \end{bmatrix} = 102{,}400 \begin{bmatrix} 2 \\ 1 \end{bmatrix} + \frac{200}{1024} \begin{bmatrix} 1 \\ 1 \end{bmatrix}$$

$$= 102{,}400 \begin{bmatrix} 2 \\ 1 \end{bmatrix} + \frac{25}{128} \begin{bmatrix} 1 \\ 1 \end{bmatrix}$$

$$= \begin{bmatrix} 204{,}800 \\ 102{,}400 \end{bmatrix} + \begin{bmatrix} \frac{25}{128} \\ \frac{25}{128} \end{bmatrix}$$

$$\approx \begin{bmatrix} 204{,}800 \\ 102{,}400 \end{bmatrix}. \qquad \blacksquare$$

Let's look at the calculation in Example 7(b) more closely. How much of the answer was due to the eigenvector $\begin{bmatrix} 2 \\ 1 \end{bmatrix}$ and how much was due to the eigenvector $\begin{bmatrix} 1 \\ 1 \end{bmatrix}$? We can see that the contribution from $\begin{bmatrix} 1 \\ 1 \end{bmatrix}$ was insignificant compared to the contribution from $\begin{bmatrix} 2 \\ 1 \end{bmatrix}$. This is because 2^n grows very fast as n grows, but $\left(\frac{1}{2} \right)^n$ does not. In general, the growth of a population is determined by the largest eigenvalue of the Leslie matrix.

The population of a species is the sum of the entries of its population vector. We define the **relative growth rate** of a population as the ratio of the population one year to its population the previous year. For example, if a population grows from 100 individuals to 150 individuals, then its relative growth rate is $\frac{150}{100} = 1.5$. We also say that the **percentage growth rate** is 50%.

THEOREM 11

Suppose that the largest eigenvalue of a Leslie matrix is r. Then the relative growth rate approaches r as the number of years approaches infinity. (Some technical conditions, which almost certainly occur for real populations, must also hold, but these are beyond the scope of this book.)

We refer to the largest eigenvalue of a matrix as the **long-term growth rate.** To find the **long-term percentage growth rate,** we subtract 1 from the long-term growth rate and write this difference as a percentage.

EXAMPLE 8   *Ovenbird Population.*   The Leslie matrix for ovenbirds in central Missouri is[15]

$$G = \begin{bmatrix} 0.728 & 1.302 \\ 0.52 & 0.62 \end{bmatrix},$$

where group 1 represents hatchlings and group 2 represents adults. Find the long-term growth rate and the long-term percentage growth rate.

Solution   We find the eigenvalues of G. The trace of G is $0.728 + 0.62 = 1.348$, and the determinant is $0.728(0.62) - 1.302(0.52) \approx -0.2257$. The characteristic equation is

$$r^2 - 1.348r - 0.2257 = 0.$$

We use the quadratic formula to solve for r.

$$r = \frac{1.348 \pm \sqrt{(1.348)^2 - 4(1)(-0.2257)}}{2(1)}$$

$$\approx \frac{1.348 \pm 1.6492}{2} = 0.674 \pm 0.8246.$$

The two eigenvalues are approximately $r \approx -0.1506$ and $r \approx 1.4986$. The larger eigenvalue $r \approx 1.4986$ is the long-term growth rate, and the long-term percentage growth rate is approximately 49.86%. That is, after several years, the total population grows at a rate of approximately 49.86% per yr.   ■

---

[15]P. A. Porneluzi and J. Faaborg, "Season-Long Fecundity, Survival, and Viability of Ovenbirds in Fragmented and Unfragmented Landscapes, *Conservation Biology,* Vol. 13, No. 2, pp. 1151–1161 (1999).

Exercise Set 6.4

Multiply the matrix and the vector to determine if the vector is an eigenvector. If so, what is the eigenvalue?

1. $\begin{bmatrix} 2 & 0 \\ 0 & 3 \end{bmatrix}, \begin{bmatrix} 1 \\ 0 \end{bmatrix}$

2. $\begin{bmatrix} 2 & 0 \\ 0 & 3 \end{bmatrix}, \begin{bmatrix} 0 \\ 1 \end{bmatrix}$

3. $\begin{bmatrix} 3 & 2 \\ 0 & 4 \end{bmatrix}, \begin{bmatrix} 0 \\ 1 \end{bmatrix}$

4. $\begin{bmatrix} 5 & 0 \\ 3 & -2 \end{bmatrix}, \begin{bmatrix} 1 \\ 0 \end{bmatrix}$

5. $\begin{bmatrix} 5 & 0 \\ 3 & -2 \end{bmatrix}, \begin{bmatrix} 0 \\ 1 \end{bmatrix}$

6. $\begin{bmatrix} 10 & 0 \\ 42 & -4 \end{bmatrix}, \begin{bmatrix} 1 \\ 3 \end{bmatrix}$

7. $\begin{bmatrix} -8.5 & -4.5 \\ 21 & 11 \end{bmatrix}, \begin{bmatrix} 2 \\ -7 \end{bmatrix}$

8. $\begin{bmatrix} 5 & 0 & 0 \\ 0 & 3 & 0 \\ 0 & 0 & -2 \end{bmatrix}, \begin{bmatrix} 1 \\ 0 \\ 0 \end{bmatrix}$

9. $\begin{bmatrix} 5 & 0 & 0 \\ 0 & 3 & 0 \\ 0 & 0 & -2 \end{bmatrix}, \begin{bmatrix} 0 \\ 1 \\ 0 \end{bmatrix}$

10. $\begin{bmatrix} 5 & 0 & 0 \\ 0 & 3 & 0 \\ 0 & 0 & -2 \end{bmatrix}, \begin{bmatrix} 0 \\ 0 \\ 1 \end{bmatrix}$

11. $\begin{bmatrix} -25 & 40 & 39 \\ -32 & 47 & 39 \\ 16 & -20 & -1 \end{bmatrix}, \begin{bmatrix} -13 \\ -13 \\ 4 \end{bmatrix}$

12. $\begin{bmatrix} -25 & 40 & 39 \\ -32 & 47 & 39 \\ 16 & -20 & -1 \end{bmatrix}, \begin{bmatrix} 3 \\ 7 \\ -2 \end{bmatrix}$

Write the vector **v** as a linear combination of the vectors **w** and **u**.

13. $\mathbf{v} = \begin{bmatrix} 3 \\ 4 \end{bmatrix}, \mathbf{w} = \begin{bmatrix} 1 \\ 0 \end{bmatrix}, \mathbf{u} = \begin{bmatrix} 0 \\ 1 \end{bmatrix}$

14. $\mathbf{v} = \begin{bmatrix} 6 \\ 4 \end{bmatrix}, \mathbf{w} = \begin{bmatrix} -2 \\ 0 \end{bmatrix}, \mathbf{u} = \begin{bmatrix} 0 \\ 3 \end{bmatrix}$

15. $\mathbf{v} = \begin{bmatrix} 3 \\ 5 \end{bmatrix}, \mathbf{w} = \begin{bmatrix} 1 \\ 1 \end{bmatrix}, \mathbf{u} = \begin{bmatrix} 1 \\ 3 \end{bmatrix}$

16. $\mathbf{v} = \begin{bmatrix} -16 \\ 27 \end{bmatrix}, \mathbf{w} = \begin{bmatrix} -2 \\ 7 \end{bmatrix}, \mathbf{u} = \begin{bmatrix} 3 \\ 4 \end{bmatrix}$

17. $\mathbf{v} = \begin{bmatrix} 5 \\ 5 \end{bmatrix}, \mathbf{w} = \begin{bmatrix} 4 \\ 1 \end{bmatrix}, \mathbf{u} = \begin{bmatrix} 3 \\ 2 \end{bmatrix}$

18. $\mathbf{v} = \begin{bmatrix} 8 \\ 10 \end{bmatrix}, \mathbf{w} = \begin{bmatrix} 0 \\ 1 \end{bmatrix}, \mathbf{u} = \begin{bmatrix} 2 \\ 3 \end{bmatrix}$

Find all the eigenvalues and the corresponding eigenvectors for the following matrices.

19. $\begin{bmatrix} 1 & 0 \\ -1 & 2 \end{bmatrix}$

20. $\begin{bmatrix} 2 & 0 \\ 1 & 3 \end{bmatrix}$

21. $\begin{bmatrix} 5 & 2 \\ -24 & -9 \end{bmatrix}$

22. $\begin{bmatrix} -16 & -20 \\ 15 & 19 \end{bmatrix}$

23. $\begin{bmatrix} -7.5 & -15.75 \\ 6 & 12 \end{bmatrix}$

24. $\begin{bmatrix} \frac{51}{13} & -\frac{20}{13} \\ -\frac{24}{13} & \frac{79}{13} \end{bmatrix}$

25. $\begin{bmatrix} 9.5 & -4.5 \\ 15 & -7 \end{bmatrix}$

26. $\begin{bmatrix} -18 & -52.5 \\ 7 & 20.5 \end{bmatrix}$

27. $\begin{bmatrix} 10 & -4 & 15 \\ 8 & -2 & 15 \\ -4 & 2 & -5 \end{bmatrix}$

28. $\begin{bmatrix} -3 & 0 & 0 \\ 9 & 6 & 30 \\ -1 & -1 & -5 \end{bmatrix}$

29. $\begin{bmatrix} -5 & 8 & 2 \\ -15 & 18 & 4 \\ 45 & -48 & -10 \end{bmatrix}$

30. $\begin{bmatrix} -8 & -2 & 0 \\ 15 & 3 & 0 \\ 45 & 12 & -1 \end{bmatrix}$

31. $\begin{bmatrix} -3 & 1 & -1 \\ 2 & 2 & -1 \\ 4 & 2 & -1 \end{bmatrix}$

32. $\begin{bmatrix} 1 & 0 & -1 \\ -2 & -2 & -6 \\ 2 & 0 & 4 \end{bmatrix}$

33. $\begin{bmatrix} 12 & -16 & 0 \\ 8 & -12 & 0 \\ 6 & -12 & 2 \end{bmatrix}$

34. $\begin{bmatrix} 8 & 0 & 12 \\ -8 & 0 & -12 \\ -2 & 0 & -2 \end{bmatrix}$

35. $\begin{bmatrix} -3 & 8 & 6 \\ -4 & 9 & 6 \\ 3 & -6 & -4 \end{bmatrix}$

36. $\begin{bmatrix} 5 & -6 & -3 \\ 6 & -6 & -2 \\ -6 & 4 & 0 \end{bmatrix}$

The matrix **A** has eigenvalues r_1 and r_2 with corresponding eigenvectors $\mathbf{v}_1$ and $\mathbf{v}_2$, respectively. Compute $\mathbf{A}^n\mathbf{w}$.

37. $r_1 = 2, r_2 = 1, \mathbf{v}_1 = \begin{bmatrix} 1 \\ 0 \end{bmatrix}, \mathbf{v}_2 = \begin{bmatrix} 0 \\ 1 \end{bmatrix}, n = 10,$

$\mathbf{w} = \begin{bmatrix} 2 \\ 3 \end{bmatrix}$

38. $r_1 = 1, r_2 = -1, \mathbf{v}_1 = \begin{bmatrix} 1 \\ 1 \end{bmatrix}, \mathbf{v}_2 = \begin{bmatrix} 0 \\ 1 \end{bmatrix},$

$n = 100, \mathbf{w} = \begin{bmatrix} 2 \\ 3 \end{bmatrix}$

39. $r_1 = -2, r_2 = 0, \mathbf{v}_1 = \begin{bmatrix} 1 \\ 1 \end{bmatrix}, \mathbf{v}_2 = \begin{bmatrix} 0 \\ 1 \end{bmatrix}, n = 10,$

$\mathbf{w} = \begin{bmatrix} 2 \\ 3 \end{bmatrix}$

40. $r_1 = 3, r_2 = 0, \mathbf{v}_1 = \begin{bmatrix} 1 \\ -1 \end{bmatrix}, \mathbf{v}_2 = \begin{bmatrix} 1 \\ 1 \end{bmatrix}, n = 4,$

$\mathbf{w} = \begin{bmatrix} 2 \\ 4 \end{bmatrix}$

A 2×2 Leslie matrix has eigenvalues r_1 and r_2. Find the long-term growth rate and the long-term percentage growth rate.

41. $r_1 = 0, r_2 = 1.5$

42. $r_1 = 1.01, r_2 = 0.5$

43. $r_1 = 1.2, r_2 = 2.1$

44. $r_1 = 3, r_2 = 4$

APPLICATIONS

45. *Bird Population.* The Leslie matrix for a bird population of hatchlings and adults is $\begin{bmatrix} 0.5 & 2 \\ 0.5 & 0.5 \end{bmatrix}$. Determine the long-term growth rate for this population.

 46. *Cerulean Warblers.* A population of cerulean warblers is grouped into hatchlings (H), second year (S), and after second year (A). Their Leslie matrix for one location is given below, showing estimates of the survivability and fecundity.[16]

$$G = \begin{bmatrix} 0 & 0.75 & 0.95 \\ 0.3 & 0 & 0 \\ 0 & 0.7 & 0.95 \end{bmatrix}$$

Determine the long-term growth rate.

47. *Dinophilus Gyrociliatus.* **Dinophilus gyrociliatus** is a small species that lives in the fouling community of harbor environments. The female population is grouped into those 6 wk old or less and those that are more than 6 wk old. The Leslie matrix is given by[17]

$$G = \begin{bmatrix} 15 & 7.5 \\ 0.8 & 0 \end{bmatrix}.$$

Determine the long-term percentage growth rate per time period.

 48. *Fairy Shrimp.* The San Diego fairy shrimp live in ponds that fill and dry many times during a year. While the ponds are dry, the fairy shrimp survive as cysts. The population is divided into three groups. Cysts that survive one dry period are in group 1, cysts that survive two dry periods are in group 2, and cysts that survive three or more dry periods are in group 3. The Leslie matrix representing the sur-

vivability and fecundity is given below.[18] In the third dry spell, there are 8365, 1095, and 310 individuals in groups 1, 2, and 3, respectively.

$$G = \begin{bmatrix} 3.6 & 0.98 & 0.65 \\ 0.5 & 0 & 0 \\ 0 & 0.5 & 0.49 \end{bmatrix}$$

Compute the long-term percentage growth rate between dry periods.

SYNTHESIS

49. An initial population vector is $\mathbf{p} = \mathbf{v} + \mathbf{u}$, where $\mathbf{v} = \begin{bmatrix} 1 \\ 1 \end{bmatrix}$ and $\mathbf{u} = \begin{bmatrix} 2 \\ 3 \end{bmatrix}$ are eigenvectors with eigenvalues 1.1 and 0.8, respectively.

  a) Use Theorem 10 to compute $\lim_{n \to \infty} \left(\dfrac{1}{1.1} \right)^n G^n \mathbf{p}$, where G is the Leslie matrix.

  tw b) Based on part (a), explain why the long-term growth rate is 1.1.

50. An initial population vector is $\mathbf{p} = \mathbf{v} + \mathbf{u}$, where $\mathbf{v} = \begin{bmatrix} 3 \\ 5 \end{bmatrix}$ and $\mathbf{u} = \begin{bmatrix} 1 \\ 2 \end{bmatrix}$ are eigenvectors with eigenvalues 1.4 and 0.1, respectively.

  a) Use Theorem 10 to compute $\lim_{n \to \infty} \left(\dfrac{1}{1.4} \right)^n G^n \mathbf{p}$, where G is the Leslie matrix.

  tw b) Based on part (a), explain why the long-term growth rate is 1.4.

51. Let $A = \begin{bmatrix} 1 & 0 \\ -1 & 2 \end{bmatrix}$ and $B = \begin{bmatrix} 2 & 5 \\ 1 & 3 \end{bmatrix}$.

  a) Compute $B^{-1}AB$.

  b) Compute the characteristic polynomial for A and $B^{-1}AB$. Compare your answers.

  tw c) How do the eigenvalues of A and $B^{-1}AB$ compare? Explain. You do not need to compute the eigenvalues to answer the question.

52. Let $A = \begin{bmatrix} 2 & 3 \\ 0 & 4 \end{bmatrix}$ and $B = \begin{bmatrix} 2 & 3 \\ 5 & 3 \end{bmatrix}$.

  a) Compute $B^{-1}AB$.

  b) Compute the characteristic polynomial for A and $B^{-1}AB$. Compare your answers.

[16]Jones, J., Barg, J.J., Sillett, T.S., Veit, M.L., and Robertson, R.J., "Season-Long Fecundity, Survival, and Viability of Ovenbirds in Fragmented and Unfragmented Landscapes," *Conservation Biology*, Vol. 13, No. 2, pp. 1151–1161 (1999).

[17]D. Prevedelli and R. Zunarelli Vandini, "Survival, fecundity and sex ratio of *Dinophilus gyrociliatus* (Polychaeta: Dinophilidae) under different dietary conditions," *Marine Biology*, Vol. 133, pp. 231–236 (1999).

[18]B.J. Ripley, J. Holtz, and M. Simovich, "Cyst bank life-history model for a fairy shrimp from ephemeral ponds," *Freshwater Biology*, Vol. 49, pp. 221–231 (2004).

tW c) How do the eigenvalues of A and $B^{-1}AB$ compare? Explain. You do not need to compute the eigenvalues to answer the question.

53. Suppose that v is an eigenvector of both $n \times n$ matrices A and B.

   a) Show that v is an eigenvector of AB.
   b) If the eigenvalue corresponding to the eigenvector v is r_1 and r_2 for the matrices A and B respectively, what is the eigenvalue corresponding to v for the matrix AB?

tW 54. The $m \times m$ Leslie matrix G has m different positive eigenvalues, and r is the largest. If $p \neq 0$ is a population vector, is it necessarily true that the long-term population growth for that initial population is r?

55. This exercise illustrates the Cayley–Hamilton theorem, which states that a matrix satisfies its characteristic polynomial. Let $A = \begin{bmatrix} 2 & 3 \\ 1 & -5 \end{bmatrix}$.

   a) Show that the characteristic polynomial for A is $r^2 + 3r - 13$.
   b) Compute $A^2 + 3A - 13I$.
   c) Write another square matrix B and compute its characteristic polynomial. Replace the variable r with the matrix B, just as we did in part (b), and compute the matrix.

6.5 Solving Difference Equations

OBJECTIVES

■ Find solutions to difference equations.
■ Determine asymptotic behavior of solutions.

Eigenvalues are used to solve a variety of problems in science and mathematics. In this section, we will illustrate their use in solving difference equations.

Solving Difference Equations

Annual plants live for only one season. At the end of the season, they produce seeds before their death. Some of the seeds germinate to continue the life cycle the following spring, some die over the winter, and others do not germinate until a later spring. The germination rate and the number of seeds each plant produces depend on numerous environmental conditions, such as the weather during the growing season and the number of hungry birds.

Suppose that no plants are in the field in year 0 and that 100 plants are introduced in year 1. We let x_n be the number of plants in year n, so $x_0 = 0$ and $x_1 = 100$.

For simplicity, let's suppose that on average, a plant produces five seeds. Of all the seeds, 15% germinate the following spring, 6% germinate the second spring, and the rest die. On average, each plant produces $(0.15)(5) = 0.75$ seeds that germinate the first spring and $(0.06)(5) = 0.3$ seeds that germinate the second spring. The number of plants in successive years satisfies the equation

$$x_{n+1} = 0.75x_n + 0.3x_{n-1}, \tag{1}$$

since on average each plant in year n contributes 0.75 seeds that grow to a plant in year $n + 1$, and each plant in year $n - 1$ contributes 0.3 seeds that become a plant in year $n + 1$. Equation (1) provides us with a way of estimating the number of plants each year.[19] The first few values of x_n are given by

$$x_0 = 0,$$
$$x_1 = 100,$$
$$x_2 = 0.75x_1 + 0.3x_0 = 0.75(100) + 0.3(0) = 75,$$

---

[19]L. Edelstein-Keshet, *Mathematical Models in Biology* (New York: Random House, 1988).

$$x_3 = 0.75x_2 + 0.3x_1 = 0.75(75) + 0.3(100) = 86.25,$$
$$x_4 = 0.75x_3 + 0.3x_2 \approx 87.19,$$
$$x_5 = 0.75x_4 + 0.3x_3 \approx 91.27.$$

Equations that define a sequence of numbers by specifying how each number is related to the previous numbers are called **recursion relations.** A *solution* to a recursion relation is a function that gives the *n*th number of the sequence without using the previous numbers in the sequence.

In this section, we will restrict our attention to special recursion relations given in the following definition.

DEFINITION

A **second-order linear difference equation** is a recursion relation of the form

$$x_{n+1} = ax_n + bx_{n-1} + g(n),$$

where a and b are constants and $g(n)$ is a function of n. If $g(n) = 0$, then the difference equation is **homogeneous.**

In the above definition, the term *second-order* is used to indicate that x_{n+1} depends on the two previous values x_n and x_{n-1}.

EXAMPLE 1   Are the following recursion relations second-order linear difference equations? If so, are they homogeneous?

a) $x_{n+1} = x_n + x_{n-1}$

b) $x_{n+1} = 2x_n - x_{n-1} + 3e^n$

c) $x_{n+1} = \sin x_n + 5 \cos x_{n-1}$

Solution

a) The recursion relation $x_{n+1} = x_n + x_{n-1}$ is a second-order linear difference equation with $a = 1$ and $b = 1$. It is also homogeneous since $g(n) = 0$.

b) The recursion relation $x_{n+1} = 2x_n - x_{n-1} + 3e^n$ is a second-order linear difference equation with $a = 2$ and $b = -1$. It is not homogeneous since $g(n) = 3e^n \neq 0$.

c) The recursion relation $x_{n+1} = \sin x_n + 5 \cos x_{n-1}$ cannot be rewritten in the proper form since it involves the sine and cosine functions of x_n and x_{n-1}. It is not a second-order linear difference equation.   ■

Homogeneous Difference Equations

Let's first solve homogeneous linear difference equations. The difference equation

$$x_{n+1} = ax_n + bx_{n-1}$$

can be written in matrix form as

$$\begin{bmatrix} x_{n+1} \\ x_n \end{bmatrix} = \begin{bmatrix} a & b \\ 1 & 0 \end{bmatrix} \begin{bmatrix} x_n \\ x_{n-1} \end{bmatrix}.$$

We can think of the matrix

$$A = \begin{bmatrix} a & b \\ 1 & 0 \end{bmatrix}$$

as a Leslie matrix. Its characteristic polynomial is

$$\det(A - rI) = \det\begin{bmatrix} a - r & b \\ 1 & -r \end{bmatrix} = (a - r)(-r) - b = r^2 - ar - b.$$

If the characteristic polynomial has two distinct real roots r_1 and r_2, then we can use Theorem 10 to write any solution to the difference equation in the form

$$x_n = c_1 r_1^n + c_2 r_2^n,$$

where c_1 and c_2 are constants. This is called the **general solution** to the difference equation. If the values of x_0 and x_1 are specified, the constants c_1 and c_2 may be found. The solution is then called the **particular solution.**

EXAMPLE 2 Let $x_{n+1} = -x_n + 2x_{n-1}$.

a) Find the general solution.

b) Find the particular solution assuming that $x_0 = 3$ and $x_1 = 0$.

Solution

a) In this difference equation, we have $a = -1$ and $b = 2$. We solve the characteristic equation:

$$r^2 - ar - b = 0$$
$$r^2 - (-1)r - 2 = 0$$
$$r^2 + r - 2 = 0$$
$$(r + 2)(r - 1) = 0.$$

The eigenvalues are $r = -2$ and $r = 1$. We conclude that the general solution of the difference equation is

$$x_n = c_1(-2)^n + c_2(1)^n = c_1(-2)^n + c_2.$$

b) In order to find the particular solution, we solve the system of equations that arises from the initial values $x_0 = 3$ and $x_1 = 0$:

$$c_1(-2)^0 + c_2 = \quad c_1 + c_2 = 3$$
$$c_1(-2)^1 + c_2 = -2c_1 + c_2 = 0.$$

We use an augmented matrix to solve for c_1 and c_2.

$$\begin{bmatrix} 1 & 1 & | & 3 \\ -2 & 1 & | & 0 \end{bmatrix}$$

$$\begin{bmatrix} 1 & 1 & | & 3 \\ 0 & 3 & | & 6 \end{bmatrix} \qquad \text{Add 2 times row 1 to row 2.}$$

$$\begin{bmatrix} 1 & 1 & | & 3 \\ 0 & 1 & | & 2 \end{bmatrix} \qquad \text{Multiply row 2 by 1/3.}$$

We can see from the second row that $c_2 = 2$, so $c_1 = 1$ from the first row. The particular solution is

$$x_n = (-2)^n + 2.$$

We can check the first few terms to verify that we did not make a mistake in our calculations. We see that

$$
\begin{aligned}
x_0 &= & 3 &= (-2)^0 + 2 \\
x_1 &= & 0 &= (-2)^1 + 2 \\
x_2 &= -x_1 + 2x_0 = & -0 + 2(3) = \quad 6 &= (-2)^2 + 2 \\
x_3 &= -x_2 + 2x_1 = & -6 + 0 = -6 &= (-2)^3 + 2 \\
x_4 &= -x_3 + 2x_2 = -(-6) + 2(6) = & 18 &= (-2)^4 + 2.
\end{aligned}
$$
∎

We can use the same technique to solve the difference equation that describes the population of annual plants discussed on page 486.

EXAMPLE 3   *Plant Propagation.* A population of 100 plants is introduced into an empty area in year 1. Each plant on average produces five seeds. Each year, 15% of the seeds produced germinate the second year, 6% of the seeds germinate the third year, and all the other seeds die. Find a formula for the number of plants in year n.

Solution   As described on page 486, the number of plants each year is given by the recursion relation

$$x_{n+1} = 0.75x_n + 0.3x_{n-1},$$

and the first two terms are $x_0 = 0$ and $x_1 = 100$.

We find the general solution by solving the characteristic equation

$$r^2 - 0.75r - 0.3 = 0.$$

We use the quadratic formula to find the solutions.

$$r = \frac{-(-0.75) \pm \sqrt{(-0.75)^2 - 4(1)(-0.3)}}{2(1)} = 0.375 \pm \frac{\sqrt{1.7625}}{2}.$$

We can use a calculator to approximate the two roots: $r_1 \approx 1.0388$ and $r_2 \approx -0.2888$. The general solution is

$$x_n = c_1(1.0388)^n + c_2(-0.2888)^n.$$

To find the particular solution, we use the values $x_0 = 0$ and $x_1 = 100$ to write a system of linear equations:

$$
\begin{aligned}
c_1(1.0388)^0 + c_2(-0.2888)^0 &= & c_1 + & \quad c_2 = & 0 \\
c_1(1.0388)^1 + c_2(-0.2888)^1 &= 1.0388c_1 - & 0.2888c_2 = & 100.
\end{aligned}
$$

We simplify and write as an augmented matrix to solve.

$$\left[\begin{array}{cc|c} 1 & 1 & 0 \\ 1.0388 & -0.2888 & 100 \end{array}\right]$$

$$\left[\begin{array}{cc|c} 1 & 1 & 0 \\ 0 & -1.3276 & 100 \end{array}\right] \qquad \text{Add } -1.0388 \text{ times row 1 to row 2.}$$

$$\left[\begin{array}{cc|c} 1 & 1 & 0 \\ 0 & 1 & -75.324 \end{array}\right] \qquad \text{Multiply row 2 by } 1/(-1.3276).$$

From the second row, we see that $c_2 = -75.324$. Substituting into the first row gives $c_1 = 75.324$. The particular solution is

$$x_n = 75.324(1.0388)^n - 75.324(-0.2888)^n.$$

To verify that we did not make a mistake, we note that

$$x_5 = 75.324(1.0388)^5 - 75.324(-0.2888)^5 \approx 91.27.$$

This agrees with the value we computed on page 487 by directly using the recursion relation.   ■

Nonhomogeneous Difference Equations

Let's investigate how to solve nonhomogeneous linear difference equations of the form

$$x_{n+1} = ax_n + bx_{n-1} + K,$$

where K is a constant. A reasonable guess for a solution is a constant $x_n = c$ for each n. Since a, b, and c are constants, we can solve an equation to determine c.

EXAMPLE 4   Find a solution to the nonhomogeneous difference equation $x_{n+1} = 5x_n - 6x_{n-1} + 2$.

Solution   We substitute the constant c for x_{n+1}, x_n, and x_{n-1} and then solve for c.

$$c = 5c - 6c + 2$$
$$2c = 2$$
$$c = 1$$

One solution is the constant $x_n = 1$.   ■

Given the nonhomogeneous linear difference equation

$$x_{n+1} = ax_n + bx_{n-1} + g(n),$$

the **associated homogeneous difference equation** is

$$x_{n+1} - ax_n + bx_{n-1}.$$

> **THEOREM 12**
>
> The general solution of a nonhomogeneous linear difference equation is the general solution of the associated homogeneous equation plus any particular solution to the nonhomogeneous equation.

Theorem 12 provides a way to find the general solution to a nonhomogeneous linear difference equation if we can both solve the homogeneous equation and find one solution to the nonhomogeneous equation. We illustrate this procedure in the next example.

EXAMPLE 5

a) Find the general solution to $x_{n+1} = 5x_n - 6x_{n-1} + 2$.

b) Find the particular solution with $x_0 = 1$ and $x_1 = 2$.

Solution

a) We begin by solving the homogeneous linear difference equation

$$x_{n+1} = 5x_n - 6x_{n-1}.$$

We find the eigenvalues by solving the characteristic equation.

$$r^2 - 5r + 6 = 0$$
$$(r - 3)(r - 2) = 0.$$

Since the eigenvalues are 3 and 2, the associated homogeneous equation has general solution

$$x_n = c_1(3)^n + c_2(2)^n.$$

The general solution of the nonhomogeneous difference equation is found by adding this general solution to a particular solution. In Example 4, we showed that a particular solution is $x_n = 1$. Therefore, the general solution is

$$x_n = c_1(3)^n + c_2(2)^n + 1.$$

b) To find the particular solution with $x_0 = 1$ and $x_2 = 2$, we solve the system

$$c_1(3)^0 + c_2(2)^0 + 1 = c_1 + c_2 + 1 = 1$$
$$c_1(3)^1 + c_2(2)^1 + 1 = 3c_1 + 2c_2 + 1 = 2.$$

Simplifying and writing as augmented matrices, we find

$$\begin{bmatrix} 1 & 1 & 0 \\ 3 & 2 & 1 \end{bmatrix}$$

$$\begin{bmatrix} 1 & 1 & 0 \\ 0 & -1 & 1 \end{bmatrix}. \qquad \text{Add } -3 \text{ times row 1 to row 2.}$$

We see that $c_2 = -1$; substituting into the first equation gives $c_1 = 1$. The particular solution is

$$x_n = 3^n - 2^n + 1. \qquad \blacksquare$$

Asymptotic Behavior of Solutions

In Example 3, we considered a population that grows according to a linear difference equation. Each generation contributes positively to the population for the next two years. That is, the population x_n satisfies the linear difference equation

$$x_{n+1} = ax_n + bx_{n-1},$$

where a and b are positive. The general solution is

$$x_n = c_1 r_1^n + c_2 r_2^n,$$

where $r_1 \geq r_2$ are the eigenvalues. For any solution with $c_1 \neq 0$ (a virtual certainty for real populations), there are three possible cases. These cases describe the **asymptotic** or long-term growth of the population.

THEOREM 13

Let $x_{n+1} = ax_n + bx_{n-1}$ where a and b are positive numbers. Then one of three cases holds.

1. If $a + b < 1$, then $\lim_{n\to\infty} x_n = 0$. In this case, the population decreases exponentially.

2. If $a + b = 1$, then $\lim_{n\to\infty} x_n = c_1 > 0$. In this case, the population approaches a nonzero constant.

3. If $a + b > 1$, then $\lim_{n\to\infty} x_n = \infty$. In this case, the population grows exponentially.

The proof of Theorem 13 can be found in the exercises. Here, we illustrate its use with some examples.

EXAMPLE 6   *Plant Propagation.* Without computing the solution, determine the asymptotic growth of a population of annual plants satisfying the recursion relation

a) $x_{n+1} = 0.82x_n + 0.19x_{n-1}$.

b) $x_{n+1} = 0.54x_n + 0.36x_{n-1}$.

c) $x_{n+1} = 0.5x_n + 0.5x_{n-1}$.

Solution

a) In this case, $a + b = 1.01 > 1$. The population grows exponentially.

b) In this case, $a + b = 0.9 < 1$. The population decreases to 0 exponentially.

c) In this case, $a + b = 1$, so the population approaches a constant.   ■

In the previous example, we were able to give a qualitative assessment of the growth of the population without actually finding the eigenvalues. Theorem 13 is very useful when we do not need the details of exactly how quickly a population grows or declines but only wish to determine if the population will be successful.

In this section, we studied how to solve linear difference equations in the case of distinct real eigenvalues. There are linear difference equations that have repeated eigenvalues and some with complex eigenvalues. These can also be solved using very similar techniques. In the exercises, we discuss how to solve second-order linear difference equations with a repeated eigenvalue. More details on solving linear difference equations with repeated or complex roots, orders other than 2, and nonconstant $g(n)$ can be found in more advanced books.[20]

---

[20]L. Edelstein-Keshet, *Mathematical Models in Biology* (New York: Random House, 1988).

Exercise Set 6.5

Which of the following recursive relations are linear difference equations? Which are homogeneous linear difference equations?

1. $x_{n+1} = x_n^2 + x_{n-1}^2$

2. $x_{n+1} = 5x_n - 3x_{n-1} + 2$

3. $x_{n+1} = -13x_n + 14x_{n-1} + 2x_{n-2}$

4. $x_{n+1} = 2x_n - x_{n-1} + 2$

5. $x_{n+1} = x_n + 2x_{n-1}$

6. $x_{n+1} = e^{x_n} - x_{n-1}$

Find the general solution.

7. $x_{n+1} = 2x_n + 3x_{n-1}$

8. $x_{n+1} = 8x_n - 15x_{n-1}$

9. $x_{n+1} = -x_n + 6x_{n-1}$

10. $x_{n+1} = 5x_n - 6x_{n-1}$

11. $x_{n+1} = x_n + 2x_{n-1}$

12. $x_{n+1} = 3x_n + 4x_{n-1}$

Find the particular solution.

13. $x_{n+1} = x_n + 2x_{n-1}; x_0 = 2, x_1 = 1$

14. $x_{n+1} = 2x_n + 8x_{n-1}; x_0 = 1, x_1 = 4$

15. $x_{n+1} = \frac{5}{2}x_n - x_{n-1}; x_0 = 0, x_1 = -3/2$

16. $x_{n+1} = 2x_n + 3x_{n-1}; x_0 = 7, x_1 = 10$

17. $x_{n+1} = 2x_n + 8x_{n-1}; x_0 = 1, x_1 = -2$

18. $x_{n+1} = 4x_n + 5x_{n-1}; x_0 = 7, x_1 = 17$

19. $x_{n+1} = -\frac{3}{2}x_n + x_{n-1}; x_0 = 5, x_1 = -\frac{5}{2}$

20. $x_{n+1} = 5x_n + 6x_{n-1}; x_0 = 17, x_1 = 32$

Find the general solution.

21. $x_{n+1} = 5x_n - 6x_{n-1} + 8$

22. $x_{n+1} = x_n + 2x_{n-1} - 12$

23. $x_{n+1} = 9x_n - 20x_{n-1} + 12$

24. $x_{n+1} = -5x_n - 6x_{n-1} + 108$

Find the particular solution.

25. $x_{n+1} = 7x_n - 12x_{n-1} + 12; x_0 = 2, x_1 = 3$

26. $x_{n+1} = x_n + 6x_{n-1} - 36; x_0 = 8, x_1 = 7$

27. $x_{n+1} = -2x_n + 8x_{n-1} - 45; x_0 = 11, x_1 = -5$

28. $x_{n+1} = 3x_n + 28x_{n-1} - 150; x_0 = 13, x_1 = -49$

29. $x_{n+1} = 9x_n - 18x_{n-1} + 20; x_0 = 7, x_1 = 26$

30. $x_{n+1} = x_n + 2x_{n-1} - 2; x_0 = 1, x_1 = -2$

Without solving the difference equation, determine the asymptotic behavior of the general solution.

31. $x_{n+1} = 0.35x_n + 0.45x_{n-1}$

32. $x_{n+1} = 0.1x_n + 0.95x_{n-1}$

33. $x_{n+1} = 2.3x_n + 1.5x_{n-1}$

34. $x_{n+1} = 0.92x_n + 0.2x_{n-1}$

35. $x_{n+1} = 0.01x_n + 1.5x_{n-1}$

36. $x_{n+1} = 0.2x_n + 0.5x_{n-1}$

APPLICATIONS

37. *Plant Propagation.* The number of blue bonnets (an annual flower) in a particular field satisfies the recursion relation

$$x_{n+1} = 0.92x_n + 0.15x_{n-1}.$$

In year 1, 50 blue bonnet plants are introduced into the flowerless field.

a) Solve the difference equation to predict the number of blue bonnet plants in subsequent years.

tw b) Describe the asymptotic growth of the population. Will the population grow, approach a constant, or die?

38. *Plant Propagation.* The number of blue bonnets in a particular field satisfies the recursion relation

$$x_{n+1} = 0.81x_n + 0.13x_{n-1}.$$

In year 1, 50 blue bonnet plants are introduced into the field.

a) Solve the difference equation to predict the number of blue bonnet plants in subsequent years.

tw b) Describe the asymptotic growth of the population. Will the population grow, approach a limit, or die?

Fox Populations. The number of red foxes in a habitat in year n can be modeled by

$$x_{n+1} = a(x_n + z_n), \qquad (1)$$

where x_n is the number of resident foxes in year n, z_n is the number of immigrant foxes in year n, and a is a constant.[21] We further assume that the number of immigrant foxes z_n is proportional to $M - x_{n-1}$, so that

$$z_n = b(M - x_{n-1}). \qquad (2)$$

Use this model to solve Exercises 39–41.

39. Use equations (1) and (2) to find a linear difference equation for x_n that does not involve z_n.

40. Suppose that $a = 1.15$, $b = 0.01$, and $M = 1000$.

a) Find the general solution to the difference equation for x_n.

b) Find the particular solution for x_n, given that $x_0 = 100$ and $x_1 = 125$.

tw c) Determine $\lim_{n \to \infty} x_n$. Interpret the meaning of this limit.

41. Suppose that $a = 0.95$, $b = 0.1$, and $M = 1000$.

a) Find the general solution to the difference equation for x_n.

b) Find the particular solution for x_n, given that $x_0 = 50$ and $x_1 = 130$.

tw c) Find $\lim_{n \to \infty} x_n$. Interpret the meaning of this limit.

[21]E. H. Harding, D. F. Doak, J. D. Albertson, "Evaluating the effectiveness of predator control: The non-native red fox as a case study," *Conservation Biology*, Vol. 15, No. 4, pp. 1114–1122 (2001).

SYNTHESIS

42. The sequence $x_0 = 0$, $x_1 = 1$, $x_2 = 0$, $x_3 = 1$, $x_4 = 0$, $x_5 = 1,\ldots$ alternates between 0 and 1.

a) Find a second-order linear difference equation that this sequence satisfies. (*Hint:* A coefficient can be 0 in the recursion relation.)

b) Solve the difference equation you found in part (a) with the initial conditions $x_0 = 0$ and $x_1 = 1$ to find a formula for the sequence that alternates between 0 and 1.

43. The sequence $x_0 = 2$, $x_1 = 4$, $x_2 = 2$, $x_3 = 4$, $x_4 = 2$, $x_5 = 4,\ldots$ alternates between 2 and 4.

a) Find a second-order linear difference equation that this sequence satisfies.

b) Solve the difference equation you found in part (a) with the initial conditions $x_0 = 2$ and $x_1 = 4$ to find a formula for the sequence that alternates between 2 and 4.

c) Use part (b) to write a formula for the weight for the kth term used in Simpson's rule, where $1 \le k \le n - 1$.

44. *Fibonacci Numbers.* The Fibonacci numbers satisfy the linear difference equation

$$x_{n+1} = x_n + x_{n-1}$$

with $x_0 = 0$ and $x_1 = 1$.

a) Find the general solution to this difference equation. Do not approximate your answer with decimals.

b) Find the particular solution.

c) Use a calculator to compute x_3 and x_{20} using part (b).

tw d) The number of digits in the number a is approximately $\log_{10} a$. Explain why the nth Fibonacci number has approximately $n/5$ digits.

Repeated Roots. If the characteristic equation for a second-order linear difference equation has a double root r, then the general solution is of the form

$$x_n = c_1 r^n + c_2 n r^n.$$

For example, if the characteristic polynomial in factored form is $(r - 7)^2$, then 7 is a double root and the general solution of the difference equation is

$$x_n = c_1(7)^n + c_2 n(7)^n.$$

45. Find the general solution of the difference equation $x_{n+1} = 4x_n - 4x_{n-1}$.

46. Find the general solution of the difference equation $x_{n+1} = 2x - x_{n-1}$.

47. Find the general solution of $x_{n+1} = -2x_n - x_{n-1} + 12$.

48. Find the particular solution of
$$x_{n+1} = 6x_n - 9x_{n-1}, x_0 = 7, x_1 = 21.$$

In Exercises 49–52, assume that $a > 0$ and $b > 0$ and consider the characteristic equation $r^2 - ar - b = 0$ with roots r_1 and r_2. Use Exercise 49 to complete Exercises 50–52.

49. Use the quadratic formula to write the solutions of the characteristic equation. You will need this formula for the rest of the problem.

a) Explain why there are exactly two real roots. (*Hint:* Are we taking the square root of a positive number, 0, or a negative number?)
b) Explain why one root is positive and the other is negative.
c) Let r_1 be the positive root and r_2 be the negative root. Explain why $r_1 > |r_2|$.

50. Assume that $a + b = 1$.

a) Show that $r_1 = 1$.
b) Show that if $x_n = c_1 r_1^n + c_2 r_2^n$ is a solution of $x_{n+1} = ax_n + bx_{n-1}$, then $\lim_{n \to \infty} x_n = c_1$.

51. Assume that $a + b < 1$.

a) Show that $r_1 < 1$.
b) Show that if $x_n = c_1 r_1^n + c_2 r_2^n$ is a solution of $x_{n+1} = ax_n + bx_{n-1}$, then $\lim_{n \to \infty} x_n = 0$.

52. Assume that $a + b > 1$.

a) Show that $r_1 > 1$.
b) Show that if $x_n = c_1 r_1^n + c_2 r_2^n$ is a solution of $x_{n+1} = ax_n + bx_{n-1}$, then $\lim_{n \to \infty} x_n = \infty$.

tw 53. Give a biological explanation of why the three conditions of Theorem 13 are reasonable.

54. Show that if x_n and x'_n are two solutions to the linear difference equation
$$x_{n+1} = ax_n + bx_{n-1},$$
then $x_n + x'_n$ and cx_n are also solutions, where c is a constant.

 Technology Connection

Plant Mass. In the absence of competitors and herbivores, plant growth can be modeled by the recursion relation

$$M_{n+1} = \frac{(1 + \rho)M_n}{1 + \theta M_n},$$

where M_n is the total plant mass after $3n$ days, ρ is the maximum growth rate, and θ is a constant.[22] Use this model to solve Exercises 55–57.

55. For a plant, $\rho = 0.3$, $\theta = 0.001$, and the starting mass is $M_0 = 1$ g.

a) Use the recursion relation to compute the total mass of the plant for days 3, 6, 9, 12, and 15 ($n = 1, 2, 3, 4, 5$).
tw b) Compare your answers to $(1.3), (1.3)^2, (1.3)^3, (1.3)^4,$ and $(1.3)^5,$ respectively. Explain.

56. For a plant, $\rho = 0.3$, $\theta = 0.001$, and the starting mass is $M_0 = 295$ g.

a) Solve the equation
$$M = \frac{(1 + \rho)M}{1 + \theta M}$$
to determine a nonzero equilibrium value for the mass of the plant.
b) Use the recursion relation to compute the total mass of the plant on days 3, 6, 9, 12, and 15 ($n = 1, 2, 3, 4, 5$).
tw c) Explain why your answers to part (b) are expected.

57. For a plant, $\rho = 0.3$, $\theta = 0.001$, and the starting mass is $M_0 = 1$ g.

a) Plot the points (n, x_n) for $n = 0, 1, 2, 3, \dots, 40$.
tw b) Does this graph look familiar?

---

[22]A. E. Weis and M. E. Hochberg, "The diverse effects of intraspecific competition on the selective advantage to resistance: A model and its predictions," *The American Naturalist,* Vol. 156, No. 3, pp. 276–292 (Sept. 2000).

Chapter 6 Summary and Review

Terms to Know

Review Exercises

These review exercises are for test preparation. They can also be used as a lengthened practice test. Answers are at the back of the book. The answers also contain bracketed section references, which tell you where to restudy if your answer is incorrect.

Let $A = \begin{bmatrix} 2 & -4 \\ 7 & 3 \end{bmatrix}$, $B = \begin{bmatrix} 1 & -3 & 2 \\ -4 & 7 & 5 \end{bmatrix}$,

$C = \begin{bmatrix} 2 & 8 \\ -2 & 1 \\ 7 & 4 \end{bmatrix}$, $D = \begin{bmatrix} 5 & 6 & -2 \\ -9 & 3 & 0 \\ 1 & -3 & 5 \end{bmatrix}$, and $v = \begin{bmatrix} 3 \\ 5 \end{bmatrix}$.

Compute the following.

1. 3B

2. AC

3. BC

4. A + B

5. B + C

6. CB

7. 2A + 3BC

8. BD

9. ABD

10. Av

11. A^3v

12. D^2Cv

Solve the following systems of linear equations using Gaussian elimination.

13. $\begin{aligned} 2x + 3y &= 11 \\ 5x - 2y &= -39 \end{aligned}$

14. $\begin{aligned} x + y + z &= 4 \\ 2x - y - 3z &= -5 \\ 3x - 3y - 3z &= -6 \end{aligned}$

15. $4x + 2y + 3z = 9$
    $2x - y - 3z = -1$
    $3x - 3y - z = 15$

Find the determinants. Is the matrix invertible?

16. $\det[-3]$

17. $\det[0]$

18. $\begin{vmatrix} -3 & 6 \\ -5 & 2 \end{vmatrix}$

19. $\begin{vmatrix} 0.5 & 5 \\ 0.2 & 2 \end{vmatrix}$

20. $\begin{vmatrix} 1 & -2 & -1 \\ 0 & 3 & -4 \\ 2 & 0 & 1 \end{vmatrix}$

21. $\begin{vmatrix} 7 & -3 & 5 \\ 2 & 6 & 1 \\ 16 & 0 & 11 \end{vmatrix}$

Find A^{-1}.

22. $A = \begin{bmatrix} 3 & 0 \\ 1 & -1 \end{bmatrix}$

23. $A = \begin{bmatrix} 1 & 3 \\ -2 & -1 \end{bmatrix}$

24. $A = \begin{bmatrix} -5 & 1 & 7 \\ -4 & 2 & 6 \\ -1 & 1 & 1 \end{bmatrix}$

25. $A = \begin{bmatrix} -20 & 7 & 10 \\ -27 & 10 & 13 \\ -14 & 5 & 7 \end{bmatrix}$

Find all the eigenvalues and the corresponding eigenvectors.

26. $\begin{bmatrix} 3 & 0 \\ 2 & 4 \end{bmatrix}$

27. $\begin{bmatrix} 4 & 2 \\ 0 & -1 \end{bmatrix}$

28. $\begin{bmatrix} 15 & 28 \\ -6 & -11 \end{bmatrix}$

29. $\begin{bmatrix} 15 & -42 \\ 4 & -11 \end{bmatrix}$

30. $\begin{bmatrix} 5 & 0 & 0 \\ 0 & -2 & 4 \\ 0 & 0 & -3 \end{bmatrix}$

31. $\begin{bmatrix} -7 & 0 & 0 \\ 3 & 2 & 0 \\ 4 & 6 & -1 \end{bmatrix}$

32. $\begin{bmatrix} 4 & 2 & 0 \\ -1 & 1 & 0 \\ -5 & -10 & -3 \end{bmatrix}$

33. $\begin{bmatrix} 3 & 0 & 2 \\ -4 & 0 & -3 \\ -4 & 0 & -3 \end{bmatrix}$

34. $\begin{bmatrix} 4 & 1 & -1 \\ 6 & -9 & 9 \\ 12 & -18 & 18 \end{bmatrix}$

Find the particular solution to the difference equation.

35. $x_{n+1} = -2x_n + 8x_{n-1}; x_0 = 3, x_1 = 0$

36. $x_{n+1} = -\dfrac{3}{2}x_n + x_{n-1} + \dfrac{3}{2}; x_0 = 0, x_1 = 2$

APPLICATIONS

37. *Equilibrium.* Two mg of glactosyl human serum albumin (Tc-GSA) is injected into the blood stream. Let B and L be the amount of Tc-GSA in the blood

and liver, respectively. The quantities B and L are related by[23]

$$2 = B + L$$
$$\frac{dB}{dt} = -0.06B + 0.03L$$
$$\frac{dL}{dt} = 0.06B - 0.03L.$$

At equilibrium, the quantities B and L are constant. Find the equilibrium amount of Tc-GSA in the blood and the liver.

38. *Bird Population.* The Leslie diagram for a population of birds is given below.

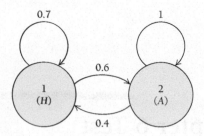

The population vector in year 1 is

$$\mathbf{p} = \begin{bmatrix} \text{hatchlings} \\ \text{adults} \end{bmatrix} = \begin{bmatrix} 60 \\ 40 \end{bmatrix}.$$

a) Write the Leslie matrix.
b) Estimate the population vector for year 2.
c) Estimate the population vector for year 3.
d) Find the long-term growth rate.

39. *Bird Population.* The Leslie matrix for a population of birds is

$$G = \begin{bmatrix} 0.8 & 2 \\ 0.5 & 0.4 \end{bmatrix}.$$

The population vector in year 1 is

$$\mathbf{p} = \begin{bmatrix} \text{hatchlings} \\ \text{adults} \end{bmatrix} = \begin{bmatrix} 100 \\ 60 \end{bmatrix}.$$

a) Estimate the population vector for year 2.
b) Estimate the population vector for year 3.
c) Find the long-term growth rate.

[23]K. Kinoshita, M. Ukikusa, K. Iwaisako, A. Arimoto, N. Fujisawa, T. Ozaki, H. Tanaka, S. Seo, M. Naitoh, A. Nomura, T. Inomoto, T. Kitai, K. Ino, H. Higashiyama, T. Hanafusa and Y. Nakajima, "Preoperative assessment of heptic function: Utility of a new convenient two-compartment model analysis using galactosyl human serum albumin scintigraphy," *Journal of Gastroenterology and Hepatology*, Vol. 18, pp. 99–104 (2003).

SYNTHESIS

tw **40.** Explain the connections between a matrix equation and a system of linear equations.

 Technology Connection

41. Let $A = \begin{bmatrix} 1 & 0 \\ 3 & 0.5 \end{bmatrix}$.

   a) Find the eigenvalues.
   b) For each eigenvalue, find the corresponding eigenvectors.
   c) Compute A^{100}.

42. Let $A = \begin{bmatrix} 1 & 0 \\ 4 & 0.9 \end{bmatrix}$.

   a) Find the eigenvalues.
   b) For each eigenvalue, find the corresponding eigenvectors.
   c) Compute A^{100}.

Chapter 6 Test

Compute.

1. $\begin{bmatrix} 7 & -2 \\ 5 & 3 \end{bmatrix} \begin{bmatrix} 1 & 6 & -3 \\ 5 & -1 & 4 \end{bmatrix}$

2. $\begin{bmatrix} 10 & -4 \\ 2 & -5 \\ 7 & 2 \end{bmatrix} \begin{bmatrix} -5 & 6 & 0 \\ -2 & 1 & 8 \end{bmatrix}$

3. $\begin{bmatrix} 1 & 6 & -3 \\ 5 & -1 & 4 \end{bmatrix} \begin{bmatrix} 10 & -4 \\ 2 & -5 \\ 7 & 2 \end{bmatrix} - \begin{bmatrix} 7 & -2 \\ 5 & 3 \end{bmatrix}$

Solve the following systems of linear equations using Gaussian elimination.

4. $2x + 3y = -1$
   $3x + 2y = 6$

5. $-x + 3y + 2z = -3$
   $3x - 4y + 2z = 26$
   $4x + 2y - 3z = -10$

Find the determinant of the matrix and determine if the matrix is invertible.

6. $\begin{bmatrix} 2 & 8 \\ 1 & 4 \end{bmatrix}$

7. $\begin{bmatrix} -3 & 6 \\ -1 & 2 \end{bmatrix}$

8. $\begin{bmatrix} 0.5 & 3 & 1.2 \\ 2.4 & 0.5 & 0.2 \\ 1.3 & 3.4 & 7.3 \end{bmatrix}$

9. $\begin{bmatrix} 2 & -5 & 0 \\ 3 & 1 & 8 \\ 1 & -1 & 4 \end{bmatrix}$

Find the inverse matrix.

10. $\begin{bmatrix} 3 & 4 \\ 1 & 1 \end{bmatrix}$

11. $\begin{bmatrix} 0 & -1 & -2 \\ 1 & 1 & 2 \\ 0 & 1 & 1 \end{bmatrix}$

Find the eigenvalues and the corresponding eigenvectors.

12. $\begin{bmatrix} 11 & 20 \\ -6 & -11 \end{bmatrix}$

13. $\begin{bmatrix} 15 & 4 \\ -42 & -11 \end{bmatrix}$

14. $\begin{bmatrix} -5 & -4 & -8 \\ -2 & 5 & 4 \\ 2 & -2 & -1 \end{bmatrix}$

15. $\begin{bmatrix} -1 & 0 & -8 \\ -6 & -7 & 8 \\ 4 & 0 & 11 \end{bmatrix}$

16. $\begin{bmatrix} -6 & -11 & -22 \\ 0 & 7 & 2 \\ 0 & -1 & 4 \end{bmatrix}$

Find the particular solution to the difference equation.

17. $x_{n+1} = 3x_n + 10x_{n-1}; x_0 = -1, x_1 = 16$

18. $x_{n+1} = 2x_n + 8x_{n-1} - 27; x_0 = 1, x_1 = 1$

APPLICATIONS

19. *Home Air Quality.* Chemicals enter a house's basement air. Let $F(t)$ and $B(t)$ be the amount of the chemical (in mg/m^3) in the first-floor air and the basement air after t minutes, respectively. The rate of change of F and B are given by the equations[24]

$$B' = -0.015B + 0.09 \quad \text{and}$$
$$F' = 0.01B - 0.025F.$$

Find the equilibrium values for B and F.

20. *Bird Population.* The Leslie matrix for a population of birds is

$$G = \begin{bmatrix} 0.6 & 3 \\ 0.4 & 0.2 \end{bmatrix}.$$

The population vector for year 1 is

$$\mathbf{p} = \begin{bmatrix} \text{hatchlings} \\ \text{adults} \end{bmatrix} = \begin{bmatrix} 200 \\ 180 \end{bmatrix}.$$

a) Estimate the population vector for year 2.
b) Estimate the population vector for year 3.

21. *Bird Population.* The 2×2 Leslie matrix for a population of birds has eigenvalues $r = 1.1$ and $r = 0.9$. Find the long-term growth rate for the population.

---

[24]D. A. Olson and R. L. Corsi, "Characterizing exposure to chemicals from soil vapor intrusion using a two-compartment model," *Atmospheric Environment*, Vol. 35, pp. 4201–4209 (2001).

SYNTHESIS

22. Let $\mathbf{A} = \begin{bmatrix} 1 & 6 \\ 5 & 1 \end{bmatrix}$ and $\mathbf{B} = \begin{bmatrix} 1 & 3 \\ 4 & 1 \end{bmatrix}$.

   a) Compute $\mathbf{A}^{-1}$.
   b) Compute $\mathbf{B}^{-1}$.
   c) Compute $\mathbf{A}^{-1}\mathbf{B}^{-1}$.
   d) Compute $(\mathbf{AB})^{-1}$.
   e) Compute $(\mathbf{BA})^{-1}$.
 f) Compare your answers to parts (c), (d), and (e). Explain.

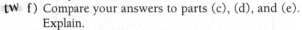 *Technology Connection*

23. *Bird Population.* The Leslie matrix for a population of birds is

$$G = \begin{bmatrix} 0.5 & 2.6 \\ 0.6 & 0.3 \end{bmatrix}.$$

The population vector for year 1 is

$$\mathbf{p} = \begin{bmatrix} \text{hatchlings} \\ \text{adults} \end{bmatrix} = \begin{bmatrix} 150 \\ 140 \end{bmatrix}.$$

a) Estimate the population vector for year 2.
b) Estimate the population vector for year 3.
c) Estimate the population vector for year 20.
d) Find the long-term growth rate.

Extended Life Science Connection

POPULATION GROWTH

The exponential growth model provides a convenient way to approximate the growth of a population. However, Leslie matrices often provide a better model by taking into account the different stages in an organism's life cycle. The purpose of this Extended Life Science Connection is to analyze the growth or decay of a population based on the fecundity and survivability of the species.

We consider an organism whose life cycle has two stages. *Biennial* plants such as carrots provide examples of organisms of this type. We make the assumption that all organisms die at the end of the second time period. The Leslie diagram for this two-stage model is given on the following page.

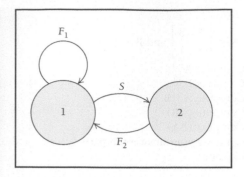

The number F_1 is the fecundity for organisms in the first stage of life, and F_2 is the fecundity for organisms in the second stage of life. The proportion of the organisms that survives from the first stage to the second stage is S.

We investigate which values of the parameters F_1, F_2, and S make the population grow, shrink, or stay at equilibrium. According to Theorem 11, the largest eigenvalue determines the long-term growth rate. The long-term growth rate is greater than 1 for a growing population, less than 1 for a shrinking population, and equal to 1 for a population at equilibrium. In the case of a shrinking population, the species will eventually become extinct unless the parameters change.

EXERCISES

tw **1.** Explain why $0 \le S \le 1$. What does it mean if $S = 0$? What does it mean if $S = 1$? We assume that $0 < S < 1$ for the remainder of the exercises.

tw **2.** Explain why it is reasonable to assume that $0 \le F_1$ and $0 \le F_2$. We make these assumptions for the remaining exercises.

3. Write the Leslie matrix G for the two-stage model.

4. Find the long-term growth rate, assuming that $F_1 = 0.5$, $F_2 = 2$, and $S = 0.25$.

5. Assume that $F_2 = 2$ and $S = 0.25$. Determine if the long-term growth rate is greater than 1 or less than 1 assuming $F_1 < 0.5$, and then assuming $F_1 > 0.5$.

6. Find the characteristic polynomial for G. Your answer will involve F_1, F_2, and S.

After its first contact with humans in 1598, the docile dodo was extinct within 83 years. How did its survivability change in 1598?

7. Show that the two eigenvalues are
$$r_1 = \frac{F_1}{2} + \frac{\sqrt{F_1^2 + 4F_2 S}}{2} \text{ and}$$
$$r_2 = \frac{F_1}{2} - \frac{\sqrt{F_1^2 + 4F_2 S}}{2}.$$

tw **8.** Explain why r_1 is positive and $r_1 > |r_2|$.

9. Show that if $F_1 = 1 - F_2 S$, then $r_1 = 1$.

10. Show that if $F_1 > 1 - F_2 S$, then $r_1 > 1$.

11. Show that if $F_1 < 1 - F_2 S$, then $r_1 < 1$.

12. Explain how the number $F_1 + F_2 S$ determines the long-term growth rate. Is this condition consistent with Exercises 4 and 5?

7

Functions of Several Variables

INTRODUCTION *Functions that have more than one input are called functions of several variables. We introduce these functions in this chapter and learn to differentiate them to find* partial derivatives. *Then we use these functions and their partial derivatives to solve maximum–minimum problems. Finally, we consider the integration of such functions.*

AN APPLICATION

The birth weight w of a baby (in grams) may be predicted using the function[1]

$$w(x, s, h, m, r, p) = x(9.38 + 0.264s + 0.000233hm + 4.62r[p + 1]).$$

In this function, x is the gestation age (in days), s is the sex of the child (1 for male, -1 for female), h is the maternal height (in centimeters), m is the maternal weight at the 26th week of pregnancy (in kilograms), r is the maternal daily weight gain (in kg/day), and p is the number of the mother's previous children. Find $w(275, 1, 160, 71, 0.068, 0)$ and $w(282, -1, 171, 76, 0.085, 3)$.

This is an example of a *function of several variables*. In this chapter, we will learn how the concepts of calculus extend to such functions.

This problem appears as Exercise 11 in Section 7.1.

[1]G. G. Nahum and H. Stanislaw, "Validation of a Birth Weight Prediction Equation Based on Maternal Characteristics," *Journal of Reproductive Medicine*, Vol. 47, 752–760 (2002).

7.1 Functions of Several Variables

OBJECTIVE

■ Find a function value for a function of several variables.

Suppose a child who is 42 in. (or 3.5 ft) tall weighs x pounds. Then her *body mass index B* is given by

$$B(x) = \frac{703x}{(42)^2}.$$

This is a function of one variable.

As the child grows, both her height and her weight change. Suppose her weight is x pounds when her height is y inches. Then her body mass index is given by

$$B(x, y) = \frac{703x}{y^2}.$$

This function assigns the input pair (x, y) to a unique output number $703x/y^2$.

DEFINITION

A **function of two variables** assigns to each input pair (x, y) exactly one output number $f(x, y)$.

We can also think of a function of two variables as a machine that has two inputs. The domain of a function of two variables is a set of pairs (x, y) in the plane. Unless otherwise restricted, when such a function is given by a formula, the domain consists of all ordered pairs (x, y) that are meaningful replacements in the formula.

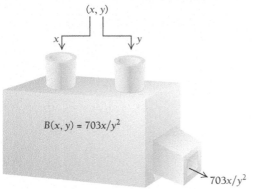

EXAMPLE 1   For $B(x, y) = 703x/y^2$, find $B(125, 63)$.

Solution $B(125, 63)$ is defined to be the value of the function found by substituting 125 for x and 63 for y:

$$B(125, 63) = \frac{703(125)}{(63)^2}$$
$$\approx 22.14.$$

Doctors use the body mass index B to determine if a patient is overweight. A person with a body mass index over 25 is considered overweight, while a body mass index over 30 corresponds to someone who is obese.[2]

The following are examples of **functions of several variables,** that is, functions of two or more variables. If there are n variables, then there are n inputs for such a function.

[2]National Center for Emergency Medicine Informatics.

EXAMPLE 2   *Fish Growth.* The relative growth rate G (in percent per day) of juvenile blue tilapia (*Oreochromis aureus*) may be modeled by the function $G(m, t) = 10^{g(m,t)} - 1$, where

$$g(m, t) = -78.2911 + 0.4655 \log m + 104.9699 \log t$$
$$- 34.3378(\log t)^2 - 0.3271 \log m\,(\log t)^2.$$

In this formula, m is the body mass (in milligrams) and t is the ambient temperature (in °C). Find the relative growth rate of a 100-mg juvenile blue tilapia that is reared in a 30°C environment.[3]

Solution   First, we substitute 100 for m and 30 for t in the function g:

$$g(100, 30) = -78.2911 + 0.4655 \log 100 + 104.9699 \log 30$$
$$- 34.3378(\log 30)^2 - 0.3271 \log 100(\log 30)^2$$
$$\approx -78.2911 + 0.4655(2) + 104.9699(1.47712)$$
$$- 34.3378(1.47712)^2 - 0.3271(2)(1.47712)^2$$
$$\approx 1.34457.$$

The relative growth rate is therefore

$$G(100, 30) = 10^{g(100,30)} - 1 \approx 10^{1.34457} - 1 \approx 21.1092.$$

The fish grow at a rate of about 21.1% per day.   ◼

This NASA photograph shows an overhead view of San Francisco and Oakland. Sociologists say that as two cities merge, the communication between them increases.

Technology Connection

EXERCISE

1. The population of San Francisco in 2000 was 777,000, while the population of Oakland was 399,000. The distance between the cities is 8 miles. What is the average number of phone calls in a day between these two cities?
Source: U.S. Bureau of the Census.

EXAMPLE 3   *Communications.* The average number of telephone calls in a day between two cities can be estimated by

$$N(d, P_1, P_2) = \frac{2.8 P_1 P_2}{d^{2.4}},$$

where d is the distance, in miles, between the cities and P_1 and P_2 are their populations.   ◼

A constant can also be thought of as a function of several variables.

EXAMPLE 4   The constant function f is given by

$$f(x, y) = -3 \quad \text{for all inputs } x \text{ and } y.$$

Find $f(5, 7)$ and $f(-2, 0)$.

---

[3] E. Baras, A. Mpo'n'tcha, H. Driouch, C. Prignon, and C. Mélard, "Ontogenetic variations of thermal optimum for growth, and its implication on thermolabile sex determination in blue tilapia," *Journal of Fish Biology,* Vol. 61, pp. 645–660 (2002).

Solution   Since this is a constant function, it has the value -3 for any x and y. Thus,

$$f(5, 7) = -3 \quad \text{and} \quad f(-2, 0) = -3.$$

Geometric Interpretations

Previously, we considered functions of one variable, represented by $y = f(x)$. To graph such functions, we needed two dimensions: one for the inputs (the x-axis) and one for the outputs (the y-axis).

Now consider a function of two variables

$$z = f(x, y).$$

As a mapping, a function of two variables can be thought of as mapping a point (x_1, y_1) in an xy-plane onto a point z_1 on a number line.

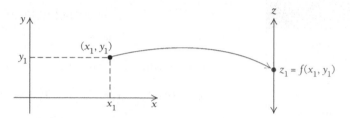

To graph a function of two variables, we need a three-dimensional coordinate system. Two dimensions are needed for the inputs (the x-axis and the y-axis), and another dimension is needed for the outputs (the z-axis). The axes are generally placed as shown below. The line z, called the z-axis, is placed perpendicular to the xy-plane at the origin.

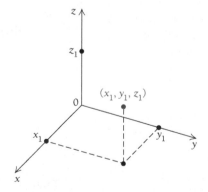

To help visualize this, think of looking into the corner of a room, where the floor is the xy-plane and the z-axis is the intersection of the two walls. To plot a point (x_1, y_1, z_1), we locate the point (x_1, y_1) in the xy-plane and move up or down in space according to the value of z_1.

EXAMPLE 5   Plot these points:

$$P_1(2, 3, 5), \quad P_2(2, -2, -4), \quad P_3(0, 5, 2), \quad \text{and} \quad P_4(2, 3, 0).$$

Solution   The solution is shown below.

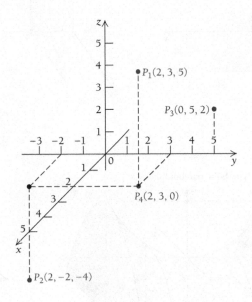

The *graph* of a function of two variables

$$z = f(x, y)$$

consists of ordered triples (x_1, y_1, z_1), where $z_1 = f(x_1, y_1)$. The domain of f is a region D in the xy-plane, and the graph of f is a surface S.

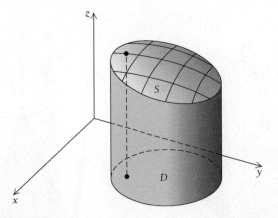

The following graphs have been generated from their equations by a computer graphics program called *Mathematica*. There are many excellent graphics programs for generating graphs of functions of two variables. The graphs can be generated with very few keystrokes. Seeing one of these graphs often reveals more insight into the behavior of the function than does its formula.

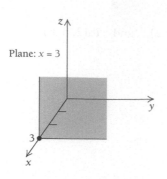

Plane: $x = 3$

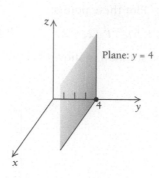

Plane: $y = 4$

Elliptic paraboloid: $z = x^2 + y^2$

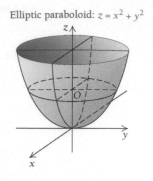

Hyperbolic paraboloid: $z = x^2 - y^2$

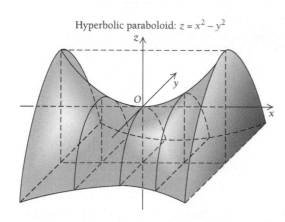

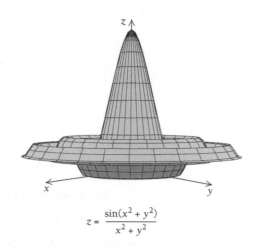

$$z = \frac{\sin(x^2 + y^2)}{x^2 + y^2}$$

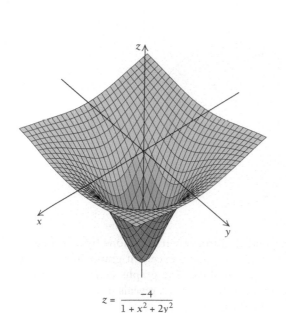

$$z = \frac{-4}{1 + x^2 + 2y^2}$$

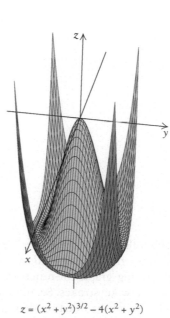

$$z = (x^2 + y^2)^{3/2} - 4(x^2 + y^2)$$

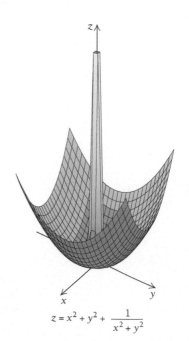

$$z = x^2 + y^2 + \frac{1}{x^2 + y^2}$$

Technology Connection

There are many 3D graphers capable of generating graphs of functions of two variables. Use one to verify the graphs shown here. What problems do you have with viewing windows?

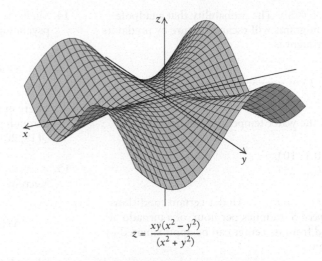

$$z = \frac{xy(x^2 - y^2)}{(x^2 + y^2)}$$

Exercise Set 7.1

1. For $f(x, y) = x^2 - 2xy$, find $f(0, -2)$, $f(2, 3)$, and $f(10, -5)$.

2. For $f(x, y) = (y^2 + 3xy)^3$, find $f(-2, 0)$, $f(3, 2)$, and $f(-5, 10)$.

3. For $f(x, y) = 3^x + 7xy$, find $f(0, -2)$, $f(-2, 1)$, and $f(2, 1)$.

4. For $f(x, y) = \log(x + y) + 3x^2$, find $f(3, 7)$, $f(1, 99)$, and $f(2, -1)$.

5. For $f(x, y) = \sin x \tan y$, find $f(\pi/2, 0)$, $f(3\pi/4, 2\pi/3)$, and $f(\pi/6, \pi/4)$.

6. For $f(x, y) = \sin(x \cos y)$, find $f(\pi, \pi/3)$, $f(\pi/3, 0)$, and $f(7\pi/6, \pi)$.

7. For $f(x, y, z) = x^2 - y^2 + z^2$, find $f(-1, 2, 3)$ and $f(2, -1, 3)$.

8. For $f(x, y, z) = 2^x + 5zy - x$, find $f(0, 1, -3)$ and $f(1, 0, -3)$.

APPLICATIONS

9. *Body Surface Area.* The Mosteller formula for approximating the surface area S in m^2 of a human is given by

$$S(h, w) = \sqrt{hw}/60,$$

where h is the height in centimeters and w is the weight in kilograms of the person. Use the Mosteller approximation to estimate the surface area of a person whose height is 165 cm and whose weight is 80 kg.[4]

10. *Body Surface Area.* The Haycock formula for approximating the surface area S in m^2 of a human is given by

$$S(h, w) = 0.024265h^{0.3964}w^{0.5378},$$

where h is the height in centimeters and w is the weight in kilograms of the person. Use the Haycock approximation to estimate the surface area of a person whose height is 165 cm and whose weight is 80 kg.[5]

11. *Birth Weights.* The birth weight w of a baby (in grams) may be predicted using the function

$$w(x, s, h, m, r, p) = x(9.38 + 0.264s$$
$$+ 0.000233hm$$
$$+ 4.62r[p + 1]).$$

In this function, x is the gestation age (in days), s is the sex of the child (1 for male, -1 for female), h is the maternal height (in centimeters), m is the maternal weight at the 26th week of pregnancy (in kilograms), r is the maternal daily weight gain (in kg/day), and p is the number of the mother's previous children.[6]

   a) Find $w(275, 1, 160, 71, 0.068, 0)$.
   b) Find $w(282, -1, 171, 76, 0.085, 3)$.

[4] http://www.halls.md/body-surface-area/refs.htm

[5] http://www.halls.md/body-surface-area/refs.htm
[6] G. G. Nahum and H. Stanislaw, "Validation of a Birth Weight Prediction Equation Based on Maternal Characteristics," *Journal of Reproductive Medicine*, Vol. 47, 752–760 (2002).

12. *Capture Probability.* The probability that a tadpole with mass m grams will escape capture by predators in an experiment is

$$P(m, T) = \frac{1}{1 + e^{3.222 - 31.669m + 0.083T}},$$

where T is the water temperature in degrees Celsius.[7]

a) Find $P(0.1, 10)$.
b) Find $P(0.2, 25)$.

13. *Wind Speed of a Tornado.* Under certain conditions, the *wind speed S*, in miles per hour, of a tornado at a distance d from its center can be approximated by the function

$$S = \frac{aV}{0.51d^2},$$

where a is a constant that depends on certain atmospheric conditions and V is the approximate volume of the tornado, in cubic feet. Approximate the wind speed 100 ft from the center of a tornado when its volume is 1,600,000 ft³ and $a = 0.78$.

[7]M. T. Anderson, J. M. Kiesecker, D. P. Chivers, and A. R. Blaustein, "The direct and indirect effects of temperature on a predator-prey relationship," *Canadian Journal of Zoology,* Vol. 79, pp. 1834–1841 (2001).

14. *Intelligence Quotient.* The *intelligence quotient* in psychology is given by

$$Q(m, c) = 100 \cdot \frac{m}{c},$$

where m is a person's mental age and c is his or her chronological, or actual, age. Find $Q(21, 20)$ and $Q(19, 20)$.

15. *Poiseuille's Law.* The speed of blood in a vessel is given by

$$V(L, p, R, r, v) = \frac{p}{4Lv}(R^2 - r^2),$$

where R is the radius of the vessel, r is the distance of the blood from the center of the vessel, L is the length of the blood vessel, p is the pressure, and v is the viscosity. Find $V(1, 100, 0.0075, 0.0025, 0.05)$.

SYNTHESIS

tw 16. Explain the difference between a function of two variables and a function of one variable.

tw 17. Find some examples of functions of several variables not considered in the text, even some that may not have formulas.

 Technology Connection

Wind Chill Temperature. Because wind speed enhances the loss of heat from the skin, we feel colder when there is wind than when there is not. The *wind chill temperature* is what the temperature would have to be with no wind in order to give the same chilling effect. The wind chill temperature W is given by

$$W(v, T) = 91.4 - \frac{(10.45 + 6.68\sqrt{v} - 0.447v)(457 - 5T)}{110},$$

where T is the actual temperature as given by a thermometer, in degrees Fahrenheit, and v is the speed of the wind, in miles per hour. Find the wind chill temperature in each case. Round to the nearest one degree.

18. $T = 20°F, v = 20$ mph

19. $T = 20°F, v = 40$ mph

20. $T = -10°F, v = 30$ mph

Use a 3D grapher to generate the graph of the function.

21. $f(x, y) = y^2$

22. $f(x, y) = x^2 + y^2$

23. $f(x, y) = (x^4 - 16x^2)e^{-y^2}$

24. $f(x, y) = 4(x^2 + y^2) - (x^2 + y^2)^2$

25. $f(x, y) = x^3 - 3xy^2$

26. $f(x, y) = \dfrac{1}{x^2 + 4y^2}$

7.2 Partial Derivatives

OBJECTIVES

■ Compute partial derivatives.
■ Approximate functions using partial derivatives.
■ Find the four second-order partial derivatives of a function in two variables.
■ Find the Jacobian of two functions of two variables.

Finding Partial Derivatives

Consider the function f given by

$$z = f(x, y) = x^2y^3 + xy + 4y^2.$$

Suppose for the moment that we fix y at 3. Then

$$f(x, 3) = x^2(3^3) + x(3) + 4(3^2) = 27x^2 + 3x + 36.$$

Note that we now have a function of only one variable. Taking the first derivative with respect to x, we have

$$54x + 3.$$

In general, without replacing y with a specific number, let's consider y fixed. Then f becomes a function of x alone and we can calculate its derivative with respect to x. This derivative is called the *partial derivative of f with respect to x.* Notation for this partial derivative is

$$\frac{\partial f}{\partial x} \quad \text{or} \quad \frac{\partial z}{\partial x}.$$

Thus let's again consider the function

$$z = f(x, y) = x^2y^3 + xy + 4y^2.$$

The color blue indicates the variable x when we fix y and treat it as a constant. The expressions y^3, y, and y^2 are then constants. We have

$$\frac{\partial f}{\partial x} = \frac{\partial z}{\partial x}$$

$$= 2xy^3 + y.$$

Similarly, we find $\partial f/\partial y$ or $\partial z/\partial y$ by fixing x (treating it as a constant) and calculating the derivative with respect to y. From

$$z = f(x, y) = x^2y^3 + xy + 4y^2, \qquad \text{The color blue indicates the variable.}$$

we get *the partial derivative with respect to y*:

$$\frac{\partial f}{\partial y} = \frac{\partial z}{\partial y}$$

$$= 3x^2y^2 + x + 8y.$$

A definition of **partial derivatives** is as follows.

DEFINITION

For $z = f(x, y)$,

$$\frac{\partial z}{\partial x} = \lim_{h \to 0} \frac{f(x + h, y) - f(x, y)}{h},$$

$$\frac{\partial z}{\partial y} = \lim_{h \to 0} \frac{f(x, y + h) - f(x, y)}{h}.$$

We can find partial derivatives of functions of any number of variables.

EXAMPLE 1   For $w = x^2 - xy + y^2 + 2yz + 2z^2 + z$, find

$$\frac{\partial w}{\partial x}, \quad \frac{\partial w}{\partial y}, \quad \text{and} \quad \frac{\partial w}{\partial z}.$$

Solution   In order to find $\partial w/\partial x$, we consider x the variable and the other letters the constants. From

$$w = x^2 - xy + y^2 + 2yz + 2z^2 + z,$$

we get

$$\frac{\partial w}{\partial x} = 2x - y.$$

From

$$w = x^2 - xy + y^2 + 2yz + 2z^2 + z,$$

we get

$$\frac{\partial w}{\partial y} = -x + 2y + 2z;$$

and from

$$w = x^2 - xy + y^2 + 2yz + 2z^2 + z,$$

we get

$$\frac{\partial w}{\partial z} = 2y + 4z + 1. \qquad \blacksquare$$

We will often make use of a simpler notation f_x for the partial derivative of f with respect to x and f_y for the partial derivative of f with respect to y. If $z = f(x, y)$, then z_x represents the partial derivative of z with respect to x and z_y represents the partial derivative of z with respect to y.

EXAMPLE 2   For $f(x, y) = 3x^2y + xy$, find f_x and f_y.

Solution   We have

$$f_x = 6xy + y,$$
$$f_y = 3x^2 + x.$$

For the function in the preceding example, let's evaluate f_x at $(2, -3)$:

$$f_x(2, -3) = 6 \cdot 2 \cdot (-3) + (-3)$$
$$= -39.$$

If we use the notation $\partial f / \partial x = 6xy + y$, where $f(x, y) = 3x^2y + xy$, the value of the partial derivative at $(2, -3)$ is given by

$$\left. \frac{\partial f}{\partial x} \right|_{(2, -3)} = 6 \cdot 2 \cdot (-3) + (-3)$$
$$= -39,$$

but this notation is not as convenient as $f_x(2, -3)$.

EXAMPLE 3   For $f(x, y) = e^{xy} + y \ln x$, find f_x and f_y.

Solution

$$f_x = y \cdot e^{xy} + y \cdot \frac{1}{x}$$
$$= ye^{xy} + \frac{y}{x},$$
$$f_y = x \cdot e^{xy} + 1 \cdot \ln x$$
$$= xe^{xy} + \ln x.$$

Technology Connection

Consider finding values of a partial derivative of $f(x, y) = 3x^3y + 2xy$ using a grapher that finds derivatives of functions of one variable. How can you find $f_x(-4, 1)$? Then how can you find $f_y(2, 6)$?

The Geometric Interpretation of Partial Derivatives

The *graph* of a function of two variables $z = f(x, y)$ is a surface S that might have a graph similar to the one shown below, where each input pair (x, y) has only one output $z = f(x, y)$.

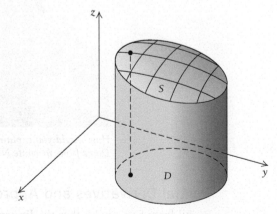

Now suppose that we hold x fixed, say, at the value a. The set of all points for which $x = a$ is a plane parallel to the yz-plane, so when x is fixed at a, y and z vary along the plane, as shown below. The plane in the figure cuts the surface in some curve C_1. The partial derivative f_y gives the slope of tangent lines to this curve.

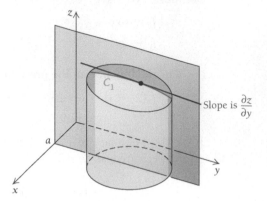

Similarly, if we hold y fixed, say, at the value b, we obtain a curve C_2, as shown. The partial derivative f_x gives the slope of tangent lines to this curve.

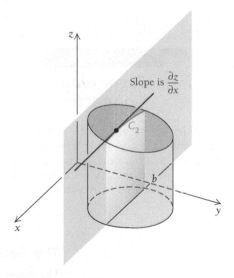

How would you explain partial derivatives using Half Dome from Yosemite National Park?

Partial Derivatives and Approximation

Recall from Section 3.6 that the **linearization**

$$f(x) \approx f(a) + f'(a)(x - a)$$

could be used to estimate the value of $f(x)$ given the value of $f(a)$. Linearization can also be used to approximate functions of two or more variables. For a function $f(x, y)$ of two variables, the linearization is

$$f(x, y) = f(a, b) + f_x(a, b)(x - a) + f_y(a, b)(y - b).$$

To use this formula, we need to evaluate $f, f_x,$ and f_y at the point (a, b).

EXAMPLE 4   *Body Mass Index.* The body mass index is defined by $B(x, y) = 703x/y^2$, where x is weight in pounds and y is height in inches.

a) A child weighs 78 lb and is 48 in. tall. Find the child's body mass index.

b) Over the next 6 mo, the child grows to 82 lb and 48.5 in. Use linearization to estimate the child's new body mass index.

Solution

a) $B(78, 48) = (703)(78)/(48)^2 \approx 23.80$

b) For this problem, we take $(x, y) = (82, 48.5)$, which is near the original point $(a, b) = (78, 48)$. Next, we find the partial derivatives.

$$B_x = \frac{703}{y^2} \quad \text{and} \quad B_y = -\frac{1406x}{y^3}$$

Therefore,

$$B_x(78, 48) = \frac{703}{(48)^2} \approx 0.3051$$

and

$$B_y(78, 48) = -\frac{1406(78)}{(48)^3} \approx -0.9916.$$

The new body mass index may be approximated as

$$B(82, 48.5) \approx B(78, 48) + B_x(78, 48)(82 - 78)$$
$$+ B_y(78, 48)(48.5 - 48)$$
$$\approx 23.80 + (0.3051)(4) + (-0.9916)(0.5)$$
$$\approx 24.52.$$

This is very close to the exact answer, $B(82, 48.5) \approx 24.51$.   ■

In Section 9.5, this two-variable linearization will be used to study the population growth of two interacting species.

Jacobian Matrix

An important matrix that is used to study *pairs* of functions of two variables (for example, the population growth of two interacting species) is called the **Jacobian.**

DEFINITION

Suppose that $f(x, y)$ and $g(x, y)$ are differentiable functions of x and y. Then the *Jacobian* of f and g is the matrix

$$\begin{bmatrix} \dfrac{\partial f}{\partial x} & \dfrac{\partial f}{\partial y} \\[2ex] \dfrac{\partial g}{\partial x} & \dfrac{\partial g}{\partial y} \end{bmatrix}.$$

EXAMPLE 5   Find the Jacobian of $f(x, y) = 2x - 3y^2$ and $g(x, y) = 3xy^3$.

Solution   Using the definition,

$$\begin{bmatrix} \dfrac{\partial f}{\partial x} & \dfrac{\partial f}{\partial y} \\[2ex] \dfrac{\partial g}{\partial x} & \dfrac{\partial g}{\partial y} \end{bmatrix} = \begin{bmatrix} 2 & -6y \\ 3y^3 & 9xy^2 \end{bmatrix}.$$

In Section 9.5, we will see how the eigenvalues of the Jacobian are used to characterize population growth.

Higher-Order Partial Derivatives

Consider

$$z = f(x, y) = 3xy^2 + 2xy + x^2.$$

Then

$$\frac{\partial z}{\partial x} = \frac{\partial f}{\partial x} = 3y^2 + 2y + 2x.$$

Suppose that we continue and find the partial derivative of $\partial z/\partial x$ with respect to y. This will be a **second-order partial derivative** of the original function z. Its notation is as follows:

$$\frac{\partial}{\partial y}\left(\frac{\partial z}{\partial x}\right) = \frac{\partial}{\partial y}\left(\frac{\partial f}{\partial x}\right)$$
$$= \frac{\partial^2 z}{\partial y\, \partial x}$$
$$= \frac{\partial^2 f}{\partial y\, \partial x}$$
$$= 6y + 2.$$

We could also denote the preceding partial derivative using the notation f_{xy}. Then

$$f_{xy} = 6y + 2.$$

Note that in the notation f_{xy}, x and y are in the order (left to right) in which the differentiation is done, but in

$$\frac{\partial^2 f}{\partial y\, \partial x},$$

the order of x and y is reversed. In both notations, the differentiation with respect to x is done first, followed by differentiation with respect to y.

Notation for the four second-order partial derivatives is as follows.

DEFINITION

1. $\dfrac{\partial^2 z}{\partial x^2} = \dfrac{\partial^2 f}{\partial x^2} = f_{xx} = z_{xx}$ Take the partial with respect to x, and then with respect to x again.

2. $\dfrac{\partial^2 z}{\partial y\, \partial x} = \dfrac{\partial^2 f}{\partial y\, \partial x} = f_{xy} = z_{xy}$ Take the partial with respect to x, and then with respect to y.

3. $\dfrac{\partial^2 z}{\partial x\, \partial y} = \dfrac{\partial^2 f}{\partial x\, \partial y} = f_{yx} = z_{yx}$ Take the partial with respect to y, and then with respect to x.

4. $\dfrac{\partial^2 z}{\partial y^2} = \dfrac{\partial^2 f}{\partial y^2} = f_{yy} = z_{yy}$ Take the partial with respect to y, and then with respect to y again.

EXAMPLE 6   For

$$z = f(x, y) = x^2 y^3 + x^4 y + x e^y,$$

find the four second-order partial derivatives.

Solution

a) $\dfrac{\partial^2 f}{\partial x^2} = f_{xx} = \dfrac{\partial}{\partial x}(2xy^3 + 4x^3 y + e^y)$ Differentiate twice with respect to x.

$\qquad\qquad = 2y^3 + 12x^2 y$

b) $\dfrac{\partial^2 f}{\partial y\, \partial x} = f_{xy} = \dfrac{\partial}{\partial y}(2xy^3 + 4x^3 y + e^y)$ Differentiate with respect to x, and then with respect to y.

$\qquad\qquad = 6xy^2 + 4x^3 + e^y$

c) $\dfrac{\partial^2 f}{\partial x\, \partial y} = f_{yx} = \dfrac{\partial}{\partial x}(3x^2 y^2 + x^4 + x e^y)$ Differentiate with respect to y, and then with respect to x.

$\qquad\qquad = 6xy^2 + 4x^3 + e^y$

d) $\dfrac{\partial^2 f}{\partial y^2} = f_{yy} = \dfrac{\partial}{\partial y}(3x^2 y^2 + x^4 + x e^y)$ Differentiate twice with respect to y.

$\qquad\qquad = 6x^2 y + x e^y$ ■

We see by comparing parts (b) and (c) in Example 6 that

$$\frac{\partial^2 f}{\partial y\, \partial x} = \frac{\partial^2 f}{\partial x\, \partial y}; \quad \text{that is,} \quad f_{xy} = f_{yx}.$$

It turns out that $f_{xy} = f_{yx}$ for virtually all functions f that arise in practice, and there are mathematical theorems that explain this phenomenon. (An example where this equality fails is given in Exercise 71 of Exercise Set 7.2.)

EXAMPLE 7   Let $f(x, y) = x^3y^3 + 2xy^2$. Compute $\dfrac{\partial^2 f}{\partial x^2} + \dfrac{\partial^2 f}{\partial y^2}$.

Solution   We begin by computing $\dfrac{\partial^2 f}{\partial x^2}$ and $\dfrac{\partial^2 f}{\partial y^2}$:

$$\frac{\partial^2 f}{\partial x^2} = \frac{\partial}{\partial x}[3x^2y^3 + 2y^2] = 6xy^3,$$

$$\frac{\partial^2 f}{\partial y^2} = \frac{\partial}{\partial y}[3x^3y^2 + 4xy] = 6x^3y + 4x.$$

Adding, we find $\dfrac{\partial^2 f}{\partial x^2} + \dfrac{\partial^2 f}{\partial y^2} = 6xy^3 + 6x^3y + 4x.$ ∎

In many biological applications, it is necessary to find a function f so that an expression such as $\dfrac{\partial^2 f}{\partial x^2} + \dfrac{\partial^2 f}{\partial y^2}$ is equal to some predetermined right-hand side. Techniques for solving these *partial differential equations* lie beyond the scope of this text.

Exercise Set 7.2

Find $\dfrac{\partial z}{\partial x}, \dfrac{\partial z}{\partial y}, \dfrac{\partial z}{\partial x}\bigg|_{(-2, -3)}$ and $\dfrac{\partial z}{\partial y}\bigg|_{(0, -5)}$

1. $z = 2x - 3xy$ **2.** $z = (x - y)^3$

3. $z = 3x^2 - 2xy + y$ **4.** $z = 2x^3 + 3xy - x$

Find $f_x, f_y, f_x(-2, 4)$, and $f_y(4, -3)$.

5. $f(x, y) = 2x - 3y$ **6.** $f(x, y) = 5x + 7y$

Find $f_x, f_y, f_x(-2, 1)$, and $f_y(-3, -2)$.

7. $f(x, y) = \sqrt{x^2 + y^2}$ **8.** $f(x, y) = \sqrt{x^2 - y^2}$

Find f_x and f_y.

9. $f(x, y) = 2x - 3y$ **10.** $f(x, y) = e^{2x-y}$

11. $f(x, y) = \sqrt{x} + \sin(xy)$ **12.** $f(x, y) = e^{2xy}$

13. $f(x, y) = x \ln y$ **14.** $f(x, y) = \dfrac{x}{y} - \dfrac{y}{x}$

15. $f(x, y) = x^3 - 4xy + y^2$

16. $f(x, y) = x^5 - 4x^2y^2 + 5xy^3 - 2y$

17. $f(x, y) = (x^2 + 2y + 2)^4$

18. $f(x, y) = y \ln(x^2 + y)$ **19.** $f(x, y) = \sin(e^{x+y})$

20. $f(x, y) = x\sqrt{2x + y^3}$ **21.** $f(x, y) = \dfrac{e^x}{y^2 + 1}$

22. $f(x, y) = \dfrac{\cos(xy)}{x - y}$

23. $f(x, y) = [x^5 + \tan(y^2)]^4$

24. $f(x, y) = \sqrt{\sin(2x + 3y)}$

25. $f(x, y) = \dfrac{y \ln x}{y^3 - 1}$ **26.** $f(x, y) = \dfrac{e^{x+2y}}{x + 2y}$

Find $\dfrac{\partial f}{\partial b}$ and $\dfrac{\partial f}{\partial m}$.

27. $f(b, m) = (m + b - 4)^2 +$
          $(2m + b - 5)^2 + (3m + b - 6)^2$

28. $f(b, m) = (m + b - 6)^2 +$
          $(2m + b - 8)^2 + (3m + b - 9)^2$

Find z_x and z_t.

29. $z = \dfrac{x^2 + t^2}{x^2 - t^2}$ **30.** $z = \dfrac{x^2 - t}{x^3 + t}$

31. $z = \dfrac{2\sqrt{x} - 2\sqrt{t}}{1 + 2\sqrt{t}}$ **32.** $z = \left(\dfrac{x^2 + t^2}{x^2 - t^2}\right)^5$

33. $z = \sqrt[4]{x^3t^5}$

34. $z = 6x^{2/3} - 8x^{1/4}t^{1/2} - 12x^{-1/2}t^{3/2}$

Compute the Jacobian of f and g.

35. $f(x, y) = x + 3y, g(x, y) = x - 2y$

36. $f(x, y) = xy^2, g(x, y) = x^2/y$

37. $f(x, y) = \sqrt{x + 3y}, g(x, y) = e^{-x-y}$

38. $f(x, y) = \ln(2x^2y^3), g(x, y) = (x + \cos y)^2$

39.–46. For Exercises 9–16, find f_{xx}, f_{xy}, f_{yx}, and f_{yy}.

Find $f_x, f_y,$ and f_z.

47. $f(x, y, z) = x^2 y^3 z^4$ **48.** $f(x, y, z) = \dfrac{xy^3}{z^4}$

49. $f(x, y, z) = e^{x+y^2+z^3}$

50. $f(x, y, z) = \ln(xy^3 z^5)$

For the following functions, use linearization and the values of $z, z_x,$ and z_y at the point (a, b) to estimate $z(x, y)$.

51. $z = f(x, y) = xy^2;\ a = 2, b = 3,$
$x = 2.01, y = 3.02$

52. $z = f(x, y) = x^2/y;\ a = 4, b = 2,$
$x = 4.02, y = 1.97$

53. $z = f(x, y) = x\sin(xy);\ a = 1, b = 0,$
$x = 0.99, y = 0.02$

54. $z = f(x, y) = y^2 e^x;\ a = 0, b = 4,$
$x = -0.02, y = 3.99$

APPLICATIONS

55. *Body Surface Area.* The Mosteller formula for approximating the surface area S in m^2 of a human is given by

$$S(h, w) = \sqrt{hw}/60,$$

where h is the height in centimeters and w is the weight in kilograms of the person.[8]

a) Use the Mosteller approximation to estimate the surface area of a child whose height is 1 m and whose weight is 28 kg.

b) Over the next 6 mo, the child grows 2 cm and gains 2 kg. Use linearization to estimate the child's surface area.

56. *Wind Speed of a Tornado.* Under certain conditions, the *wind speed S* (in miles per hour) of a tornado at a distance d (in feet) from its center can be approximated by the function

$$S = \frac{aV}{0.51 d^2},$$

where a is a constant that depends on certain atmospheric conditions and V is the approximate volume of the tornado, in cubic feet. Suppose that $a = 0.78$.

a) An observer spots a tornado about 300 ft away and estimates its volume to be 1,000,000 ft^3. Find the wind speed at the observer's location.

b) Suppose the observer underestimated the distance by 5 ft and the volume by 50,000 ft^3. Use linearization to approximate the wind speed.

57. *Birth Weights.* The birth weight w of a baby (in grams) may be predicted using the function

$$w(x, s, h, m, r, p) = x(9.38 + 0.264s$$
$$+ 0.000233hm$$
$$+ 4.62r[p + 1]).$$

In this function, x is the gestation age (in days), s is the sex of the child (1 for male, -1 for female), h is the maternal height (in centimeters), m is the maternal weight at the 26th week of pregnancy (in kilograms), r is the maternal daily weight gain (in kg/day), and p is the number of the mother's previous children.[9]

a) A mother goes to her doctor during the seventh month of her first pregnancy. Suppose the doctor estimates that the baby boy will have a gestation age of 280 days. The mother's height is 150 cm, her weight was 65 kg in her 26th week of pregnancy, and she is gaining about 0.08 kg per day during her third trimester. Estimate the birth weight of the baby.

b) Suppose the gestation age turns out to be 276 days and that the mother gains 0.081 kg per day during her third trimester. Use linearization to approximate the baby's predicted birth weight.

58. *Wind Chill Temperature.* The wind chill temperature is calculated by using the formula

$$W(v, T) = 91.4$$
$$- \frac{(10.45 + 6.68\sqrt{v} - 0.447v)(457 - 51)}{110},$$

where T is temperature in degrees Fahrenheit and v is the speed of the wind in miles per hour.

a) Find the wind chill temperature if $T = 25°F$ and $v = 20$ mph.

b) Suppose over the next hour the temperature drops by 1°F and the wind speed increases by 1 mph. Use linearization to approximate the new wind chill temperature.

59. *Capture Probability.* The probability that a tadpole with mass m grams will escape capture by predators in an experiment is

$$P(m, T) = \frac{1}{1 + e^{3.222 - 31.669m + 0.083T}},$$

---

[8]http://www.halls.md/body-surface-area/refs.htm

[9]G. G. Nahum and H. Stanislaw, "Validation of a Birth Weight Prediction Equation Based on Maternal Characteristics," *Journal of Reproductive Medicine,* Vol. 47, 752–760 (2002).

where T is the water temperature in degrees Celsius.[10]

a) The mass of a tadpole is measured to be 0.15 g, and the water temperature is measured to be 20°C. Find the probability of escaping capture.

b) Suppose that the mass of the tadpole is really 0.155 g and the water temperature is 19.5°C. Use linearization to approximate the new probability of escaping capture.

Temperature–Humidity Heat Index. In the summer, humidity affects the actual temperature, making a person feel hotter due to a reduced heat loss from the skin caused by higher humidity. The *temperature–humidity index,* T_h, is what the temperature would have to be with no humidity in order to give the same heat effect. One index often used is given by

$$T_h = 1.98T - 1.09(1 - H)(T - 58) - 56.9,$$

where T is the air temperature, in degrees Fahrenheit, and H is the relative humidity, which is the ratio of the amount of water vapor in the air to the maximum amount of water vapor possible in the air at that temperature. H is usually expressed as a percentage. Find the temperature–humidity index in each case. Round to the nearest tenth of a degree.

60. $T = 90°$ and $H = 90\%$

61. $T = 90°$ and $H = 100\%$

62. $T = 78°$ and $H = 100\%$

tw 63. Find $\dfrac{\partial T_h}{\partial H}$ and interpret its meaning.

[10]M. T. Anderson, J. M. Kiesecker, D. P. Chivers, and A. R. Blaustein, "The direct and indirect effects of temperature on a predator-prey relationship," *Canadian Journal of Zoology,* Vol. 79, pp. 1834–1841 (2001).

tw 64. Find $\dfrac{\partial T_h}{\partial T}$ and interpret its meaning.

Reading Ease. The following formula is used by psychologists and educators to predict the *reading ease* E of a passage of words:

$$E = 206.835 - 0.846w - 1.015s,$$

where w is the number of syllables in a 100-word section and s is the average number of words per sentence. Find the reading ease in each case.

65. $w = 146$ and $s = 5$

66. $w = 180$ and $s = 6$

67. Find $\dfrac{\partial E}{\partial w}$.

68. Find $\dfrac{\partial E}{\partial s}$.

SYNTHESIS

69. Consider $f(x, y) = \ln (x^2 + y^2)$. Show that f is a solution to the partial differential equation

$$\frac{\partial^2 f}{\partial x^2} + \frac{\partial^2 f}{\partial y^2} = 0.$$

70. Consider $f(x, y) = x^3 - 5xy^2$. Show that f is a solution to the partial differential equation

$$xf_{xy} - f_y = 0.$$

71. Consider the function f defined as follows:

$$f(x, y) = \begin{cases} \dfrac{xy(x^2 - y^2)}{x^2 + y^2}, & \text{for } (x, y) \neq (0, 0), \\ 0, & \text{for } (x, y) = (0, 0). \end{cases}$$

a) Find $f_x(0, y)$ by evaluating the limit

$$\lim_{h \to 0} \frac{f(h, y) - f(0, y)}{h}.$$

b) Find $f_y(x, 0)$ by evaluating the limit

$$\lim_{h \to 0} \frac{f(x, h) - f(x, 0)}{h}.$$

c) Now find and compare $f_{yx}(0, 0)$ and $f_{xy}(0, 0)$.

tw 72. Explain the meaning of the first partial derivatives of a function of two variables in terms of slopes of tangent lines.

7.3   Maximum-Minimum Problems

OBJECTIVE

■ Find relative extrema of a function of two variables.

We will now find maximum and minimum values of functions of two variables.

DEFINITION

A function f of two variables:

1. has a **relative maximum** at (a, b) if

   $$f(x, y) \leq f(a, b)$$

   for all points (x, y) in a rectangular region containing (a, b);

2. has a **relative minimum** at (a, b) if

   $$f(x, y) \geq f(a, b)$$

   for all points (x, y) in a rectangular region containing (a, b).

This definition is illustrated in Figs. 1 and 2. A relative maximum (or minimum) may not be an "absolute" maximum (minimum), as illustrated in Fig. 3.

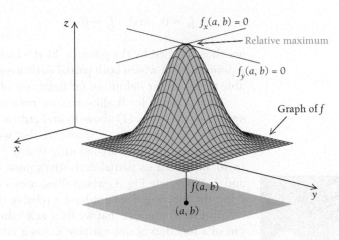

Figure 1

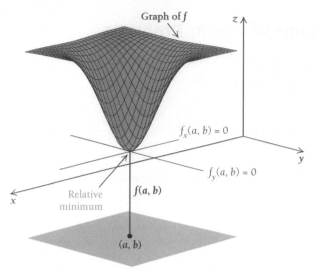

Graph of f

$f_x(a, b) = 0$

$f_y(a, b) = 0$

$f(a, b)$

Relative
minimum

(a, b)

Figure 2

Absolute
maximum

Relative maximum

Relative
minimum

Figure 3

Determining Maximum and Minimum Values

Suppose that a function f assumes a relative maximum (or minimum) value at some point (a, b) inside its domain. We assume that f and its partial derivatives exist and are "continuous" inside its domain, though we will not take the space to define continuity. If we hold y fixed at the value b, then $f(x, b)$ is the output of a function f of one variable x, and the resulting function has its relative maximum value at $x = a$. Thus its derivative must be 0 there—that is, $f_x = 0$ at the point (a, b). Similarly, $f_y = 0$ at (a, b). The equations

$$f_x = 0 \quad \text{and} \quad f_y = 0 \tag{1}$$

are thus satisfied by the point (a, b) at which the relative maximum occurs. We call a point (a, b) at which both partial derivatives are 0 a **critical point.** This is comparable to the earlier definition for functions of one variable.

One strategy for finding relative maximum or minimum values is to solve the system of equations (1) above to find critical points. Just as for functions of one variable, this strategy does *not* guarantee that we will have a relative maximum or minimum value. We have argued only that *if* f has a maximum or minimum value at (a, b), *then* both its partial derivatives must be 0 at that point. Look back at Figs. 1 and 2. Then note Fig. 4, which illustrates a case in which the partial derivatives are 0 but the function does not have a relative maximum or minimum value at (a, b).

In Fig. 4, suppose that we fix y at a value b. Then $f(x, b)$, considered as the output of a function of one variable x, has a minimum at a, but f does not. Similarly, if we fix x at a, then $f(a, y)$, considered as the output of a function of one variable y, has a maximum at b, but f does not. The point $f(a, b)$ is called a **saddle point.** In other words, $f_x(a, b) = 0$ and $f_y(a, b) = 0$ [the point (a, b) is a critical point], but f does not attain a relative maximum or minimum value at (a, b). A saddle point for a function of two variables is comparable to a point of inflection (which is simultaneously a critical point) for a function of one variable.

Where is the saddle point?

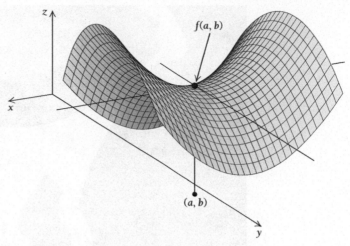

Figure 4

A test for finding relative maximum and minimum values that involves the use of first- and second-order partial derivatives is stated below. We will not prove this theorem.

THEOREM 1

The *D*-test

To find the relative maximum and minimum values of f:

1. Find f_x, f_y, f_{xx}, f_{yy}, and f_{xy}.
2. Find the critical points by solving the system of equations $f_x = 0$, $f_y = 0$. Let (a, b) represent a solution.
3. Evaluate D, where $D = f_{xx}(a, b) \cdot f_{yy}(a, b) - [f_{xy}(a, b)]^2$. Then:

   a) f has a maximum at (a, b) if $D > 0$ and $f_{xx}(a, b) < 0$.

   b) f has a minimum at (a, b) if $D > 0$ and $f_{xx}(a, b) > 0$.

   c) f has neither a maximum nor a minimum at (a, b) if $D < 0$. The function has a **saddle point** at (a, b). See Fig. 4.

   d) This test is not applicable if $D = 0$.

The **D-test** is somewhat analogous to the Second-Derivative Test (Section 3.2) for functions of one variable. Saddle points are analogous to critical points where concavity changes and that are not relative maximum or minimum values.

A relative maximum or minimum may not be an *absolute* maximum or minimum value. Tests for absolute maximum or minimum values are rather complicated. We will restrict our attention to finding *relative* maximum or minimum values. Fortunately, in most of our applications, relative maximum or minimum values turn out to be absolute as well.

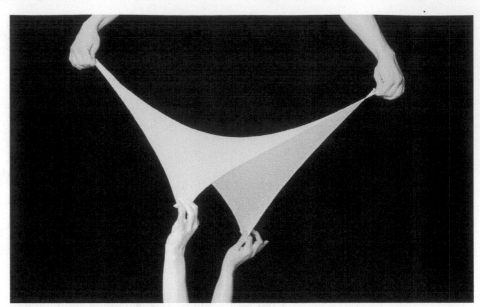

The shape of a perfect tent. To give a tent roof the maximum strength possible, designers draw the fabric into a series of three-dimensional shapes that, viewed in profile, resemble a horse's saddle and that mathematicians call an anticlastic curve. Two people with a stretchy piece of fabric such as Lycra Spandex can duplicate the shape, as shown above. One person pulls up and out on two diagonal corners; the other person pulls down and out on the other two corners. The opposing tensions draw each point of the fabric's surface into rigid equilibrium. The more pronounced the curve, the stiffer the surface.

EXAMPLE 1 Find the relative maximum or minimum values of

$$f(x, y) = x^2 + xy + y^2 - 3x.$$

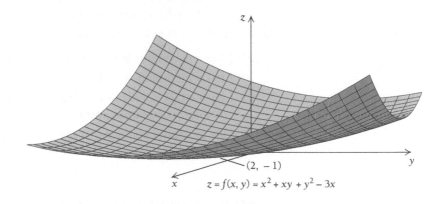

$z = f(x, y) = x^2 + xy + y^2 - 3x$

Solution

1. Find f_x, f_y, f_{xx}, f_{yy}, and f_{xy}:

$$f_x = 2x + y - 3, \qquad f_y = x + 2y,$$
$$f_{xx} = 2, \qquad\qquad\quad f_{yy} = 2,$$
$$f_{xy} = 1.$$

2. Solve the system of equations $f_x = 0, f_y = 0$:

$$2x + y - 3 = 0, \tag{1}$$
$$x + 2y = 0. \tag{2}$$

Solving equation (2) for x, we get $x = -2y$. Substituting $-2y$ for x in equation (1) and solving, we get

$$2(-2y) + y - 3 = 0$$
$$-4y + y - 3 = 0$$
$$-3y = 3$$
$$y = -1.$$

To find x when $y = -1$, we substitute -1 for y in either equation (1) or equation (2). We choose equation (2):

$$x + 2(-1) = 0$$
$$x = 2.$$

Thus, $(2, -1)$ is the only critical point and $f(2, -1)$ is our candidate for a maximum or minimum value.

3. We must check to see whether $f(2, -1)$ is a maximum or minimum value:

$$D = f_{xx}(2, -1) \cdot f_{yy}(2, -1) - [f_{xy}(2, -1)]^2$$
$$= 2 \cdot 2 - [1]^2$$
$$= 3.$$

Thus, $D = 3$ and $f_{xx}(2, -1) = 2$. Since $D > 0$ and $f_{xx}(2, -1) > 0$, it follows that f has a relative minimum at $(2, -1)$. The minimum value is by substitution:

$$f(2, -1) = 2^2 + 2(-1) + (-1)^2 - 3 \cdot 2$$
$$= 4 - 2 + 1 - 6$$
$$= -3.$$ ∎

EXAMPLE 2   Find the relative maximum and minimum values of

$$f(x, y) = xy - x^3 - y^2.$$

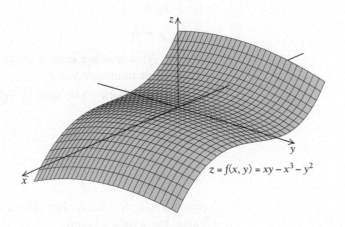

$z = f(x, y) = xy - x^3 - y^2$

Solution

1. Find f_x, f_y, f_{xx}, f_{yy}, and f_{xy}:

$$f_x = y - 3x^2, \qquad f_y = x - 2y,$$
$$f_{xx} = -6x; \qquad f_{yy} = -2;$$
$$f_{xy} = 1.$$

2. Solve the system of equations $f_x = 0, f_y = 0$:

$$y - 3x^2 = 0, \tag{1}$$
$$x - 2y = 0. \tag{2}$$

Solving equation (1) for y, we get $y = 3x^2$. Substituting $3x^2$ for y in equation (2) and solving, we get

$$x - 2(3x^2) = 0$$
$$x - 6x^2 = 0$$
$$x(1 - 6x) = 0. \qquad \text{Factoring}$$

Setting each factor equal to 0 and solving, we have

$$x = 0 \quad or \quad 1 - 6x = 0$$
$$x = \tfrac{1}{6}.$$

a) To find y when $x = 0$, we substitute 0 for x in either equation (1) or equation (2). We choose equation (2):

$$0 - 2y = 0$$
$$-2y = 0$$
$$y = 0.$$

Thus, $(0, 0)$ is a critical point and $f(0, 0)$ is one candidate for a maximum or minimum value.

b) To find the other critical point, we substitute $\tfrac{1}{6}$ for x in either equation (1) or equation (2). We choose equation (2):

$$\tfrac{1}{6} - 2y = 0$$
$$-2y = -\tfrac{1}{6}$$
$$y = \tfrac{1}{12}.$$

Thus, $\left(\tfrac{1}{6}, \tfrac{1}{12}\right)$ is another critical point and $f\left(\tfrac{1}{6}, \tfrac{1}{12}\right)$ is another candidate for a maximum or minimum value.

3. We must check both $(0, 0)$ and $\left(\tfrac{1}{6}, \tfrac{1}{12}\right)$ to see whether they yield maximum or minimum values:

a) For $(0, 0)$: $D = f_{xx}(0, 0) \cdot f_{yy}(0, 0) - [f_{xy}(0, 0)]^2$
$$= (-6 \cdot 0) \cdot (-2) - [1]^2$$
$$= -1.$$

Since $D < 0$, it follows that $f(0, 0)$ is neither a maximum nor a minimum value, but a saddle point.

b) For $\left(\frac{1}{6},\frac{1}{12}\right)$: $D = f_{xx}\left(\frac{1}{6},\frac{1}{12}\right) \cdot f_{yy}\left(\frac{1}{6},\frac{1}{12}\right) - \left[f_{xy}\left(\frac{1}{6},\frac{1}{12}\right)\right]^2$

$$= \left(-6 \cdot \frac{1}{6}\right) \cdot (-2) - [1]^2$$

$$= -1(-2) - 1$$

$$= 1.$$

Thus, $D = 1$ and $f_{xx}\left(\frac{1}{6},\frac{1}{12}\right) = -1$. Since $D > 0$ and $f_{xx}\left(\frac{1}{6},\frac{1}{12}\right) < 0$, it follows that f has a relative maximum at $\left(\frac{1}{6},\frac{1}{12}\right)$ and that maximum value is found as follows:

$$f\left(\frac{1}{6},\frac{1}{12}\right) = \frac{1}{6} \cdot \frac{1}{12} - \left(\frac{1}{6}\right)^3 - \left(\frac{1}{12}\right)^2$$

$$= \frac{1}{72} - \frac{1}{216} - \frac{1}{144} = \frac{1}{432}.$$

Exercise Set 7.3

Use the D-test to identify where relative extrema and/or saddle points occur.

1. $f(x, y) = x^2 + xy + y^2 - y$

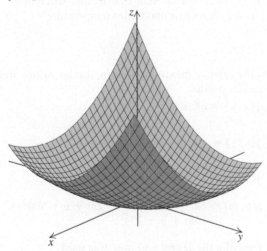

2. $f(x, y) = x^2 + xy + y^2 - 5y$

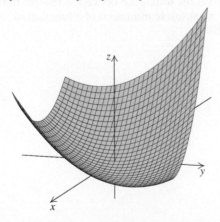

3. $f(x, y) = 2xy - x^3 - y^2$

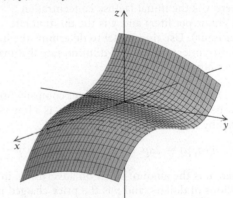

4. $f(x, y) = 4xy - x^3 - y^2$

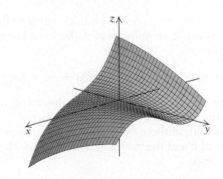

5. $f(x, y) = x^3 + y^3 - 3xy$

6. $f(x, y) = x^3 + y^3 - 6xy$

7. $f(x, y) = x^2 + y^2 - 2x + 4y - 2$

8. $f(x, y) = x^2 + 2xy + 2y^2 - 6y + 2$

9. $f(x, y) = x^2 + y^2 + 2x - 4y$

10. $f(x, y) = 4y + 6x - x^2 - y^2$

11. $f(x, y) = 4x^2 - y^2$

12. $f(x, y) = x^2 - y^2$

13. $f(x, y) = e^{x^2+y^2+1}$

14. $f(x, y) = e^{x^2-2x+y^2-4y+2}$

APPLICATIONS

15. *Production of Ethanol.* When certain substances ferment, they produce ethanol. The ethanol productivity by *Kluyveromyces lactis* (in grams per liter per hour) may be modeled by the function

$$P(x, y) = -2.23 + 0.0345x + 25.6y$$
$$- 0.000115x^2 + 0.109xy - 63.2y^2,$$

where x is the initial lactose concentration (in grams per liter) and y is the dilution rate (per hour). Use this model to determine the initial lactose concentration and dilution rate that maximizes ethanol productivity.[11]

16. *Maximizing Profit.* A one-product company finds that its profit, in millions of dollars, is a function P given by

$$P(a, p) = 2ap + 80p - 15p^2 - \tfrac{1}{10}a^2p - 100,$$

where a is the amount spent on advertising, in millions of dollars, and p is the price charged per item of the product, in dollars. Find the maximum value of P and the values of a and p at which it is attained.

17. *Maximizing Profit.* A one-product company finds that its profit, in millions of dollars, is a function P given by

$$P(a, n) = -5a^2 - 3n^2 + 48a - 4n + 2an + 300,$$

where a is the amount spent on advertising, in millions of dollars, and n is the number of items sold, in thousands. Find the maximum value of P and the values of a and n at which it is attained.

18. *Minimizing the Cost of a Container.* A trash company is designing an open-top, rectangular container that will have a volume of 320 ft^3. The cost of making the bottom of the container is $5 per square foot, and the cost of the sides is $4 per square foot. Find the dimensions of the container that will minimize total cost. (*Hint:* Make a substitution using the formula for volume.)

19. *Temperature.* A flat metal plate is located on a coordinate plane. The temperature of the plate, in degrees Fahrenheit, at point (x, y) is given by

$$T(x, y) = x^2 + 2y^2 - 8x + 4y.$$

Find the minimum temperature and where it occurs. Is there a maximum temperature?

20. *Temperature.* A flat metal plate is located on a coordinate plane. The temperature of the plate, in degrees Fahrenheit, at a point (x, y) is given by

$$T(x, y) = 2x^2 + 3y^2 - 2x + 6y.$$

Find the minimum temperature and where it occurs. Is there a maximum temperature?

SYNTHESIS

Find the relative maximum and minimum values and the saddle points.

21. $f(x, y) = e^x + e^y - e^{x+y}$

22. $f(x, y) = xy + \dfrac{2}{x} + \dfrac{4}{y}$

23. $f(x, y) = 2y^2 + x^2 - x^2y$

24. $S(b, m) = (m + b - 72)^2 + (2m + b - 73)^2 + (3m + b - 75)^2$

tw 25. Describe the D-test and how it is used.

tw 26. Explain the difference between a relative minimum and an absolute minimum of a function of two variables.

[11]S. F. Deriase, L. M. Farahat, and A. I. El-Batal, "Optimization of process parameters for the continuous ethanol production by *Kluyveromyces lactis* immobilized cells in hydrogel copolymer carrier," *Acta Microbiologica Polonica*, Vol. 50, pp. 45–51 (2001).

Solar Radiation. The annual radiation (in megajoules per square centimeter) for north-facing land in the northern hemisphere may be modeled with the equation

$$R(l, s) = e^{-1.236 + 1.35 \cos l \cos s - 1.707 \sin l \sin s}.$$

In this equation, l is the latitude (between $0°$ and $90°$) and s is the steepness of the ground (between $0°$ and $60°$).[12]

---

[12]B. McCune and D. Keon, "Equations for potential annual direct incident radiation and heat load," *Journal of Vegetation Science,* Vol. 13, pp. 603–606 (2002).

27. a) Compute $\partial R / \partial l$.
    b) Show that $\partial R / \partial l$ must be negative over the permissible values of l, s, and a. (*Hint:* Are the trigonometric functions positive or negative in the first quadrant?)
    c) Conclude that the solar radiation is maximized if $l = 0°$, corresponding to land at the equator.
    tw d) Explain why this conclusion is reasonable.

28. a) Compute $\partial R / \partial s$.
    b) Show that $\partial R / \partial s$ must be negative over the permissible values of l, s, and a.
    c) Conclude that the solar radiation is maximized if $s = 0°$, or if the land is perfectly flat.
    tw d) Explain why this conclusion is reasonable.

 Technology Connection

Use a 3D grapher to graph each of the following functions. Then estimate any relative extrema.

29. $f(x, y) = \dfrac{-5}{x^2 + 2y^2 + 1}$

30. $f(x, y) = \dfrac{y + x^2 y^2 - 8x}{xy}$

7.4   An Application: The Method of Least Squares

OBJECTIVES

- Find a regression line.
- Solve applied problems involving regression lines.

The problem of fitting an equation to a set of data occurs frequently. We considered this in Section 1.2. Such an equation provides a model of the phenomena from which predictions can be made. For example, in business, one might want to predict future sales on the basis of past data. In ecology, one might want to predict future demands for natural gas on the basis of past need. Suppose that we are trying to find a linear equation

$$y = mx + b$$

to fit the data. To determine this equation is to determine the values of m and b. But how? Let's consider some factual data.

The table below shows the average age of mothers who had their first child in the given year. Suppose that we plot these points and try to draw a line through them that fits. Note that there are several ways in which this might be done (see Figs. 1 and 2). Each would give a different predicted average age of mothers who have their first child in 2005 or 2010.[13]

Year, x	Average Age of New Mothers, y
1. 1985	23.7
2. 1990	24.2
3. 1995	24.5
4. 2000	24.9

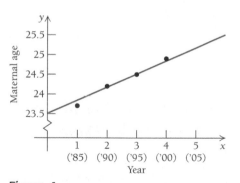

Figure 1

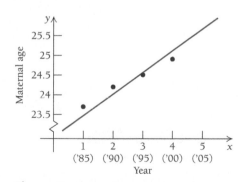

Figure 2

Note that the time is given in increments of five years, making computations easier.

---

[13]National Vital Statistics Reports, Vol. 51, No. 1 (2002).

Consider the data points $(1, 23.7)$, $(2, 24.2)$, $(3, 24.5)$, and $(4, 24.9)$, as plotted in Fig. 3. We will try to fit these data with a line

$$y = mx + b$$

by determining the values of m and b. Note the y-errors, or y-deviations—$y_1 - 23.7$, $y_2 - 24.2$, $y_3 - 24.5$, and $y_4 - 24.9$—between the observed points $(1, 23.7)$, $(2, 24.2)$, $(3, 24.5)$, and $(4, 24.9)$ and the points $(1, y_1)$, $(2, y_2)$, $(3, y_3)$, and $(4, y_4)$ on the line. We would like, somehow, to minimize these deviations in order to have a good fit. One way of minimizing the deviations is based on the **method of least squares.**

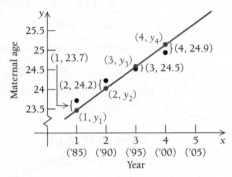

Figure 3

The Method of Least Squares

The line of best fit is the line for which the sum of the squares of the y-deviations is a minimum. This is called the **regression line.**

Using the method of least squares for the maternal data, we would minimize

$$S = (y_1 - 23.7)^2 + (y_2 - 24.2)^2 + (y_3 - 24.5)^2 + (y_4 - 24.9)^2. \qquad (1)$$

Since the points $(1, y_1)$, $(2, y_2)$, $(3, y_3)$, and $(4, y_4)$ must be solutions of $y = mx + b$, it follows that

$$y_1 = m(1) + b = m + b,$$
$$y_2 = m(2) + b = 2m + b,$$
$$y_3 = m(3) + b = 3m + b,$$
$$y_4 = m(4) + b = 4m + b.$$

Substituting $m + b$ for y_1, $2m + b$ for y_2, $3m + b$ for y_3, and $4m + b$ for y_4 in equation (1), we have

$$S = (m + b - 23.7)^2 + (2m + b - 24.2)^2 + (3m + b - 24.5)^2$$
$$+ (4m + b - 24.9)^2. \qquad (2)$$

Thus, to find the regression line for the given set of data, we must find the values of m and b that minimize the function S given by the sum in equation (2).

To apply the *D*-test, we first find the partial derivatives $\partial S/\partial b$ and $\partial S/\partial m$:

$$\frac{\partial S}{\partial b} = 2(m + b - 23.7) + 2(2m + b - 24.2) + 2(3m + b - 24.5)$$
$$+ 2(4m + b - 24.9)$$
$$= 20m + 8b - 194.6,$$

and

$$\frac{\partial S}{\partial m} = 2(m + b - 23.7) + 2(2m + b - 24.2)2 + 2(3m + b - 24.5)3$$
$$+ 2(4m + b - 24.9)4$$
$$= 60m + 20b - 490.4.$$

We set these derivatives equal to 0 and solve the resulting system:

$$20m + 8b - 194.6 = 0$$
$$60m + 20b - 490.4 = 0$$

or

$$5m + 2b = 48.65$$
$$15m + 5b = 122.6.$$

Using the techniques of Chapter 6, the solution of this system is

$$b = 23.35, \qquad m = 0.39.$$

We leave it to the student to complete the *D*-test to verify that (23.35, 0.39) does, in fact, yield the minimum of *S*.

We need not bother to compute $S(23.35, 0.39)$. The values of *m* and *b* are all we need to determine $y = mx + b$. The regression line is

$$y = 0.39x + 23.35.$$

We see the graph of the "best-fit" regression line together with the data in Fig. 4. Compare it to Figs. 1 and 2.

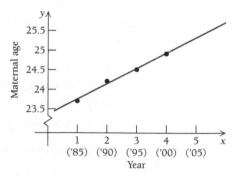

Figure 4

Extrapolating, we predict that the average age of first-time mothers in 2010 will be

$$y = 0.39(6) + 23.35 = 25.69 \text{ yr.}$$

The method of least squares is a statistical process illustrated here with only four data points in order to simplify the explanation. Most statistical researchers would warn that many more than four data points should be used to get a "good" regression line. Furthermore, making predictions too far into the future from any linear model may not be valid. It can be done, but the further into the future the prediction is made, the more dubious one should be about the prediction.

*The Regression Line for an Arbitrary Collection of Data Points $(c_1, d_1), (c_2, d_2), \ldots, (c_n, d_n)$

Look again at the regression line

$$y = 0.39x + 23.35$$

for the data points $(1, 23.7), (2, 24.2), (3, 24.5)$, and $(4, 24.9)$. Let's consider the arithmetic averages, or means, of the x-coordinates, denoted by $\bar{x}$, and the y-coordinates, denoted by $\bar{y}$:

$$\bar{x} = \frac{1 + 2 + 3 + 4}{4} = 2.5,$$

$$\bar{y} = \frac{23.7 + 24.2 + 24.5 + 24.9}{4} = 24.325.$$

It turns out that the point $(\bar{x}, \bar{y})$, or $(2.5, 24.325)$, is on the regression line since

$$24.325 = 0.39(2.5) + 23.35.$$

Thus the regression line is

$$y - \bar{y} = m(x - \bar{x}),$$

or

$$y - 24.325 = m(x - 2.5).$$

All that remains, in general, is to determine m.

Suppose that we want to find the regression line for an arbitrary number of points $(c_1, d_1), (c_2, d_2), \ldots, (c_n, d_n)$. To do so, we find the values m and b that minimize the function S given by

$$S(b, m) = (y_1 - d_1)^2 + (y_2 - d_2)^2 + \cdots + (y_n - d_n)^2$$

$$= \sum_{i=1}^{n} (y_i - d_i)^2.$$

Using a procedure like the one used earlier, the values of b and m that minimize $S(b, m)$ may be found. We state this result as a theorem.

---

*This part is considered optional and can be omitted without loss of continuity.

THEOREM 2

The regression line for the points $(c_1, d_1), (c_2, d_2), \ldots, (c_n, d_n)$ has the form

$$y - \bar{y} = m(x - \bar{x}),$$

where

$$\bar{x} = \frac{\sum\limits_{i=1}^{n} c_i}{n}, \qquad \bar{y} = \frac{\sum\limits_{i=1}^{n} d_i}{n},$$

and

$$m = \frac{\sum\limits_{i=1}^{n}(c_i - \bar{x})(d_i - \bar{y})}{\sum\limits_{i=1}^{n}(c_i - \bar{x})^2}.$$

Let's see how this works out for the maternal age example used previously.

	c_i	d_i	$(c_i - \bar{x})$	$(c_i - \bar{x})^2$	$(d_i - \bar{y})$	$(c_i - \bar{x})(d_i - \bar{y})$
	1	23.7	-1.5	2.25	-0.625	0.9375
	2	24.2	-0.5	0.25	-0.125	0.0625
	3	24.5	0.5	0.25	0.175	0.0875
	4	24.9	1.5	2.25	0.575	0.8625
Total	10	97.3		5		1.95
	$\bar{x} = \dfrac{10}{4} = 2.5$		$\bar{y} = \dfrac{97.3}{4} = 24.325$		$m = \dfrac{1.95}{5} = 0.39$	

Thus the regression line is

$$y - 24.325 = 0.39(x - 2.5),$$

which simplifies to

$$y = 0.39x + 23.35.$$

Technology Connection

As we have seen in Section 1.2 and in other parts of the book, graphers can perform linear regression, as well as other kinds of regression such as quadratic, exponential, and logarithmic. Use such a grapher to fit a linear equation to the maternal age data.

With some graphers, you will also obtain a number r, called the *coefficient of correlation*. Although we cannot develop that concept in detail in this text, keep in mind that r is used to describe the strength of the linear relationship between x and y. The closer r is to either -1 or 1, the better the correlation.

For the maternal age data, $r \approx 0.995$, which indicates a tight linear relationship. However, remember that a high linear correlation does not necessarily indicate a "cause-and-effect" connection between the variables.

Exercise Set 7.4

 All of the following exercises can be done with a grapher or spreadsheet if your instructor so directs. The grapher can also be used to check your work.

APPLICATIONS

1. *Hockey Ticket Prices.* Ticket prices for NHL hockey games have experienced steady growth, as shown in the following table.[14]

Years, x, since 1994–95 Season	Average Ticket Price (dollars)
0. 1994–1995	33.49
1. 1995–1996	34.72
2. 1996–1997	38.04
3. 1997–1998	42.15
4. 1998–1999	42.78
5. 1999–2000	46.38
6. 2000–2001	47.70

a) Find the regression line $y = mx + b$.
b) Use the regression line to predict the average ticket price of an NHL game in 2010–2011 and in 2015–2016.

2. *Strain Measurements.* A machine for measuring the stress on stab-resistant body armor is calibrated using electrical signals. Representative data from this calibration is shown below.[15]

Load, x (newtons)	Electrical Output, y (volts)
500	0.77
1000	0.98
1500	1.27
2000	1.45
2500	1.70

a) Find the regression line $y = mx + b$.
b) Use the regression line to predict the voltage generated by a stress of 4000 newtons.

3. *Life Expectancy of Women.* Consider the data below, which show the average life expectancy of women in various years.[16]

Year, x	Life Expectancy of Women, y (years)
0. 1950	71.1
10. 1960	73.1
20. 1970	74.7
30. 1980	77.4
40. 1990	78.8
50. 2000	79.5

a) Find the regression line $y = mx + b$.
b) Use the regression line to predict the life expectancy of women in 2010 and 2015.

4. *Life Expectancy of Men.* Consider the data below, which show the average life expectancy of men in various years.[17]

Year, x	Life Expectancy of Men, y (years)
0. 1950	65.6
10. 1960	66.6
20. 1970	67.1
30. 1980	70.0
40. 1990	71.8
50. 2000	74.1

a) Find the regression line $y = mx + b$.
b) Use the regression line to predict the life expectancy of men in 2010 and 2015.

[14]Amusement Business.
[15]E. K. J. Chadwick, A. C. Nicol, S. Floyd, and T. G. F. Gray, "A telemetry-based device to determine the force-displacement behaviour of materials in high impact loading situations," *Journal of Biomechanics,* Vol. 33, pp. 361–365 (2000).

[16]Centers for Disease Control.
[17]Centers for Disease Control.

5. *Teen Pregnancy.* Consider the data below, which show the birthrate of teenagers aged 18–19 years.[18]

Year, x	Birthrate per 1000 teens, y
0. 1996	86.0
1. 1997	83.6
2. 1998	82.0
3. 1999	80.3
4. 2000	79.5

a) Find the regression line $y = mx + b$.
b) Use the regression line to predict the birthrate for 18- and 19-year-olds in 2010.

6. *Teen Pregnancy.* Consider the data below, which show the birthrate of teenagers aged 15–17 years.[19]

Year, x	Birthrate per 1000 teens, y
0. 1996	33.8
1. 1997	32.1
2. 1998	30.4
3. 1999	28.7
4. 2000	27.5

a) Find the regression line $y = mx + b$.
b) Use the regression line to predict the birthrate for 15-, 16-, and 17-year-olds in 2010.

7. *Grade Predictions.* A professor wants to predict students' final examination scores on the basis of their midterm test scores. An equation was determined on the basis of data on the scores of three students who took the same course with the same instructor the previous semester (see the following table).

Midterm Score, x	Final Exam Score, y
70%	75%
60	62
85	89

a) Find the regression line $y = mx + b$.

---

[18]National Vital Statistics Report, Vol. 49, No. 10.
[19]National Vital Statistics Report, Vol. 49, No. 10.

b) The midterm score of a student was 81. Use the regression line to predict the student's final exam score.

8. *Predicting the World Record in the High Jump.* On July 29, 1989, Javier Sotomayer of Cuba set an astounding world record of 8 ft in the high jump. It has been established that most world records in track and field can be modeled by a linear function. The following table shows world records for various years.

Year, x (Use the actual year for x)	World Record in High Jump, y (in inches)
1912 (George Horne)	78.0
1956 (Charles Dumas)	84.5
1973 (Dwight Stones)	90.5
1989 (Javier Sotomayer)	96.0

a) Find the regression line $y = mx + b$.
b) Use the regression line to predict the world record in the high jump in 2010 and in 2050.

tw c) Does your answer in part (b) for 2050 seem realistic? Explain why extrapolating so far into the future could be a problem.

tw **9.** Explain the concept of linear regression to a friend.
tw **10.** Discuss the idea of linear regression with an expert in your major. Write a brief report.

SYNTHESIS

Log-Log Plots. Linear regression can also be used to find a relationship of the form $y = Ax^c$ between two quantities, as discussed in Section 4.2. To do so, we first use the conversions $X = \log x$ and $Y = \log y$. Regression is then used to find a linear relationship between X

and Y. Finally, this relationship between X and Y is converted to one between x and y using properties of logarithms.

11. *Species Richness.* The table below shows the number of bird species found in some North American land areas.[20]

Land Area x (acres)	Bird Species Count, y
30	25
200	30
20,000	80
25,000,000	170
1,000,000,000	250

a) Make a table for $X = \log x$ and $Y = \log y$.
b) Find the regression line $Y = mX + b$.
c) Use properties of logarithms to find a relationship $y = Ax^c$.
d) Predict the number of bird species in a land area that covers 1,000,000 acres.

12. *Species Richness.* The table below shows the number of plant species on some islands near the Galápagos Islands.[21]

Land Area, x (sq miles)	Plant Species Count, y
2	15
2	30
9	50
17	110
180	175
2500	270

a) Make a table for $X = \log x$ and $Y = \log y$.
b) Find the regression line $Y = mX + b$.
c) Use properties of logarithms to find a relationship $y = Ax^c$.
d) Predict the number of plant species on an island with an area of 5 sq mi.

 Technology Connection

13. *Predicting the World Record in the One-Mile Run.*

a) Find the regression line $y = mx + b$ that fits the set of data in the table. (*Hint:* Convert each time to decimal notation; for instance, $4{:}24.5 = 4\frac{24.5}{60} \approx 4.4083$.)
b) Use the regression line to predict the world record in the mile in 2010 and in 2015.
c) In July 1999, Hicham El Guerrouj set a new world record of 3:43.13 for the mile. How does this compare with what can be predicted by the regression?

Year, x (Use the actual year for x)	World Record in Mile (min:sec), y
1875 (Walter Slade)	4:24.5
1894 (Fred Bacon)	4:18.2
1923 (Paavo Nurmi)	4:10.4
1937 (Sidney Wooderson)	4:06.4
1942 (Gunder Hägg)	4:06.2
1945 (Gunder Hägg)	4:01.4
1954 (Roger Bannister)	3:59.6
1964 (Peter Snell)	3:54.1
1967 (Jim Ryun)	3:51.1
1975 (John Walker)	3:49.4
1979 (Sebastian Coe)	3:49.0
1980 (Steve Ovett)	3:48.40
1985 (Steve Cram)	3:46.31
1993 (Noureddine Morceli)	3:44.39

---

[20]N. A. Campbell and J. B. Reece, *Biology* 6th ed. (San Francisco: Benjamin Cummings, 2002).
[21]N. A. Campbell and J. B. Reece, *Biology* 6th ed. (San Francisco: Benjamin Cummings, 2002).

7.5 Multiple Integration

OBJECTIVE

■ Evaluate a multiple integral.

The following is an example of a **double integral:**

$$\int_3^6 \int_{-1}^2 10xy^2 \, dx \, dy, \quad \text{or} \quad \int_3^6 \left(\int_{-1}^2 10xy^2 \, dx \right) dy.$$

Evaluating a double integral is somewhat similar to "undoing" a second partial derivative. We first evaluate the inside x-integral, treating y as a constant:

$$\int_{-1}^2 10xy^2 \, dx = 10y^2 \left[\frac{x^2}{2} \right]_{-1}^2 = 5y^2[x^2]_{-1}^2 = 5y^2[2^2 - (-1)^2] = 15y^2.$$

Color indicates the variable. All else is constant.

Then we evaluate the outside y-integral:

$$\int_3^6 15y^2 \, dy = 15 \left[\frac{y^3}{3} \right]_3^6 = 5[y^3]_3^6 = 5(6^3 - 3^3) = 945.$$

If the dx and dy and the limits of integration are interchanged, as follows,

$$\int_{-1}^2 \int_3^6 10xy^2 \, dy \, dx,$$

we first evaluate the inside y-integral, treating x as a constant:

$$\int_3^6 10xy^2 \, dy = 10x \left[\frac{y^3}{3} \right]_3^6 = \frac{10x}{3} \left[y^3 \right]_3^6 = \frac{10}{3} x(6^3 - 3^3) = 630x.$$

Then we evaluate the outside x-integral:

$$\int_{-1}^2 630x \, dx = 630 \left[\frac{x^2}{2} \right]_{-1}^2 = 315[x^2]_{-1}^2 = 315[2^2 - (-1)^2] = 945.$$

Note that we get the same result. This is not always true, but will be for the types of functions that we consider and for almost all functions that arise in applications.

Sometimes variables occur as limits of integration.

EXAMPLE 1 Evaluate

$$\int_0^1 \int_{x^2}^x xy^2 \, dy \, dx.$$

Solution We first evaluate the y-integral, treating x as a constant:

$$\int_{x^2}^x xy^2 \, dy = x \left[\frac{y^3}{3} \right]_{x^2}^x = \frac{1}{3} x[x^3 - (x^2)^3] = \frac{1}{3}(x^4 - x^7).$$

Then we evaluate the outside integral:

$$\int_0^1 \frac{1}{3}(x^4 - x^7) \, dx = \frac{1}{3} \left[\frac{x^5}{5} - \frac{x^8}{8} \right]_0^1$$

$$= \frac{1}{3} \left[\left(\frac{1^5}{5} - \frac{1^8}{8} \right) - \left(\frac{0^5}{5} - \frac{0^8}{8} \right) \right] = \frac{1}{40}.$$

Thus,

$$\int_0^1 \int_{x^2}^x xy^2 \, dy \, dx = \frac{1}{40}.$$

■

The Geometric Interpretation of Multiple Integrals

Suppose that the region G in the xy-plane is bounded by the functions $y_1 = g(x)$ and $y_2 = h(x)$ and the lines $x_1 = a$ and $x_2 = b$. We want the volume V of the solid above G and under the surface $z = f(x, y)$. Recall from Section 5.8 that the volume of a solid can be found by integrating the cross-sectional area. If the cross-sectional area at x is given by $A(x)$ (shown in blue in Fig. 1), then

$$V = \int_a^b A(x) \, dx. \tag{1}$$

For a given value of x, the cross-sectional area $A(x)$ is an area under the curve $z = f(x, y)$ between $y_1 = g(x)$ and $y_2 = h(x)$. This area can be found by integrating with respect to y:

$$A(x) = \int_{y_1}^{y_2} f(x, y) \, dy = \int_{g(x)}^{h(x)} f(x, y) \, dy.$$

Substituting into (1), we obtain the entire volume:

$$V = \int_a^b A(x) \, dx$$

$$= \int_a^b \left[\int_{g(x)}^{h(x)} f(x, y) \, dy \right] dx,$$

or

$$V = \int_a^b \int_{g(x)}^{h(x)} f(x, y) \, dy \, dx.$$

We sometimes write this integral as

$$\iint_G z \, dy \, dx \quad \text{or} \quad \iint_G f(x, y) \, dy \, dx$$

for brevity.

We saw in Section 5.2 that finding an area and finding an integral are not exactly identical: Curves that lie below the x-axis will have negative integrals. In the same way, the double integral of a function that is negative (thus lying below the xy-plane) will also be negative.

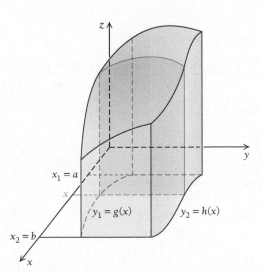

Figure 1

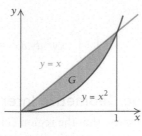

Figure 2

In Example 1, the region of integration G is the plane region between the graphs of $y = x^2$ and $y = x$, as shown in Fig. 2.

EXAMPLE 2   Find $\iint_G \dfrac{1}{xy}\,dy\,dx$, where the region G is shown in the figure below.

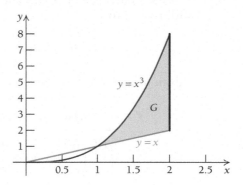

Solution   The region G lies between $x = 1$ and $x = 2$. Also, at any value of x, the region G lies between $y = x$ and $y = x^3$. Therefore, the integral becomes

$$\int_1^2 \int_x^{x^3} \frac{1}{xy}\,dy\,dx.$$

We first evaluate the inside y-integral, treating x as a constant:

$$\int_x^{x^3} \frac{1}{xy}\,dy = \left[\frac{1}{x}\ln y\right]_x^{x^3} = \frac{\ln(x^3) - \ln(x)}{x} = \frac{3\ln x - \ln x}{x} = \frac{2\ln x}{x}.$$

We now evaluate the outside x-integral, which is

$$\int_1^2 \frac{2\ln x}{x}\,dx.$$

We evaluate this integral using substitution. The indefinite integral is

$$\int \frac{2\ln x}{x}\,dx = \int 2u\,du \qquad \underline{\text{Substitution}} \quad \begin{aligned} u &= \ln x \\ du &= dx/x \end{aligned}$$

$$= u^2 + C$$

$$= (\ln x)^2 + C.$$

The outside x-integral is therefore

$$\int_1^2 \frac{2\ln x}{x}\,dx = [(\ln x)^2]_1^2 = (\ln 2)^2 - (\ln 1)^2 = (\ln 2)^2.$$

Thus,

$$\int_1^2 \int_x^{x^3} \frac{1}{xy}\,dy\,dx = (\ln 2)^2.$$ ∎

Exercise Set 7.5

Evaluate.

1. $\displaystyle\int_0^1 \int_0^1 2y\,dx\,dy$

2. $\displaystyle\int_0^1 \int_0^1 2x\,dx\,dy$

3. $\displaystyle\int_{-1}^1 \int_x^1 xy\,dy\,dx$

4. $\displaystyle\int_{-1}^1 \int_x^2 (x+y)\,dy\,dx$

5. $\displaystyle\int_0^1 \int_{-1}^3 (x+y)\,dy\,dx$

6. $\displaystyle\int_0^1 \int_{-1}^1 (x+y)\,dy\,dx$

7. $\displaystyle\int_0^1 \int_{x^2}^x (x+y)\,dy\,dx$

8. $\displaystyle\int_0^2 \int_0^x e^{x+y}\,dy\,dx$

9. $\displaystyle\int_0^1 \int_1^{e^x} \frac{1}{y}\,dy\,dx$

10. $\displaystyle\int_0^1 \int_{-1}^x (x^2+y^2)\,dy\,dx$ **11.** $\displaystyle\int_0^2 \int_0^x (x+y^2)\,dy\,dx$

12. $\displaystyle\int_1^3 \int_0^x 2e^{x^2}\,dy\,dx$

13. Find the volume of the solid capped by the surface $z = 1 - y - x^2$ over the region bounded above and below by $y = 1 - x^2$ and $y = 0$ and left and right by $x = 0$ and $x = 1$, by evaluating the integral

$$\int_0^1 \int_0^{1-x^2} (1 - y - x^2)\,dy\,dx.$$

14. Find the volume of the solid capped by the surface $z = x + y$ over the region bounded above and below by $y = 1 - x$ and $y = 0$ and left and right by $x = 0$ and $x = 1$, by evaluating the integral

$$\int_0^1 \int_0^{1-x} (x + y)\,dy\,dx.$$

Evaluate $\iint_G f(x, y)\,dy\,dx$ for the given region and function.

15. $f(x, y) = ye^{x^3}$

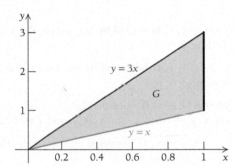

16. $f(x, y) = \sqrt{xy}$

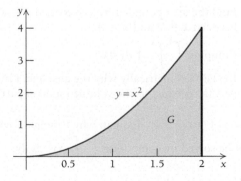

17. $f(x, y) = (7 - 2x)y$

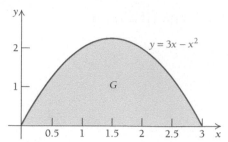

18. $f(x, y) = \dfrac{\ln (x + y)}{x(x + y)}$

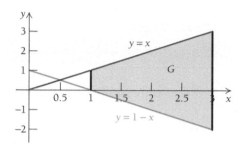

19. $f(x, y) = x^2y$, G bounded by the x-axis, the y-axis, and the line $y = 1 - x$

20. $f(x, y) = x + y$, G bounded by the x-axis and $y = x - x^2$

21. $f(x, y) = y^2 \cos x$, G bounded by $y = \sin x$, $y = -\sin x$, the y-axis, and $x = \pi/2$

22. $f(x, y) = \sqrt{x + y}$, G bounded by the x-axis, the y-axis, $x = 16$, and $y = 9$

SYNTHESIS

23. *Area and Double Integrals.*

a) Find the area bounded by $y = x^3$ and $y = x$ between $x = 2$ and $x = 4$.

b) Compute $\displaystyle\int_2^4 \int_x^{x^3} 1 \, dy \, dx$.

tw c) Explain geometrically why the area found in part (a) is equal to the volume found in part (b).

tw 24. *Order of Integration.* Explain why Example 1 could not be solved by computing $\displaystyle\int_{x^2}^x \int_0^1 xy^2 \, dx \, dy$.

A *triple integral* such as

$$\int_r^s \int_c^d \int_a^b f(x, y, z) \, dx \, dy \, dz$$

is evaluated in much the same way as the double integral. We first evaluate the inside x-integral, treating y and z as constants. Then we evaluate the middle y-integral, treating z as a constant. Finally, we evaluate the outside z-integral. Evaluate these triple integrals.

25. $\displaystyle\int_0^1 \int_1^3 \int_{-1}^2 (2x + 3y - z) \, dx \, dy \, dz$

26. $\displaystyle\int_0^2 \int_1^4 \int_{-1}^2 (8x - 2y + z) \, dx \, dy \, dz$

27. $\displaystyle\int_0^1 \int_0^{1-x} \int_0^{2-x} xyz \, dz \, dy \, dx$

28. $\displaystyle\int_0^2 \int_{2-y}^{6-2y} \int_0^{\sqrt{4-y^2}} z \, dz \, dx \, dy$

tw 29. Describe the geometric meaning of the multiple integral of a function of two variables.

 Technology Connection

30. Use a grapher that does multiple integration to evaluate some double integrals found in this exercise set.

Evaluate. Use Simpson's rule with $n = 10$ to approximate the outer integral.

31. $\displaystyle\int_1^2 \int_0^x \dfrac{2x^2y}{y^2 + 1} \, dy \, dx$

32. $\displaystyle\int_1^2 \int_1^2 e^{xy} \, dy \, dx$

33. $\displaystyle\int_0^1 \int_0^{x^2} \sqrt{x + y} \, dy \, dx$

34. $\displaystyle\int_\pi^{2\pi} \int_0^x \cos(xy) \, dy \, dx$

Chapter 7   Summary and Review

Terms to Know

Function of two variables, p. 502
Function of several variables, p. 502
Partial derivative, p. 510
Linearization, pp. 512–513
Jacobian, p. 513

Second-order partial derivative, p. 514
Relative maximum, p. 519
Relative minimum, p. 519
Critical point, p. 520

Saddle point, p. 520
D-test, p. 521
Method of least squares, p. 529
Regression line, p. 529
Double integral, p. 536

Review Exercises

These review exercises are for test preparation. They can also be used as a lengthened practice test. Answers are at the back of the book. The answers also contain bracketed section references, which tell you where to restudy if your answer is incorrect.

Given $f(x, y) = e^y + 3xy^3 + 2y$, find each of the following.

1. $f(2, 0)$ **2.** f_x **3.** f_y

4. f_{xy} **5.** f_{yx} **6.** f_{xx}

7. f_{yy}

Given $z = \ln(2x^3 + y) + \sin(xy^2)$, find each of the following.

8. $\dfrac{\partial z}{\partial x}$ **9.** $\dfrac{\partial z}{\partial y}$ **10.** $\dfrac{\partial^2 z}{\partial x\, \partial y}$

11. $\dfrac{\partial^2 z}{\partial x^2}$ **12.** $\dfrac{\partial^2 z}{\partial y^2}$

13. Find the Jacobian of $f(x, y) = x + x^2 y^3$ and $g(x, y) = e^{x-5y}$.

For the following functions, use linearization and the values of z, z_x, and z_y at (a, b) to estimate $z(x, y)$.

14. $z = f(x, y) = xe^y$; $a = 1, b = 0, x = 1.01$, $y = 0.02$

15. $z = f(x, y) = \sqrt{xy^3}$; $a = 2, b = 2, x = 1.99$, $y = 2.02$

Find the relative maximum and minimum values.

16. $f(x, y) = x^3 - 6xy + y^2 + 6x + 3y - \frac{1}{5}$

17. $f(x, y) = x^2 - xy + y^2 - 2x + 4y$

18. $f(x, y) = 3x - 6y - x^2 - y^2$

19. $f(x, y) = x^4 + y^4 + 4x - 32y + 80$

20. *NFL Salaries.* Consider the data in the following table regarding the average salary of an NFL player over a recent 10-year period.[22]

Year, x	Average Salary of an NFL Player, y (in millions)
1. 1992	0.48
2. 1993	0.67
3. 1994	0.63
4. 1995	0.72
5. 1996	0.79
6. 1997	0.74
7. 1998	0.99
8. 1999	1.06
9. 2000	1.12
10. 2001	1.10

a) Find the regression line $y = mx + b$.
b) Use the regression line to predict the average NFL player salary in 2007 and in 2010.

---

[22]http://www.sportsfansofamerica.com

21. Consider the data in the following table regarding enrollment in colleges and universities during a recent three-year period.

Year, x	Enrollment, y (in millions)
1	7.2
2	8.0
3	8.4

a) Find the regression line $y = mx + b$.
b) Use the regression line to predict enrollment in the fourth year.

Evaluate.

22. $\int_0^1 \int_{x^2}^{3x} (x^3 + 2y)\, dy\, dx$ 23. $\int_0^1 \int_{x^2}^{x} (x - y)\, dy\, dx$

SYNTHESIS

24. Evaluate

$$\int_0^2 \int_{1-2x}^{1-x} \int_0^{\sqrt{2-x^2}} z\, dz\, dy\, dx.$$

 Technology Connection

25. Use a 3D grapher to graph

$$f(x, y) = x^2 + 4y^2.$$

Chapter 7 Test

Given $f(x, y) = \cos(e^{2y}x^2)$, find each of the following.

1. $f(-1, 2)$ 2. $\dfrac{\partial f}{\partial x}$ 3. $\dfrac{\partial f}{\partial y}$

4. $\dfrac{\partial^2 f}{\partial x^2}$ 5. $\dfrac{\partial^2 f}{\partial x\, \partial y}$ 6. $\dfrac{\partial^2 f}{\partial y^2}$

For the following functions, use linearization and the values of z, z_x, and z_y, at (a, b) to estimate $z(x, y)$.

7. $z = f(x, y) = x^3/y;\ a = 2, b = 4,$
   $x = 2.02, y = 4.01$

8. $z = f(x, y) = x\sqrt{y};\ a = 3, b = 9,$
   $x = 2.99, y = 9.03$

9. Find the Jacobian of $f(x, y) = x/(y^2 + 1)$ and $g(x, y) = \cos(e^x y)$.

Find the relative maximum and minimum values.

10. $f(x, y) = x^2 - xy + y^3 - x$

11. $f(x, y) = 4y^2 - x^2$

12. *Breast Cancer.* Consider the data in the following table regarding the death rate due to breast cancer in women ages 55–64 in the United States.[23]

Year, x	Death Rate per 100,000 Women, y
1. 1990	78.6
2. 1995	69.8
3. 2000	59.7

a) Find the regression line $y = mx + b$.
b) Use the regression line to predict the death rate due to breast cancer in 2010.

13. Evaluate

$$\int_0^2 \int_1^x (x^2 - y)\, dy\, dx.$$

SYNTHESIS

14. Evaluate

$$\int_0^1 \int_0^x \int_0^{x+y} (x + y + z)\, dz\, dy\, dx$$

 Technology Connection

15. Use a 3D grapher to graph

$$f(x, y) = x - \tfrac{1}{2}y^2 - \tfrac{1}{3}x^3.$$

---

[23]U.S. Bureau of the Census.

Extended Life Science Connection

STOCKING FISH

To restock a natural habitat, a large number of fish are raised in a hatchery and then released near the shore. The fish swim along the coast in constantly changing directions. If the fish move independently of each other (and not in schools), we say that the swimming patterns of each fish may be described by *Brownian motion*. In this Extended Life Science Connection, we will learn how fish spread throughout a bay.[24]

DIFFUSION

Let N be the number of fish that are released into the bay. After t days, the density of fish x meters away (along the shore) is given by $f(x, t)$, a function of two variables. Suppose for the moment that there is no river current or wind that would tend to push the fish in a certain direction. Then f satisfies the *partial differential equation*

$$\frac{\partial f}{\partial t} = \frac{\sigma^2}{2} \frac{\partial^2 f}{\partial x^2}, \qquad (1)$$

where σ is a constant. This equation is called the *diffusion equation*.[25] Many biological and

---

[24]C. R. Sparrevohn, A. Nielsen, and J. G. Støttrup, "Diffusion of fish from a single release point," *Canadian Journal of Fisheries and Aquatic Sciences,* Vol. 59, pp. 844–853 (2002).

[25]The derivation of the diffusion equation is beyond the scope of this text but may be found in *Random Walks in Biology,* by H. C. Berg (Princeton, NJ: Princeton University Press, 1983).

ecological phenomena may be accurately modeled by (1) and its variants, including the spread of pollutants, the movement of molecules in the body by passive transport, a room's scent after an aerosol can is sprayed, the uptake of nutrients in plants, the drug release from medications, and the temperature of a narrow room when either heated or cooled.

In Chapters 8 and 9, we will study ordinary differential equations; however, a full treatment of partial differential equations lies beyond the scope of this text. It can be shown that a solution of (1) is given by

$$f(x, t) = \frac{N}{2\pi\sigma\sqrt{t}} e^{-x^2/(2\sigma^2 t)}. \qquad (2)$$

As we will see in Chapter 10, (2) is also related to the probability density function of the normal distribution.

EXERCISES

1. Use (2) to compute $\dfrac{\partial f}{\partial t}$ and $\dfrac{\partial^2 f}{\partial x^2}$. Simplify your answers to verify that (1) holds.

2. Suppose that $N = 1000$ and $\sigma = 100$.

   a) What is the density of the fish 100 m away from the release point after 3 days?

   b) Use a grapher to plot the graph of $g(x) = f(x, 1)$ for $-600 \le x \le 600$.

   c) Repeat part (b) for the graph of $g(x) = f(x, 3)$.

   d) Repeat part (b) for the graph of $g(x) = f(x, 5)$.

   tw e) Explain the behavior of these graphs. What happens to the fish as the time t increases?

3. Suppose that $N = 1000$ and $\sigma = 100$.

   a) Use a grapher to plot the graph of $h(t) = f(100, t)$ for $0 \le t \le 10$. At what time does the density of the fish appear to be the largest at this location?

b) Use the graph from part (a) to evaluate
$$\lim_{t \to 0^+} h(t).$$

c) Repeat part (a) for $h(t) = f(200, t)$.

d) Repeat part (a) for $h(t) = f(300, t)$.

e) Find when the fish density is largest at position x. (*Hint:* Use Exercise 1 to solve the equation $\dfrac{\partial f}{\partial t} = 0$ for t.)

tw f) Repeat part (a) for $h(t) = f(-100, t)$. Why is the density of fish at $x = -100$ the same as the density of fish at $x = 100$?

RIVER CURRENT

The diffusion equation (1) does not completely model the movement of the fish since the river current will also affect their movement. Suppose that the river current tends to move the fish downstream at rate α, a constant. Then equation (1) should be modified as

$$\frac{\partial f}{\partial t} = \frac{\sigma^2}{2} \frac{\partial^2 f}{\partial x^2} - \alpha \frac{\partial f}{\partial x}. \tag{3}$$

This equation is called the *advection-diffusion equation*, or the *convection-diffusion equation*.

EXERCISES

4. Verify by substitution that the function

$$f(x, t) = \frac{N}{2\pi\sigma\sqrt{t}} e^{-(x-\alpha t)^2/(2\sigma^2 t)} \tag{4}$$

satisfies (3).

5. Suppose that $N = 1000$, $\sigma = 100$, and $\alpha = 50$.

a) Use a grapher to plot the graph of $g(x) = f(x, 1)$ for $-600 \le x \le 600$.

b) Repeat part (a) for the graph of $g(x) = f(x, 3)$.

c) Repeat part (a) for the graph of $g(x) = f(x, 5)$.

tw d) Explain the behavior of these graphs. What happens to the fish as the time t increases? Why do the peaks of these graphs move to the right?

e) Use (4) to compute $\dfrac{\partial f}{\partial x}$. Solve the equation

$$\frac{\partial f}{\partial x} = 0$$

for x to determine where the fish density is highest after t days. Show that $x = \alpha t$.

tw f) Why is the density of fish highest at $x = \alpha t$ at time t?

6. Suppose that $N = 1000$, $\sigma = 100$, and $\alpha = 50$.

a) Use a grapher to plot the graph of $h(t) = f(100, t)$ for $0 \le t \le 10$. At what time does the density of the fish appear to be the largest at this location?

b) Repeat part (a) for $h(t) = f(200, t)$.

c) Repeat part (a) for $h(t) = f(300, t)$.

tw d) Repeat part (a) for $h(t) = f(-100, t)$. Why is the density of fish at $x = -100$ different from the density of fish at $x = 100$?

8

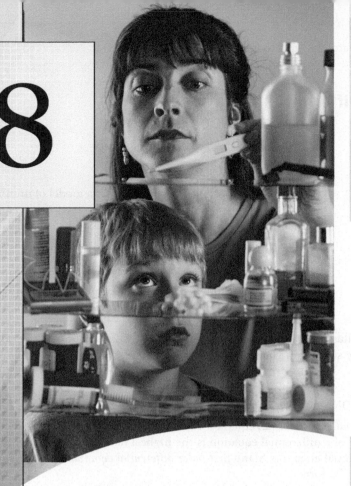

First-Order Differential Equations

AN APPLICATION   A person with a cold ingests a spherical tablet. Let $Q(t)$ be the total amount of the drug released up to time t. Then Q satisfies the differential equation

$$\frac{dQ}{dt} = 3kQ_0^{1/3}(Q_0 - Q)^{2/3},$$

where k is a constant and Q_0 is the total amount of drug contained in the tablet. Use the initial condition $Q(0) = 0$ to show that[1]

$$Q(t) = Q_0[1 - (1 - kt)^3].$$

This problem appears as Exercise 37 in Section 8.4.

---

[1]T. Koizumi, G. C. Ritthidej, and T. Phaechamud, "Mechanistic modeling of drug release from chitosan coated tablets," *Journal of Controlled Release*, Vol. 70, pp. 277–284 (2001).

INTRODUCTION   *Differential equations, equations that involve derivatives, naturally arise in biological and physical problems. In this chapter, we will learn many ways of solving differential equations. We will also discuss the application of differential equations to the life sciences.*

8.1 Differential Equations and Initial-Value Problems

OBJECTIVES

■ Solve certain differential equations.
■ Solve certain initial-value problems.
■ Sketch solutions of a differential equation using a direction field.
■ Verify that a given function is a solution of a differential equation.

In Chapter 4, we studied the equation

$$y'(x) = ky(x),$$

where k is a constant. This equation is a model of uninhibited population growth if k is positive. However, if k is negative, then this is a model for population decay. Its solution is

$$y(x) = y_0 e^{kx},$$

where the constant y_0 is the initial size of the population.

The equation $y' = ky$ is an example of a **differential equation.** It is called a *first-order* differential equation because the highest derivative contained in the equation is y'. As this one equation illustrates, differential equations are rich in applications. Other applications to the life sciences will be presented throughout this chapter.

DEFINITION

A *differential equation* is an equation containing one or more derivatives. The **order** of a differential equation is the highest derivative contained in the differential equation. Many *first-order differential equations* may be written in the form

$$y' = f(x, y),$$

where the right-hand side is some function of x and y.

In this book, the kinds of differential equations we consider are similar to the following:

$$y' = 2x + 3y, \qquad y' = e^{x+y}, \quad \text{and} \quad x^2 y' + xy = y^2.$$

For all of these equations, it is possible to rewrite the equation so that y' is alone on one side, while the other side is some function of x and y. We will not consider differential equations like $y' + \ln y' = x$, where we cannot solve explicitly for y'.

A *solution* of the differential equation $y' = f(x, y)$ is a function such that, when substituted for y in the equation, makes the equation true.

Differential Equations and Indefinite Integrals

If the right-hand side does not depend on y, as in $y' = 2x$ or $y' = x^2 + 1$, then the differential equation has the form

$$y' = f(x).$$

As we saw in Chapter 5, the function y may be obtained by integrating:

$$y = \int f(x)\, dx.$$

EXAMPLE 1   Solve $y' = x/2$.

Solution   Integrating, we find

$$y = \int \frac{x}{2}\, dx = \frac{x^2}{4} + C.$$ ■

Note the constant of integration in the solution to Example 1. This solution is called a **general solution** because taking different values of C gives *all* the solutions. Taking specific values of C gives **particular solutions.** For example, the following are particular solutions of $y' = x/2$:

$$y = \frac{1}{4} x^2 + 1, \qquad y = \frac{1}{4} x^2, \qquad y = \frac{1}{4} x^2 - 2.$$

The general solution can be envisioned as the set of all particular solutions.

Direction Fields

The solutions of differential equations can be visualized using a graph called a **direction field.** To draw the direction field for $y' = x/2$, we draw a short arrow with slope $x/2$ at various points (x, y). For example, the direction field at $(1, -1)$ should have slope $1/2$. This is shown by the green arrow in the figure. Drawing an accurate direction field by hand can be somewhat laborious; however, computer software can be used to construct direction fields.

Also included in the graph are the three particular solutions given earlier. Notice that the direction field is tangent to these solutions at all points of contact. This occurs because the slopes of the tangent lines are given by y', which is equal to $x/2$ for this direction field. Direction fields may be visualized as "current" through which the solutions of differential equations flow.

We notice that different solutions do not intersect. In other words, requiring a solution to pass through a certain point uniquely determines the solution. Knowing the value of a function at a particular point is called an **initial condition.** A first-order differential equation together with an initial condition is called a first-order **initial-value problem.**

Most first-order initial-value problems that arise in practice have unique solutions.

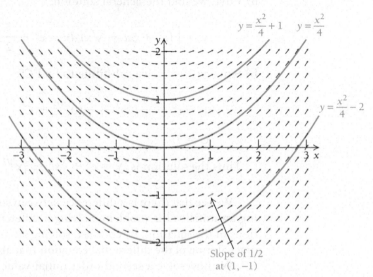

> **THEOREM 1**
>
> Consider the initial-value problem
>
> $$y' = f(x, y), \qquad y(x_0) = y_0.$$
>
> Assume that both f and $\partial f/\partial y$ are continuous near the point (x_0, y_0). Then the initial-value problem has a unique solution on some interval including x_0. As a consequence, different solutions of the differential equation $y' = f(x, y)$ cannot intersect.

The proof of Theorem 1 is beyond the scope of this text. Unless stated otherwise, these continuity conditions on f will be assumed throughout this chapter. As discussed in the exercises, it is possible for an initial-value problem to have multiple solutions if f violates either continuity condition.

EXAMPLE 2 Verify that the initial-value problem $y' = 0.03y, y(0) = 2$ has a unique solution.

Solution In this example, $f(x, y) = 0.03y$, and so $\partial f/\partial y = 0.03$. Since both functions are continuous, the initial-value problem has a unique solution in some interval containing 0. However, Theorem 1 does not give any information about the full domain of the solution. ∎

Initial-Value Problems

As discussed above, an initial condition specifies one solution of a first-order differential equation. Algebraically, the initial condition is used to determine the constant C.

EXAMPLE 3 Solve $y' = e^x + 5x - \sqrt{x}$ given that $y(0) = 8$.

Solution

a) First, we find the general solution:

$$y = \int (e^x + 5x - \sqrt{x}) \, dx = e^x + \frac{5}{2}x^2 - \frac{2}{3}x^{3/2} + C.$$

b) Since $y(0) = 8$, we substitute to find C:

$$8 = e^0 + \frac{5}{2}(0)^2 - \frac{2}{3}(0)^{3/2} + C$$

$$C = 7.$$

Thus the solution is $y = e^x + \frac{5}{2}x^2 - \frac{2}{3}x^{3/2} + 7$. ∎

We see that there is a difference between solving a differential equation and solving an initial-value problem. Solving a differential equation means finding all solutions of the equation. By contrast, solving an initial-value problem means finding the one solution of the differential equation that also satisfies the given initial condition.

We now solve a second-order initial-value problem.

EXAMPLE 4   Solve $f''(x) = x^2 - x$ given that $f(1) = 0$ and $f'(1) = -1$.

Solution

a) First, we find the general solution for $f'(x)$:

$$f'(x) = \int f''(x)\,dx = \frac{x^3}{3} - \frac{x^2}{2} + C_1.$$

b) Since $f'(1) = -1$, we substitute to find C_1:

$$-1 = \frac{(1)^3}{3} - \frac{(1)^2}{2} + C_1$$

$$C_1 = -\frac{5}{6}.$$

Thus we have $f'(x) = x^3/3 - x^2/2 - 5/6$.

c) We integrate again to find the general solution for $f(x)$:

$$f(x) = \int f'(x)\,dx = \frac{x^4}{12} - \frac{x^3}{6} - \frac{5x}{6} + C_2.$$

d) Since $f(1) = 0$, we substitute to find C_2:

$$0 = \frac{(1)^4}{12} - \frac{(1)^3}{6} - \frac{5(1)}{6} + C_2$$

$$C_2 = \frac{11}{12}.$$

The solution is therefore $f(x) = x^4/12 - x^3/6 - 5x/6 + 11/12$.   ∎

Notice that, for this second-order initial-value problem, two initial conditions are needed to uniquely specify the solution. In general, to specify a unique solution, the number of initial conditions must match the order of the differential equation.

In the next chapter, we will discuss solutions of second- and higher-order differential equations in greater detail.

Verifying Solutions

To verify that a function is a solution of a differential equation, we find the necessary derivatives and substitute.

EXAMPLE 5   Show that $y = 4e^x + 5e^{3x}$ is a solution of $y'' - 4y' + 3y = 0$.

Solution   We first find y' and y'':

$$y' = 4e^x + 15e^{3x} \quad \text{and} \quad y'' = 4e^x + 45e^{3x}.$$

Then we substitute into the differential equation:

$$y'' - 4y' + 3y = (4e^x + 45e^{3x}) - 4(4e^x + 15e^{3x}) + 3(4e^x + 5e^{3x})$$

$$= 4e^x + 45e^{3x} - 16e^x - 60e^{3x} + 12e^x + 15e^{3x}$$

$$= 0.$$

Since $y'' - 4y' + 3y = 0$ for all values of x, we know that $y = 4e^x + 5e^{3x}$ is one solution of the differential equation.   ∎

Exercise Set 8.1

Find the general solution. You may need to use substitution, integration by parts, or the table of integrals.

1. $y' = 4x^3$

2. $y' = \sqrt{2x - 3}$

3. $y' = \dfrac{3}{x} - x^2 + x^5$

4. $y' = 5 \sin x - 4$

5. $y' = 4e^{3x} + \sqrt{x}$

6. $y' = \dfrac{2}{\sqrt{x}} + \sec^2 x$

7. $y' = x^2\sqrt{3x^3 - 5}$

8. $y' = \dfrac{\ln x}{x}$

9. $y' = \dfrac{\sin 2x}{(4 + \cos 2x)^3}$

10. $y' = x \cos x$

11. $y' = \dfrac{1}{1 - x^2}$

12. $y' = \dfrac{x}{(3x + 2)^2}$

Solve the initial-value problem.

13. $y' = x^2 + 2x - 3;\, y(0) = 4$

14. $y' = 2x;\, y(1) = 7$

15. $y' = e^{3x} + 1;\, y(0) = 2$

16. $y' = \sin x - x;\, y(0) = 3$

17. $f'(x) = x^{2/3} - x;\, f(1) = -6$

18. $f'(x) = 1/x - 2x + x^{1/2};\, f(4) = 2$

19. $y' = x\sqrt{x^2 + 1};\, y(0) = 3$

20. $y' = xe^{x^2};\, y(0) = 1$

21. $y' = x^3/(x^4 + 1)^2;\, y(1) = -2$

22. $y' = \sin x/(2 + \cos x);\, y(\pi) = 3$

23. $y' = xe^x;\, y(0) = 2$

24. $y' = x \sin x;\, y(0) = 3$

25. $y' = \ln x;\, y(1) = 2$

26. $y' = x \ln x;\, y(1) = 0$

27. $y' = x \sin(x^2);\, y(0) = 3$

28. $y' = e^x\sqrt{e^x + 3};\, y(0) = 1$

Solve the higher-order initial-value problem.

29. $f''(x) = 2;\, f(0) = 3$ and $f'(0) = 4$

30. $f''(x) = 12x;\, f(1) = 2$ and $f'(1) = -2$

31. $f''(x) = x + 1/x^3;\, f(2) = 1$ and $f'(2) = 0$

32. $f''(x) = e^{2x};\, f(0) = 4$ and $f'(0) = 2$

33. $f''(x) = \sin 3x;\, f(\pi) = -2$ and $f'(\pi) = -3$

34. $f'''(x) = x;\, f(0) = 2, f'(0) = -3$, and $f''(0) = 1/2$

For the given direction fields, **(a)** find the slope of the direction field at $(-2, 1)$, and **(b)** sketch the solution that passes through the given point P for $-3 \le x \le 3$.

35. $y' = x^2 - 1$

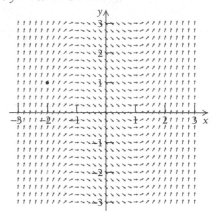

36. $y' = y/2$

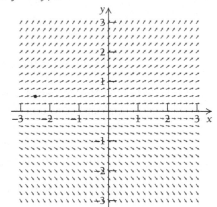

37. $y' = 2x/3 + y$

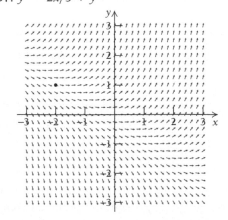

38. $y' = xy$

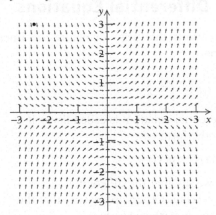

Verify that the given function is a solution of the differential equation.

39. $y'' - 1/x = 0;\ y = x \ln x + 3x - 2$

40. $y'' - 1/x = 0;\ y = x \ln x - 5x + 7$

41. $y'' - 2y' + y = 0;\ y = e^x + 3xe^x$

42. $y'' - 2y' + y = 0;\ y = -2e^x + xe^x$

43. $y'' - 4y' + 3y = 0;\ y = 2e^x - 7e^{3x}$

44. $y' - 2y = e^{2x};\ y = xe^{2x}$

45. $y'' - 2y' + 5y = 0;\ y = e^x \sin 2x$

46. $y'' + 4y = 4 \sin 2x;\ y = (1 - x) \cos 2x$

APPLICATIONS

47. *Weber–Fechner Law.* In psychology, the Weber–Fechner model of stimulus-response asserts that the rate of change dR/dS of the reaction R with respect to a stimulus S is inversely proportional to the stimulus. That is,

$$\frac{dR}{dS} = \frac{k}{S},$$

where k is some positive constant. We also assume that $S > 0$. Let S_0 be the detection threshold value, so that $R(S_0) = 0$. Solve for R as a function of S. Your answer will involve k and S_0.

48. *DNA.* A repressor molecule lands on a strand of DNA with length L a distance x away from a promoter on one end of the strand. It moves along the DNA until it is captured by the promoter. The *expected capture time* T of such a molecule may be modeled by the differential equation

$$T''(x) = -\frac{1}{D}, \quad \text{given} \quad T(0) = 0 \quad \text{and} \quad T'(L) = 0,$$

where D is the diffusion constant. Solve for $T(x)$. Your answer will involve L and D.[2]

SYNTHESIS

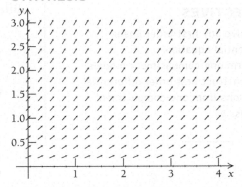

49. *Multiple Solutions.* Consider the differential equation $y' = \sqrt{y}$. The direction field for this differential equation is shown above.

 a) Show that $y(x) = 0$ is a solution of this initial-value problem.

 b) Show that $y(x) = x^2/4$ is a solution of this initial-value problem for $x > 0$.

 c) Draw these two solutions on the direction field. Verify that the direction field is tangent to both solutions at all points of contact.

tw d) Verify that these solutions cross at the origin. Does the initial-value problem
   $y' = \sqrt{y},\ y(0) = 0$ have a unique solution?

 e) Let $f(x, y) = \sqrt{y}$. Compute $\partial f/\partial y$.

tw f) Does $f(x, y) = \sqrt{y}$ satisfy the continuity criteria of Theorem 1 near $(0, 0)$?

50. *DNA.* Refer back to Exercise 48. Find the average of the function $T(x)$ on the interval $[0, L]$. Your answer will involve L and D. This is called the *average time to capture* for molecules that arrive randomly on a strand of DNA.[3]

51.–54. For the direction fields in Exercises 35–38:

 a) Use the differential equation to determine the points (x, y) where $y' = 0$.

tw b) If $y' = 0$ at a point (x, y), explain why the direction field must be horizontal at (x, y).

tw c) Confirm that the direction field is indeed horizontal at the points found in part (a).

---

[2]H. C. Berg, *Random Walks in Biology* (Princeton, NJ: Princeton University Press, 1983).

[3]H. C. Berg, *Random Walks in Biology* (Princeton, NJ: Princeton University Press, 1983).

8.2 Linear First-Order Differential Equations

OBJECTIVES

■ Solve any first-order linear differential equation.
■ Form one-compartment models.
■ Construct first-order differential equations from one-compartment models.

Many differential equations that arise in nature are **linear differential equations**.

DEFINITION

A first-order differential equation is called *linear* if it can be written in the form

$$y' + p(x)y = q(x).$$

If $q(x) = 0$, then we say that the differential equation is *homogeneous*. Otherwise, we say that the differential equation is *nonhomogeneous*.

For example, the equation

$$(x^2 + 1)y' - xy = e^x$$

may be written as a linear differential equation by dividing by $(x^2 + 1)$:

$$y' - \frac{x}{x^2 + 1}y = \frac{e^x}{x^2 + 1}.$$

We see that $p(x) = -x/(x^2 + 1)$ and $q(x) = e^x/(x^2 + 1)$. On the other hand, $y' + xy^2 = e^x$ is *not* a linear equation because the y-term is squared.

Theorem 1 states that, under certain continuity conditions, a first-order initial-value problem is assured to have a unique solution. However, Theorem 1 gave no information about the full domain of this solution. Much more about the domain can be ascertained if the differential equation is known to be linear.

THEOREM 2

Consider the initial-value problem

$$y' + p(x)y = q(x), \qquad y(x_0) = y_0.$$

Suppose that p and q are both continuous on an interval I that contains x_0. Then there is a unique solution to this initial-value problem defined for every point in I.

As a consequence, if p and q are both continuous for all real numbers, then the solution of this initial-value problem will be defined for all real numbers.

EXAMPLE 1 Find an interval where the solution of

$$y' - \frac{x}{x^2 + 1}y = \frac{e^x}{x^2 + 1}, \qquad y(0) = 4$$

is defined.

Solution   The functions $p(x) = -x/(x^2 + 1)$ and $q(x) = e^x/(x^2 + 1)$ are both continuous for all real numbers. Therefore, this initial-value problem has a unique solution defined on the entire real line.   ∎

EXAMPLE 2   Find an interval where the solution of

$$xy' + y = x^3, \qquad y(2) = -3$$

is defined.

Solution   This differential equation may be rewritten as

$$y' + \frac{1}{x}y = x^2.$$

The function $q(x) = x^2$ is continuous for all real numbers. However, the function $p(x) = 1/x$ is undefined at $x = 0$ but is continuous on both $(0, \infty)$ and $(-\infty, 0)$. Therefore, the differential equation has solutions defined on $(0, \infty)$ and on $(-\infty, 0)$. Since $x = 2$ lies in the interval $(0, \infty)$, we conclude that the solution is defined on $(0, \infty)$.   ∎

In Example 2, our conclusion is that the domain *includes* the interval $(0, \infty)$. Notice that we do not say that the domain *is* the interval $(0, \infty)$. As discussed in the exercises, it is possible to choose the initial condition $y(2)$ so that the solution is defined for all real numbers. However, for most choices of the initial condition $y(2)$, the domain will be $(0, \infty)$, and the solution will not extend to nonpositive values of x.

In Example 2, if the initial condition had been given at a negative value of x, then the solution would have been defined on the interval $(-\infty, 0)$.

Solving Linear Differential Equations

To solve the linear first-order differential equation

$$y' + p(x)y = q(x), \tag{1}$$

we let $F(x)$ be any specific antiderivative of $p(x)$, so that $F'(x) = p(x)$. Also, let $G(x) = e^{F(x)}$. Using the Chain Rule, we have

$$G'(x) = e^{F(x)}F'(x) = G(x)p(x).$$

The function $G(x)$ is called an **integrating factor;** its usefulness will become apparent shortly.

To solve for y, we first multiply both sides of (1) by $G(x)$:

$$G(x)y' + p(x)G(x)y = q(x)G(x)$$
$$G(x)y' + G'(x)y = q(x)G(x). \qquad \text{\small Since } G'(x) = p(x)G(x)$$

According to the Product Rule, the left-hand side is the derivative of $G(x)y$. Therefore,

$$[G(x)y]' = q(x)G(x).$$

This equation may be integrated to solve for y. We state this result as a theorem.

> ### THEOREM 3
>
> The general solution of the linear first-order differential equation
>
> $$y' + p(x)y = q(x)$$
>
> may be found by solving
>
> $$[G(x)y]' = q(x)G(x),$$
>
> where $G(x) = e^{F(x)}$ and $F(x)$ is any antiderivative of $p(x)$.

Using this theorem, we can solve any first-order linear differential equation as long as we can compute the integrals. We illustrate this theorem with several examples.

EXAMPLE 3   Solve $y' - 3x^2y = x^2$.

Solution

a) We see that $p(x) = -3x^2$ and $q(x) = x^2$. (Notice that the negative sign is included in the definition of p.) Since both p and q are continuous on the entire real line, we know that all solutions of this differential equation will be defined on the entire real line.

b) We now find the indefinite integral of $p(x)$:

$$\int p(x)\,dx = \int (-3x^2)\,dx = -x^3 + C.$$

Since $F(x)$ can be any antiderivative of $p(x)$, we choose $F(x) = -x^3$ for simplicity. (Notice that the arbitrary constant does not have to be included when determining F.)

c) Next, we determine $G(x)$:

$$G(x) = e^{F(x)} = e^{-x^3}.$$

If we wish to verify that this is the correct integrating factor, we may multiply both sides of the differential equation by $G(x)$:

$$e^{-x^3}y' - 3x^2e^{-x^3}y = x^2e^{-x^3}$$
$$(e^{-x^3}y)' = x^2e^{-x^3}.$$

Notice that the left-hand side is the derivative of $e^{-x^3}y$, according to the Product Rule.

d) Finally, we use Theorem 3 to solve the differential equation:

$$[G(x)y]' = q(x)G(x)$$
$$(e^{-x^3}y)' = x^2e^{-x^3}$$
$$e^{-x^3}y = \int x^2e^{-x^3}\,dx$$
$$e^{-x^3}y = \int \frac{1}{3}e^{-u}\,du \qquad \underline{\text{Substitution}} \quad \begin{aligned} u &= x^3 \\ du &= 3x^2\,dx \end{aligned}$$
$$e^{-x^3}y = -\frac{1}{3}e^{-u} + C$$

$$e^{-x^3}y = -\frac{1}{3}e^{-x^3} + C$$

$$y = -\frac{1}{3} + Ce^{x^3}.$$

Here, C is an arbitrary constant. As expected, there are infinitely many solutions to this differential equation. ∎

CAUTION When solving a first-order differential equation, the unknown constant in the general solution may appear as a coefficient or in some other way. This is in contrast with indefinite integrals, where the constant of integration is always added to form the general solution.

The constant C may be identified by specifying an initial condition.

EXAMPLE 4   Solve $xy' + 2y = 5x^3$ given that $y(1) = 5/4$.

Solution

a) To use Theorem 3, we first divide through by x:

$$y' + \frac{2}{x}y = 5x^2.$$

We see that $p(x) = 2/x$ and $q(x) = 5x^2$. The function q is continuous on the entire real line. However, p is continuous on $(0, \infty)$ and $(-\infty, 0)$ but is discontinuous at $x = 0$. Since the initial condition is given at $x = 1$, we know that the domain of the solution will include $(0, \infty)$.

CAUTION Before using Theorem 3, make sure that the coefficient of y' is 1.

b) We now find the indefinite integral of $p(x)$:

$$\int p(x)\, dx = \int \frac{2}{x}\, dx = 2\ln|x| + C.$$

We choose $F(x) = 2\ln|x|$.

c) Next, we determine $G(x)$. Using properties of logarithms, we find

$$\begin{aligned}
G(x) &= e^{F(x)} \\
&= e^{2\ln|x|} \\
&= e^{\ln|x|^2} \qquad &&\text{Property P3 of logarithms: } k\ln a = \ln a^k \\
&= |x|^2 \qquad &&\text{Property P7 of logarithms: } e^{\ln a} = a \\
&= x^2. \qquad &&\text{Since } x^2 \text{ is always nonnegative}
\end{aligned}$$

d) We use Theorem 3 to solve the differential equation:

$$\begin{aligned}
[G(x)y]' &= q(x)G(x) \\
(x^2 y)' &= (5x^2)(x^2) \\
x^2 y &= \int 5x^4\, dx \\
x^2 y &= x^5 + C \\
y &= x^3 + \frac{C}{x^2}.
\end{aligned}$$

e) We now use the initial condition $y(1) = 5/4$ to determine C:

$$y(1) = (1)^3 + \frac{C}{(1)^2}$$

$$\frac{5}{4} = 1 + C$$

$$\frac{1}{4} = C.$$

Therefore, our solution is $y = x^3 + \dfrac{1}{4x^2}$. As expected, this solution is valid for $x > 0$ but is undefined at $x = 0$. ∎

The direction field for the initial-value problem of Example 4 is shown below; the solution is also shown. This direction field was obtained by writing the differential equation as

$$y' = \frac{-2y}{x} + 5x^2.$$

Notice that the direction field becomes quite steep as x approaches 0. Based on the direction field, we shouldn't expect a solution that passes through the point $(1, 5/4)$ to extend to the negative real axis. For this reason, we say that the domain of the solution is $(0, \infty)$ even though $x^3 + 1/(4x^2)$ is defined for negative values of x.

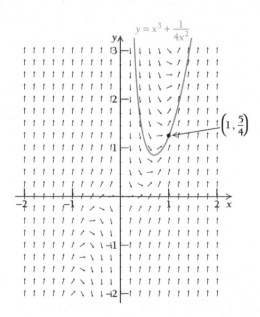

One-Compartment Models

Many biological and ecological processes may be modeled using differential equations. In the next examples, we illustrate how linear first-order differential equations may be used to predict concentrations of hazardous chemicals.

EXAMPLE 5 *Biohazards.* A mussel is placed into water polluted with polychlorinated biphenyls (PCBs). Let $Q(t)$ be the concentration of PCB in the mussel in micrograms (per gram of tissue) after t days. For low concentrations of pollution, the mussel absorbs PCBs at the rate of 12 micrograms of PCB per gram of tissue per day. Also, the elimination rate of PCBs from the mussel is $0.18Q$ micrograms per gram of tissue per day.[4] Construct a differential equation for Q.

Solution   There are two different effects on Q. First, there is an elimination rate of PCBs from the mussel:

Elimination rate $= -0.18Q$ micrograms per day (per gram of tissue).

The negative sign indicates that PCBs are being eliminated from the mussel. Notice that this rate is proportional to Q.

Second, the mussel absorbs PCBs at a rate determined by the outside environment. This rate is *not* proportional to Q.

Absorption rate $= 12$ micrograms per day (per gram of tissue).

Adding these two effects gives the rate of change of Q:

$$Q' = -0.18Q + 12.$$

In the previous example, the two effects on the concentration's rate of change may be visualized using a **one-compartment model** (or a *single-compartment model*), as shown in the margin. The compartment represents the concentration Q. The absorption rate is represented by the incoming arrow, and the elimination rate is represented by the outgoing arrow.

The elimination rate is called a **relative rate** since it is proportional to the current concentration Q. (This rate is represented by the solid arrow.) On the other hand, the absorption rate is not proportional to Q. (This rate is represented by the dashed arrow.)[5] We call this kind of rate an **uptake rate.**

---

[4]K. N. Yu, P. K. S. Lam, C. C. C. Cheung, and C. W. Y. Yip, "Mathematical modeling of PCB bioaccumulation in *Perna viridis*," *Marine Pollution Bulletin*, Vol. 45, pp. 332–338 (2002).

[5]While compartment models appear throughout the scientific literature, the labeling convention for the arrows is unfortunately not uniform. For example, solid arrows are used by some authors to represent both uptake and relative rates. While the labeling convention in this book distinguishes between these two kinds of rates, students should not expect them to be distinguished when reading the literature.

EXAMPLE 6   *Biohazards.* Let $Q(t)$ be as in Example 5.

a) Solve for Q if no PCB is initially present in the mussel.

b) For large values of t, what value does $Q(t)$ approach?

Solution

a) As discussed in Example 5, the differential equation for Q is

$$Q' = -0.18Q + 12, \quad \text{or} \quad Q' + 0.18Q = 12.$$

Also, since no PCBs are initially present, the initial condition is $Q(0) = 0$.

To solve this differential equation, we observe that $p(t) = 0.18$ and $q(t) = 12$. Therefore, we may choose $F(t) = 0.18t$ and hence $G(t) = e^{0.18t}$. Using Theorem 3, we see that

$$[G(t)Q(t)]' = q(t)G(t)$$
$$(e^{0.18t}Q(t))' = 12e^{0.18t}$$

$$e^{0.18t}Q(t) = \int 12e^{0.18t}\, dt$$

$$e^{0.18t}Q(t) = \frac{200}{3}e^{0.18t} + C$$

$$Q(t) = \frac{200}{3} + Ce^{-0.18t}.$$

To solve for the constant C, we use the initial condition $Q(0) = 0$. Letting $t = 0$, we get

$$Q(0) = \frac{200}{3} + Ce^{-0.18(0)}$$

$$0 = \frac{200}{3} + Ce^{0}$$

$$-\frac{200}{3} = C.$$

Substituting for C,

$$Q(t) = \frac{200}{3} - \frac{200}{3}e^{-0.18t} = \frac{200}{3}(1 - e^{-0.18t}).$$

b) As t tends to infinity, $e^{-0.18t}$ tends to 0. Therefore,

$$\lim_{t\to\infty} Q(t) = \lim_{t\to\infty} \frac{200}{3}(1 - e^{-0.18t}) = \frac{200}{3}(1 - 0) = \frac{200}{3}.$$

In other words, the mussel should eventually have very close to $66\frac{2}{3}$ micrograms of PCB per gram of tissue.   ■

To interpret this result further, we observe that the concentration level of PCBs within the mussel should reach some *equilibrium* value over a long period of time. At this equilibrium value, Q does not change and hence $Q' = 0$. Substituting into the original differential equation, we see that

$$0 + 0.18Q = 12$$

$$Q = \frac{200}{3} \quad \text{at equilibrium.}$$

This straightforward calculation of the equilibrium value agrees with the limit $\lim_{t \to \infty} Q(t)$. In Section 8.3, we will discuss the equilibrium values of differential equations in greater detail.

Relative, Uptake, and Downtake Rates

There are two qualitatively different kinds of rates that may appear in a one-compartment model. First, there may be a *relative rate,* which is proportional to the quantity studied. Examples of this kind of rate are growth rates (for population growth), decay rates (for population decay), and elimination rates. This type of rate is expressed as a *number* per time. In Example 5, there were $0.18Q$ micrograms of PCB eliminated per day (per gram of tissue). This is more succinctly written as an elimination rate of 0.18 per day.

Second, there may be an *uptake rate* or a **downtake rate,** which is either constant or a function of t. However, neither an uptake rate nor a downtake rate is proportional to the quantity studied. This type of rate is expressed as some number of *units* per time. In Example 5, the uptake rate of PCB into the mussel was 12 micrograms of PCB per day (per gram of tissue).

To properly construct compartment models, these different effects on the instantaneous rate of change must be correctly distinguished.

EXAMPLE 7 *Mixing Chemicals.* A tank contains 100 gal of brine whose concentration is 2.5 lb of salt per gal. Brine containing 2 lb of salt per gal runs into the tank at a rate of 5 gal per min, and the mixture runs out of the tank at the same rate. The concentration of salt within the tank is kept uniform by stirring. Let $S(t)$ be the amount of salt in the tank in pounds at time t.

a) Construct a one-compartment model for S.

b) Construct a differential equation for S.

c) Determine how much salt is in the tank at time t.

d) For large t, how much salt is in the tank?

Solution

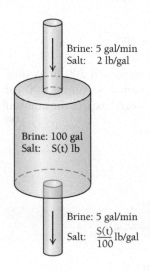

Brine: 5 gal/min
Salt: 2 lb/gal

Brine: 100 gal
Salt: S(t) lb

Brine: 5 gal/min
Salt: $\frac{S(t)}{100}$ lb/gal

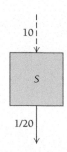

10

S

1/20

a) There are two effects on the amount of salt in the tank. First, since brine enters the tank at 5 gal per min and each gallon of this brine contains 2 lb of salt, we see that salt enters the tank at the rate of 10 lb per min. This rate is independent of how much salt is currently in the tank and hence is an uptake rate.

The second effect is the elimination of salt from the tank. Every minute, 5 of the 100 gal are removed from the tank. In other words, 5% of the brine in the tank is removed every minute, so 5% of the salt in the tank is removed every minute. This rate is proportional to the amount of salt currently in the tank. Therefore, the elimination rate may be expressed as a relative rate of 5% per min, or 1/20 per min.

These two rates are shown in the compartment model.

b) Based on the one-compartment model, the differential equation for S is

$$S' = -\frac{1}{20}S + 10$$

$$S' + \frac{1}{20}S = 10.$$

c) To solve this differential equation, we observe that $p(t) = 1/20$ and $q(t) = 10$. Therefore, we may choose $F(t) = t/20$ and hence $G(t) = e^{t/20}$. Using Theorem 3, we see that

$$[G(t)S(t)]' = q(t)G(t)$$
$$(e^{t/20}S(t))' = 10e^{t/20}$$
$$e^{t/20}S(t) = \int 10e^{t/20}\, dt$$
$$e^{t/20}S(t) = 200e^{t/20} + C$$
$$S(t) = 200 + Ce^{-t/20}.$$

To solve for the constant C, we observe that the tank initially contains $(2.5)(100) = 250$ lb of salt. The initial condition is thus $S(0) = 250$. Solving for C, we obtain

$$S(0) = 200 + Ce^{-0/20}$$
$$250 = 200 + Ce^0$$
$$50 = C.$$

Substituting for C, we conclude that $S(t) = 200 + 50e^{-t/20}$.

d) As t tends to infinity, $e^{-t/20}$ tends to 0. Therefore,

$$\lim_{t\to\infty} S(t) = \lim_{t\to\infty}(200 + 50e^{-t/20}) = (200 + 0) = 200.$$

In other words, the amount of salt in the tank should approach 200 lb as time increases. This also makes sense intuitively; since the tank has volume 100 gal and the incoming brine contains 2 lb of salt per gal, we would expect the amount of salt in the tank to approach 200 lb. ∎

Exercise Set 8.2

Determine an interval on which a unique solution of the initial-value problem will exist. Do not actually find the solution.

1. $y' - x^2y = x^3$; $y(1) = -2$

2. $y' + \dfrac{1}{x-2}y = \sin x$; $y(-2) = 4$

3. $y' + y\tan x = x^2 + 1$; $y(0) = 3$

4. $(x-2)^2y' + 4y = \dfrac{1}{x+1}$; $y(1) = 2$

5. $(t-2)^2y' + 4y = \dfrac{1}{t+1}$; $y(4) = -6$

6. $(e^t - 1)y' + e^t = \dfrac{1}{t^2 - t - 6}$; $y(1) = 8$

Solve the differential equation.

7. $y' - 3y = 0$

8. $y' + 4y = 2$

9. $y' + y\cos 2x = \cos 2x$

10. $xy' + 3xy = 6x^3 - 4x$

11. $y' - 2ty = 2t$

12. $(t^2 + 1)y' + 2ty = t^3 + t$

13. $y' - y = e^t$

14. $y' + 4t^3y = 4t^4e^{-t^4}$

15. $x^2y' - 4xy = x^7e^{x^2} + 3x^5 - 6$

16. $xy' + y = x\sin x$

17. $xy' + 3y = x\ln x$

18. $y' + y = x$

19. $y' + y = e^{-t} \ln t$

20. $y' \cos 2t - y \sin 2t = \sin 2t(3 + \cos 2t)$

Solve the initial-value problem. State an interval on which the solution exists.

21. $y' + 4y = 6; y(0) = 2$

22. $xy' + y = 0; y(1) = 3$

23. $(1 + \sin x)y' + y \cos x = \cos x + \sin x \cos x; y(\pi/2) = -1$

24. $y' + y \sin x = \sin x; y(\pi/2) = 4$

25. $t^2 y' + ty = t^4; y(2) = 5$

26. $(t^3 + 1)y' + 3t^2 y = t^3 + 1; y(0) = 2$

27. $y' - 5y = x + 1; y(0) = 1$

28. $y' + e^{2x}y = e^{2x}; y(0) = 0$

29. $(e^x - 2)y' + e^x y = 2e^{-2x} - e^{-3x}; y(0) = 3$

30. $y' + y \cot x = \cos x; y(\pi/2) = -1$

31. $y' + 3t^2 y = t^2 e^{t^3}; y(0) = 3$

32. $t^2 y' + y = 1; y(1) = 4$

33. $y' + 4ty = t; y(0) = 3$

34. $y' + y \tan t = \cos t; y(\pi/3) = \pi$

Find the general solution for the following one-compartment models.

35.

36.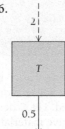

37.

38.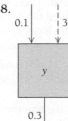

APPLICATIONS

39. *Uninhibited Population Growth.* Use Theorem 3 to solve the differential equation $P' = kP$ given the initial condition $P(0) = P_0$.

40. *Mixing Chemicals.* A tank initially contains 180 lb of salt dissolved in 500 gal of water. Saltwater containing 0.5 lb of salt per gal enters the tank at the rate of 3 gal per min. The mixture (kept uniform by stirring) is removed at the same rate. How many pounds of salt are in the tank after an hour?

41. *Mixing Chemicals.* A tank initially contains 100 lb of salt dissolved in 800 gal of water. Saltwater containing 1 lb of salt per gal enters the tank at the rate of 4 gal per min. The mixture (kept uniform by stirring) is removed at the same rate. How many pounds of salt are in the tank after 2 hr?

42. *Crop Yield.* A farmer is growing soybeans in a field. The more fertilizer he spreads, the greater his yield Y, up to the limiting value of 60 bushels per acre. If he spreads x pounds of fertilizer per acre, the yield follows the differential equation

$$\frac{dY}{dx} = k(60 - Y),$$

where k is a constant.

a) Write this first-order linear differential equation in standard form.

b) Solve for $Y(x)$ given that $Y(0) = 15$.

c) Suppose that $Y(10) = 21$. Use this to determine k.

d) Find $Y(5)$.

e) How many pounds of fertilizer, per acre, must be spread to yield 42 bushels per acre?

Note: This model is not completely valid since over-fertilizing the field will tend to kill the crop.

43. *Newton's Law of Cooling.* An object has an initial temperature of T_0 and is placed into a surrounding medium with a lower temperature C. The temperature T of the cooling object drops at a rate that is proportional to the difference $T - C$. That is,

$$\frac{dT}{dt} = -k(T - C),$$

where k is a positive constant and t is time.

a) Use the techniques of this section and the condition $T(0) = T_0$ to solve for $T(t)$.

b) A metal object that has been heated to 143°F is placed into a room that is kept at a constant 70°F. After 30 min, it is observed to cool to 117°F. How long will it take the object to cool to 90°F?

SYNTHESIS

44. *Equilibrium.* In Example 7, determine the equilibrium value for S by setting $S' = 0$ in the differential

equation. Does your answer agree with the equilibrium value found in Example 7?

45. *Domain of Solution.* Consider the initial-value problem $xy' + 2y = 5x^3$, $y(1) = 1$. (This is the differential equation used in Example 4 but with a different initial condition.)

   a) According to Theorem 3, the solution is guaranteed to be valid over what interval?

   b) Solve the initial-value problem.

   tw c) Show that the solution of part (b) is defined for all real numbers, unlike the solution found in Example 4.

   tw d) Sketch this solution on the direction field shown with Example 4. Based on the direction field, why does this solution include negative values of x?

46. *Domain of Solution.* Consider the differential equation $xy' + y = x^3$, which was considered in Example 2.

   a) Find the general solution of this differential equation.

   b) Find the value of the initial condition $y(2)$ that makes the solution defined for all real numbers.

47. *Mixing Chemicals.* A 2000-gal tank initially contains 200 lb of salt dissolved in 500 gal of water. Pure water is pumped into the tank at the rate of 5 gal per min, while the well-stirred mixture is drawn off at the rate of 2 gal per min. The process stops when the tank is full. How much salt is left in the tank when the tank is full?

Anesthesia. As a patient receives anesthesia through a gas mask, the pressure $P(t)$ of the anesthesia in one section of the body increases at a rate proportional to the difference of the arterial pressure $f(t)$ and the current sectional pressure. In other words,

$$\frac{dP}{dt} = k[f(t) - P],$$

where k is a constant. Assume that no anesthesia is initially present.[6]

48. Suppose that $k = 1.2$ and the gas mask dispenses anesthesia so that $f(t) = 1$. Solve for $P(t)$.

49. Solve for $P(t)$ if $k = 1.2$ and $f(t) = 1 - e^{-kt}$.

50. Solve for $P(t)$ if $k = 1.2$ and $f(t) = \sin t$.

One-Compartment Models. In the following problems, a, b, and k are constants. Some of your answers will contain these constants.

51. *Pollution.* Rainbow fish are exposed to elevated copper concentrations due to a nearby copper mine. Let $Q(t)$ be the concentration of copper per gram of fish after t hours. Suppose that the fish intake copper at the rate of a nanograms of copper per gram of fish per hour. The elimination rate of copper from the fish is b per hour.[7]

   a) Draw a one-compartment model for Q.

   b) Solve for $Q(t)$ given the initial condition $Q(0) = 0$.

   c) Find the equilibrium value for Q.

52. *Radioactive Waste.* Lichen are placed into an environment where radioactive lead-210 is constantly produced. Let C denote the concentration of radioactive material present in 1 sq m of lichen. We assume that the lichen absorb a atoms of lead-210 per second. There are two ways that lead-210 can be removed from the lichen: radioactive decay (with rate k per second) and natural elimination (with rate b per second).[8]

   a) Draw a one-compartment model for C.

   b) Solve for $C(t)$ given the initial condition $C(0) = 0$.

   c) Find the amount of lead-210 present at equilibrium.

[6] J. P. Whiteley, D. J. Gavaghan, and C. E. W. Hahn, "Modelling inert gas exchange in tissue and mixed-venous blood return to the lungs," *Journal of Theoretical Biology,* Vol. 209, pp. 431–443 (2001).

[7] S. A. Gale, S. V. Smith, R. P. Lim, R. A. Jeffree, and P. Petocz, "Insights into the mechanisms of copper tolerance of a population of black-banded rainbowfish (*Melanotaenia nigrans*) (Richardson) exposed to mine leachate, using $^{64/67}$Cu," *Aquatic Toxicology,* Vol. 62, pp. 135–153 (2003).

[8] G. Kirchner and O. Daillant, "The potential of lichens as long-term biomonitors of natural and artificial radionuclides," *Environmental Pollution,* Vol. 120, pp. 145–150 (2002).

8.3 Autonomous Differential Equations and Stability

OBJECTIVES

■ Find equilibrium values of autonomous differential equations.
■ Assess the stability of equilibrium values.
■ Sketch solutions of autonomous differential equations.

Recall that a first-order differential equation has the form $y' = f(x, y)$, where the right-hand side is some function of x and y. In Section 8.1, we saw several examples where the right-hand side was a function of only x. These differential equations could be solved using indefinite integrals.

In an **autonomous differential equation,** the right-hand side is a function of only y. For example, the differential equation $y' = 6y^2 - y^3$ is autonomous, but the differential equation $y' = x - 6y$ is not autonomous.

Autonomous differential equations are often used to model population growth. If y represents the size of a population, it seems reasonable to assume that the rate of population growth should be completely determined by the current population size and not dependent on time.

Some autonomous differential equations can be solved exactly using techniques discussed in Section 8.4. In this section, we discuss techniques of understanding the qualitative behavior of solutions without explicitly solving the differential equation.

Equilibrium Values and Stability

As discussed in Sections 4.3 and 4.4, one model of population growth (or decay) sets the *relative* (or *per capita*) rate of growth as a constant, so that the population satisfies the autonomous differential equation $y' = ky$ for some constant k. The solution of this differential equation is $y = y_0 e^{hx}$, where y_0 is the population at time 0.

If $y_0 = 0$, then we obtain a constant solution $y = 0$. A constant solution to an autonomous differential equation is called an **equilibrium value.**

DEFINITION

An *equilibrium value* of the autonomous differential equation $y' = f(y)$ is a constant solution $y = c$. Since the derivative of a constant is 0, equilibrium values may be found by solving the equation $f(y) = 0$.

A typical direction field for $y' = ky$ is shown below if $k > 0$. As before, the solutions are tangent to the direction field at all points of contact. The constant solution $y = 0$ corresponds to the horizontal arrows along the x-axis.

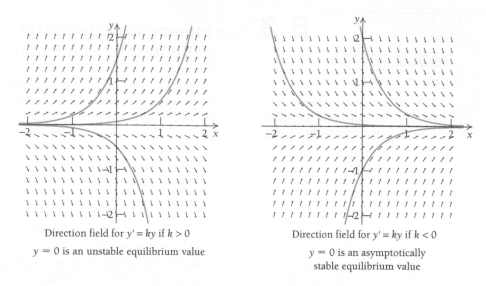

Direction field for $y' = ky$ if $k > 0$

$y = 0$ is an unstable equilibrium value

Direction field for $y' = ky$ if $k < 0$

$y = 0$ is an asymptotically
stable equilibrium value

If $k > 0$, then nonzero solutions to $y' = ky$ exhibit exponential growth and hence diverge from 0 for large values of x. We say that the equilibrium value of $y = 0$ is **unstable.**

However, if $k < 0$, then the direction field points toward the x-axis. In this case, nonzero solutions get closer and closer to the equilibrium value of $y = 0$ for large values of x. We say that the equilibrium value of $y = 0$ is **asymptotically stable.**

These characterizations of stability apply generally to autonomous differential equations.

DEFINITION

Let c be an equilibrium value of the autonomous differential equation $y' = f(y)$. Also, consider a solution y of this differential equation with an initial value $y(0) = y_0$ that is close (but not equal) to c.

1. If y diverges from c for large values of x, we say that c is *unstable*.
2. If y converges to c for large values of x, we say that c is *asymptotically stable*.
3. Suppose that either of the following occurs:
    a) Solutions slightly above $y = c$ diverge from $y = c$, while solutions slightly below $y = c$ converge to $y = c$, or
    b) Solutions slightly above $y = c$ converge to $y = c$, while solutions slightly below $y = c$ diverge from $y = c$.

Then we say that c is **semistable.**

Assessing Stability Using the Sign of y'

To assess the stability of an equilibrium value of $y' = f(y)$, we will first determine where nonconstant solutions increase and decrease. In other words, we need to determine where y' is positive and negative.

EXAMPLE 1 Find the equilibrium values of the differential equation $y' = y^2 - y - 2$, and assess the stability of each.

Solution   The equilibrium values may be found by setting $y' = 0$:

$$y^2 - y - 2 = 0$$
$$(y - 2)(y + 1) = 0.$$

We see that there are two equilibrium values: $y = 2$ and $y = -1$. (Notice that these equilibrium values were computed without solving the differential equation.) We use these points to divide the y-axis into three intervals: $(-\infty, -1)$, $(-1, 2)$, and $(2, \infty)$. In a manner similar to the curve-sketching techniques of Chapter 3, we choose a test value for y in each interval to determine the sign of y'.

$$(-\infty, -1):\quad \text{Test } -2,\quad y' = (-2)^2 - (-2) - 2 = 4 > 0;$$
$$(-1, 2):\quad \text{Test } 0,\quad y' = (0)^2 - (0) - 2 = -2 < 0;$$
$$(2, \infty):\quad \text{Test } 3,\quad y' = (3)^2 - (3) - 2 = 4 > 0.$$

These test values are used to determine the behavior of the direction field. In the table below, these intervals are arranged in reverse numerical order to simulate the y-axis. If the arrows point toward an equilibrium value, then that equilibrium value is asymptotically stable. If the arrows point away from an equilibrium value, then that equilibrium value is unstable. Finally, if one arrow points toward the equilibrium value but the other points away, then that equilibrium value is semistable.

Interval	Test Value	Sign of y'	Result	Stability
$(2, \infty)$	$y = 3$	$y' > 0$	↗	$y = 2$ is unstable
$(-1, 2)$	$y = 0$	$y' < 0$	↘	
$(-\infty, -1)$	$y = -2$	$y' > 0$	↗	$y = -1$ is asymptotically stable

We conclude that $y = 2$ is unstable and $y = -1$ is asymptotically stable. This information can be used to sketch sample solutions of the differential equation, as shown in the figure. (Since different solutions of this differential equation cannot intersect, the nonconstant solutions which converge to $y = -1$ do not actually intersect this line.)

Notice that the x-axis is unmarked in this sketch: in the above analysis, we did not determine how quickly solutions either converge to or diverge from the equilibrium values.   ■

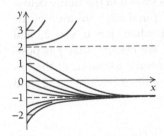

EXAMPLE 2 Determine the equilibrium values of $y' = y^4 - 4y^2$, and assess the stability of each.

Solution   We set $y' = 0$ to find the equilibrium values:

$$y^4 - 4y^2 = 0$$
$$y^2(y^2 - 4) = 0$$
$$y^2(y - 2)(y + 2) = 0.$$

We see that there are three equilibrium values: $y = 0, y = 2$, and $y = -2$. We use these points to divide the y-axis into four intervals: $(-\infty, -2), (-2, 0), (0, 2)$, and

$(2, \infty)$. In a manner similar to the curve-sketching techniques of Chapter 3, we choose a test value of y in each interval to determine the sign of y'.

$$(-\infty, -2): \quad \text{Test } -3, \quad y' = (-3)^4 - 4(-3)^2 = 45 > 0;$$
$$(-2, 0): \quad \text{Test } -1, \quad y' = (-1)^4 - 4(-1)^2 = -3 < 0;$$
$$(0, 2): \quad \text{Test } 1, \quad y' = (1)^4 - 4(1)^2 = -3 < 0;$$
$$(2, \infty): \quad \text{Test } 3, \quad y' = (3)^4 - 4(3)^2 = 45 > 0.$$

Interval	Test Value	Sign of y'	Result	Stability
$(2, \infty)$	$y = 3$	$y' > 0$	↗	$y = 2$ is unstable
$(0, 2)$	$y = 1$	$y' < 0$	↘	$y = 0$ is semistable
$(-2, 0)$	$y = -1$	$y' < 0$	↘	$y = -2$ is asymptotically stable
$(-\infty, -2)$	$y = -3$	$y' > 0$	↗	

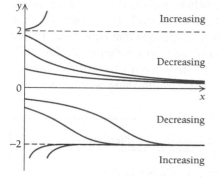

Increasing

Decreasing

Decreasing

Increasing

We conclude that $y = 2$ is unstable, $y = -2$ is asymptotically stable, and $y = 0$ is semistable. This information can be used to sketch sample solutions of the differential equation, as shown in the figure.

In Example 2, $y = 0$ was semistable because solutions either converged to or diverged from 0 depending on whether the initial value was slightly above 0 or slightly below 0. If it had turned out that solutions slightly above 0 diverged from 0 while those slightly below 0 converged to 0, we would still have classified $y = 0$ as semistable.

Using the Sign of $f'(y)$

Let's take a closer look at the equilibrium values of $y' = f(y) = y^4 - 4y^2$, the differential equation studied in Example 2. The graph of f is shown in the figure below. (It should be emphasized that this graph is the graph of f and *not* a solution of the differential equation.) This function is equal to zero when $y = 0, y = 2$, and $y = -2$, corresponding to the three equilibrium values. We observe that f is positive on the intervals $(-\infty, -2)$ and $(2, \infty)$. On these intervals, solutions of the differential equation $y' = f(y)$ increase. Also, f is negative on the intervals $(-2, 0)$ and $(0, 2)$. On these intervals, solutions of $y' = f(y)$ decrease.

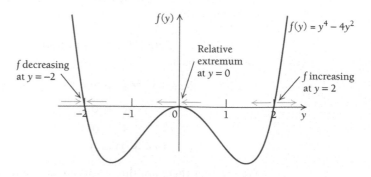

The blue arrows in the figure indicate where the solutions increase and decrease, as found in Example 2. Arrows point to the left where $f(y)$ is negative, while arrows point to the right where $f(y)$ is positive. The arrows also indicate the stability of the equilibrium values. First, the blue arrows point together at $y = -2$, corresponding to an asymptotically stable equilibrium value. At this point, the function f is decreasing, which we can directly verify by computing $f'(-2)$:

$$f'(y) = 4y^3 - 8y$$
$$f'(-2) = 4(-2)^3 - 8(-2) = -16.$$

Since $f'(-2) < 0$, we see that f is decreasing when $y = -2$.

Second, the blue arrows point away from $y = 2$, so this equilibrium value is unstable. At this point, f is increasing since $f'(2)$ is positive:

$$f'(2) = 4(2)^3 - 8(2) = 16 > 0.$$

Third, one blue arrow points away from $y = 0$, while the other arrow points toward $y = 0$. Accordingly, $y = 0$ is a semistable equilibrium value. At this equilibrium value, y attains a relative extremum (in this case, a relative maximum).

The following theorem summarizes how the sign of f' can be used to assess the stability of equilibrium values.

THEOREM 4

Let $y = c$ be an equilibrium point of the autonomous differential equation $y' = f(y)$, where f is differentiable.

1. If $f'(c) < 0$, then $y = c$ is asymptotically stable.
2. If $f'(c) > 0$, then $y = c$ is unstable.
3. If $f'(c) = 0$ and f has a relative extremum at c, then $y = c$ is semistable.

We have seen that the graph of f may be used to assess the stability of equilibrium values. The graph of f also may be used to determine where solutions have points of inflection. We recall from Chapter 3 that points of inflection of y are identified by setting $y'' = 0$. If $y' = f(y)$, we may compute y'' using the Chain Rule:

$$y'' = \frac{d}{dx}(y') = \frac{d}{dx}f(y) = f'(y)y' = f'(y)f(y).$$

If y is not an equilibrium value, then $f(y) \neq 0$. Therefore, points of inflection are identified by solving the equation $f'(y) = 0$.

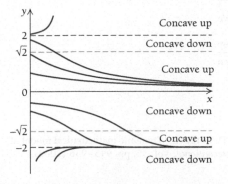

EXAMPLE 3   Determine the points of inflection for solutions of $y' = y^4 - 4y^2$.

Solution   In this example, $f(y) = y^4 - 4y^2$. To identify points of inflection, we solve the equation $f'(y) = 0$.

$$f'(y) = 4y^3 - 8y = 0$$
$$4y(y^2 - 2) = 0$$

There are two nonequilibrium solutions: $y = \sqrt{2}$ and $y = -\sqrt{2}$. In the margin, we add two blue horizontal lines to the sketch made in Example 1. Solutions change concavity if they cross either blue line. ∎

EXAMPLE 4 *Fat Crystallization.* Let $y(t)$ be the proportion of crystallizable milk fat in a sample after t hours. Then y satisfies the differential equation[9]

$$y' = 8(y^5 - y).$$

Determine the equilibrium values and assess the stability of each. Use this information and the points of inflection to sketch solutions of the differential equation.

Solution   In this differential equation, $f(y) = 8(y^5 - y)$. We set $f(y) = 0$ to find the equilibrium values:

$$8(y^5 - y) = 0$$
$$8y(y^4 - 1) = 0$$
$$8y(y^2 - 1)(y^2 + 1) = 0$$
$$8y(y - 1)(y + 1)(y^2 + 1) = 0.$$

We see that there are three equilibrium values: $y = 0$, $y = 1$, and $y = -1$.

Next, we use f' to assess stability:

$$f'(y) = 8(5y^4 - 1);$$
$$f'(0) = 8(5[0]^4 - 1) = -8 < 0; \qquad \text{Asymptotically stable}$$
$$f'(1) = 8(5[1]^4 - 1) = 32 > 0; \qquad \text{Unstable}$$
$$f'(-1) = 8(5[-1]^4 - 1) = 32 > 0. \qquad \text{Unstable}$$

We conclude that $y = 0$ is asymptotically stable while $y = 1$ and $y = -1$ are unstable. To find the points of inflection for nonconstant solutions, we solve the equation $f'(y) = 0$:

$$f'(y) = 8(5y^4 - 1) = 0$$
$$5y^4 - 1 = 0$$
$$y^4 = \frac{1}{5}$$
$$y = \pm\frac{1}{\sqrt[4]{5}} \approx \pm 0.669.$$

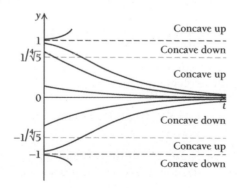

[9]I. Foubert, P. A. Vanrolleghem, B. Vanhoutte, and K. Dewettinck, "Dynamic mathematical model of the crystallization kinetics of fats," *Food Research International*, Vol. 35, pp. 945–956 (2002).

Solutions of the differential equation are sketched in the figure. These solutions have a simple physical interpretation. If the milk fat has no impurities and is completely crystallizable ($y = 1$), then this equilibrium condition will continue for all time. However, if even a small number of impurities are initially present, then the proportion of the sample that is crystallizable will eventually approach 0 as time increases. (A technique for actually computing y as a function of time is presented in the exercises.)

We note in passing that, since y is a proportion, the solutions that are either always greater than 1 or always less than 0 are physically impossible. ∎

Logistic Growth Model

A population that grows with a constant *per capita* rate k satisfies the uninhibited growth model $y' = ky$. We have seen that nonzero solutions of this differential equation grow exponentially. For real populations, however, such unbounded growth is physically unrealistic over all timescales. There may be some compelling reason, such as a limitation on food, living space, or other natural resources, that decreases the per capita growth rate as the population increases.

We will construct a model of inhibited growth by appending an additional factor to the uninhibited growth model. One such simple model is called the **logistic growth model** and is defined by

$$y' = ky\left(1 - \frac{y}{L}\right),$$

where k and L are positive constants. The per capita growth rate is $k(1 - y/L)$, which decreases from k to 0 as y increases from 0 to L. The physical significance of L will become apparent shortly.

EXAMPLE 5 Determine the equilibrium values of the logistic growth model and assess the stability of each.

Solution   Setting $y' = 0$, we see that the equilibrium levels are $y = 0$ and $y = L$. Next, we compute $f'(y)$:

$$f(y) = ky - \frac{k}{L}y^2$$

$$f'(y) = k - \frac{2k}{L}y.$$

We then compute f' at the equilibrium values. We recall that k is assumed to be a positive constant.

$$f'(0) = k - \frac{2k}{L}(0) = k > 0; \qquad \text{Unstable}$$

$$f'(L) = k - \frac{2k}{L}(L) = -k < 0. \qquad \text{Asymptotically stable}$$

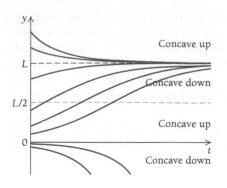

We conclude that $y = 0$ is unstable, while $y = L$ is asymptotically stable.

Next, we find the point of inflection by solving $f'(y) = 0$:

$$f'(y) = k - \frac{2k}{L}y = 0$$

$$y = \frac{L}{2}.$$

Solutions of the logistic growth model using the equilibrium values and this point of inflection are shown in the figure. ∎

In the logistic growth model, L represents the *limiting* population. If the initial population is less than L but is nonzero, then the population will increase to L. On the other hand, if the initial population is greater than L, then the population will decrease to L. Negative solutions of the logistic growth model are of course physically unrealistic.

Using techniques that will be developed in Section 8.4, it can be shown that all nonconstant solutions of the logistic growth model can be written as

$$y(t) = \frac{Ly_0}{y_0 + (L - y_0)e^{-kt}},$$

where y_0 is the initial value $y(0)$. This is the logistic equation, which was first introduced in Section 4.3.

EXAMPLE 6 *Bacteria Growth.* Many bacteria strains are used by the dairy industry to produce different types of fermented milks and yogurts. During a fermentation experiment, the population y (in millions per mL) of bacteria *Lactobacillus fermentum* after t hours in a wheat medium satisfied the differential equation

$$y' = 0.532y\left(1 - \frac{y}{1900}\right).$$

Initially, 8 million bacteria per mL were present. Determine the number of bacteria per mL after 5 hr.[10]

Solution The population y satisfies the logistic growth model with $k = 0.532$, $L = 1900$, and initial condition $y_0 = 8$. Substituting into the solution of the logistic growth model, we have

$$y(t) = \frac{(8)(1900)}{8 + (1900 - 8)e^{-(0.532)t}} = \frac{15{,}200}{8 + 1892e^{-0.532t}}.$$

Substituting 5 for t, we find

$$y(5) = \frac{15{,}200}{8 + 1892e^{-(0.532)(5)}} = \frac{15{,}200}{8 + 1892e^{-2.66}} \approx 108.3.$$

After 5 hr, there are approximately 108.3 million bacteria per mL. ∎

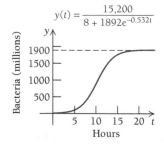

$$y(t) = \frac{15{,}200}{8 + 1892e^{-0.532t}}$$

---

[10]D. Charalampopoulos, S. S. Pandiella, and C. Webb, "Growth studies of potentially probiotic lactic acid bacteria in cereal-based substrates," *Journal of Applied Microbiology,* Vol. 92, pp. 851–859 (2002).

Logistic Growth Model with a Threshold

The logistic model is not the only differential equation that models inhibited growth. Another model of inhibited growth incorporates a *threshold level T* as follows:

$$y' = ky\left(1 - \frac{y}{L}\right)\left(\frac{y}{T} - 1\right).$$

Notice the extra factor of $(y/T - 1)$ on the right-hand side. We will assume that k is positive and that $0 < T < L$.

In the exercises, we will show that $y = 0$ and $y = L$ are asymptotically stable equilibrium values, while $y = T$ is an unstable equilibrium value. Solutions of the logistic growth model with a threshold are sketched in the margin.

If the initial population is greater than T, then the limiting size of the population is L, and the long-term survival of the population is assured. On the other hand, if the initial population is less than the threshold T, then the population experiences the so-called Allee effect and declines toward 0. Populations that experience the Allee effect eventually become extinct.

The Allee effect describes the fate of the passenger pigeon. Though billions of passenger pigeons were present in North America at the beginning of the 19th century, their population was greatly reduced by hunting. After the population dropped below a threshold level in the 1880s, the passenger pigeon was doomed to extinction. The last passenger pigeon died in 1914.

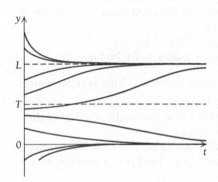

Exercise Set 8.3

For Exercises 1–10, (a) find the equilibrium value(s) of the differential equation, (b) assess the stability of each equilibrium value, (c) determine the point(s) of inflection, and (d) sketch sample solutions of the differential equation.

1. $y' = 2 - y$
2. $y' = 5y + 4$
3. $y' = y^2 - 5y + 4$
4. $y' = (1 - y)^2$
5. $y' = y^3 - 2y^2$
6. $y' = -y^3 + 4y^2 - 4y$
7. $y' = y^3 + 8y^2 + 15y$
8. $y' = 5 - e^y$
9. $y' = e^{2y} - e^y$
10. $y' = e^{2y} - 5e^y + 6$

APPLICATIONS

11. *Bacteria Growth.* During a fermentation experiment, the population $P(t)$ (in millions per mL) of bacteria *Lactobacillus plantarum* after t hours in a malt medium satisfied the differential equation

$$\frac{dP}{dt} = 0.4P\left(1 - \frac{P}{12,500}\right).$$

Initially, 9 million bacteria per mL were present.[11]

a) Determine the number of bacteria per mL after 7 hr.

b) Determine when 900 million bacteria per mL will be present.

12. *Bacteria Growth.* During a fermentation experiment, the population $P(t)$ (in millions per mL) of bacteria *Lactobacillus reuteri* after t hours in a barley medium satisfied the differential equation

$$\frac{dP}{dt} = 0.14P\left(1 - \frac{P}{20}\right).$$

Initially, 1.4 million bacteria per mL were present.[12]

[11]D. Charalampopoulos, S. S. Pandiella, and C. Webb, "Growth studies of potentially probiotic lactic acid bacteria in cereal-based substrates," *Journal of Applied Microbiology*, Vol. 92, pp. 851–859 (2002).

[12]D. Charalampopoulos, S. S. Pandiella, and C. Webb, "Growth studies of potentially probiotic lactic acid bacteria in cereal-based substrates," *Journal of Applied Microbiology*, Vol. 92, pp. 851–859 (2002).

a) Determine the number of bacteria per mL after 8 hr.

b) Determine when 8 million bacteria per mL will be present.

13. *Theta-Logistic Growth Model.* A generalization of the logistic growth model is the theta-logistic growth model[13]

$$y' = ky\left(1 - \left[\frac{y}{L}\right]^{\theta}\right),$$

where $k > 0$, $\theta > 0$, and $L > 0$.

a) Verify that this differential equation reduces to the logistic growth model if $\theta = 1$.

b) Determine the per capita growth rate.

c) Find the nonnegative equilibrium values, and assess the stability of each.

d) Sketch sample solutions of the differential equation.

14. *Gompertz Equation.* Another model of inhibited growth is the Gompertz equation[14]

$$y' = ky \ln (L/y),$$

where $k > 0$ and y is assumed to be positive.

a) Determine the per capita growth rate.

b) Find the equilibrium values, and assess the stability of each.

c) Sketch sample solutions of the differential equation.

15. *Disease.* Let $y(t)$ be the number of people in a confined population who have contracted a contagious but nonfatal disease. Then y may be modeled by the differential equation

$$y' = ky(P - y),$$

where P is the population.

a) Find the equilibrium values, and assess the stability of each.

b) Sketch sample solutions of the differential equation.

c) Suppose only one or two members of a large confined population initially have the disease. What eventually happens to the population?

Harvesting. Some populations, like fisheries, are harvested as they grow. The rate of harvesting may be assumed to be proportional to the population: The larger the population, the greater the harvest. The rate of growth may thus be modeled by the differential equation

$$y' = ky\left(1 - \frac{y}{L}\right) - sy,$$

obtained by subtracting sy from the logistic growth model. This differential equation is called the Schaefer model.

16. *Equilibrium Values.*

a) Show that $y = 0$ is an equilibrium value.

b) Find the other equilibrium value $y = c$ in terms of k, L, and s.

c) Show that $c > 0$ if $s < k$.

tw d) Explain the inequality $s < k$ biologically.

17. *Maximum Sustainable Harvest.* This exercise continues Exercise 16.

a) Assuming that $s < k$, assess the stability of the equilibrium values.

b) At the equilibrium value $y = c$, the harvesting rate H is equal to sc. Find H as a function of s, k, and L.

c) Determine the value of s (in terms of k and L) that maximizes H.

18. *Harvesting and the Allee Effect.* A certain harvested population satisfies the differential equation

$$y' = 2y\left(1 - \frac{y}{20}\right)\left(\frac{y}{5} - 1\right) - \frac{7y}{25}.$$

tw a) Explain how the numbers 2, 20, 5, and 7/25 affect the size of the population.

b) Find the equilibrium values of the differential equation, and assess the stability of each.

tw c) Is the asymptotically stable equilibrium value less than or greater than 20? Explain this result biologically.

tw d) Is the unstable equilibrium value less than or greater than 5? Explain this result biologically.

SYNTHESIS

19. Let $y(t)$ be the solution of $y' = y^2 - 6y$ with initial value $y(0) = 5$. Find $\lim_{t \to \infty} y(t)$.

20. Let $y(t)$ be the solution of $y' = y^2 + y - 20$ with initial value $y(0) = -2$. Find $\lim_{t \to \infty} y(t)$.

---

[13]O. H. Diserud and S. Engen, "A General and Dynamic Species Abundance Model, Embracing the Lognormal and the Gamma Models," *American Naturalist,* Vol. 155, pp. 497–511 (2000).

[14]O. H. Diserud and S. Engen, "A General and Dynamic Species Abundance Model, Embracing the Lognormal and the Gamma Models," *American Naturalist,* Vol. 155, pp. 497–511 (2000).

Semistable Equilibrium Values. **In Exercises 21–22,** let $y = c$ be an equilibrium value of the differential equation $y' = f(y)$.

21. Suppose that $f(c) = 0$, $f'(c) = 0$, and $f''(c) > 0$.

tw a) Show that $y = c$ is semistable.

tw b) Show that solutions slightly above $y = c$ diverge from $y = c$, while solutions slightly below $y = c$ converge to $y = c$.

22. Suppose that $f(c) = 0$, $f'(c) = 0$, and $f''(c) < 0$.

tw a) Show that $y = c$ is semistable.

tw b) Show that solutions slightly above $y = c$ converge to $y = c$, while solutions slightly below $y = c$ diverge from $y = c$.

Logistic Growth Model with a Threshold. **Consider the** differential equation

$$y' = ky\left(1 - \frac{y}{L}\right)\left(\frac{y}{T} - 1\right),$$

where $0 < T < L$ and $K > 0$.

23. Verify that $y = 0$ and $y = L$ are asymptotically stable equilibrium values.

24. Verify that $y = T$ is an unstable equilibrium value.

25. Find the points of inflection.

Harvesting. **A population is harvested at a constant** rate, regardless of the size of the population. The population y satisfies the differential equation

$$y' = ky\left(1 - \frac{y}{L}\right) - s.$$

26. Assume that $s < kL/4$.

a) Determine the two equilibrium values in terms of k, L, and s.

b) Show that the smaller of the two equilibrium values is unstable.

c) Show that the greater of the two equilibrium values is asymptotically stable.

tw d) What happens to the population if the initial population is less than both equilibrium values?

27. Assume that $s > kL/4$.

tw a) Show that there are no equilibrium values.

tw b) Show that the right-hand side is always less than or equal to $-(s - kL/4)$.

tw c) Explain why the population must decline to 0.

28. Assume that $s = kL/4$. Show that $y = L/2$ is a semistable equilibrium value.

Fat Crystallization. **Let** $y(t)$ be the proportion of crystallizable fat in a sample after t hours. Then y satisfies the differential equation

$$y' = k(y^n - y),$$

where k is a constant and n is the Avrami exponent for decrystallization reactions. In practice, n is an integer greater than 1 that is computed from the time dependence of nucleation and the number of dimensions in which crystal growth occurs.[15] Use this differential equation to solve Exercises 29–32.

29. Show that $y = 0$ is an asymptotically stable equilibrium point.

30. Show that $y = 1$ is an unstable equilibrium point.

31. Determine the point of inflection between $y = 0$ and $y = 1$.

32. Although this differential equation for y is nonlinear, we will solve it using Theorem 3 after an initial modification.

a) Divide both sides of the differential equation by y^n. Show that

$$y^{-n}\frac{dy}{dt} + ky^{1-n} = k.$$

b) Let $z(t) = [y(t)]^{1-n}$. Use the Chain Rule to find z' in terms of y and y'.

c) Use the result of part (b) to show that

$$\frac{dz}{dt} + k(1 - n)z = k(1 - n).$$

d) Find the general solution for $z(t)$.

e) Find the general solution for $y(t)$.

f) Suppose that the initial condition is $y(0) = p$, where p is the initial proportion of crystallizable fat. Show that

$$y(t) = [1 + (p^{1-n} - 1)e^{(n-1)kt}]^{1/(1-n)}.$$

 Technology Connection

Fat Crystallization. **The following exercises are a continuation of Exercise 32.**

33. For a sample of cocoa butter, $p = 0.999998$, $k = 5$, and $n = 4$.

a) Use a grapher to plot the graph of $y(t)$ for $0 \le t \le 2$.

[15]I. Foubert, P. A. Vanrolleghem, B. Vanhoutte, and K. Dewettinck, "Dynamic mathematical model of the crystallization kinetics of fats," *Food Research International*, Vol. 35, pp. 945–956 (2002).

b) Calculate $y(1)$.

c) Determine when half of the sample will be crystallized.

34. For a sample of milk fat, $p = 0.999993$, $k = 8$, and $n = 5$.

a) Use a grapher to plot the graph of $y(t)$ for $0 \leq t \leq 2$. Does the rate of crystallization appear to be faster or slower than in the previous exercise?

b) Calculate $y(1)$.

c) Determine when half of the sample will be crystallized.

8.4 Separable Differential Equations

OBJECTIVE

■ Solve certain differential equations by separating the variables.

Consider the first-order initial-value problem

$$3y^2 \frac{dy}{dx} + x = 0, \quad y(4) = 2.$$

This differential equation is neither linear nor autonomous and hence cannot be solved using previous techniques. To solve this differential equation, we subtract x from both sides of the equation and integrate to obtain

$$3y^2 \frac{dy}{dx} = -x \tag{1}$$

$$\int 3y^2 \frac{dy}{dx}\, dx = \int -x\, dx.$$

Recall from Section 5.5, where we discussed integration by substitution, that we may write

$$\int 3y^2 \frac{dy}{dx}\, dx \quad \text{as} \quad \int 3y^2\, dy.$$

Therefore,

$$\int 3y^2\, dy = \int -x\, dx$$

$$y^3 = -\frac{x^2}{2} + C$$

$$y = \sqrt[3]{C - \frac{x^2}{2}}.$$

We then use the initial condition $y(4) = 2$ to solve for C:

$$2 = \sqrt[3]{C - \frac{(4)^2}{2}}$$

$$8 = C - 8$$

$$16 = C.$$

The solution of the initial-value problem is therefore

$$y = \sqrt[3]{16 - \frac{x^2}{2}}.$$

The above solution may be shortened somewhat by formally treating dy/dx as a quotient, as we did in Chapter 5 when we discussed integration by substitution. If we "multiply" both sides of (1) by dx, we obtain

$$3y^2 \, dy = -x \, dx. \tag{2}$$

We say that we have **separated the variables,** meaning that all the expressions involving y are on one side and all those involving x are on the other. We say that the differential equation is *separable*. The solution is then found by integrating both sides of (2), as shown above.

EXAMPLE 1   Solve the differential equation $\dfrac{dy}{dx} = 3x^2 e^{2y}$.

Solution   We first separate the variables:

$$\frac{dy}{e^{2y}} = 3x^2 \, dx$$

$$e^{-2y} \, dy = 3x^2 \, dx.$$

We then integrate both sides:

$$\int e^{-2y} \, dy = \int 3x^2 \, dx$$

$$-\frac{1}{2} e^{-2y} = x^3 + C$$

$$e^{-2y} = -2x^3 - 2C$$

$$-2y = \ln(-2x^3 - 2C)$$

$$y = -\frac{1}{2} \ln(-2x^3 - 2C)$$

$$y = -\frac{1}{2} \ln(C_1 - 2x^3). \qquad \text{Let } C_1 = -2C$$

The solution is $y = -\dfrac{1}{2} \ln(C_1 - 2x^3)$. ■

EXAMPLE 2   Solve the initial-value problem $\dfrac{dy}{dx} = \dfrac{x}{y}$ with initial condition $y(5) = 3$.

Solution   We first separate the variables:

$$y \, dy = x \, dx.$$

We then integrate both sides:

$$\int y \, dy = \int x \, dx$$

$$\frac{y^2}{2} = \frac{x^2}{2} + C$$

$$y^2 = x^2 + 2C$$

$$y^2 = x^2 + C_1. \qquad \text{Let } C_1 = 2C$$

We thus obtain the solutions

$$y = \sqrt{x^2 + C_1} \quad \text{and} \quad y = -\sqrt{x^2 + C_1}.$$

We now use the initial condition to specify the solution. Since $y(5) = 3$ is positive, we use the positive solution $y = \sqrt{x^2 + C_1}$. Substituting, we find

$$3 = \sqrt{(5)^2 + C_1}$$

$$C_1 = -16.$$

The solution of the initial-value problem is $y = \sqrt{x^2 - 16}$.

Domain of Solutions

Solutions of linear and nonlinear differential equations may have significantly different properties. First, the method of determining the domain of the solution is different. For the linear initial-value problem

$$y' + p(x)y = q(x), \qquad y(x_0) = y_0,$$

we recall that Theorem 2 stated that the solution was defined on any interval including x_0, for which both p and q were continuous. This property held for any value of the initial condition y_0.

In the nonlinear case, however, it is often impossible to predetermine the domain of the solution of an initial-value problem without explicitly finding the solution. In Example 2, the solution of the initial-value problem $y' = x/y$, $y(5) = 3$ was found to be $y = \sqrt{x^2 - 16}$. This solution is shown below in red on the direction field for $y' = x/y$. Notice that the arrows do not permit the red curve to be extended to the left of $x = 4$. Algebraically, the function $y = \sqrt{x^2 - 16}$ is undefined if $-4 < x < 4$. Since the initial condition was given at $x = 5$, we see that the domain of the solution is the interval $[4, \infty)$.

Furthermore, the domain of the solution may change with different initial conditions. The solution in green was found using the initial condition $y(5) = \sqrt{21}$. We see that this solution has domain $[2, \infty)$. The solution in black was found using the initial condition $y(5) = \sqrt{29}$. The domain of this solution is the entire real line. The computations of these domains are discussed in the exercises.

Summarizing, Theorem 1 guarantees that a first-order initial-value problem has a unique solution in some interval. Precisely determining that interval is generally harder for nonlinear differential equations than for linear differential equations.

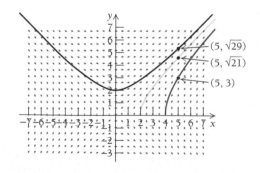

Implicitly Defined Solutions

Another difference between linear and nonlinear differential equations is the form of the solutions. We recall that solutions of a linear differential equation may always be written in the form $y = Cf(x) + g(x)$ and thus can always be explicitly determined as a function of x.

By contrast, solutions of first-order nonlinear differential equations may exhibit a variety of forms. In fact, it may be impossible to explicitly solve for y as a function of x.

EXAMPLE 3   Find the solution of the initial-value problem

$$y' = \frac{2x}{y + e^{5y}}, \qquad y(2) = 0.$$

Solution   We begin by separating the variables, replacing y' by dy/dx:

$$\frac{dy}{dx} = \frac{2x}{y + e^{5y}}$$

$$(y + e^{5y})\, dy = 2x\, dx.$$

We then integrate both sides:

$$\int (y + e^{5y})\, dy = \int 2x\, dx$$

$$\frac{y^2}{2} + \frac{e^{5y}}{5} = x^2 + C.$$

This equation defines y *implicitly* as a function of x. However, it is impossible to isolate y on one side to find the solution as a function of x.

Nevertheless, we may use the initial condition $y(2) = 0$ to find the constant C:

$$\frac{(0)^2}{2} + \frac{e^{5(0)}}{5} = (2)^2 + C$$

$$\frac{1}{5} = 4 + C$$

$$-\frac{19}{5} = C.$$

The solution of the initial-value problem is implicitly given by

$$\frac{y^2}{2} + \frac{e^{5y}}{5} = x^2 - \frac{19}{5}.$$

Even though we cannot find y as a function of x explicitly, we can graph the solution using a grapher. We can also use other analytical methods to interpret the solution.   ■

EXAMPLE 4   Solve $\dfrac{dy}{dt} = ty(y + 2)^2$.

Solution Separating variables, we find

$$\frac{dy}{y(y+2)^2} = t\,dt$$

$$\int \frac{dy}{y(y+2)^2} = \int t\,dt$$

The integral on the left-hand side may be found by using *Formula 33* from the table of integrals:

$$\int \frac{dx}{x(ax+b)^2} = \frac{1}{b(ax+b)} + \frac{1}{b^2}\ln\left|\frac{x}{ax+b}\right| + C.$$

Replacing *x* by *y*, *a* by 1, and *b* by 2, we find

$$\int \frac{dy}{y(y+2)^2} = \frac{1}{2(y+2)} + \frac{1}{4}\ln\left|\frac{y}{y+2}\right| + C.$$

Continuing the solution of the differential equation, we find

$$\int \frac{dy}{y(y+2)^2} = \int t\,dt$$

$$\frac{1}{2(y+2)} + \frac{1}{4}\ln\left|\frac{y}{y+2}\right| = \frac{t^2}{2} + C.$$

This equation defines *y* implicitly as a function of *t*. In this example, we cannot solve for *y* explicity. ∎

Since nonlinear differential equations are generally difficult to solve, alternative methods of analyzing the solutions of nonlinear differential equations have been developed. We have already seen in Section 8.3 how equilibrium values can be used to sketch the solutions of autonomous differential equations. In Section 8.5, we will develop numerical methods of solving initial-value problems.

Despite these differences between linear and nonlinear differential equations, one important similarity remains. Under the assumptions of Theorem 1, initial-value problems have unique solutions, so different solutions never intersect. As shown below, this important property will be useful in identifying all solutions of separable differential equations.

Identifying Constant Solutions

Division is sometimes required when separating the variables. When this happens, some care is needed to find all solutions.

EXAMPLE 5 Solve $y' - y^2 t\sin(t^2) = 0$.

Solution We rewrite the differential equation as

$$\frac{dy}{dt} = y^2 t\sin(t^2).$$

If we assume that y is never equal to 0, we may separate the variables and integrate.

$$\frac{dy}{y^2} = t \sin(t^2) \, dt$$

$$\int \frac{dy}{y^2} = \int t \sin(t^2) \, dt$$

$$\int \frac{dy}{y^2} = \frac{1}{2} \int \sin u \, du \qquad \underline{\text{Substitution}} \quad \begin{aligned} u &= t^2 \\ du &= 2t \, dt \end{aligned}$$

$$-\frac{1}{y} = -\frac{1}{2} \cos u + C$$

$$-\frac{1}{y} = -\frac{1}{2} \cos(t^2) + C$$

$$\frac{1}{y} = \frac{1}{2} \cos(t^2) - C$$

$$y = \frac{1}{\frac{1}{2} \cos(t^2) - C}$$

$$y = \frac{2}{\cos(t^2) - 2C}$$

$$y = \frac{2}{\cos(t^2) + C_1} \qquad \text{Let } C_1 = -2C$$

In the above solution, we assumed that y was never equal to 0 so that we could divide both sides of the differential equation by y^2. Indeed, the solution found will be nonzero for any value of t in the domain.

Nevertheless, we may directly verify that the constant function $y = 0$ is a solution of the differential equation:

$$y' - y^2 t \sin(t^2) = (0)' - (0)^2 t \sin(t^2) = 0.$$

Therefore, $y = 0$ is also a solution of the differential equation.

Finally, no function that equals 0 for some but not all values of t can be a solution of the differential equation. Such a function would intersect the solution $y = 0$. However, since the differential equation satisfies the conditions of Theorem 1, different solutions cannot intersect.

We conclude that the solutions of the differential equation are

$$y = \frac{2}{\cos(t^2) + C_1} \quad \text{and} \quad y = 0. \qquad \blacksquare$$

In this example, there is no constant C_1 that would reduce the form of the nonconstant solutions to the constant solution. Nevertheless, some separable differential equations have constant solutions that may be thought of as special cases of the nonconstant solutions.

EXAMPLE 6   Solve $y' = x + xy$.

Solution   We separate variables, replacing y' by dy/dx:

$$\frac{dy}{dx} = x(1 + y).$$

If y never equals -1, then $1 + y$ is never equal to 0. We may then separate the variables:

$$\frac{dy}{1 + y} = x \, dx.$$

Then we integrate both sides.

$$\int \frac{dy}{1 + y} = \int x \, dx$$

$$\ln|1 + y| = \frac{x^2}{2} + C$$

$$|1 + y| = e^{x^2/2 + C}$$

$$|1 + y| = e^C e^{x^2/2}$$

$$1 + y = \pm e^C e^{x^2/2} \qquad \text{Remove the absolute value signs}$$

$$1 + y = C_1 e^{x^2/2} \qquad \qquad \text{Let } C_1 = \pm e^C$$

$$y = C_1 e^{x^2/2} - 1.$$

In the above calculation, we assumed that y was never equal to -1 so that we could divide by $1 + y$. This suggests that the constant function $y = -1$ may also be a solution of the differential equation. To check, we substitute into the two sides of the differential equation:

$$y' = (-1)' = 0 \qquad \text{and} \qquad x(1 + y) = x(1 + [-1]) = 0.$$

Since the two sides of the differential equation agree, we see that $y = -1$ is also a solution of the differential equation.

Summarizing, if y is a solution of the differential equation, then

$$y = C_1 e^{x^2/2} - 1 \qquad \text{or} \qquad y = -1.$$

However, in this case, the solutions of the differential equation may be written more succinctly. The constant solution $y = -1$ may be obtained by setting $C_1 = 0$. Therefore, listing $y = -1$ as a separate solution is redundant. We may write the solutions of the differential equation as

$$y = C_1 e^{x^2/2} - 1,$$

where C_1 may be any constant, including 0.   ■

In Example 6, the constant solution could be thought of as a special case of the other solutions. By contrast, in Example 5, the constant solution was *not* a special case of the other solutions, and so we separately listed the constant solution in addition to the other solutions.

Exercise Set 8.4

Solve the differential equation. Be sure to check for possible constant solutions. If necessary, write your answer implicitly.

1. $\dfrac{dy}{dx} = 4x^3 y$

2. $3y^2 \dfrac{dy}{dx} = 5x$

3. $y' = \dfrac{x}{2y}$

4. $y' = x^2 y^3$

5. $y' = \dfrac{x\sqrt{y^2 + 1}}{y}$

6. $y' = \dfrac{e^y + 1}{e^y}$

7. $\dfrac{dy}{dx} = \dfrac{x^2 \sec y}{(x^3 + 1)^3}$

8. $\dfrac{dy}{dx} = e^x y \sqrt{e^x + 4}$

9. $\dfrac{dy}{dt} = y + \dfrac{1}{y^2}$

10. $\dfrac{dy}{dt} = t^2 e^{t^3} y^3$

11. $y' = x \cos^2 y$

12. $y' = \dfrac{x}{y + y^5}$

13. $y' = \dfrac{\sqrt{x}}{\sin y + \cos y}$

14. $y' = \dfrac{e^{x+2y}}{e^y + 1}$

15. $\dfrac{dy}{dt} = \dfrac{t}{(t^2 + 1)(y^4 + 1)}$

16. $\dfrac{dy}{dt} = \dfrac{\sin^2 y}{t}$

17. $\dfrac{dy}{dx} = 3x^2(y - 2)^2$

18. $\dfrac{dy}{dx} = y^2 - 1$

Solve the initial-value problem. If necessary, write your answer implicitly.

19. $3y^2 \dfrac{dy}{dx} - 2x = 0;\ y(2) = 5$

20. $\dfrac{dy}{dx} = 2x + xy;\ y(0) = 1$

21. $\dfrac{dy}{dt} = e^{2t} \sin^2 y;\ y(0) = \pi/4$

22. $\dfrac{dy}{dt} = y + \dfrac{4}{y};\ y(0) = 3$

23. $y' = \dfrac{y^2 \ln t}{t};\ y(1) = -4$

24. $y' = \dfrac{y \cos x}{1 + y^2};\ y(\pi/2) = 1$

25. $\dfrac{dy}{dx} = \dfrac{x(y^3 + 2)^2}{3y^2};\ y(1) = 2$

26. $\dfrac{dy}{dx} = xe^{2y};\ y(0) = 4$

27. $y' = \dfrac{t^2(2 + \sin y)^2}{\cos y};\ y(-1) = \pi/6$

28. $y' = e^{2t+y};\ y(1) = -2$

29. $y' = \dfrac{\sqrt{t}}{e^{4y} + e^{5y}};\ y(1) = 0$

30. $y' = \tan y;\ y(0) = \pi/6$

31. Solve the initial-value problem $y' = x/y$, $y(5) = \sqrt{21}$. Verify that the domain of the solution is $[2, \infty)$.

32. Solve the initial-value problem $y' = x/y$, $y(5) = \sqrt{29}$. Verify that the solution is defined for all real numbers.

APPLICATIONS

33. *Uninhibited Growth Model*

   a) Separate the variables to solve the initial-value problem $dy/dt = ky$ with initial condition $y(0) = y_0$.

tw **b)** Verify that your answer is identical to the uninhibited growth model of Section 4.3.

34. *Chemostat.* Let $P(t)$ be the population of organisms in a chemostat. While the organisms reproduce with relative growth rate k per hour, water is drained from the chemostat at a rate of r liters per hour, while fresh water is added at the same rate. The volume of the chemostat is V liters.

   a) Draw a one-compartment model for P.
   b) Construct a differential equation for P.
   c) Solve for k in terms of r and V if the population remains constant.

35. *Measuring a Decay Rate.* Iodine-125 is a radioactive element used to treat cancer. A sample of 5 g of iodine-125 has a mass of 4.404 g after 11 days. Find the decay rate.

36. *Newton's Law of Cooling.* An object has an initial temperature of T_0 and is placed into a surrounding medium with a lower temperature C. The temperature T of the cooling object drops at a rate that is proportional to the difference $T - C$. That is,

$$\frac{dT}{dt} = -k(T - C),$$

where k is a positive constant and t is time.

   a) Use the condition $T(0) = T_0$ to solve for $T(t)$.
   b) Verify that your answer is identical to the answer given in Section 4.4.

37. *Drug Release.* A person with a cold ingests a spherical tablet. Let $Q(t)$ be the total amount of the drug released up to time t. Then Q satisfies the differential equation

$$\frac{dQ}{dt} = 3kQ_0^{1/3}(Q_0 - Q)^{2/3},$$

where k is a constant and Q_0 is the total amount of drug contained in the tablet. Use separation of variables and the initial condition $Q(0) = 0$ to show that[16]

$$Q(t) = Q_0[1 - (1 - kt)^3].$$

38. *Population Growth.* According to one theory of population growth,[17] the world's human population $N(t)$, in millions, during the last millennium satisfied the differential equation

$$\frac{dN}{dt} = kN^2,$$

where k is a constant and t is the number of years since A.D. 1000.

a) Use separation of variables and the initial condition $N(0) = N_0$ to show that

$$N(t) = \frac{N_0}{1 - kN_0 t}.$$

b) Let $k = 5.59 \times 10^{-6}$ and $N_0 = 1/(1024k)$. Compute $\lim\limits_{t \to 1024^-} N(t)$.

tw c) Interpret the answer to part (b) in a sentence. Does this theory of population growth appear to be realistic?

39. *Population Growth.* Another theory of population growth says that, during the last millennium, the world's human population $N(t)$, in millions, satisfied the differential equation[18]

$$\frac{dN}{dt} = ke^{at} N,$$

[16]T. Koizumi, G. C. Ritthidej, and T. Phaechamud, "Mechanistic modeling of drug release from chitosan coated tablets," *Journal of Controlled Release*, Vol. 70, pp. 277–284 (2001).

[17]S. D. Varfolomeyev and K. G. Gurevich, "The hyperexponential growth of the human population on a macrohistorical scale," *Journal of Theoretical Biology*, Vol. 212, pp. 367–372 (2001).

[18]S. D. Varfolomeyev and K. G. Gurevich, "The hyperexponential growth of the human population on a macrohistorical scale," *Journal of Theoretical Biology*, Vol. 212, pp. 367–372 (2001).

where a and k are constants and t is the number of years since A.D. 1000.

a) Use separation of variables and the initial condition $N(0) = N_0$ to show that

$$N(t) = N_0 e^{k(e^{at} - 1)/a}.$$

b) Suppose $N_0 = 375.6$, $a = 0.00734$, and $k = 0.00185a$. Use this model to predict the world population in 2010 and in 2015.

c) Using the constants in part (b), predict when the world population will reach 10 billion.

SYNTHESIS

40. *Domain of Solutions.* The solution of a nonlinear initial-value problem may have a domain that depends on the initial condition.

a) Solve the initial-value problem $y' = y^2$, $y(0) = 1$. Find the domain of the solution.

b) Change the initial condition in part (a) to $y(0) = 2$. Find the domain of the solution.

c) Change the initial condition to $y(0) = k$, where k is not equal to 0. Find the domain of the solution.

d) Change the initial condition to $y(0) = 0$. Find the domain of the solution.

41. *Logistic Growth Model.* Consider the logistic growth model

$$y' = ky\left(1 - \frac{y}{L}\right)$$

with initial condition $y(0) = y_0$.

a) Using Formula 32 from the table of integrals, show that a nonconstant solution of the logistic growth model may be written as

$$y = \frac{C L e^{kt}}{L + C e^{kt}},$$

where C is a constant.

b) Use the initial condition to show that $C = (Ly_0)/(L - y_0)$.

c) Show that

$$y(t) = \frac{Ly_0}{y_0 + (L - y_0)e^{-kt}}.$$

d) Verify that the equilibrium values are special cases of this equation.

42. *Logistic Growth Model.* This is a continuation of Exercise 41.

a) Suppose that $y_0 \geq 0$. Show that the solution is defined for all real numbers t.

b) Suppose that $y_0 < 0$. Show that the domain of the solution is defined only if

$$t < \frac{1}{k} \ln \left(1 - \frac{K}{y_0} \right).$$

43. *Plant Growth.* Let $M(t)$ be the mass of a plant grown in a stand after t days, and let $R(t) = M'(t)$ denote its rate of growth. Then R may be modeled with the differential equation

$$\frac{dR}{dt} = kR \left(1 - \frac{R}{L} \right),$$

where k is a constant and L is the limiting growth rate.[19]

a) Solve for $R(t)$ given the initial condition $R(0) = R_0$.

---

[19]J. L. Monteith, "Fundamental equations for growth in uniform stands of vegetation," *Agriculture and Forest Meteorology*, Vol. 104, pp. 5–11 (2000).

b) Integrate your answer to show that

$$M(t) = M_0 + \frac{L}{k} \ln \left(\frac{L - R_0 + R_0 e^{kt}}{L} \right),$$

where M_0 is the initial mass of the plant. (*Hint:* To begin, multiply the numerator and denominator by e^{kt}. Then integrate using substitution.)

 Technology Connection

44. *Plant Growth.* This is a continuation of Exercise 43.

a) Let $L = 20$, $R_0 = 0.06$, $k = 0.01$, and $M_0 = 1$. Use a grapher to plot $M(t)$ for $0 \le t \le 8$.

tw b) Use the logistic model of inhibited growth to explain the slope of M for large t.

8.5 Numerical Solutions of Differential Equations

OBJECTIVES

■ Find an approximate solution to a first-order initial-value problem using Euler's method.

■ Find an approximate solution to a first-order initial-value problem using the Runge–Kutta method.

So far, we have considered various methods for finding exact solutions of first-order initial-value problems. Unfortunately, many applications give rise to differential equations that cannot be solved exactly. In these cases, it is necessary to use methods for generating approximate solutions for the given equation.

This situation is analogous to the evaluation of definite integrals. Recall that, while we have studied several techniques for computing definite integrals exactly, there are important definite integrals that cannot be exactly evaluated using any technique. Consequently, we developed numerical methods in Section 5.7 to approximate the values of such integrals. We also saw that some techniques are generally more computationally efficient than others.

In this section, we will develop methods of numerically solving the initial-value problem

$$y' = f(x, y), \qquad y(x_0) = y_0. \tag{1}$$

These techniques are called *Euler's method* and the *Runge–Kutta method*.

Euler's Method

The first approximation technique that we consider is called **Euler's method.** We first choose a distance Δx, called the *step size*. Then we define the sequence of points

$$x_1 = x_0 + \Delta x,$$
$$x_2 = x_1 + \Delta x = x_0 + 2\Delta x,$$
$$x_3 = x_2 + \Delta x = x_0 + 3\Delta x,$$

and so on. (Recall that a similar sequence of points was used in Section 5.7 for numerical integration.) Euler's method will approximate the exact solution only at these selected points.

Next, we approximate the exact solution by the tangent line to y at $x = x_0$. In other words, we use the linearization

$$y(x) \approx y(x_0) + y'(x_0)(x - x_0).$$

The right-hand side is the equation of the tangent line T_0, as shown in the figure.

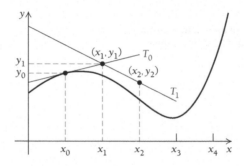

Using (1), we know that

$$y'(x_0) = f(x_0, y(x_0)) = f(x_0, y_0).$$

Substituting into the above approximation, we have

$$y(x) \approx y(x_0) + f(x_0, y_0)(x - x_0).$$

If we let $x = x_1$, then

$$y(x_1) \approx y(x_0) + f(x_0, y_0)(x_1 - x_0) = y(x_0) + f(x_0, y_0)\,\Delta x.$$

We denote this last expression by y_1:

$$y_1 = y(x_0) + f(x_0, y_0)\,\Delta x.$$

This is our approximation for the exact solution at x_1, so that $y(x_1) \approx y_1$.

Now we can repeat the whole process. We use the tangent line to the exact solution at $x = x_1$ to obtain

$$\begin{aligned}
y(x) &\approx y(x_1) + y'(x_1)(x - x_1) \\
&= y(x_1) + f(x_1, y(x_1))(x - x_1) &&\text{Using (1)} \\
&\approx y_1 + f(x_1, y_1)(x - x_1). &&\text{Using } y_1 \approx y(x_1)
\end{aligned}$$

This line is depicted as T_1 in the figure. If we let $x = x_2$, then

$$y(x_2) \approx y_1 + f(x_1, y_1)(x_2 - x_1) = y_1 + f(x_1, y_1)\,\Delta x.$$

We define y_2 to be the last expression; this approximates the value $y(x_2)$.

Euler's method is derived by continuing in this way.

> **DEFINITION**
>
> To approximate the solution of (1) using Euler's method, we use the formula
>
> $$y_{n+1} = y_n + f(x_n, y_n)\,\Delta x,$$
>
> where $y_n \approx y(x_n)$.

We now illustrate Euler's method with an example.

EXAMPLE 1 Use Euler's method with $\Delta x = 0.2$ to approximate the solution of $y' = (4x - 1)y, y(0) = 2$ on the interval $[0, 1]$.

Solution Before applying Euler's method, we note that the exact solution is $y = 2e^{2x^2 - x}$; this may be found by using either separation of variables or Theorem 3. We have deliberately chosen an initial-value problem that can be solved exactly so that we can compare the approximations from Euler's method with this exact solution.

Since the initial condition is $y(0) = 2$, $x_0 = 0$ and $y_0 = 2$. Successively adding $\Delta x = 0.2$ to $x_0 = 0$, we have $x_1 = 0.2, x_2 = 0.4, x_3 = 0.6, x_4 = 0.8$, and $x_5 = 1$.

We first use Euler's method to find y_1:

$$\begin{aligned}
y_1 &= y_0 + f(x_0, y_0)\,\Delta x \\
&= y_0 + [4x_0 - 1]y_0\,\Delta x \\
&= 2 + [4(0) - 1](2)(0.2) \\
&= 1.6.
\end{aligned}$$

Next,

$$\begin{aligned}
y_2 &= y_1 + f(x_1, y_1)\,\Delta x \\
&= y_1 + [4x_1 - 1]y_1\,\Delta x \\
&= 1.6 + [4(0.2) - 1](1.6)(0.2) \\
&= 1.536.
\end{aligned}$$

Continuing,

$$\begin{aligned}
y_3 &= 1.536 + [4(0.4) - 1](1.536)(0.2) = 1.72032 \\
y_4 &= 1.72032 + [4(0.6) - 1](1.72032)(0.2) \approx 2.20201 \\
y_5 &= 2.20201 + [4(0.8) - 1](2.20201)(0.2) \approx 3.17089.
\end{aligned}$$

In the figure on the next page, these points are shown as red points; the point $(x_0, y_0) = (0, 2)$ is also shown for completeness. (The points are connected by red lines to guide the eye.) Also, the exact solution is shown in black. Notice that the first couple of points are somewhat accurate, but when we reach $x_5 = 1$, the approximation $y_5 = 3.17089$ bears little resemblance to the actual value,

$$y(1) = 2e^{2(1)^2 - 1} \approx 5.43656.$$

∎

x	Δx 0.1	0.01	0.001	Exact
0.1	1.800	1.842	1.846	1.846
0.2	1.692	1.766	1.773	1.774
0.3	1.658	1.762	1.773	1.774
0.4	1.691	1.831	1.845	1.846
0.5	1.793	1.978	1.998	2.000
0.6	1.972	2.225	2.252	2.255
0.7	2.248	2.602	2.642	2.646
0.8	2.653	3.166	3.225	3.232
0.9	3.236	4.005	4.098	4.109
1	4.078	5.269	5.419	5.437

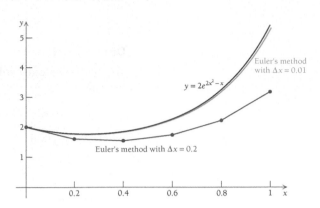

Technology Connection

Euler's Method

Euler's method is easily implemented using a spreadsheet such as Microsoft *Excel*. To repeat Example 1 with $\Delta x = 0.1$, first type the title **Xn** into cell A1 and the title **Yn** into cell B1. The initial conditions 0 and 2 are entered into cells A2 and B2, respectively. Euler's method is then implemented by typing **=A2+0.1** into cell A3 and **=B2+0.1*((4*A2-1)*B2)** into cell B3. Finally, fill down to repeat Euler's method. The first column of the table above was generated by this method.

To compute the exact solution at these points, first type the title **Exact** into cell C1. Then type **=2*EXP(2*A2^2-A2)** into cell C2 and fill down to cell C12.

	A	B	C
1	Xn	Yn	Exact
2	0	2	2
3	0.1	1.8	1.846233
4	0.2	1.692	1.773841
5	0.3	1.65816	1.773841
6	0.4	1.691323	1.846233
7	0.5	1.792803	2
8	0.6	1.972083	2.254994
9	0.7	2.248174	2.64626
10	0.8	2.652846	3.232149
11	0.9	3.236472	4.108866
12	1	4.077955	5.436564

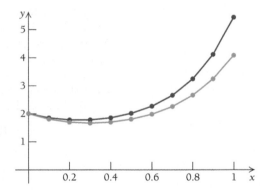

Spreadsheets are also able to graph this data. To do so, you should highlight the box from cell A2 to cell C12. Then select **Insert** and then **Chart** from the menu bar. From the options, choose **XY (Scatter)**, the type of plot you prefer (with or without lines and points), and then follow the instructions. Since the spreadsheet only connects the dots, the graph generated by column C is not quite the same as the graph of the exact solution above.

EXERCISES

1. Use a spreadsheet with $\Delta x = 0.01$ to replicate the second column of the table above.

2. Graph your solution and the exact solution. Your graph should match the black and blue lines in the figure above.

3. Find the error in the estimates of $y(1)$ for these three step sizes. As the step size is cut by a factor of ten, the error decreases by approximately what factor?

More accurate approximations may be obtained by using a smaller Δx in Euler's method. This is shown in the table as Δx decreases to 0.1, 0.01, and 0.001. We see that with $\Delta x = 0.001$, we obtain the approximation $y(1) \approx 5.419$, which is significantly closer to the correct answer. The approximation using $\Delta x = 0.01$ is shown in blue in the figure; with such a small step size, these points look like a solid curve. (The approximation using $\Delta x = 0.001$ is indistinguishable from the exact solution on the scale of this figure.)

While taking a smaller step size yields more accurate results, the disadvantage is that the computations of Euler's method must be repeated more frequently. In the above table, choosing $\Delta x = 0.001$ meant that Euler's method had to be repeated 1000 times to approximate $y(1)$. Fortunately, Euler's method can be easily implemented on any standard spreadsheet (like Microsoft *Excel*), as described in the Technology Connection. Some advanced graphers also have built-in spreadsheets. The method can also be easily programmed on a programmable calculator or a computer.

There is a second difficulty with choosing a small step size Δx besides the computational burden of repeating Euler's method many times. If Δx is very small (on the order of 10^{-9}), then the accuracy of Euler's method is compromised by the errors inherent in the floating-point arithmetic used by computers. In other words, we simply cannot reduce the step size indefinitely in Euler's method to obtain more accurate results.

Because of these difficulties, Euler's method is rarely implemented in practice. Instead, more accurate methods are used, as we now discuss.

Runge–Kutta Method

Euler's method may be modified to obtain more accurate algorithms for approximating the solutions of initial-value problems. Because of their accuracy, much larger step sizes can be confidently used. One of the best algorithms is called the **Runge–Kutta method,** which we now describe. It should be thought of as an enhanced version of Euler's method.

DEFINITION

To approximate the solution of the initial-value problem

$$y' = f(x, y), \qquad y(x_0) = y_0$$

using the Runge–Kutta method, we use the following five formulas in order:

$$a_n = f(x_n, y_n)$$

$$b_n = f\left(x_n + \frac{\Delta x}{2}, y_n + \frac{a_n \Delta x}{2}\right)$$

$$c_n = f\left(x_n + \frac{\Delta x}{2}, y_n + \frac{b_n \Delta x}{2}\right)$$

$$d_n = f(x_n + \Delta x, y_n + c_n \Delta x)$$

$$y_{n+1} = y_n + \frac{\Delta x}{6}(a_n + 2b_n + 2c_n + d_n).$$

Once again, Δx is the step size and y_n approximates $y(x_n)$.

The Runge–Kutta method replaces the term $f(x_n, y_n)$ in Euler's method with the term $(a_n + 2b_n + 2c_n + d_n)/6$; this term is a weighted average over four values of $f(x, y)$. The motivation for this method is beyond the scope of this text.

EXAMPLE 2   Use the Runge–Kutta method with $\Delta x = 0.2$ to approximate the solution of $y' = (4x - 1)y$, $y(0) = 2$ on the interval $[0, 1]$.

Solution   Successively adding $\Delta x = 0.2$ to $x_0 = 0$, we have $x_1 = 0.2$, $x_2 = 0.4$, $x_3 = 0.6$, $x_4 = 0.8$, and $x_5 = 1$. Also, from the initial condition, we know that $y_0 = 2$.
For the function $f(x, y) = (4x - 1)y$ and the step size $\Delta x = 0.2$, the Runge–Kutta method becomes

$$a_n = [4x_n - 1]y_n,$$
$$b_n = [4(x_n + 0.1) - 1][y_n + 0.1a_n],$$
$$c_n = [4(x_n + 0.1) - 1][y_n + 0.1b_n],$$
$$d_n = [4(x_n + 0.2) - 1][y_n + 0.2c_n],$$
$$y_{n+1} = y_n + \frac{0.2}{6}[a_n + 2b_n + 2c_n + d_n].$$

For example, if $n = 0$, we have

$$a_0 = [4x_0 - 1]y_0 = [4(0) - 1][2] = -2$$

$$b_0 = [4(x_0 + 0.1) - 1][y_0 + 0.1a_0]$$
$$= [4(0 + 0.1) - 1][2 + 0.1(-2)] = -1.08$$

$$c_0 = [4(x_0 + 0.1) - 1][y_0 + 0.1b_0]$$
$$= [4(0 + 0.1) - 1][2 + 0.1(-1.08)] = -1.1352$$

$$d_0 = [4(x_0 + 0.2) - 1][y_0 + 0.2c_0]$$
$$= [4(0 + 0.2) - 1][2 + 0.2(-1.1352)] = -0.354592$$

$$y_1 = y_0 + \frac{0.2}{6}[a_0 + 2b_0 + 2c_0 + d_0]$$

$$= 2 + \frac{0.2}{6}[(-2) + 2(-1.08) + 2(-1.1352) + (-0.354592)]$$

$$= 1.7738336.$$

We then compute $a_1, b_1, c_1,$ and d_1 in order to obtain y_2, and so on. The results, rounded to five decimal places, are shown in the table to the left, along with values of the exact solution $y = 2e^{2x^2 - x}$.   ■

x	Runge–Kutta with $\Delta x = 0.2$	Exact
0.2	1.77383	1.77384
0.4	1.84622	1.84623
0.6	2.25496	2.25499
0.8	3.23190	3.23215
1	5.43479	5.43656

This Runge–Kutta estimate of $y(1)$ is much more accurate than the Euler's method estimates seen earlier. Using Euler's method, a step size of approximately $\Delta x = 0.0001$ would be needed to estimate $y(1)$ as accurately as the Runge–Kutta method in this example.

The above Runge–Kutta estimate of $y(1)$ requires 25 separate calculations, which is five times as many steps as Euler's method with the same step size. However, to obtain the same level of accuracy, Euler's method would require a very small step size and hence 10,000 calculations. While each step of the Runge–Kutta method requires more work than Euler's method, the great advantage of the Runge–Kutta method is the reduced overall computation.

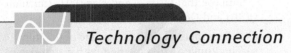

Technology Connection

Runge-Kutta Method

The Runge–Kutta method may also be implemented using a spreadsheet. To repeat Example 2, we type the titles **Xn**, **Yn**, **An**, **Bn**, **Cn**, and **Dn** into cells A1 through F1. The initial conditions 0 and 2 are entered into cells A2 and B2, respectively. We then enter formulas into cells C2 through F2, A3 and B3, as follows:

Cell	Formula
C2	=(4*A2-1)*B2
D2	=(4*(A2+0.1)-1)*(B2+0.1*C2)
E2	=(4*(A2+0.1)-1)*(B2+0.1*D2)
F2	=(4*(A2+0.2)-1)*(B2+0.2*E2)
A3	=A2+0.2
B3	=B2+0.2/6*(C2+2*D2+2*E2+F2)

We then fill down to obtain the approximations to y, which are listed in cells B2 through B7.

	A	B	C	D	E	F
1	Xn	Yn	An	Bn	Cn	Dn
2	0	2	-2	-1.08	-1.1352	-0.35459
3	0.2	1.773834	-0.35477	0.347671	0.36172	1.107707
4	0.4	1.846224	1.107735	1.956998	2.041924	3.156453
5	0.6	2.254959	3.156942	4.627175	4.891817	7.113309
6	0.8	3.2319	7.11018	10.25159	11.06835	16.33671
7	1	5.434792				
8						

EXERCISES

1. Find the Runge–Kutta approximation to $y(1)$ using a step size of $\Delta x = 0.1$.

2. In Example 2, the error on the estimate of $y(1)$ was $5.43656 - 5.43479 = 0.0177$. What is the error using a step size of $\Delta x = 0.1$?

3. By about what factor does the error decrease when the step size is cut in half?

Exercise Set 8.5

To facilitate calculation, use a spreadsheet for all problems in this section.

For the following exercises,

  a) Find the solution to the initial-value problem using Euler's method on the given interval with the indicated step size Δx.
  b) Repeat using the Runge–Kutta method.
  c) Find the exact solution.
  d) Compare the exact value at the interval's right endpoint with the approximations derived in parts (a) and (b).

1. $y' = 2x$, $y(0) = -2$, $\Delta x = 0.2$, on $[0, 1]$

2. $y' = -y$, $y(0) = 4$, $\Delta x = 0.2$, on $[0, 2]$

3. $y' = 2xy$, $y(1) = 2$, $\Delta x = 0.2$, on $[1, 2]$

4. $y' = x\sqrt{y}$, $y(1) = 1$, $\Delta x = 0.2$, on $[1, 3]$

5. $y' = 2x + y$, $y(-1) = 2$, $\Delta x = 0.1$, on $[-1, 2]$

6. $y' = \dfrac{x\sqrt{y^2 + 1}}{y}$, $y(0) = \sqrt{8}$, $\Delta x = 0.02$, on $[0, 1]$

7. $y' = x^2 y$, $y(0) = 1$, $\Delta x = 0.1$, on $[0, 2]$

8. $y' = \dfrac{2}{x} y = x$, $y(1) = 2$, $\Delta x = 0.1$, on $[1, 3]$

APPLICATIONS

9. *Bacteria Growth.* During a fermentation experiment, the population P (in millions per mL) of bacteria *Lactobacillus fermentum* after t hours in a wheat medium satisfied the differential equation

$$\frac{dP}{dt} = 0.532P\left(1 - \frac{P}{1900}\right).$$

Initially, 8 million bacteria per mL were present.[20] Calculate $P(5)$ and $P(8)$ using

  a) Euler's method with a step size of 0.1 hr.
  b) The Runge–Kutta method with a step size of 0.1 hr.
  c) The exact solution which is given in Example 5 of Section 8.3.

10. *Drug Release.* A person with a cold ingests a spherical tablet. Let $Q(t)$ be the total amount of the drug released up to t minutes. Then Q satisfies the differential equation

$$\frac{dQ}{dt} = 3kQ_0^{1/3}(Q_0 - Q)^{2/3},$$

where k is a constant and Q_0 is the total amount of drug contained in the tablet. Also, $Q(0) = 0$ since no medicine is initially released.[21]

Suppose that $Q_0 = 1$ and $k = 0.01$. Calculate $Q(30)$ and $Q(90)$ using

  a) Euler's method with a step size of 0.5 min.
  b) The Runge–Kutta method with a step size of 0.5 min.
  c) The exact solution, which is given in Exercise 37 of Section 8.4.

11. *Anesthesia.* The pressure $P(t)$ of anesthesia in one section of the body after t minutes may be modeled with the differential equation[22]

$$\frac{dP}{dt} + kP = kf(t),$$

where k is a constant and $f(t)$ is the arterial pressure. Assume that no anesthesia is initially present, so that $P(0) = 0$.

Suppose that $k = 1.2$ and $f(t) = 1 - e^{-kt}$. Calculate $P(2)$ and $P(4)$ using

  a) Euler's method with a step size of 0.2 min.
  b) The Runge–Kutta method with a step size of 0.2 min.
  c) The exact solution, which is given in Exercise 49 of Section 8.2.

[20]D. Charalampopoulos, S. S. Pandiella, and C. Webb, "Growth studies of potentially probiotic lactic acid bacteria in cereal-based substrates," *Journal of Applied Microbiology*, Vol. 92, pp. 851–859 (2002).

[21]T. Koizumi, G. C. Ritthidej, and T. Phaechamud, "Mechanistic modeling of drug release from chitosan coated tablets," *Journal of Controlled Release*, Vol. 70, pp. 277–284 (2001).

[22]J. P. Whiteley, D. J. Gavaghan, and C. E. W. Hahn, "Modelling inert gas exchange in tissue and mixed-venous blood return to the lungs," *Journal of Theoretical Biology*, Vol. 209, pp. 431–443 (2001).

12. *Fat Crystallization.* The fraction $h(t)$ of crystallizable fat in a sample after t hours satisfies the differential equation[23]

$$\frac{dh}{dt} = k(h^n - h).$$

For a sample of cocoa butter, $h(0) = 0.999998$, $k = 5$, and $n = 4$. Calculate $h(1)$ and $h(2)$ using

a) Euler's method with a step size of 0.05 min.

b) The Runge–Kutta method with a step size of 0.05 min.

c) The exact solution, which is given in Exercise 33 of Section 8.3.

13. *Biohazards.* A mussel is placed into polluted water containing a high concentration of polychlorinated biphenyls (PCBs). Let $Q(t)$ be the concentration of PCB in the mussel (in micrograms of PCB per gram of tissue) after t days. Then Q satisfies the differential equation[24]

$$\frac{dQ}{dt} = \frac{75Q}{Q + 200} - 0.18Q.$$

Assuming that $Q(0) = 1$, find $Q(6)$ using

a) Euler's method with a step size of 0.25 day.

b) The Runge–Kutta method with a step size of 0.25 day.

SYNTHESIS

14. *Biohazards.* This exercise continues Exercise 13.

a) Use the Runge–Kutta method with a step size of 0.25 day to estimate $Q(20)$, $Q(40)$, $Q(60)$, $Q(80)$, and $Q(100)$.

b) Use the techniques of Section 8.3 to determine the asymptotically stable equilibrium value of the differential equation $Q' = 75Q/(Q + 200) - 0.18Q$. Do the solutions of part (a) approach this value as t increases?

c) Approximately how many days will it take for $Q(t)$ to be within 10 of its asymptotically stable equilibrium?

tw 15. *Choosing a Step Size.* Suppose that the initial-value problem $y' = f(x, y)$, $y(0) = y_0$ is approximated numerically using either Euler's method or the Runge–Kutta method.

a) Is it possible to use either method to directly approximate $y(1)$ if $\Delta x = 0.07$?

b) Repeat part (a) if $\Delta x = 0.06$.

c) Repeat part (a) if $\Delta x = 0.05$.

d) What is the condition on Δx that allows a direct approximation of $y(1)$?

e) Suppose that we wish to approximate $y(a)$ instead of $y(1)$. What is the relationship between Δx and a that allows a direct approximation?

 Technology Connection

Accuracy and Step Size. For both Euler's method and the Runge–Kutta method, we expect to get more accurate solutions as the step size decreases. In Exercises 16–17, we will quantify how the accuracy increases.

16. *Euler's Method.* Consider the linear initial-value problem $y' = e^x - 2y$, $y(0) = 1$.

a) Use Euler's method with $\Delta x = 0.1$ to approximate $y(2)$. Find your answer to six decimal places.

b) Repeat part (a) with $\Delta x = 0.05$.

c) Use the exact solution of the initial-value problem to find $y(2)$.

d) Find the errors in the approximations of parts (a) and (b).

e) In Euler's method, by approximately what factor does the error decrease when the step size is cut in half?

17. *Runge–Kutta Method.* Consider the initial-value problem in Exercise 16.

a) Use the Runge–Kutta method with $\Delta x = 0.1$ to approximate $y(2)$. Find your answer to eight decimal places.

b) Repeat part (a) with $\Delta x = 0.05$.

c) Use the exact solution of the initial-value problem to find $y(2)$.

d) Find the errors in the approximations of parts (a) and (b).

e) In the Runge–Kutta method, by approximately what factor does the error decrease when the step size is cut in half?

[23]I. Foubert, P. A. Vanrolleghem, B. Vanhoutte, and K. Dewettinck, "Dynamic mathematical model of the crystallization kinetics of fats," *Food Research International,* Vol. 35, pp. 945–956 (2002).

[24]K. N. Yu, P. K. S. Lam, C. C. C. Cheung, and C. W. Y. Yip, "Mathematical modeling of PCB bioaccumulation in *Perna viridis,*" *Marine Pollution Bulletin,* Vol. 45, pp. 332–338 (2002).

Chapter 8   Summary and Review

Terms to Know

Differential equation, p. 546
Order, p. 546
General solution, p. 547
Particular solution, p. 547
Direction field, p. 547
Initial condition, p. 547
Initial-value problem, p. 547
Linear differential equation, p. 552

Integrating factor, p. 553
One-compartment model, p. 557
Relative rate, p. 557
Uptake rate, p. 557
Downtake rate, p. 559
Autonomous differential equation,
   p. 563
Equilibrium value, p. 563

Unstable, p. 564
Asymptotically stable, p. 564
Semistable, p. 564
Logistic growth model, p. 569
Separating variables, p. 575
Euler's method, p. 584
Runge–Kutta method, p. 587

Review Exercises

These review exercises are for test preparation. They can also be used as a lengthened practice test. Answers are at the back of the book. The answers also contain bracketed section references, which tell you where to restudy if your answer is incorrect.

Solve the differential equation.

1. $y' = 6x^2$

2. $y' = \dfrac{\ln x}{x}$

3. $y' = xe^x$

4. $y' = 5x^2 - 6x^2 y$

5. $xy' + 2y = x^3 + x$

6. $y' + x^2 y = x^2 e^{x^3}$

7. $y' = -x/y$

8. $y' = x \csc y$

9. $y' = \dfrac{x^3 - 3x + 1}{y + \sqrt{y}}$

Solve the initial-value problem.

10. $y' = 4x - 5;\ y(-1) = 3$

11. $y' = \cos x \sqrt{4 - 3 \sin x};\ y(\pi/2) = -1$

12. $y' = 5(x - y);\ y(0) = 2$

13. $x^2 y' + 4xy = 8x;\ y(1) = 4$

14. $y' = \dfrac{x^3}{1 + 4y^7};\ y(2) = -1$

15. $y' = \dfrac{3x^2}{y};\ y(1) = -2$

For Exercises 16–17, (a) find the equilibrium value(s) of the differential equation, (b) assess the stability of each equilibrium value, (c) determine the point(s) of inflection, and (d) sketch sample solutions of the differential equation.

16. $y' = y^2 - 2y - 24$

17. $y' = y^3 - 6y^2 + 9y$

APPLICATIONS

18. *Bacteria Growth.* During a fermentation experiment, the population $P(t)$ (in millions per mL) of bacteria *Lactobacillus fermentum* after t hours in a malt medium satisfied the differential equation

$$\frac{dP}{dt} = 0.62P\left(1 - \frac{P}{4800}\right).$$

Initially, 7 million bacteria per mL were present.[25]

a) Determine the number of bacteria per mL after 4 hr.

b) Determine when 500 million bacteria per mL will be present.

19. *Acceptance of a New Medicine.* The proportion P of doctors who are aware of a new medicine after t months may be modeled by the differential equation

$$\frac{dP}{dt} = k(1 - P),$$

where k is a constant.

a) Solve for $P(t)$ given that $P(0) = 0$.

---

[25]D. Charalampopoulos, S. S. Pandiella, and C. Webb, "Growth studies of potentially probiotic lactic acid bacteria in cereal-based substrates," *Journal of Applied Microbiology*, Vol. 92, pp. 851–859 (2002).

b) Suppose that $P(4) = 0.5$. Use this to determine k.

c) How many months will it take for 90% of doctors to become aware of the new medicine?

20. *Carbon Decomposition.* Let C denote the quantity of organic carbon in a sample of soil. Let a denote the decay rate of organic carbon per year, and let b denote the uptake rate of organic carbon into the sample per year.[26]

a) Draw a one-compartment model for C.

b) Show that $C' = -aC + b$.

c) Solve for $C(t)$ given the initial condition $C(0) = C_0$.

d) Find the amount of organic carbon present at equilibrium.

SYNTHESIS

21. *Creaky Voicing.* Let $x(t)$ denote the lateral displacement of the lower vocal folds at time t when someone with a creaky voice (or pressed voice) speaks. Then x satisfies the differential equation

$$r\frac{dx}{dt} + \frac{1}{c}(x - x_0) = pd$$

as speech begins. In this equation, r is the resistance of the lower vocal folds per unit length, c is the mechanical compliance per unit length, x_0 is the resting position, p is the subglottal pressure, and d is the length of the lower vocal folds.[27]

a) Find the general solution of this differential equation.

b) Assume that the vocal folds are at rest before the outward displacement begins, so that $x(0) = 0$. Use this initial condition to solve for $x(t)$.

22. *Creaky Voicing.* After some amount of time T has passed, the lower vocal folds of a creaky-voiced speaker displaces by a certain critical distance, which we denote by a. At this time, the subglottal pressure drops to 0. The equation of motion for the lower vocal fold is then

$$r\frac{dx}{dt} + \frac{1}{c}(x - x_0) = 0,$$

where r, c, and x_0 are as defined in Exercise 21.[28]

a) Find the general solution of this differential equation.

b) Use the initial condition $x(T) = a$ to solve for $x(t)$.

 Technology Connection

For the initial-value problem, use (a) Euler's method and (b) the Runge–Kutta method with the given step size Δx to approximate $y(2)$.

23. $y' = x - y$, $y(0) = 3$, $\Delta x = 0.2$

24. $y' = \sin y/(x^2 + 4)$, $y(0) = 1$, $\Delta x = 0.1$

---

[26]X. Yang, M. Wang, Y. Huang, and Y. Wang, "A one-compartment model to study soil carbon decomposition rate at equilibrium situation," *Ecological Modelling*, Vol. 151, pp. 63–73 (2002).

[27]K. N. Stevens, *Acoustic Phonetics* (Cambridge, MA: MIT Press, 1998).

---

[28]K. N. Stevens, *Acoustic Phonetics* (Cambridge, MA: MIT Press, 1998).

Chapter 8 Test

Solve the differential equation.

1. $y' = 2x^2 - 4$

2. $y' = e^{2x}(e^{2x} - 2)^4$

3. $y' = x \sin x$

4. $y' + 4y = 6$

5. $x^3y' + 2x^2y = 1$

6. $y' + xy = 3x$

7. $y' = e^{3x-2y}$

8. $y' = xy^3$

9. $y' = \dfrac{x^2}{y^3 - 2y}$

Solve the initial-value problem.

10. $y' = x^2 + 3x + 5; y(1) = 7$

11. $y' = x^3 \sin(x^4 + \pi); y(0) = -1$

12. $xy' + 3y = 0; y(4) = 1$

13. $y' - y \tan x = 1; y(\pi/4) = 2$

14. $y' = \dfrac{x + 1}{y}; y(2) = 4$

15. $y' = \dfrac{1}{e^y + 1}; y(0) = -1$

For Exercises 16–17, **(a)** find the equilibrium value(s) of the differential equation, **(b)** assess the stability of each equilibrium value, **(c)** determine the point(s) of inflection, and **(d)** sketch sample solutions of the differential equation.

16. $y' = 2y^2 + 6y$

17. $y' = y^2 - 9$

APPLICATIONS

18. *Hullian Model of Learning.* According to the Hullian model of learning, the probability P of mastering a certain concept after t learning trials may be modeled by the differential equation

$$\frac{dP}{dt} = k(1 - P),$$

where k is a constant.

a) Write this first-order linear differential equation in standard form.
b) Solve for $P(t)$ given that $P(0) = 0$.
c) Suppose that $P(5) = 0.6$. Use this to determine k.
d) Find $P(10)$.
e) Find t so that $P(t) = 0.9$. In other words, how many trials are necessary in order for the probability of mastering the concept to be 0.90?

19. *Mixing Chemicals.* A tank initially contains 100 lb of salt dissolved in 400 gal of water. Saltwater containing 2 lb of salt per gal enters the tank at the rate of 4 gal per min. The mixture (kept uniform by stirring) is removed at the same rate. How many pounds of salt are in the tank after an hour?

20. *Dye Degradation.* A dye is placed into a mixture of acetone and triethylamine solution. Let $D(t)$ be the concentration of the dye after t seconds. The acetone causes the dye to degrade with a decay rate of

0.226 per sec, while the triethylamine causes a downtake rate of 3×10^{-8} molars per second.[29]

a) Draw a one-compartment model for D.
b) Show that $D' = -0.226D - (3 \times 10^{-8})$.
c) Solve for $D(t)$ given the initial condition $D(0) = D_0$.

21. *Bacteria Growth.* During a fermentation experiment, the population $P(t)$ (in millions per mL) of bacteria *Lactobacillus reuteri* after t hours in a wheat medium satisfied the differential equation

$$\frac{dP}{dt} = 0.13P\left(1 - \frac{P}{16}\right).$$

Initially, 1.7 million bacteria per mL were present.[30]

a) Determine the number of bacteria per mL after 6 hr.
b) Determine when 10 million bacteria per mL will be present.

SYNTHESIS

22. *Yeast Growth.* Let $P(t)$ be the number of yeast cells in a laboratory experiment. Suppose that $P(t)$ follows a logistic model of inhibited growth, given by

$$\frac{dP}{dt} = kP\left(1 - \frac{P}{700}\right).$$

Find k if $P(0) = 10$ and $P(8) = 300$.

 Technology Connection

For the initial-value problem, use **(a)** Euler's method and **(b)** the Runge–Kutta method with the given step size Δx to approximate $y(2)$.

23. $y' = y + 2, y(0) = 3, \Delta x = 0.2$

24. $y' = \sin(ty), y(0) = 1, \Delta x = 0.1$

[29]W. Chu and S. M. Tsui, "Modeling of photodecoloration of azo dye in a cocktail photolysis system," *Water Research,* Vol. 36, pp. 3350–3358 (2002).
[30]D. Charalampopoulos, S. S. Pandiella, and C. Webb, "Growth studies of potentially probiotic lactic acid bacteria in cereal-based substrates," *Journal of Applied Microbiology,* Vol. 92, pp. 851–859 (2002).

Extended Life Science Connection

LARVAE AND FOREST DEFOLIATION

Suppose that larvae are placed on a tree leaf with an initial mass F_0. After T days, the leaf's mass has decreased to some smaller size F_1. How much of the leaf did the larvae eat?

At first glance, it seems that the obvious answer is the difference $F_0 - F_1$. However, there are actually two sources of leaf decay: the digestion by the larvae and also natural moisture loss due to the environment. In this Extended Life Science Connection, we will see how differential equations can be used to isolate the component of leaf decay due to food consumption. This application of differential equations was used to study forest defoliation due to the native paropsine chrysomelid leaf beetle *Chrysophtharta bimaculata* in Tasmania, Australia.

We first describe the experiment that we will model. Two leaves are used, which are called the *feed leaf* and the *control leaf*. We denote by F_0 and C_0 the initial masses of the feed leaf and control leaf, respectively. Larvae are placed on the feed leaf, while the control leaf is placed into a protective bag. The control leaf will be subject only to the effect of moisture loss to the environment, while the feed leaf will be subject to both moisture loss and larval feeding.

Both leaves are placed on the same petri dish and inserted into an incubator for T days, usually between 1 and 2. After removing the leaves from the incubator, the final masses F_1 and C_1 of the feed and control leaves are measured. Based on the masses F_0, F_1, C_0, and C_1, we seek to measure the amount of the feed leaf that is eaten by the larvae.

We assume that the rate of moisture loss is proportional to the leaf mass with relative rate b. We also assume that the feeding rate per larva is constant over the incubation period. This is a reasonable assumption; the incubation period is chosen to be short enough so that larval growth should be insignificant.

EXERCISES

1. We begin our analysis by considering moisture loss in the control leaf.

   a) Let $c(t)$ be the mass of the control leaf after t days. Draw a one-compartment model for c.

   b) Explain why
   $$\frac{dc}{dt} = -bc,$$
   where b is a constant.

   c) Solve for $c(t)$ using the initial condition $c(0) = C_0$.

   d) Substitute T for t in your equation to show that
   $$C_1 = C_0 e^{-bT}.$$

   e) Let $p = C_1/C_0$, or the mass of the control leaf after incubation as a proportion of its initial mass. Show that
   $$p = e^{-bT} \quad \text{and} \quad b = -\frac{\ln p}{T}.$$

2. We now turn to the feed leaf. Let b be the moisture constant as before, N the total number of larvae on the leaf, and a the average mass eaten per larva per day.

   a) Let $f(t)$ be the mass of the feed leaf after t days. Draw a one-compartment model for f.

b) Explain why

$$\frac{df}{dt} = -aN - bf.$$

c) Solve for $f(t)$ using the initial condition $f(0) = F_0$.

d) Substitute T for t in your equation to show that

$$F_1 = -\frac{aN}{b}(1 - e^{-bT}) + F_0 e^{-bT}.$$

e) Simplify this expression by using the results of Exercise 1(e).

3. Finally, we develop an equation for the mass of leaf that is consumed by the larvae.

a) Let M be the total mass of leaf that is consumed by all larvae during the incubation. Explain why $M = aNT$.

b) Use the result of Exercise 2(d) to solve for the consumption rate a.

c) Show that

$$M = \frac{-(pF_0 - F_1)\ln p}{1 - p}.$$

4. In an experiment, 20 first-instar larvae of *Chrysophtharta bimaculata* are placed on a feed leaf from *Eucalyptus nitens*. The control leaf had an initial mass of 0.352 g, which reduced to 0.324 g after 2 days of incubation. The feed leaf had an initial mass of 0.346 g, which reduced to 0.307 g after incubation.

a) Find the total mass of leaf consumed by the larvae.

b) Express the total consumed mass as a percentage of the initial mass of the feed leaf.

c) Find the average daily mass consumed by each larva.

5. In a second experiment, 30 fourth-instar larvae of *Chrysophtharta bimaculata* are placed on a feed leaf from *Eucalyptus regnans*. The control leaf had an initial mass of 1.282 g, which reduced to 1.085 g after 30 hr of incubation. The feed leaf had an initial mass of 1.165 g, which reduced to 0.365 g after incubation.

a) Find the total mass of leaf consumed by the larvae.

b) Express the total consumed mass as a percentage of the initial mass of the feed leaf.

c) Find the average daily mass consumed by each larva.

REFERENCES

1. S. G. Candy and S. C. Baker, "Calculating food consumption in the laboratory: A formula to adjust for natural mass loss," *Australian Journal of Entomology,* Vol. 41, pp. 170–173 (2002).

2. S. C. Baker, J. A. Elak, and S. G. Candy, "Comparison of feeding efficiency, development time and survival of Tasmanian eucalyptus leaf beetle larvae *Chrysophtharta bimaculata* (Olivier) (Coleoptera: Chrysomelidae) on two hosts," *Australian Journal of Entomology,* Vol. 41, pp. 174–181 (2002).

9

Higher-Order and Systems of Differential Equations

9.1 Higher-Order Homogeneous Differential Equations

9.2 Higher-Order Nonhomogeneous Differential Equations

9.3 Systems of Linear Differential Equations

9.4 Matrices and Trajectories

9.5 Models of Population Biology

9.6 Numerical Methods

AN APPLICATION   Chemicals enter a house's basement air at the rate of 0.1 mg per min. Both the first floor and the basement are initially uncontaminated and have volumes of 200 m³. Air flows from the basement into the first floor at the rate of 2 m³ per min, while air flows through the first floor to the outside at the rate of 4 m³ per min.[1] Show that the equilibrium values of the chemical in the first-floor and basement air are 10 mg and 5 mg, respectively.

This problem appears as Exercise 32 in Section 9.3.

INTRODUCTION   *Systems of differential equations model many biological systems, including the populations of species that share a habitat. In this chapter, we will develop two techniques for solving linear systems of differential equations. The first technique reduces a system to a higher-order differential equation, while the second technique uses the eigenvalues and eigenvectors of a matrix. We will also discuss qualitative techniques of graphing solutions of nonlinear systems.*

[1]D. A. Olson and R. L. Corsi, "Characterizing exposure to chemicals from soil vapor intrusion using a two-compartment model," *Atmospheric Environment*, Vol. 35, pp. 4201–4209 (2001).

9.1 Higher-Order Homogeneous Differential Equations

OBJECTIVES

■ Find the general solution of higher-order homogeneous linear differential equations with constant coefficients.

■ Use initial conditions to solve initial-value problems.

In the previous chapter, we developed exact, graphical, and numerical techniques for solving a first-order differential equation. For example, we saw that the uninhibited growth model

$$y' - ky = 0 \tag{1}$$

has general solution $y = Ce^{kx}$. This general solution may be found either by using Theorem 3 of Chapter 8 or by separating the variables.

In this section, we will consider the solution of *linear nth-order homogeneous differential equations,* which have the form

$$a_n(x)y^{(n)} + a_{n-1}(x)y^{(n-1)} + \cdots + a_1(x)y' + a_0(x)y = 0,$$

where the coefficients $a_0(x), a_1(x), \ldots, a_n(x)$ are functions of x. Unfortunately, not all of these differential equations can be solved explicitly. For the moment, we'll consider second-order linear homogeneous equations with constant coefficients. These have the form

$$ay'' + by' + cy = 0, \tag{2}$$

where $a, b,$ and c are constants and $a \neq 0$.

The first-order linear differential equation (1) also has constant coefficients, and its general solution is based upon an exponential function. This suggests that we attempt to find a solution of (2) of the form e^{rx}, where r is a constant. If we let $y = e^{rx}$ in (2) above, then, since $y' = re^{rx}$ and $y'' = r^2 e^{rx}$, we get

$$a(r^2 e^{rx}) + b(re^{rx}) + c(e^{rx}) = 0.$$

Noting that each term in the equation above has the common factor e^{rx}, we write

$$e^{rx}(ar^2 + br + c) = 0.$$

Now e^{rx} is a positive number for all choices of real numbers r and x. Dividing by e^{rx}, we get

$$ar^2 + br + c = 0.$$

The roots of this quadratic equation are the values of r. These roots are then used to find solutions of the form e^{rx}.

EXAMPLE 1   Solve $y'' + 5y' + 6y = 0$.

Solution   We try $y = e^{rx}$. Then $y' = re^{rx}$ and $y'' = r^2 e^{rx}$. Substituting these quantities into the equation gives

$$r^2 e^{rx} + 5re^{rx} + 6e^{rx} = 0$$
$$e^{rx}(r^2 + 5r + 6) = 0$$
$$r^2 + 5r + 6 = 0$$
$$(r + 2)(r + 3) = 0.$$

We have two solutions, or roots, -2 and -3. This gives us the two solutions e^{-2x} and e^{-3x} of the differential equation. These two solutions are called the **fundamental solutions** of the differential equation.

It turns out that the general solution to the equation is given by $y = C_1 e^{-2x} + C_2 e^{-3x}$, where C_1 and C_2 are arbitrary constants. We can verify that y is a solution by finding y' and y'' and substituting y, y', and y'' into the original equation. We get

$$y = C_1 e^{-2x} + C_2 e^{-3x},$$
$$y' = -2C_1 e^{-2x} - 3C_2 e^{-3x},$$
$$y'' = 4C_1 e^{-2x} + 9C_2 e^{-3x}.$$

Substituting, we get

$$
\begin{aligned}
y'' + 5y' + 6y &= (4C_1 e^{-2x} + 9C_2 e^{-3x}) + 5(-2C_1 e^{2x} - 3C_2 e^{3x}) \\
&\quad + 6(C_1 e^{-2x} + C_2 e^{-3x}) \\
&= 4C_1 e^{-2x} + 9C_2 e^{-3x} - 10C_1 e^{-2x} - 15C_2 e^{-3x} \\
&\quad + 6C_1 e^{-2x} + 6C_2 e^{-3x} \\
&= 0.
\end{aligned}
$$

In Example 1, we used the fact that if $f(x)$ and $g(x)$ are solutions to a linear homogeneous differential equation, then so is $C_1 f(x) + C_2 g(x)$. We state this important principle, called the *principle of superposition*, in the form of a theorem. The proof of this theorem is verified in the exercises.

THEOREM 1

If $f_1(x), f_2(x), \ldots, f_n(x)$ are solutions to a linear nth order homogeneous differential equation, then so is

$$\sum_{i=1}^{n} C_i f_i(x) = C_1 f_1(x) + C_2 f_2(x) + \cdots + C_n f_n(x),$$

where $C_1, C_2, \ldots, C_n$ are arbitrary constants.

In Example 1, we obtained the expression $y = C_1 e^{-2x} + C_2 e^{-3x}$ for the solution of $y'' + 5y' + 6y = 0$. It can be shown that this expression is called the **general solution** to the equation. This means that every function of this form is a solution to the equation and that every solution to the equation can be written in this form. (Proving that this expression is the general solution is beyond the scope of this text.)

For the equation $y'' + 5y' + 6y = 0$, the general solution contained two arbitrary constants. In all of the equations that we'll consider, the number of constants in the general solution will be equal to the order of the equation.

EXAMPLE 2   Find the general solution to $y''' + y'' - 6y' = 0$.

Solution   As in Example 1, we let $y = e^{rx}$. We obtain $y' = re^{rx}, y'' = r^2 e^{rx}$, and $y''' = r^3 e^{rx}$. Substituting into the equation, we obtain

$$
\begin{aligned}
r^3 e^{rx} + r^2 e^{rx} - 6r e^{rx} &= 0 \\
r(r^2 + r - 6)e^{rx} &= 0 \\
r(r^2 + r - 6) &= 0 \qquad \text{Since } e^{rx} \text{ is never } 0 \\
r(r + 3)(r - 2) &= 0.
\end{aligned}
$$

So e^{-3x}, e^{2x}, and $e^{0x} = 1$ are the three fundamental solutions of this equation. The general solution is

$$y = C_1e^{-3x} + C_2e^{2x} + C_3(1) = C_1e^{-3x} + C_2e^{2x} + C_3,$$

which may be verified by substituting into the original differential equation.   ■

In each of Examples 1 and 2, we made the substitution $y = e^{rx}$, reduced the given differential equation to a polynomial equation in the variable r, and then solved this polynomial equation for r. The resulting values for r determined the fundamental solutions and then the general solution.

But there's no need to repeat the substitution $y = e^{rx}$ and the subsequent algebra each time. We can bypass this work, going directly from the given differential equation to the associated polynomial equation, called the **auxiliary equation.**

EXAMPLE 3   Find the auxiliary equation of $y'' - 7y' + 10y = 0$.

Solution   Replacing y'' by r^2, y' by r, and y by 1 gives

$$r^2 - 7r + 10 = 0.$$ ■

EXAMPLE 4   Solve $y'' - 7y' + 10y = 0$.

Solution   As discussed above, the auxiliary equation is $r^2 - 7r + 10 = 0$. Factoring, we get $(r - 2)(r - 5) = 0$, so its roots are 2 and 5. The fundamental solutions are e^{2x} and e^{5x}, so the general solution is

$$y = C_1e^{2x} + C_2e^{5x}.$$ ■

Repeated Roots

The technique that we have developed works as long as the auxiliary equation has distinct real roots. It must be modified if the real roots are repeated.

EXAMPLE 5   Solve $y'' - 2y' + y = 0$.

Solution   The auxiliary equation is $r^2 - 2r + 1 = 0$. Factoring yields $(r - 1)^2 = 0$. The roots are 1 and 1, or simply 1. Sure enough, $r = 1$ yields the solution $y = e^x$. We might try $y = C_1e^x + C_2e^x$ as the general solution. But

$$C_1e^x + C_2e^x = (C_1 + C_2)e^x = Ke^x,$$

where $K = C_1 + C_2$ is a constant. In other words, simply repeating the e^x does not give the general solution because it only yields one of the two fundamental solutions.

To obtain the general solution, let's substitute $y = ve^x$ into the original equation, where v is a function to be determined. Then, by the Product Rule,

$$y' = v'e^x + ve^x$$

and

$$y'' = (v''e^x + v'e^x) + (v'e^x + ve^x) = v''e^x + 2v'e^x + ve^x.$$

Substituting all of this into $y'' - 2y' + y = 0$ yields

$$(v''e^x + 2v'e^x + ve^x) - 2(v'e^x + ve^x) + ve^x = 0$$

$$v''e^x = 0$$

$$v'' = 0. \quad \text{Since } e^x \text{ is never 0}$$

Integrating twice yields $v = C_1 + C_2 x$. Finally, replacing v in the expression $y = ve^x$ gives

$$y = (C_1 + C_2 x)e^x = C_1 e^x + C_2 x e^x.$$

This is the general solution of $y'' - 2y' + y = 0$, which can be verified by substituting into the original differential equation. Not surprisingly, the solution e^x found earlier is a special case of this general solution; we obtain e^x by substituting $C_1 = 1$ and $C_2 = 0$.   ■

The calculations in Example 5 weren't easy. Rather than repeat them each time a double root r occurs in the auxiliary equation, we point out that the corresponding fundamental solutions of the differential equation are e^{rx} and xe^{rx}. This fact will be verified in the exercises for second-order equations.

If some roots are distinct and some are repeated, we simply combine the fundamental solutions.

EXAMPLE 6   Solve $y''' + 4y'' + 4y' = 0$.

Solution   The auxiliary equation is $r^3 + 4r^2 + 4r = 0$, or $r(r + 2)^2 = 0$. So $0, -2$, and -2 are the roots. The corresponding fundamental solutions are $e^{0x} = 1$, e^{-2x}, and xe^{-2x}. The general solution is

$$y = C_1 + C_2 e^{-2x} + C_3 x e^{-2x}.$$ ■

If a root r appears three times, then we use e^{rx}, xe^{rx}, and $x^2 e^{rx}$ as the three fundamental solutions.

EXAMPLE 7   Solve $y''' - 3y'' + 3y' - y = 0$.

Solution   The auxiliary equation is $r^3 - 3r^2 + 3r - 1 = 0$, or $(r - 1)^3 = 0$. So the roots are 1, 1, and 1. The three fundamental solutions are e^x, xe^x, and $x^2 e^x$. The general solution is

$$y = C_1 e^x + C_2 x e^x + C_3 x^2 e^x.$$ ■

Complex Roots

We have considered examples for which the roots of the auxiliary equation are real numbers. Now consider the equation

$$y'' + y = 0.$$

The auxiliary equation is $r^2 + 1 = 0$. The roots of this equation are $r = \pm\sqrt{-1} = \pm i$, where i denotes $\sqrt{-1}$, as discussed in Chapter 1.

Although the auxiliary equation has two complex roots, it's easy to solve $y'' + y = 0$. If $y = \sin x$, then $y' = \cos x$ and $y'' = -\sin x = -y$. In other words, $y = \sin x$ is one solution. A similar calculation, which is left to the student, reveals that $y = \cos x$ is another solution. The general solution is $y = C_1 \sin x + C_2 \cos x$.

This example illustrates the general case. As discussed in the exercises, if the roots of the auxiliary equation are the complex numbers $a \pm bi$, then the two corresponding fundamental solutions are $e^{ax} \sin bx$ and $e^{ax} \cos bx$.

EXAMPLE 8   Solve $y'' + 16y = 0$.

Solution   The auxiliary equation is $r^2 + 16 = 0$. Solving directly yields

$$r = \pm\sqrt{-16} = \pm 4\sqrt{-1} = \pm 4i = 0 \pm 4i.$$

In this case, we have two complex roots with $a = 0$ and $b = 4$. The two fundamental solutions are $e^{0x} \sin 4x = \sin 4x$ and $e^{0x} \cos 4x = \cos 4x$. The general solution is

$$y = C_1 \sin 4x + C_2 \cos 4x. \qquad\blacksquare$$

EXAMPLE 9   Solve $y'' - 2y' + 10y = 0$.

Solution   The auxiliary equation is $r^2 - 2r + 10 = 0$. Applying the quadratic formula yields

$$r = \frac{-(-2) \pm \sqrt{(-2)^2 - 4(1)(10)}}{2 \cdot 1} = \frac{2 \pm \sqrt{-36}}{2} = \frac{2 \pm 6i}{2} = 1 \pm 3i.$$

So $a = 1$ and $b = 3$. The corresponding fundamental solutions are $e^x \sin 3x$ and $e^x \cos 3x$. The general solution is

$$y = C_1 e^x \sin 3x + C_2 e^x \cos 3x. \qquad\blacksquare$$

If some of the roots are real and some are complex, we simply combine our earlier techniques.

EXAMPLE 10   Solve $y''' - 2y'' + 4y' - 8y = 0$.

Solution   The auxiliary equation is $r^3 - 2r^2 + 4r - 8 = 0$. Using grouping, this factors as $(r - 2)(r^2 + 4) = 0$. The roots are 2, $2i$, and $-2i$. The corresponding fundamental solutions are e^{2x}, $\sin 2x$, and $\cos 2x$. The general solution is

$$y = C_1 e^{2x} + C_2 \sin 2x + C_3 \cos 2x. \qquad\blacksquare$$

If complex roots are repeated, then we modify the solution just as we did in the case of repeated real roots.

EXAMPLE 11   Solve $y^{(4)} + 2y'' + y = 0$.

Solution   The auxiliary equation is $r^4 + 2r^2 + 1 = 0$, or $(r^2 + 1)^2 = 0$. The roots are $\pm i$ and $\pm i$. The fundamental solutions corresponding to $\pm i$ are $\sin x$ and $\cos x$. Two more fundamental solutions corresponding to the second pair $\pm i$ are $x \sin x$ and $x \cos x$. The general solution is

$$y = C_1 \sin x + C_2 \cos x + C_3 x \sin x + C_4 x \cos x. \qquad\blacksquare$$

Initial-Value Problems

In Chapter 8, we saw that an initial condition specified the solution of a first-order initial-value problem. In general, n initial conditions are required to find the unique solution y of an nth-order initial-value problem. These initial conditions specify the values of $y, y', \ldots, y^{(n-1)}$ at a specified value of x.

Technology Connection

Exploratory

For Examples 1–11, use a grapher to plot the general solution for various choices of the constants. Some of these solutions will converge to 0 as x gets large, while others will diverge from 0. What condition on the roots of the auxiliary equation is needed to guarantee that the solutions must converge to 0?

EXAMPLE 12   Solve the initial-value problem

$$y'' - 2y' - 35y = 0, \qquad y(0) = 4, \qquad y'(0) = -2.$$

Solution   The auxiliary equation is $r^2 - 2r - 35 = 0$. Factoring, we get $(r - 7)(r + 5) = 0$, so that its roots are 7 and -5. The fundamental solutions are e^{7x} and e^{-5x}, so the general solution is

$$y = C_1 e^{7x} + C_2 e^{-5x}.$$

To find the unique solution of the initial-value problem, we first find y':

$$y' = 7C_1 e^{7x} - 5C_2 e^{-5x}.$$

We then substitute the initial conditions into the formulas for y and y':

$$\begin{array}{ll} y(0) = C_1 e^{7(0)} + C_2 e^{-5(0)} & \quad 4 = C_1 + C_2 \\ & \text{or} \\ y'(0) = 7C_1 e^{7(0)} - 5C_2 e^{-5(0)}, & \quad -2 = 7C_1 - 5C_2. \end{array}$$

This system of equations may be solved by using any of the techniques developed in Chapter 6. Multiplying the first equation by 5 and adding to the second equation, we obtain

$$18 = 12C_1, \quad \text{or} \quad C_1 = \frac{3}{2}.$$

Substituting into the first equation, we find

$$4 = \frac{3}{2} + C_2, \quad \text{or} \quad C_2 = \frac{5}{2}.$$

The solution of the initial-value problem is therefore $y = \frac{3}{2} e^{7x} + \frac{5}{2} e^{-5x}$.   ■

EXAMPLE 13   Solve the initial-value problem

$$y'' + 4y' + 29y = 0, \qquad y(0) = 2, \qquad y'(0) = -3.$$

Solution   The auxiliary equation is $r^2 + 4r + 29 = 0$. Using the quadratic formula, we find

$$r = \frac{-4 \pm \sqrt{(4)^2 - 4(1)(29)}}{2} = \frac{-4 \pm \sqrt{-100}}{2} = \frac{-4 \pm 10i}{2} = -2 \pm 5i.$$

The corresponding fundamental solutions are $e^{-2x} \sin 5x$ and $e^{-2x} \cos 5x$. The general solution is

$$y = C_1 e^{-2x} \sin 5x + C_2 e^{-2x} \cos 5x.$$

To find the unique solution of the initial-value problem, we first find y':

$$\begin{aligned} y' &= (-2C_1 e^{-2x} \sin 5x + 5C_1 e^{-2x} \cos 5x) \\ &\quad + (-2C_2 e^{-2x} \cos 5x - 5C_2 e^{-2x} \sin 5x) \\ &= (-2C_1 - 5C_2) e^{-2x} \sin 5x + (5C_1 - 2C_2) e^{-2x} \cos 5x. \end{aligned}$$

We then substitute the initial conditions into the formulas for y and y':

$$y(0) = C_1 e^{-2(0)} \sin(5[0]) + C_2 e^{-2(0)} \cos(5[0])$$

$$y'(0) = (-2C_1 - 5C_2)e^{-2(0)} \sin(5[0]) + (5C_1 - 2C_2)e^{-2(0)} \cos(5[0]),$$

or, since $\sin 0 = 0$ and $\cos 0 = 1$,

$$2 = C_2 \qquad \text{Initial condition: } y(0) = 2$$

$$-3 = 5C_1 - 2C_2. \qquad \text{Initial condition: } y'(0) = -3$$

The first equation gives $C_2 = 2$. Substituting into the second equation, we find

$$-3 = 5C_1 - 2(2), \quad \text{or} \quad C_1 = \frac{1}{5}.$$

The solution of the initial-value problem is therefore $y = \frac{1}{5}e^{-2x} \sin 5x + 2e^{-2x} \cos 5x$. ■

We end this section by summarizing the techniques that we have seen thus far.

To solve the linear nth-order homogeneous differential equation

$$a_n y^{(n)} + \cdots + a_2 y'' + a_1 y' + a_0 y = 0$$

with constant coefficients, use the following steps.

1. Construct the auxiliary equation.

2. Solve the auxiliary equation to determine the n roots (including possible repeated roots).

3. List the fundamental solutions, using the table below. The total number of fundamental solutions is the same as the order of the differential equation.

4. Use the principle of superposition to construct the general solution.

Distinct Real Root, r 1 solution	Distinct Pair of Complex Roots, $r = a \pm bi$ 2 solutions
e^{rx}	$e^{ax} \sin bx$ and $e^{ax} \cos bx$
Real root r that is repeated k times k solutions	Pair of complex roots $r = a \pm bi$ that are repeated k times $2k$ solutions
$e^{rx}, xe^{rx}, \ldots, x^{k-1}e^{rx}$	$e^{ax} \sin bx, xe^{ax} \sin bx, \ldots, x^{k-1}e^{ax} \sin bx,$ and $e^{ax} \cos bx, xe^{ax} \cos bx, \ldots, x^{k-1}e^{ax} \cos bx$

Exercise Set 9.1

Solve.

1. $y'' - 6y' + 5y = 0$
2. $y'' - 8y' + 15y = 0$
3. $y'' - y' - 2y = 0$
4. $y'' + 2y' - 3y = 0$
5. $y'' + 3y' + 2y = 0$
6. $y'' + 5y' + 6y = 0$
7. $2y'' - 5y' + 2y = 0$
8. $2y'' - y' - y = 0$
9. $y'' - 9y = 0$
10. $y'' - 3y = 0$
11. $y'' + 10y' + 25y = 0$
12. $y'' - 8y' + 16y = 0$
13. $4y'' + 12y' + 9y = 0$
14. $y''' + 8y'' + 16y' = 0$
15. $y''' + y'' + 4y' + 4y = 0$
16. $y''' + 8y = 0$
17. $y''' + 6y'' + 12y' + 8y = 0$
18. $y^{(4)} + 3y''' + 3y'' + y' = 0$
19. $y''' - 6y'' + 3y' - 18y = 0$
20. $y^{(4)} - 4y''' = 0$
21. $y^{(4)} - 5y''' + 4y'' = 0$
22. $y^{(4)} + 3y''' - y'' - 3y' = 0$
23. $y'' + 36y = 0$
24. $y'' + 4y' + 13y = 0$
25. $y'' + 8y' + 41y = 0$
26. $y'' + y' + y = 0$
27. $y''' + 2y'' + 5y' = 0$
28. $y''' - 2y'' + 4y' - 8y = 0$
29. $y''' - y = 0$
30. $y^{(4)} + 8y'' + 16y = 0$
31. $2y'' + 2y' - 5y = 0$
32. $2y'' - 5y' + y = 0$
33. $3y'' - 2y' + 10y = 0$
34. $2y'' + 6y' + 5y = 0$

Solve the initial-value problem.

35. $y'' - y' = 0,\ y(0) = 0,\ y'(0) = -1$
36. $y'' + y = 0,\ y(\pi) = 4,\ y'(\pi) = 3$
37. $y'' - y = 0,\ y(0) = 1,\ y'(0) = 3$
38. $y'' - 2y' + y = 0,\ y(0) = 1,\ y'(0) = -2$
39. $y'' + 4y' + 4y = 0,\ y(0) = 3,\ y'(0) = 2$
40. $y'' - 2y' + 2y = 0,\ y(0) = 1,\ y'(0) = -1$
41. $y''' + y' = 0,\ y(\pi) = 1,\ y'(\pi) = 8,\ y''(\pi) = 4$
42. $y''' + 6y'' + 9y' = 0,\ y(0) = 2,\ y'(0) = -2,$
    $y''(0) = 3$

APPLICATIONS

43. *Population Growth.* **The** estimated number $N(t)$ of bacteria (in thousands) present t hours after the beginning of an experiment satisfies the differential equation

$$N'' - 3N' + 2N = 0.$$

Find the approximate number of bacteria present after 4 hr if $N(0) = 5$ and $N'(0) = 15$.

Leukocytes. Blood leukocytes, or white blood cells, may either be circulating or marginated. Let c be the rate at which leukocytes change from marginated to circulating, and let m be the rate at which they change from circulating to marginated. Also, let a be the rate at which leukocytes are absorbed by surrounding tissue. Then the difference $N(t)$ of circulating leukocytes from its equilibrium value follows the differential equation

$$N'' + (c + m + 2a)N' + a(c + m + a)N = 0,$$

where t is measured in hours.[2]

tw 44. Set up the auxiliary equation. Show that its two roots are always negative. (*Hint:* After using the quadratic formula, the part under the square root should look familiar after simplification.)

45. Under normal conditions, $c = 2.7$, $m = 3.94$, and $a = 0.0576$. Find the general solution of $N(t)$.

46. In the presence of a tumor, $c = 1.64$, $m = 2.2$, and $a = 0.0576$. Find the general solution of $N(t)$.

47. During long, sustained aerobic exercise, $c = 0.31$, $m = 1.64$, and $a = 0.51$. Find the general solution of $N(t)$.

SYNTHESIS

48. You are told that the roots of the auxiliary equation of a certain sixth-order linear homogeneous differential equation with constant coefficients are 2, 2, 3, 3, 3, and 3. Find the general solution to the equation.

49. You are told that the roots of the auxiliary equation of a certain sixth-order linear homogeneous differential equation with constant coefficients are 7, 9, $2 \pm 4i$, and $2 \pm 4i$. Find the general solution to the equation.

---

[2]K. Iadocicco, L. H. A. Monteiro, and J. G. Chaui-Berlinck, "A theoretical model for estimating the margination constant of leukocytes," *BMC Physiology*, Vol. 3, article 2 (2002).

50. *Phonetics.* The shape of the vocal tract tends to promote certain sound frequencies. For example, to produce the first vowel (called a *schwa*) in the word *about,* the vocal tract opens widely. The cross-sectional area throughout the vocal tract is approximately the same and may be modeled by a cylinder with one end open (the lips) and the other end closed (the glottis).

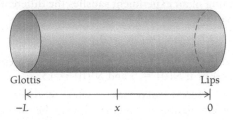

Glottis   Lips

$-L$ x 0

Let $p(x)$ denote the sound pressure at position x within the cylinder. Then $p(x)$ satisfies the *boundary-value problem*

$$p'' + \left(\frac{2\pi k}{c}\right)^2 p = 0, \qquad p(0) = 0, \qquad p'(-L) = 0,$$

where k is the frequency of the sound, c is the speed of sound, and L is the length of the vocal tract.[3]

a) Find the general solution of this differential equation.

b) Use the condition $p(0) = 0$ to show that $p(x) = C \sin(2\pi k x/c)$, where C is a constant.

---

[3]K. N. Stevens, *Acoustic Phonetics* (Cambridge, MA: MIT Press, 1998).

c) Use the condition $p'(-L) = 0$ to show that $k = [(2n - 1)c]/(4L)$ for some integer n.

Notice that a boundary-value problem need not have a unique solution, unlike an initial-value problem.

51. *Superposition.* Let y_1 and y_2 both be solutions to the differential equation

$$ay'' + by' + cy = 0.$$

Verify that $C_1 y_1 + C_2 y_2$ is also a solution to the equation, where C_1 and C_2 are constants.

52. *Superposition.* Show that the conclusion of the previous exercise holds for any linear nth-order homogeneous differential equation.

53. *Repeated Roots.* Suppose that the equation $ax^2 + bx + c = 0$ has only one root, $x = r$.

a) Use the quadratic formula to show that $r = -b/2a$.

b) Explain why $ar^2 + br + c = 0$.

tw c) Explain why $2ar + b = 0$.

d) Verify that $y = xe^{rx}$ is a solution of the differential equation $ay'' + by' + cy = 0$.

54. *Complex Roots.* Suppose that the roots of the equation $ar^2 + br + c = 0$ are the complex numbers $p \pm qi$, where $q \neq 0$.

a) Explain why $a(p^2 - q^2) + bp + c = 0$ and $2apq + bq = 0$.

b) Verify that $y = e^{px} \sin qx$ is a solution of the differential equation $ay'' + by' + cy = 0$.

c) Verify that $y = e^{px} \cos qx$ is also a solution.

9.2 Higher-Order Nonhomogeneous Differential Equations

OBJECTIVE

■ Solve a linear nonhomogeneous differential equation with constant coefficients if the forcing function is a polynomial.

In the previous section, all of the equations considered were linear homogeneous differential equations with constant coefficients. These had the form

$$a_n y^n + a_{n-1} y^{(n-1)} + \cdots + a_1 y' + a_0 y = 0. \tag{1}$$

In this section, we consider *nonhomogeneous* equations of the form

$$a_n y^{(n)} + a_{n-1} y^{(n-1)} + \cdots + a_1 y' + a_0 y = q(x). \tag{2}$$

That is, we will no longer require that the equation be homogeneous. The function q is sometimes called the **forcing function**. Since the left-hand sides of (1) and (2) are the same, we will say that (1) is the **associated homogeneous equation** of (2).

To solve nonhomogeneous equations, we will use the following theorem.

THEOREM 2

Suppose y_p is any solution of (2); such a solution is called a **particular solution** of (2). Suppose also that y_h is the general solution of (1), the associated homogeneous equation. Then the general solution of (2) is given by $y = y_p + y_h$.

Proof   We first verify that $y = y_p + y_h$ is a solution of (2):

$$a_n y^{(n)} + \cdots + a_1 y' + a_0 y = a_n(y_p + y_h)^{(n)} + \cdots + a_1(y_p + y_h)' + a_0(y_p + y_h)$$
$$= [a_n y_p^{(n)} + \cdots + a_1 y_p' + a_0 y_p] + [a_n y_h^{(n)} + \cdots + a_1 y_h' + a_0 y_h]$$
$$= q(x) + 0 \qquad \text{Since } y_p \text{ satisfies (2) and } y_h \text{ satisfies (1)}$$
$$= q(x).$$

Therefore, y is a solution of (2).

We have shown that, if y_h is a solution of (1), then $y = y_p + y_h$ is a solution of (2). We must now show that every solution of (2) may be written as the sum of y_p and some solution of (1).

Suppose that y is a solution of (2). Then, as we show below, the function $y - y_p$ is a solution of the associated homogeneous equation:

$$a_n(y - y_p)^{(n)} + \cdots + a_1(y - y_p)' + a_0(y - y_p)$$
$$= [a_n y^{(n)} + \cdots + a_1 y' + a_0 y] - [a_n y_p^{(n)} + \cdots + a_1 y_p' + a_0 y_p]$$
$$= q(x) - q(x) \qquad \text{Since both } y \text{ and } y_p \text{ satisfy (2)}$$
$$= 0.$$

Therefore, we may write $y_h = y - y_p$, where y_h is a solution of (1). In other words, y is the sum of y_p and a solution of the associated homogeneous equation. ∎

Theorem 2 has a simple, practical application. Suppose that the general solution y_h of the associated homogeneous differential equation (2) is known. If we can find a single, particular solution y_p, then we may add y_p and y_h to find the general solution of (1).

The Method of Undetermined Coefficients

In Section 9.1, we discussed how to find the general solution y_h of a homogeneous differential equation. We now consider a technique, called the **method of undetermined coefficients,** for finding a particular solution y_p. This technique can be applied whenever the function $q(x)$ is a polynomial, including a constant. We state this technique as a theorem but omit its proof.

> **THEOREM 3**
>
> Suppose that $q(x)$ is a polynomial of degree n, where $q(x)$ is as in (2).
>
> 1. If 0 is *not* a root of the auxiliary equation of the associated homogeneous equation, then a particular solution y_p will be a polynomial of degree n.
>
> 2. If 0 appears k times as a root of the auxiliary equation, a particular solution y_p will be the product of x^k and a polynomial of degree n.
>
> The coefficients of the polynomial may be found by solving a system of equations.

We illustrate the method of undetermined coefficients with a few examples.

EXAMPLE 1   Solve $y'' - 6y' + 8y = 2x^2 - 1$.

Solution

a) We must first solve the associated homogeneous equation $y'' - 6y' + 8y = 0$. The auxiliary equation is $r^2 - 6r + 8 = 0$, and its roots are 2 and 4. The two corresponding fundamental solutions are e^{2x} and e^{4x}. Therefore, the general solution of the associated homogeneous equation is

$$y_h = C_1 e^{2x} + C_2 e^{4x}.$$

b) Since 0 is not a root of the auxiliary equation and $q(x) = 2x^2 - 1$ is quadratic, the method of undetermined coefficients says that we should try

$$y_p = Ax^2 + Bx + C,$$

where A, B, and C are constants to be determined.

To find these constants, we observe that

$$y_p' = 2Ax + B \quad \text{and} \quad y_p'' = 2A.$$

Substituting into the original nonhomogeneous equation, we find

$$y_p'' - 6y_p' + 8y_p = 2x^2 - 1$$
$$2A - 6(2Ax + B) + 8(Ax^2 + Bx + C) = 2x^2 - 1$$
$$8Ax^2 + (-12A + 8B)x + (2A - 6B + 8C) = 2x^2 - 1.$$

In order for these two expressions to be equal, the coefficients of corresponding terms must be equal. So

$$8A = 2, \qquad \text{x^2-coefficients are equal.}$$
$$-12A + 8B = 0, \qquad \text{x-coefficients are equal.}$$
$$2A - 6B + 8C = -1. \qquad \text{Constant coefficients are equal.}$$

This is a system of three equations in three unknowns, which may be solved using any of the techniques of Chapter 6. This particular system of equations can be solved by using back-substitution. The first equation gives $A = 1/4$. Substituting into the second equation, we find

$$-12\left(\frac{1}{4}\right) + 8B = 0, \quad \text{or} \quad B = \frac{3}{8}.$$

Substituting both A and B into the third equation gives

$$2\left(\frac{1}{4}\right) - 6\left(\frac{3}{8}\right) + 8C = -1, \quad \text{or} \quad C = \frac{3}{32}.$$

In summary, a particular solution is given by $y_p = x^2/4 + 3x/8 + 3/32$.

c) We add y_h and y_p to find the general solution of the nonhomogeneous equation:

$$y = C_1 e^{2x} + C_2 e^{4x} + \frac{1}{4}x^2 + \frac{3}{8}x + \frac{3}{32}. \qquad \blacksquare$$

In the previous examples, 0 was not a root of the auxiliary equation. We now discuss finding a particular solution when 0 appears in the roots of the auxiliary equation.

EXAMPLE 2   Solve $y'' - 2y' = 6$.

Solution

a) The associated homogeneous equation is $y'' - 2y' = 0$, and its auxiliary equation is $r^2 - 2r = 0$. Since the roots are 0 and 2, the two corresponding fundamental solutions are $e^{0x} = 1$ and e^{2x}. Therefore, the general solution of the associated homogeneous equation is

$$y_h = C_1 + C_2 e^{2x}.$$

b) Since 0 is a root of the auxiliary equation and $q(x) = 6$ is constant, the method of undetermined coefficients says that we should try

$$y_p = Cx.$$

To find C, we observe that

$$y_p' = C \quad \text{and} \quad y_p'' = 0.$$

Substituting into the original nonhomogeneous equation, we find

$$y_p'' - 2y_p' = 6$$
$$0 - 2(C) = 6$$
$$C = -3.$$

In other words, a particular solution is given by $y_p = -3x$.

c) We add y_h and y_p to find the general solution of the nonhomogeneous equation:

$$y = C_1 + C_2 e^{2x} - 3x.$$

We notice that multiplying by x to find the particular solution was important. If we had erroneously tried $y_p = C$, then $y_p' = 0$ and $y_p'' = 0$. Substituting back into the nonhomogeneous equation would then have yielded $0 + 0 = 6$, which is clearly impossible. $\qquad \blacksquare$

EXAMPLE 3   Solve the initial-value problem $y''' - y'' = 2x - 3$ given that $y(0) = 1$, $y'(0) = 0$, and $y''(0) = 3$.

Solution

a) The associated homogeneous equation is $y''' - y'' = 0$, and the auxiliary equation is $r^3 - r^2 = 0$. Since its roots are 0, 0, and 1, the three corresponding fundamental solutions are $e^{0x} = 1$, $xe^{0x} = x$, and e^x. Therefore, the general solution of the associated homogeneous equation is

$$y_h = C_1 + C_2 x + C_3 e^x.$$

b) Since 0 appears twice in the roots of the auxiliary equation and $q(x) = 2x - 3$ is linear, we should try

$$y_p = x^2(Ax + B) = Ax^3 + Bx^2$$

according to the method of undetermined coefficients. To find the constants A and B, we observe that

$$y_p' = 3Ax^2 + 2Bx, \quad y_p'' = 6Ax + 2B, \quad \text{and} \quad y_p''' = 6A.$$

Substituting into the original nonhomogeneous equation, we find

$$y_p''' - y_p'' = 2x - 3$$
$$6A - (6Ax + 2B) = 2x - 3$$
$$-6Ax + (6A - 2B) = 2x - 3.$$

Equating corresponding coefficients of x yields

$$-6A = 2 \quad \text{and} \quad 6A - 2B = -3.$$

The first equation gives $A = -1/3$. Substituting into the second equation, we find

$$6\left(-\frac{1}{3}\right) - 2B = -3, \quad \text{or} \quad B = \frac{1}{2}.$$

In summary, a particular solution is given by $y_p = -x^3/3 + x^2/2$.

c) We add y_h and y_p to find the general solution of the nonhomogeneous equation:

$$y = C_1 + C_2 x + C_3 e^x - \frac{1}{3}x^3 + \frac{1}{2}x^2.$$

d) It remains for us to use the initial conditions $y(0) = 1$, $y'(0) = 0$, and $y''(0) = 3$ to compute C_1, C_2, and C_3. To do this, we observe that

$$y' = C_2 + C_3 e^x - x^2 + x,$$
$$y'' = C_3 e^x - 2x + 1.$$

Substituting $x = 0$ into y, y', and y'', we find

$$y(0) = C_1 + C_2(0) + C_3 e^0 - \frac{1}{3}(0)^3 + \frac{1}{2}(0)^2 \qquad 1 = C_1 + C_3$$

$$y'(0) = C_2 + C_3 e^0 - (0)^2 + (0) \qquad \text{or} \quad 0 = C_2 + C_3$$

$$y''(0) = C_3 e^0 - 2(0) + 1, \qquad\qquad\qquad 3 = C_3 + 1.$$

Solving the third equation, we see that $C_3 = 2$. Substituting back into the first two equations, we obtain $C_1 = -1$ and $C_2 = -2$.

In conclusion, the solution of this initial-value problem is

$$y = -1 - 2x + 2e^x - \frac{1}{3}x^3 + \frac{1}{2}x^2.$$

Once again, notice that three initial conditions are needed to determine the three constants in the general solution. ∎

In the previous example, notice that the constants A and B were determined by the method of undetermined coefficients. By contrast, the constants C_1, C_2, and C_3 were found by using the initial conditions and *after* finding A and B.

The method of undetermined coefficients described in this section applies to forcing functions that are polynomials. This technique can be modified if the forcing function is an exponential function, a sine function, or a cosine function. It can also be used if the forcing function is the sum of any of these types of functions. However, if the forcing function does not fit any of these forms, other methods exist for finding a particular solution, including a technique called *variation of parameters*. These generalizations for finding particular solutions lie beyond the scope of this text.

Exercise Set 9.2

Solve.

1. $y'' + y = 7$

2. $y'' - 3y' + 2y = 4$

3. $y'' - 2y' + y = 3$

4. $y'' - 7y' + 10y = 10x - 27$

5. $y'' + 4y' + 4y = 8 - 12x$

6. $y'' - 4y' + 5y = 5x + 1$

7. $y'' + 4y' + 3y = 6x^2 - 4$

8. $y'' + 4y = 8x^2 - 12x$

9. $y'' - y' - 2y = x^3 - 1$

10. $y'' + 2y' + y = x^3 + 2x - 1$

11. $y'' - 3y' = 4$

12. $y''' + y' = -2$

13. $y''' + y'' = -2$

14. $y'' - 2y' = -4x$

15. $y''' + 4y'' + 20y' = 40x - 12$

16. $y'' - y' = 3x^2 - 8x + 5$

Solve the initial-value problem.

17. $y'' - y' - 2y = 2x - 1$, $y(0) = 6$, $y'(0) = 0$

18. $y'' + 9y = 9 - 9x$, $y(0) = 3$, $y'(0) = 2$

19. $y'' + 2y' + y = x^2$, $y(0) = 1$, $y'(0) = 0$

20. $y'' - 5y' + 6y = 18x^2 + 12x + 1$, $y(0) = 3$, $y'(0) = 2$

21. $y'' + 4y' = 16x$, $y(0) = 2$, $y'(0) = -3$

22. $y''' + 3y'' = 18x - 18$, $y(0) = 3$, $y'(0) = -1$, $y''(0) = 10$

23. $y'' + 2y' + 2y = 2$, $y(0) = 2$, $y'(0) = 1$

24. $y'' + 4y' + 13y = 26$, $y(0) = 1$, $y'(0) = 0$

25. $y''' + 4y'' + 5y' = 25x - 5$, $y(0) = 1$, $y'(0) = 0$, $y''(0) = 1$

26. $y''' - 6y'' + 9y' = 27x^2$, $y(0) = 2$, $y'(0) = -3$, $y''(0) = 1$

APPLICATIONS

Spring-Mass Systems. The displacement $x(t)$ of a spring from its rest position after t seconds follows the differential equation

$$mx'' + \gamma x' + kx = q(t),$$

where m is the mass of the object attached to the spring, $q(t)$ is the forcing function, and γ and k are the stiffness and damping coefficients, respectively. Suppose that the spring starts at rest, so that $x(0) = 0$ and $x'(0) = 0$. Solve for $x(t)$ given the following conditions.

27. $m = 1$, $k = 16$, $\gamma = 0$, $q(t) = 1$

28. $m = 1$, $k = 4$, $\gamma = 0$, $q(t) = 2$

29. $m = 1$, $k = 5$, $\gamma = 2$, $q(t) = 10$

30. $m = 2$, $k = 50$, $\gamma = 20$, $q(t) = 25t$

31. *Vocal Folds.* When our vocal folds vibrate under normal speech, their initial movement is governed by a second-order differential equation. Let $x(t)$ denote the lateral displacement of the lower vocal folds at time t. Then the equation of motion for the lower vocal folds is

$$mx'' + \frac{1}{c}(x - x_0) = pd.$$

In this equation, m is the mass of the lower vocal folds per unit length, c is the mechanical compliance per unit length, x_0, is the resting position, p is the subglottal pressure, and d is the length of the lower vocal folds.[4]

a) Find the general solution of this differential equation.

b) Assume that the vocal folds are at rest before the outward displacement begins, so that $x(0) = 0$ and $x'(0) = 0$. Solve this initial-value problem.

Parallel Reactions. Suppose a reactant has two isomeric forms A and B. The reactant initially has concentration q and is entirely in form A, but may freely convert from form A to form B and back again. Also, once in form A, the reactant may produce two products C and D. The concentration $F(t)$ of product C after t minutes satisfies the second-order differential equation

$$F'' + (a + b + c + d)F' + b(c + d)F = bcq,$$

where the constants a, b, c, and d describe the rates that A, B, C, and D are produced. Also, $F(0) = 0$ and $F'(0) = cq$ initially.[5]

32. Suppose that $a = 1, b = 1, c = 1, d = 0.5$, and $q = 1$. Find $F(t)$.

33. Suppose that $a = 0, b = 1, c = 0.05, d = 0$, and $q = 1$. Find $F(t)$.

34. Suppose that $a = 1, b = 1, c = 0.1, d = 2$, and $q = 87$. Find $F(t)$.

SYNTHESIS

35. Solve the first-order differential equation $y' + 2y = x^2$ using Theorem 3 of Chapter 8.

36. Repeat Exercise 35 using the method of undetermined coefficients.

---

[4]K. N. Stevens, *Acoustic Phonetics* (Cambridge, MA: MIT Press, 1998).

[5]E. T. Rakitzis and P. Papandreou, "Kinetic analysis of parallel reactions, the initial reactant of which presents with two interconvertible isomeric forms (hydrolysis and hydroxylaminolysis of 6-phosphogluconolactone)," *Journal of Theoretical Biology*, Vol. 200, pp. 427–434 (1999).

9.3 Systems of Linear Differential Equations

OBJECTIVES

■ Solve a system of first-order linear homogeneous differential equations.

■ Solve a first-order nonhomogeneous system with a polynominal forcing function.

In Sections 9.1 and 9.2, we developed methods of solving a differential equation for a single variable y. However, in many applications, we are interested in finding the values of two (or more) quantities, say, x and y, that are functions of yet another variable t (usually thought of as time). For example, if x and y represent the populations of two competing species that share a common habitat, then it is reasonable to assume that the rates of change dx/dt and dy/dt both depend on the current populations x and y.

First-Order Homogeneous Systems

If a, b, c, and d are constants, then we say that the two equations

$$\frac{dx}{dt} = ax + by, \tag{1}$$

$$\frac{dy}{dt} = cx + dy, \tag{2}$$

form a **first-order linear system** of differential equations. Such a system may be solved by using the following procedure.

To solve the system of equations given by (1) and (2), we may use the **reduction method:**

a)   Find x'' by differentiating both sides of (1) with respect to t.
b)   Eliminate the y'-term by using (2).
c)   Eliminate the y-term by using (1).
d)   The result of the previous three steps will be a second-order differential equation for x. This equation may be solved using techniques described in Section 9.1.
e)   Substitute the solution for x into (1) to find y.

We illustrate this procedure with several examples.

EXAMPLE 1   Solve the first-order linear system

$$\frac{dx}{dt} = x + 2y,$$

$$\frac{dy}{dt} = 4x - y.$$

Solution

a) We begin by differentiating both sides of the first equation with respect to t:

$$x'' = x' + 2y'.$$

b) To eliminate the y'-term, we use the second equation:

$$x'' = x' + 2y' = x' + 2(4x - y) = x' + 8x - 2y. \qquad (3)$$

c) To eliminate the y-term, we solve for y by using the first equation again:

$$x' = x + 2y, \quad \text{or} \quad y = \frac{1}{2}(x' - x).$$

Substituting into (3), we find

$$x'' = x' + 8x - 2\left[\frac{1}{2}(x' - x)\right]$$

$$x'' = x' + 8x - x' + x$$

$$x'' - 9x = 0.$$

d) We solve this second-order differential equation using the techniques presented in Section 9.1. The auxiliary equation is $r^2 - 9 = 0$, or $(r - 3)(r + 3) = 0$. The roots are 3 and -3, so that the general solution is

$$x(t) = C_1 e^{3t} + C_2 e^{-3t}. \qquad (4)$$

e) Finally, to find y, we again use (1). After differentiating both sides of (4), we obtain

$$x'(t) = 3C_1e^{3t} - 3C_2e^{-3t}. \tag{5}$$

Substituting (4) and (5) into (1) yields

$$x' = x + 2y$$
$$3C_1e^{3t} - 3C_2e^{-3t} = C_1e^{3t} + C_2e^{-3t} + 2y$$
$$2C_1e^{3t} - 4C_2e^{-3t} = 2y$$
$$C_1e^{3t} - 2C_2e^{-3t} = y(t).$$

In conclusion, the general solution of this system of equations is

$$x(t) = C_1e^{3t} + C_2e^{-3t} \quad \text{and} \quad y(t) = C_1e^{3t} - 2C_2e^{-3t}.$$

This may also be verified by substituting into the system of equations.

Note that, although there is an infinite number of solutions, the solutions for x and y are *coupled*. That is, once x is determined by specifying C_1 and C_2, y is also determined by the same constants.   ∎

EXAMPLE 2   Solve the first-order linear system

$$\frac{dx}{dt} = -x - 3y$$

$$\frac{dy}{dt} = 3x - y.$$

Solution

a) We differentiate the first equation with respect to t, obtaining $x'' = -x' - 3y'$.

b) We substitute the second equation into the result from part (a):

$$x'' = -x' - 3(3x - y) = -x' - 9x + 3y.$$

c) To eliminate the y-term, we use the first equation again to obtain $3y = -x' - x$. Substituting into the result from part (b), we get

$$x'' = -x' - 9x + (-x' - x), \quad \text{or} \quad x'' + 2x' + 10x = 0.$$

d) The auxiliary equation of this second-order equation is $r^2 + 2r + 10 = 0$. We find the roots using the quadratic formula:

$$r = \frac{-2 \pm \sqrt{4 - 4(1)(10)}}{2} = \frac{-2 \pm \sqrt{-36}}{2} = \frac{-2 \pm 6i}{2} = -1 \pm 3i.$$

We conclude that $x(t) = C_1e^{-t}\sin 3t + C_2e^{-t}\cos 3t$.

e) To find y, we first differentiate x:

$$x'(t) = -C_1e^{-t}\sin 3t + 3C_1e^{-t}\cos 3t - C_2e^{-t}\cos 3t - 3C_2e^{-t}\sin 3t$$
$$= (-C_1 - 3C_2)e^{-t}\sin 3t + (3C_1 - C_2)e^{-t}\cos 3t.$$

Substitution into the first equation yields

$$(-C_1 - 3C_2)e^{-t}\sin 3t + (3C_1 - C_2)e^{-t}\cos 3t = -(C_1e^{-t}\sin 3t + C_2e^{-t}\cos 3t) - 3y(t)$$
$$-3C_2e^{-t}\sin 3t + 3C_1e^{-t}\cos 3t = -3y(t)$$
$$C_2e^{-t}\sin 3t - C_1e^{-t}\cos 3t = y(t).$$

The general solution of this system of equations is therefore

$$x(t) = C_1 e^{-t} \sin 3t + C_2 e^{-t} \cos 3t \quad \text{and} \quad y(t) = C_2 e^{-t} \sin 3t - C_1 e^{-t} \cos 3t.$$

As before, these solutions are coupled. ■

EXAMPLE 3   Solve the initial-value problem

$$x' = y, \qquad y' = -x + 2y$$

given the initial conditions $x(0) = 1$ and $y(0) = 3$.

Solution

a) We differentiate the first equation with respect to t, obtaining $x'' = y'$.

b) Substituting the second equation, we find $x'' = -x + 2y$.

c) Since $y = x'$, we arrive at the second-order equation for x:

$$x'' = -x + 2x', \quad \text{or} \quad x'' - 2x' + x = 0.$$

d) The auxiliary equation of this second-order equation is $r^2 - 2r + 1 = 0$, or $(r - 1)^2 = 0$. Since the roots are 1 and 1, the general solution is

$$x(t) = C_1 e^t + C_2 t e^t.$$

e) Using the first equation,

$$y(t) = x'(t) = C_1 e^t + (C_2 e^t + C_2 t e^t) = (C_1 + C_2) e^t + C_2 t e^t.$$

The general solution of this system of equations is therefore

$$x(t) = C_1 e^t + C_2 t e^t \quad \text{and} \quad y(t) = (C_1 + C_2) e^t + C_2 t e^t.$$

f) Finally, we use the initial conditions $x(0) = 1$ and $y(0) = 3$ to solve for C_1 and C_2. Substituting 0 into the general solution, we find

$$x(0) = C_1 e^0 + C_2 (0) e^0 \qquad \qquad 1 = C_1$$
$$\qquad \qquad \qquad \qquad \qquad \qquad \text{or}$$
$$y(0) = (C_1 + C_2) e^0 + C_2 (0) e^0, \qquad 3 = C_1 + C_2.$$

Therefore, $C_1 = 1$ and hence $C_2 = 2$. The solution of this initial-value problem is

$$x(t) = e^t + 2t e^t \quad \text{and} \quad y(t) = 3 e^t + 2t e^t.$$

The graphs of $x(t)$ and $y(t)$ are shown in the margin. ■

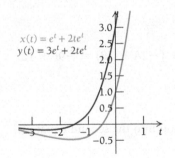

$x(t) = e^t + 2te^t$
$y(t) = 3e^t + 2te^t$

First-Order Nonhomogeneous Systems

If $f(t)$ and $g(t)$ are polynomials in t, then the preceding techniques can be used to solve nonhomogeneous systems that have the form

$$\frac{dx}{dt} = ax + by + f(t),$$

$$\frac{dy}{dt} = cx + dy + g(t).$$

Once again, a, b, c, and d are constants.

EXAMPLE 4   Solve the first-order linear system

$$\frac{dx}{dt} = x + y + 9t,$$

$$\frac{dy}{dt} = 4x + y + 3.$$

Solution

a) We differentiate the first equation with respect to t, obtaining $x'' = x' + y' + 9$.

b) We substitute the second equation into the result from part (a):

$$x'' = x' + (4x + y + 3) + 9 = x' + 4x + y + 12.$$

c) To eliminate the y-term, we use the first equation again to obtain $y = x' - x - 9t$. Substituting into the result from part (b), we get

$$x'' = x' + 4x + (x' - x - 9t) + 12, \quad \text{or} \quad x'' - 2x' - 3x = -9t + 12. \tag{6}$$

d) The second-order equation for x is nonhomogeneous, but it can be solved using the method of undetermined coefficients. We first solve the associated homogeneous differential equation. The auxiliary equation is $r^2 - 2r - 3 = 0$, or $(r - 3)(r + 1) = 0$. Since the roots are 3 and -1, the general solution of the associated homogeneous equation is

$$x_h(t) = C_1 e^{3t} + C_2 e^{-t}.$$

Next, since the forcing function is linear, we try a particular solution of the form

$$x_p = At + B$$

using the method of undetermined coefficients. To determine A and B, we notice that

$$x_p' = A \quad \text{and} \quad x_p'' = 0.$$

Substituting into (6), we find that

$$(0) - 2(A) - 3(At + B) = -9t + 12, \quad \text{or} \quad -3At + (-2A - 3B) = -9t + 12.$$

Matching corresponding coefficients, we obtain

$$-3A = -9 \quad \text{and} \quad -2A - 3B = 12.$$

The first equation implies that $A = 3$, and substitution into the second equation gives $B = -6$. The general solution for x is

$$x(t) = x_h + x_p = C_1 e^{3t} + C_2 e^{-t} + 3t - 6.$$

e) To find y, we first differentiate x:

$$x'(t) = 3C_1 e^{3t} - C_2 e^{-t} + 3.$$

Substitution into the first equation yields

$$3C_1 e^{3t} - C_2 e^{-t} + 3 = (C_1 e^{3t} + C_2 e^{-t} + 3t - 6) + y + 9t$$

$$2C_1 e^{3t} - 2C_2 e^{-t} - 12t + 9 = y(t).$$

In summary, the general solution is

$$x(t) = C_1 e^{3t} + C_2 e^{-t} + 3t - 6 \quad \text{and} \quad y(t) = 2C_1 e^{3t} - 2C_2 e^{-t} - 12t + 9. \quad ■$$

Two-Compartment Models

Systems of differential equations naturally arise in **two-compartment models.** As with a one-compartment model, a two-compartment model may have an uptake rate or a downtake rate that is not proportional to the quantities studied. This model may have relative rates, like a growth rate, decay rate, or elimination rate. It may also have another kind of relative rate, called a **transfer rate,** that governs movement between compartments.

EXAMPLE 5   *Liver Function.* To measure liver function among patients with liver cancer, 3 mg of glactosyl human serum albumin (Tc-GSA) is injected into the bloodstream. Let $B(t)$ and $L(t)$ denote the total amount (in mg) of Tc-GSA present in the blood and liver after t minutes, respectively. After this injection, we assume that Tc-GSA is transferred from the blood into the liver at the rate of 6% per min, and that Tc-GSA is transferred from the liver into the blood at the rate of 3% per min. Also, we assume that no Tc-GSA is eliminated from the body after the injection.[6]

a) Draw a two-compartment model for B and L.

b) Construct a system of differential equations for B and L.

c) Solve for $B(t)$ and $L(t)$.

d) Calculate $\lim_{t \to \infty} B(t)$ and $\lim_{t \to \infty} L(t)$.

Solution

a) The two-compartment model for B and L is shown in the margin.

b) Using the two-compartment model, B and L satisfy the system of equations

$$B' = -0.06B + 0.03L \quad \text{and} \quad L' = 0.06B - 0.03L.$$

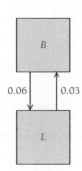

c) We solve this system using the reduction method. Differentiating the first equation with respect to t, we obtain $B'' = -0.06B' + 0.03L'$. Substituting the second equation, we find

$$B'' = -0.06B' + 0.03(0.06B - 0.03L), \quad \text{or} \quad B'' = -0.06B' + 0.0018B - 0.0009L.$$

To eliminate the L-term, we use the first equation again to obtain

$$L = \frac{B' + 0.06B}{0.03}.$$

Substituting, we obtain

$$B'' = -0.06B' + 0.0018B - \frac{0.0009}{0.03}(B' + 0.06B)$$

$$B'' = -0.06B' + 0.0018B - 0.03B' - 0.0018B$$

$$B'' + 0.09B' = 0.$$

---

[6]K. Kinoshita, M. Ukikusa, K. Iwaisako, A. Arimoto, N. Fujisawa, T. Ozaki, H. Tanaka, S. Seo, M. Naitoh, A. Nomura, T. Inomoto, T. Kitai, K. Ino, H. Higashiyama, T. Hanafusa, and Y. Nakajima, "Preoperative assessment of hepatic function: Utility of a new convenient two-compartment model analysis using galactosyl human serum albumin scintigraphy," *Journal of Gastroenterology and Hepatology*, Vol. 18, pp. 99–104 (2003). The technique described in this example may be used to calculate the volume of the liver, as discussed in the exercises.

How are relative rates computed? In Chapter 8, we saw several examples of one-compartment models whose rate constants were measurable using empirical data. Another example of measuring relative rates was presented in the Extended Life Science Connection of Chapter 8.

For an n-compartment model, inferring these rates is somewhat more complicated. Assuming that all uptake and downtake rates are constants and that the auxiliary equation only has distinct, real, nonzero roots, then the general solution will be the sum of n exponential terms and a constant. After the solution is physically measured, regression analysis is used to approximately evaluate the exponential terms. The relative rates may be computed using these exponential terms.

While this method works in principle, the inevitability of measurement error may greatly affect the accuracy of the regression analysis, especially if the exponents differ by a factor of less than two.

The auxiliary equation of this second-order equation is $r^2 + 0.09r = 0$, or $r(r + 0.09) = 0$. Since the roots are 0 and -0.09, the general solution is

$$B(t) = C_1 + C_2 e^{-0.09t}.$$

To find L, we first differentiate B:

$$B'(t) = -0.09C_2 e^{-0.09t}.$$

Substitution into the first equation yields

$$B'(t) = -0.06B(t) + 0.03L(t)$$
$$-0.09C_2 e^{-0.09t} = -0.06(C_1 + C_2 e^{-0.09t}) + 0.03L(t)$$
$$0.06C_1 - 0.03C_2 e^{-0.09t} = 0.03L(t)$$
$$2C_1 - C_2 e^{-0.09t} = L(t).$$

The general solution of this system of equations is therefore

$$B(t) = C_1 + C_2 e^{-0.09t} \quad \text{and} \quad L(t) = 2C_1 - C_2 e^{-0.09t}.$$

To solve for C_1 and C_2, we use the initial conditions $B(0) = 3$ and $L(0) = 0$. Substituting $t = 0$ into the general solution, we obtain

$$B(0) = C_1 + C_2 e^{-0.09(0)} \qquad 3 = C_1 + C_2$$
$$\text{or}$$
$$L(0) = 2C_1 - C_2 e^{-0.09(0)}, \qquad 0 = 2C_1 - C_2.$$

From the second equation, we see that $C_2 = 2C_1$. Substituting into the first equation, we find

$$3 = C_1 + 2C_1, \quad \text{or} \quad C_1 = 1.$$

Substituting back, we find $C_2 = 2$.

In summary, the amounts of Tc-GSA in the blood and liver are given by

$$B(t) = 1 + 2e^{-0.09t} \quad \text{and} \quad L(t) = 2 - 2e^{-0.09t},$$

respectively.

d) As t tends to infinity, the term $e^{-0.09t}$ in both $B(t)$ and $L(t)$ tends to 0. Therefore,

$$\lim_{t \to \infty} B(t) = 1 \quad \text{and} \quad \lim_{t \to \infty} L(t) = 2. \qquad \blacksquare$$

In Example 5, we notice that

$$B(t) + L(t) = (1 + 2e^{-0.09t}) + (2 - 2e^{-0.09t}) = 3,$$

the total amount of Tc-GSA that is injected. In other words, $L = 3 - B$. This makes sense; after the injection, we are assuming that the Tc-GSA is not eliminated from the body during the course of the clinical trial. Since the initial conditions were $(B(0), L(0)) = (3, 0)$ and the limiting point is $(1, 2)$, we conclude that the graph of $(B(t), L(t))$ lies on the line segment shown in the figure. The arrow indicates the direction of time. At all times, the pair $(B(t), L(t))$ lies somewhere on this line segment. This graph is called a *trajectory*.

The point $(1, 2)$ is called an *equilibrium point* of the system. This equilibrium point may be found directly from the original system of differential equations. At equilibrium, $dB/dt = 0$ and $dL/dt = 0$. Substituting into the first equation of the system,

$$0 = -0.06B + 0.03L, \quad \text{or} \quad L = 2B.$$

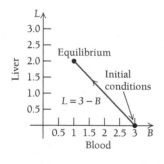

Since the total amount of Tc-GSA is 3 mg, we conclude that $B = 1$ and $L = 2$.
We will discuss equilibrium points and trajectories more in the rest of this chapter.

Exercise Set 9.3

Find the general solution of the system of equations.

1. $x' = y, y' = -2x + 3y$

2. $x' = x + 2y, y' = x + 2y$

3. $x' = 2x + y, y' = 3x + 4y$

4. $x' = 3x - 4y, y' = 2x - 3y$

5. $x' = -0.5x + y, y' = 0.5x$

6. $x' = -x - 3y, y' = -y$

7. $x' = 4x + y, y' = -x + 2y$

8. $x' = 2x - y, y' = 4x - 2y$

9. $x' = 2y, y' = -18x$

10. $x' = x - y, y' = 5x - y$

11. $x' = 3x - 5y, y' = x - y$

12. $x' = -3x + 5y, y' = -x + y$

13. $x' = 5x - y - 5, y' = 2x + 2y + 2$

14. $x' = 2x + y + 1, y' = -2y - 22$

15. $x' = y + 2t + 3, y' = -x + 4t - 2$

16. $x' = -x + y + 2t + 2, y' = 2x + 2t - 1$

Solve the initial-value problem.

17. $x' = 2x - y, y' = 2x + 5y, x(0) = 3, y(0) = -5$

18. $x' = x + y, y' = x + y, x(0) = 2, y(0) = 0$

19. $x' = 2x + 3y, y' = -3x + 8y, x(0) = 1, y(0) = 1$

20. $x' = 2x + 3y, y' = -3x + 8y, x(0) = 1, y(0) = 2$

21. $x' = y, y' = -4x, x(0) = 1, y(0) = 2$

22. $x' = x + 5y, y' = -x - 3y, x(0) = 10, y(0) = -3$

23. $x' = 2x + y + 3, y' = 5x - 2y + 12, x(0) = 6, y(0) = -3$

24. $x' = y + 5, y' = -x + 2y + 10, x(0) = 1, y(0) = -2$

Two-Compartment Models. Find the general solution of the following two-compartment models.

25.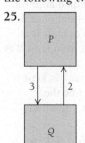

26.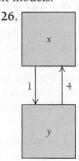

27.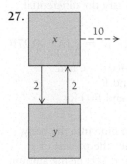

28.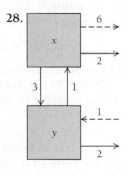

APPLICATIONS

29. *Mixing Chemicals.* Tank A contains 2000 lb of salt dissolved in 1000 gal of water. Tank B contains 1000 lb of salt dissolved in 1000 gal of water. The mixture from tank A is pumped to tank B at the rate of 500 gal per hr, while that from tank B is pumped to tank A at the same rate. Assume that the mixture in each tank is kept uniform by stirring. Let $A(t)$ and $B(t)$ be the amount of salt in tanks A and B after t hours, respectively.

a) Determine the salt transfer rates from tank A to tank B and from tank B to tank A. (*Hint:* If $B(t)$ pounds of salt are dissolved in 1000 gal and 500 gal are pumped to tank A, how many pounds of salt get pumped into tank A?)

b) Draw a two-compartment model for $A(t)$ and $B(t)$.

c) Show that $A(t)$ and $B(t)$ satisfy the differential equations

$$A' = -0.5A + 0.5B \quad \text{and} \quad B' = 0.5A - 0.5B.$$

d) Use the initial conditions $A(0) = 2000$ and $B(0) = 1000$ to solve for A and B.

e) Use a grapher to plot $A(t)$ and $B(t)$ for $0 \le t \le 4$.

f) What are the equilibrium values of A and B?

30. *Mixing Chemicals.* Tank A contains 100 gal of pure water. Tank B contains 33 lb of salt dissolved in 50 gal of water. Pure water is poured into tank B at the rate of 3.5 gal per min while an equal amount of the mixture is drained from the bottom of tank B. The mixture from tank A is pumped to tank B at the rate of 10 gal per min, while that from tank B is pumped to tank A at the same rate.

   Assume that the mixture in each tank is kept uniform by stirring. Let $A(t)$ and $B(t)$ be the amount of salt in tanks A and B after t minutes, respectively.

a) Draw a two-compartment model for $A(t)$ and $B(t)$.

b) Show that $A(t)$ and $B(t)$ satisfy the differential equations

$$A' = -0.1A + 0.2B \quad \text{and} \quad B' = 0.1A - 0.27B.$$

c) Use the initial conditions $A(0) = 0$ and $B(0) = 33$ to solve for A and B.

d) Use a grapher to plot $A(t)$ and $B(t)$ for $0 \le t \le 50$.

31. *Zinc Depletion.* After intake of zinc into the body, the zinc may be found in either the plasma or a portion of the liver. Let $P(t)$ and $L(t)$ be the amount of zinc in the plasma and liver after t days, respectively. The transfer rate of zinc from the plasma to the liver is 3 per day. The transfer rate from the liver to the plasma is 0.6 per day. Finally, the zinc is removed from the body via plasma at a rate of 2.24 per day.[7]

a) Draw a two-compartment model for L and P.

b) Find a system of differential equations satisfied by L and P.

c) Solve for $P(t)$ and $L(t)$ given that $P(0) = 0$ and $L(0) = 241/15$.

32. *Home Air Quality.* Chemicals enter a house's basement air at the rate of 0.1 mg per min. Let $F(t)$ and $B(t)$ denote the total amount of chemical present in the first-story air and the basement air after t minutes, respectively. Both the first floor and the basement have volumes of 200 m^3. Air flows from the basement into the first floor at the rate of 2 m^3 per min, while air flows through the first floor to the outside at the rate of 4 m^3 per min. Uncontaminated air from the outside replenishes the air that flows out of the house.[8]

a) Draw a two-compartment model for B and F.

b) Show that $B(t)$ and $F(t)$ satisfy the system of equations

$$B' = -0.01B + 0.1 \quad \text{and} \quad F' = 0.01B - 0.02F.$$

c) Suppose that no chemicals are initially present in either floor. Solve for $B(t)$ and $F(t)$.

d) Show that the equilibrium values of B and F are 10 mg and 5 mg, respectively.

SYNTHESIS

33. *Home Air Quality.* This is a continuation of Exercise 32.

a) Assume that the first story and basement are initially uncontaminated. Show that $F(t) = [B(t)]^2/20$.

b) Verify that the initial conditions and the equilibrium values of B and F satisfy this equation.

c) Make a sketch of the trajectory of B and F.

[7]K. Pinna, L. R. Woodhouse, B. Sutherland, D. M. Shames, and J. C. King, "Exchangable zinc pool masses and turnover are maintained in healthy men with low zinc intakes," *Journal of Nutrition*, Vol. 131, pp. 2288–2294 (2001).

[8]D. A. Olson and R. L. Corsi, "Characterizing exposure to chemicals from soil vapor intrusion using a two-compartment model," *Atmospheric Environment*, Vol. 35, pp. 4201–4209 (2001).

Leukocytes. Blood leukocytes may either be circulating or marginated. Let $L(t)$ and $M(t)$ be the number of circulating and marginated leukocytes after t hours, respectively. Circulating leukocytes are created by bone marrow with an uptake rate of p. Let c be the transfer rate at which leukocytes change from marginated to circulating, and let m be the transfer rate at which they change from circulating to marginated. Also, let a be the elimination rate at which both circulating and marginated leukocytes are absorbed by surrounding tissue.[9]

34. Draw a two-compartment model for L and M.

35. Determine a nonhomogeneous system of differential equations for L and M.

36. Show that L satisfies the second-order differential equation

$$L'' + (c + m + 2a)L' + a(c + m + a)L = (c + a)p.$$

37. Show that a particular solution of this nonhomogeneous equation is the constant

$$L_p = \frac{(c + a)p}{a(c + m + a)}.$$

38. Show that this particular solution is the equilibrium value of L. (*Hint:* What are L' and L'' at equilibrium?)

39. Let $N(t)$ be the difference of the number of circulating leukocytes from its equilibrium value. That is,

$$N(t) = L(t) - \frac{(c + a)p}{a(c + m + a)}.$$

Show that N is a solution of the homogeneous differential equation given in Exercise 44 of Exercise Set 9.1.

40. *Liver Function.* In this exercise, we present an alternate way of solving Example 5. With this technique, B may be found by solving a *first*-order differential equation.

a) Use the differential equations to show that $B + L$ is a constant. (*Hint:* What is $B' + L'$?)

b) Explain why $B(t) + L(t) = 3$.

c) Substitute $L = 3 - B$ into the differential equation for B. Show that $B' + 0.09B = 0.09$.

d) Use the initial condition $B(0) = 3$ to solve for B. Then use part (b) to solve for L. Your answers should match those found in Example 5.

Liver Function. The following exercises are an extension of Example 5. Suppose that an initial dosage of D mg of Tc-GSA is injected into the blood, while no Tc-GSA is present in the liver. Let p denote the transfer rate of Tc-GSA from the blood into the liver, and let q denote the transfer rate of Tc-GSA from the liver into the blood. As shown below, the volume of the liver and its ability to function can be measured using laboratory data using a two-compartment model.

41. Draw a two-compartment model for $B(t)$ and $L(t)$.

42. Solve for $B(t)$ and $L(t)$.

43. Let $b(t)$ be the *concentration* of Tc-GSA in the bloodstream. Thus, $b(t) = B(t)/V_B$, where V_B is the total blood volume (in cm^3). Show that $b(t)$ has the form

$$b(t) = C_1 + C_2 e^{-rt},$$

where C_1, C_2, and r are constants.

44. Solve for C_1, C_2, and r in terms of p, q, D, and V_B.

45. Show that

$$p = \frac{C_2 r}{C_1 + C_2}, \quad q = \frac{C_1 r}{C_1 + C_2}, \quad \text{and} \quad V_B = \frac{D}{C_1 + C_2}.$$

In other words, the transfer rates can be measured based on laboratory data.

46. Let V_L denote the volume of the liver. At equilibrium, $V_B/V_L = q/p$. Show that[10]

$$V_L = \frac{C_2 D}{C_1(C_1 + C_2)}.$$

---

[9]K. Iadocicco, L. H. A. Monteiro, and J. G. Chaui-Berlinck, "A theoretical model for estimating the margination constant of leukocytes," *BMC Physiology,* Vol. 3, article 2 (2002).

[10]K. Kinoshita, M. Ukikusa, K. Iwaisako, A. Arimoto, N. Fujisawa, T. Ozaki, H. Tanaka, S. Seo, M. Naitoh, A. Nomura, T. Inomoto, T. Kitai, K. Ino, H. Higashiyama, T. Hanafusa, and Y. Nakajima, "Preoperative assessment of hepatic function: Utility of a new convenient two-compartment model analysis using galactosyl human serum albumin scintigraphy," *Journal of Gastroenterology and Hepatology,* Vol. 18, pp. 99–104 (2003). Laboratory technicians can approximate C_1, C_2, and r based on blood tests taken over time. Therefore, by using the result of Example 5, the volume of the liver can be measured based on blood tests. Furthermore, the efficiency of the liver, which is measured by p and q, may also be measured using the result of Exercise 46.

9.4 Matrices and Trajectories

OBJECTIVES

■ Use eigenvalues and eigenvectors to solve systems of differential equations.
■ Sketch the trajectories of solutions.

In Section 9.3, we developed the reduction method of solving the system of linear differential equations

$$x' = ax + by,$$
$$y' = cx + dy.$$

This system of equations may also be solved by using matrices. Notice that the equations correspond to the two rows of the matrix equation

$$\begin{bmatrix} x \\ y \end{bmatrix}' = \begin{bmatrix} a & b \\ c & d \end{bmatrix} \begin{bmatrix} x \\ y \end{bmatrix}$$

by using the rules of matrix multiplication. The derivative symbol on the left side indicates taking the derivative of the entries within the matrix. This system may be succinctly written as

$$z' = Az, \quad \text{where } z = \begin{bmatrix} x \\ y \end{bmatrix} \quad \text{and} \quad A = \begin{bmatrix} a & b \\ c & d \end{bmatrix}.$$

This is called the *matrix form* of a system of differential equations.

EXAMPLE 1   Verify that the functions $x(t) = e^t, y(t) = 4e^t - e^{2t}$ solve the system of differential equations

$$\begin{bmatrix} x \\ y \end{bmatrix}' = \begin{bmatrix} 1 & 0 \\ -4 & 2 \end{bmatrix} \begin{bmatrix} x \\ y \end{bmatrix}.$$

Solution   The left-hand side of the equation is

$$\begin{bmatrix} x' \\ y' \end{bmatrix} = \begin{bmatrix} (e^t)' \\ (4e^t - e^{2t})' \end{bmatrix} = \begin{bmatrix} e^t \\ 4e^t - 2e^{2t} \end{bmatrix}.$$

Using matrix multiplication, the right-hand side is

$$\begin{bmatrix} 1 & 0 \\ -4 & 2 \end{bmatrix} \begin{bmatrix} e^t \\ 4e^t - e^{2t} \end{bmatrix} = \begin{bmatrix} (1)(e^t) + (0)(4e^t - e^{2t}) \\ (-4)(e^t) + (2)(4e^t - e^{2t}) \end{bmatrix} = \begin{bmatrix} e^t \\ 4e^t - 2e^{2t} \end{bmatrix}.$$

The two sides agree, so x and y solve this system of differential equations.   ■

Eigenvalues and Eigenvectors

We have seen that the general solution of $y' = ky$ is $y = Ce^{kx}$. Therefore, to solve the system of differential equations $z' = Az$, let's attempt to find a solution of the form

$$z = \begin{bmatrix} ae^{rt} \\ be^{rt} \end{bmatrix} = e^{rt}v, \quad \text{where } v = \begin{bmatrix} a \\ b \end{bmatrix} \neq \begin{bmatrix} 0 \\ 0 \end{bmatrix}.$$

Taking the derivative, we find

$$z' = \begin{bmatrix} are^{rt} \\ bre^{rt} \end{bmatrix} = re^{rt} \begin{bmatrix} a \\ b \end{bmatrix} = re^{rt}v.$$

Therefore, if z is a solution of the system of differential equations, then

$$re^{rt}\mathbf{v} = \mathbf{A}e^{rt}\mathbf{v}$$

$$r\mathbf{v} = \mathbf{A}\mathbf{v}. \qquad \text{Since } e^{rt} \text{ is never } 0$$

This is the same equation that we used in Chapter 6 to find the eigenvalues and eigenvectors of the matrix $\mathbf{A}$. In other words, if r is an eigenvalue of $\mathbf{A}$ with associated eigenvector $\mathbf{v}$, then $z = \mathbf{v}e^{rt}$ is a solution of the system of differential equations. This observation is the basis of the **matrix method** of solving systems of differential equations.

THEOREM 4

Suppose that $r_1, r_2, \ldots, r_n$ are distinct real eigenvalues of the n-by-n matrix $\mathbf{A}$ with associated eigenvectors $\mathbf{v}_1, \mathbf{v}_2, \ldots, \mathbf{v}_n$, respectively. Then the general solution of the system of differential equations $z' = \mathbf{A}z$ is given by

$$z = C_1 e^{r_1 t}\mathbf{v}_1 + C_2 e^{r_2 t}\mathbf{v}_2 + \cdots + C_n e^{r_n t}\mathbf{v}_n.$$

We will see later how to use the matrix method if there are repeated or complex eigenvalues.

EXAMPLE 2   Solve the first-order linear system

$$x' = -4x + y, \qquad y' = 2x - 3y.$$

Solution   This system may be written in matrix form as

$$\begin{bmatrix} x \\ y \end{bmatrix}' = \begin{bmatrix} -4 & 1 \\ 2 & -3 \end{bmatrix}\begin{bmatrix} x \\ y \end{bmatrix}, \quad \text{so that} \quad \mathbf{A} = \begin{bmatrix} -4 & 1 \\ 2 & -3 \end{bmatrix}.$$

We use Theorem 9 in Chapter 6 to find the eigenvalues of $\mathbf{A}$. The trace of $\mathbf{A}$ is $-4 - 3 = -7$, while the determinant is $(-4)(-3) - (1)(2) = 10$. Therefore, we must solve

$$r^2 - (-7)r + (10) = 0, \quad \text{or} \quad (r + 2)(r + 5) = 0.$$

The eigenvalues are therefore -2 and -5.

Using Theorem 9 of Chapter 6, the eigenvectors associated with -2 are nonzero multiples of $\begin{bmatrix} 1 \\ (-2) - (-4) \end{bmatrix} = \begin{bmatrix} 1 \\ 2 \end{bmatrix}$. Also, the eigenvectors associated with -5 are nonzero multiples of $\begin{bmatrix} 1 \\ (-5) - (-4) \end{bmatrix} = \begin{bmatrix} 1 \\ -1 \end{bmatrix}$. Therefore, by Theorem 4, the general solution of this system is

$$z = C_1 e^{-2t}\begin{bmatrix} 1 \\ 2 \end{bmatrix} + C_2 e^{-5t}\begin{bmatrix} 1 \\ -1 \end{bmatrix}, \quad \text{or} \quad \begin{bmatrix} x \\ y \end{bmatrix} = \begin{bmatrix} C_1 e^{-2t} + C_2 e^{-5t} \\ 2C_1 e^{-2t} - C_2 e^{-5t} \end{bmatrix}.$$

Therefore, $x(t) = C_1 e^{-2t} + C_2 e^{-5t}$ and $y(t) = 2C_1 e^{-2t} - C_2 e^{-5t}$.   ∎

Relationships Between the Reduction and Matrix Methods

The reduction method and the matrix method may both be used to solve systems of differential equations. Not surprisingly, the two methods give the same solution.

EXAMPLE 3   Solve the first-order linear system

$$x' = x + 2y, \qquad y' = 4x - y.$$

This system was originally considered in Example 1 of Section 9.3.

Solution   This system may be written in matrix form as

$$\begin{bmatrix} x \\ y \end{bmatrix}' = \begin{bmatrix} 1 & 2 \\ 4 & -1 \end{bmatrix} \begin{bmatrix} x \\ y \end{bmatrix}, \quad \text{so that } A = \begin{bmatrix} 1 & 2 \\ 4 & -1 \end{bmatrix}.$$

The trace of A is $1 + (-1) = 0$, while the determinant is $(1)(-1) - (2)(4) = -9$. Using Theorem X of Chapter 6, the eigenvalues are found by solving

$$r^2 - (0)r + (-9) = 0, \quad \text{or} \quad (r-3)(r+3) = 0.$$

Therefore, the eigenvalues are 3 and -3.

To find the eigenvectors, we again use Theorem 9 of Chapter 6. The eigenvectors associated with 3 are nonzero multiples of $\begin{bmatrix} 2 \\ 3-1 \end{bmatrix} = \begin{bmatrix} 2 \\ 2 \end{bmatrix}$, while the eigenvectors associated with -3 are nonzero multiples of $\begin{bmatrix} 2 \\ -3-1 \end{bmatrix} = \begin{bmatrix} 2 \\ -4 \end{bmatrix}$. For simplicity, we choose $\begin{bmatrix} 1 \\ 1 \end{bmatrix}$ and $\begin{bmatrix} 1 \\ -2 \end{bmatrix}$, respectively, as representative eigenvectors.

Using Theorem 4, the general solution of this system is

$$z = C_1 e^{3t} \begin{bmatrix} 1 \\ 1 \end{bmatrix} + C_2 e^{-3t} \begin{bmatrix} 1 \\ -2 \end{bmatrix}, \quad \text{or} \quad \begin{bmatrix} x \\ y \end{bmatrix} = \begin{bmatrix} C_1 e^{3t} + C_2 e^{-3t} \\ C_1 e^{3t} - 2C_2 e^{-3t} \end{bmatrix}.$$

Therefore, $x(t) = C_1 e^{3t} + C_2 e^{-3t}$ and $y(t) = C_1 e^{3t} - 2C_2 e^{-3t}$.   ∎

Looking back, we see that this solution agrees with the solution found in Example 1 of Section 9.3, where the reduction method was used. To see the relationship between the reduction and matrix methods, let's take a closer look at these two different approaches to the solution. Using the reduction method, we found that the auxiliary equation was $r^2 - 9 = 0$. This was also the characteristic equation when we used the matrix method. This equivalence between the two methods generally holds for systems of differential equations.

THEOREM 5

When solving a first-order linear system, the auxiliary equation found using the reduction method is the same as the characteristic equation found using the matrix method.

Theorem 5 allows us to solve first-order linear systems even if the eigenvalues are either repeated or complex.

EXAMPLE 4   Solve the first-order linear system

$$x' = 4x + 2y, \qquad y' = -10x - 4y.$$

Solution   The trace of $\begin{bmatrix} 4 & 2 \\ -10 & -4 \end{bmatrix}$ is $4 + (-4) = 0$, while the determinant is $(4)(-4) - (2)(-10) = 4$. We must therefore solve the equation $r^2 + 4 = 0$. We find

$$r = \pm\sqrt{-4} = \pm 2i.$$

Since the eigenvalues are complex, we cannot directly use Theorem 4 to solve this system.

Nevertheless, according to Theorem 5 and the discussion from Section 9.3, we know that the general solution of x must be

$$x(t) = C_1 \sin 2t + C_2 \cos 2t.$$

Therefore, we may solve for y using the first equation of the system. To begin, we find that $x'(t) = 2C_1 \cos 2t - 2C_2 \sin 2t$. Substituting into the first equation yields

$$x' = 4x + 2y$$
$$2C_1 \cos 2t - 2C_2 \sin 2t = 4(C_1 \sin 2t + C_2 \cos 2t) + 2y$$
$$(-4C_1 - 2C_2) \sin 2t + (2C_1 - 4C_2) \cos 2t = 2y$$
$$(-2C_1 - C_2) \sin 2t + (C_1 - 2C_2) \cos 2t = y(t). \qquad\blacksquare$$

In this example, we see that direct computation of the eigenvalues may replace steps (a) through (c) of the reduction method.

Trajectories

Let's return to the system $x' = x + 2y,\, y' = 4x - y$, which we considered in Example 3. The general solution was found to be

$$x(t) = C_1 e^{3t} + C_2 e^{-3t} \quad \text{and} \quad y(t) = C_1 e^{3t} - 2C_2 e^{-3t}.$$

For example, if $C_1 = 1$ and $C_2 = 2$, then one solution of this system is $x(t) = e^{3t} + 2e^{-3t}$ and $y(t) = e^{3t} - 4e^{-3t}$. The graphs of these functions are shown below. Also shown is a graph of the points $(x(t), y(t))$. For example, since $x(0) = 3$ and $y(0) = -3$, the point $(3, -3)$ is plotted. The set of all points $(x(t), y(t))$ is called the **trajectory** of the solution.

t	$x(t) = e^{3t} + 2e^{-2t}$	$y(t) = e^{3t} - 4e^{-3t}$
-0.4	4.752	-12.979
-0.2	3.532	-6.740
0	3.000	-3.000
0.2	3.163	-0.373
0.4	4.219	2.115
0.6	6.652	5.388
0.8	11.427	10.660

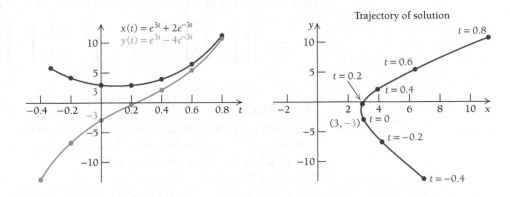

Technology Connection

Plotting Trajectories

To plot the trajectory of $x(t) = e^{3t} + 2e^{-3t}$ and $y(t) = e^{3t} - 4e^{-3t}$, we first set the grapher in parametric mode. We then enter the window settings; these specify the minimum and maximum values for x, y, and t. We then enter x and y as functions of t. The plot generated by the grapher is shown below.

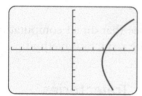

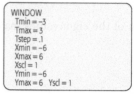

EXERCISES

1. Use a grapher to plot the trajectory $x(t) = e^{3t}$, $y(t) = e^{3t}$.

2. Use a grapher to plot the trajectory $x(t) = e^{-3t}$, $y(t) = -2e^{-3t}$.

3. Choose several pairs of C_1 and C_2 to plot several trajectories of

$$x(t) = C_1 e^{3t} + C_2 e^{-3t},$$
$$y(t) = C_1 e^{3t} - 2C_2 e^{-3t},$$

on the same graph. Your graph should be similar to the one on page 627.

EXAMPLE 5 Find the trajectory if $C_1 = 0$ and $C_2 = 0$.

Solution In this case, $x(t) = 0$ and $y(t) = 0$. The trajectory would simply be the origin. ∎

If a system of differential equations has a constant solution, the trajectory of that solution is called an **equilibrium point**. It's easy to verify that the origin is always an equilibrium point of the linear system of equations $z' = Az$:

$$z' = \begin{bmatrix} 0 \\ 0 \end{bmatrix}' = \begin{bmatrix} 0 \\ 0 \end{bmatrix}, \quad \text{and} \quad Az = \begin{bmatrix} a & b \\ c & d \end{bmatrix} \begin{bmatrix} 0 \\ 0 \end{bmatrix} = \begin{bmatrix} 0 \\ 0 \end{bmatrix}.$$

THEOREM 6

The origin is an equilibrium point of the linear system of differential equations $z' = Az$. Furthermore, if A only has nonzero eigenvalues, then the origin is the only equilibrium point.

Stability of the Origin

In Section 8.3, we developed techniques for assessing the stability of equilibrium values of an autonomous differential equation. We were also able to sketch solutions using this stability information.

A similar analysis may be conducted for the first-order linear system $z' = Az$ using the eigenvalues and eigenvectors of A. The stability of the equilibrium point at the origin depends on the nature of the eigenvalues of A. We illustrate this by reexamining previous examples.

A Positive Eigenvalue and a Negative Eigenvalue

We begin with the system $x' = x + 2y$, $y' = 4x - y$, which we solved in Example 3. In matrix form, the general solution of this system is

$$\begin{bmatrix} x \\ y \end{bmatrix} = C_1 e^{3t} \begin{bmatrix} 1 \\ 1 \end{bmatrix} + C_2 e^{-3t} \begin{bmatrix} 1 \\ -2 \end{bmatrix}.$$

Saddle point unstable

If we set $C_2 = 0$, we obtain a solution that is a multiple of the eigenvector $\begin{bmatrix} 1 \\ 1 \end{bmatrix}$. The trajectories of such solutions are depicted by the blue line in the figure in the margin. Notice that the arrows point *away* from the origin since e^{3t} grows exponentially.

On the other hand, if we set $C_1 = 0$, we obtain a solution that is a multiple of the eigenvector $\begin{bmatrix} 1 \\ -2 \end{bmatrix}$. Such solutions are depicted by the green line in the figure. Notice that the arrows point *toward* the origin; since e^{-3t} decays exponentially, both x and y approach 0 as t increases.

The other trajectories are hyperbolic in shape and are shown with red curves. For large positive values of t, the e^{-3t}-term converges to 0, and so the solution will get closer and closer to the blue line. On the other hand, for large negative values of t, the e^{3t}-term converges to 0, and the solution will get closer and closer to the green line.

Notice that the trajectories can be sketched by hand once the eigenvalues and eigenvectors are computed. We also observe that different trajectories do not cross. As a consequence, the red trajectories do not actually touch either the blue or green lines.

If the eigenvalues have opposite signs, the origin is called a **saddle point.** Also, since almost all trajectories diverge from the origin as t increases, we say that the origin is an **unstable** equilibrium.

Two Distinct Positive Eigenvalues or Two Distinct Negative Eigenvalues

Second, we consider the system $x' = -4x + y$, $y' = 2x - 3y$, which was solved in Example 2. The general solution was found to be

$$\begin{bmatrix} x \\ y \end{bmatrix} = C_1 e^{-2t} \begin{bmatrix} 1 \\ 2 \end{bmatrix} + C_2 e^{-5t} \begin{bmatrix} 1 \\ -1 \end{bmatrix}.$$

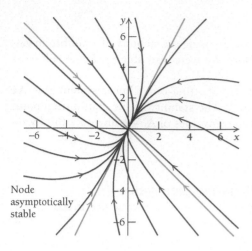

Node
asymptotically
stable

Lines corresponding to $\begin{bmatrix} 1 \\ 2 \end{bmatrix}$ and $\begin{bmatrix} 1 \\ -1 \end{bmatrix}$ are shown by the green and blue lines, respectively. Since both -2 and -5 are negative, the arrows along these lines point toward the origin.

Other trajectories are shown in red. These trajectories bend approximately along the green line close to the origin. To understand why this occurs, we recall that, for large positive values of t, e^{-2t} will be larger than e^{-5t}. Therefore, as t increases, the term involving e^{-5t} decreases quickly to 0. Accordingly, the solutions will approach the green line as trajectories get close to the origin.

Far away from the origin, the trajectories are approximately parallel with the blue line. For large negative values of t, the term with e^{-5t} will be the larger of the two terms and hence will dominate the behavior of the trajectory. Once again, after the eigenvalues and eigenvectors are determined, a reasonable sketch of the trajectories can be made by hand and without the assistance of a grapher.

Since the origin describes the limiting behavior of the trajectories, no nonconstant trajectory reaches the origin for finite values of t. Therefore, different trajectories do not intersect, even though they appear to intersect at the origin.

If the eigenvalues are distinct but have the same sign, the origin is called a **node**. In this example, since all trajectories point toward the origin, we say that the origin is an **asymptotically stable** equilibrium. On the other hand, if the eigenvalues had been both positive, the trajectories would point outward instead of inward. In this case, the origin would have been an unstable equilibrium point.

Two Complex Eigenvalues

Third, we consider the system $x' = -x - 3y$, $y' = 3x - y$. We showed in Example 2 of Section 9.3 that the roots of the auxiliary equation (and hence the eigenvalues) are the complex numbers $-1 \pm 3i$. We also showed that the general solution is

$$x(t) = C_1 e^{-t} \sin 3t + C_2 e^{-t} \cos 3t \quad \text{and} \quad y(t) = C_2 e^{-t} \sin 3t - C_1 e^{-t} \cos 3t.$$

The term e^{-t} indicates exponential decay. Also, the terms $\sin 3t$ and $\cos 3t$ indicate rotational behavior, like the motion of a point around a circle. It is not too surprising, then, that the trajectories are inward spirals, as shown in the figure below. In this case, the origin is called a **spiral point**. Since the spirals point inward, the origin is an asymptotically stable equilibrium point.

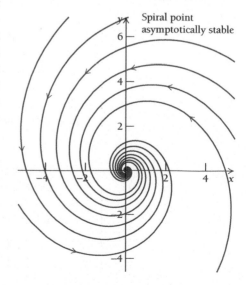

Spiral point
asymptotically stable

The spirals may be either circular or elliptical in shape. For this reason, it may be difficult to draw an accurate sketch by hand if the eigenvalues are complex. Nevertheless, we may determine if the rotation is clockwise or counterclockwise by substituting $x = 1$ and $y = 0$ into the system of differential equations:

$$x' = -(1) - 3(0) = -1 < 0, \quad \text{and} \quad y' = 3(1) - (0) = 3 > 0.$$

Therefore, at the point $(1, 0)$, the x-coordinate of the trajectory is decreasing while the y-coordinate is increasing. For this example, this corresponds to counterclockwise rotation. (For other examples, it may be necessary to substitute other values of x and y to determine the direction of the rotation.)

If the eigenvalues are complex with positive real part, then the trajectories will be *outward* spirals. The origin then would be an unstable equilibrium point.

Two Purely Imaginary Eigenvalues

A fourth type of behavior occurs for systems with purely imaginary values. To see this, we consider the system $x' = 4x + 2y, y' = -10x - 4y$. In Example 4, we showed that the general solution of this system is

$$x(t) = C_1 \sin 2t + C_2 \cos 2t \quad \text{and} \quad y(t) = (-2C_1 - C_2) \sin 2t + (C_1 - 2C_2) \cos 2t.$$

The terms $\sin 2t$ and $\cos 2t$ indicate rotational behavior; however, there is no exponential term to indicate either growth or decay. In this case, the trajectories are ellipses centered at the origin, which in this case is called a **center** and is a **stable** equilibrium point. (Notice that "stable" is not synonymous with "asymptotically stable.")

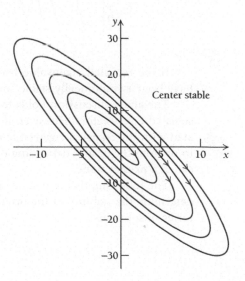

It is possible for the trajectories corresponding to a center to be circles instead of ellipses. Furthermore, the direction of the rotation may be determined by substituting $x = 1$ and $y = 0$ as before:

$$x' = 4(1) + 2(0) = 4 > 0, \quad \text{and} \quad y' = -10(1) - 4(0) = -10 < 0.$$

At the point $(1, 0)$, the x-coordinate is increasing while the y-coordinate is decreasing. For this example, this corresponds to clockwise rotation.

One Repeated Eigenvalue

Finally, we consider the first-order linear system $x' = y, y' = -x + 2y$. We showed in Example 3 of Section 9.3 that the general solution is

$$x(t) = C_1 e^t + C_2 t e^t \quad \text{and} \quad y(t) = (C_1 + C_2)e^t + C_2 t e^t.$$

In the exercises, we will show that 1 is a repeated eigenvalue and has one direction of eigenvectors. In this case, the origin is called an **improper node.** Since the eigenvalue is positive, this improper node is unstable. The trajectories are hard to draw without the aid of a grapher.

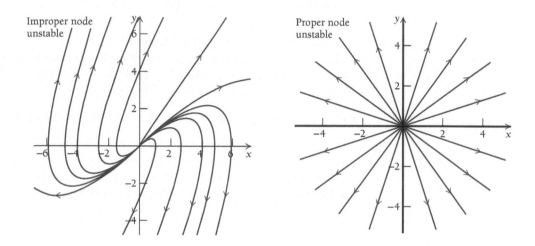

If the repeated eigenvalue had been negative, then the improper node would have been asymptotically stable, and the trajectories would point toward the origin.

The above discussion holds for a repeated eigenvalue with one direction of associated eigenvectors. However, if $\mathbf{A} = r\mathbf{I}$, then *all* vectors are eigenvectors associated with the repeated eigenvalue r. The trajectories would be lines pointing either to or from the origin, depending on the sign of r. In this case, the origin would be called a **proper node.**

The following theorem summarizes how the eigenvalues and eigenvectors of $\mathbf{A}$ determine the stability of the origin, the classification of the origin, and the trajectories of solutions.

THEOREM 7

Suppose that the eigenvalues r_1 and r_2 of **A** are nonzero and have associated eigenvectors $\mathbf{v}_1$ and $\mathbf{v}_2$, respectively. Then the equilibrium point at the origin is unstable if at least one eigenvalue is positive or is complex with positive real part. On the other hand, the origin is asymptotically stable if both eigenvalues are negative or are complex with negative real part.

The classification of the origin and the trajectories of the solutions may be described as follows:

Eigenvalues	Stability	Origin	Trajectories
$0 < r_2 < r_1$	Unstable	Node	Along $\mathbf{v}_2$ near the origin and nearly parallel to $\mathbf{v}_1$ far from the origin
$0 < r_1 = r_2$, $\mathbf{A} \neq r_1\mathbf{I}$	Unstable	Improper node	Difficult to draw by hand
$0 < r_1 = r_2$, $\mathbf{A} = r_1\mathbf{I}$	Unstable	Proper node	Lines pointing away from the origin
$r_2 < 0 < r_1$	Unstable	Saddle point	Hyperbolic in shape, pointing toward the line corresponding to $\mathbf{v}_1$
$r_2 < r_1 < 0$	Asymptotically stable	Node	Along $\mathbf{v}_1$ near the origin and nearly parallel to $\mathbf{v}_2$ far from the origin
$r_1 = r_2 < 0$, $\mathbf{A} \neq r_1\mathbf{I}$	Asymptotically stable	Improper node	Difficult to draw by hand
$r_1 = r_2 < 0$, $\mathbf{A} = r_1\mathbf{I}$	Asymptotically stable	Proper node	Lines pointing toward the origin
$r_1, r_2 = a \pm bi$, $a > 0$	Unstable	Spiral point	Outward spirals, either clockwise or counterclockwise
$r_1, r_2 = a \pm bi$, $a < 0$	Asymptotically stable	Spiral point	Inward spirals, either clockwise or counterclockwise
$r_1, r_2 = \pm bi$	Stable	Center	Concentric ellipses or circles, either clockwise or counterclockwise

Exercise Set 9.4

Rewrite the system of differential equations into matrix form.

1. $x' = x - y, y' = 3x - 2y$
2. $x' = y, y' = 2x$
3. $x' = 4x + 2y, y' = y$
4. $x' = -x + 3y, y' = 3x + y$

Write as two differential equations.

5. $\begin{bmatrix} x \\ y \end{bmatrix}' = \begin{bmatrix} 1 & 3 \\ 5 & 7 \end{bmatrix} \begin{bmatrix} x \\ y \end{bmatrix}$

6. $\begin{bmatrix} x \\ y \end{bmatrix}' = \begin{bmatrix} 2 & -3 \\ -1 & 5 \end{bmatrix} \begin{bmatrix} x \\ y \end{bmatrix}$

7. $\begin{bmatrix} x \\ y \end{bmatrix}' = \begin{bmatrix} 0 & 3 \\ 1 & -2 \end{bmatrix} \begin{bmatrix} x \\ y \end{bmatrix}$

8. $\begin{bmatrix} x \\ y \end{bmatrix}' = \begin{bmatrix} 4 & 0 \\ 2 & -3 \end{bmatrix} \begin{bmatrix} x \\ y \end{bmatrix}$

Verify by substitution that the given functions solve the system of differential equations.

9. $\begin{bmatrix} x \\ y \end{bmatrix}' = \begin{bmatrix} 4 & 1 \\ -2 & 1 \end{bmatrix} \begin{bmatrix} x \\ y \end{bmatrix}$;
$x = 2e^{3t} - e^{2t}, y = -2e^{3t} + 2e^{2t}$

10. $\begin{bmatrix} x \\ y \end{bmatrix}' = \begin{bmatrix} -4 & -3 \\ 6 & 5 \end{bmatrix} \begin{bmatrix} x \\ y \end{bmatrix}$;
$x = -e^{2t} + 2e^{-t}, y = -2e^{2t} - 2e^{-t}$

11. $\begin{bmatrix} x \\ y \end{bmatrix}' = \begin{bmatrix} -5 & -4 \\ 10 & 7 \end{bmatrix} \begin{bmatrix} x \\ y \end{bmatrix}$;
$x = -2e^t \sin 2t, y = 3e^t \sin 2t + e^t \cos 2t$

12. $\begin{bmatrix} x \\ y \end{bmatrix}' = \begin{bmatrix} -1 & -2 \\ 1 & 1 \end{bmatrix} \begin{bmatrix} x \\ y \end{bmatrix}$;
$x = -3 \sin t + \cos t, y = 2 \sin t + \cos t$

Determine the eigenvalues for the system of differential equations. If the eigenvalues are real and distinct, find the general solution by determining the associated eigenvectors. If the eigenvalues are complex or repeated, solve using the reduction method.

13. $x' = -2y, y' = x + 3y$
14. $x' = 4x + y, y' = -2x + y$
15. $x' = 2x + 4y, y' = 3x - 2y$
16. $x' = x - 2y, y' = -4x - y$
17. $x' = -2x + 4y, y' = -x - 7y$
18. $x' = -10x + 4y, y' = -3x - 2y$

19. $x' = -5x + 10y, y' = -4x + 7y$
20. $x' = -2x - 2y, y' = x$
21. $x' = -y, y' = x + 2y$
22. $x' = x + y, y' = -9x - 5y$

Sketch trajectories of the solutions of $z' = Az$ given the eigenvalues and eigenvectors of A. Classify the origin and assess its stability.

23. $r_1 = 3, r_2 = -2, v_1 = \begin{bmatrix} 1 \\ 2 \end{bmatrix}, v_2 = \begin{bmatrix} -3 \\ 1 \end{bmatrix}$

24. $r_1 = 1, r_2 = -1, v_1 = \begin{bmatrix} 1 \\ 0 \end{bmatrix}, v_2 = \begin{bmatrix} -2 \\ 3 \end{bmatrix}$

25. $r_1 = 3, r_2 = 1, v_1 = \begin{bmatrix} 1 \\ -1 \end{bmatrix}, v_2 = \begin{bmatrix} 2 \\ 1 \end{bmatrix}$

26. $r_1 = 4, r_2 = 3, v_1 = \begin{bmatrix} 2 \\ 1 \end{bmatrix}, v_2 = \begin{bmatrix} 1 \\ 2 \end{bmatrix}$

27. $r_1 = -2, r_2 = -3, v_1 = \begin{bmatrix} 0 \\ 1 \end{bmatrix}, v_2 = \begin{bmatrix} 3 \\ 1 \end{bmatrix}$

28. $r_1 = -1, r_2 = -4, v_1 = \begin{bmatrix} 2 \\ -1 \end{bmatrix}, v_2 = \begin{bmatrix} 1 \\ 1 \end{bmatrix}$

29.–38. For Exercises 13–22, sketch trajectories of the solutions. Classify the origin and assess its stability.

SYNTHESIS

39. *Improper Node.*
    a) Show that 1 is a repeated eigenvalue for the system $x' = y, y' = -x + 2y$, and find an associated eigenvector.
    tw b) How is this eigenvector reflected in the graph on page 630?

40. *Proper Node.* Consider the system of differential equations $x' = rx, y' = ry$, where r is a constant.
    a) Show that r is a repeated eigenvalue and that every vector is an associated eigenvector.
    b) Solve this system of differential equations.
    c) Show that either $x = 0$ or $y = mx$ for some constant m.
    d) Verify that the graph on page 630 shows the trajectories if $r > 0$.
    e) Sketch a graph of the trajectories if $r < 0$.

41. *Auxiliary Equation.* Let r be a root of the auxiliary equation after the linear system
$$x' = ax + by, \qquad y' = cx + dy$$

is reduced to a second-order equation. Show that r is a solution of the equation

$$r^2 - (a + d)r + (ad - bc) = 0,$$

which is the characteristic equation in Theorem 9 of Chapter 6.

42. *Uniqueness.* Assume that **A** only has nonzero eigenvalues. Show that the origin is the only equilibrium point of $z' = \mathbf{A}z$.

43. *Trajectories if 0 is an Eigenvalue.*

   a) Show that 0 is an eigenvalue of the system

   $$x' = -2x + 6y, \quad y' = 2x - 6y.$$

   b) Find the general solution.
   c) Show that $y = b - x$ for some constant b.
   d) Show that all points on the line $y = x/3$ are equilibrium points for this system.
   e) Draw sample trajectories for this system of equations.

Larger Systems. The matrix method may also be used for systems of three or more functions. For Exercises 44–49, find the general solution.

44. $x' = -y - 3z, y' = 2x + 3y + 3z,$
    $z' = -2x + y + z$

45. $x' = x - 3y, y' = -3x - 3y - 4z,$
    $z' = 3x + 5y + 6z$

46. $x' = -3x + 12y + 6z, y' = -2x - 9y - 6z,$
    $z' = 4x + 3z$

47. $x' = 3x + 2z, y' = -x + 2y - 2z, z' = -x$

48. $x' = 2x + 7y + 7z, y' = x + 2y + z,$
    $z' = -x - 7y - 6z$

49. $x' = 2x + 6z, y' = x + y, z' = -x + y - 4z$

tw 50. Describe the relative merits of the reduction method and the matrix method.

Soil Organic Matter. Let S, B, and H denote the amount of carbon-14 (^{14}C) present in the soil, microbial biomass, and humus, respectively. The movement of ^{14}C between these compartments may be described by the model below. In this figure, a denotes the absorption rate of ^{14}C into the biomass from the soil, b denotes the rate ^{14}C is eliminated from the biomass as carbon dioxide, and c and d are transfer rates.[11]

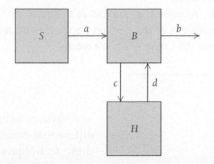

51. Find a system of differential equations for S, B, and H.

52. Compute the eigenvalues if $a = 5, b = 4, c = 7/2,$ and $d = 1$.

53. Assess the stability of the origin.

tw 54. Explain the stability of the origin biologically. (*Hint:* Where does all the ^{14}C eventually go?)

55. Solve the system of differential equations if $S(0) = 1, B(0) = 0,$ and $H(0) = 0$.

---

[11]A. Parshotam, S. Saggar, K. Tate, and R. Parfitt, "Modelling organic matter dynamics in New Zealand soil," *Environment International,* Vol. 27, pp. 111–119 (2001).

9.5 Models of Population Biology

OBJECTIVES

■ Use the Jacobian to assess the stability of equilibrium points of nonlinear systems.

■ Analyze the qualitative behavior of solutions of the competing species model and the Lotka–Volterra model.

In the previous two sections, we developed techniques for solving a linear system of equations

$$x' = ax + by, \qquad y' = cx + dy.$$

Many biological phenomena, however, are better described by autonomous systems of *nonlinear* differential equations. These have the form

$$x' = F(x, y) \quad \text{and} \quad y' = G(x, y), \tag{1}$$

for some functions F and G. Notice that the right-hand sides do not explicitly depend on t.

Various techniques for exactly and approximately solving systems of nonlinear differential equations are used in biology and the life sciences; a full treatment of these techniques may be found in a more advanced textbook. Here we present a method (described in Theorem 8) for assessing the stability of equilibrium points. We will then apply this method to various models that arise in population biology.

Stability of Equilibrium Points

To find the equilibrium points, we set $x' = 0$ and $y' = 0$, obtaining $F(x, y) = 0$ and $G(x, y) = 0$. This nonlinear system may not be easily solved. Nevertheless, suppose that we are able to determine that (x_0, y_0) is an equilibrium point. To assess its stability, we recall from Section 7.2 that linearization may be used to approximate F and G near (x_0, y_0):

$$x' = F(x, y) \approx F(x_0, y_0) + \left[\frac{\partial F}{\partial x}(x_0, y_0)\right](x - x_0) + \left[\frac{\partial F}{\partial y}(x_0, y_0)\right](y - y_0),$$

$$y' = G(x, y) \approx G(x_0, y_0) + \left[\frac{\partial G}{\partial x}(x_0, y_0)\right](x - x_0) + \left[\frac{\partial G}{\partial y}(x_0, y_0)\right](y - y_0).$$

Since (x_0, y_0) is an equilibrium point, we know that $F(x_0, y_0) = 0$ and $G(x_0, y_0) = 0$. We may further simplify these equations by letting $u(t) = x(t) - x_0$ and $v(t) = y(t) - y_0$. Since x_0 and y_0 are constants, we know that $u' = x'$ and $v' = y'$. Therefore, the system of differential equations may be rewritten as

$$u' \approx \left[\frac{\partial F}{\partial x}(x_0, y_0)\right]u + \left[\frac{\partial F}{\partial y}(x_0, y_0)\right]v,$$

$$v' \approx \left[\frac{\partial G}{\partial x}(x_0, y_0)\right]u + \left[\frac{\partial G}{\partial y}(x_0, y_0)\right]v. \tag{2}$$

The terms in brackets are constants and may be evaluated by computing the partial derivatives of F and G at the equilibrium point. Therefore, in the vicinity of an equilibrium point (x_0, y_0), we may approximate the nonlinear system (1) with the *linear* system (2). The stability of the equilibrium point may thus be assessed by finding the eigenvalues of (2).

The above discussion is the basis of the following theorem.

THEOREM 8

Suppose that F and G have continuous second derivatives. Then the stability of an equilibrium point of (1) may be assessed by finding the eigenvalues of the Jacobian matrix $\begin{bmatrix} \partial F/\partial x & \partial F/\partial y \\ \partial G/\partial x & \partial G/\partial y \end{bmatrix}$ evaluated at (x_0, y_0). The stability may be described as follows:

Eigenvalues	Stability	Equilibrium Point
Both positive	Unstable	Node (or possibly a spiral point for a repeated eigenvalue)
Opposite signs	Unstable	Saddle point
Both negative	Asymptotically stable	Node (or possibly a spiral point for a repeated eigenvalue)
$a \pm bi, a > 0$	Unstable	Spiral point
$a \pm bi, a < 0$	Asymptotically stable	Spiral point
$\pm bi$	Indeterminant	Center or spiral point

As with linear systems, the equilibrium value is asymptotically stable if the eigenvalues are both negative or are complex with negative real part. Also, the equilibrium value is unstable if at least one eigenvalue is positive or the eigenvalues are complex with positive real part. However, if the eigenvalues are purely imaginary in a nonlinear system, we cannot determine the stability of the equilibrium point without further analysis. This is unlike the case of linear systems, where we could conclude that the equilibrium point is a center.

EXAMPLE 1 Find the equilibrium points for the nonlinear system

$$x' = xy, \qquad y' = x + 2y - 8$$

and assess the stability of each.

Solution

a) We begin by finding the equilibrium points. At an equilibrium point, we must have

$$xy = 0 \quad \text{and} \quad x + 2y - 8 = 0.$$

Since this is a nonlinear system of equations, the techniques of Chapter 6 cannot be used to solve for x and y.

To find the equilibrium points, we observe that the first equation implies that either $x = 0$ or $y = 0$. If $x = 0$, then we may solve for y using the second equation:

$$0 + 2y - 8 = 0, \quad \text{or} \quad y = 4.$$

Therefore, $(0, 4)$ is an equilibrium point.

On the other hand, if $y = 0$, then we may solve for x using the second equation:

$$x + 2(0) - 8 = 0, \quad \text{or} \quad x = 8.$$

Therefore, $(8, 0)$ is also an equilibrium point. (We note in passing that it is impossible for a linear system of equations to have exactly two distinct solutions.)

b) Next, we find the Jacobian of $F(x, y) = xy$ and $G(x, y) = x + 2y - 8$:

$$\begin{bmatrix} \partial F/\partial x & \partial F/\partial y \\ \partial G/\partial x & \partial G/\partial y \end{bmatrix} = \begin{bmatrix} y & x \\ 1 & 2 \end{bmatrix}.$$

c) To assess the stability of the equilibrium points, we substitute the equilibrium points into the Jacobian and find the eigenvalues of the resulting matrix. First, at the equilibrium point $(0, 4)$, this matrix is equal to $\begin{bmatrix} 4 & 0 \\ 1 & 2 \end{bmatrix}$. Since the upper-right entry is 0, we see that the eigenvalues are 4 and 2. Since both eigenvalues are positive, we conclude that $(0, 4)$ is a node and an unstable equilibrium point.

Second, at the equilibrium point $(8, 0)$, the Jacobian is equal to $\begin{bmatrix} 0 & 8 \\ 1 & 2 \end{bmatrix}$. The trace of this matrix is $0 + 2 = 2$, while the determinant is $(0)(2) - (8)(1) = -8$. Therefore, the eigenvalues satisfy

$$r^2 - 2r - 8 = 0, \quad \text{or} \quad (r - 4)(r + 2) = 0.$$

Since the eigenvalues 4 and -2 have opposite signs, we conclude that $(8, 0)$ is a saddle point and an unstable equilibrium. ∎

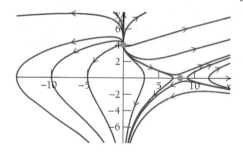

Actual trajectories of this system are shown in the figure; these were computed using the numerical techniques discussed in the next section. We notice that $(0, 4)$ is a node while $(8, 0)$ is a saddle point. Since both are unstable equilibrium points, trajectories diverge away from both points. However, in the above stability analysis, we did not determine precisely how the trajectories behave far away from the equilibrium points.

Competing Species

In Chapter 8, we studied the logistic growth model, which models inhibited population growth due to exhaustable natural resources. This differential equation may be written as

$$y' = ky\left(1 - \frac{y}{L}\right) = y\left(k - \frac{ky}{L}\right) = y(k - ay),$$

where $a = k/L$. The magnitude of a describes how the per capita growth rate decreases as the population increases.

The inhibited growth model is appropriate for a single species in an environment with limited resources. However, if two noninteracting but **competing species** share the same environment and compete for the same limited resources, then it is natural to assume that the presence of one species will inhibit the growth of the other species. If x and y represent the populations of the two species, then we may extend the logistic growth model to the following nonlinear system:

$$x' = x(k_1 - a_1 x - b_1 y), \qquad y' = y(k_2 - b_2 x - a_2 y).$$

The coefficients a_1 and a_2 give the *intraspecies* effect on the per capita growth rates, while the coefficients b_1 and b_2 give the *interspecies* effect on the per capita growth rates.

This system of differential equations cannot be solved in general. Nevertheless, we may find the equilibrium points of this system and assess their stability. For this system, it turns out that we can sketch trajectories without resorting to numerical techniques.

EXAMPLE 2   Let x and y be the populations (in thousands) of two competing species. Suppose that x and y satisfy the system of differential equations

$$x' = x(0.2 - 0.05x - 0.02y), \qquad y' = y(0.1 - 0.01x - 0.02y).$$

Find the equilibrium points and assess the stability of each. Use this stability information to sketch the trajectories of the solutions.

Solution

a) To find the equilibrium points, we set $x' = 0$ and $y' = 0$:

$$0 = x(0.2 - 0.05x - 0.02y) \quad \text{and} \quad 0 = y(0.1 - 0.01x - 0.02y).$$

From the first equation, we see that either $x = 0$ or $0.05x + 0.02y = 0.2$, which may be rewritten as $5x + 2y = 20$. From the second equation, we see that either $y = 0$ or $0.01x + 0.02y = 0.1$, which may be rewritten as $x + 2y = 10$.

There are four equilibrium points that arise from these equations. First, if $x = 0$ and $y = 0$, then we obtain the equilibrium point $(0, 0)$. Second, if $x = 0$ and $x + 2y = 10$, then $y = 5$, yielding the equilibrium point $(0, 5)$. Third, if $5x + 2y = 20$ and $y = 0$, then $x = 4$, yielding the equilibrium point $(4, 0)$. Fourth, if

$$5x + 2y = 20 \quad \text{and} \quad x + 2y = 10,$$

then we have a system of equations that can be solved using any of the techniques from Chapter 6. Subtracting the two equations, we find that

$$4x = 10, \quad \text{or} \quad x = 2.5.$$

Substituting into the second equation, we find

$$2.5 + 2y = 10, \quad \text{or} \quad y = 3.75.$$

Therefore, the fourth equilibrium point is $(2.5, 3.75)$.

Equations	Equilibrium Point	Biological Interpretation
$x = 0, \quad y = 0$	$(0, 0)$	Neither species is present
$x = 0, \quad x + 2y = 10$	$(0, 5)$	Only the second species is present
$5x + 2y = 20, \quad y = 0$	$(4, 0)$	Only the first species is present
$5x + 2y = 20, \quad x + 2y = 10$	$(2.5, 3.75)$	Coexistence

b) Next, we find the Jacobian of

$$F(x, y) = x(0.2 - 0.05x - 0.02y) \quad \text{and} \quad G(x, y) = y(0.1 - 0.01x - 0.02y).$$

The partial derivatives of F and G are

$$\frac{\partial F}{\partial x} = 0.2 - 0.1x - 0.02y, \qquad \frac{\partial F}{\partial y} = -0.02x,$$

$$\frac{\partial G}{\partial x} = -0.01y, \qquad \frac{\partial G}{\partial y} = 0.1 - 0.01x - 0.04y.$$

Therefore, the Jacobian is

$$\begin{bmatrix} 0.2 - 0.1x - 0.02y & -0.02x \\ -0.01y & 0.1 - 0.01x - 0.04y \end{bmatrix}.$$

c) We now use the eigenvalues of the Jacobian to assess the stability of the equilibrium points. First, at $(0, 0)$, the Jacobian is

$$\begin{bmatrix} 0.2 - 0.1(0) - 0.02(0) & -0.02(0) \\ -0.01(0) & 0.1 - 0.01(0) - 0.04(0) \end{bmatrix} = \begin{bmatrix} 0.2 & 0 \\ 0 & 0.1 \end{bmatrix}.$$

We see that the eigenvalues are 0.2 and 0.1. Since both eigenvalues are positive, we see that $(0, 0)$ is a node and an unstable equilibrium.

Second, at $(0, 5)$, the Jacobian is

$$\begin{bmatrix} 0.2 - 0.1(0) - 0.02(5) & -0.02(0) \\ -0.01(5) & 0.1 - 0.01(0) - 0.04(5) \end{bmatrix} = \begin{bmatrix} 0.1 & 0 \\ -0.05 & -0.1 \end{bmatrix}.$$

We see that the eigenvalues are 0.1 and -0.1. Since the eigenvalues have opposite signs, we see that $(0, 5)$ is a saddle point and an unstable equilibrium.

Third, at $(4, 0)$, the Jacobian is

$$\begin{bmatrix} 0.2 - 0.1(4) - 0.02(0) & -0.02(4) \\ -0.01(0) & 0.1 - 0.01(4) - 0.04(0) \end{bmatrix} = \begin{bmatrix} -0.2 & -0.08 \\ 0 & 0.06 \end{bmatrix}.$$

We see that the eigenvalues are -0.2 and 0.06. Since the eigenvalues have opposite signs, we see that $(4, 0)$ is a saddle point and an unstable equilibrium.

Fourth, at $(2.5, 3.75)$, the Jacobian is

$$\begin{bmatrix} 0.2 - 0.1(2.5) - 0.02(3.75) & -0.02(2.5) \\ -0.01(3.75) & 0.1 - 0.01(2.5) - 0.04(3.75) \end{bmatrix} = \begin{bmatrix} -0.125 & -0.05 \\ -0.0375 & -0.075 \end{bmatrix}.$$

The trace of this matrix is $-0.125 - 0.075 = -0.2$, and the determinant is $(-0.125)(-0.075) - (-0.05)(-0.0375) = 0.0075$. To find the eigenvalues, we must solve the equation $r^2 + 0.2r + 0.0075 = 0$:

$$r = \frac{-0.2 \pm \sqrt{(0.2)^2 - 4(1)(0.0075)}}{2(1)} = \frac{-0.2 \pm 0.1}{2} = -0.05 \quad \text{or} \quad -0.15.$$

Since the eigenvalues are negative, we see that $(2.5, 3.75)$ is a node and an asymptotically stable equilibrium point.

The above stability analysis may be used to sketch trajectories of solutions, as shown in the figure below. Many trajectories curl around the saddle points and converge toward the asymptotically stable equilibrium point $(2.5, 3.75)$. This is true even for trajectories that begin far from the origin: For large values of x and y, both x' and y' will be negative, and so trajectories that begin far from the origin will point toward the origin. We conclude that the long-run values of x and y will be 2.5 and 3.75, respectively.

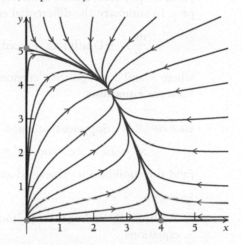

Since the equilibrium points $(4, 0)$ and $(0, 5)$ are saddle points, we expect there to be a pair of trajectories to converge to each point. Indeed, trajectories along the x-axis converge to $(4, 0)$, and trajectories along the y-axis converge to $(0, 5)$. If one species is not initially present, then the trajectory will converge to one of these points. However, if one species has even a small presence in the habitat, then the model predicts that the trajectory will converge to $(2.5, 3.75)$.

In the previous example, the only asymptotically stable equilibrium point corresponded to coexistence. However, for other choices of the model constants, the only asymptotically stable equilibrium point or points may correspond to having only one of the two species present. For such systems, sustained coexistence is not possible. Examples of such systems are presented in the exercises.

Predator-Prey Models

In Example 2, we considered two noninteracting species that competed for resources in a common habitat. We now consider a habitat where one species (the predators) prey upon the second species (the prey). Let $x(t)$ and $y(t)$ denote the number of prey and predators, respectively, present at time t.

There are two effects on the prey population growth rate. First, if we assume that plenty of food is provided for the prey and that no predators are present, then it seems reasonable that the per capita growth rate of the prey population should be constant. Second, any meeting between predator and prey will usually be beneficial for the predator and detrimental to the prey. The number of encounters between predators and prey should be proportional to xy. Therefore, we assume that the prey population death rate due to predation is proportional to xy. Combining these two effects, the differential equation for x is

$$\frac{dx}{dt} = \text{Growth rate} - \text{Predation rate} = px - qxy,$$

where p and q are positive constants.

We now consider the predator population growth rate. In the absence of their source of food, the predator per capita death rate should be constant. However, the predator population should increase at a rate proportional to xy as they feed on the prey. In summary, the differential equation for y is

$$\frac{dy}{dt} = -\text{Death rate} + \text{Feeding rate} = -ry + sxy,$$

where r and s are positive constants. These two equations are called the **Lotka–Volterra equations.**

EXAMPLE 3   Suppose that x and y satisfy the nonlinear system

$$x' = 2x - xy \quad \text{and} \quad y' = -y + 0.4xy.$$

Find the equilibrium values and assess the stability of each.

Solution

a) We begin by finding the equilibrium values. These satisfy the nonlinear system of equations

$$2x - xy = x(2 - y) = 0 \quad \text{and} \quad -y + 0.4xy = y(0.4x - 1) = 0.$$

From the first equation, we see that $x = 0$ or $y = 2$. If $x = 0$, then the second equation becomes

$$y(0.4[0] - 1) = 0, \quad \text{or} \quad y = 0.$$

Therefore $(0, 0)$ is an equilibrium point. On the other hand, if $y = 2$, then the second equation becomes

$$(2)(0.4x - 1) = 0, \quad \text{or} \quad x = 2.5.$$

Therefore, $(2.5, 2)$ is also an equilibrium point.

b) Next, we find the Jacobian of $F(x, y) = 2x - xy$ and $G(x, y) = -y + 0.4xy$:

$$\begin{bmatrix} \partial F/\partial x & \partial F/\partial y \\ \partial G/\partial x & \partial G/\partial y \end{bmatrix} = \begin{bmatrix} 2 - y & -x \\ 0.4y & -1 + 0.4x \end{bmatrix}.$$

c) We now find the eigenvalues of the Jacobian at the equilibrium points. First, at $(0, 0)$, the Jacobian is equal to $\begin{bmatrix} 2 & 0 \\ 0 & -1 \end{bmatrix}$. Since the eigenvalues $r = 2$ and $r = -1$ have opposite signs, we conclude that $(0, 0)$ is an unstable equilibrium and a saddle point.

Next, at $(2.5, 2)$, the Jacobian is equal to $\begin{bmatrix} 0 & -2.5 \\ 0.8 & 0 \end{bmatrix}$. The trace of this matrix is $0 + 0 = 0$, while the determinant is $(0)(0) - (-2.5)(0.8) = 2$. The eigenvalues are the solutions of $r^2 + 2 = 0$, which are $\pm i\sqrt{2}$.

It turns out that, for Lotka–Volterra systems, the coexistence equilibrium is *always* a center.[12] Therefore, $(2.5, 2)$ is a center, and most trajectories are closed paths that enclose $(2.5, 2)$. ■

The figure shows trajectories of x and y from Example 3. Notice that these cycles revolve around the equilibrium point $(2.5, 2)$. We also see that $(0, 0)$ is a saddle point. The outward trajectory on the x-axis corresponds to a prey population that increases without the threat of predators, while the inward trajectory on the y-axis corresponds to a predator population that decreases in the absence of its prey.

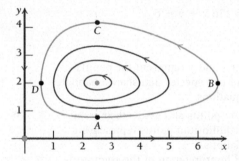

Let's examine why cyclical trajectories are ecologically reasonable. On the trajectory marked in green, there are few predators and an abundance of prey at the point A. With food abundant, the predator population increases. Also, with so few predators, the prey population increases until point B is reached. At this point, so many predators are present that the prey population decreases while the predator population increases. At point C, food becomes scarce for the predators, so the population of the predators also decreases. At point D, the predator population is low enough that the prey population begins to increase again. When point A is reached, the cycle starts anew.

Various modifications of the Lotka–Volterra equations are presented in the exercises.

---

[12]Theorem 8 states that, for general nonlinear systems, purely imaginary eigenvalues could correspond to either a center or a spiral point. By contrast, recall that purely imaginary eigenvalues always correspond to a center for systems of linear differential equations. The proof that coexistence equilibrium is always a center for all Lotka–Volterra systems lies beyond the scope of this book.

Exercise Set 9.5

Find the equilibrium points and assess the stability of each.

1. $x' = x(x + y + 4)$, $y' = y(x - 2y + 1)$

2. $x' = x(y - 2)$, $y' = y(x - 3)$

3. $x' = x^2 + y^2 - 25$, $y' = x + y - 7$

4. $x' = x^2 + y^2 - 5$, $y' = x^2 - 4x - y + 5$

5. $x' = (2 - x)\sqrt{y}$, $y' = xy - 8$

6. $x' = x + y - 3$, $y' = x^2 + 3y^2 - 7$

7. $x' = -x + y^2$, $y' = x + y^4 - 2$

8. $x' = -2x^2 - y$, $y' = x^4 + y - 8$

9. $x' = x - e^y$, $y' = x + e^{2y} - 2$

10. $x' = x - e^y$, $y' = 2 \ln x + y - 6$

APPLICATIONS

Competing Species. Let x and y represent the populations (in thousands) of two species that share a habitat. For each system of equations:

a) Find the equilibrium points and assess their stability. Solve only for equilibrium points representing nonnegative populations.

b) Give the biological interpretation of the asymptotically stable equilibrium point(s).

11. $x' = x(0.1 - 0.01x - 0.005y)$,
    $y' = y(0.05 - 0.001x - 0.002y)$

12. $x' = x(0.04 - 0.0008x - 0.0024y)$,
    $y' = y(0.02 - 0.0012x - 0.0004y)$

13. $x' = x(0.1 - 0.005x - 0.002y)$,
    $y' = y(0.05 - 0.001x - 0.002y)$

14. $x' = x(0.01 - 0.0001x - 0.0007y)$,
    $y' = y(0.02 - 0.0006x - 0.0002y)$

15. $x' = x(0.04 - 0.0004x - 0.0008y)$,
    $y' = y(0.1 - 0.002x - 0.005y)$

16. $x' = x(0.1 - 0.006x - 0.0008y)$,
    $y' = y(0.2 - 0.001x - 0.006y)$

Fishing. Let x and y represent the populations (in thousands) of two kinds of fish in a pond. Suppose that x and y satisfy the system

$$x' = x(0.1 - 0.01x - 0.005y) \quad \text{and}$$

$$y' = y(0.2 - 0.015x - 0.02y).$$

17. Show that the coexistence equilibrium is asymptotically stable.

18. Suppose a fisherman fishes for only the first species of fish from the pond, so that the first differential equation becomes

$$x' = x(0.1 - 0.01x - 0.005y) - 0.01x.$$

However, the second species of fish is unaffected by the fisherman, so the second differential equation remains the same.

a) Find the new coexistence equilibrium point, and assess its stability.

tw b) Explain why the new equilibrium point is different from that found in Exercise 17.

19. Suppose that the first species is heavily fished, so that the first differential equation becomes

$$x' = x(0.1 - 0.01x - 0.005y) - 0.08x.$$

a) Show that no state of coexistence is an equilibrium point.

b) Determine the only asymptotically stable equilibrium point.

tw c) Explain why your answer to part (b) is different from the assessment found in Exercise 18.

Predators and Prey. Let x and y represent the populations (in thousands) of prey and predators that share a habitat. For the given system of differential equations, find and classify the equilibrium points.

20. $x'(t) = 0.5x - 0.4xy$, $y'(t) = -0.4y + 0.2xy$

21. $x'(t) = 0.6x - 0.3xy$, $y'(t) = -y + 0.2xy$

22. $x'(t) = 0.8x - 0.2xy$, $y'(t) = -0.6y + 0.1xy$

23. $x'(t) = 0.5x - 0.2xy$, $y'(t) = -0.4y + 0.1xy$

Prey with Inhibited Growth. The Lotka–Volterra equations assume that, in the absence of predators, the prey will have a constant per capita growth rate. However, if there are limited resources in the habitat, it may be more reasonable to assume that the prey population follows the logistic growth model.

The systems of differential equations in Exercises 24–27 are similar to those of Exercises 20–23 except that the growth of the prey is inhibited. For the following systems, find the equilibrium points and assess their stability.

24. $x'(t) = 0.5x(1 - x/10) - 0.4xy$,
    $y'(t) = -0.4y + 0.2xy$

25. $x'(t) = 0.6x(1 - x/20) - 0.3xy$,
    $y'(t) = -y + 0.2xy$

26. $x'(t) = 0.8x(1 - x/24) - 0.2xy,$
    $y'(t) = -0.6y + 0.1xy$

27. $x'(t) = 0.5x(1 - x/10) - 0.2xy,$
    $y'(t) = -0.4y + 0.1xy$

28. Compare the coexistence equilibrium points in Exercises 24–27 to those of Exercises 20–23. What is the effect of inhibited prey growth on the equilibrium point?

Neurons. Let x and y represent the proportion of excitatory and inhibitory neurons firing per unit of time in an unstimulated cortical column. Then x and y may be modeled by the differential equations[13]

$$x' = \frac{ax - by}{\sqrt{(ax - by)^2 + 1}} \quad \text{and}$$

$$y' = -dy + \frac{cx}{\sqrt{c^2x^2 + 1}},$$

where $a, b, c,$ and d are constants.

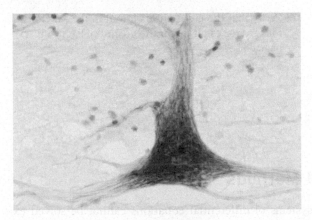

29. Let $a = 4, b = 3, c = 3,$ and $d = 1.$
    a) Show that the origin is an equilibrium point.
    b) Find the other equilibrium point. (Since x and y are proportions, they must both be positive.)
    c) Assess the stability of the origin.

30. Let $a = 1, b = 3, c = 3,$ and $d = 4.$
    a) Show that the origin is an equilibrium point.
    b) Find the other equilibrium point.
    c) Assess the stability of the origin.

[13]L. H. A. Monteiro, M. A. Bussab, and J. G. Chaui Berlinck, "Analytical Results on a Wilson-Cowan Neuronal Network Modified Model," *Journal of Theoretical Biology,* Vol. 219, pp. 83–91 (2002).

SYNTHESIS

Competing Species. In Exercises 31–33, we will examine further properties of the model for competing species.

31. *Eigenvectors.* This exercise continues Example 2.
    a) At the equilibrium point $(0, 5)$, compute the eigenvectors that correspond to the eigenvalues of the Jacobian. How are these eigenvectors reflected in the trajectories on page 637?
    b) Repeat for the equilibrium point $(4, 0)$.
    c) Repeat for the equilibrium point $(2.5, 3.75)$.

32. *Sustained Coexistence.*
    a) For Exercises 11–16, which systems had an asymptotically stable equilibrium of coexistence?
    b) Which systems had an asymptotically unstable equilibrium of coexistence?

33. *Condition for Sustained Coexistence.* Recall that the system of differential equations for competing species generally has the form

$$x' = x(k_1 - a_1x - b_1y) \quad \text{and}$$
$$y' = y(k_2 - b_2x - a_2y),$$

where a_1 and a_2 measure the intraspecies inhibition and b_1 and b_2 measure the interspecies inhibition.

    a) For the systems in Exercise 32, determine the larger of a_1a_2 and b_1b_2.
    tw b) Explain in biological terms why the relative magnitudes of a_1a_2 and b_1b_2 should determine whether sustained coexistence is asymptotically stable or is unstable.

Limited Predation. If prey are abundant, then predators cannot possibly use all of them for food. For this reason, one extension of the Lotka–Volterra equations is the system

$$x' = px\left(1 - \frac{x}{L}\right) - \frac{qxy}{1 + ax}, \quad y' = -ry + \frac{sxy}{1 + ax},$$

where a is a nonnegative constant.

tw 34. Explain the presence of the $(1 + ax)$ terms biologically.

35. Verify that the Lotka–Volterra equations are recovered if $a = 0.$

36. Assess the stability of the coexistence equilibrium point if $p = 0.2, q = 0.04, r = 0.05, s = 0.03, a = 0.1,$ and $L = 10.$

37. Repeat Exercise 36 if $L = 40.$ The change in the equilibrium point for environments capable of sustaining more prey has been called the *paradox of enrichment.*

Lotka–Volterra Equations. In the next few exercises, we analyze further the Lotka–Volterra equations.

38. *Equilibrium Point.*

a) Find the two equilibrium points of the general Lotka–Volterra system

$$x' = px - qxy, \qquad y' = -ry + sxy,$$

where $p, q, r,$ and s are positive.

b) Assess the stability of both equilibrium points.

39. *Bounds of the Trajectories.*

a) According to the Chain Rule, $dy/dt = (dy/dx)(dx/dt)$. Use this and the Lotka–Volterra equations to show that

$$\frac{dy}{dx} = \frac{y(-r + sx)}{x(p - qy)}.$$

tw b) Show that the trajectories have horizontal tangent lines when $x = r/s$.

tw c) Show that the trajectories have vertical tangent lines when $y = p/q$.

tw d) How do your answers to parts (b) and (c) compare to the coexistence equilibrium point found in Exercise 38?

40. *Separation of Variables.* Use separation of variables and Exercise 39 to show that

$$p \ln y - qy + r \ln x - sx = C$$

for some constant C.

41. Consider the Lotka–Volterra system of equations

$$x' = 2x - xy \quad \text{and} \quad y' = -y + 0.4xy,$$

which was considered in Example 3. Suppose the initial conditions are $x(0) = 5$ and $y(0) = 1$. Use the result to Exercise 40 to find an implicit equation for x and y.

 Technology Connection

Predators and Prey. Exercises 42–46 are a continuation of Exercise 41.

42. Let $y = 2$ in the implicit equation found in Exercise 41. Use Newton's method to find the two values of x that satisfy this equation.

43. Repeat Exercise 42 for $y = 1$.

44. Repeat Exercise 42 for $y = 3$.

45. Repeat Exercise 42 for $y = 4$.

46. Verify that the eight points found in the previous four exercises all lie on the green trajectory on page 641.

9.6 Numerical Methods

OBJECTIVE

■ Find an approximate solution to a first-order system using Euler's method.

Some systems of differential equations cannot be solved exactly. For such systems, **Euler's method** may be extended to find a numerical approximation of the solution. For the system

$$x' = F(x, y, t) \quad \text{and} \quad y' = G(x, y, t),$$

with initial conditions $x(t_0) = x_0$ and $y(t_0) = y_0$, Euler's method becomes

$$x_{n+1} = x_n + F(x_n, y_n, t_n)\,\Delta t \quad \text{and} \quad y_{n+1} = y_n + G(x_n, y_n, t_n)\,\Delta t.$$

Euler's method can also be extended for systems of three or more differential equations.

EXAMPLE 1 Use Euler's method with $\Delta t = 0.1$ to approximate the solution of the initial-value problem

$$x' = 2x - 4y + 3t, \qquad y' = 5x - 4y - t^2, \qquad x(0) = 2, \qquad y(0) = 3$$

for $0 \le t \le 1$. Plot the trajectory of x and y.

Solution The initial conditions are $x(0) = 2$ and $y(0) = 3$, so that $t_0 = 0, x_0 = 2$, and $y_0 = 3$. Successively adding $\Delta t = 0.1$ to $t_0 = 0$, we have $t_1 = 0.1, t_2 = 0.2$, and so on, until $t_{10} = 1$.

For this system of differential equations, we see that $F(x, y, t) = 2x - 4y + 3t$ and $G(x, y, t) = 5x - 4y - t^2$. Since $\Delta t = 0.1$, Euler's method becomes

$$x_{n+1} = x_n + 0.1(2x_n - 4y_n + 3t_n) \quad \text{and} \quad y_{n+1} = y_n + 0.1(5x_n - 4y_n - t_n^2).$$

We first use Euler's method to find x_1 and y_1.

$$x_1 = x_0 + 0.1(2x_0 - 4y_0 + 3t_0) = 2 + 0.1(2[2] - 4[3] + 3[0]) = 1.2,$$
$$y_1 = y_0 + 0.1(5x_0 - 4y_0 - t_0^2) = 3 + 0.1(5[2] - 4[3] - [0]^2) = 2.8.$$

Next,

$$x_2 = x_1 + 0.1(2x_1 - 4y_1 + 3t_1) = 1.2 + 0.1(2[1.2] - 4[2.8] + 3[0.1]) = 0.35,$$
$$y_2 = y_1 + 0.1(5x_1 - 4y_1 - t_1^2) = 2.8 + 0.1(5[1.2] - 4[2.8] - [0.1]^2) = 2.279.$$

We continue until we reach x_{10} and y_{10}; these computations may be facilitated by using a spreadsheet, as discussed in the Technology Connection. The trajectory is shown in the figure.

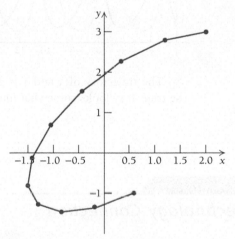

EXAMPLE 2   *Predators and Prey.* Let $x(t)$ and $y(t)$ be the populations (in thousands) of prey and predators in a closed environment, where t is measured in years. Suppose that x and y satisfy the system of differential equations

$$x' = 2x - xy \quad \text{and} \quad y' = -y + 0.4xy.$$

Also, suppose that $x(0) = 5$ and $y(0) = 1$. Use Euler's method with $\Delta t = 0.0005$ to approximate x and y for $0 \leq t \leq 15$. Plot the graphs $x(t)$ and $y(t)$, and plot the trajectory of x and y.

Solution   For this system of differential equations, we see that

$$F(x, y, t) = 2x - xy \quad \text{and} \quad G(x, y, t) = -y + 0.4xy.$$

If $\Delta t = 0.0005$, Euler's method for this system becomes

$$x_{n+1} = x_n + 0.0005(2x_n - x_n y_n),$$
$$y_{n+1} = y_n + 0.0005(-y_n + 0.4x_n y_n).$$

These calculations may be performed by a spreadsheet, as discussed in the Technology Connection.

Using a spreadsheet, the graphs of $x(t)$ and $y(t)$ are shown in the figure on the next page. As expected for solutions of the Lotka–Volterra equations, the graphs of $x(t)$ and $y(t)$ appear to be periodic functions. In other words, the predator and prey populations appear to go through cycles with a period of approximately 5 years. Using the spreadsheet, the length of this cycle can be more precisely determined. Since

$$x(4.84) \approx 5.02 \quad \text{and} \quad y(4.84) \approx 1.00,$$

we conclude that the period is about 4.84 years. Notice that the point $(5, 1)$ is not exactly repeated; this is due to the inherent numerical error in Euler's method.

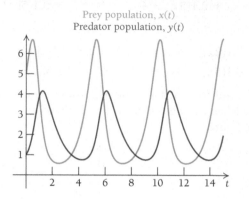

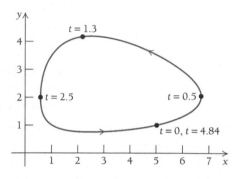

The trajectory of x and y is also shown. As discussed in the previous section, the trajectory looks somewhat like an oval that repeats itself over and over again. ■

Technology Connection

Systems of Equations

To solve Example 1 with a spreadsheet, we would type the titles **Tn**, **Xn**, and **Yn** into cells A1, B1, and C1, respectively. We type the initial conditions 0, 2, and 3 into cells A2, B2, and C2. We then use Euler's method by typing **=A2+0.1** into cell A3, **=B2+0.1*(2*B2-4*C2+3*A2)** into cell B3, and **=C2+0.1*(5*B2-4*C2-A2^2)** into cell C3. We then fill down to repeat Euler's method.

The graphs of x and y may be made by highlighting columns A, B, and C, starting with row 2. Furthermore, the trajectory may be graphed by highlighting columns B and C, starting with row 2.

	A	B	C
1	Tn	Xn	Yn
2	0	2	3
3	0.1	1.2	2.8
4	0.2	0.35	2.279
5	0.3	-0.4316	1.5384
6	0.4	-1.04328	0.69824
7			

EXERCISES

1. If $\Delta t = 0.01$, how many rows are needed to reach $t = 1$?

2. Repeat Example 1 with $\Delta t = 0.01$.

3. Use a spreadsheet to graph the trajectory of x and y if $\Delta t = 0.01$.

4. Your graph in Exercise 3 should be somewhat different from the figure on page 645. Which graph do you expect to be more accurate?

A small value of Δt was used in Example 2 to minimize numerical errors. However, even with $\Delta t = 0.0005$, the point $(5, 1)$ is not repeated exactly, due to the inaccuracy of Euler's method. Though we do not do so here, the Runge–Kutta method can be generalized for systems of equations and usually gives accurate results without requiring very small step sizes.

Exercise Set 9.6

To facilitate calculation, use a spreadsheet for all problems in this section.

Use Euler's method with the indicated value of Δt to approximate the solution to the given system of differential equations on the given interval.

1. $x'(t) = x - 4y + t,\ y'(t) = x - 3y,\ x(0) = 6,$
   $y(0) = 3,\ \Delta t = 0.2,\ \text{on } [0, 1]$

2. $x'(t) = x + y,\ y'(t) = 4x - 2y,\ x(0) = 3,$
   $y(0) = -2,\ \Delta t = 0.1,\ \text{on } [0, 1]$

3. $x'(t) = y,\ y'(t) = -2x + 3y,\ x(0) = 0,$
   $y(0) = 1,\ \Delta t = 0.05,\ \text{on } [0, 2]$

4. $x'(t) = -3x + 4y,\ y'(t) = -2x + 3y,\ x(0) = 4,$
   $y(0) = 3,\ \Delta t = 0.05,\ \text{on } [0, 2]$

5. $x'(t) = x + t,\ y'(t) = x + y + 2t,\ x(0) = 3,$
   $y(0) = -1,\ \Delta t = 0.01,\ \text{on } [0, 3]$

6. $x'(t) = y - 2t^2,\ y'(t) = 2x + y - 3,\ x(0) = 0,$
   $y(0) = 1,\ \Delta t = 0.01,\ \text{on } [0, 3]$

APPLICATIONS

Competing Species. For the systems of differential equations in Exercises 7–12, use Euler's method with $\Delta t = 2$ to

a) Plot the graphs of x and y for $0 \le t \le 500$.
b) Plot the trajectory of x and y.

7. $x' = x(0.1 - 0.01x - 0.004y),$
   $y' = y(0.05 - 0.001x - 0.0016y),\ x(0) = 5,$
   $y(0) = 4$

8. $x' = x(0.1 - 0.01x - 0.004y),$
   $y' = y(0.05 - 0.001x - 0.0016y),\ x(0) = 20,$
   $y(0) = 4$

9. $x' = x(0.04 - 0.001x - 0.0022y),$
   $y' = y(0.02 - 0.0012x - 0.0004y),\ x(0) = 5,$
   $y(0) = 7$

10. $x' = x(0.04 - 0.001x - 0.0022y),$
    $y' = y(0.02 - 0.0012x - 0.0004y),\ x(0) = 5,$
    $y(0) = 7.5$

11. $x' = x(0.1 - 0.006x - 0.002y),$
    $y' = y(0.05 - 0.001x - 0.002y),\ x(0) = 1,$
    $y(0) = 10$

12. $x' = x(0.1 - 0.006x - 0.002y),$
    $y' = y(0.05 - 0.001x - 0.002y),\ x(0) = 2,$
    $y(0) = 1$

Lotka–Volterra Equations. For the Lotka–Volterra equations in Exercises 13–16, use Euler's method with $\Delta t = 0.001$ to

a) Plot the graphs of x and y for $0 \le t \le 10$.
b) Plot the trajectory of x and y.
c) Measure (to the nearest 10th of a year) how much time is needed to complete one cycle.

13. $x' = x - xy,\ y' = -y + 0.2xy,\ x(0) = 3,\ y(0) = 3$

14. $x' = x - xy,\ y' = -y + 0.2xy,\ x(0) = 10,\ y(0) = 2$

15. $x' = 0.8x - 0.2xy,\ y' = -0.6y + 0.1xy,$
    $x(0) = 8,\ y(0) = 3$

16. $x' = 0.8x - 0.2xy,\ y' = -0.6y + 0.1xy,$
    $x(0) = 12,\ y(0) = 4$

17. *Prey with Inhibited Growth.* Consider the system of equations

$$x' = 2x\left(1 - \frac{x}{20}\right) - xy$$

$$y' = -y + 0.4xy$$

with the initial conditions $x(0) = 5$ and $y(0) = 1$. This is similar to the system in Example 2. Use $\Delta t = 0.0005$ to graph the trajectory of this system for $0 \le t \le 15$. Is the trajectory still a closed cycle?

18. *Wastewater Treatment.* Let $S(t)$ be the concentration (per liter) of volatile fatty acids in dairy farming wastewater after t days of treatment. The wastewater is treated by bacteria that have a concentration of $X(t)$ after t days. Then S and X may be modeled by the differential equations[14]

$$S' = -\frac{22SX}{30 + S}$$

$$X' = \left(\frac{0.4S}{30 + S} - 0.004\right)X.$$

The initial conditions are $S(0) = 3$ and $X(0) = 0.06$. Use Euler's method with $\Delta t = 0.1$ to find $S(10)$ and $X(10)$.

19. *Groundwater Contamination.* Let $W(t)$, $E(t)$, and $U(t)$ represent the number of radioactive cadmium-109 ($^{109}Cd^{2+}$) tracers present in river water, extractable sediments, and unextractable sediments, respectively. Assume that these functions satisfy the system[15]

$$W' = -0.059W + 0.0006E$$

$$E' = 0.059W - 0.2506E + 1.06U$$

$$U' = 0.25E - 1.06U.$$

a) Suppose the initial conditions are $W(0) = 0.003$, $E(0) = 0.84$, and $U(0) = 0.16$. Use Euler's method with $\Delta t = 0.5$ to find $W(20)$ and $E(20)$.

b) Use a spreadsheet to graph $W(t)$, $E(t)$, and $U(t)$ on the interval $[0, 100]$.

c) What do the equilibrium values appear to be?

20. *HIV Infection.* Let $T(t)$, $I(t)$, and $V(t)$ represent the concentrations (per mm^3) of healthy $CD4^+$ T cells, infected $CD4^+$ T cells, and free HIV after t days.

These functions may be modeled by the system of equations[16]

$$T' = 10 + 0.01T - (2 \times 10^{-5})T^2$$
$$- (2 \times 10^{-5})TI - (2.4 \times 10^{-5})VT$$

$$I' = (2 \times 10^{-5})VT - 0.26I$$

$$V' = 120I - (2.4 \times 10^{-5})VT - 2.4V.$$

a) Suppose that $T(0) = 1000$, $I(0) = 0$, and $V(0) = 0.001$. Use Euler's method with $\Delta t = 0.1$ to approximate $T(35)$, $I(35)$, and $V(35)$.

b) Use a spreadsheet to compute $T(t)$, $I(t)$, and $V(t)$ for t on the interval $[0, 250]$. Graph each function.

c) What do the equilibrium values appear to be?

SYNTHESIS

Competing Species. Exercises 21–26 refer to a system of differential equation given in a previous exercise. Use the techniques of Section 9.5 to assess the stability of the equilibrium points. Are any of the equilibrium points reflected in the previously found trajectories?

21. Exercises 7 and 8

22. Exercises 9 and 10

23. Exercises 11 and 12

24. Exercises 13 and 14

25. Exercises 15 and 16

26. Exercise 17

 Technology Connection

27. Use different choices of the initial conditions to sketch trajectories of the system $x' = y$, $y' = -x + 2y$. Your answer should look like the improper node graph on page 630.

[14]T. G. Müller, N. Noykova, M. Gyllenberg, and J. Timmer, "Parameter identification in dynamical models of anaerobic waste water management," *Mathematical Biosciences,* Vol. 177 & 178, pp. 147–160 (2002).

[15]W. J. F. Standring, D. H. Oughton, and B. Salbu, "Remobilisation of ^{109}Cd, ^{65}Zn and ^{54}Mn from freshwater-labelled river sediments when mixed with seawater," *Environment International,* Vol. 28, pp. 185–195 (2002).

[16]R. V. Culshaw and S. Ruan, "A delay differential-equation model of HIV infection of $CD4^+$ T-cells," *Mathematical Biosciences,* Vol. 165, pp. 27–39 (2000).

Chapter 9   Summary and Review

Terms to Know

Review Exercises

These review exercises are for test preparation. They can also be used as a lengthened practice test. Answers are at the back of the book. The answers also contain bracketed section references, which tell you where to restudy if your answer is incorrect.

Solve.

1. $y'' - 7y' + 10y = 0$

2. $y'' - 6y' + 9y = 0$

3. $y''' + 4y' = 0$

4. $y'' + 2y' + 26y = 100$

5. $y'' - y' - 20y = 40x^2 - 16x + 15$

Solve the initial-value problem.

6. $y'' + 5y' + 6y = 0, y(0) = 2, y'(0) = -3$

7. $y'' + 4y' + 4y = 0, y(0) = 3, y'(0) = 6$

8. $y'' + 8y' + 20y = 0, y(0) = -4, y'(0) = 1$

9. $y'' - 8y' + 15y = 30t + 60, y(0) = 2, y'(0) = 1$

10. $y'' - 2y' + y = x^2, y(0) = 3, y'(0) = -10$

Solve the first-order linear system.

11. $x' = 2x + 6y, y' = 4x$

12. $x' = 2x - y, y' = x + 4y$

13. $x' = x - 4y, y' = 10x + 5y$

14. $x' = 7x + 3y, y' = -9x - 5y$

Sketch trajectories of solutions of $z' = Az$ given the eigenvalues and associated eigenvectors of A. Classify the origin and assess its stability.

15. $r_1 = 2, r_2 = -3, v_1 = \begin{bmatrix} 2 \\ 1 \end{bmatrix}, v_2 = \begin{bmatrix} 1 \\ 3 \end{bmatrix}$

16. $r_1 = 5, r_2 = 3, v_1 = \begin{bmatrix} 1 \\ -1 \end{bmatrix}, v_2 = \begin{bmatrix} 0 \\ 1 \end{bmatrix}$

17. $r_1 = -2 + 3i, r_2 = -2 - 3i$

18. $r_1 = 4i, r_2 = -4i$

Solve the initial-value problem.

19. $x' = 3x + y, y' = x + 3y, x(0) = 2, y(0) = 0$

20. $x' = 4x + y, y' = -x + 2y, x(0) = 3, y(0) = 1$

21. Assess the stability of the equilibrium points of the nonlinear system $x' = x(y + 2), y' = x + y + 5$.

APPLICATIONS

22. Solve for x and y in the two-compartment model.

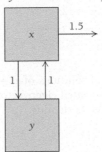

23. *Mixing Chemicals.* Tank A contains 400 lb of salt dissolved in 1000 gal of water. Tank B contains 500 lb of salt dissolved in 1000 gal of water. The mixture from tank A is pumped to tank B at the rate of 200 gal per hr, while that from tank B is pumped to tank A at the same rate. Assume that the mixture in each tank is kept uniform by stirring. Let $A(t)$ and $B(t)$ be the amount of salt in tanks A and B after t hours, respectively.

a) Draw a two-compartment model for $A(t)$ and $B(t)$.
b) Use the initial conditions $A(0) = 400$ and $B(0) = 500$ to solve for A and B.
c) What are the equilibrium values of A and B?

Let x and y represent the populations (in thousands) of two species that share a habitat. For each system of equations, find and assess the stability of the equilibrium points. Remember that, for real

populations, both x_0 and y_0 must be nonnegative for (x_0, y_0) to be an equilibrium point.

24. *Competing Species.*
$$x' = x(0.1 - 0.0025x - 0.003y),$$
$$y' = y(0.05 - 0.002x - 0.0032y)$$

25. *Predators and Prey.* $x' = x - xy,$
$$y' = -0.8y + 0.2xy$$

 Technology Connection

26. *Predators and Prey.* Use Euler's method with step size $\Delta t = 0.001$ to plot the trajectory of the Lotka–Volterra system

$$x' = 0.7x - 0.25xy, \quad y' = -0.5y + 0.1xy,$$
$$x(0) = 1, y(0) = 3,$$

for $0 \le t \le 20$.

Chapter 9 Test

Solve.

1. $y'' - 8y' + 15y = 0$
2. $y'' - 12y' + 36y = 0$
3. $y'' + 36y = 72$
4. $y'' + y' - 12y = 4x - 3$

Solve the initial-value problem.

5. $y'' - 6y' + 5y = 0, y(0) = 1, y'(0) = 0$
6. $y'' + 4y' + 5y = 0, y(0) = 2, y'(0) = -2$
7. $y'' + 6y' + 9y = 12, y(0) = 0, y'(0) = 3$
8. $y'' + 25y = 15x - 5, y(0) = 3, y'(0) = -1$

Solve the first-order linear system.

9. $x' = 2x + 5y, y' = 3x$
10. $x' = -x - 3y, y' = 2x + 4y$
11. $x' = 3x - y, y' = -2x + 4y$
12. $x' = -5x - 20y, y' = 5x + 7y$

Sketch trajectories of solutions of $z' = Az$ given the eigenvalues and associated eigenvectors of A. Classify the origin and assess its stability.

13. $r_1 = -2 + 4.4i, r_2 = -2 - 4.4i$

14. $r_1 = 4, r_2 = -1, v_1 = \begin{bmatrix} 2 \\ 3 \end{bmatrix}, v_2 = \begin{bmatrix} 1 \\ -2 \end{bmatrix}$

15. $r_1 = -1, r_2 = -2, v_1 = \begin{bmatrix} -4 \\ 1 \end{bmatrix}, v_2 = \begin{bmatrix} 1 \\ 2 \end{bmatrix}$

16. $r_1 = 2i, r_2 = -2i$

17. Assess the stability of the equilibrium points of the nonlinear system $x' = x^2 + y^2 - 13,$
$y' = x - y + 5.$

APPLICATIONS

18. Solve for x and y in the two-compartment model.

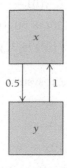

19. *Mixing Chemicals.* Tank A contains 100 lb of salt dissolved in 500 gal of water. Tank B contains 150 lb of salt dissolved in 500 gal of water. The mixture from tank A is pumped to tank B at the rate of 50 gal per hr, while that from tank B is pumped to tank A at the same rate. Furthermore, pure water enters tank A at the rate of 75 gal per hr, while brine drains from tank A at the same rate. Assume that the mixture in each tank is kept uniform by stirring. Let $A(t)$ and $B(t)$ be the amount of salt in tanks A and B after t hours, respectively.

a) Draw a two-compartment model for $A(t)$ and $B(t)$.
b) Use the initial conditions $A(0) = 100$ and $B(0) = 150$ to solve for A and B.
c) What are the equilibrium values of A and B?

Let x and y represent the populations (in thousands) of two species that share a habitat. For each system of equations, find and assess the stability of the equilibrium points. Remember that, for real populations, both x_0 and y_0 must be nonnegative for (x_0, y_0) to be an equilibrium point.

20. *Competing Species.* $x' = x(0.1 - 0.004x - 0.002y)$, $y' = y(0.06 - 0.002x - 0.003y)$

21. *Predators and Prey.* $x' = x - 0.5xy$, $y' = -y + 0.1xy$

 Technology Connection

Competing Species. For the systems of differential equations in Exercises 22–23, use Euler's method with $\Delta t = 2$ to plot the trajectory of x and y for $0 \le t \le 500$.

22. $x' = x(0.04 - 0.001x - 0.003y)$, $y' = y(0.02 - 0.001x - 0.0004y)$, $x(0) = 4, y(0) = 3$

23. $x' = x(0.04 - 0.001x - 0.003y)$, $y' = y(0.02 - 0.001x - 0.0004y)$, $x(0) = 4, y(0) = 4$

Extended Life Science Connection

EPIDEMICS

Differential equations may be used to predict the rate at which a disease spreads throughout a population. For many diseases, the population may be divided into three groups:

1. Individuals who are *susceptible* to the disease but have not yet been infected.
2. Individuals who are *infected* and able to spread the disease to susceptibles, and
3. Individuals who were infected but have either recovered or died. These individuals can neither reacquire the disease nor infect others and are *removed* from the infected group.

We will denote the populations of these three groups after t days by $S(t)$, $I(t)$, and $R(t)$, respectively. We will assume that susceptibles become infected at a rate proportional to the number of contacts between susceptible and infected individuals. We also assume that individuals are removed from the infected group at a constant relative rate.

We say that an *epidemic* occurs if the infected population increases at least once. In other words, if $I'(t) > 0$ for some value of t, then an epidemic occurs.

EXERCISES

tw **1.** Explain why the above assumptions yield the differential equations

$$S' = -aSI, \quad I' = aSI - bI, \quad \text{and}$$
$$R' = bI, \tag{1}$$

where a and b are constants.

2. Let $P = S + I + R$, the total population. Show that, under the assumptions of this model, the total population remains constant.

3. Suppose that $a = 10^{-5}$ and $b = 0.1$. Suppose also that there are initially 500 infected individuals present and that no one is initially in the recovered group. For the given population sizes, use Euler's method with $\Delta t = 0.1$ to numerically approximate $S(t)$, $I(t)$, and $R(t)$ for $0 \leq t \leq 100$. Use a spreadsheet to graph these three functions.

a) $P = 2000$

b) $P = 5000$

c) $P = 20,000$

d) $P = 30,000$

4. Based on the graphs of $I(t)$, for which populations did an epidemic occur?

The graphs of Exercise 3 illustrate two important features of the spread of a disease. First, epidemics do not occur in small populations but may occur in large populations. Second, after an epidemic has run its course, there will be susceptibles who remain uninfected. In the following exercises, we will investigate further these two properties of epidemics.

MINIMUM POPULATION SIZE

First, we investigate the conditions for an epidemic to occur. We assume there are some infected individuals present at time 0.

5. Use the differential equation for I to show that the sign of $I'(t)$ agrees with the sign of $aS(t) - b$.

6. Explain why an epidemic must occur if $P > b/a$. In other words, explain why epidemics occur only if the population is large enough.

7. Refer to the graphs of Exercise 3. Did epidemics occur for populations larger than b/a?

UNINFECTED INDIVIDUALS

We now turn to predicting the number of individuals that will be uninfected after the epidemic has run its course.

8. From the Chain Rule, $dI/dt = (dI/dS) \times (dS/dt)$. Use this to show that

$$\frac{dI}{dS} = \frac{b}{aS} - 1.$$

9. In Exercise 3, the initial conditions were $I(0) = 500$ and $S(0) = P - 500$. Integrate the result of Exercise 8 and use these initial conditions to solve for C. Show that

$$I = \frac{b}{a}(\ln S - \ln [P - 500]) - S + P.$$

10. Let x denote the number of susceptible individuals who remain after the epidemic. Explain why

$$0 = \frac{b}{a}(\ln x - \ln [P - 500]) - x + P.$$

11. Use Newton's method with initial estimate $x_0 = 10$ to approximately solve for x if $P = 20,000$.

12. Is your solution in approximate agreement with the graph of $S(t)$ made in Exercise 3? (The agreement can be improved by choosing a smaller step size in Euler's method.)

13. Repeat Exercise 11 for $P = 30,000$.

10

Probability

AN APPLICATION   On average, an earthquake large enough to cause damage occurs in Nevada approximately every 3 yr. What is the probability that at least one large earthquake will hit Nevada within the next 4 yr?

This problem appears as Exercise 57 in Exercise Set 10.6.

INTRODUCTION *Gene transmission, cell decay rates, and clinical medical trials may all be modeled with the laws of probability. This chapter introduces probability and its connections to calculus. Both discrete and continuous notions of probability are presented, including the celebrated "bell curve" of the normal distribution.*

10.1 Probability

OBJECTIVES

■ Compute probabilities using the Principle of Equally Likely Outcomes.
■ Use the Addition, Complement, and Multiplication Rules.
■ Compute conditional probabilities.

Probability

We define an *event* to be a set of possible outcomes of an experiment. For the experiment of flipping a coin, landing heads is an event. For the experiment of rolling a die, getting a three on the die is an event. If we repeat the experiment many times and compute the number of times an event occurs as a proportion of the number of experiments, then that ratio approaches the **probability** of the event. We denote the probability of the event A as $P(A)$. For example, if we toss a coin a great number of times, say, 1000, and heads comes up 503 times, then we would estimate the probability of a head coming up on another toss as

$$P(\text{head}) = \frac{503}{1000} = 0.503.$$

Face Value	Frequency
1	166
2	176
3	158
4	178
5	153
6	169
Total	1000

EXAMPLE 1 *Rolling a Die.* A die is rolled 1000 times. The table in the margin shows the number of times the die came up each possible number. Use this experiment to estimate the probability of each possible outcome.

Solution Of the 1000 rolls, 166 were ones. Based on the experiment, we estimate $P(1) = 166/1000 = 0.166$. Similarly, we would estimate $P(2) = 0.176$, $P(3) = 0.158, P(4) = 0.178, P(5) = 0.153,$ and $P(6) = 0.169$. ∎

If p is a probability of an event, then $0 \leq p \leq 1$. A probability of 0.5 indicates that the event has just as much chance of occurring as not occurring, but a probability of 0.99 indicates that we are pretty sure the event will occur. A probability of 1 means that we are certain that the event will occur.

Equally Likely Outcomes

In Example 1, we may have expected the probability of all six outcomes to be equal. They may be, but determining whether they are or not is the subject of *statistical inference.* In this chapter, we will focus not so much on experimentally determining probabilities, but on computing probabilities under reasonable theoretical assumptions.

Often, we may reasonably assume the outcomes of an experiment to be *equally likely.* In Example 1, if the die was symmetric (except for minor differences among the sides) and was evenly weighted, then we would expect the six possible outcomes to be equally likely. In this case, the probability of each side landing face up would be 1/6. If the outcomes are equally likely, then the die is called **fair.**

Often statements of probability are formulated by using *expert opinion*. For example, when financial analysts forecast how the stock market will behave in the future, their predictions are based on expert opinion. These predictions cannot be determined experimentally since each day's trading is conducted only once.

The following probabilities are more expressions of expert opinion:

"The probability that a heart attack victim will survive the week is 3/4."

"If you kiss someone who has a cold, the probability of your catching a cold is 0.57."

"There is a 60% chance of rain tomorrow."

THEOREM 1
The Principle of Equally Likely Outcomes

If an experiment is conducted with n possible outcomes, all equally likely, and A is an event with k possible outcomes, then the probability that A occurs is k/n. That is,

$$P(A) = \frac{k}{n}.$$

This fraction may be interpreted as the number of *favorable* outcomes (for the event A) divided by the number of *possible* outcomes.

EXAMPLE 2   A jar contains 7 black balls, 6 yellow balls, 4 green balls, and 3 red balls. The jar is shaken and you remove a ball without looking. What is the probability that **(a)** the ball is red? **(b)** that it is white? **(c)** that it is either black, yellow, green, or red?

Solution

a) There are 20 balls that we assume are identical except for their colors, so each is equally likely to be picked. There are 3 red balls, and so

$$P(\text{red}) = \frac{3}{20}$$

by the Principle of Equally Likely Outcomes.

This answer does not mean that if you tried the experiment 20 times (replacing the removed ball between experiments), you would remove a red ball exactly 3 times, nor does it mean that if you did the experiment 20,000,000 times that you would remove a red ball exactly 3,000,000 times. It means that the proportion of the experiments that you remove a red ball gets very close to 3/20 as you repeat the experiment many times.

b) There are no white balls, so

$$P(\text{white}) = \frac{0}{20} = 0.$$

c) The ball must be either black, yellow, green, or red, and so

$$P(\text{black or yellow or green or red}) = \frac{20}{20} = 1.$$

∎

The Addition and Complement Rules

While the Principle of Equally Likely Outcomes may be used to quantify probabilities, the task of counting the favorable and possible outcomes may be quite burdensome. We now discuss the Addition Rule, the Complement Rule, and the Multiplication Rule, which are techniques for computing probabilities without explicitly listing the outcomes.

EXAMPLE 3   *Playing Cards.* A card is drawn from a well-shuffled standard deck of 52 cards.

a) What is the probability of drawing an ace?

b) What is the probability of drawing a king?

c) What is the probability of not drawing a king?

d) What is the probability of drawing an ace or a king?

Solution

a) There are 52 possible outcomes, and each is equally likely. Since there are 4 aces in the deck, the Principle of Equally Likely Outcomes says that the probability of drawing an ace is $P(\text{ace}) = 4/52 = 1/13$.

b) Since there are also 4 kings, $P(\text{king}) = 4/52 = 1/13$.

c) Since there are 4 kings in the deck, there are 48 cards in the deck that are not kings. So $P(\text{not king}) = 48/52 = 12/13$.

d) There are a total of 8 aces and kings, so $P(\text{ace or king}) = 8/52 = 2/13$.   ■

   We say that two events are **disjoint** if they cannot occur in the same experiment. For example, drawing an ace and drawing a king are disjoint events. For these disjoint events, we saw in Example 3 that

$$P(\text{ace or king}) = P(\text{ace}) + P(\text{king}).$$

This is an example of the **Addition Rule.**

THEOREM 2

Addition Rule

Suppose that A and B are disjoint events. Then $P(A \text{ or } B) = P(A) + P(B)$.

The assumption of disjoint events is essential. For example, drawing an ace and drawing a heart are not disjoint events. In this case,

$$P(\text{ace or heart}) \neq P(\text{ace}) + P(\text{heart}).$$

In the exercises, we will discuss how the Addition Rule extends to events that are not disjoint.

CAUTION If there is a chance that A and B can occur in the same experiment, then $P(A \text{ or } B) \neq P(A) + P(B)$.

Also in Example 3, $P(\text{king}) = 12/13$ and $P(\text{not king}) = 1/13$. These disjoint events are called *complements* because exactly one of these two events must occur. As a result, these probabilities sum to 1:

$$P(\text{king}) + P(\text{not king}) = 1, \quad \text{or} \quad P(\text{not king}) = 1 - P(\text{king}).$$

This is an example of the **Complement Rule.**

> **THEOREM 3**
>
> Complement Rule
>
> Let A be an event. Then $P(\text{not } A) = 1 - P(A)$.

EXAMPLE 4 A jar contains 7 black balls, 6 yellow balls, 4 green balls, and 3 red balls. The jar is well shaken and a ball is drawn from the jar.

a) Find the probability that the ball is yellow.
b) Find the probability that the ball is red.
c) Find the probability that the ball is red or yellow.
d) Find the probability that the ball is not red.

Solution

a) There are 20 balls in the jar and 6 of them are yellow. Since each ball is equally likely to be drawn, we have $P(\text{yellow}) = 6/20 = 3/10$.

b) There are 3 red balls, so $P(\text{red}) = 3/20$.

c) Since the events of drawing a yellow ball and drawing a red ball are disjoint, we may use the Addition Rule:

$$P(\text{red or yellow}) = 3/20 + 6/20 = 9/20.$$

d) Using the Complement Rule,

$$P(\text{not red}) = 1 - 3/20 = 17/20.$$

Both parts (c) and (d) could also be done by using the Principle of Equally Likely Outcomes. The Addition and Complement Rules provide an alternate, and often faster, method of computing these probabilities. ■

The Multiplication Rule

The Addition Rule is used to find the probability that one of two disjoint events occurs. By contrast, the Multiplication Rule is used to find the probability that two events both occur.

EXAMPLE 5   A large class of students are given identical boxes containing three tickets, marked 1, 2, and 3. Students are then asked to draw two tickets from the box without placing the first ticket back into the box. A prize will be given to students who draw the 3 ticket first and the 1 ticket second.

a) About what proportion of students draw the 3 ticket first?

b) Of the students who drew the 3 ticket first, about what proportion draw the 1 ticket second?

c) About what proportion of students in the class win the prize?

Solution

a) According to the Principle of Equally Likely Outcomes, about 1/3 of the students draw the 3 ticket first.

b) For the students who drew the 3 ticket first, their box now contains the 1 ticket and the 2 ticket. On the second draw, about 1/2 of these students will draw the 1 ticket.

c) About 1/2 of 1/3 of all the students should win the prize. In other words, about 1/6 of the students in the class should win the prize. ■

   Example 5(b) asks for the probability that the 1 ticket is drawn second *given that* the 3 ticket is drawn first. This is expressed by the symbol

$$P\left(1 \text{ second} \mid 3 \text{ first}\right).$$

The vertical line should be read as "given that." Using this notation, the result of Example 5(c) may be written as

$$P\left(3 \text{ first and } 1 \text{ second}\right) = P\left(3 \text{ first}\right)P\left(1 \text{ second} \mid 3 \text{ first}\right)$$
$$= \left(\frac{1}{3}\right)\left(\frac{1}{2}\right) = \frac{1}{6}.$$

This is an example of the **Multiplication Rule.**

THEOREM 4

Multiplication Rule

Let A and B be two events. Then

$$P(A \text{ and } B) = P(A)P(B \mid A).$$

In this equation, $P(B \mid A)$, read "the probability of B given A," denotes the **conditional probability** that event B happens given that event A has already happened.

EXAMPLE 6   *Playing Cards.* Two cards are dealt from a well-shuffled deck of cards.

a) What is the probability that the first card is a heart?

b) Given that the first card is a heart, what is the probability that the second card is a heart?

c) What is the probability that both cards are hearts?

Solution   Let A be the event "the first card is a heart," and let B be the event "the second card is a heart."

a) There are 52 cards in a deck of cards, of which 13 are hearts. By the Principle of Equally Likely Outcomes, $P(A) = 13/52 = 1/4$.

b) If the first dealt card is a heart, then there are 51 remaining cards in the deck, of which 12 are hearts. Therefore, $P(B \mid A) = 12/51 = 4/17$.

c) By the Multiplication Rule,

$$P(A \text{ and } B) = P(A)P(B \mid A) = \left(\frac{13}{52}\right)\left(\frac{12}{51}\right) = \left(\frac{1}{4}\right)\left(\frac{4}{17}\right) = \frac{1}{17}.$$ ∎

In principle, Example 6 could be solved by using the Principle of Equally Likely Outcomes. This would entail listing all 2652 ways that the first two cards could be dealt. After listing all of these possible outcomes, the number of favorable outcomes containing two hearts must be counted. We see that the Multiplication Rule gives an effective shortcut for computing probabilities when listing all of the possible outcomes is unrealistic.

Independence

In many important examples, the occurrence of one event has nothing to do with the occurrence of another event. We say that such events are **independent.**

> **DEFINITION**
>
> We say that the events A and B are *independent* if
>
> $P(B \mid A) = P(B)$.
>
> In words, the event A occurring does not affect whether or not the event B occurs. For independent events, Theorem 4 reduces to
>
> $P(A \text{ and } B) = P(A)P(B)$.
>
> Events that are not independent are called *dependent*.

CAUTION   *Independent* does not mean the same thing as *disjoint*. If the events A and B are disjoint and the event A occurs, then B cannot occur. For disjoint events, $P(B \mid A) = 0$.

If two events are independent, then the probability that both happen is the product of the probabilities of each of them happening. However, blindly multiplying probabilities is inappropriate if the events are dependent.

In 1999, Sally Clark was convicted by a British court of murdering her first two infant children, even though no cause of death could be established. The defense contended that her children had died of sudden infant death syndrome (SIDS).

To counter this argument, the prosecution noted that about one British infant in 8550 dies of SIDS. Squaring this probability, the probability for two children to die of SIDS is 1 in 73 million. The prosecution thus argued that it was extremely unlikely that the two brothers died of SIDS. Influenced by this argument, the court sentenced her to two life sentences.

In 2001, the Royal Statistical Society criticized this misuse of probability. The events of two siblings dying of SIDS can be plausibly explained by genetic or environmental factors and hence are dependent. Since squaring 1/8550 was inappropriate, her conviction was overturned in 2003.

EXAMPLE 7 *Rolling a Die.* A fair die is rolled twice. What is the probability that both rolls are sixes?

Solution The probability that the first roll is a six is 1/6. Whether or not the first roll is a six, the probability that the second roll is a six is also 1/6. Because the two events are independent,

$$P(\text{first a six and second a six}) = P(\text{first a six})P(\text{second a six})$$

$$= \left(\frac{1}{6}\right)\left(\frac{1}{6}\right) = \frac{1}{36}.$$ ■

Exercise Set 10.1

A jar contains 10 black balls, 23 yellow balls, 14 green balls, and 3 red balls. The jar is shaken and you remove a ball without looking. Find the probability of the event.

1. The ball is red.

2. The ball is black.

3. The ball is green.

4. The ball is white.

5. The ball is yellow.

6. The ball is red or black.

7. The ball is green or black.

8. The ball is not red.

9. The ball is not green.

10. The ball is neither yellow nor red.

Determine if the events A and B are (a) independent or (b) disjoint.

11. A coin is flipped twice. Let A be the event "the first flip is heads," and let B be the event "the second flip is tails."

12. A coin is flipped. Let A be the event "the flip is heads," and let B be the event "the flip is tails."

13. A die is rolled once. Let A be the event "the die is a four," and let B be the event "the die is a five."

14. A die is rolled twice. Let A be the event "the first die is a four," and let B be the event "the second die is a five."

15. Two cards are dealt from a deck of cards. Let A be the event "the first card is an ace," and let B be the event "the second card is an ace."

16. A card is dealt from a deck of cards. Let A be the event "the card is a queen," and let B be the event "the card is a spade."

17. A card is dealt from a deck of cards. Let A be the event "the card is a queen," and let B be the event "the card is a king."

18. Two cards are dealt from a deck of cards. Let A be the event "the first card is a queen," and let B be the event "the second card is a king."

Two tickets are drawn from the box

The first ticket is not returned to the box before the second ticket is selected. Find the probability of the event.

19. The first ticket is **1**.

20. The first ticket is **5**.

21. The first ticket is not **2**.

22. The first ticket is not **1**.

23. The first ticket is **2** or **3**.

24. The first ticket is **1**, **2**, or **5**.

25. The first ticket is **1**, and the second ticket is **5**.

26. The first ticket is **2**, and the second ticket is **3**.

27. The first ticket is **3**, and the second ticket is not **2**.

28. The first ticket is **2**, and the second ticket is not **5**.

29.–32. Repeat Exercises 25–28 if the first ticket is returned to the box before the second ticket is drawn.

Two tickets are drawn from the box

| 1 | 1 | 1 | 2 | 2 | 3 | 3 | 3 | 3 | 4 |

The first ticket is not returned to the box before the second ticket is selected. Find the probability of the event.

33. The first ticket is **3**.

34. The first ticket is **1**.

35. The first ticket is not **4**.

36. The first ticket is not **2**.

37. The first ticket is **1** or **3**.

38. The first ticket is **2**, **3**, or **4**.

39. The first ticket is **2**, and the second ticket is **2**.

40. The first ticket is **3**, and the second ticket is **1**.

41. The first ticket is **4**, and the second ticket is not **3**.

42. The first ticket is **1**, and the second ticket is not **1**.

43.–46. Repeat Exercises 39–42 if the first ticket is returned to the box before the second ticket is drawn.

SYNTHESIS

Rolling a Die. A fair die is rolled three times.

47. Find the probability that all three rolls are sixes.

48. Find the probability that none of the rolls is a six.

49. Find the probability that not all of the rolls are sixes.

tw 50. Why are the answers to Exercises 48 and 49 different?

Flipping a Coin. A fair coin is flipped five times.

51. Find the probability that all five flips are heads.

52. Find the probability that none of the flips is heads.

53. Find the probability that not all of the flips are heads.

tw 54. Why are the answers to Exercises 52 and 53 different?

Rolling a Die. A die is rolled four times.

55. What is the probability that the rolls are all sixes?

56. What is the probability that the rolls are all the same?

57. What is the probability that the rolls are all different?

58. What is the probability that a face appears more than once?

Addition Rule for Events that are not Disjoint. If the events A and B are not disjoint, then the Addition Rule may be generalized as

$$P(A \text{ or } B) = P(A) + P(B) - P(A \text{ and } B).$$

To illustrate this, suppose a card is selected from a well-shuffled deck. Let A be the event "the card is an ace," and let B be the event "the card is a heart." Then $P(A) = 4/52$, $P(B) = 13/52$, and $P(A \text{ and } B) = 1/52$ since there is one card in the deck that is both an ace and a heart (namely, the ace of hearts). As a result,[1]

$$P(\text{ace or heart}) = P(\text{ace}) + P(\text{heart})$$
$$- P(\text{ace and heart})$$
$$= \frac{4}{52} + \frac{13}{52} - \frac{1}{52} = \frac{16}{52} = \frac{4}{13}.$$

Notice that subtracting 1/52 removes the effect of double counting the ace of hearts. In general, the Multiplication Rule is used to calculate $P(A \text{ and } B)$.

59. Two dice are rolled. Find the probability that the first die is a six or the second die is a four.

60. Two coins are flipped. Find the probability that the first coin lands heads or the second coin lands tails.

61. Two cards are dealt from a well-shuffled deck. Find the probability that the first card is an ace or the second card is a king.

62. Two dice are rolled. Find the probability that the first die is not a six or the second die is not a four.

tw 63. Give four examples of probability in daily life.

---

[1]Notice that the use of the word *or* in probability differs slightly from its usage in ordinary speech. When you hear someone say, "I'll take genetics or psychology next semester," you don't think that he'll enroll in both classes. In contrast, when we write $P(A \text{ or } B)$ in probability, it may be possible for both A and B to occur.

10.2 Multiplication Trees and Bayes' Rule

OBJECTIVES

■ Use multiplication trees.
■ Compute posterior probabilities using Bayes' Rule.

Probabilities of events determined by one random experiment may be computed using the Principle of Equally Likely Outcomes. If an event is determined by two random experiments (for example, flipping a fair coin twice), then the Addition and Multiplication Rules may be used to compute the probability. A convenient visualization of events determined by two or more random experiments is called a **multiplication tree,** as shown in the next example.

EXAMPLE 1 *Playing Cards.* Two cards are dealt from a well-shuffled deck.

a) Find the probability that 2 hearts appear.

b) Find the probability that 0 hearts appear.

c) Find the probability that 1 heart appears.

d) Find the probability that the second card is a heart.

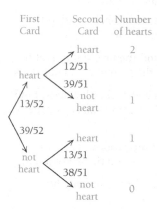

First Card · Second Card · Number of hearts

Solution The multiplication tree for this example is shown in the margin. The arrows denote the possible outcomes of the first card and the second card. The probabilities for each arrow denote the probabilities (or the conditional probabilities) of the event occurring. Two of these probabilities, 13/52 and 12/51, were computed in Example 6 of Section 10.1. The others are left to the student. The last column is a summary of each branch of the multiplication tree. For this problem, the summary is the number of hearts that appear. The probability of two events occurring may then be found by multiplying the probabilities for the respective arrows.

a) Using the multiplication tree, we see that

$$P(2 \text{ hearts}) = \left(\frac{13}{52}\right)\left(\frac{12}{51}\right) = \frac{1}{17} \approx 0.059.$$

This is the same answer obtained in Example 6(c) of Section 10.1.

b) Using the multiplication tree, we see that

$$P(0 \text{ hearts}) = \left(\frac{39}{52}\right)\left(\frac{38}{51}\right) = \frac{19}{34} \approx 0.559.$$

c) There are two ways that one heart could appear, corresponding to the second and third entries of the final column. These two ways are disjoint, so we may add the probabilities:

$$P(1 \text{ heart}) = \left(\frac{13}{52}\right)\left(\frac{39}{51}\right) + \left(\frac{39}{52}\right)\left(\frac{13}{51}\right) = \frac{13}{34} \approx 0.382.$$

Not surprisingly, the answers to parts (a), (b), and (c) add to 1.

d) There are two ways that the second card could be a heart, corresponding to the first and third entries of the final column. These two ways are disjoint, and so we may add the probabilities:

$$P(\text{second card a heart}) = \left(\frac{13}{52}\right)\left(\frac{12}{51}\right) + \left(\frac{39}{52}\right)\left(\frac{13}{51}\right) = \frac{1}{4}.$$

This answer is intuitively clear. The question asks for the *unconditional* probability that the second card is a heart without any knowledge of the first card. If the first card is dealt facedown without looking at it, there are still 52 cards of unknown suit, even though one has been physically removed from the deck. The probability that the second card is a heart (without looking at the first card) is therefore $13/52 = 1/4$. ∎

Genetics

The Addition and Multiplication Rules can be used to explain many principles of genetics. In the following paragraphs, we review some basic definitions of genetics by discussing the flower color of pea plants.[2]

Pea plants may have either purple flowers or white flowers, which are examples of *phenotypes*. Phenotypes are determined by two *alleles,* or variants of a gene. For example, if F is the allele for purple flowers and f is the allele for white flowers, then there are three possible *genotypes:* FF, Ff, and ff. The genotypes FF and ff are called *homozygous,* while the hybrid genotype Ff is called *heterozygous.*

In some cases, the alleles are *dominant* and *recessive.* For flower color, F is the dominant allele and f is the recessive allele. If at least one F allele is present (genotypes FF and Ff), then the flower is purple. If no F allele is present (genotype ff), then the flower is white.

The genotypes FF and Ff correspond to purple flowers and hence belong to the same phenotype. Nevertheless, these two genotypes may be distinguished with a *test cross,* as discussed in the exercises.

During pollination, the egg and sperm gametes each get only one of the two alleles. According to the *law of segregation,* these alleles are selected independently and with equal probability from the two alleles of the parents. These two alleles from the gametes form the genotype of the offspring.

EXAMPLE 2   *Genetics.* Suppose that pea plants of genotypes FF and ff are crossed. A sperm allele is transmitted from the FF plant, while an ovum allele is transmitted from the ff plant.

a) Find the probability that an offspring is of each of the three genotypes.

b) Find the probability that the offspring will have purple flowers.

c) Find the probability that the offspring will have white flowers.

---

[2]N. A. Campbell and J. B. Reece, *Biology,* 6th ed., (New York: Benjamin Cummings, 2002).

Solution

Sperm Ovum Genotype
Allele Allele

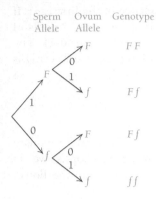

a) We begin by constructing a multiplication tree, which is shown in the margin. Notice that the sperm allele must be F, while the ovum allele must be f. Also, since these alleles are transmitted independently, the second stage of the multiplication tree has the same probabilities.

From the multiplication tree, we see that

$$P(FF) = (1)(0) = 0,$$
$$P(Ff) = (1)(1) + (0)(0) = 1, \quad \text{and}$$
$$P(ff) = (0)(1) = 0.$$

b) Since there are two genotypes for the phenotype of purple flowers, we may use the Addition Rule:

$$P(\text{purple}) = P(FF) + P(Ff) = 0 + 1 = 1.$$

c) Since there is one genotype for the phenotype of white flowers, $P(\text{white}) = P(ff) = 0$.

Not surprisingly, the offspring of this cross are guaranteed to be heterozygous and hence purple. (This calculation may also be visualized using a *Punnett square*, as discussed in the exercises.) ∎

Sperm Ovum Genotype
Allele Allele

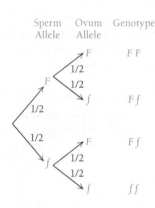

EXAMPLE 3 *Genetics.* Suppose that a pea plant of genotype Ff is self-pollinated.
a) Find the probability that an offspring is of each of the three genotypes.
b) Find the probability that the offspring will have purple flowers.
c) Find the probability that the offspring will have white flowers.

Solution

a) The multiplication tree is shown in the margin; verifying the probabilities on each branch is left to the student. From the multiplication tree, we see that

$$P(FF) = (0.5)(0.5) = 0.25,$$
$$P(Ff) = (0.5)(0.5) + (0.5)(0.5) = 0.5, \quad \text{and}$$
$$P(ff) = (0.5)(0.5) = 0.25.$$

b) Using the Addition Rule,

$$P(\text{purple}) = P(FF) + P(Ff) = 0.25 + 0.5 = 0.75.$$

c) $P(\text{white}) = P(ff) = 0.25.$

If a large number of offspring are produced, there should be three times as many purple-flowering plants as white-flowering plants. The occurrence of this 3:1 ratio for many characteristics of pea plants, including flower color, led Gregor Mendel to postulate his principles of genetics. ∎

Other applications of probability to genetics are presented in the exercises.

Bayes' Rule[3]

One important application of conditional probabilities is the clinical diagnosis of a disease based on medical studies. In these studies, researchers give patients who are previously known to either have or not have a certain disease a medical test. Ideally, the test correlates perfectly with whether or not the patient has the disease. However, medical tests are usually not completely accurate. A *false positive* occurs when a disease-free patient tests positive. A *false negative* occurs when a patient who has the disease tests negative.

Let $D+$ and $D-$ denote the events that a patient does or does not have the disease, respectively. Also, let $T+$ and $T-$ denote the events that the test is positive or negative, respectively. The probability that a patient with the disease is accurately identified by the test is called the *sensitivity* of the test. In other words,

$$\text{Sensitivity} = P(T+ \mid D+) = 1 - P(\text{false positive}).$$

Furthermore, the probability that a disease-free patient is accurately identified by the test is called the *specificity* of the test. In other words,

$$\text{Specificity} = P(T- \mid D-) = 1 - P(\text{false negative}).$$

Ideally, the sensitivity and specificity of a test are both close to 1.

EXAMPLE 4   *Coronary Artery Disease.* In a study of 101 patients, 37 do not have coronary artery disease (CAD) and 64 have CAD. All 101 patients were given a certain echocardiography test for CAD. Of the 37 patients without CAD, 34 had a negative test while 3 had a positive test. Of the 64 patients with CAD, 54 had a positive test and 10 had a negative test. Find the sensitivity and the specificity of the test.[4]

Solution   To find the sensitivity, we see that the test correctly identified 54 of the 64 patients with CAD. Therefore,

$$\text{Sensitivity} = P(T+ \mid D+) = \frac{54}{64} \approx 0.84.$$

Second, the test correctly identified 34 of the 37 patients without CAD. Therefore,

$$\text{Specificity} = P(T- \mid D-) = \frac{34}{37} \approx 0.92.$$

In this example, we take these relative frequencies as exact probabilities. ■

Medical studies measure the accuracy of a test on patients who are already known to have or not have a disease. On the other hand, when a patient visits a doctor's office, the status of the disease in the patient is unknown. Before a test is performed, the doctor estimates $P(D+)$, the probability that the patient has the disease,

---

[3]This section may be omitted without loss of continuity.
[4]W. Mathias, Jr., J. M. Tsutsui, J. L. Andrade, I. Kowatsch, P. A. Lemos, S. M. B. Leal, B. K. Khandheria, and J. F. Ramires, "Value of Rapid Beta-Blocker Injection at Peak Dobutamine-Atropine Stress Echocardiography for Detection of Coronary Heart Disease," *Journal of the American College of Cardiography*, Vol. 41, pp. 1583–1589 (2003).

based on professional expertise. After the test, the probability of having the disease is then updated based on the extra information from the test. Mathematically, this probability is either $P(D+ \mid T+)$ or $P(D+ \mid T-)$, depending on the result of the test.

Notice that the roles of the D and T events have been reversed. In medical studies, the occurrence of the disease is known, and probabilities are assigned to whether or not the test is effective. In clinical diagnosis, the result of the test is known, and probabilities for having the disease are assigned based on the test result and the doctor's previous estimate for the patient having the disease.

To calculate a probability such as $P(D+ \mid T+)$, we rewrite the Multiplication Rule. Since

$$P(D+ \text{ and } T+) = P(T+)P(D+ \mid T+),$$

we may divide both sides by $P(T+)$ to obtain

$$P(D+ \mid T+) = \frac{P(D+ \text{ and } T+)}{P(T+)}.$$

The two probabilities on the right-hand side may be computed using a multiplication tree, as illustrated in the next example.

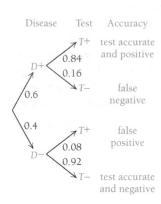

Disease   Test   Accuracy

$T+$ test accurate and positive

$T-$ false negative

$T+$ false positive

$T-$ test accurate and negative

EXAMPLE 5 *Diagnosing Coronary Artery Disease.* A man in his 50s with a family history of coronary artery disease (CAD) sees a doctor, complaining of chest pain. Because of his age and family history, the doctor estimates the probability that the patient has CAD as 0.6. The patient is then given the echocardiography test of Example 4, and the test returns positive. Find the probability that the patient has CAD given the positive test result.

Solution   Based on the multiplication tree in the margin, we see that

$$P(D+ \text{ and } T+) = (0.6)(0.84) = 0.504.$$

Also, there are two ways that the test can return positive. The test could either be accurate and positive, or the test could be a false positive. Adding the probabilities for these two disjoint events, we have

$$P(T+) = (0.6)(0.84) + (0.4)(0.08) = 0.536.$$

Therefore,

$$P(D+ \mid T+) = \frac{P(D+ \text{ and } T+)}{P(T+)} = \frac{0.504}{0.536} \approx 0.94.$$

From the result of the test, we see that there is a 94% chance that the patient has CAD and a 6% chance that the patient does not have CAD. Based on these probabilities, the doctor can be reasonably sure that the patient has CAD and can prescribe treatment accordingly. ■

Reviewing the calculations of Example 5, the answer may be written as

$$P(D+ \mid T+) = \frac{(0.6)(0.84)}{(0.6)(0.84) + (0.4)(0.08)}.$$

We see that the numerator is one of the two terms in the denominator. The term 0.6 was the doctor's estimate of $P(D+)$, so 0.4 is equal to $P(D-)$. Also, the terms

Quantifying probabilities, such as the risk of having coronary artery disease, has obvious importance in medicine, biology, and engineering. However, these probabilities cannot be measured accurately with limited empirical data. In the absence of data, they are instead estimated by expert opinion.

Firm reliance on opinion may have disastrous consequences. Prior to the destruction of the space shuttle *Challenger* in 1986, shuttle engineers estimated the probability of catastrophe as 1 in 300. However, NASA management used an unreasonable estimate of 1 in 100,000. As a result, management was willing to launch *Challenger* in subfreezing weather, even though engineers had doubts about the shuttle's reliability under those conditions.

Though the specific causes of the accident were different, the same combination of a design flaw and human error influenced by inaccurate probabilities led to the destruction of the space shuttle *Columbia* in 2003.

0.84 and 0.08 are the conditional probabilities $P(T+ \mid D+)$ and $P(T+ \mid D-)$ from Example 5. Replacing these numbers with the corresponding probabilities, we obtain **Bayes' Rule.**

THEOREM 5

Bayes' Rule

$$P(D+ \mid T+) = \frac{P(D+)P(T+ \mid D+)}{P(D+)P(T+ \mid D+) + P(D-)P(T+ \mid D-)}.$$

Similar formulas may be written for $P(D- \mid T+)$, $P(D+ \mid T-)$, and $P(D- \mid T-)$.

Bayes' Rule is used to update the probability of an event (like occurrence of a disease) given extra information (like a positive test). It is often easier to formulate Bayes' Rule using a multiplication tree, as in Example 5, than to memorize this formula.

Notice that the doctor had to make an initial estimate of $P(D+)$ before using Bayes' Rule. This initial estimate, usually determined by expert opinion, is called the **prior probability.** The conditional probability (using the test result) is called the **posterior probability.** In Example 5, the prior probability was 0.6 and the posterior probability was 0.94.

Notice that the posterior probability is dependent on the prior probability. If the prior probability changes, then the posterior probability will also change, as seen in the next example.

EXAMPLE 6   *Diagnosing Coronary Artery Disease.* A man in his 20s with no available family history sees a doctor complaining of chest pain. Because of his age, the doctor estimates the probability that the patient has CAD as 0.05. The patient is then given the echocardiography test of Example 4, and the test returns positive. Find the probability that the patient has CAD given the positive test result.

Solution   In this case, $P(D+) = 0.05$ and $P(D-) = 0.95$. Using Bayes' Rule,

$$P(D+ \mid T+) = \frac{P(D+)P(T+ \mid D+)}{P(D+)P(T+ \mid D+) + P(D-)P(T+ \mid D-)}$$

$$= \frac{(0.05)(0.84)}{(0.05)(0.84) + (0.95)(0.08)}$$

$$= \frac{0.042}{0.118} \approx 0.356.$$

While the posterior probability is larger than the prior probability, it's still more likely than not that the patient is disease free. This conclusion is different than that for the patient of Example 5 because the prior probability is very close to 0.

Example 6 may also be solved by using a multiplication tree and the definition of conditional probability, as in Example 5. This computation is left to the student. ∎

Exercise Set 10.2

Two tickets are drawn from the box

$$\boxed{1 \quad 2 \quad 3}.$$

The first ticket is not returned to the box before the second ticket is selected. Use a multiplication tree to find the probability of the event.

1. The second ticket is **2**.

2. The sum of the tickets is 3.

3. The sum of the tickets is 4.

4. The sum of the tickets is 6.

5.–8. Repeat Exercises 1–4 if the first ticket is returned to the box before the second ticket is drawn.

Four fair coins are tossed. Use a multiplication tree to find the probability of the event.

9. All four flips are heads.

10. All four flips are the same.

11. The first flip is different from the last three flips.

12. The number of heads in the first two flips is the same as the number of heads in the last two flips.

13. Exactly one of the flips is heads.

Two cards are dealt from a deck of cards. Use a multiplication tree to find the probability of the event.

14. Both cards are kings.

15. Exactly one of the cards is a spade.

16. Exactly one of the cards is an ace.

APPLICATIONS

17. *Test Cross*

a) The genotypes *Ff* and *ff* are crossed. Find the probability that the offspring are of each of the three genotypes.

tw b) Compare your answer in part (a) to the answer of Example 2. Explain how the *test cross* of a dominant phenotype with a recessive phenotype can distinguish between the *FF* and *Ff* genotypes.

Law of Independent Assortment. In Mendel's pea plants, the allele *Y* for yellow peas is dominant to the allele *y* for green peas, and the allele *R* for round peas is dominant to the allele *r* for wrinkled peas. According to the *law of independent assortment,* the transmission of the *Y* or *y*

allele is independent of the transmission of the *R* or *r* allele.[5]

18. Suppose that a dihybrid pea plant (*YyRr*) is self-pollinated.

a) Find the probability that the offspring has the genotype *YYRR*. (*Hint:* By independence, $P(YYRR) = P(YY)P(RR)$.)

b) Find the probability that the offspring has the genotype *yyRr*.

c) Find the probability that the offspring has the genotype *yyrr*.

19. Suppose that a dihybrid pea plant (*YyRr*) is self-pollinated.

a) Use a multiplication tree and the result of Example 3 to find the probability that the offspring is yellow and round.

b) Find the probability that the offspring is green and round.

c) Find the probability that the offspring is yellow and wrinkled.

d) Find the probability that the offspring is green and wrinkled.

20. *Ectopic Pregnancy.* The receptor interleukin-8 is measured as a test for ectopic (or tubal) pregnancy. For 17 women with ectopic pregnancy, 14 tested positive and 3 tested negative. For 55 women without an ectopic pregnancy, 10 tested positive and 45 tested negative.[6]

a) Find the sensitivity of the test.

b) Find the specificity of the test.

c) An obstetrician treats a woman with severe pelvic pain. Based on her professional experience, the doctor believes the probability of an ectopic pregnancy to be 0.4. The patient is tested, and the test returns positive. Find the posterior probability of an ectopic pregnancy.

21. *Grafting Blood Vessels.* Blood pressure can be used to determine whether an arteriovenous graft, implanted in a previous bypass surgery, is functioning or not. Of 42 patients with a malfunctioning graft, 37 had a positive test while 5 had a negative test.

---

[5]N. A. Campbell and J. B. Reece, *Biology,* 6th ed. (New York: Benjamin Cummings, 2002).

[6]D. Soriano, D. Hugol, N. T. Quang, and E. Darai, "Serum concentrations of interleukin-2R (IL-2R), IL-6, IL-8 and tumor necrosis factor alpha in patients with ectopic pregnancy," *Fertility and Sterility,* Vol. 79, pp. 975–980 (2003).

Of 110 patients with a functioning graft, 4 had a positive test while 106 had a negative test.[7]

a) Find the sensitivity of the test.

b) Find the specificity of the test.

c) A doctor sees a patient after bypass surgery. The doctor believes that the probability of a malfunctioning graft is 0.2. The patient is tested, and the test returns negative. Find the posterior probability of a malfunctioning graft.

22. *Down Syndrome.* In a hospital, 29 babies were born with Down syndrome. Of these 29, 12 had an absent nasal bone in sonograms taken in the second trimester of pregnancy. In 102 other fetuses who did not have Down syndrome, none had absent nasal bones.[8]

a) Find the sensitivity of the test.

b) Find the specificity of the test.

c) A pregnant woman is 40 years old. According to health statistics, the probability that her child will have Down syndrome is 0.012. A second-trimester sonogram shows a nasal bone present. Find the probability that the baby has Down syndrome.

23. *Cirrhosis.* To test for liver cirrhosis, a diagnostic test of albumin levels is performed. The test has sensitivity 67% and specificity 76%.[9]

a) For a certain patient, a doctor estimates the probability of cirrhosis as 0.1. The test for cirrhosis returns positive. Find the posterior probability of cirrhosis.

b) For a certain patient, a doctor estimates the probability of cirrhosis as 0.5. The test for cirrhosis returns positive. Find the posterior probability of cirrhosis.

SYNTHESIS

Punnett Squares. Because of independence and the equal probability of inheriting either allele, geneticists sometimes use *Punnett squares* to visualize the probabilities that the offspring will be of the three genotypes. For example, the Punnett square for the offspring of a self-pollinated pea plant heterozygous for purple flowers (Ff) would be

	F	f
F	FF	Ff
f	Ff	ff

The red F and f are the possible alleles transmitted by the sperm, and the blue F and f are the possible alleles transmitted by the ovum. If one of the four squares is selected at random, the probability that it is dominant homozygous is $1/4$, the probability that is heterozygous is $2/4$, and the probability that it is recessive homozygous is $1/4$. Comparing with Example 3, we see that these are the same probabilities obtained by using a multiplication tree.

24. Use a Punnett square to solve Example 2.

25. Use a Punnett square to solve Exercise 17(a).

26. Use a 4-by-4 Punnett square to solve Exercise 18.

27. Use a 4-by-4 Punnett square to solve Exercise 19.

Transmission of Diseases. Many human diseases, including hemophilia[10] and Tay-Sachs disease, are inherited genetically. Suppose that D is the dominant allele for not having the disease, and d is the recessive allele for having the disease. The genotype DD does not carry the disease. The genotype Dd is not diseased but is a carrier for the disease. The genotype dd is diseased.

Suppose that a diseased individual has only a 20% chance of surviving to adulthood. Individuals who are not diseased (both the DD and Dd genotypes) have a 96% chance of surviving to adulthood. Suppose further that nondiseased carriers (genotype Dd) consist of 1% of the nondiseased population.

---

[7]G. R. Sirken, C. Shah, and R. Raja, "Slow-flow venous pressure for detection of arteriovenous graft malfunction," *Kidney International,* Vol. 63, pp. 1894–1898 (2003).

[8]A. Vintzileos, C. Walters, and L. Yeo, "Absent nasal bone in the prenatal detection of fetuses with trisomy 21 in a high-risk population," *Obstetrics and Gynecology,* Vol. 101, pp. 905–908 (2003).

[9]M. Pompili, G. Addolorato, G. Pignataro, C. Rossi, C. Zuppi, M. Covino, A. Grieco, G. Gasbarrini, and G. L. Rapaccini, "Evaluation of the Albumin-γ-Glutamyltransferase Isoenzyme as a Diagnostic Marker of Hepatocellular Carcinoma-Complicating Liver Cirrhosis," *Journal of Gastroenterology and Hepatology,* Vol. 18, pp. 288–295 (2003).

---

[10]Because of the frequency of hemophilia among her male descendants, geneticists are reasonably certain that Queen Victoria of England (1819–1901) was a carrier for hemophilia. Because royalty tended to marry royalty in the 19th century, several members of Europe's royal families were hemophiliacs, including Alexis, son of the last czar of Russia. Historian Robert Massie, in his book *Nicholas and Alexandra,* argues persuasively that the mysterious Rasputin became influential with the Russian royal family because of his ability to assuage the suffering of young Alexis. More information about the genetic transmission of diseases may be found in W. S. Klug and M. R. Cummings, *Concepts of Genetics* (New York: MacMillan, 2002).

Suppose that two carriers, neither of whom is diseased, mate and have a son, Alex.

28. What is the probability that Alex is of each of the three genotypes?

29. What is the probability that Alex is of each of the three genotypes given that he survives to adulthood?

10.3 The Binomial Distribution

OBJECTIVES

■ Use random variables.
■ Find the probability of a certain number of successes in a binomial experiment.
■ Draw the histogram for the distribution of a random variable.

Many probability problems ask for the number of successes that occur when a certain trial is independently repeated a fixed number of times. In this section, we will see that these probabilities may be described by the *Binomial(n, p) distribution*. We motivate this result with the following example.

EXAMPLE 1   *Rolling a Die.* A fair die is rolled three times. Use a multiplication tree to determine the probability that **(a)** 0 sixes appear, **(b)** 1 six appears, **(c)** 2 sixes appear, and **(d)** 3 sixes appear.

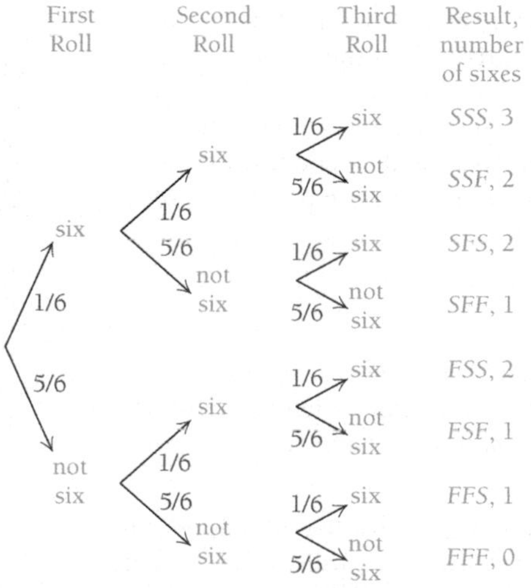

| First Roll | Second Roll | Third Roll | Result, number of sixes |

Solution   Let S denote the event "the roll is a six," and F the event "the roll is not a six." Using the independence of the rolls, the multiplication tree for the three rolls is shown above. (The sequence SFF, for example, is the event "the first roll is a six, the second roll is not a six, and the third roll is not a six.")

a) From the multiplication tree, there is one way (FFF) that 0 sixes could appear:

$$P(0 \text{ sixes appear}) = P(FFF) = \left(\frac{5}{6}\right)^3 = \frac{125}{216}.$$

b) There are three ways (SFF, FSF, FFS) that exactly 1 six could appear. Each has probability $(1/6)(5/6)^2$. Using the Addition Rule, we have

$$P(1 \text{ six appears}) = P(SFF \text{ or } FSF \text{ or } FFS)$$
$$= P(SFF) + P(FSF) + P(FFS)$$
$$= 3\left(\frac{1}{6}\right)\left(\frac{5}{6}\right)^2$$
$$= \frac{75}{216}.$$

c) There are three ways (SSF, SFS, FSS) that exactly 2 sixes could appear. Each has probability $(1/6)^2(5/6)$. Using the Addition Rule as in part (b), we have

$$P(2 \text{ sixes appear}) = P(SSF \text{ or } SFS \text{ or } FSS) = 3\left(\frac{1}{6}\right)^2\left(\frac{5}{6}\right) = \frac{15}{216}.$$

d) Finally, there is only one way (SSS) that 3 sixes could appear:

$$P(3 \text{ sixes appear}) = P(SSS) = \left(\frac{1}{6}\right)^3 = \frac{1}{216}.$$

Not surprisingly, these probabilities add to 1:

$$\frac{125}{216} + \frac{75}{216} + \frac{15}{216} + \frac{1}{216} = \frac{216}{216} = 1.$$

Counting the number of sixes on three rolls of a die is an example of a **binomial experiment.** In a binomial experiment,

1. A trial is repeated n times, where n is some fixed number.
2. The trials are independent of each other.
3. The outcome of each trial may be categorized as a success or a failure.
4. For each trial, the probability of success is some constant p, while the probability of failure is $1 - p$. We will often write $q = 1 - p$.
5. Finally, we wish to count the number of successes that occur in the n trials.

Example 1 illustrates a binomial experiment. A trial (rolling a die) is repeated three times. The rolls are independent. Each trial is either a success (the die lands six) or a failure (the die does not land six). For each trial, the probability of success is $p = 1/6$, so that the probability of failure is $q = 1 - p = 5/6$. Finally, the central problem is to find probabilities for the number of sixes that could appear.

Other examples of binomial experiments are the number of heads that appear when a coin is flipped repeatedly and the number of questions correctly guessed on a multiple-choice test.

Random Variables

In this section, we will introduce a formula for computing the probabilities of a binomial experiment without explicitly constructing the multiplication tree. Before doing so, we introduce some notation. Let X denote the number of sixes that appear on three rolls of a fair die. Then the result of Example 1 may be more succinctly written as

$$P(X = 0) = \frac{125}{216}, \qquad P(X = 1) = \frac{75}{216},$$

$$P(X = 2) = \frac{15}{216}, \quad \text{and} \quad P(X = 3) = \frac{1}{216}. \tag{1}$$

Notice that it is perfectly acceptable to use two equal signs in the same equation. For example, the second equation is read "the probability that the number of sixes is 1 is 75/216."

The variable X is called a **random variable** since the value of X may change for different rolls of the die. Furthermore, since there are finitely many possible values of X, this random variable is called **discrete.** Listing the possible values of a random variable X as well as the probabilities for each, as in equation (1), gives a **distribution** of X.

DEFINITION

A *random variable* X is a variable whose value may change with a random experiment. A *discrete random variable* is a random variable that can only assume a finite number of values. The possible values of X together with their associated probabilities form the *distribution* of X.

  We will always write random variables with capital letters to distinguish them from the usual variables that we considered throughout this book.

A random variable whose possible values form an infinite set of integers is also called discrete.

EXAMPLE 2   *Rolling a Die.*   A fair die is rolled once. If it lands six, you win $4, but you lose $1 if it does not land six. Let M be the amount of money you win. Find the distribution of M.

Solution   The possible values of M are 4 (if you win) and -1 (if you lose). The probabilities of winning and losing are 1/6 and 5/6, respectively. So the distribution of M may be written as

$$P(M = 4) = \frac{1}{6} \quad \text{and} \quad P(M = -1) = \frac{5}{6}. \qquad \blacksquare$$

The Binomial(n, p) Distribution and Binomial Coefficients

Let X denote the number of successes in a binomial experiment with n trials and success probability p. Then we say that X has the **Binomial(n, p) distribution.** For example, the number of sixes that appear when a fair die is rolled three times has the Binomial (3, 1/6) distribution. This distribution was written in full in equation (1). However, the distribution may be more succinctly written using the following theorem.

THEOREM 6

Let X denote the number of successes in a binomial experiment of n trials with success probability p. Then

$$P(X = k) = \binom{n}{k} p^k q^{n-k}, \qquad k = 0, 1, 2, \ldots, n,$$

where $q = 1 - p$. The **binomial coefficient** $\binom{n}{k}$, which is read "n choose k," is defined by

$$\binom{n}{k} = \frac{n!}{k!\,(n-k)!},$$

where $n! = 1 \cdot 2 \cdot 3 \cdot \cdots \cdot n$ if $n \geq 1$ and $0! = 1$.

This theorem may be proven by using a multiplication tree similar to that used in Example 1.

The probability formula of the Binomial(n, p) distribution contain the binomial coefficient $\binom{n}{k}$. We illustrate how factorials are used to calculate these coefficients in the next example. Many calculators have a special function for finding these coefficients, as discussed in the Technology Connection on page 674.

EXAMPLE 3 Use factorials to compute the following binomial coefficients:
(a) $\binom{3}{1}$, (b) $\binom{6}{2}$, (c) $\binom{18}{3}$, and (d) $\binom{4}{0}$.

Solution

a) By definition, $3! = (1)(2)(3) = 6$, $2! = (1)(2) = 2$, and $1! = 1$. Therefore,

$$\binom{3}{1} = \frac{3!}{1!\,(3-1)!} = \frac{3!}{(1!)(2!)} = \frac{6}{(1)(2)} = 3.$$

b) $\binom{6}{2} = \frac{6!}{2!\,(6-2)!} = \frac{6!}{(2!)(4!)} = \frac{(6)(5)(4)(3)(2)(1)}{(2)(1)(4)(3)(2)(1)} = \frac{(6)(5)}{(2)(1)} = 15.$

We see that it is sometimes easier to compute a binomial coefficient by canceling factors instead of multiplying out the factorials. Notice that the 6 tells us where to start the numerator, while the 2 tells us how many terms are in both the numerator and denominator. The 2 also tells us where to start the denominator.

c) Both the numerator and denominator will have three terms. The numerator starts at 18, while the denominator starts at 3. Therefore,

$$\binom{18}{3} = \frac{(18)(17)(16)}{(3)(2)(1)} = 816.$$

Technology Connection

Binomial Coefficients

Many calculators can directly compute binomial coefficients, but the sequence of keys to press varies from calculator to calculator. On TI calculators, the command for finding binomial coefficients is **NCR**, which may be found under the **PRB** menu after pressing the **MATH** button.

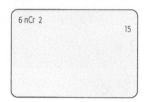

EXERCISES
Use a calculator to compute (if possible) the following binomial coefficients.

1. $\begin{pmatrix} 6 \\ 0 \end{pmatrix}$ **2.** $\begin{pmatrix} 14 \\ 8 \end{pmatrix}$

3. $\begin{pmatrix} 4 \\ 15 \end{pmatrix}$ **4.** $\begin{pmatrix} 7 \\ -1 \end{pmatrix}$

d) Since $0! = 1$, we have

$$\begin{pmatrix} 4 \\ 0 \end{pmatrix} = \frac{4!}{(0!)(4!)} = \frac{24}{(1)(24)} = 1.$$

Notice that we divide by 1, and not 0, in this example.   ■

CAUTION The binomial coefficient $\begin{pmatrix} n \\ k \end{pmatrix}$ is generally not equal to the fraction $\frac{n}{k}$.

The binomial coefficients may also be found using *Pascal's triangle*, as discussed in the exercises.

Applications of the Binomial(*n*, *p*) Distribution

EXAMPLE 4   *Rolling a Die.* A fair die is rolled three times. Use the Binomial(*n*, *p*) distribution to find the probability that exactly 1 six appears.

Solution   Let X be the number of sixes that appear. As discussed above, X has the Binomial(3, 1/6) distribution. Using Theorem 6 with $n = 3$, $k = 1$, and $p = 1/6$, we find

$$P(X = 1) = \begin{pmatrix} n \\ k \end{pmatrix} p^k q^{n-k}$$

$$= \begin{pmatrix} 3 \\ 1 \end{pmatrix} \left(\frac{1}{6}\right)^1 \left(\frac{5}{6}\right)^{3-1} \quad \text{Since } q = 1 - p = 5/6$$

$$= (3)\left(\frac{1}{6}\right)\left(\frac{25}{36}\right) \quad \text{From example 3(a)}$$

$$= \frac{75}{216}.$$

This agrees with the answer found in Example 1. Referring back to Example 1, we also see the significance of the binomial coefficient $\begin{pmatrix} 3 \\ 1 \end{pmatrix} = 3$. This factor is necessary because there are three ways (*SFF*, *FSF*, and *FFS*) of having exactly one success in three trials.   ■

EXAMPLE 5   *Guessing on a Test.* A multiple-choice test offers five possible answers for every question, of which only one is the correct answer. After answering all the questions that she knows, a student guesses on the remaining six questions. What is the probability that she gets exactly two of these guessed questions correct?

Solution   First, we verify that this is a binomial experiment. There are a fixed number of trials (6). Each trial can be characterized as a success (she gets the answer correct) or as a failure (she gets the answer wrong). The answers for each question are guessed independently. The probability of success for each trial is the same (0.2).

Technology Connection

Binomial(n, p) Distribution

Some calculators can directly compute the probabilities of the Binomial(*n*, *p*) distribution. On TI calculators, this is done by selecting the **BINOMPDF** function, which is found under the **DISTR** menu. We show below the result of Example 4.

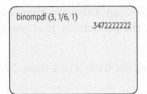

binompdf (3, 1/6, 1)
                              .3472222222

EXERCISE

1. Repeat Example 1 using the **BINOMPDF** function.

Finally, the central problem is counting the number of successes that occur. We may therefore use the Binomial(6, 0.2) distribution to compute this probability.

Let *X* be the number of questions that are correctly guessed. Then

$$P(X = 2) = \binom{n}{k} p^k q^{n-k}$$

$$= \binom{6}{2}(0.2)^2(0.8)^{6-2} \quad \text{Since } n = 6, k = 2, p = 0.2, \text{ and } q = 1 - p = 0.8$$

$$= (15)(0.04)(0.4096) \quad \text{Since } \binom{6}{2} = 15 \text{ from Example 3(b)}$$

$$= 0.24576. \qquad ∎$$

CAUTION In the previous example, *p* was 1/5, the probability of success on any one trial. This probability was not 2/6, even though we sought the probability of 2 successes in 6 trials.

Some calculators have a function for directly computing the probabilities of a Binomial(*n*, *p*) distribution, as discussed in the Technology Connection.

EXAMPLE 6 *Drawing Tickets from a Box.* Six tickets are drawn from the box

| 0 | 0 | 0 | 0 | 1 |

The drawings are made at random **with replacement.** That is, after a ticket is drawn and observed, it is returned to the box before the next ticket is drawn. Find the probability that the sum of the six tickets is equal to 2.

Solution First, we verify that this is a binomial experiment. There are a fixed number of trials (6). Each trial can be characterized as a success (a **1** is drawn) or as a failure (a **0** is drawn). The tickets are drawn independently. The probability of success for each trial is the same (0.2). Finally, the central problem is counting the number of successes that occur, since the sum of the tickets is 2 only if exactly 2 **1** tickets are drawn.

Let *X* be the number of **1**s that are drawn. Then *X* has the Binomial (6, 1/5) distribution. Using the result of Example 5, $P(X = 2) = 0.24576$. ∎

A very important principle is illustrated with the last example. A Binomial(*n*, *p*) distribution may be interpreted as the sum of tickets drawn at random with replacement from a box of **0**s and **1**s, where the proportion of **1**s in the box is equal to *p*. This interpretation is called the **box model** and will be used throughout this chapter.

Random Sampling

One important application of the Binomial(*n*, *p*) distribution occurs in **random sampling** from a population. A common example of random sampling occurs when a polling organization asks likely voters who they favor in the next election.

As before, random sampling can be thought of as drawing tickets from a box of
0s and 1s. However, with random sampling, the drawings are made **without replacement.** That is, once somebody is selected to be in the sample, that person cannot be reselected. In terms of the box model, the proportions of 0s and 1s in the box
will change as tickets are drawn from the box. When drawing without replacement,
the draws are not independent.

However, if the number of tickets in the box is much larger than the number of
draws, then the proportions of 0s and 1s will change by only a very small amount
as tickets are drawn and removed. As a result, the events are *effectively* independent,
and so the Binomial(n, p) distribution can still be used with high accuracy.

EXAMPLE 7   *Polling.* According to the 2000 U.S. Census, the city of Phoenix, Arizona, has a population of 1,321,045, of whom 449,972 are Latino or Hispanic. If 50
citizens of Phoenix are selected at random, what is the probability that precisely 16
of them are Latino or Hispanic?

Solution   We may represent the population of Phoenix with a box model containing 1,321,045 tickets, as follows:

$$\boxed{449{,}972\ \text{1s}\qquad 871{,}073\ \text{0s}}\ .$$

The 1 ticket represents selecting a Latino or Hispanic, while the 0 ticket represents
selecting somebody else. The number of Latinos or Hispanics in the sample may be
thought of as the number of 1s that appear after drawing from this box without replacement 50 times.

Let X denote the number of Latinos or Hispanics selected. We have 50 trials, and
the proportion of 1s in the box is

$$\frac{449{,}972}{1{,}321{,}045} \approx 0.340618.$$

So X has the Binomial(50, 0.340618) distribution, and we have

$$P(X = 16) = \binom{50}{16}(0.340618)^{16}(1 - 0.340618)^{34} \approx 0.114613,$$

using a calculator for approximation.[11]   ■

Application to Genetics

Many principles of genetics may be explained using the Binomial(n, p) distribution.

EXAMPLE 8   *Heterozygous Cross.* Two heterozygous pea plants with purple flowers (Ff) are crossed. Find the probability that the offspring will be heterozygous.

---

[11]The exact answer, accounting for the dependence of the draws, agrees with the answer shown in
Example 7 to five decimal places. This confirms that the dependence of the draws is a mere tech-
nicality for this problem; a larger error would occur if the U.S. Bureau of the Census had overes-
timated the population of Phoenix by 3.

Solution   Let X be the number of F alleles inherited by the offspring. The probability that each gamete will be an F allele is $1/2$. Since the two gametes are selected from the two parents independently, X has a Binomial($2, 1/2$) distribution. Therefore,

$$P(X = 1) = \binom{2}{1}\left(\frac{1}{2}\right)^1\left(\frac{1}{2}\right)^1$$

$$P(\text{offspring is } Ff) = \frac{1}{2}.$$

This agrees with the answer found using multiplication trees (Example 3 of Section 10.2) and using Punnett squares (on page XXX).   ■

EXAMPLE 9   *Hardy–Weinberg Law.* Suppose that a trait has two alleles, A and a. Suppose that the frequency of A alleles for both sperm and egg gametes in the population is p, so that the frequency of a alleles in the gametes is $q = 1 - p$. A sperm and an egg are randomly selected and crossed. Find the probabilities that the offspring will contain 0 A alleles, 1 A allele, and 2 A alleles.

Solution   Let X denote the number of A alleles in the offspring. Then X corresponds to the number of successes (A alleles, with probability p) in two independent trials. Therefore, X has a Binomial($2, p$) distribution, and so

$$P(\text{offspring is } aa) = P(X = 0) = \binom{2}{0}p^0q^2 = q^2,$$

$$P(\text{offspring is } Aa) = P(X = 1) = \binom{2}{1}p^1q^1 = 2pq,$$

$$P(\text{offspring is } AA) = P(X = 2) = \binom{2}{2}p^2q^0 = p^2.$$ ■

If the proportions of AA, Aa, and aa genotypes in a population are p^2, $2pq$, and q^2, respectively, then the population is said to satisfy the *Hardy–Weinberg Law*. In the exercises, we will show that such a population is in *equilibrium*, meaning that we would expect these proportions to remain the same from generation to generation.

EXAMPLE 10   *Albinism.* Albinism, a pigmentation deficiency, is a recessively inherited trait with an incidence rate of about 1 out of every 10,000 in a given population. Assuming that the genotypes for albinism follow the Hardy–Weinberg Law, find the proportion of heterozygotes and dominant homozygotes in the population.[12]

Solution   Using the Hardy–Weinberg Law, we are given that $q^2 = 1/10,000 = 0.0001$. Therefore, $q = 0.01$ and $p = 1 - q = 0.99$. The proportion of homozygous dominant individuals in the population is thus

$$p^2 = (0.99)^2 = 0.9801.$$

---

[12]W. S. Klug and M. R. Cummings, *Concepts of Genetics* (New York: MacMillan, 2002).

Also, the proportion of heterozygous individuals is

$$2pq = 2(0.99)(0.01) = 0.0198.$$

We see that the allele for albinism is present in nearly 2% of the population, even though albinism is a rare trait. These heterozygotes are called *carriers,* since their genotype contains the allele for albinism even though the symptoms of albinism are not physically apparent.   ■

Polygenic Traits

Thus far, we have considered genetic characteristics that are determined by a single gene pair. Under the *multiple-gene hypothesis,* some genetic characteristics are *polygenic* since they are determined by several gene pairs instead of a single gene pair. For example, three gene pairs determine the grain color of certain types of wheat. That is, the genotype *AABBCC* represents dark red wheat, while the genotype *aabbcc* represents white. Any other genotype, such as *AaBBCc,* would be seen as an *intermediate* shade of red. The depth of the redness is controlled by the number of dominant alleles, so that the genotypes *AaBBCc* and *AAbbCC* would belong to the same phenotype.[13]

EXAMPLE 11   *Polygenic Traits.* Suppose that two trihybrid individuals with genotype *AaBbCc* are crossed. The law of independent assortment states that different gene alleles are inherited independently (see p. XXX). Find the probability that the offspring will contain exactly four dominant alleles.

Solution   Let X denote the number of dominant alleles inherited by the offspring. From both parents, the offspring receives an A allele with probability $1/2$, a B allele with probability $1/2$, and a C allele with probability $1/2$. Each of these six alleles (three from both parents) is independently inherited, so X has a Binomial$(6, 1/2)$ distribution. Therefore,

$$P(X = 4) = \binom{6}{4}\left(\frac{1}{2}\right)^2\left(\frac{1}{2}\right)^4 = (15)\left(\frac{1}{4}\right)\left(\frac{1}{16}\right) = \frac{15}{64}.$$ ■

Histograms

If X has a Binomial$(6, 1/2)$ distribution, as in Example 11, then we may directly compute

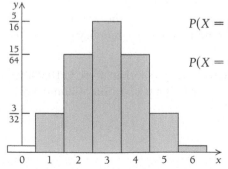

$$P(X = 0) = \frac{1}{64}, \qquad P(X = 1) = \frac{3}{32}, \qquad P(X = 2) = \frac{15}{64}, \qquad P(X = 3) = \frac{5}{16},$$

$$P(X = 4) = \frac{15}{64}, \qquad P(X = 5) = \frac{3}{32}, \quad \text{and} \quad P(X = 6) = \frac{1}{64}.$$

These probabilities may be graphically represented with a bar chart called a **histogram.** Rectangles of width 1 are centered on the numbers $0, 1, \dots, 6$ on the x-axis. The heights of the rectangles are given by the probabilities. The histogram for the distribution of X is shown in the margin.

[13]W. S. Klug and M. R. Cummings, *Concepts of Genetics* (New York: MacMillan, 2002).

The value with the highest rectangle, 3, is called the **mode** of the distribution since it is the most likely value. It is possible for a distribution to have more than one mode.

Since the width of each rectangle is 1, we notice that

$$\text{Area of rectangle} = (\text{Height}) \times (\text{Width})$$
$$= (\text{Probability})(1) = \text{Probability}.$$

In other words, the areas of the rectangles give the probabilities. Since the probabilities add to 1, we conclude that *the total area of the histogram is equal to 1.*

This important fact about histograms will be extended when we study continuous probability distributions in Section 10.5.

Exercise Set 10.3

Compute the binomial coefficients, if possible.

1. $\binom{4}{2}$ 2. $\binom{5}{3}$ 3. $\binom{3}{3}$

4. $\binom{2}{2}$ 5. $\binom{8}{0}$ 6. $\binom{4}{5}$

7. $\binom{5}{6}$ 8. $\binom{6}{0}$ 9. $\binom{9}{-4}$

10. $\binom{12}{5}$ 11. $\binom{16}{10}$ 12. $\binom{3}{-1}$

Suppose that X has the Binomial(n, p) distribution. Find the probability given the values of n and p.

13. $P(X = 2)$ if $n = 3, p = 1/3$

14. $P(X = 4)$ if $n = 6, p = 3/4$

15. $P(X = 3)$ if $n = 4, p = 0.8$

16. $P(X = 1)$ if $n = 8, p = 0.2$

17. $P(X = 5)$ if $n = 6, p = 0.9$

18. $P(X = 4)$ if $n = 6, p = 0.9$

19. $P(X = 10)$ if $n = 20, p = 0.6$

20. $P(X = 8)$ if $n = 15, p = 0.7$

21. $P(X = 5)$ if $n = 4, p = 0.3$

22. $P(X = 6)$ if $n = 3, p = 0.6$

Draw the histograms of the Binomial(n, p) distribution for the following values of n and p.

23. $n = 3, p = 0.4$ 24. $n = 3, p = 0.7$

25. $n = 4, p = 0.6$ 26. $n = 5, p = 0.3$

27. Suppose that $P(M = 0) = 0.2$, $P(M = 1) = 0.5$, and $P(M = 2) = 0.3$. Draw the histogram for the distribution of M.

28. Suppose that $P(M = 1) = 0.25$, $P(M = 3) = 0.4$, and $P(M = 6) = 0.35$. Draw the histogram for the distribution of M.

APPLICATIONS

29. *Coin Flipping.* Eight fair coins are flipped. Find the probability that exactly five of the coins land heads.

30. *Coin Flipping.* Three fair coins are flipped. Find the probability that at least one of the coins lands tails.

31. *Coin Flipping.* Six fair coins are flipped. Find the probability that fewer than two of the coins land heads.

32. *Coin Flipping.* Five fair coins are flipped. Find the probability that at least three of the coins land tails.

33. *Rolling Dice.* Five dice are rolled. Find the probability that exactly two of the dice are fours.

34. *Rolling Dice.* Two dice are rolled. Find the probability that at least one of the dice is a one.

35. *Rolling Dice.* Ten dice are rolled. Find the probability that at least one of the dice is a two.

36. *Rolling Dice.* Six dice are rolled. Find the probability that at most one die is a six.

37. *Rolling Dice.* Nine dice are rolled. Find the probability that more than six dice are fives.

38. *Guessing on a Test.* A multiple-choice test offers four possible answers for every question, of which only one is the correct answer. After completing all the questions that she knows, a student guesses on

the remaining seven questions. What is the probability that she gets exactly two of these guessed answers correct?

39. *Random Sampling.* According to the 2000 U.S. Census, the city of Fort Worth, Texas, has a population of 534,694, of whom 319,159 are white. If 40 residents of Fort Worth are selected at random, what is the probability that exactly 30 of them are white?

40. *Random Sampling.* According to the 2000 U.S. Census, the city of Miami, Florida, has a population of 362,470, of whom 238,351 are Latino or Hispanic. If 30 residents of Miami are selected at random, what is the probability that exactly 20 of them are Latino or Hispanic?

Amino Acids. A synthetically constructed RNA heteropolymer is formed by placing two nucleotides in a random sequence. Experiments using these synthetic strands were used to first determine the base composition assignments on mRNA for amino acids.[14]

41. Suppose that a nucleotide mixture contains 20% adenine (A) and 80% guanine (G). Find the probability that a randomly chosen sequence of three nucleotides on the resulting RNA strand contains exactly two guanines.

42. Suppose that a nucleotide mixture contains 40% cytosine (C) and 60% uracil (U). Find the probability that a randomly chosen sequence of three nucleotides on the resulting RNA strand are all cytosines.

Evaluating Habitats. Sites in the Nadgee Nature Reserve in Australia are rated with a *habitat complexity score,* which evaluates their fitness as habitats for ground-dwelling mammals. These scores range from 0 (impossible) to 12 (ideal). The habitat complexity score for a randomly selected site may be modeled by a Binomial(12, p) distribution.[15]

43. Immediately after a fire, $p = 0.346$.

   a) Find the probability that a randomly selected site would be ranked as a 2 or less.

   b) Find the probability that a randomly selected site would be ranked as at least a 9.

44. Eighteen years after the fire, the nature reserve recovered significantly and p increased to 0.638.

   a) Find the probability that a randomly selected site would be ranked as a 2 or less.

b) Find the probability that a randomly selected site would be ranked as at least a 9.

Hardy–Weinberg Law. To assess genetic susceptibility to various diseases, the frequencies of the *C* and *G* alleles for the gene encoding interleukin-6 was determined using DNA testing among North Carolina women.[16]

45. Among African-American women, the frequencies of the *C* and *G* alleles were measured to be 0.08 and 0.92, respectively. Use the Hardy–Weinberg law to find the expected proportions of the *CC*, *CG*, and *GG* genotypes among African-American women.

46. Among Caucasian women, the frequencies of the *C* and *G* alleles were measured to be 0.41 and 0.59, respectively. Use the Hardy–Weinberg law to find the expected proportions of the *CC*, *CG*, and *GG* genotypes among Caucasian women.

Prevalence of Carriers. Assume that the genotypes for the following hereditary diseases occur in the given populations according to the Hardy–Weinberg Law. For all of these diseases, the allele for having the disease is recessive to the allele for not having the disease.[17]

47. *Cystic Fibrosis.* One in every 2500 whites of European descent has cystic fibrosis. Find the proportion of whites of European descent who are carriers (that is, have the allele for the disease but are not diseased) for cystic fibrosis.

48. *Tay-Sachs Disease.* Among Ashkenazic Jews (Jewish people whose ancestors lived in central Europe), about one newborn in 3600 has Tay-Sachs disease. Find the proportion of Ashkenazic Jews who are carriers for Tay-Sachs disease.

49. *Tay-Sachs Disease.* Among non-Jews, about one newborn in 360,000 has Tay-Sachs disease. Find the proportion of non-Jews who are carriers for Tay-Sachs disease.

50. *Sickle-Cell Disease.* One in 400 African-Americans has sickle-cell disease. Find the proportion of African-Americans who are carriers for sickle-cell disease.[18]

---

[14]W. S. Klug and M. R. Cummings, *Concepts of Genetics* (New York: MacMillan, 2002).

[15]N. C. Coops and P. C. Catling, "Estimating forest habitat complexity in relation to time since fire," *Austral Ecology,* Vol. 25, pp. 344–351 (2000).

[16]M. I. Hassan, Y. Aschner, C. H. Manning, J. Xn, and J. L. Aschner, "Racial differences in selected cytokine allelic and genotypic frequencies among healthy, pregnant women in North Carolina," *Cytokine,* Vol. 21, pp. 10–16 (2003).

[17]N. A. Campbell and J. B. Reece, *Biology* (New York: Benjamin Cummings, 2002).

[18]For this disease, the sickle-cell allele is incompletely dominated by the non-sickle-cell allele. As a result, it is possible for a carrier to exhibit some symptoms of the disease.

SYNTHESIS

51. *Basketball.* At a college basketball promotion, a student shoots five three-pointers. The student wins a prize if he makes at least three shots. Assume the probability that any one shot is made to be 0.4 and that the shots are independent. Find the probability that he wins a prize.

Distributions. For Exercises 52–55, find the distribution of X by (a) listing the possible values for X and (b) finding the probability associated with each value. Then (c) draw the histogram for the distribution.

52. A fair die is rolled once. Let X be the number of spots that appear.

53. Two cards are dealt from a deck of cards. Let X be the number of these cards that are hearts. (*Hint:* See Example 1 of Section 10.2.)

54. A ticket is drawn at random from the box ⎡0 1 1⎤. Let X be the ticket that is drawn.

55. Two tickets are drawn at random without replacement from the box ⎡0 1 1⎤. Let X be the number of 1s that are drawn.

Pascal's Triangle. For small values of n, the binomial coefficient $\binom{n}{k}$ may be quickly found by using *Pascal's triangle*. To form Pascal's triangle, we start with a 1 and place 1's diagonally down to the left and to the right. Entries inside of Pascal's triangle are found by adding the two entries immediately above it. For example, in Pascal's triangle below, $2 + 1 = 3$ and $5 + 10 = 15$.

Row 0						1						
Row 1					1		1					
Row 2				1		2		1				
Row 3			1		3		3		1			
Row 4		1		4		6		4		1		
Row 5	1		5		10		10		5		1	
Row 6	1	6		15		20		15		6		1

Both the row numbering and column numbering start from 0. For example, the green 1 is in row 4, column 0, and the red 15 is in row 6, column 2.

56. The blue 2 is located in the second row, first column of Pascal's triangle. Verify that $\binom{2}{1} = 2$.

57. The blue 3 is located in the third row, second column of Pascal's triangle. Verify that $\binom{3}{2} = 3$.

58. The red 15 is located in the sixth row, second column of Pascal's triangle. Verify that $\binom{6}{2} = 15$.

59. The green 1 is located in the fourth row, 0th column of Pascal's triangle. Verify that $\binom{4}{0} = 1$.

60. a) Use Pascal's triangle to compute $\binom{7}{3}$.

  b) Verify your answer to part (a) using factorials.

61. a) Use Pascal's triangle to compute $\binom{8}{5}$.

  b) Verify your answer to part (a) using factorials.

The following exercises explain why Pascal's triangle generates the binomial coefficients.

62. Use factorials to compute $\binom{n}{0}$ for any n.

63. Use factorials to compute $\binom{n}{n}$ for any n.

64. Suppose that $0 < k < n$. Use factorials to show that

$$\binom{n-1}{k-1} + \binom{n-1}{k} = \binom{n}{k}.$$

tw 65. In words, explain the results of Exercises 62–64 in terms of Pascal's triangle.

Combinatories. In the multiplication tree of Example 1, we listed in terms of S and F the eight possible outcomes of a binomial experiment with three trials.

66. Suppose a binomial experiment has two trials.

  a) In terms of S and F, list the four possible outcomes.

  b) Of these four possible outcomes, how many correspond to exactly one success?

  c) Verify that your answer to part (b) is equal to $\binom{2}{1}$.

67. Suppose a binomial experiment has four trials.

  a) In terms of S and F, list the 16 possible outcomes.

  b) Of these 16 possible outcomes, how many correspond to exactly two successes?

  c) Of these 16 possible outcomes, how many correspond to exactly three successes?

tw d) Verify that your answer to part (b) is equal to $\binom{4}{2}$.

tw e) Verify that your answer to part (c) is equal to $\dbinom{4}{3}$.

68. *Hardy–Weinberg Law.* In Example 9, we showed that the probabilities that the offspring is of each of the three genotypes are q^2, $2pq$, and p^2, respectively. Show that the sum of these probabilities is equal to 1. (Use the fact that $p + q = 1$.)

Multinomial Distribution. Some experiments have more than two possible outcomes for each trial but otherwise satisfy the conditions for a binomial experiment. For example, suppose that a population satisfies the Hardy–Weinberg Law for a certain trait with two alleles, F and f, and suppose that the frequencies of these alleles in the population are p and q. If an individual is chosen at random from the population, then the individual could have one of three possible genotypes (FF, Ff, and ff). As a result, classifying the genotypes of a random sample from this population is not an example of a binomial experiment.

However, probabilities where the genotype can be one of three possibilities can be computed using the *multinomial distribution*. If a random sample of size n is taken from the population, then the probability that i of them have genotype FF, j of them have genotype Ff, and k of them have genotype ff is given by

$$P(i, j, k) = \frac{n!}{i!\,j!\,k!}(p^2)^i\,(2pq)^j\,(q^2)^k.$$

(In this formula, we assume that $i + j + k = n$, the total size of the sample.)

For example, among Caucasian women, the frequencies of the C and G alleles encoding interleukin-6 were measured to be 0.41 and 0.59, respectively.[19]

69. Suppose a random sample of 10 Caucasian women is taken. Find the probability that 2 of the women have genotype CC, 5 have genotype CG, and 3 have genotype GG.

70. Suppose a random sample of 12 Caucasian women is taken. Find the probability that 4 of the women have genotype CC, 3 have genotype CG, and 5 have genotype GG.

71. *Hardy–Weinberg Law.* Suppose that an individual is randomly selected from a population that satisfies the Hardy–Weinberg Law. Let p and q denote the proportions of A and a alleles in the population.

a) The allele inherited by the offspring may be visualized with the following multiplication tree. Fill in the probabilities of the unmarked arrows.

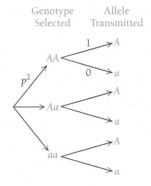

tw b) In the multiplication tree, there are three ways corresponding to the transmission of the A allele. Show that the probability that the A allele is transmitted is p.

tw c) Show that the probability that the a allele is transmitted is q.

tw d) Use the result of Example 9 to show that, if one generation satisfies the Hardy–Weinberg Law and survives to mate, then the offspring generation should also satisfy it.

The Hardy–Weinberg Law is called an *equilibrium state* for the population. The stability of this equilibrium state, assuming that different genotypes have different probabilities of living long enough to mate, is discussed in the Extended Life Science Connection of Chapter 3.

---

[19]M. I. Hassan, Y. Aschner, C. H. Manning, J. Xu, and J. L. Aschner, "Racial differences in selected cytokine allelic and genotypic frequencies among healthy, pregnant women in North Carolina," *Cytokine,* Vol. 21, pp. 10–16 (2003).

10.4 Expected Value and Standard Deviation for Discrete Random Variables

OBJECTIVES

■ Compute the mean and standard deviation of random variables.
■ Compute the mean and standard deviation of the Binomial(n, p) distribution.
■ Convert values into standard units.

In Section 10.3, we mostly used the Binomial(n, p) distribution to calculate probabilities of the form $P(X = a)$. Instead of computing all of the probabilities, qualitative information about a random variable may be obtained from its *expected value* and *standard deviation*. We discuss the computations of these in this section.

Expected Value

EXAMPLE 1 *Rolling a Die.* (This example continues Example 1 of Section 10.1.) A die is rolled 1000 times. The table in the margin shows the number of times the die came up each possible number. Use this experiment to find the average value of the 1000 rolls.

Face Value	Frequency
1	166
2	176
3	158
4	178
5	153
6	169
Total	1000

Solution　The sum of the 1000 rolls is equal to

$$(1)(166) + (2)(176) + (3)(158) + (4)(178) + (5)(153) + (6)(169) = 3483.$$

The average of the 1000 rolls is therefore $3483/1000 = 3.483$. ∎

Notice that the solution of Example 1 may be written as

$$\text{Average} = \frac{(1)(166) + (2)(176) + (3)(158) + (4)(178) + (5)(153) + (6)(169)}{1000}$$

$$= (1)\left(\frac{166}{1000}\right) + (2)\left(\frac{176}{1000}\right) + (3)\left(\frac{158}{1000}\right)$$

$$+ (4)\left(\frac{178}{1000}\right) + (5)\left(\frac{153}{1000}\right) + (6)\left(\frac{169}{1000}\right)$$

$$= (1)(0.166) + (2)(0.176) + (3)(0.158)$$
$$+ (4)(0.178) + (5)(0.153) + (6)(0.169).$$

To find the average, the possible values $(1, 2, \ldots, 6)$ are respectively multiplied by the experimentally derived probabilities for each value (see Example 1 of Section 10.1). This motivates the definition of the *expected value* for a random variable.

DEFINITION

The **expected value** of a discrete random variable X is defined by

$$E(X) = \sum kP(X = k),$$

where the sum is taken over all possible values of X. This is also called the **mean** of X or the *average* of X. The symbol μ is often used to denote $E(X)$.

　(The range for the index k is not written since this depends on the random variable X.)

EXAMPLE 2   *Rolling a Die.* Let X be the number of spots that appear when a fair die is rolled. Use the Principle of Equally Likely Outcomes to find $\mu = E(X)$.

Solution   According to the Principle of Equally Likely Outcomes,

$$P(X = 1) = P(X = 2) = P(X = 3) = P(X = 4) = P(X = 5) = P(X = 6) = \frac{1}{6}.$$

Using the definition, we find

$$\mu = E(X) = \sum_{k=1}^{6} kP(X = k) = 1\left(\frac{1}{6}\right) + 2\left(\frac{1}{6}\right) + 3\left(\frac{1}{6}\right) + 4\left(\frac{1}{6}\right) + 5\left(\frac{1}{6}\right) + 6\left(\frac{1}{6}\right) = 3.5.$$

Since the probabilities are all equally likely, the mean of X is the same as the average of the six numbers 1, 2, 3, 4, 5, and 6.   ∎

Notice that the mean of 3.5 is approximately equal to the *long-run average* of 1000 rolls, calculated in Example 1. Of course, when doing an experiment, the average of these 1000 rolls is not necessarily *equal* to 3.5. Indeed, the *law of averages* dictates that the long-run average will get closer and closer to the mean with more and more rolls.

It is possible for $E(X)$ to be equal to a value that cannot be attained by X. For example, let X be the number of people living in a randomly selected American family. According to the 2000 U.S. Census, $E(X) = 3.17$. This fact may be expressed by the statement "The average American family has 3.17 people." Obviously, no family actually consists of 3.17 people.

EXAMPLE 3   *Rolling a Die.* A fair die is rolled once. If it lands a six, you win \$4, but you lose \$1 if it does not land six. Let M be the amount of money you win. Find $\mu = E(M)$.

Solution   In Example 2 of Section 10.3, we showed that $P(M = 4) = 1/6$ and $P(M = -1) = 5/6$. Therefore,

$$\mu = E(M) = 4\left(\frac{1}{6}\right) + (-1)\left(\frac{5}{6}\right) = -\frac{1}{6} \approx -\$0.167.$$ ∎

There is a natural physical interpretation of the expected value. The histogram rests on the x-axis as solid blocks on a thin rod. The expected value may be viewed as the "fulcrum" of a histogram, or where the histogram would be balanced if you were to place your finger under the x-axis.

The histograms for the random variables X and M from Examples 2 and 3 are shown in the figures below. For X, the histogram is perfectly symmetric, and so the expected value is simply the average of the six possible outcomes, or 3.5. For M, placing the fulcrum at the origin would cause the histogram to tip over to the left. However, the histogram may be balanced by placing the fulcrum a little to the left of the origin.

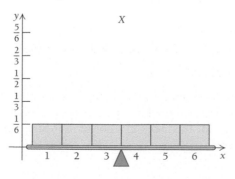

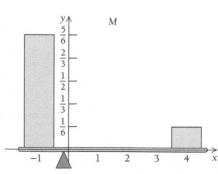

Standard Deviation and Variance

In the figures above, the histograms are "spread" about the mean. Also, it appears visually that the histogram for M is more spread out than the histogram for X. The spread of a random variable is measured with the *standard deviation*, which we now define.

DEFINITION

1. Let X be a discrete random variable whose expected value is $\mu = E(X)$. Then we define the **variance** of X as

$$Var(X) = E([X - \mu]^2) = \sum (k - \mu)^2 P(X = k)$$

and the sum is taken over all possible values of X.

2. The **standard deviation** of X is the square root of the variance of X:

$$SD(X) = \sqrt{Var(X)}.$$

The symbol σ is often used to denote $SD(X)$, so that $\sigma^2 = Var(X)$.

The positive term $(k - \mu)^2$ in the definition of $Var(X)$ is the square of the distance between k and the mean μ.

EXAMPLE 4 *Rolling a Die.* Let X be the number of spots that appear when a fair die is rolled. Find $Var(X)$ and $SD(X)$.

Solution Let μ be the mean of X; we showed that $\mu = 7/2$ in Example 2. Using the definition of variance and the probabilities for X, we find

$$Var(X) = \sum_{k=1}^{6} (k - \mu)^2 P(X = k)$$

$$= \left(1 - \frac{7}{2}\right)^2\left(\frac{1}{6}\right) + \left(2 - \frac{7}{2}\right)^2\left(\frac{1}{6}\right) + \left(3 - \frac{7}{2}\right)^2\left(\frac{1}{6}\right)$$

$$+ \left(4 - \frac{7}{2}\right)^2\left(\frac{1}{6}\right) + \left(5 - \frac{7}{2}\right)^2\left(\frac{1}{6}\right) + \left(6 - \frac{7}{2}\right)^2\left(\frac{1}{6}\right)$$

$$= \frac{25}{24} + \frac{9}{24} + \frac{1}{24} + \frac{1}{24} + \frac{9}{24} + \frac{25}{24}$$

$$= \frac{35}{12}.$$

Taking the square root of both sides, the standard deviation of X is $\sigma = SD(X) = \sqrt{35/12} \approx 1.708$. ∎

The computation of $Var(X)$ may be facilitated by using the following theorem.

THEOREM 7

Let X be a discrete random variable whose expected value is $\mu = E(X)$. Then the variance of X is given by

$$\text{Var}(X) = E(X^2) - \mu^2,$$

where

$$E(X^2) = \sum k^2 P(X = k)$$

and the sum is taken over all possible values of X.

EXAMPLE 5 *Rolling a Die.* Let X be the number of spots that appear when a fair die is rolled. Use Theorem 7 to find $\text{Var}(X)$ and $\text{SD}(X)$.

Solution Using the probabilities for X, we find

$$E(X^2) = \sum_{k=1}^{6} k^2 P(X = k)$$

$$= (1)^2\left(\frac{1}{6}\right) + (2)^2\left(\frac{1}{6}\right) + (3)^2\left(\frac{1}{6}\right) + (4)^2\left(\frac{1}{6}\right) + (5)^2\left(\frac{1}{6}\right) + (6)^2\left(\frac{1}{6}\right) = \frac{91}{6}.$$

In Example 2, we showed that $\mu = 7/2$. Therefore, the variance is

$$\text{Var}(X) = E(X^2) - \mu^2 = \frac{91}{6} - \left(\frac{7}{2}\right)^2 = \frac{35}{12}.$$

Once again, $\text{SD}(X) = \sqrt{35/12} \approx 1.708$. Notice that this computation of $\text{SD}(X)$ is computationally less burdensome than that of Example 4. ∎

EXAMPLE 6 *Rolling a Die.* A fair die is rolled once. If it lands a six, you win $4, but you lose $1 if it does not land six. Let M be the amount of money you win. Find $\sigma = \text{SD}(M)$.

Solution Let μ be the mean of M; we showed that $\mu = -1/6$ in Example 3. Using the probabilities for M, we find

$$E(M^2) = (-1)^2\left(\frac{5}{6}\right) + (4)^2\left(\frac{1}{6}\right) = \frac{7}{2}.$$

Therefore, the variance is

$$\text{Var}(X) = E(X^2) - \mu^2 = \frac{7}{2} - \left(-\frac{1}{6}\right)^2 = \frac{125}{36}.$$

Taking the square root of both sides, the standard deviation of M is $\sigma = \text{SD}(M) = \sqrt{125/36} = 5\sqrt{5}/6 \approx 1.863$.

Notice that, for both X and M, the difference between the largest possible value and the smallest possible value is 5. However, the standard deviation of M is larger than the standard deviation of X. This occurs because the histogram of M is more "spread out" than the histogram of X. ∎

Binomial(n, p) Distribution

The mean and standard deviation of random variables that follow the Binomial(n, p) distribution may also be calculated.

EXAMPLE 7   *Rolling Dice.* Let X be the number of sixes that appear when three dice are rolled. Find $\mu = E(X)$ and $\sigma = SD(X)$.

Solution   Notice that X has the Binomial(3, 1/6) distribution. The possible values of X are 0, 1, 2, and 3. Using the result of Example 1 of Section 10.3, we have

$$\mu = E(X) = \sum_{k=0}^{3} kP(X = k) = 0\left(\frac{125}{216}\right) + 1\left(\frac{75}{216}\right) + 2\left(\frac{15}{216}\right) + 3\left(\frac{1}{216}\right) = \frac{1}{2}.$$

As expected, this answer is three times larger than 1/6, the expected number of sixes on one die, which we found in Example 2.

To find $SD(X)$, we first compute $E(X^2)$ using these same probabilities:

$$E(X^2) = \sum_{k=0}^{3} k^2 P(X = k) = (0)^2\left(\frac{125}{216}\right) + (1)^2\left(\frac{75}{216}\right) + (2)^2\left(\frac{15}{216}\right) + (3)^2\left(\frac{1}{216}\right) = \frac{2}{3}.$$

Therefore,

$$Var(X) = E(X^2) - \mu^2 = \frac{2}{3} - \left(\frac{1}{2}\right)^2 = \frac{5}{12},$$

and so $\sigma = SD(X) = \sqrt{5/12}$.   ◼

The following theorem gives an easier way of computing the mean and standard deviation of a Binomial(n, p) random variable instead of explicitly adding many terms.

THEOREM 8

Let X have a Binomial(n, p) distribution, and let $q = 1 - p$. Then $E(X) = np$, $Var(X) = npq$, and $SD(X) = \sqrt{npq}$.

EXAMPLE 8   *Rolling Dice.* Let X be the number of sixes that appear when three dice are rolled. Find $\mu = E(X)$ and $\sigma = SD(X)$ using Theorem 8.

Solution   Since X has the Binomial(3, 1/6) distribution, we may use Theorem 8 with $n = 3$, $p = 1/6$, and $q = 5/6$:

$$\mu = E(X) = np = (3)\left(\frac{1}{6}\right) = \frac{1}{2},$$

$$\sigma = SD(X) = \sqrt{npq} = \sqrt{(3)\left(\frac{1}{6}\right)\left(\frac{5}{6}\right)} = \sqrt{\frac{5}{12}}.$$

These agree with the answers found in Example 7, where we explicitly used the definitions of $E(X)$ and $SD(X)$. Since X is a binomial random variable, using Theorem 8 significantly shortens the computational effort required to find $E(X)$ and $SD(X)$.   ◼

EXAMPLE 9   *Genetics.* From a heterozygous cross (see Example 3 of Section 10.2), 192 seeds are planted. Find the expected number of recessive homozygous offspring, and find the standard deviation.

Solution   Let X denote the number of recessive homozygous offspring. From Example 3 of Section 10.2, the success probability is $1/4$. Since this trial is independently repeated 192 times. X has a Binomial(192, 1/4) distribution. Therefore, $E(X) = (192)(1/4) = 48$ and $SD(X) = \sqrt{(192)(1/4)(3/4)} = 6$.   ■

The solution for the expected value in Example 9 is intuitively clear. If 192 offspring are produced and there is one chance in four for an offspring to be recessive homozygous, we would expect about 48 of the offspring to be recessive homozygous.

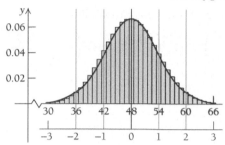

The histogram for a Binomial(192, 1/4) distribution is shown as the rectangles in the figure to the left. Notice that the mode is 48, which happens to be equal to the expected value for the distribution. In general, the mode of a Binomial(n, p) will either equal or be very close to the expected value.

While it is possible for a random variable with a Binomial(192, 1/4) distribution to be less than 30 or greater than 66, both of these events are extremely unlikely. As a result, the figure only shows the histogram between 30 and 66.

Standard Units

The standard deviation measures the spread about the mean. In the above figure, the blue vertical lines are found by adding and subtracting 6, or one standard deviation, to the mean of 48. The green vertical lines are found by adding and subtracting 12 (or two standard deviations) to the mean.

Since 60 is two standard deviations above the mean, we say that 60 converts to 2 **standard units**. Also, since 30 is three standard deviations below the mean, we say that 30 converts to -3 standard units. The mean itself converts to 0 standard units. These values in standard units are depicted in the figure by the lower horizontal axis. Standard units measure how far away a value is from the mean relative to the size of the standard deviation.

> **DEFINITION**
>
> To convert a value x into *standard units*, we use the formula
>
> $$z = \frac{x - \mu}{\sigma},$$
>
> where μ and σ are the mean and standard deviation of the distribution, respectively.

EXAMPLE 10   Suppose that $E(X) = 15$ and $SD(X) = 4$. Convert 17 into standard units.

Solution   Using the definition,

$$z = \frac{17 - 15}{4} = 0.5.$$

The value of 17 converts to 0.5 standard units. In other words, 17 is half of a standard deviation above the mean.   ∎

In the above example, z was positive because the value was larger than the mean. Negative values of z correspond to values that are less than the mean. Finally, $z = 0$ corresponds to the mean.

We observe that the histogram on page 688 appears to be bell-shaped. We will discuss the implications of this observation in Section 10.7.

Exercise Set 10.4

1. Suppose that $P(X = 0) = 0.2$, $P(X = 1) = 0.3$, and $P(X = 2) = 0.5$. Find $E(X)$.

2. Suppose that $P(X = 1) = 0.4$, $P(X = 3) = 0.4$, and $P(X = 5) = 0.2$. Find $E(X)$.

3. Suppose that $P(X = 0) = 0.2$, $P(X = 1) = 0.3$, and $P(X = 2) = 0.5$. Find $Var(X)$ and $SD(X)$.

4. Suppose that $P(X = 1) = 0.4$, $P(X = 3) = 0.4$, and $P(X = 5) = 0.2$. Find $Var(X)$ and $SD(X)$.

Find $E(X)$ for the Binomial(n, p) distribution using **(a)** the definition and **(b)** Theorem 8. These two answers should be the same.

5. $n = 2$, $p = 0.25$ 6. $n = 2$, $p = 1/3$

7. $n = 3$, $p = 0.5$ 8. $n = 3$, $p = 0.25$

Find $SD(X)$ for the following Binomial(n, p) distribution using **(a)** the definition, **(b)** Theorem 7, and **(c)** Theorem 8. These three answers should be the same.

9. $n = 2$, $p = 0.25$ 10. $n = 2$, $p = 1/3$

11. $n = 3$, $p = 0.5$ 12. $n = 3$, $p = 0.25$

Let X have the Binomial(n, p) distribution for the given values of n and p. Find $E(X)$ and $SD(X)$.

13. $n = 6$, $p = 0.2$ 14. $n = 10$, $p = 0.5$

15. $n = 20$, $p = 0.1$ 16. $n = 25$, $p = 0.36$

17. $n = 50$, $p = 0.4$ 18. $n = 100$, $p = 0.7$

Convert the given value into standard units given the mean and standard deviation of X.

19. $x = 20$, $E(X) = 16$, $SD(X) = 2$

20. $x = 10$, $E(X) = 12$, $SD(X) = 5$

21. $x = 13.1$, $E(X) = 13.5$, $SD(X) = 0.24$

22. $x = -1.2$, $E(X) = -1.2$, $SD(X) = 0.9$

23. $x = 29.3$, $E(X) = 20.3$, $SD(X) = 4.5$

24. $x = 0.14$, $E(X) = 0.25$, $SD(X) = 0.04$

APPLICATIONS

25. *Seagrass.* Let X denote the number of tertiary strands generated by the seagrass *Posidonia oceanica*

by each growth unit in a year. Then X follows a Binomial(12, 0.73) distribution.[20]

a) Find $E(X)$ and $SD(X)$.
b) Find $P(X = 8)$.

Evaluating Habitats. Sites in the Nadgee Nature Reserve in Australia are rated with a *habitat complexity score,* which evaluates their fitness as habitats for ground-dwelling mammals. Let X denote the habitat complexity score of a randomly chosen site; these scores range from 0 (impossible) to 12 (ideal). Then X may be modeled by a Binomial(12, p) distribution.[21]

26. Immediately after a fire, $p = 0.346$. Find $E(X)$ and $SD(X)$.

---

[20]H. Molenaar, D. Barthélémy, P. de Reffye, A. Meinesz, and I. Mialet, "Modeling architecture and growth patterns of *Posidonia oceanica,*" Vol. 66, pp. 85–99 (2000).
[21]N. C. Coops and P. C. Catling, "Estimating forest habitat complexity in relation to time since fire." *Austral Ecology,* Vol. 25, pp. 344–351 (2000).

27. Eighteen years after the fire, the nature reserve recovered significantly and p increased to 0.638. Find $E(X)$ and $SD(X)$.

SYNTHESIS

Properties of Mean and Standard Deviation. In the following exercises, we derive several properties about the mean and standard deviation.

tw 28. If a is a constant, we define $E(aX)$ to be

$$E(aX) = \sum akP(X = k).$$

Show that $E(aX) = aE(X)$. (*Hint:* Factor the a from the right side.)

tw 29. Show that $Var(aX) = a^2 Var(X)$. (*Hint:* $Var(aX) = E(a^2X^2) - [E(aX)^2].$)

tw 30. Use Exercise 29 to show that $SD(aX) = |a|SD(X)$.

10.5 Continuous Random Variables

OBJECTIVES

■ Define probability density functions.
■ Compute probabilities for uniformly distributed random variables.
■ Evaluate the mean, variance, and standard deviation of continuous random variables.

Suppose we throw a dart at a number line in such a way that it always lands in the interval $[1, 3]$. Let X be the number that the dart hits. There are an infinite number of possibilities for X. Note that X is a random variable whose possible values consist of an entire interval of real numbers. Such a random variable is called a **continuous random variable.**

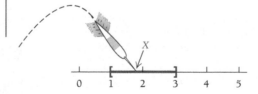

What is the probability that the dart lands at the point $\sqrt{2}$ exactly? It is very unlikely. In fact, the probability is zero that the dart lands at any one particular point. One important difference between a discrete and a continuous random variable is that each possible value for a discrete random variable has a nonzero probability, but for a continuous random variable, the probability of any particular value is zero.

Let's consider some other examples of continuous random variables.

EXAMPLE 1 Suppose that X is the corn acreage of a randomly chosen farm in North America. Is it reasonable to model X as a continuous random variable?

Solution Technically, there are only finitely many farms, and each has a fixed acreage of corn. Therefore, the random variable X is discrete. However, the number of farms is very large and it seems reasonable to assume that for any number x between 0 and the highest acreage a, there is at least one farm with corn acreage very near x. Therefore, it is reasonable to model X as a continuous random variable. If we know the value of a, then we could say that X is *distributed* over the interval $[0, a]$. Otherwise, we could say that X is distributed over the interval $[0, \infty)$. ■

Suppose X is a discrete random variable. As we saw in Section 10.3, the probability

$$P(c \le X \le d) = \sum_{k=c}^{d} p(k)$$

may be seen as the area of rectangles. For a continuous random variable, probability is seen as the area under the graph of a curve. As we learned in Chapter 5, these areas may be computed by integration. The function to be integrated is called a *probability density function* or *density function*

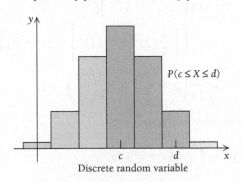

Discrete random variable

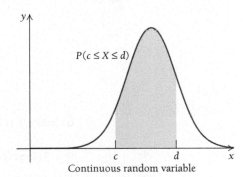

Continuous random variable

DEFINITION

A continuous random variable X distributed over an interval $[a, b]$ has a **density function** f that satisfies the following three conditions:

1. f is nonnegative over $[a, b]$; that is, $f(x) \ge 0$ for all x in $[a, b]$.

2. $\displaystyle\int_a^b f(x)\, dx = 1$.

3. $P(c \le X \le d) = \displaystyle\int_c^d f(x)\, dx$ for any subinterval $[c, d]$ of $[a, b]$.

In the above definition, the first condition is necessary since probabilities are nonnegative. Also, the second condition says $P(a \le X \le b) = 1$. This is true since X is assumed to be between a and b. If a function f satisfies these first two conditions, then f is a density function for some continuous random variable.

Although the definition of a density function was stated for a closed interval $[a, b]$, the definition extends in a natural way if the random variable is distributed over $(-\infty, b], [a, \infty)$, or $(-\infty, \infty)$.

EXAMPLE 2 Let X be a random variable with density function

$$f(x) = \frac{3}{14}(x + x^2) \quad \text{over} \quad [0, 2].$$

a) Verify that $f(x)$ is a density function.

b) Compute $P(0 \le X \le 1)$.

c) Compute $P(X \le 1)$.

Solution

a) We first check that $f(x) \ge 0$. Since we are only considering values of x between 0 and 2, $f(x) = \frac{3}{14}(x + x^2) \ge 0$.

 We next verify that the integral is 1.

$$\int_0^2 f(x)\, dx = \int_0^2 \frac{3}{14}(x + x^2)\, dx$$

$$= \frac{3}{14}\left[\frac{1}{2}x^2 + \frac{1}{3}x^3\right]_0^2$$

$$= \frac{3}{14}\left(\left[\frac{1}{2}(2)^2 + \frac{1}{3}(2)^3\right] - [0 + 0]\right)$$

$$= \frac{3}{14}\left(2 + \frac{8}{3}\right) = \frac{3}{14}\left(\frac{14}{3}\right) = 1.$$

b) Since f is the density function for X,

$$P(0 \le X \le 1) = \int_0^1 \frac{3}{14}(x + x^2)\, dx$$

$$= \frac{3}{14}\left[\frac{1}{2}x^2 + \frac{1}{3}x^3\right]_0^1$$

$$= \frac{3}{14}\left(\left[\frac{1}{2}(1)^2 + \frac{1}{3}(1)^3\right] - [0 + 0]\right) = \frac{3}{14}\left(\frac{1}{2} + \frac{1}{3}\right) = \frac{5}{28}.$$

c) Since the domain of the density function f contains no negative numbers, $P(X < 0) = 0$. Therefore, $P(X \le 1) = P(0 \le X \le 1) = \frac{5}{28}$. ■

EXAMPLE 3 A 3 cc solution contains exactly one bacterium. The solution is slowly transferred into a flask using a pipette. The random variable X is the amount of solution already transferred into the flask at the instant the bacterium enters the flask. Then X is a continuous random variable distributed over the interval $[0, 3]$. Suppose that the density function for X is

$$f(x) = \frac{1}{3} \quad \text{over} \quad [0, 3].$$

Find the probability that X is in the interval

a) $0 \le X \le 1$.

b) $1 \le X \le 2$.

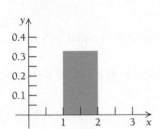

Solution

a) The probability is given by integrating $f(x)$ between 0 and 1.

$$P(0 \le X \le 1) = \int_0^1 \frac{1}{3} \, dx = \left[\frac{1}{3} x \right]_0^1 = \frac{1}{3} - 0 = \frac{1}{3}.$$

We see that the integral gives the red area shown in the figure in the margin.

b) We integrate between 1 and 2.

$$P(1 \le X \le 2) = \int_1^2 \frac{1}{3} \, dx = \left[\frac{1}{3} x \right]_1^2 = \frac{1}{3}(2 - 1) = \frac{1}{3}.$$

We see that the integral gives the blue area shown in the figure in the margin.

EXAMPLE 4 In Example 3, the density function was $f(x) = \frac{1}{3}$. Suppose instead that the density function for X is

$$f(x) = \frac{1}{9} x^2 \quad \text{over} \quad [0, 3].$$

a) Verify that $f(x)$ is a density function.

b) Compute the probability that the bacterium enters the flask with the first cc of transferred solution.

c) Compute the probability that the bacterium enters the flask with the last cc of transferred solution.

Solution

a) We see that $f(x) \ge 0$ since x^2 is always at least 0. Also,

$$\int_0^3 f(x) \, dx = \int_0^3 \frac{1}{9} x^2 \, dx = \frac{1}{27}[x^3]_0^3 = \frac{1}{27}(3^3 - 0^3) = 1.$$

b) The probability that the bacterium enters the flask with the first cc of transferred solution is given by the red area.

$$P(0 \le X \le 1) = \int_0^1 f(x) \, dx = \int_0^1 \frac{1}{9} x^2 \, dx = \frac{1}{27}[x^3]_0^1 = \frac{1}{27}(1^3 - 0^3) = \frac{1}{27} \approx 0.0370.$$

c) The probability that the bacterium enters the flask with the last cc of transferred solution is given by the blue area.

$$P(2 \le X \le 3) = \int_2^3 f(x) \, dx = \int_2^3 \frac{1}{9} x^2 \, dx = \frac{1}{27}[x^3]_2^3 = \frac{1}{27}(3^3 - 2^3) = \frac{19}{27} \approx 0.7037.$$

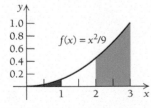

It makes sense that the probability is much higher that the bacterium enters the flask with the last cc, since the area under the curve between 2 and 3 looks to be much greater than the area between 0 and 1.

Observed Significance Levels

Random variables that arise in statistical applications are often assumed to have a certain probability distribution. Based on this assumption, researchers often compute the probability that the random variable is at least as large as some observed value. In statistics, this probability is called the **observed level of significance.**

EXAMPLE 5 A random variable X is assumed to have probability density function

$$f(x) = \frac{1}{x^2} \quad \text{over} \quad [1, \infty).$$

a) Find the observed level of significance if the random variable is equal to 5 in an experiment.

b) Find the observed level of significance if the random variable is equal to 100 in a second experiment.

Solution

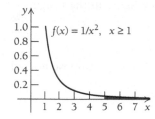

a) In this example, the observed level of significance is the probability $P(X \geq 5)$. We compute this probability using integration.

$$P(X \geq 5) = \int_5^\infty \frac{dx}{x^2}$$

$$= \lim_{b \to \infty} \int_5^b \frac{dx}{x^2}$$

$$= \lim_{b \to \infty} \left[-\frac{1}{x} \right]_5^b$$

$$= \lim_{b \to \infty} \left[-\frac{1}{b} - \left(-\frac{1}{5} \right) \right]$$

$$= \frac{1}{5}.$$

The observed level of significance is $1/5$, or 0.2.

b) The observed level of significance for the second experiment is $P(X \geq 100)$.

$$P(X \geq 100) = \int_{100}^\infty \frac{dx}{x^2}$$

$$= \lim_{b \to \infty} \int_{100}^b \frac{dx}{x^2}$$

$$= \lim_{b \to \infty} \left[-\frac{1}{x} \right]_{100}^b$$

$$= \lim_{b \to \infty} \left[-\frac{1}{b} - \left(-\frac{1}{100} \right) \right]$$

$$= \frac{1}{100}.$$

The observed level of significance is $1/100$, or 0.01.

In the previous example, the observed levels of significance corresponded to areas under right tails of the probability density function. As we will see in Section 10.7, it is also possible to define a left-tailed observed level of significance.

A small observed level of significance means that it is very unlikely that a random variable would be at least as large as the value that was observed in an experiment. For this reason, researchers often interpret a small observed level of significance as evidence against the assumed probability density function of the random variable. More about the process of statistical inference can be found in a textbook on statistics.

Critical Values

As well as the observed level of significance, scientists often compute the **critical value** that corresponds to some predetermined **significance level,** which is usually denoted by α. The significance level sets a desired probability, or an area under a tail of the probability density function. The critical value is the value that corresponds to this probability.

EXAMPLE 6   The random variable X has probability density function

$$f(x) = \frac{2}{x^3} \quad \text{over} \quad [1, \infty).$$

Find the critical value if the significance level is $\alpha = 0.01$.

Solution   To begin, we find the right-tail probability $P(X \geq c)$, where c is the critical value to be found.

$$P(X \geq c) = \int_c^\infty \frac{2}{x^3}\, dx$$

$$= \lim_{b \to \infty} \int_c^b \frac{2}{x^3}\, dx$$

$$= \lim_{b \to \infty} \left[-\frac{1}{x^2} \right]_c^b$$

$$= \lim_{b \to \infty} \left[-\frac{1}{b^2} - \left(-\frac{1}{c^2} \right) \right]$$

$$= \frac{1}{c^2}.$$

We then find the value of c so that this probability is equal to the significance level of $\alpha = 0.01$.

$$\frac{1}{c^2} = 0.01$$

$$c = 10. \qquad \text{Since } c \geq 1$$

The critical value is therefore 10. ∎

EXAMPLE 7   The random variable X has probability density function

$$f(x) = \frac{1}{x^2} + \frac{4}{x^3} \quad \text{over} \quad [2, \infty).$$

Find the critical value if the significance level is $\alpha = 0.05$.

Solution   We begin by computing $P(X \geq c)$.

$$P(X \geq c) = \int_c^\infty \left(\frac{1}{x^2} + \frac{4}{x^3} \right) dx$$

$$= \lim_{b \to \infty} \int_c^b \left(\frac{1}{x^2} + \frac{4}{x^3} \right) dx$$

$$= \lim_{b \to \infty} \left[-\frac{1}{x} - \frac{2}{x^2} \right]_c^b$$

$$= \lim_{b \to \infty} \left[\left(-\frac{1}{b} - \frac{2}{b^2} \right) - \left(-\frac{1}{c} - \frac{2}{c^2} \right) \right]$$

$$= \frac{1}{c} + \frac{2}{c^2}.$$

We then find the value of c so that this probability is equal to the significance level of $\alpha = 0.05$.

$$\frac{1}{c} + \frac{2}{c^2} = 0.05$$

$$0 = 0.05 - \frac{1}{c} - \frac{2}{c^2}$$

$$0 = 0.05c^2 - c - 2.$$

We then use the quadratic formula to solve for c.

$$c = \frac{1 \pm \sqrt{1 - 4(0.05)(-2)}}{2(0.05)}$$

$$= \frac{1 \pm \sqrt{1.4}}{0.1}$$

$$c \approx 21.832 \quad \text{or} \quad c \approx -1.832.$$

Since -1.832 lies outside of the domain of the probability density function, we conclude that the critical value is approximately 21.832.   ∎

In statistical applications, both critical values and observed levels of significance are used to draw conclusions from the results of an experiment, and both are often published in scientific articles. In this context, the significance level α is interpreted as the probability of making a certain kind of error in interpreting the results of the experiment. More about this use of critical values may be found in a statistics textbook.

Constructing Probability Density Functions

Suppose that we have a nonnegative function $f(x)$ whose definite integral over some interval $[a, b]$ is $A > 0$. Then

$$\int_a^b f(x)\, dx = A.$$

If we wish to find a number k to make $kf(x)$ a density function, then we must have

$$1 = \int_a^b kf(x)\, dx = k \int_a^b f(x)\, dx = kA.$$

Solving $1 = kA$ for k gives

$$k = \frac{1}{A}.$$

EXAMPLE 8   Find k such that

$$f(x) = kx^2$$

is a probability density function over the interval $[2, 5]$. Then write the probability density function.

Solution   We have

$$\int_2^5 x^2\, dx = \left[\frac{x^3}{3} \right]_2^5$$

$$= \frac{5^3}{3} - \frac{2^3}{3} = \frac{125}{3} - \frac{8}{3} = \frac{117}{3}.$$

Thus,

$$k = \frac{1}{\frac{117}{3}} = \frac{3}{117},$$

and the probability density function is

$$f(x) = \frac{3}{117}x^2 \quad \text{for } 2 \le x \le 5.$$ ∎

EXAMPLE 9   Find a k such that

$$f(x) = kx^3$$

is a probability density function over the interval $[-1, 2]$.

Solution   The function $f(x) = x^3$ is negative for $x < 0$ and positive for $x > 0$. Since a density function cannot be negative, we conclude that we cannot make a density function of the form $f(x) = kx^3$ over the interval $[-1, 2]$. ∎

Uniform Distributions

Suppose a bacterium is in a 3 cc solution. If it is just as likely to be in the first 1 cc transferred to a flask as in any other 1 cc volume of the solution, then the graph of the density function is a horizontal line.

EXAMPLE 10 The graph of the density function $f(x)$ is a horizontal line for $a \leq x \leq b$. Find $f(x)$.

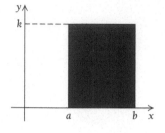

Solution Since the graph of $f(x)$ is a horizontal line, $f(x) = k$ for some constant k. Since f is a density function, the area of the rectangle shown in the margin is 1. Therefore,

$$k(b - a) = 1, \quad \text{or} \quad k = \frac{1}{b - a}.$$

The density function is

$$f(x) = \frac{1}{b - a} \quad \text{for } a \leq x \leq b. \qquad \blacksquare$$

DEFINITION

A continuous random variable X is said to be *uniformly distributed* over an interval $[a, b]$ if its density function f is given by

$$f(x) = \frac{1}{b - a} \quad \text{for } a \leq x \leq b.$$

If X is uniformly distributed over the interval $[a, b]$, we will sometimes say that X has a **Uniform(a, b) distribution,** or that X is Uniform(a, b).

EXAMPLE 11 The loudness V of a bird's song is uniformly distributed between 30 and 50 decibels. Find the probability that the loudness of the bird's song is between 35 and 45 decibels.

Solution The probability density function for V is

$$f(v) = \frac{1}{50 - 30} = \frac{1}{20}, \quad \text{for } 30 \leq v \leq 50.$$

The probability is

$$P(35 \leq V \leq 45) = \int_{35}^{45} \frac{1}{20} \, dv = \frac{1}{20}[v]_{35}^{45} = \frac{1}{20}(45 - 35) = \frac{10}{20} = \frac{1}{2}. \qquad \blacksquare$$

Expected Value and Standard Deviation

In Section 10.4, we defined the expected value or mean of a discrete random variable to be $\sum kP(X = k)$, where the sum is taken over all possible values k of the random variable X. We define the expected value of a continuous random variable analogously, using an integral instead of a sum.

DEFINITION

Let X be a continuous random variable distributed over $[a, b]$ with density function f.

a) The *expected value* (or *mean*) of X is

$$\mu = E(X) = \int_a^b xf(x)\,dx.$$

b) The *variance* of X is

$$\sigma^2 = \mathrm{Var}(X) = \int_a^b (x - \mu)^2 f(x)\,dx.$$

As with discrete random variables, we also have

$$\mathrm{Var}(X) = E(X^2) - \mu^2,$$

where $E(X^2) = \displaystyle\int_a^b x^2 f(x)\,dx.$

c) The *standard deviation* of X is the square root of the variance:

$$\sigma = \mathrm{SD}(X) = \sqrt{\mathrm{Var}(X)}.$$

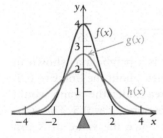

For both discrete and continuous random variables, the mean measures the balance point of the distribution, and the standard deviation is a measure of how much a random variable is likely to deviate from the mean. The graph in the margin shows density functions $f(x)$, $g(x)$, and $h(x)$ for three random variables, all with mean 0. Note that $f(x)$, the red curve, is concentrated closer to 0, so a random variable with density function $f(x)$ has a small standard deviation. The density function $h(x)$, the green curve, is spread out the most. This indicates a random variable with a high standard deviation. The blue curve has a standard deviation somewhere between the other two.

EXAMPLE 12   Let X be a random variable with density function $f(x) = 6x - 6x^2$ over the interval $0 \le x \le 1$.

a) Compute the expected value of X.

b) Compute the variance of X.

c) Compute the standard deviation of X.

Solution

a) We compute the expected value of X using the definition.

$$\mu = \int_0^1 xf(x)\,dx$$

$$= \int_0^1 x(6x - 6x^2)\,dx$$

$$= \int_0^1 (6x^2 - 6x^3)\,dx$$

$$= \left[2x^3 - \frac{3}{2}x^4\right]_0^1 = \left(2 - \frac{3}{2}\right) - (0 - 0) = \frac{1}{2} = 0.5$$

b) We compute $E(X^2)$ using the density function for X.

$$E(X^2) = \int_0^1 x^2 f(x)\,dx$$

$$= \int_0^1 x^2(6x - 6x^2)\,dx$$

$$= \int_0^1 (6x^3 - 6x^4)\,dx$$

$$= \left[\frac{6}{4}x^4 - \frac{6}{5}x^5\right]_0^1$$

$$= \left(\frac{6}{4} - \frac{6}{5}\right) - (0 - 0)$$

$$= 0.3$$

We now compute the variance:

$$\mathrm{Var}(X) = E(X^2) - \mu^2 = 0.3 - (0.5)^2 = 0.05.$$

c) Since the variance is 0.05, the standard deviation is $\sqrt{0.05} \approx 0.2236$.

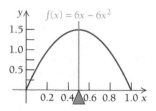

The graph of the density function $f(x) = 6x - 6x^2$ is a parabola as shown in the figure in the margin. Since the graph of $f(x)$ is symmetric about the line $x = 0.5$, if a fulcrum were placed on the x-axis at $x = 0.5$, the area under the graph would just balance. Therefore, the mean should be at the balance point, $x = 0.5$. The standard deviation measures the spread of the distribution. Since the value of X can never be more than 0.5 away from the mean 0.5, certainly the spread should be no more than 0.5. Furthermore, since most of the area under the parabola $f(x) = 6x - 6x^2$ is near the mean, we would expect the standard deviation to be closer to zero than to 0.5. Based on the graph, a standard deviation of approximately 0.2236 is therefore reasonable. ■

For special distributions, such as the uniform distribution, there are formulas that allow us to determine the mean and standard deviation.

> **THEOREM 9**
>
> If X has a Uniform(a, b) distribution, then
>
> 1. $E(X) = \dfrac{a + b}{2}$,
>
> 2. $\text{Var}(X) = \dfrac{(b - a)^2}{12}$, and
>
> 3. $\text{SD}(X) = \sqrt{\text{Var}(X)} = \dfrac{b - a}{\sqrt{12}}$.

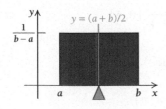

We leave the proof to the exercises, but the formulas seem reasonable. For a uniform distribution, the expected value is the average of the endpoints. This makes sense as the distribution function is symmetric about the line $y = \dfrac{a + b}{2}$. Furthermore, the standard deviation is proportional to the range of the distribution $b - a$.

EXAMPLE 13   Compute the mean, variance, and standard deviation of a Uniform$(1, 7)$ random variable.

Solution   We use Theorem 9 with $a = 1$ and $b = 7$. The mean is

$$E(X) = \mu = \frac{1 + 7}{2} = 4.$$

The variance is

$$\text{Var}(X) = \frac{(7 - 1)^2}{12} = 3.$$

The standard deviation is the square root of the variance:

$$\text{SD}(X) = \sqrt{3}.$$

Exercise Set 10.5

Verify the two properties of a probability density function over the given interval.

1. $f(x) = 2x$, $[0, 1]$

2. $f(x) = \frac{1}{3}$, $[4, 7]$

3. $f(x) = \frac{3}{26}x^2$, $[1, 3]$

4. $f(x) = \dfrac{1}{x + 1}$, $[0, e - 1]$

Use the density function to compute the probability.

5. $f(x) = 4x^3$ over $[0, 1]$, $P(1/4 \le X \le 3/4)$

6. $f(x) = 12x^2 (1 - x)$ over $[0, 1]$, $P(X \le 1/2)$

7. $f(x) = 20x(1 - x)^3$ over $[0, 1]$, $P(1/4 \le X)$

8. $f(x) = \frac{1}{2}\sin x$ over $[0, \pi]$, $P(0 \le X \le \pi/4)$

9. $f(x) = \sec^2 x$ over $[0, \pi/4]$, $P(X \ge \pi/6)$

10. $f(x) = \dfrac{1}{\sqrt{2x + 3}}$ over $[3, 13/2], P(X \le 4)$

11. $f(x) = \dfrac{e^x}{e - 1}$ over $[0, 1], P(0 \le X \le 1/2)$

12. $f(x) = 2xe^{-x^2}$ over $[0, \infty), P(X \le 1)$

In Exercises 13–18, a random variable is assumed to have the given probability density function. Find the observed significance level if the random variable is equal to the given value in an experiment.

13. $f(x) = 1/x^2$ over $[1, \infty)$; observed value of 16

14. $f(x) = 2/x^3$ over $[1, \infty)$; observed value of 20

15. $f(x) = e^{-x}$ over $[0, \infty)$; observed value of 3.4

16. $f(x) = xe^{-x}$ over $[0, \infty)$; observed value of 5

17. $f(x) = 2x/(x^2 + 1)^2$ over $[0, \infty)$; observed value of 4

18. $f(x) = 3x^2e^{-x^3}$ over $[0, \infty)$; observed value of 2

In Exercises 19–24, the probability density function of a random variable and a significance level α are given. Find the critical value.

19. $f(x) = 1/x^2$ over $[1, \infty)$; $\alpha = 0.05$

20. $f(x) = 3/x^4$ over $[1, \infty)$; $\alpha = 0.01$

21. $f(x) = e^{-x}$ over $[0, \infty)$; $\alpha = 0.05$

22. $f(x) = 2xe^{-x^2}$ over $[0, \infty)$; $\alpha = 0.01$

23. $f(x) = 3.8/x^3 + 33.6/x^5$ over $[2, \infty)$; $\alpha = 0.05$

24. $f(x) = 0.32e^{-x} + 1.36e^{-2x}$ over $[0, \infty)$; $\alpha = 0.01$

In Exercises 25–30, find k such that the function is a probability density function over the given interval. Then write the probability density function. If there is no k that makes the function a probability density function, state why.

25. $f(x) = kx, [1, 3]$

26. $f(x) = kx, [-1, 4]$

27. $f(x) = kx^2, [-1, 1]$

28. $f(x) = kx^3, [-2, 2]$

29. $f(x) = k(x^2 - x), [0, 2]$

30. $f(x) = kx^2(x - 2)^2, [-2, 2]$

In Exercises 31–37:

a) Find k such that the function is a probability density function over the given interval. Then write the probability density function.

b) Find the probability if X has the given density function.

31. a) $f(x) = k(2 - x)^3, [0, 2]$
    b) $P(X \le 1)$

32. a) $f(x) = \dfrac{k}{x}, [1, 4]$
    b) $P(X \le 3)$

33. a) $f(x) = k \sin x, [\pi/6, \pi/2]$
    b) $P\left(X \ge \dfrac{\pi}{3}\right)$

34. a) $f(x) = k \cos^2 x, [0, \pi]$
    b) $P\left(X \ge \dfrac{\pi}{4}\right)$

35. a) $f(x) = k\dfrac{2x + 3}{x^2 + 3x + 4}, [0, 5]$
    b) $P(2 \le X \le 4)$

36. a) $f(x) = k(x - \sin x)e^{-x}, [0, \infty)$
    b) $P(X \ge 3)$

37. a) $f(x) = kx \sin x, [-\pi, \pi]$
    b) $P(X \le \pi/2)$

Compute the probability using the given distribution.

38. $P(0 \le X \le 3), X$ is Uniform$(0, 7)$

39. $P(3 \le X \le 4), X$ is Uniform$(2, 4)$

40. $P(10 \le X \le 30), X$ is Uniform$(-10, 70)$

41. $P(200 \le X \le 350), X$ is Uniform$(100, 500)$

For each probability density function, find $E(X)$, Var(X), and SD(X).

42. $f(x) = \dfrac{1}{3}, [2, 5]$

43. $f(x) = \dfrac{2}{9}x, [0, 3]$

44. $f(x) = \dfrac{1}{3}x^2, [-2, 1]$

45. $f(x) = \dfrac{1}{x \ln 2}, [1, 2]$

46. $f(x) = \dfrac{1}{96}(1 + x + x^2), [0, 6]$

47. $f(x) = \dfrac{4}{21}(x + x^3), [1, 2]$

48. $f(x) = \dfrac{12}{17}(\sqrt{x} + \sqrt[3]{x}), [0, 1]$

49. $f(x) = \dfrac{1}{2} \sin x, [0, \pi]$

50. $f(x) = \cos x, \left[0, \dfrac{\pi}{2}\right]$

51. $f(x) = \dfrac{3}{14}\sqrt{x+1}, [0, 3]$

52. $f(x) = \ln x, [1, e]$

53. $f(x) = 3e^{-3x}, [0, \infty)$

54. $f(x) = 3e^{-3(x-1)}, [1, \infty)$

Find the mean, variance, and standard deviation for a random variable with the given distribution.

55. Uniform$(3, 9)$

56. Uniform$(-2, 5)$

57. Uniform$(10, 20)$

58. Uniform$(0.001, 0.002)$

APPLICATIONS

59. *Bacterium Transfer.* A pipette contains exactly one bacterium in 5 cc of a well-mixed solution. Find the probability that the bacterium is in the first 3 cc of solution transferred to a flask.

60. *Bacterium Transfer.* A pipette contains exactly one bacterium in 4 cc of a well-mixed solution. Find the probability that the bacterium is in the last 3 cc of solution transferred to a flask.

61. *Bird Calls.* The loudness, measured in decibels, of a bird's call has a Uniform$(60, 70)$ distribution.

    a) Find the probability that the bird's call is between 60 and 65 decibels.
    b) Find the probability that the bird's call is between 68 and 70 decibels.

62. *Bird Calls.* The loudness, measured in decibels, of a bird's call has a Uniform$(55, 73)$ distribution.

    a) Find the probability that the bird's call is between 55 and 61 decibels.
    b) Find the probability that the bird's call is between 67 and 73 decibels.

SYNTHESIS

63. The function $f(x) = x^3$ is a probability density function over $[0, b]$. What is b?

64. The function $f(x) = 12x^2$ is a probability density function over $[-a, a]$. What is a?

65. Let X have a Uniform(a, b) distribution.

    tW   a) Use the definition of expected value to prove part 1 of Theorem 9.

b) Compute $E(X^2)$.

tW   c) Use the definition of variance to prove part 2 of Theorem 9.

66. Let X be a random variable with density function $f(x)$ over $[a, b]$ and expected value $E(X) = \mu$. The definition of the variance of X is

$$\text{Var}(X) = \int_a^b (x - \mu)^2 f(x)\, dx.$$

Simplify the integral to show

$$\text{Var}(X) = E(X^2) - \mu^2,$$

where $E(X^2) = \displaystyle\int_a^b x^2 f(x)\, dx.$

67. *Mortality Rate.* The number of people in the United States of any age x between 1 and 85 who died in 2000 can be approximated by the function[22]

$$f(x) = 1152.9e^{0.051476x}.$$

    a) Find a value k to make the function $kf(x)$ a probability density function.
    b) Use your answer in part (a) to approximate the mean age among all who died in 2000 with age between 1 and 85.
    c) Find the standard deviation of the ages of people who died in 2000 whose age was between 1 and 85.

68. *Habitat for Humanity.* The relative number of household incomes in the United States can be approximated by the graph of the function[23]

$$f(x) = k(0.00394 - 0.00247x$$
$$+ 0.00000910x^2 + 0.0365 \ln x),$$

where x is the income measured in thousands of dollars and $2 \le x \le 125$.

    a) Find the k that makes $f(x)$ a density function.
    b) To qualify for a Habitat for Humanity house, the family income must be between \$14,000 and \$30,000. If a family is picked at random from those of the population whose income is between \$2000 and \$125,000, find the probability that the family's income qualifies for a Habitat for Humanity house.

---

[22]Centers for Disease Control.
[23]U.S. Bureau of the Census.

69. *Tree Cavities.* Cavities in trees are necessary for certain birds and other animals to survive. For a plot of woods in Missouri, let X represent the number of cavities per hectare. The density function for X can be approximated by[24]

$$f(x) = \frac{a}{b}\left(\frac{x-c}{b}\right)^{a-1} e^{-(x-c)^a/b^a},$$

where a, b, and c are constants, and $x \geq c$.

tw a) Verify property 2 of a density function for $f(x)$. (*Hint:* Try substituting $u = (x-c)/b$.)

b) For plots of trees no more than 30 yr old, it was found that $a \approx 0.68$, $b \approx 0.89$, and $c \approx 0.0$. Compute $P(X \leq 2)$.

tw **70.** Explain the idea of a probability density function.

 Technology Connection

Student t-Distribution. A common distribution that arises in statistics is the *Student t-distribution with n degrees of freedom*, or the "t_n distribution" for short. The probability density functions of the t_n distribution for some small values of n are given below.

t_2: $\quad f(x) = \dfrac{1}{(x^2 + 2)^{3/2}} \qquad$ t_3: $\quad f(x) = \dfrac{6\sqrt{3}}{\pi(x^2 + 3)^2}$

t_5: $\quad f(x) = \dfrac{200\sqrt{5}}{3\pi(x^2 + 5)^3} \qquad$ t_7: $\quad f(x) = \dfrac{5488\sqrt{7}}{5\pi(x^2 + 7)^4}$

All of these probability density functions are over the interval $(-\infty, \infty)$.

Use numerical integration to find the observed significance level for the t_n distribution for the given observed value from an experiment.[25]

71. t_2; observed value of 4. (*Hint:* Since the Student t-distribution is symmetric about 0, this observed level of significance is equal to $0.5 - P(0 \leq X \leq 4)$.)

72. t_3; observed value of 2

73. t_5; observed value of 2.8

74. t_7; observed value of 2.8

Chi-Squared Distribution. Another common distribution that arises in statistics is the *chi-squared distribution with n degrees of freedom,* or the "$\chi^2(n)$ distribution" for short. The probability density function of the $\chi^2(n)$ distribution for some small values of n are given below.

$$\chi^2(3): \quad f(x) = \sqrt{\frac{x}{2\pi}}\, e^{-x/2}$$

$$\chi^2(4): \quad f(x) = \frac{x}{4}\, e^{-x/2}$$

$$\chi^2(6): \quad f(x) = \frac{x^2}{16}\, e^{-x/2}$$

$$\chi^2(8): \quad f(x) = \frac{x^3}{96}\, e^{-x/2}$$

All of these probability density functions are over the interval $[0, \infty)$.

Use numerical integration to find the observed significance level for the $\chi^2(n)$ distribution for the given observed value from an experiment.

75. $\chi^2(4)$; observed value of 15

76. $\chi^2(6)$; observed value of 16

77. $\chi^2(8)$; observed value of 10

78. $\chi^2(3)$; observed value of 5

Critical Values of the $\chi^2(n)$ Distribution. Use integration by parts to determine the right-tail probability $P(X \geq c)$. Then use Newton's method to determine the critical value for the $\chi^2(n)$ distribution given significance level α. Use a starting value of 5 to initiate Newton's method. (Critical values such as these are often found in statistical tables.)

79. $\chi^2(4)$; $\alpha = 0.05$

80. $\chi^2(4)$; $\alpha = 0.01$

81. $\chi^2(6)$; $\alpha = 0.05$

82. $\chi^2(6)$; $\alpha = 0.01$

---

[24]Z. Fan, D. R. Larsen, S. R. Shifley, and F. R. Thompson, *Forest Ecology and Management*, Vol. 179, pp. 231–242 (2003).

[25]Though we do not do so here, it is also possible to compute observed significance levels using a left tail of the t_n distribution.

10.6 The Poisson Process

OBJECTIVES

■ Compute probabilities of exponential random variables.

■ Compute probabilities of Poisson random variables.

Exponential Distributions

The duration of a phone call, the time until a cell divides, and the amount of time required to learn a task are all examples of *exponentially distributed* random variables.

> **DEFINITION**
>
> A continuous random variable is *exponentially distributed* if it has a probability density function given by
>
> $$f(x) = \lambda e^{-\lambda x} \quad \text{over the interval } [0, \infty),$$
>
> where $\lambda > 0$ is a constant.

If X is exponentially distributed with parameter λ, we will sometimes say that X has an **Exponential(λ) distribution.**

The function $f(x) = 2e^{-2x}$ is the density function for an Exponential(2) distribution. The integral

$$\int_0^\infty 2e^{-2x}\, dx = 1$$

was computed in Section 5.9. The general case

$$\int_0^\infty \lambda e^{-\lambda x}\, dx = 1$$

can be verified in a similar way. This is left as an exercise.

EXAMPLE 1 It is often assumed that the *cell cycle time,* the time T in minutes until a single cell divides, is an Exponential(λ) random variable. The parameter λ is equal to $\dfrac{1}{a}$, where a is the average cell cycle time. Assuming that the average cell cycle time is 100 min, find the probability that a given cell divides in 60 min or less.

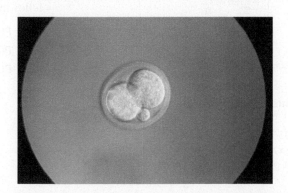

Solution   We first determine λ:

$$\lambda = \frac{1}{a} = \frac{1}{100} = 0.01.$$

The probability density function for T is

$$f(t) = \lambda e^{-\lambda t} = 0.01 e^{-0.01t} \quad \text{for } t \geq 0.$$

Now we can compute the probability that the cell divides in 60 min or less.

$$
\begin{aligned}
P(0 \leq T \leq 60) &= \int_0^{60} 0.01 e^{-0.01t} \, dt \\
&= \left[\frac{0.01}{-0.01} e^{-0.01t} \right]_0^{60} \\
&= \left[-e^{0.01t} \right]_0^{60} \\
&= \left(-e^{0.01 \cdot 60} \right) - \left(-e^{-0.01 \cdot 0} \right) \\
&= -e^{-0.6} + 1 \\
&\approx 0.4512. \qquad \text{Using a scientific calculator for approximation}
\end{aligned}
$$

Technology Connection

Exponential Distribution

Graphers may or may not have the exponential density function built in. The TI-83 does not. Even so, we can use the TI-83 to help us visualize and compute probabilities involving an Exponential distribution. Let's visualize and compute $P(0.2 \leq X \leq 0.8)$, where X has an Exponential(5) distribution. First we press the $\boxed{Y=}$ key and enter the density function $f(x) = 5e^{-5x}$.

We now enter a reasonable window, in this case, $[0, 1, 0, 5]$. We can use the integration capabilities of the TI-83 by pressing the $\boxed{\text{CALC}}$ button and selecting $\int f(x) \, dx$. When prompted, we enter the lower limit of 0.2 and the upper limit of 0.8.

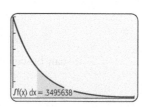

We see that $P(0.2 \leq X \leq 0.8) \approx 0.3496$.

EXERCISES
Use a grapher to compute the probabilities of the following Exponential(λ) random variables.

1. $P(0.3 \leq X \leq 0.9), \lambda = 5$

2. $P(0 \leq X \leq 2), \lambda = 4$

3. $P(1 \leq X \leq 3), \lambda = 0.5$

Poisson Process

How can we determine that a random variable has an exponential distribution? There are some conditions that, if satisfied, imply that X has an exponential distribution. For example, a radioactive material will set off a counter every time there is a decay. Although the decays occur at random times, the process seems to satisfy three conditions. Loosely stated, the three conditions are

1. For a very short time interval, if we double the length of time, the probability of a decay during the time interval will approximately double.
2. In a short time interval, we are very unlikely to observe two or more decays.
3. Just because we detected a decay during a 1-sec time interval, we do not expect it to be any more or less likely that we will detect another decay in the next 1-sec time interval.

A *process* is an experiment where events occur at random times. If a process satisfies these three conditions, then we call it a **Poisson process.**

Events like mutations in a population and earthquakes can be modeled as a Poisson process. However, a Poisson process may not be an appropriate model for other events. For instance, cars on a highway tend to cluster behind a slowly moving car. If an event is the passing of a car, then condition (3) is not satisfied. This is because if a car just passed, then it is likely that more passes from the other cars in the cluster will occur in the near future.

Time Between Events

Given a Poisson process, we can define two different random variables. The first random variable X is the length of time until the next event. The random variable X has an exponential distribution with density function

$$f(x) = \lambda e^{-\lambda x},$$

over the interval $[0, \infty)$. The value of λ is the average number of events per unit of time. Equivalently, λ can be determined by the formula

$$\lambda = \frac{1}{a},$$

where a is the average length of time between events.

EXAMPLE 2   *Mutations* In the Luria–Delbrück mutation model, it is assumed that the occurrences of mutations follow a Poisson process whose value of λ depends on the population size.[26] Suppose that on average, the number of mutations per hour in a population is 0.25. Let X be the number of hours between mutations. Compute the probability that

a) at least 4 hr pass after a mutation occurs until the next mutation occurs.

b) at most 8 hr pass after a mutation occurs until the next mutation occurs.

c) at least 4 hr, but not more than 8 hr pass after a mutation until the next one occurs.

---

[26]Q. Zheng, "Progress of a half century in the study of the Luria–Delbrück distribution," *Mathematical Biosciences,* Vol. 162, pp. 1–32 (1999).

Solution Since there is an average of 0.25 mutations per hour, $\lambda = 0.25$. Therefore, the probability density function for X is

$$f(x) = 0.25e^{-0.25x},$$

where $x \geq 0$.

a) We are to compute $P(X \geq 4)$.

$$\begin{aligned}
P(X \geq 4) &= \int_{4}^{\infty} 0.25e^{-0.25x}\,dx \\
&= \lim_{b \to \infty} \left[-e^{-0.25x}\right]_{4}^{b} \\
&= \lim_{b \to \infty} \left(-e^{-0.25b}\right) - \left(-e^{-0.25(4)}\right) \\
&= 0 + e^{-1} = \frac{1}{e} \approx 0.3679
\end{aligned}$$

b) In this case we need to compute $P(X \leq 8)$.

$$\begin{aligned}
P(X \leq 8) &= P(0 \leq X \leq 8) \qquad \text{Since the domain of } f(x) \text{ is } x \geq 0 \\
&= \int_{0}^{8} 0.25e^{-0.25x}\,dx \\
&= \left[-e^{-0.25x}\right]_{0}^{8} \\
&= -e^{-0.25(8)} - \left(-e^{0.25(0)}\right) = -e^{-2} + 1 \approx 0.8647
\end{aligned}$$

c) We need to compute $P(4 \leq X \leq 8) = \int_{4}^{8} 0.25e^{-0.25x}\,dx$. Following parts (a) and (b), we see that

$$\int_{4}^{8} 0.25e^{-0.25x}\,dx = -e^{-0.25(8)} - \left(e^{-0.25(4)}\right) = -e^{-2} + e^{-1} \approx 0.2325. \quad \blacksquare$$

Counting Events

The second random variable we can define with a Poisson process is the number of events observed in a time interval of length h. If Y denotes the number of events that occur, then Y is a discrete random variable. If we let $\mu = \lambda h$, where λ is as above, then the random variable Y has a *Poisson(μ)* distribution as defined below.

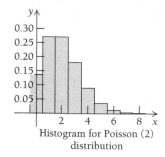

Histogram for Poisson (2) distribution

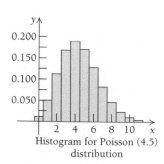

Histogram for Poisson (4.5) distribution

DEFINITION

If Y is a discrete random variable that satisfies

$$P(Y = k) = \frac{\mu^k}{k!}e^{-\mu} \qquad k = 0, 1, 2, \ldots$$

for $\mu > 0$, then we say that Y has a **Poisson(μ) distribution.**

EXAMPLE 3 *Mutations* From Example 2, let Y be the number of mutations that occur in a 2-hr period. For a fixed 2-hr period, compute the probability that

a) no mutations occur.

b) exactly one mutation occurs.

c) at least one mutation occurs.

Solution   From Example 2, we know that $\lambda = 0.25$. We are also given that the length of the time interval is $h = 2$. Since $\lambda h = 0.25(2) = 0.5$, the random variable Y has a Poisson(0.5) distribution.

a) If no mutations occur, then $Y = 0$.

$$P(Y = 0) = \frac{\mu^k}{k!}e^{-\mu} = \frac{(0.5)^0}{0!}e^{-0.5} = \frac{1}{1}e^{-0.5} = e^{-0.5} \approx 0.6065.$$

b) If exactly one mutation occurs, then $Y = 1$.

$$P(Y = 1) = \frac{\mu^k}{k!}e^{-\mu} = \frac{(0.5)^1}{1!}e^{-0.5} = \frac{0.5}{1}e^{-0.5} \approx 0.3033.$$

c) If at least one mutation occurs, then $Y \geq 1$ or, equivalently, Y is not 0. We have

$$
\begin{aligned}
P(Y \geq 1) &= P(Y \neq 0) \\
&= 1 - P(Y = 0) \\
&\approx 1 - 0.6065 \qquad \text{By part (a)} \\
&= 0.3935.
\end{aligned}
$$

Technology Connection

Poisson Distribution

We can solve Example 3 using a grapher. Press the DISTR button and select **POISSONPDF**. Then enter the value of μ (in this case, 0.5) followed by the value of the random variable. For part (a), $Y = 0$.

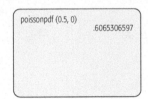

poissonpdf (0.5, 0)
.6065306597

EXERCISES

Suppose that Y has a Poisson(μ) distribution. Compute the probabilities.

1. $P(Y = 1)$, $\mu = 1$

2. $P(Y = 2)$, $\mu = 1$

3. $P(Y = 3)$, $\mu = 1$

4. $P(Y = 0)$, $\mu = 5$

5. $P(Y = 1)$, $\mu = 5$

6. $P(Y = 2)$, $\mu = 5$

Mixed Solutions

Suppose that a few bacteria are in a well-mixed solution and a portion of the solution is transferred into another container. We can use a Poisson distribution to analyze the number of bacteria that were transferred. If we let b be the number of bacteria per unit volume and V be the volume of the transferred liquid, then the number of bacteria transferred has a Poisson distribution with parameter $\mu = bV$.

EXAMPLE 4   A well-mixed solution contains 0.5 bacteria per cc. A volume of 2 cc is transferred into a test tube.

a) What is the probability that no bacteria were transferred?

b) What is the probability that exactly one bacterium was transferred?

c) What is the probability that at least one bacterium was transferred?

Solution   Let X be the number of bacteria transferred. Then X has a Poisson distribution with $\mu = 0.5(2) = 1$.

a) $P(X = 0) = \dfrac{1^0}{0!}e^{-1} = \dfrac{1}{1} \cdot \dfrac{1}{e} \approx 0.3679.$

b) $P(X = 1) = \dfrac{1^1}{1!}e^{-1} = \dfrac{1}{1} \cdot \dfrac{1}{e} \approx 0.3679.$

c) From part (a) we know the probability that no bacteria were transferred is 0.3679. Therefore, the probability that at least one transferred is

$$P(X \geq 1) = 1 - P(X = 0) \approx 1 - 0.3679 = 0.6321.$$

Mean and Standard Deviation

We next state the mean and variance of an Exponential(λ) distribution and a Poisson(μ) distribution.

THEOREM 10

Let X be an Exponential(λ) random variable. Then

$$E(X) = \frac{1}{\lambda} \quad \text{and} \quad SD(X) = \frac{1}{\lambda}.$$

Let Y be a Poisson(μ) random variable. Then

$$E(Y) = \mu \quad \text{and} \quad SD(Y) = \sqrt{\mu}.$$

If the expected value of an exponential random variable is a, then Theorem 10 says $\lambda = \dfrac{1}{a}$. Recall that in Example 1 we used $\lambda = \dfrac{1}{a}$, where the average cell cycle time was $a = 100$. Theorem 10 justifies this choice of parameter since $a = \dfrac{1}{\lambda}$, or equivalently, $\lambda = \dfrac{1}{a} = \dfrac{1}{100}$.

EXAMPLE 5   Compute the mean, variance, and standard deviation of a random variable X that has an Exponential(4) distribution.

Solution   According to Theorem 10, the mean is

$$E(X) = \mu = \frac{1}{\lambda} = \frac{1}{4}.$$

Also, the standard deviation is the same as the mean, so

$$SD(X) = \frac{1}{4}.$$

The variance of X is

$$\text{Var}(X) = \left(\frac{1}{4}\right)^2 = \frac{1}{16}.$$ ■

EXAMPLE 6   Find $E(Y)$ and $SD(Y)$ if Y has a Poisson(5) distribution.

Solution   Using Theorem 10, we see that

$$E(Y) = 5 \quad \text{and} \quad SD(Y) = \sqrt{5}.$$ ■

Exercise Set 10.6

Suppose that X has a Poisson(μ) distribution. Compute the following quantities.

1. $P(X = 0)$, if $\mu = 2.2$
2. $P(X = 3)$, if $\mu = 3.5$
3. $P(X = 1)$, if $\mu = 1.3$
4. $P(X = 9)$, if $\mu = 7$
5. $P(X = 7)$, if $\mu = 4$
6. $P(X = 6)$, if $\mu = 5.2$
7. $P(X \leq 2)$, if $\mu = 3$
8. $P(X \leq 3)$, if $\mu = 1$
9. $P(X \geq 3)$, if $\mu = 1.5$
10. $P(X \geq 2)$, if $\mu = 4.1$
11. $P(3 \leq X \leq 5)$, if $\mu = 1.2$
12. $P(2 \leq X \leq 5)$, if $\mu = 2$
13. $E(X)$, if $\mu = 2.3$
14. $E(X)$, if $\mu = 4$
15. $SD(X)$, if $\mu = 4$
16. $SD(X)$, if $\mu = 5.7$

Suppose that X has an Exponential(λ) distribution. Compute the following quantities.

17. $P(0 \leq X \leq 2)$, if $\lambda = 1$
18. $P(2 \leq X \leq 3)$, if $\lambda = 1$
19. $P(0 \leq X \leq 2)$, if $\lambda = 3$
20. $P(2 \leq X \leq 3)$, if $\lambda = 3$
21. $P(2 \leq X)$, if $\lambda = 2$
22. $P(3 \leq X)$, if $\lambda = 2$
23. $P(X \leq 0.5)$, if $\lambda = 2.5$
24. $P(X \leq 1)$, if $\lambda = 2.5$
25. $E(X)$, if $\lambda = 4$
26. $E(X)$, if $\lambda = 10$
27. $SD(X)$, if $\lambda = 4$
28. $SD(X)$, if $\lambda = 10$

 29.–40. Use a grapher to do Exercises 1–12.

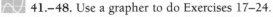 41.–48. Use a grapher to do Exercises 17–24.

Find the mean, variance, and standard deviation for a random variable with the given distribution.

49. Poisson(4)
50. Poisson(3.5)
51. Exponential(3)
52. Exponential(5)

APPLICATIONS

Cell Cycle Time. The cycle time of a cell may be modeled as an exponential distribution. Use this distribution to solve Exercises 53 and 54.

53. *E. Coli*, a bacteria present in a healthy human intestine, has an average cell cycle time of approximately 22 min at 40°C. Find the probability that an *E. Coli* cell divides in less than 20 min.[27]

54. *Giardia lambia,* the protozoa most dreaded by backpackers, has an average cell cycle time of approximately 18 hr at 37°C. Find the probability that a *Giardia lambia* cell divides in less than 15 hr.[28]

55. *Mutations.* In a stable population, on average, three mutations occur every minute.

   a) Find the probability that exactly one mutation occurs in 2 min.
   b) Find the probability that exactly five mutations occur in 3 min.
   c) Find the probability that no mutation occurs in a minute.

56. *Mutations.* In a stable population, on average, four mutations occur every hour.

   a) Find the probability that exactly two mutations occur in an hour.
   b) Find the probability that exactly 10 mutations occur in 2 hr.
   c) Find the probability that fewer than three mutations occur in 2 hr.

Predicting Earthquakes. In Exercises 57 and 58, assume that the occurrence of earthquakes in the given region follows a Poisson process.

57. On average, a large earthquake (large enough to cause damage) occurs in Nevada approximately every 3 yr.[29]

   a) What is the probability that at least one large earthquake will hit Nevada within the next 4 yr?
   b) What is the probability that the time between the next two large earthquakes in Nevada will be at least 2 yr?
   c) What is the probability that there will be no large earthquakes in Nevada in the next year?
   d) What is the probability that there will be exactly one large earthquake in Nevada in the next year?
   e) What is the probability that there will be at least two large earthquakes in Nevada in the next year?

58. On average, a large earthquake occurs near Reno, Nevada, approximately every 10 yr.[30]

   a) What is the probability that at least one serious earthquake will hit near Reno within the next 5 yr?

---

[27]Prescott, Harley, Klein, *Microbiology* (New York: McGraw Hill, 2002).

[28]Prescott, Harley, Klein, *Microbiology* (New York: McGraw Hill, 2002).
[29]http://www.seismo.unr.edu/Perminfo/nevada.html
[30]http://www.seismo.unr.edu/Perminfo/nevada.html

b) What is the probability that the time between the next two serious earthquakes near Reno will be at least 8 yr?

c) What is the probability that there will be no serious earthquakes near Reno in the next 20 yr?

d) What is the probability that there will be exactly one serious earthquake near Reno in the next 25 yr?

e) What is the probability that there will be at least two serious earthquakes near Reno in the next 15 yr?

59. *E. Coli.* Let X denote the number of E. *coli* coliforms found in a randomly selected 5-cm^2 sample of minced beef before processing. Suppose that X has a Poisson(6) distribution.[31]

a) Find $E(X)$ and $SD(X)$.

b) Find the probability that the sample contains no E. *coli*.

60. *Seagrass.* Let X denote the annual number of principal strands generated by a specimen of the seagrass *Posidonia oceanica*. Then X follows a Poisson(13.1) distribution.[32]

a) Find $E(X)$ and $SD(X)$.

b) Find $P(X = 10)$.

SYNTHESIS

61. Use the definition of a density function to verify that

$$f(x) = \lambda e^{-\lambda x}, \quad \text{for } x \geq 0,$$

is a probability density function for any positive value of λ.

62. *Memoryless Property.* Let X be an Exponential(λ) distributed random variable.

a) Compute $P(X > c)$.

b) Use the formula $P(X > a + b$ and $X > a) = P(X > a + b|X > a)P(X > a)$ and part (a) to show $P(X > a + b|X > a) = P(X > b)$.

tw c) Use part (b) to explain why the past does not affect when the next event will occur if the time until an event has an exponential distribution.

---

[31]R. D. Reinders, R. De Jonge, and E. G. Evers, "A statistical method to determine whether micro-organisms are randomly distributed in a food matrix, applied to coliforms and *Escherichia coli* O157 in minced beef." *Food Microbiology,* Vol. 20, pp. 297–303 (2003).

[32]H. Molenaar, D. Barthélémy, P. de Reffye, A. Meinesz, and I. Mialet, "Modelling architecture and growth patterns of *Posidonia oceanica,* Vol. 66, pp. 85–99 (2000).

Medians. Let X be a continuous random variable over $[a, b]$ with density function f. Then the median of X is the number m for which

$$\int_a^m f(x) \, dx = \frac{1}{2}.$$

63. Compute the median of X if the density function is $f(x) = \frac{1}{2}x$ over $[0, 2]$.

64. Compute the median of X if the density function is $f(x) = \frac{3}{2}x^2$ over $[-1, 1]$.

tw 65. X has a Uniform(0, 3) distribution. How does the median compare to the mean?

tw 66. X has a Uniform(a, b) distribution. How does the median compare to the mean?

67. Suppose that X has an Exponential(1) distribution.

a) Compute the median for X.

tw b) How does the median compare to the mean? Does this make sense based on the graph of the density function? Explain.

68. Suppose X has an Exponential(λ) distribution.

a) Compute the median for X.

tw b) How does the median compare with the mean? Does this make sense based on the graph of the density function? Explain.

69. Let X have an Exponential(λ) distribution.

a) Use the definition of an exponential distribution on page 705 and the definition of mean on page 699 to compute $E(X)$.

b) Use the definition of variance on page 699 to compute $Var(X)$.

 Technology Connection

Comparison of Poisson and Binomial Random Variables. If n is large and μ is small, then the Binomial(n, μ) distribution is very close to the Poisson(np) distribution. Compute the following probabilities using (a) the Binomial(n, p) distribution and (b) the Poisson(μ) distribution, where $\mu = np$. Are the two answers very different?

70. $P(X = 3)$ if $n = 1000, p = 0.002$

71. $P(X = 4)$ if $n = 1000, p = 0.003$

72. $P(X = 5)$ if $n = 5000, p = 0.001$

73. $P(X = 2)$ if $n = 5000, p = 0.0005$

74. $P(X = 3)$ if $n = 10{,}000, p = 0.0005$

75. $P(X = 2)$ if $n = 10{,}000, p = 0.0001$

10.7 The Normal Distribution

OBJECTIVES

■ Evaluate normal distribution
probabilities using a table.
■ Approximate binomial distributions
with normal distributions.

The Standard Normal Distribution

The **standard normal curve** is the graph of the function

$$f(x) = \frac{1}{\sqrt{2\pi}}\, e^{-x^2/2} \quad \text{over } (-\infty, \infty).$$

Although it is beyond the scope of this book, it is possible to verify that $f(x)$ is a probability density function. If X is a continuous random variable with this density function, we say that X has a *standard normal distribution*.

From the graph in the margin, we see that the normal curve is symmetric about the line $x = 0$, and so the mean of a standard normal random variable is 0. Furthermore, it can be shown that the standard deviation is 1. In order to compute probabilities of a normal random variable, we must integrate the above density function. That is, we need to compute the area under the normal curve.

By symmetry, the area to the right of 0 is 0.5, and the area to the left of 0 is 0.5. However, the areas under the normal curve for other intervals rarely can be computed exactly. Because of the normal curve's importance, tables that approximate areas under this curve have been computed.

Table 2 in the back of the book is one such table. It shows areas under the normal curve between 0 and z. We illustrate its use with a few examples. (These areas may also be computed on some calculators, as discussed in the Technology Connection on page 715.)

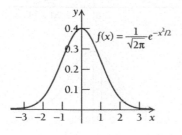

EXAMPLE 1   The random variable Z has a standard normal distribution.

a) Find $P(0 \le Z \le 1.68)$.

b) Find $P(-0.97 \le Z \le 0)$.

c) Find $P(-2.43 \le Z \le 1.01)$.

d) Find $P(1.90 \le Z \le 2.74)$.

e) Find $P(-2.98 \le Z \le -0.42)$.

f) Find $P(Z \ge 0.61)$.

Solution

a) The answer is

$$P(0 \le Z \le 1.68) = \int_0^{1.68} \frac{1}{\sqrt{2\pi}}\, e^{-x^2/2}\, dx,$$

but it is impossible to compute this integral exactly using techniques of integration. We therefore use Table 2 to find the area under the normal curve between 0 and 1.68.

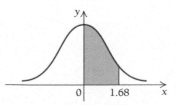

We look up this area in Table 2 by going down the left column to 1.6, then moving to the right to the column headed 0.08. There we read 0.4535. The area between 0 and 1.68 is 0.4535.

b) Because of the symmetry of the graph, the area between −0.97 and 0 is the same as the area between 0 and 0.97. Both areas are equal to 0.3340.

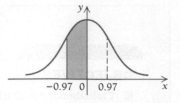

c) This area may be found by adding the area between −2.43 and 0 to the area between 0 and 1.01:

Area = 0.4925 + 0.3438 = 0.8363.

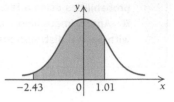

d) This area may be found by subtracting the area between 0 and 1.90 from the area between 0 and 2.74:

Area = 0.4969 − 0.4713 = 0.0256.

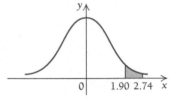

e) This area may be found by subtracting the area between −0.42 and 0 from the area between −2.98 and 0:

Area = 0.4986 − 0.1628 = 0.3358.

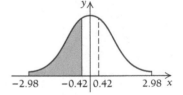

f) The area to the right of 0 is 0.5000. So the area to the right of 0.61 may be found by subtracting the area between 0 and 0.61 from 0.5000:

Area = 0.5000 − 0.2291 = 0.2709.

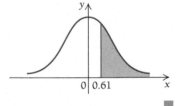

We see that the area under the normal curve for any interval may be found by using Table 2 and possibly addition or subtraction.

Some random variables have a distribution that is very similar to the standard normal distribution, but they may have a different spread or a different mean than the standard normal distribution.

DEFINITION

If X is a random variable with density function given by

$$f(x) = \frac{1}{\sigma\sqrt{2\pi}} e^{-\left(\frac{x-\mu}{\sigma}\right)^2} \quad \text{over } (-\infty, \infty),$$

where μ and σ are constants with $\sigma > 0$, then we say that X has a **Normal(μ, σ) distribution**.[33]

---

[33]Some textbooks refer to this as the Normal(μ, σ^2) distribution. In this text, we choose the second parameter to be the standard deviation instead of the variance.

Technology Connection

Areas under the Normal Curve

On a TI calculator, the area under the normal curve between two endpoints may be computed using the function **NORMALCDF**, which is found under the **DISTR** menu. For example, to solve Example 1(a), we would input **NORMALCDF(0,1.68)**. If the region has no right endpoint, as in Example 1(f), we input **1E99** as the right endpoint. The TI calculator treats this large number as infinity. (For a region with no left endpoint, we would input −**1E99** as the left endpoint.)

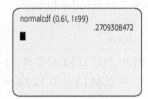

Areas under the normal curve may be visualized by using the function **ShadeNorm**, which is found under the **DRAW** submenu after pressing **DISTR**. Before using this command, however, the graphing window must be chosen. For the standard normal curve, a good graphing window is [−**3.5, 3.5, −0.1, 0.4**].

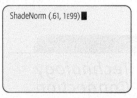

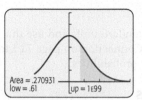

Note: The letters *cdf* are an abbreviation for *cumulative distribution function.*

EXERCISES

1. Use a calculator to solve Example 1(b)–(e).

2. Some of these answers may differ in the fourth decimal place. Why does this occur?

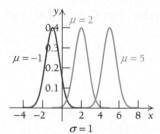

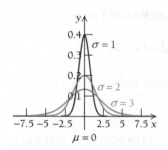

We will verify in the exercises that the $f(x)$ given in the definition is a distribution function. The figures in the margin illustrate that the value of μ determines where the distribution is centered, and the value of σ determines the spread of the distribution. That is, the mean is μ and the standard deviation is σ. The standard normal distribution may also be referred to as the Normal(0, 1) distribution.

EXAMPLE 2 Suppose that X has a Normal(3, 5) distribution. Write the density function for X.

Solution We are given that $\mu = 3$ and $\sigma = 5$. Therefore, the density function is

$$f(x) = \frac{1}{5\sqrt{2\pi}}\, e^{-\left(\frac{x-3}{5}\right)^2} \quad \text{over } (-\infty, \infty).$$

If X has a Normal(μ, σ) distribution, we will rarely use the formula for its density function. Instead, when we compute probabilities, we will use Table 2. We convert to standard units by applying the formula

$$Z = \frac{X - \mu}{\sigma}.$$

This change of variable makes Z a Normal(0, 1) random variable, allowing us to use Table 2 to compute probabilities for X.

The variable Z is often used to represent a Normal(0, 1) random variable. We will follow this convention throughout the rest of the book.

Technology Connection

The Normal Distribution

Convert to standard units and use the Technology Connection on page 715 to compute the probabilities.

EXERCISES

1. The weight in pounds of the students in a calculus class have a Normal(150, 25) distribution.

   a) What is the probability that a student's weight is between 125 lb and 170 lb?

   b) What is the probability that a student's weight is greater than 200 lb?

2. *SAT Scores.* The distribution of the math SAT test scores is approximately Normal(519, 115).[34]

   a) What percentage of the scores is between 250 and 450?

   b) What percentage of the scores is above 500?

EXAMPLE 3   Find $P(2 \leq X \leq 4)$ given that the distribution for X is Normal(1, 3).

Solution   The random variable X has a normal distribution with mean 1 and standard deviation 3. We first convert the boundaries for X to standard units.

$$2 \Rightarrow \frac{2-1}{3} = \frac{1}{3} \approx 0.3333 \text{ in standard units};$$

$$4 \Rightarrow \frac{4-1}{3} = 1 \text{ in standard units.}$$

Using Table 2, we have

$$P(2 \leq X \leq 4) = P(0.3333 \leq Z \leq 1)$$
$$\approx 0.3413 - 0.1293 = 0.2120.$$

(The more accurate answer of 0.2108 is obtained if a grapher is used instead of Table 2.)   ■

Many naturally occurring phenomena follow a normal distribution.

EXAMPLE 4   *Population Height.* Let X be the height in inches of a randomly picked male adult in a certain city. Suppose that we can model the random variable X as Normal(69, 2.5). Find the probability that a randomly picked adult male in that city has a height between 70 and 75 in.

Solution   We first convert 70 and 75 to standard units.

$$70 \Rightarrow \frac{70-69}{2.5} = 0.4 \text{ in standard units};$$

$$75 \Rightarrow \frac{75-69}{2.5} = 2.4 \text{ in standard units.}$$

We have

$$P(70 \leq X \leq 75) = P(0.4 \leq Z \leq 2.4) \approx 0.4918 - 0.1554 = 0.3364.$$

This may also be solved by using the Technology Connection.   ■

---

[34]The College Board.

Approximating a Binomial(n, p) Distribution with a Normal Curve

In Section 10.3, we saw that the histogram for a Binomial(n, p) random variable looks very much like a normal curve. If n is at least 30 and both np and nq are at least 10, then the Binomial(n, p) distribution may be reasonably approximated by the normal curve. This approximation provides a useful way of estimating probabilities without explicitly adding the areas of rectangles in a histogram, as we did in Section 10.3.

We illustrate the use of the **normal approximation** in the following example.

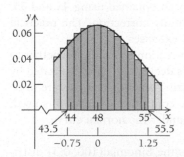

EXAMPLE 5   *Genetics.* From a heterozygous cross, 192 seeds are planted. Use the normal curve to approximate the probability that between 44 and 55 (inclusive) of the seeds are recessive homozygous.

Solution   Let X denote the number of recessive homozygous offspring. Using Example 9 on page XXX X has the Binomial(192, $1/4$) distribution, so $n = 192$, $p = 1/4$, and $q = 3/4$. Since $np = 48 \geq 10$ and $nq = 144 \geq 10$, the normal curve may be used to approximate this probability. We also see that the mean of X is $\mu = np = 48$ and the standard deviation of X is $\sigma = \sqrt{npq} = 6$.

A portion of the histogram for X is shown in the figure. The probability that is sought is $P(44 \leq X \leq 55)$, which corresponds to the areas of the 12 rectangles shown in the figure. The left endpoint of these rectangles is at 43.5, while the right endpoint is at 55.5. Before approximating the probability in question, we convert 43.5 and 55.5 into standard units.

$$43.5 \Rightarrow \frac{43.5 - \mu}{\sigma} = \frac{43.5 - 48}{6} = -0.75 \text{ in standard units;}$$

$$55.5 \Rightarrow \frac{55.5 - \mu}{\sigma} = \frac{55.5 - 48}{6} = 1.25 \text{ in standard units.}$$

The values in standard units are depicted in the figure by the lower horizontal axis.

Under 0.75 and 1.25 in Table 2, we find 0.2734 and 0.3944, respectively. Adding these areas, we obtain

$$P(44 \leq X \leq 55) \approx 0.2734 + 0.3944 = 0.6678.$$

A more accurate answer of 0.6677 is obtained by using the **NORMALCDF** function on a TI calculator, as discussed in the Technology Connection. ■

Technology Connection

Areas Under the Normal Curve

Example 5 may be done on a TI calculator *without* first converting to standard units by entering

NORMALCDF(43.5,55.5,48,6).

The third and fourth arguments are the mean μ and standard deviation σ, respectively.

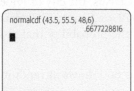

To calculate this probability *without* the normal approximation, we would have to explicitly add 12 probabilities using the Binomial(192, $1/4$) distribution:

$$P(44 \leq X \leq 55) = P(X = 44) + P(X = 45) + \cdots + P(X = 55)$$

$$= \binom{192}{44}\left(\frac{1}{4}\right)^{44}\left(\frac{3}{4}\right)^{148} + \binom{192}{45}\left(\frac{1}{4}\right)^{45}\left(\frac{3}{4}\right)^{147}$$

$$+ \cdots + \binom{192}{55}\left(\frac{1}{4}\right)^{55}\left(\frac{3}{4}\right)^{137}.$$

To four decimal places, this sum is equal to 0.6646. We see that the area under the normal curve provides an excellent approximation to the true probability while simultaneously avoiding cumbersome calculations with binomial probabilities.

In Example 5, we did not convert 44 and 55 into standard units. If we had, we would have omitted half of the rectangle over 44 and half of the rectangle over 55. Therefore, to obtain a more accurate approximation, we used 43.5 and 55.5 as the left and right endpoints, respectively. Correcting by increasing or decreasing the endpoints by 0.5 is called the **continuity correction.**

While an adequate approximation would have been obtained using 44 and 55, a better approximation results from using the continuity correction. The practical effect of the continuity correction is discussed in the exercises.

The primary difficulty in using the continuity correction is deciding whether to increase an endpoint by 0.5 or decrease it by 0.5. This decision may be facilitated by drawing a rough sketch of the histogram and visualizing which rectangles are to be included. In the next example, we illustrate how the continuity correction is used; we leave to the student the computation of the area under the normal curve.

EXAMPLE 6 *Continuity Correction.* Suppose X has the Binomial$(100, 0.3)$ distribution. If the normal approximation is used, what should be converted to standard units to find (a) $P(24 \leq X \leq 30)$, (b) $P(31 < X < 38)$, (c) $P(X < 25)$, and (d) $P(X \geq 37)$?

Solution

a) We need to find the area of the histogram between 24 and 30, including both rectangles at 24 and 30. So we would convert 23.5 and 30.5 into standard units.

b) We need to find the area between 31 and 38, without including the rectangles at 31 and 38. So we would convert 31.5 and 37.5 into standard units.

c) We need to find the area to the left of 25, without including the rectangle at 25. So we would convert 24.5 into standard units.

d) We need to find the area to the right of 37, including the rectangle at 37. So we would convert 36.5 into standard units. ∎

EXAMPLE 7 *Polling.* According to the 2000 U.S. Census, the city of Phoenix, Arizona, has a population of 1,321,045, of whom 449,972 are Latino or Hispanic. If 800 citizens of Phoenix are selected at random, what is the probability that at least 250 of them will be Latino or Hispanic?

Solution We may represent this random sample as 800 draws at random without replacement from the box

449,972 **1**s	871,073 **0**s

Let X denote the number of **1**s that are drawn. Even though the draws are made without replacement, the box is sufficiently large that X may be well approximated with a Binomial(n, p) distribution, where

$$n = 800 \quad \text{and} \quad p \approx \frac{449,972}{1,321,045} \approx 0.3406.$$

Since $q = 1 - p = 0.6594$, we use Theorem 8 to obtain

$$\mu = E(X) = np \approx (800)(0.3406) = 272.48$$

The normal approximation of the Binomial(n, p) distribution partially explains the prevalence of the normal curve for many biological phenomena. Suppose that a polygenic trait is controlled by g gene pairs. (The case of $g = 3$ was considered in Example 11 of Section 10.3.) If two heterozygotes for all g gene pairs are crossed, the number of dominant alleles inherited by the offspring follows a Binomial$(2g, 1/2)$ distribution.

For traits such as skin color or height, it is thought that g is quite large. Therefore, the number of inherited dominant alleles will approximately follow a normal curve. One result of this approximation is that the heights of adult men and women tend to follow a normal distribution.

and

$$\sigma = \text{SD}(X) = \sqrt{npq} \approx \sqrt{(800)(0.3406)(0.6594)} \approx 13.4.$$

To find the probability that at least 250 1s are chosen, we convert 249.5 into standard units:

$$249.5 \Rightarrow \frac{249.5 - 272.48}{13.4} \approx -1.7149 \text{ in standard units.}$$

Looking up 1.71 in Table 2, we find 0.4564. This represents the area between -1.71 and 0 under the normal curve. To this we add 0.5, the area to the right of 0. Our answer is therefore

$$P(X \geq 250) \approx 0.4564 + 0.5 = 0.9564.$$

A more accurate approximation, obtained by using a TI calculator, is 0.9568. ∎

Technology Connection

Normal Table

How are the areas under the normal curve, listed in Table 2, computed? Each entry is the area under the normal curve between 0 and a number z. So to compute an entry, we need to compute

$$\int_0^z \frac{1}{\sqrt{2\pi}} e^{\frac{-x^2}{2}} \, dx.$$

Although we cannot use techniques of integration to compute an exact value for this integral, we can use the methods of numerical integration that we learned in Section 5.7 to approximate the integral. The TI-83 is capable of performing numerical integration. Let's use the TI-83's capabilities to verify some values in Table 2.

We first compute the area between 0 and 1. To do this, we enter the normal density function by pressing the key $\boxed{\text{Y=}}$ Next to Y1= we paste in the normal density function by pressing $\boxed{\text{2nd}}$ $\boxed{\text{DISTR}}$. We select **NORMALPDF** and fill in the parameters as shown in the screen below. Alternatively, we could have entered the formula for the normal density function,

$$Y_1 = \frac{1}{\sqrt{2\pi}} e^{\frac{-x^2}{2}}.$$

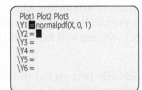

The TI-83 will graph the function before it computes the integral, so set the window to $[-3, 3, 0, 0.5]$ with Xscl $= 1$, Yscl $= 0.1$, and Xres $= 1$. Now we use the $\boxed{\text{CALC}}$ menu and select $\int f(x)\, dx$. We are prompted to enter the lower value, 0. We are then prompted to enter the upper limit, 1. Our answer is 0.34134475, which agrees with Table 2.

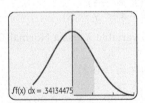

EXERCISES

Verify the values in Table 2 for the value of z indicated.

1. $z = 0.1$ **2.** $z = 0.5$

3. $z = 0.75$ **4.** $z = 1.5$

5. $z = 2$ **6.** $z = 3$

7. Reset the window to $[-1, 6, 0, 0.5]$. Compute the areas between 0 and z that are not given in Table 2.

   a) $z = 4$

   b) $z = 5$

   c) $z = 6$

tw d) Explain why your answers to parts (a), (b), and (c) indicate that we are integrating a density function.

Exercise Set 10.7

Let Z be a Normal$(0, 1)$ random variable. Find the probability that Z is in the interval.

1. $[-2, 2]$
2. $[-1.45, 1.45]$
3. $[-0.26, 0.7]$
4. $[-1.29, 2.04]$
5. $[1.26, 1.43]$
6. $[1.94, 2.93]$
7. $[-2.47, -0.38]$
8. $[-1.65, -1.08]$
9. $[0, \infty)$
10. $(-\infty, 0]$
11. $[1.17, \infty)$
12. $(-\infty, 3.06]$

13. A random variable X has a Normal$(2, 5)$ distribution.
   a) What is the mean?
   b) What is the standard deviation?
   c) Find $P(X \geq 4)$.
   d) Find $P(X \leq 3)$.
   e) Find $P(-8 \leq X \leq 7)$.

14. A random variable X has a Normal$(-3, 3)$ distribution.
   a) What is the mean?
   b) What is the standard deviation?
   c) Find $P(X \geq -5)$.
   d) Find $P(X \leq -4)$.
   e) Find $P(-4 \leq X \leq 5)$.

15. A random variable X has a Normal$(0, 0.1)$ distribution.
   a) Find $P(X \geq -0.2)$.
   b) Find $P(X \leq -0.05)$.
   c) Find $P(-0.08 \leq X \leq 0.09)$.

16. A random variable X has a Normal$(0.1, 1)$ distribution.
   a) Find $P(X \geq 2)$.
   b) Find $P(X \leq 1)$.
   c) Find $P(-1.1 \leq X \leq -0.1)$.

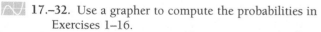 **17.–32.** Use a grapher to compute the probabilities in Exercises 1–16.

Suppose X has the Binomial(n, p) distribution. Use the normal approximation to estimate the given probability

33. $P(X \geq 18)$ if $n = 100, p = 0.2$

34. $P(35 \leq X \leq 48)$ if $n = 400, p = 0.1$

35. $P(X < 30)$ if $n = 64, p = 1/2$

36. $P(X > 85)$ if $n = 100, p = 0.9$

37. $P(280 < X < 300)$ if $n = 1000, p = 0.3$

38. $P(X \leq 158)$ if $n = 500, p = 0.34$

APPLICATIONS

39. *Fish Mass.* The mass (measured in kilograms) of a pike caught at random in a certain lake has a Normal$(4, 1)$ distribution.
   a) Find the probability that the fish has a mass of at least 5 kg.
   b) Find the probability that the fish has a mass of at least 2 kg.
   c) Suppose that all fish of mass less than 1 kg are thrown back. What is the probability that a fish will be thrown back because of its mass?

40. *Fish Length.* The reported length (measured in inches) of a pike caught at random in a certain lake has a Normal$(13, 3)$ distribution.
   a) Find the probability that the fish has length at least 6 in.
   b) Find the probability that the fish has a length at most 8 in.
   c) Suppose that all fish of length less than 5 in. are thrown back. What is the probability that a fish will be thrown back because of its length?

41. *Genetics.* From a heterozygous cross, 300 offspring are produced. Find the probability that more than 80 of the offspring are dominant homozygous.

42. *Genetics.* From a heterozygous cross, 190 offspring are produced. Find the probability that between 80 and 100 (inclusive) of the offspring are heterozygous.

43. *Genetics.* From a heterozygous cross, 70 offspring are produced. Find the probability that fewer than 20 of the offspring are recessive homozygous.

44. *Random Sampling.* According to the 2000 U.S. Census, the city of Fort Worth, Texas, has a population of 534,694, of whom 319,159 are white. If 400 citizens of Fort Worth are selected at random, what is the probability that more than 260 of them will be white?

45. *Random Sampling.* According to the 2000 U.S. Census, the city of Miami, Florida, has a population of 362,470, of whom 238,351 are Latino or Hispanic. If 120 citizens of Miami are selected at random, what is the probability that between 80 and 90 (inclusive) of them will be Latino or Hispanic?

Roulette. A gambler repeatedly bets on red in roulette. The chance of winning on one play is $p = 9/19$. In order to come out ahead, the gambler has to win more bets than he loses.

46. Suppose the gambler plays 100 times. In order to come out ahead, the gambler has to win at least 51 times. Use the normal approximation to find the probability that the gambler comes out ahead.

47. Suppose the gambler plays 1000 times. In order to come out ahead, the gambler has to win at least 501 times.

    a) Use the normal approximation to find the probability that the gambler comes out ahead.

    tw b) Compare your answer to that of Exercise 46. Which probability is larger? Does this make sense?

Percentiles. Let X be a continuous random variable. The *p*th *percentile* of its distribution is defined to be the number t so that the probability that x is less than or equal to t is $p\%$:

$$P(X \leq t) = \frac{p}{100}.$$

For example, if X has the standard normal density, then

$$P(X \leq 1.28) \approx 90\%$$

using Table 2. Therefore, we say that 1.28 is the 90th percentile for a standard normal distribution.

48. Find the following percentiles for a standard normal distribution.

    a) 30th percentile
    b) 50th percentile
    c) 95th percentile

49. *SAT Scores.* SAT verbal test scores are normally distributed with mean 507 and standard deviation 111. Find the following percentiles for SAT scores.[35]

    a) 35th percentile
    b) 60th percentile
    c) 92nd percentile

SYNTHESIS

Observed Levels of Significance and Critical Values. Observed levels of significance, which we discussed in Section 10.5, may be computed for the standard normal distribution by using the normal table. However, since the normal distribution is defined for all real numbers, the observed level of significance could correspond to the area under either a left tail or a right tail.[36]

Usually, if the observed value is greater than the mean, then a right tail is used. On the other hand, if the observed value is less than the mean, then a left tail is used. (Recall that we only used right tails to compute observed levels of significance in Section 10.5.)

In Exercises 50–55, a random variable X is assumed to have a standard normal distribution. Find the observed significance level if the random variable is equal to the given value in an experiment.

50. Observed value of 2.13

51. Observed value of 0.93

52. Observed value of -1.96

53. Observed value of -2.33

54. Observed value of 0.19

55. Observed value of -1.24

In Exercises 56–61, a significance level α and a tail of the standard normal distribution are given. Use the normal table to approximately determine the critical value.

56. $\alpha = 0.05$, right tail

57. $\alpha = 0.01$, right tail

58. $\alpha = 0.005$, right tail

59. $\alpha = 0.05$, left tail

60. $\alpha = 0.01$, left tail

61. $\alpha = 0.005$, left tail

In Exercises 62–71, assume that the random variable X is normally distributed. Use the given information to find the unknown parameter or parameters of the distribution.

62. If $E(X) = 3$ and $P(3 \leq X \leq 5) = 0.4332$, find $SD(X)$.

63. If $E(X) = 10$ and $P(9 \leq X \leq 10) = 0.4922$, find $SD(X)$.

64. If $E(X) = 4$ and $P(3 \leq X \leq 5) = 0.7620$, find $SD(X)$.

65. If $E(X) = -3$ and $P(-6 \leq X \leq 0) = 0.3108$, find $Var(X)$.

66. If $SD(X) = 3$ and $P(X \geq 2) = 0.6293$, find $E(X)$.

67. If $SD(X) = 2$ and $P(X \geq -1) = 0.409$, find $E(X)$.

68. If $P(X \geq 4.5) = 0.5$ and $P(X \geq 7) = 0.008$, find $E(X)$ and $SD(X)$.

69. If $P(X \geq 2.8) = 0.5$ and $P(X \leq 10.3) = 0.8944$, find $E(X)$ and $SD(X)$.

70. If $P(X \geq 10) = 0.3936$ and $P(X \leq 6) = 0.3936$, find $E(X)$ and $SD(X)$.

71. If $P(X \geq -3.1) = 0.8461$ and $P(X \leq 1.4) = 0.8461$, find $E(X)$ and $SD(X)$.

---

[35]The College Board.

[36]Under certain circumstances, statisticians will use the areas of both tails to compute observed significance levels and critical values.

Use the definition of median given on page 712 to solve Exercises 72 and 73.

tw 72. Explain why the median of a Normal(0, 1) distribution is 0.

tw 73. Explain why the median of a Normal(μ, σ) distribution is μ.

 Technology Connection

Continuity Correction. In Exercises 74–77, we investigate the effect of continuity correlation for the Binomial(n, p) distribution for small and large values of n.

74. Suppose X has the Binomial(100, 0.2) distribution.

a) Find $P(15 \leq X \leq 20)$ using the continuity correction.

b) Repeat part (a) *without* using the continuity correction. That is, convert 15 and 20 into standard units before using the normal curve.

c) By how much do your answers to parts (a) and (b) differ?

75. Suppose X has the Binomial(10,000, 0.2) distribution.

a) Find $P(1950 \leq X \leq 2000)$ using the continuity correction.

b) Repeat part (a) *without* using the continuity correction.

c) By how much do your answers to parts (a) and (b) differ?

76. Suppose X has the Binomial(1,000,000, 0.2) distribution.

a) Find $P(199{,}500 \leq X \leq 200{,}000)$ using the continuity correction.

b) Repeat part (a) *without* using the continuity correction.

c) By how much do your answers to parts (a) and (b) differ?

tw 77. Compare your answers to Exercises 74(c)–77(c). If n is very large, is the continuity correction important or unimportant?

78. Use the Trapezoidal Rule to estimate the integral
$$\frac{1}{\sqrt{2\pi}} \int_0^1 e^{-x^2/2} \, dx$$ with $n = 2, 4, 6, 8$ subintervals. Compare your answers to the appropriate entry in Table 2.

79. Use Simpson's rule to estimate the integral
$$\frac{1}{\sqrt{2\pi}} \int_0^1 e^{-x^2/2} \, dx$$ with $n = 2, 4, 6, 8$ subintervals. Compare your answers to the appropriate entry in Table 2.

Chapter 10   Summary and Review

Terms to Know

Probability, p. 654
Fair, p. 654
Principle of Equally Likely Outcomes, p. 655
Disjoint, p. 656
Addition Rule, p. 656
Complement Rule, p. 657
Multiplication Rule, p. 658
Conditional probability, p. 658
Independent, p. 659
Multiplication tree, p. 662
Bayes' Rule, p. 667
Prior probability, p. 667
Posterior probability, p. 667
Binomial experiment, p. 671
Random variable, p. 672

Discrete random variable, p. 672
Distribution, p. 672
Binomial(n, p) distribution, p. 672
Binomial coefficient, p. 673
Drawing with replacement, p. 675
Box model, p. 675
Random sampling, p. 675
Drawing without replacement, p. 676
Histogram, p. 678
Mode, p. 679
Expected value, p. 683, 699
Standard deviation, p. 685, 699
Mean, p. 683
Variance, p. 685, 699
Standard units, p. 688

Continuous random variable, p. 700
Density function, p. 691
Observed level of significance, p. 694
Critical value, p. 695
Significance level, p. 695
Uniform(a, b) distribution, p. 698
Exponential(λ) distribution, p. 705
Poisson process, p. 707
Poisson(μ) distribution, p. 708
Normal curve, p. 713
Normal(μ, σ) distribution, p. 714
Normal approximation, p. 717
Continuity correction, p. 718

Review Exercises

These review exercises are for test preparation. They can also be used as a lengthened practice test. Answers are at the back of the book. The answers also contain bracketed section references, which tell you where to restudy if your answer is incorrect.

Two tickets are drawn at random with replacement from the box

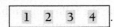

Find the probability of each event.

1. The first ticket is 3.
2. The first ticket is not 2.
3. The first ticket is 1 or 2.
4. The first ticket is 3 and the second ticket is 2.

Find the probability for the given random variable.

5. $P(X = 2)$; X is Binomial$(3, 0.4)$.
6. $P(X = 4)$; X is Binomial$(6, 0.3)$.
7. $P(1 \le X \le 5)$; X is Uniform$(0, 10)$.
8. $P(X \ge 0)$; X is Uniform$(-4, 10)$.
9. $P(2 \le Y \le 10)$; Y is Normal$(4, 5)$.
10. $P(Y \ge 7)$; Y is Normal$(8.2, 1.3)$.
11. $P(1.5 \le T \le 3)$, T is Exponential(0.5).
12. $P(T \le 5)$; T is Exponential(0.1).
13. $P(X = 3)$; X is Poisson(2.5).
14. $P(X \le 2)$; X is Poisson(4).

Find $E(X)$, $Var(X)$, and $SD(X)$ for the given random variable.

15. X is Binomial$(6, 0.12)$.
16. X is Binomial$(14, 0.8)$.
17. X is Uniform$(20, 36)$.
18. X is Uniform$(-3, 5)$.
19. X is standard normal.
20. X is Normal$(-5, 4)$.
21. X is Exponential(4).
22. X is Exponential(0.4).
23. X is Poisson(4).
24. X is Poisson(3).

Determine the probability given the probability density function.

25. $P(X \ge 3)$ if $f(x) = \dfrac{1}{60}x^3$ over $[2, 4]$

26. $P(1 \le X \le 2)$ if $f(x) = 3x^2 e^{-x^3}$ over $[0, \infty)$
27. A random variable X is assumed to have probability density function $f(x) = 1/x^2$ over $[1, \infty)$. In an experiment, the random variable has an observed value of 500. Find the observed significance level.
28. A random variable X has the Exponential(4) distribution. Find the critical value that corresponds to the significance level $\alpha = 0.05$.

Find a number k such that $kf(x)$ is a density function over the indicated interval.

29. $f(x) = 3x^2 + 3$ over $[0, 4]$
30. $f(x) = \dfrac{1}{x}$ over $[1, 10]$
31. Compute $E(X)$, $Var(X)$, and $SD(X)$ if the probability density function of X is $f(x) = x^3/60$ over $[2, 4]$.

APPLICATIONS

32. *Pea Plants.* Suppose a dihybrid pea plant for pea color and pea shape ($YyRr$) is self-pollinated. Find the probability that the offspring has genotype $YyRr$.
33. *Testing for a Disease.* A diagnostic test for a certain disease has a sensitivity of 0.82 and a specificity of 0.9. A doctor estimates the probability that a patient has the disease to be 0.4. The diagnostic test is given and returns positive. What is the posterior probability that the patient has the disease?
34. *Rolling Dice.* Five dice are rolled. Find the probability that exactly two of the dice are threes.
35. *Random Sampling.* According to the 2000 U.S. Census, the city of Miami, Florida, has a population of 362,470, of whom 238,351 are Latino or Hispanic. If 500 citizens of Miami are selected at random, what is the probability that between 300 and 340 (inclusive) of them will be Latino or Hispanic?
36. *Random Number Generator.* A random number generator is designed to output numbers with the Uniform$(0, 10)$ distribution. What is the probability that the output is between 1.25 and 3.75?
37. *Exam Scores.* The grades on an exam followed a Normal$(75, 8)$ distribution. A student who took the exam is picked at random.

   a) What is the probability that the student's score is above 90?

b) What is the probability that the student's score is between 80 and 90?

c) What is the probability that the student's score is below 60?

38. *Error Analysis.* The error in measuring the diameter of a large tree in inches is Normal$(-0.25, 0.5)$. Find the probability that the error is between -1 and 1 in.

39. *Cell Cycle Time.* The average cell cycle time for a certain bacteria is 50 min. Find the probability that a cell divides in under 40 min.

40. *Earthquake Prediction.* In a certain part of the world, a major earthquake occurs on average once every 10 yr. Find the probability that an earthquake occurs there in the next 7 yr.

41. A fair coin is flipped 100 times. Is it more likely that heads comes up exactly 50 times or that heads comes up more than 60 times?

 Technology Connection

42. Table 2 stops at three standard units. Use numerical integration to determine the following probabilities for Z, a standard normal distribution.

a) $P(0 \leq Z \leq 3)$. Compare with Table 2.
b) $P(3 \leq Z \leq 4)$
c) $P(4 \leq Z \leq 5)$
d) Find $P(Z > 5)$ using parts (a), (b), and (c).

Chapter 10 Test

Two cards are dealt without replacement from a well-shuffled deck of cards. Find the probability of each event.

1. The first card is a heart.

2. The first card is not an ace.

3. The first card is a club or a diamond.

4. The first card is a heart and the second card is a diamond.

Find the probability for the given random variable.

5. $P(X = 8)$; X is Binomial$(10, 0.7)$.

6. $P(N = 4)$; N is Poisson(1.5).

7. $P(2.3 \leq X \leq 4.1)$; X is Uniform$(1, 10)$.

8. $P(2 \leq T \leq 5)$; T is Exponential(0.5).

9. $P(-1 \leq X \leq 4)$; X is Normal$(3, 2)$.

Find $E(X)$, Var(X), and SD(X) for the given random variable.

10. X is Binomial$(9, 0.4)$.

11. X is Poisson(2.4).

12. X is Uniform$(20, 40)$.

13. X is Exponential(5).

14. X is Normal$(3.2, 0.5)$.

The probability density function of X is $f(x) = 24/x^4$ over $[2, \infty)$. Compute the following quantities for Exercises 15–20.

15. $P(X \leq 4)$ 16. $P(5 \leq X \leq 10)$

17. $E(X)$ 18. SD(X)

19. The observed level of significance if a value of 8 is observed

20. The critical value corresponding to a level of significance of $\alpha = 0.05$

21. *Testing for a Disease.* A diagnostic test for a certain disease has a sensitivity of 0.74 and a specificity of 0.85. The probability that a patient has the disease is estimated as 0.3. The diagnostic test is given and returns positive. What is the posterior probability that the patient has the disease?

22. *Coin Flipping.* Five coins are flipped. Find the probability that exactly four of the coins land heads.

23. *Random Sampling.* According to the 2000 U.S. Census, the city of Fort Worth, Texas, has a population of 534,694, of whom 319,159 are white. If 200 citizens of Fort Worth are selected at random, what is the probability that fewer than 110 of them will be white?

24. *Bird Calls.* The loudness, measured in decibels, of a bird call has a Uniform(42, 54) distribution. Find the probability that the loudness is between 45 and 48 decibels.

25. *Cell Cycle Time.* The average cell cycle time for a bacterium is 60 min. Find the probability that the bacterium divides in under 45 min.

26. *Height of Men.* In a certain village, the height of men in inches follows the Normal(70, 3) distribution. Find the probability that a randomly selected man from the village is at least 75 in. tall.

SYNTHESIS

The random variable X has probability density function $f(x) = 25xe^{-5x}$ over $[0, \infty)$. Compute the following quantities.

27. $P(X > 0.5)$

28. $E(X)$

29. $SD(X)$

 Technology Connection

30. *SAT Scores.* The distribution of the math SAT test scores is approximately Normal(519, 115).[37]

    a) What percentage of the scores is below 400?

    b) What percentage of the scores is above 600?

———————
[37]The College Board.

Extended Life Science Connection

AXENIC CULTURES

The UTEX Culture Collection of Algae at the University of Texas at Austin contains over 2000 different strains of living algae. Samples of these cultures are used by experimental botanists around the world. For some experiments, such as measurements of metabolic functions or DNA sequencing, it is critical that the algae culture is not contaminated with bacteria or other organisms. Such a culture is called *axenic*.

For aqueous cultures, a common technique for creating an axenic culture from a contaminated culture is the *dilution method*. Since the dilution method is routinely used at UTEX and other labs to create and maintain axenic cultures, it is important to maximize its success rate. In this Extended Life Science Connection, we will analyze the dilution method of creating an axenic culture from one that is contaminated with bacteria.

To perform the dilution method, a botanist may first dilute and stir the contaminated culture so that, on average, there is one alga per cubic centimeter (cc) and b bacteria per cc. We assume that the bacteria and alga neither attract nor repel each other. A small sample of the culture,

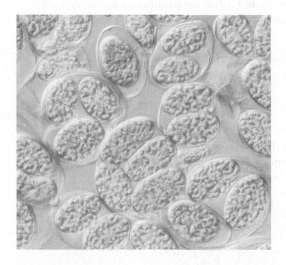

with volume v cc, is transferred into a sterile container. The dilution method is *successful* if the sample contains some algae but no bacteria. Determining whether or not the sample is axenic may be accomplished by microscopic examination after allowing the sample to grow.

DETERMINING THE OPTIMAL SAMPLE VOLUME

Using the dilution method does not guarantee that the sample is axenic. To maximize the probability that the sample is axenic, we must choose the best sample volume v. If v is too large, then the sample is likely to be contaminated with bacteria. On the other hand, if v is too small, with the intention of not collecting any bacteria, then it is likely that the sample won't contain any algae either.

EXERCISES

1. Let A be the number of algae in the sample, and let B be the number of bacteria in the sample.

   tw a) Explain why it is reasonable to model A and B as independent random variables.

   tw b) Explain why it is reasonable to model both A and B as Poisson random variables.

   tw c) Let p be the probability that the sample is axenic. Explain why
   $$p = P(A \geq 1 \text{ and } B = 0).$$

2. Let λ_A and λ_B be the parameters for the Poisson random variables A and B, respectively.

   tw a) Show that $\lambda_A = v$ and $\lambda_B = vb$.

   b) Compute $P(A \geq 1)$.

   c) Compute $P(B = 0)$.

   d) Use the independence of A and B to show that
   $$p = e^{-vb} - e^{-v(b+1)}.$$

3. Let $x = e^{-v}$. Show that $p = x^b - x^{b+1}$.

4. In this exercise, we will choose the sample volume to maximize the probability that the sample is axenic.
   a) Use Maximum–Minimum Principle 2 to show that p is maximized if $x = b/(b + 1)$.
   b) For this maximum, show that
   $$v = \ln\left(\frac{b + 1}{b}\right) \quad \text{and} \quad p = \frac{b^b}{(b + 1)^{b+1}}.$$

5. Compute v and p for the following values of b:
   a) $b = 0.5$.
   b) $b = 2$.
   c) $b = 4$.

FEASIBILITY

We have just determined that if we transfer $\ln\left(\dfrac{b + 1}{b}\right)$ cc to a single sample, then the probability for an axenic sample is maximized. In practice, botanists withdraw not one but several samples from a very large aqueous culture, so that each sample is a small part of the whole culture. This is analogous to surveying a large population, or drawing without replacement from a very large box of tickets.

The dilution method is called *feasible* if 20 samples are taken and the probability that at least one of these 20 samples is axenic is 95% or higher.

EXERCISES

tw 6. Explain why it is reasonable to assume that the successes of the 20 samples are independent events.

7. Let q be the probability that none of the 20 samples is a success. Show that
   $$q = (1 - p)^{20} = \left(1 - \frac{b^b}{(b + 1)^{b+1}}\right)^{20}.$$

8. Use a calculator to find the probability that all 20 samples fail for the given value of b.
   a) $b = 0.5$
   b) $b = 2$
   c) $b = 4$

9. Based on Exercise 8, find the probability that at least one sample is axenic for the given value of b.
   a) $b = 0.5$
   b) $b = 2$
   c) $b = 4$
   d) For which of these values of b is the dilution method feasible?

10. We now determine the feasibility of this technique.

    a) Use a grapher to solve the equation
    $$0.05 = \left(1 - \frac{b^b}{(b + 1)^{b+1}}\right)^{20}.$$

    b) Explain why the solution is the maximum value of b for which the procedure is feasible.

    tw c) For which values of b is it feasible to use the dilution method for producing an axenic culture? Explain.

REFERENCES

1. http://www.bio.utexas.edu/research/utex/

Appendix A

Review of Basic Algebra

This appendix covers most of the algebraic topics essential to a study of calculus. It might be used in conjunction with Chapter 1 or as the need for certain skills arises throughout the book.

Exponential Notation

Let's review the meaning of an expression

$$a^n,$$

where a is any real number and n is an integer; that is, n is a number in the set $\{\ldots, -3, -2, -1, 0, 1, 2, 3, \ldots\}$. The number a is called the **base** and n is called the **exponent**. When n is greater than 1, then

$$a^n = \underbrace{a \cdot a \cdot a \cdot \cdots \cdot a}_{n \text{ factors}}.$$

In other words, a^n is the product of n **factors**, each of which is a.

EXAMPLE 1 Express without exponents.

a) $4^3 = 4 \cdot 4 \cdot 4 = 64$
b) $(-2)^5 = (-2)(-2)(-2)(-2)(-2) = -32$
c) $(-2)^4 = (-2)(-2)(-2)(-2) = 16$
d) $-2^4 = -(2^4) = -(2)(2)(2)(2) = -16$
e) $(1.08)^2 = 1.08 \times 1.08 = 1.1664$
f) $\left(\dfrac{1}{2}\right)^3 = \dfrac{1}{2} \cdot \dfrac{1}{2} \cdot \dfrac{1}{2} = \dfrac{1}{8}$

■

We define an exponent of 1 as follows:

$$a^1 = a, \quad \text{for any real number } a.$$

OBJECTIVES

■ Manipulate exponential expressions.
■ Multiply and factor algebraic expressions.
■ Solve equations, inequalities, and applied problems.

In other words, any real number to the first power is that number itself.

We define an exponent of 0 as follows:

$$a^0 = 1, \quad \text{for any nonzero real number } a.$$

That is, any nonzero real number a to the zero power is 1.

EXAMPLE 2 Express without exponents.

a) $(-2x)^0 = 1$ b) $(-2x)^1 = -2x$

c) $\left(\dfrac{1}{2}\right)^0 = 1$ d) $e^0 = 1$

e) $e^1 = e$ f) $\left(\dfrac{1}{2}\right)^1 = \dfrac{1}{2}$

The meaning of a negative integer as an exponent is as follows:

$$a^{-n} = \frac{1}{a^n}, \quad \text{for any nonzero real number } a.$$

That is, any nonzero real number a to the $-n$ power is the reciprocal of a^n.

EXAMPLE 3 Express without negative exponents.

a) $2^{-5} = \dfrac{1}{2^5} = \dfrac{1}{2 \cdot 2 \cdot 2 \cdot 2 \cdot 2} = \dfrac{1}{32}$

b) $10^{-3} = \dfrac{1}{10^3} = \dfrac{1}{10 \cdot 10 \cdot 10} = \dfrac{1}{1000}, \quad \text{or} \quad 0.001$

c) $\left(\dfrac{1}{4}\right)^{-2} = \dfrac{1}{\left(\dfrac{1}{4}\right)^2} = \dfrac{1}{\dfrac{1}{4} \cdot \dfrac{1}{4}} = \dfrac{1}{\dfrac{1}{16}} = 1 \cdot \dfrac{16}{1} = 16$

d) $x^{-5} = \dfrac{1}{x^5}$

e) $e^{-k} = \dfrac{1}{e^k}$

f) $t^{-1} = \dfrac{1}{t^1} = \dfrac{1}{t}$

Properties of Exponents

Note the following:

$$b^5 \cdot b^{-3} = (b \cdot b \cdot b \cdot b \cdot b) \cdot \frac{1}{b \cdot b \cdot b}$$

$$= \frac{b \cdot b \cdot b}{b \cdot b \cdot b} \cdot b \cdot b$$

$$= 1 \cdot b \cdot b = b^2.$$

We could have obtained the same result by adding the exponents. This is true in general.

> ### THEOREM 1
>
> For any nonzero real number a and any integers n and m,
>
> $$a^n \cdot a^m = a^{n+m}.$$
>
> (To multiply when the bases are the same, add the exponents.)

EXAMPLE 4 Multiply.

a) $x^5 \cdot x^6 = x^{5+6} = x^{11}$

b) $x^{-5} \cdot x^6 = x^{-5+6} = x$

c) $2x^{-3} \cdot 5x^{-4} = 10x^{-3+(-4)} = 10x^{-7}, \quad \text{or} \quad \dfrac{10}{x^7}$

d) $r^2 \cdot r = r^{2+1} = r^3$

Note the following:

$$b^5 \div b^2 = \frac{b^5}{b^2} = \frac{b \cdot b \cdot b \cdot b \cdot b}{b \cdot b}$$

$$= \frac{b \cdot b}{b \cdot b} \cdot b \cdot b \cdot b$$

$$= 1 \cdot b \cdot b \cdot b = b^3.$$

We could have obtained the same result by subtracting the exponents. This is true in general.

> ### THEOREM 2
>
> For any nonzero real number a and any integers n and m,
>
> $$\frac{a^n}{a^m} = a^{n-m}.$$
>
> (To divide when the bases are the same, subtract the exponent in the denominator from the exponent in the numerator.)

EXAMPLE 5 Divide.

a) $\dfrac{a^3}{a^2} = a^{3-2} = a^1 = a$

b) $\dfrac{x^7}{x^7} = x^{7-7} = x^0 = 1$

c) $\dfrac{e^3}{e^{-4}} = e^{3-(-4)} = e^{3+4} = e^7$

d) $\dfrac{e^{-4}}{e^{-1}} = e^{-4-(-1)} = e^{-4+1} = e^{-3}, \quad \text{or} \quad \dfrac{1}{e^3}$

Note the following:

$$(b^2)^3 = b^2 \cdot b^2 \cdot b^2 = b^{2+2+2} = b^6.$$

We could have obtained the same result by multiplying the exponents.

THEOREM 3

For any nonzero real numbers a and b, and any integers n and m,

$$(a^n)^m = a^{nm}, \quad (ab)^n = a^n b^n, \quad \text{and} \quad \left(\frac{a}{b}\right)^n = \frac{a^n}{b^n}.$$

EXAMPLE 6 Simplify.

a) $(x^{-2})^3 = x^{-2 \cdot 3} = x^{-6}$, or $\dfrac{1}{x^6}$

b) $(e^x)^2 = e^{2x}$

c) $(2x^4 y^{-5} z^3)^{-3} = 2^{-3}(x^4)^{-3}(y^{-5})^{-3}(z^3)^{-3}$

$$= \frac{1}{2^3} x^{-12} y^{15} z^{-9}, \quad \text{or} \quad \frac{y^{15}}{8x^{12} z^9}$$

d) $\left(\dfrac{x^2}{p^4 q^5}\right)^3 = \dfrac{(x^2)^3}{(p^4 q^5)^3} = \dfrac{x^6}{(p^4)^3 (q^5)^3} = \dfrac{x^6}{p^{12} q^{15}}$ ■

Multiplication

The distributive laws are important in multiplying. These laws are as follows.

The Distributive Laws

For any numbers A, B, and C,

$$A(B + C) = AB + AC \quad \text{and} \quad A(B - C) = AB - AC.$$

EXAMPLE 7 Multiply.

a) $3(x - 5) = 3 \cdot x - 3 \cdot 5 = 3x - 15$

b) $P(1 + i) = P \cdot 1 + P \cdot i = P + Pi$

c) $(x - 5)(x + 3) = (x - 5)x + (x - 5)3$

$$= x \cdot x - 5x + 3x - 5 \cdot 3$$
$$= x^2 - 2x - 15$$

d) $(a + b)(a + b) = (a + b)a + (a + b)b$

$$= a \cdot a + ba + ab + b \cdot b$$
$$= a^2 + 2ab + b^2$$ ■

The following formulas, which are obtained using the distributive laws, are also useful in multiplying.

$$(A + B)^2 = A^2 + 2AB + B^2 \tag{1}$$
$$(A - B)^2 = A^2 - 2AB + B^2 \tag{2}$$
$$(A - B)(A + B) = A^2 - B^2 \tag{3}$$

EXAMPLE 8   Multiply.

a) $(x + h)^2 = x^2 + 2xh + h^2$

b) $(2x - t)^2 = (2x)^2 - 2(2x)t + t^2 = 4x^2 - 4xt + t^2$

c) $(3c + d)(3c - d) = (3c)^2 - d^2 = 9c^2 - d^2$ ■

Factoring

Factoring is the reverse of multiplication. That is, to factor an expression, we find an equivalent expression that is a product. Always remember to look first for a common factor.

EXAMPLE 9   Factor.

a) $P + Pi = P \cdot 1 + P \cdot i = P(1 + i)$ We used a distributive law.

b) $2xh + h^2 = h(2x + h)$

c) $x^2 - 6xy + 9y^2 = (x - 3y)^2$

d) $x^2 - 5x - 14 = (x - 7)(x + 2)$ We looked for factors of -14 whose sum is -5.

e) $6x^2 + 7x - 5 = (2x - 1)(3x + 5)$ We first considered ways of factoring the first coefficient—for example, $(2x\ \ )(3x\ \ )$. Then we looked for factors of -5 such that when we multiply, we obtain the given expression.

f) $x^2 - 9t^2 = (x - 3t)(x + 3t)$ We used the formula $(A - B)(A + B) = A^2 - B^2$.   ■

Some expressions with four terms can be factored by first looking for a common binomial factor. This is called **factoring by grouping.**

EXAMPLE 10   Factor.

a) $t^3 + 6t^2 - 2t - 12 = t^2(t + 6) - 2(t + 6)$ Factoring the first two terms and then the second two terms

$$= (t^2 - 2)(t + 6)$$ Factoring out the common binomial factor, $t + 6$

b) $x^3 - 7x^2 - 4x + 28 = x^2(x - 7) - 4(x - 7)$ Factoring the first two terms and then the second two terms

$$= (x^2 - 4)(x - 7)$$ Factoring out the common binomial factor, $x - 7$

$$= (x - 2)(x + 2)(x - 7)$$ ■

Solving Equations

Basic to the solution of many equations are the Addition Principle and the Multiplication Principle. We can add (or subtract) the same number on both sides of an equation and obtain an equivalent equation; that is, a new equation that has the same solutions as the original equation. We can also multiply (or divide) by a nonzero number on both sides of an equation and obtain an equivalent equation.

The Addition Principle

For any real numbers a, b, and c,

$$a = b \quad \text{is equivalent to} \quad a + c = b + c.$$

The Multiplication Principle

For any real numbers a, b, and c (where $c \neq 0$),

$$a = b \quad \text{is equivalent to} \quad a \cdot c = b \cdot c.$$

When solving an equation, we use these equation-solving principles and other properties of real numbers to get the variable alone on one side. Then it is easy to determine the solution.

EXAMPLE 11 Solve: $-\frac{5}{6}x + 10 = \frac{1}{2}x + 2$.

Solution We first multiply by 6 on both sides to clear the fractions:

$$6\left(-\tfrac{5}{6}x + 10\right) = 6\left(\tfrac{1}{2}x + 2\right) \qquad \text{Using the Multiplication Principle}$$

$$6\left(-\tfrac{5}{6}x\right) + 6 \cdot 10 = 6\left(\tfrac{1}{2}x\right) + 6 \cdot 2 \qquad \text{Using the Distributive Law}$$

$$-5x + 60 = 3x + 12 \qquad \text{Simplifying}$$

$$60 = 8x + 12 \qquad \begin{array}{l}\text{Using the Addition Principle:}\\ \text{We add } 5x \text{ on both sides.}\end{array}$$

$$48 = 8x \qquad \text{Adding } -12 \text{ on both sides}$$

$$\tfrac{1}{8} \cdot 48 = \tfrac{1}{8} \cdot 8x \qquad \text{Multiplying by } \tfrac{1}{8} \text{ on both sides}$$

$$6 = x.$$

The variable is now alone, and we see that 6 is the solution. We can check by substituting 6 into the original equation.

The third principle for solving equations is the *Principle of Zero Products*.

The Principle of Zero Products

For any numbers a and b, if $ab = 0$, then $a = 0$ or $b = 0$; and if $a = 0$ or $b = 0$, then $ab = 0$.

An equation being solved by this principle must have a 0 on one side and a product on the other. The solutions are then obtained by setting each factor equal to 0 and solving the resulting equations.

EXAMPLE 12 Solve: $3x(x - 2)(5x + 4) = 0$.

Solution We have

$$3x(x - 2)(5x + 4) = 0$$

$$3x = 0 \quad or \quad x - 2 = 0 \quad or \quad 5x + 4 = 0 \qquad \text{Using the Principle of Zero Products}$$

$$\tfrac{1}{3} \cdot 3x = \tfrac{1}{3} \cdot 0 \quad or \qquad x = 2 \quad or \qquad 5x = -4 \qquad \text{Solving each separately}$$

$$x = 0 \qquad or \qquad x = 2 \quad or \qquad x = -\tfrac{4}{5}.$$

The solutions are 0, 2, and $-\tfrac{4}{5}$.

Note that the Principle of Zero Products can be applied *only* when a product is 0. For example, although we may know that $ab = 8$, we *do not know* that $a = 8$ or $b = 8$.

EXAMPLE 13 Solve: $4x^3 = x$.

Solution We have

$$4x^3 = x$$

$$4x^3 - x = 0 \qquad \text{Adding } -x$$

$$x(4x^2 - 1) = 0$$

$$x(2x - 1)(2x + 1) = 0 \qquad \text{Factoring}$$

$$x = 0 \quad or \quad 2x - 1 = 0 \quad or \quad 2x + 1 = 0 \qquad \text{Using the Principle of Zero Products}$$

$$x = 0 \quad or \qquad 2x = 1 \quad or \qquad 2x = -1$$

$$x = 0 \quad or \qquad x = \tfrac{1}{2} \quad or \qquad x = -\tfrac{1}{2}.$$

The solutions are $0, \tfrac{1}{2}$, and $-\tfrac{1}{2}$.

Rational Equations

Expressions like the following are polynomials in one variable:

$$x^2 - 4, \qquad x^3 + 7x^2 - 8x + 9, \qquad t - 19.$$

The **least common multiple, LCM,** of two polynomials is found by factoring and using each factor the greatest number of times that it occurs in any one factorization.

EXAMPLE 14 Find the LCM: $x^2 + 2x + 1$, $5x^2 - 5x$, and $x^2 - 1$.

Solution

$$\left.\begin{array}{l} x^2 + 2x + 1 = (x + 1)(x + 1); \\ 5x^2 - 5x = 5x(x - 1); \\ x^2 - 1 = (x + 1)(x - 1) \end{array}\right\} \quad \text{Factoring}$$

$$\text{LCM} = 5x(x + 1)(x + 1)(x - 1)$$

A **rational expression** is a ratio of polynomials. Each of the following is a rational expression:

$$\frac{x^2 - 6x + 9}{x^2 - 4}, \qquad \frac{x - 2}{x - 3}, \qquad \frac{a + 7}{a^2 - 16}, \qquad \frac{5}{5t - 15}.$$

A **rational equation** is an equation containing one or more rational expressions. Here are some examples:

$$\frac{2}{3} - \frac{5}{6} = \frac{1}{x}, \qquad x + \frac{6}{x} = 5, \qquad \frac{2x}{x - 3} - \frac{6}{x} = \frac{18}{x^2 - 3x}.$$

To solve a rational equation, we first clear the equation of fractions by multiplying on both sides by the LCM of all the denominators. The resulting equation might have solutions that are *not* solutions of the original equation. Thus we must check all possible solutions in the original equation.

EXAMPLE 15   Solve: $\dfrac{2x}{x - 3} - \dfrac{6}{x} = \dfrac{18}{x^2 - 3x}$.

Solution   The LCM of the denominators is $x(x - 3)$. We multiply by $x(x - 3)$.

$$x(x - 3)\left(\frac{2x}{x - 3} - \frac{6}{x}\right) = x(x - 3)\left(\frac{18}{x^2 - 3x}\right) \qquad \text{Multiplying by the LCM on both sides}$$

$$x(x - 3) \cdot \frac{2x}{x - 3} - x(x - 3) \cdot \frac{6}{x} = x(x - 3)\left(\frac{18}{x^2 - 3x}\right) \qquad \text{Multiplying to remove parentheses}$$

$$2x^2 - 6(x - 3) = 18 \qquad \text{Simplifying}$$

$$2x^2 - 6x + 18 = 18$$

$$2x^2 - 6x = 0$$

$$2x(x - 3) = 0$$

$$2x = 0 \quad or \quad x - 3 = 0$$

$$x = 0 \quad or \qquad x = 3$$

The numbers 0 and 3 are possible solutions. We look at the original equation and see that each makes a denominator 0. We can also carry out a check, as follows.

Check:

For 0:

$$\frac{2x}{x - 3} - \frac{6}{x} = \frac{18}{x^2 - 3x}$$

$$\frac{2(0)}{0 - 3} - \frac{6}{0} \overset{?}{=} \frac{18}{0^2 - 3(0)}$$

$$0 - \frac{6}{0} \;\Big|\; \frac{18}{0} \qquad \text{UNDEFINED; FALSE}$$

For 3:

$$\frac{2x}{x - 3} - \frac{6}{x} = \frac{18}{x^2 - 3x}$$

$$\frac{2(3)}{3 - 3} - \frac{6}{3} \overset{?}{=} \frac{18}{3^2 - 3(3)}$$

$$\frac{6}{0} - 2 \;\Big|\; \frac{18}{0} \qquad \text{UNDEFINED; FALSE}$$

The equation has *no solution*.   ∎

EXAMPLE 16   Solve: $\dfrac{x^2}{x - 2} = \dfrac{4}{x - 2}$.

Solution   The LCM of the denominators is $x - 2$. We multiply by $x - 2$.

$$(x - 2) \cdot \frac{x^2}{x - 2} = (x - 2) \cdot \frac{4}{x - 2}$$

$$x^2 = 4 \qquad \text{Simplifying}$$

$$x^2 - 4 = 0$$

$$(x + 2)(x - 2) = 0$$

$$x = -2 \quad or \quad x = 2 \qquad \text{Using the Principle of Zero Products}$$

Check:

For 2:

$$\frac{x^2}{x - 2} = \frac{4}{x - 2}$$

$$\frac{2^2}{2 - 2} \; ? \; \frac{4}{2 - 2}$$

$$\frac{4}{0} \; \Big| \; \frac{4}{0} \qquad \text{UNDEFINED;}$$

$$\text{FALSE}$$

For -2:

$$\frac{x^2}{x - 2} = \frac{4}{x - 2}$$

$$\frac{(-2)^2}{-2 - 2} \; ? \; \frac{4}{-2 - 2}$$

$$\frac{4}{-4} \; \Big| \; \frac{4}{-4}$$

$$-1 \; \Big| \; -1 \qquad \text{TRUE}$$

The number -2 is a solution, but 2 is not (it results in division by 0). ■

Solving Inequalities

Two inequalities are **equivalent** if they have the same solutions. For example, the inequalities $x > 4$ and $4 < x$ are equivalent. Principles for solving inequalities are similar to those for solving equations. We can add the same number on both sides of an inequality. We can also multiply on both sides by the same nonzero number, but if that number is negative, we must reverse the inequality sign. The following are the inequality-solving principles.

The Inequality-Solving Principles

For any real numbers $a, b,$ and $c,$

$$a < b \quad \text{is equivalent to} \quad a + c < b + c.$$

For any real numbers $a, b,$ and any *positive* number $c,$

$$a < b \quad \text{is equivalent to} \quad ac < bc.$$

For any real numbers $a, b,$ and any *negative* number $c,$

$$a < b \quad \text{is equivalent to} \quad ac > bc.$$

Similar statements hold for $\leq$ and $\geq$.

EXAMPLE 17 Solve: $17 - 8x \geq 5x - 4$.

Solution We have

$$17 - 8x \geq 5x - 4$$
$$-8x \geq 5x - 21 \qquad \text{Adding } -17$$
$$-13x \geq -21 \qquad \text{Adding } -5x$$
$$-\frac{1}{13}(-13x) \leq -\frac{1}{13}(-21) \qquad \text{Multiplying by } -\frac{1}{13} \text{ and}$$
$$\qquad\qquad\qquad\qquad \textit{reversing the inequality sign}$$
$$x \leq \frac{21}{13}.$$

Any number less than or equal to $\frac{21}{13}$ is a solution.

Applications

To solve applied problems, we first translate to mathematical language, usually an equation. Then we solve the equation and check to see whether the solution to the equation is a solution to the problem.

EXAMPLE 18 *Weight Gain.* After a 5% gain in weight, an animal weighs 693 lb. What was its original weight?

Solution We first translate to an equation:

$$\underbrace{(Original\ weight)}_{w} + 5\%\underbrace{(Original\ weight)}_{w} = 693$$

Now we solve the equation:

$$w + 5\%w = 693$$
$$1 \cdot w + 0.05w = 693$$
$$(1 + 0.05)w = 693$$
$$1.05w = 693$$
$$w = \frac{693}{1.05} = 660.$$

Check: $600 + 5\% \times 660 = 660 + 0.05 \times 660 = 660 + 33 = 693$.

The original weight of the animal was 660 lb.

EXAMPLE 19 *Total Sales.* Raggs, Ltd., a clothing firm, determines that its total revenue, in dollars, from the sale of x suits is given by

$$200x + 50.$$

Determine the number of suits that the firm must sell to ensure that its total revenue will be more than $70,050.

Solution We translate to an inequality and solve:

$$200x + 50 > 70{,}050$$
$$200x > 70{,}000 \qquad \text{Adding } -50$$
$$x > 350. \qquad \text{Multiplying by } \frac{1}{200}$$

Thus the company's total revenue will exceed $70,050 when it sells more than 350 suits.

Exercise Set A

Express without exponents.

1. 5^3
2. 7^2
3. $(-7)^2$
4. $(-5)^3$
5. $(1.01)^2$
6. $(1.01)^3$
7. $\left(\dfrac{1}{2}\right)^4$
8. $\left(\dfrac{1}{4}\right)^3$
9. $(6x)^0$
10. $(6x)^1$
11. t^1
12. t^0
13. $\left(\dfrac{1}{3}\right)^0$
14. $\left(\dfrac{1}{3}\right)^1$

Express without negative exponents.

15. 3^{-2}
16. 4^{-2}
17. $\left(\dfrac{1}{2}\right)^{-3}$
18. $\left(\dfrac{1}{2}\right)^{-2}$
19. 10^{-1}
20. 10^{-4}
21. e^{-b}
22. t^{-k}
23. b^{-1}
24. h^{-1}

Multiply.

25. $x^2 \cdot x^3$
26. $t^3 \cdot t^4$
27. $x^{-7} \cdot x$
28. $x^5 \cdot x$
29. $5x^2 \cdot 7x^3$
30. $4t^3 \cdot 2t^4$
31. $x^{-4} \cdot x^7 \cdot x$
32. $x^{-3} \cdot x \cdot x^3$
33. $e^{-t} \cdot e^t$
34. $e^k \cdot e^{-k}$

Divide.

35. $\dfrac{x^5}{x^2}$
36. $\dfrac{x^7}{x^3}$
37. $\dfrac{x^2}{x^5}$
38. $\dfrac{x^3}{x^7}$
39. $\dfrac{e^k}{e^k}$
40. $\dfrac{t^k}{t^k}$
41. $\dfrac{e^t}{e^4}$
42. $\dfrac{e^k}{e^3}$
43. $\dfrac{t^6}{t^{-8}}$
44. $\dfrac{t^5}{t^{-7}}$
45. $\dfrac{t^{-9}}{t^{-11}}$
46. $\dfrac{t^{-11}}{t^{-7}}$
47. $\dfrac{ab(a^2b)^3}{ab^{-1}}$
48. $\dfrac{x^2y^3(xy^3)^2}{x^{-3}y^2}$

Simplify.

49. $(t^{-2})^3$
50. $(t^{-3})^4$
51. $(e^x)^4$
52. $(e^x)^5$
53. $(2x^2y^4)^3$
54. $(2x^2y^4)^5$
55. $(3x^{-2}y^{-5}z^4)^{-4}$
56. $(5x^3y^{-7}z^{-5})^{-3}$
57. $(-3x^{-8}y^7z^2)^2$
58. $(-5x^4y^{-5}z^{-3})^4$
59. $\left(\dfrac{cd^3}{2q^2}\right)^4$
60. $\left(\dfrac{4x^2y}{a^3b^3}\right)^3$

Multiply.

61. $5(x - 7)$
62. $x(1 + t)$
63. $(x - 5)(x - 2)$
64. $(x - 4)(x - 3)$
65. $(a - b)(a^2 + ab + b^2)$
66. $(x^2 - xy + y^2)(x + y)$
67. $(2x + 5)(x - 1)$
68. $(3x + 4)(x - 1)$
69. $(a - 2)(a + 2)$
70. $(3x - 1)(3x + 1)$
71. $(5x + 2)(5x - 2)$
72. $(t - 1)(t + 1)$
73. $(a - h)^2$
74. $(a + h)^2$
75. $(5x + t)^2$
76. $(7a - c)^2$
77. $5x(x^2 + 3)^2$
78. $-3x^2(x^2 - 4)(x^2 + 4)$

Use the following equation (equation 5) for Exercises 79–81.

$$
\begin{aligned}
(x + h)^3 &= (x + h)(x + h)^2 \\
&= (x + h)(x^2 + 2xh + h^2) \\
&= (x + h)x^2 + (x + h)2xh + (x + h)h^2 \\
&= x^3 + x^2h + 2x^2h + 2xh^2 + xh^2 + h^3 \\
&= x^3 + 3x^2h + 3xh^2 + h^3 \qquad (5)
\end{aligned}
$$

79. $(a + b)^3$
80. $(a - b)^3$
81. $(x - 5)^3$
82. $(2x + 3)^3$

Factor.

83. $x - xt$
84. $x + xh$
85. $x^2 + 6xy + 9y^2$
86. $x^2 - 10xy + 25y^2$
87. $x^2 - 2x - 15$
88. $x^2 + 8x + 15$
89. $x^2 - x - 20$
90. $x^2 - 9x - 10$
91. $49x^2 - t^2$
92. $9x^2 - b^2$
93. $36t^2 - 16m^2$
94. $25y^2 - 9z^2$
95. $a^3b - 16ab^3$
96. $2x^4 - 32$
97. $a^8 - b^8$
98. $36y^2 + 12y - 35$
99. $10a^2x - 40b^2x$
100. $x^3y - 25xy^3$
101. $2 - 32x^4$
102. $2xy^2 - 50x$
103. $9x^2 + 17x - 2$
104. $6x^2 - 23x + 20$
105. $x^3 + 8$
   (*Hint*: See Exercise 66.)
106. $a^3 - 27b^3$
   (*Hint*: See Exercise 65.)
107. $y^3 - 64t^3$
108. $m^3 + 1000p^3$
109. $3x^3 - 6x^2 - x + 2$

110. $5y^3 + 2y^2 - 10y - 4$

111. $x^3 - 5x^2 - 9x + 45$

112. $t^3 + 3t^2 - 25t - 75$

Solve.

113. $-7x + 10 = 5x - 11$

114. $-8x + 9 = 4x - 70$

115. $5x - 17 - 2x = 6x - 1 - x$

116. $5x - 2 + 3x = 2x + 6 - 4x$

117. $x + 0.8x = 216$ 118. $x + 0.5x = 210$

119. $x + 0.08x = 216$ 120. $x + 0.05x = 210$

121. $2x(x + 3)(5x - 4) = 0$

122. $7x(x - 2)(2x + 3) = 0$

123. $x^2 + 1 = 2x + 1$ 124. $2t^2 = 9 + t^2$

125. $t^2 - 2t = t$ 126. $6x - x^2 = x$

127. $6x - x^2 = -x$ 128. $2x - x^2 = -x$

129. $9x^3 = x$ 130. $16x^3 = x$

131. $(x - 3)^2 = x^2 + 2x + 1$

132. $(x - 5)^2 = x^2 + x + 3$

133. $\dfrac{4x}{x + 5} + \dfrac{20}{x} = \dfrac{100}{x^2 + 5x}$

134. $\dfrac{x}{x + 1} + \dfrac{3x + 5}{x^2 + 4x + 3} = \dfrac{2}{x + 3}$

135. $\dfrac{50}{x} - \dfrac{50}{x - 2} = \dfrac{4}{x}$ 136. $\dfrac{60}{x} = \dfrac{60}{x - 5} + \dfrac{2}{x}$

137. $0 = 2x - \dfrac{250}{x^2}$ 138. $5 - \dfrac{35}{x^2} = 0$

139. $3 - x \le 4x + 7$ 140. $x + 6 \le 5x - 6$

141. $5x - 5 + x > 2 - 6x - 8$

142. $3x - 3 + 3x > 1 - 7x - 9$

143. $-7x < 4$ 144. $-5x \ge 6$

145. $5x + 2x \le -21$ 146. $9x + 3x \ge -24$

147. $2x - 7 < 5x - 9$ 148. $10x - 3 \ge 13x - 8$

149. $8x - 9 < 3x - 11$ 150. $11x - 2 \ge 15x - 7$

151. $8 < 3x + 2 < 14$ 152. $2 < 5x - 8 \le 12$

153. $3 \le 4x - 3 \le 19$ 154. $9 \le 5x + 3 < 19$

155. $-7 \le 5x - 2 \le 12$

156. $-11 \le 2x - 1 < -5$

APPLICATIONS

157. *Investment increase.* An investment is made at $8\frac{1}{2}\%$, compounded annually. It grows to $705.25 at the end of 1 yr. How much was invested originally?

158. *Investment increase.* An investment is made at 7%, compounded annually. It grows to $856 at the end of 1 yr. How much was invested originally?

159. *Total revenue.* A firm determines that the total revenue, in dollars, from the sale of x units of a product is

$$3x + 1000.$$

Determine the number of units that must be sold so that its total revenue will be more than $22,000.

160. *Total revenue.* A firm determines that the total revenue, in dollars, from the sale of x units of a product is

$$5x + 1000.$$

Determine the number of units that must be sold so that its total revenue will be more than $22,000.

161. *Weight gain.* After a 6% gain in weight, an animal weighs 508.8 lb. What was its original weight?

162. *Weight gain.* After a 7% gain in weight, an animal weighs 363.8 lb. What was its original weight?

163. *Population increase.* After a 2% increase, the population of a city is 826,200. What was the former population?

164. *Population increase.* After a 3% increase, the population of a city is 741,600. What was the former population?

165. *Grade average.* To get a B in a course, a student's average must be greater than or equal to 80% (at least 80%) and less than 90%. On the first three tests, the student scores 78%, 90%, and 92%. Determine the scores on the fourth test that will guarantee a B.

166. *Grade average.* To get a C in a course, a student's average must be greater than or equal to 70% and less than 80%. On the first three tests, the student scores 65%, 83%, and 82%. Determine the scores on the fourth test that will guarantee a C.

Appendix B

Functions

Ordered Pairs and Graphs

Each point in the plane corresponds to an **ordered pair** of numbers that are referred to as the point's **coordinates**. The first number in the ordered pair is the x-coordinate, and the second number is the y-coordinate. The x-coordinate measures the position along the x-axis while the y-coordinate measures the position along the y-axis, as seen in the graph.

A solution to an equation in two variables is an ordered pair of numbers that, when substituted for the variables, gives a true statement. For example, the ordered pair $(-1, 2)$ indicates that $x = -1$ and $y = 2$. The ordered pair is a solution to $3x^2 + y = 5$ since $3(-1)^2 + 2 = 5$. We could easily check that the ordered pairs $(0, 5)$ and $(2, -7)$ are also solutions to the equation $3x + y = 5$. Unlike equations with only one variable, we expect many solutions to equations with two variables. To better understand all the solutions to an equation, we can make a drawing that represents all the solutions to the equation. This drawing is called the **graph of the equation.**

EXAMPLE 1 Graph: $y - x^2 + 1 = 0$.

Solution To graph this equation, we plot several solutions to the equation and look for patterns. The graph could be a line, a curve, or more than one curve. In order to make it easier to find points, we solve for y.

$$y - x^2 + 1 = 0$$
$$y - x^2 + 1 + x^2 - 1 = 0 + x^2 - 1 \qquad \text{Add } x^2 - 1$$
$$y = x^2 - 1$$

OBJECTIVES

- Identify functions.
- Graph functions.
- Find the domain and range of functions.

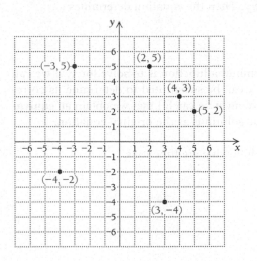

We can easily see that if we substitute $x = 0$ into the equation $y = x^2 - 1$, we get $y = 0^2 - 1 = -1$. One point on the graph has coordinates $(0, -1)$. We make a table to determine other points on the graph.

x	y	(x, y)
-2	3	$(-2, 3)$
-1	0	$(-1, 0)$
0	-1	$(0, -1)$
1	0	$(1, 0)$
2	3	$(2, 3)$

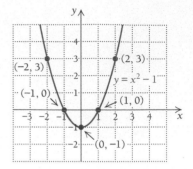

By connecting the points, we obtain the graph.   ■

Functions

If we can solve an equation for y, then we can compute the value of y that corresponds to any value for x. This makes it much easier to graph enough points to see the pattern. Equations of this sort are referred to as **functions** and are usually easier to graph than equations that are not functions.

> **DEFINITION**
>
> Given an equation of x and y, suppose that each possible value of x corresponds to exactly one value of y. Then the equation determines a *function*.

One way to see if an equation determines a function is to solve for y. If we can solve for y, then each possible value of x can be substituted to determine its corresponding y-value. When solving for y, we must be careful that we obtain an **equivalent equation,** that is, one with the same solutions as the original equation.

EXAMPLE 2 Determine which of the following are functions.

a) $3y + 2x = 6$

b) $y^2 = x$

Solution

a) We solve for y.

$$3y + 2x = 6$$

$$3y + 2x - 2x = 6 - 2x \qquad \text{Subtracting } 2x$$

$$3y = 6 - 2x \qquad \text{Simplifying}$$

$$\frac{1}{3} \cdot (3y) = \frac{1}{3} \cdot (6 - 2x) \qquad \text{Multiplying by } \frac{1}{3}$$

$$y = \frac{1}{3} \cdot (6) - \frac{1}{3} \cdot (2x)$$ Distributive law

$$y = 2 - \frac{2}{3}x$$

The equation $3y + 2x = 6$ is equivalent to the equation $y = 2 - \frac{2}{3}x$. It is easy to see that for any value of x, there is exactly one value of y.

b) We can see that this is not a function since both $(4, 2)$ and $(4, -2)$ are solutions. If we attempt to solve for y, we must take square roots, remembering that there are two.

$$y^2 = x$$
$$y = \pm\sqrt{x}$$

We need to include the possibility that y could be negative in order to have an equation equivalent to $y^2 = x$. We can see from the equation $y = \pm\sqrt{x}$ that we have two possible values of y (one positive and one negative) for any positive value of x. This does not determine a function. ■

We usually write a function as an equation where we have solved for y. Using function notation, we could write $y = f(x)$. For example, instead of writing $y = x^2 + 3x - 5$, we may write $f(x) = x^2 + 3x - 5$. Instead of saying the value of y is -5 when x is 0, we may use the function notation to simply write $f(0) = -5$. Similarly, $f(2) = 2^2 + 3(2) - 5 = 5$.

It helps to think of a function as a machine. Think of $f(2) = 5$ as putting the **input** number 2 into a function machine; the **output** is the number 5. The collection of input numbers, or possible values for x, is called the **domain** and the collection of output numbers, or possible values for y, is called the **range.**

CAUTION The notation $f(x)$ does not mean "f times x."

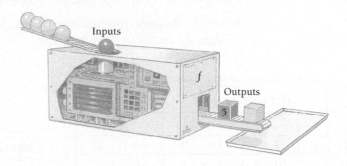

EXAMPLE 3 Let $f(x) = x + \dfrac{1}{x}$.

a) Determine $f(2)$.

b) Determine $f(-1)$.

c) Determine $f(0)$.

d) Determine $f(x + h)$.

Solution

a) $f(2) = 2 + \dfrac{1}{2}$ Substituting 2 for x

$\quad\quad = \dfrac{5}{2}$

b) $f(-1) = -1 + \dfrac{1}{-1}$ Substituting -1 for x

$\quad\quad = -1 - 1 = -2$

c) The number 0 is not part of the domain of f since we cannot divide by 0. Therefore, $f(0)$ is undefined.

d) Here we are given an expression instead of a number. In parts (a) and (b) we replaced x with 2 and -1, respectively. If we wish to evaluate $f(x + h)$, we must replace x with $(x + h)$.

$$f(x + h) = (x + h) + \frac{1}{(x + h)}$$

$$= \frac{(x + h)^2}{x + h} + \frac{1}{x + h}\quad\quad \text{Common denominator}$$

$$= \frac{(x + h)^2 + 1}{x + h}\quad\quad \text{Adding the fractions}$$

$$= \frac{x^2 + 2xh + h^2 + 1}{x + h}$$

Graphs of Functions

Graphing a function is no different from graphing an equation. If we wish to graph $f(x) = x^2 - 1$, we graph the equation $y = x^2 - 1$, just as we did in Example 1.

EXAMPLE 4   Graph the function $f(x) = x^3 + x - 1$.

Solution   To graph this function, we first find several points on the graph. Then we connect the points to form the graph.

x	$f(x) = x^3 + x - 1$	$(x, f(x))$
-2	$(-2)^3 + (-2) - 1 = -11$	$(-2, -11)$
-1	$(-1)^3 + (-1) - 1 = -3$	$(-1, -3)$
0	$(0)^3 + 0 - 1 = -1$	$(0, -1)$
1	$(1)^3 + 1 - 1 = 1$	$(1, 1)$
2	$(2)^3 + 2 - 1 = 9$	$(2, 9)$

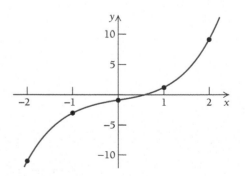

If $y = f(x)$ is a function, then for each value of x in the domain, there is only one value of y. Geometrically, this means that any vertical line intersects the graph of a function in at most one point. This is called the **vertical line test.**

The two graphs below represent functions since each vertical line intersects the graph at most once.

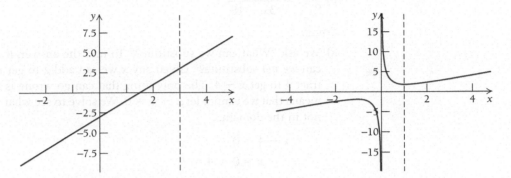

Even though the line $x = 0$ does not intersect the second graph, it is still a graph of a function since no vertical line intersects the graph in two or more points.

The graph below does not represent a function since the vertical line $x = 1$ intersects the graph in two points.

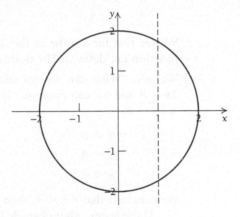

Finding Domain and Range

The domain of a function $y = f(x)$ consists of all possible values of x, while the range consists of all possible y-values. For example, if x is a square's side length and $y = f(x)$ is the area of the square, then $f(x) = x^2$. What is the domain of this function? Since x represents a square's side length, x must be positive. We would say the domain is all positive values of x.

Sometimes, we are not given an interpretation of a function. For example, suppose that we were just given the function $f(x) = x^2$, but we are not interpreting x as a length. In this case, we assume that the domain is all values of x where $f(x)$ is defined. Since we can square any number, the domain in this case is all real numbers. In general, if we are only given the formula for $y = f(x)$, we assume the domain to be all the values of x where the formula makes sense.

EXAMPLE 5 Determine the domain of each function.

a) $f(x) = \dfrac{x + 2}{x - 4}$

b) $g(x) = \sqrt{x - 3}$

c) $g(x) = \dfrac{\sqrt{2x - 8}}{3x - 18}$

Solution

a) We ask "What can we substitute?" To find the answer, it is easier to ask, "What can we *not* substitute?" Given any x we can add 2 to get $x + 2$ and we can subtract 4 to get $x - 4$. The only thing that can go wrong is if we divide by 0. This means that we cannot let $x - 4 = 0$. We solve to see what value or values of x are not in the domain.

$$x - 4 = 0$$
$$x = 0 + 4 = 4$$

We conclude that the domain consists of all values of x except $x = 4$.

b) We ask, "What can we *not* substitute?" Given any x we can subtract 3 to obtain $x - 3$. However, we cannot take the square root of negative numbers. This time we solve an inequality.

$$x - 3 < 0$$
$$x < 3$$

We see that for x to be in the domain, $x \geq 3$ since only for values $x < 3$ is the function not defined. The domain consists of all values of x with $x \geq 3$.

c) We ask, "What can we *not* substitute?" For any value of x, we can compute $2x - 8$ and we can compute $3x - 18$. If $2x - 8 < 0$, then we cannot compute $\sqrt{2x - 8}$.

$$2x - 8 < 0$$
$$2x < 8$$
$$x < 4$$

We conclude that if $x < 4$, then x is not in the domain.

There is one other obstacle for x to be in the domain. We cannot divide by 0. We find the value(s) of x that force division by 0.

$$3x - 18 = 0$$
$$3x = 18$$
$$x = 6$$

We have just determined that $x = 6$ is also not in the domain. We conclude that the domain consists of all values of x with $x \geq 4$ except $x = 6$. ■

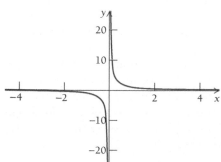

How do we determine the range of a function? If we have the graph of the function, then we can see which values of y are included in the graph. In the graph in the margin, we see that the range consists of all real values of y except $y = 0$.

Exercise Set B

Graph the points.

1. $(0, 3)$ **2.** $(3, 0)$ **3.** $(-1, 2)$

4. $(2, -3)$ **5.** $(5, 2)$ **6.** $(-3, -2)$

7. Let $g(x) = x^2 + 2x + 10$. Compute

  a) $g(0)$.   **b)** $g(3)$.   **c)** $g(-3)$.
  d) $g(5)$.   **e)** $g(-5)$.   **f)** $g(x + 1)$.
  g) $g(x + h)$.

8. Let $f(x) = \dfrac{1}{x + 1}$. Compute

  a) $f(0)$.   **b)** $f(3)$.   **c)** $f(-3)$.
  d) $f(1)$.   **e)** $f(-1)$.   **f)** $f(x - 1)$.
  g) $f(x + h)$.

Graph the equations.

9. $y + x = 2$ **10.** $2y + 3x = 12$

11. $y - x^2 = 4$ **12.** $y + 2x^2 = 1$

13. Identify which are graphs of functions.

  a)

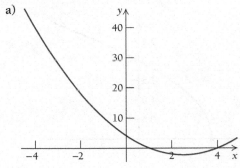

  b)

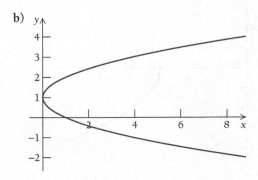

c)

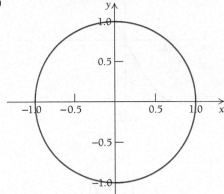

d)

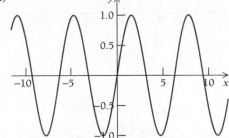

14. Identify which are graphs of functions.

  a)

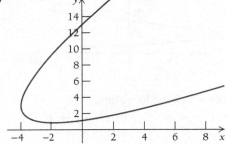

b)

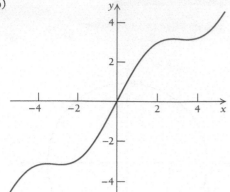

Find the domain and range from the graph.

25.

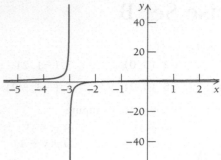

c)

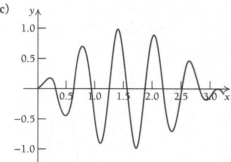

26.

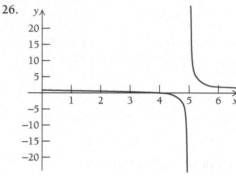

d)

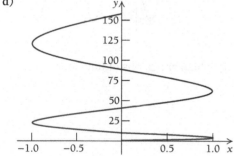

27.

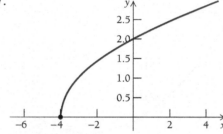

28.

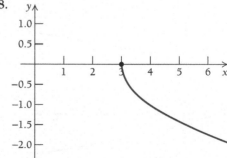

15. Identify which are functions.

  a) $x^2 + y^2 = 4$ b) $x^2 + y = 4$
  c) $x + y^2 = 4$ d) $x + y = 4$

16. Identify which are functions.

  a) $x^3 + y = 12$ b) $2x + 3 - y^4 = 0$
  c) $2x + 3y^2 = 5$ d) $x + y^3 = 10$

Graph the functions.

17. $f(x) = 3x + 4$ **18.** $f(x) = -2x + 7$

19. $g(x) = 0.5x - 2.5$ **20.** $g(x) = 1.5x + 3$

21. $f(x) = x^3$ **22.** $f(x) = x^4$

23. $r(x) = x^2 + x + 1$ **24.** $r(x) = 2x^2 - x + 3$

Find the domain.

29. $f(x) = x^2 + 3x - 9$

30. $f(x) = 2x^3 + 3x - 7$

31. $f(x) = \sqrt{x + 1}$ **32.** $f(x) = \sqrt{x - 1}$

33. $g(x) = \dfrac{1}{x + 2}$

34. $g(x) = \dfrac{x - 1}{x + 5}$

37. $p(x) = \dfrac{\sqrt[3]{x + 1}}{\sqrt[4]{x - 1}}$

38. $p(x) = \dfrac{x^2}{\sqrt[5]{x}}$

35. $r(x) = \dfrac{x}{\sqrt{x + 1}}$

36. $r(x) = \dfrac{\sqrt[4]{x}}{x - 1}$

Tables

TABLE 1

Integration Formulas

1. $\displaystyle\int x^n\,dx = \frac{x^{n+1}}{n+1} + C, n \neq -1$

2. $\displaystyle\int \frac{dx}{x} = \ln|x| + C$

3. $\displaystyle\int u\,dv = uv - \int v\,du$

4. $\displaystyle\int e^x\,dx = e^x + C$

5. $\displaystyle\int e^{ax}\,dx = \frac{1}{a} \cdot e^{ax} + C$

6. $\displaystyle\int xe^{ax}\,dx = \frac{1}{a^2} \cdot e^{ax}(ax - 1) + C$

7. $\displaystyle\int x^n e^{ax}\,dx = \frac{x^n e^{ax}}{a} - \frac{n}{a}\int x^{n-1} e^{ax}\,dx + C$

8. $\displaystyle\int \ln x\,dx = x\ln x - x + C$

9. $\displaystyle\int (\ln x)^n\,dx = x(\ln x)^n - n\int (\ln x)^{n-1}\,dx + C, n \neq -1$

10. $\displaystyle\int x^n \ln x\,dx = x^{n+1}\left[\frac{\ln x}{n+1} - \frac{1}{(n+1)^2}\right] + C, n \neq -1$

11. $\displaystyle\int a^x\,dx = \frac{a^x}{\ln a} + C, a > 0, a \neq 1$

12. $\displaystyle\int \sin ax\,dx = -\frac{1}{a}\cos ax + C$

(continued)

TABLE 1
Integration Formulas (*continued*)

13. $\int \cos ax \, dx = \dfrac{1}{a} \sin ax + C$

14. $\int \tan ax \, dx = -\dfrac{1}{a} \ln |\cos ax| + C$

15. $\int \cot ax \, dx = \dfrac{1}{a} \ln |\sin ax| + C$

16. $\int \sec ax \, dx = \dfrac{1}{a} \ln |\sec ax + \tan ax| + C$

17. $\int \csc ax \, dx = -\dfrac{1}{a} \ln |\csc ax + \cot ax| + C$

18. $\int \sec ax \tan ax \, dx = \dfrac{1}{a} \sec ax + C$

19. $\int \csc ax \cot ax \, dx = -\dfrac{1}{a} \csc ax + C$

20. $\int \sec^2 ax \, dx = \dfrac{1}{a} \tan ax + C$

21. $\int \csc^2 ax \, dx = -\dfrac{1}{a} \cot ax + C$

22. $\int x^n \sin x \, dx = -x^n \cos x + n \int x^{n-1} \cos x \, dx$

23. $\int x^n \cos x \, dx = x^n \sin x - n \int x^{n-1} \sin x \, dx$

24. $\int \dfrac{1}{\sqrt{x^2 + a^2}} \, dx = \ln |x + \sqrt{x^2 + a^2}| + C$

25. $\int \dfrac{1}{\sqrt{x^2 - a^2}} \, dx = \ln |x + \sqrt{x^2 - a^2}| + C$

26. $\int \dfrac{1}{x^2 - a^2} \, dx = \dfrac{1}{2a} \ln \left| \dfrac{x - a}{x + a} \right| + C$

27. $\int \dfrac{1}{a^2 - x^2} \, dx = \dfrac{1}{2a} \ln \left| \dfrac{a + x}{a - x} \right| + C$

28. $\int \dfrac{1}{x\sqrt{a^2 + x^2}} \, dx = -\dfrac{1}{a} \ln \left| \dfrac{a + \sqrt{a^2 + x^2}}{x} \right| + C$

29. $\int \dfrac{1}{x\sqrt{a^2 - x^2}} \, dx = -\dfrac{1}{a} \ln \left| \dfrac{a + \sqrt{a^2 - x^2}}{x} \right| + C, \, 0 < x < a$

30. $\int \dfrac{x}{ax + b} \, dx = \dfrac{x}{a} - \dfrac{b}{a^2} \ln |ax + b| + C$

31. $\int \dfrac{x}{(ax + b)^2} \, dx = \dfrac{b}{a^2(ax + b)} + \dfrac{1}{a^2} \ln |ax + b| + C$

TABLE 1
Integration Formulas (*continued*)

32. $\displaystyle\int \frac{1}{x(ax + b)}\, dx = \frac{1}{b} \ln \left| \frac{x}{ax + b} \right| + C$

33. $\displaystyle\int \frac{1}{x(ax + b)^2}\, dx = \frac{1}{b(ax + b)} + \frac{1}{b^2} \ln \left| \frac{x}{ax + b} \right| + C$

34. $\displaystyle\int \sqrt{x^2 \pm a^2}\, dx = \frac{1}{2}\left[x\sqrt{x^2 \pm a^2} \pm a^2 \ln \left| x + \sqrt{x^2 \pm a^2} \right| \right] + C$

35. $\displaystyle\int x\sqrt{a + bx}\, dx = \frac{2}{15b^2}(3bx - 2a)(a + bx)^{3/2} + C$

36. $\displaystyle\int x^2\sqrt{a + bx}\, dx = \frac{2}{105b^3}(15b^2 x^2 - 12abx + 8a^2)(a + bx)^{3/2} + C$

37. $\displaystyle\int \frac{x\, dx}{\sqrt{a + bx}} = \frac{2}{3b^2}(bx - 2a)\sqrt{a + bx} + C$

38. $\displaystyle\int \frac{x^2\, dx}{\sqrt{a + bx}} = \frac{2}{15b^3}(3b^2 x^2 - 4abx + 8a^2)\sqrt{a + bx} + C$

TABLE 2
Areas for a Standard Normal Distribution

Entries in the table represent area under the curve between $t = 0$ and a positive value of t. Because of the symmetry of the curve, area under the curve between $t = 0$ and a negative value of t would be found in a similar manner.

Area = Probability
$$= P(0 \leq x \leq t)$$
$$= \int_0^t \frac{1}{\sqrt{2\pi}} e^{-x^2/2}\, dx$$

t	0.00	0.01	0.02	0.03	0.04	0.05	0.06	0.07	0.08	0.09
0.0	.0000	.0040	.0080	.0120	.0160	.0199	.0239	.0279	.0319	.0359
0.1	.0398	.0438	.0478	.0517	.0557	.0596	.0636	.0675	.0714	.0753
0.2	.0793	.0832	.0871	.0910	.0948	.0987	.1026	.1064	.1103	.1141
0.3	.1179	.1217	.1255	.1293	.1331	.1368	.1406	.1443	.1480	.1517
0.4	.1554	.1591	.1628	.1664	.1700	.1736	.1772	.1808	.1844	.1879
0.5	.1915	.1950	.1985	.2019	.2054	.2088	.2123	.2157	.2190	.2224
0.6	.2257	.2291	.2324	.2357	.2389	.2422	.2454	.2486	.2517	.2549
0.7	.2580	.2611	.2642	.2673	.2704	.2734	.2764	.2794	.2823	.2852
0.8	.2881	.2910	.2939	.2967	.2995	.3023	.3051	.3078	.3106	.3133
0.9	.3159	.3186	.3212	.3238	.3264	.3289	.3315	.3340	.3365	.3389
1.0	.3413	.3438	.3461	.3485	.3508	.3531	.3554	.3577	.3599	.3621
1.1	.3643	.3665	.3686	.3708	.3729	.3749	.3770	.3790	.3810	.3830
1.2	.3849	.3869	.3888	.3907	.3925	.3944	.3962	.3980	.3997	.4015
1.3	.4032	.4049	.4066	.4082	.4099	.4115	.4131	.4147	.4162	.4177
1.4	.4192	.4207	.4222	.4236	.4251	.4265	.4279	.4292	.4306	.4319
1.5	.4332	.4345	.4357	.4370	.4382	.4394	.4406	.4418	.4429	.4441
1.6	.4452	.4463	.4474	.4484	.4495	.4505	.4515	.4525	.4535	.4545
1.7	.4554	.4564	.4573	.4582	.4591	.4599	.4608	.4616	.4625	.4633
1.8	.4641	.4649	.4656	.4664	.4671	.4678	.4686	.4693	.4699	.4706
1.9	.4713	.4719	.4726	.4732	.4738	.4744	.4750	.4756	.4761	.4767
2.0	.4772	.4778	.4783	.4788	.4793	.4798	.4803	.4808	.4812	.4817
2.1	.4821	.4826	.4830	.4834	.4838	.4842	.4846	.4850	.4854	.4857
2.2	.4861	.4864	.4868	.4871	.4875	.4878	.4881	.4884	.4887	.4890
2.3	.4893	.4896	.4898	.4901	.4904	.4906	.4909	.4911	.4913	.4916
2.4	.4918	.4920	.4922	.4925	.4927	.4929	.4931	.4932	.4934	.4936
2.5	.4938	.4940	.4941	.4943	.4945	.4946	.4948	.4949	.4951	.4952
2.6	.4953	.4955	.4956	.4957	.4959	.4960	.4961	.4962	.4963	.4964
2.7	.4965	.4966	.4967	.4968	.4969	.4970	.4971	.4972	.4973	.4974
2.8	.4974	.4975	.4976	.4977	.4977	.4978	.4979	.4979	.4980	.4981
2.9	.4981	.4982	.4982	.4983	.4984	.4984	.4985	.4985	.4986	.4986
3.0	.4987	.4987	.4987	.4988	.4988	.4989	.4989	.4989	.4990	.4990

Photo Credits

1, PhotoDisc   2, PhotoDisc   8, James P. Blair, ©2000 PhotoDisc, Inc.   12, Emma Lee/LifeFile, ©2000 PhotoDisc, Inc.   15, ©James L. Amos/Corbis   17, PhotoLink/ ©PhotoDisc, Inc.   69, ©Owen Franken/Corbis   70, Courtesy Nancy Bertler/AP Images 77, Corbis   98 (both), Beth Anderson   125 (both), Rick Haston, Latent Images 126, Corbis RF   130, ©Michael Dwyer, Stock Boston   132, ©Paul McCormick, The Image Bank   161, PhotoDisc   193, Don Smetzer, ©Tony Stone, Worldwide 212, Icon SMI/Corbis   224, PhotoDisc   232, PhotoDisc   237, Kaz Cuba/©2000 PhotoDisc, Inc   252, Jonathan S. Blair, National Geographic/Getty   253, PhotoDisc 254, The Science Picture Company/Alamy   260, ©Tom Haseltine/Getty   263, ©Reuters/ Corbis   267, PhotoDisc PP   277, David Cole/Alamy   296, ©Patrick Robert/Corbis 298, PhotoLink/©PhotoDisc, Inc.   300, Bruce Coleman, Inc.   304, Bob Daemmrich, Stock Boston   306, PhotoDisc   317, SuperStock   322, Corbis/Digital Stock 323, Photo Disc PP   327, PhotoDisc   328, ©Peter Menzel, Stock Boston   333, PhotoLink/©2000 PhotoDisc   334, Andy and Sandy Carey/©2000 PhotoDisc, Inc. 335, Tom Walker, Stock Boston   337, ©Royalty-Free Corbis   346 (both) ©Royalty- Free Corbis   348, Michael Grecco, Stock Boston   379, ©Lester V. Bergman/Corbis 386, ©Raymond Gehman/National Geographic/Getty Images   409, David Simson, Stock Boston   421, ©Bettmann/Corbis   425, Dimair/Shutterstock   429, ©Giff Beaton 434, ©Joe McDonald/Corbis   442, ©Robert Pickett/Corbis   481, PhotoDisc 493, ©Royalty-Free Corbis   500, The Bridgeman Art Library/Getty Images 501, Stockbyte/Getty   503, NASA Public Domain   508, Corbis   512, ©Royalty- Free/Corbis   518, Jules Frazier/©2000 PhotoDisc   520, Rick Haston, Latent Images 522, Rick Haston, Latent Images   527, PhotoDisc   528, Digital Vision 534, Reuters/Philippe Wojazer/Archive Photos   535, Lonely Planet Images/Getty Images   543, PhotoDisc   545, PhotoDisc   557, ©Raymond Gehman/Corbis 562, Digital Vision/Getty   593, PhotoDisc   597, Digital Vision/Getty   597, ©James Marshall/Corbis   639, PhotoDisc   653, Everett Collection/Alamy   656, Oleg Sayatoslavsky, LifeFile/©2000 PhotoDisc, Inc.   663, ©Royalty-Free Corbis 689, ©W. Wisniewski/zefa/Corbis   704, ©Tim Davis/Corbis   705, ©Clouds Hill Imaging, Ltd./Corbis   725, The Culture Collection of Algae at the University of Texas at Austin

Answers

SECTION 1.1

1. $y = -4$

3. $x = -4.5$

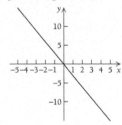

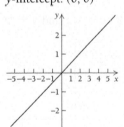

5. $m = -3$,
y-intercept: $(0, 0)$

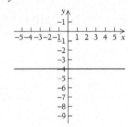

7. $m = 0.5$,
y-intercept: $(0, 0)$

9. $m = -2$,
y-intercept: $(0, 3)$

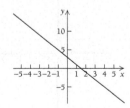

11. $m = -1$,
y-intercept: $(0, -2)$

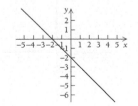

13. $m = -2$, y-intercept: $(0, 2)$ **15.** $m = -1$,
y-intercept: $(0, -\frac{5}{2})$ **17.** $m = \frac{1}{2}$, y-intercept: $(0, -4)$
19. $y = -5x$ **21.** $y = -2x + 7$ **23.** $y = 2x - 6$
25. $y = \frac{1}{2}x - 6$ **27.** $y = 3$ **29.** $\frac{3}{2}$ **31.** $-\frac{3}{34}$
33. No slope **35.** 0 **37.** 3 **39.** 2 **41.** $y = \frac{3}{2}x + 4$
43. $y = -\frac{3}{34}x + \frac{91}{170}$ **45.** $x = 3$
47. $y = 3$ **49.** $y = 3x$ **51.** $y = 2x + 3$ **53.** $\frac{2}{25}$ or 8%
55. 0.035 or 3.5% **57.** 0.16 year per year
59. (a) $R = 4.17T$ **(b)** 25.02 **61. (a)** $B = 0.025W$
(b) $B = 2.5\%W$. The brain weight is 2.5% of the total
body weight. **(c)** 3 lb **63. (a)** $D(0°) = 115$ ft,
$D(-20°) = 75$ ft, $D(10°) = 135$ ft, $D(32°) = 179$ ft
(b) **tw** **65. (a)** 145.78 cm **(b)** 142.98 cm
67. (a) $A(0) = 19.7$; $A(1) = 19.78$; $A(10) = 20.5$;
$A(30) = 22.1$; $A(50) = 23.7$ **(b)** 24.5
(c)

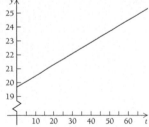

69. (a) $f(x) = x^2 + 3x + 1$, $y = x + 1$

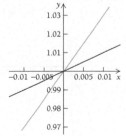

(b) Answers will vary. **(c)** $y = 3x + 1$

SECTION 1.2

1. $y = \frac{1}{2}x^2$ and $y = -\frac{1}{2}x^2$

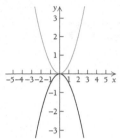

3. $y = x^2$ and $y = (x - 1)^2$

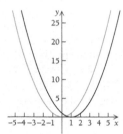

5. $y = x^2$ and $y = (x + 1)^2$

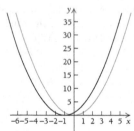

7. $y = x^3$ and $y = x^3 + 1$

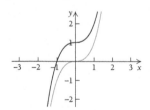

9. Parabola, $(-2, -11)$ **11.** Not a parabola
13. $y = x^2 - 4x + 3$ **15.** $y = -x^2 + 2x - 1$

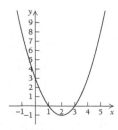
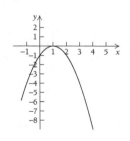

17. $y = 2x^2 + 4x - 7$ **19.** $y = \frac{1}{2}x^2 + 3x - 5$

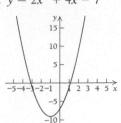
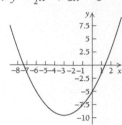

21. $1 \pm \sqrt{3}$ **23.** $-\frac{4}{3} \pm \frac{\sqrt{10}}{3}$ **25.** $1 \pm 3i$

27. $-3 \pm \sqrt{10}$ **29.** $-2 \pm 2i$ **31.** $\frac{1 \pm \sqrt{2}}{2}$

33. $f(7) = 84$, $f(10) = 220$, $f(12) = 364$
35. 2012   **37.** tw   **39.** tw   **41.** $0, 1, -1$
43. $-1.83, -0.856, 3.19$ **45.** $-10.2, -1.87, -0.821,$
$-0.303, 0.0985, 0.535, 1.22, 3.30$ **47.** $-0.279x + 4.04$
49. $0.942x^2 - 2.65x - 27.9$
51. $0.237x^4 - 0.885x^3 - 29.2x^2 + 165.2x - 210.1$

SECTION 1.3

1. $y = |x|$ and $y = |x + 3|$

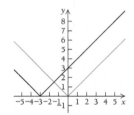

3. $y = \sqrt{x}$ and $y = \sqrt{x + 1}$

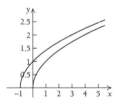

5. $y = \frac{2}{x}$ **7.** $y = \frac{-2}{x}$

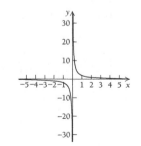
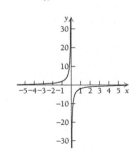

9. $y = \dfrac{1}{x^2}$

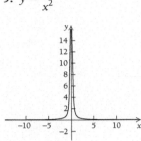

11. $y = \sqrt[3]{x}$

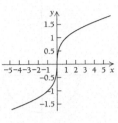

b)

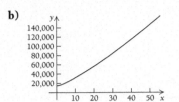

65. tw   **67.** ± 2.646 **69.** No solution

SECTION 1.4

1. $\dfrac{2\pi}{3}$

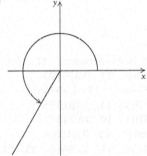

13. $y = \dfrac{x^2 - 9}{x + 3}$

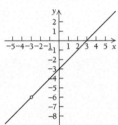

15. $y = \dfrac{x^2 - 1}{x - 1}$

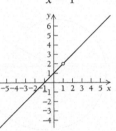

17. $x^{3/2}$ **19.** $a^{3/5}$ **21.** $t^{1/7}$ **23.** $t^{-4/3}$ **25.** $t^{-1/2}$

27. $(x^2 + 7)^{-1/2}$ **29.** $\sqrt[5]{x}$ **31.** $\sqrt[3]{y^2}$ **33.** $\dfrac{1}{\sqrt[5]{t^2}}$

35. $\dfrac{1}{\sqrt[3]{b}}$ **37.** $\dfrac{1}{\sqrt[6]{e^{17}}}$ **39.** $\dfrac{1}{\sqrt{x^2 - 3}}$ **41.** $\dfrac{1}{\sqrt[3]{t^2}}$

43. 27   **45.** 16   **47.** 8   **49.** All real numbers except 5

51. All real numbers except 2 and 3   **53.** $[-\frac{4}{5}, \infty)$

55.

W	0	10	20	30	40	50	100	150
T	0	20	51	86	126	168	417	709

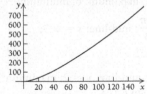

3. $\dfrac{4\pi}{3}$

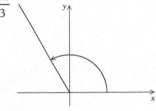

57. a) 1.93196 m^2 **b)** 1.87753 m^2

c)

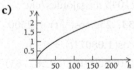

5. 3π

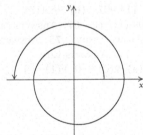

59. 3461.54 cm^2 **61.** $-\dfrac{7}{2} \pm \dfrac{\sqrt{13}}{2}$

63. a) 105,254, 114,595, and 146,957 particles per cc

7. 135 degrees

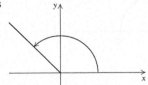

9. 270 degrees

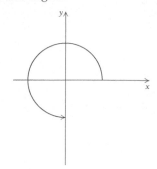

11. −60 degrees

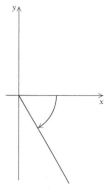

13. Not coterminal   **15.** Not coterminal   **17.** Not coterminal   **19.** Coterminal   **21.** 0.559193
23. 0.978148   **25.** 0.0874887   **27.** 1.48256
29. 1.08636   **31.** 0.587785   **33.** 0.481575
35. 2.61313   **37.** 0.745705   **39.** 26.7437°
41. 70.1231°   **43.** 66.8605°   **45.** 0.631059
47. 0.927295   **49.** 0.10956   **51.** 33.5468   **53.** 23.3359

55. 48.1897°   **57.** 62.6761°   **59.** $\dfrac{-1 + \sqrt{3}}{2\sqrt{2}}$

61. 1841.57 ft   **63. (a)** 114.907   **(b)** 96.4181
(c) 310.268   **65.** 4.55494 cm/sec   **67.** Answers vary.
69. Answers vary.   **71.** Answers vary.   **73.** Answers
vary.   **75.** Answers vary.   **77.** Answers vary.
79. Answers vary.   **81. (a)** $V(0) = 0, V(1) = 1$
(b) tw   **83.** 0.821936

SECTION 1.5

1.

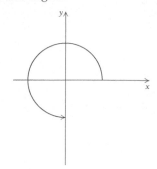

3.

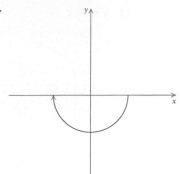

5.

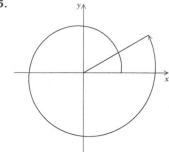

7. 0   **9.** $-\frac{1}{2}$ **11.** −1   **13.** $-\sqrt{3}$ **15.** −0.573576
17. −0.8391   **19.** 3.62796   **21.** −0.587785

23. −0.332736   **25.** $\dfrac{\pi}{6} + 2\pi n, \dfrac{5\pi}{6} + 2\pi n$

27. $\dfrac{\pi}{2} n$ **29.** $\dfrac{5\pi}{36} + \dfrac{2\pi}{3} n, -\dfrac{11\pi}{36} + \dfrac{2\pi}{3} n$ **31.** $\dfrac{2\pi}{3} n$

33. $-\dfrac{\pi}{6} + 2\pi n, \dfrac{7\pi}{6} + 2\pi n$

35. No solution   **37.** amplitude: 2; period: π; mid-line:
$y = 4$; maximum: 6; minimum: 2   **39.** amplitude: 5;
period: 4π; mid-line: $y = 1$; maximum: 6; minimum: −4

41. amplitude: $\frac{1}{2}$; period: $\dfrac{2\pi}{3}$; mid-line: $y = -3$;

maximum: $-\frac{5}{2}$; minimum: $-\frac{7}{2}$

43. amplitude: 4; period: 2; mid-line: $y = 2$; maximum: 6;
minimum: −2   **45.** $y = 7 \sin t + 3$

47. $y = 2 \cos(\frac{1}{2}t) - 1$ **49.** 0.571045 megajoules/cm²
51. 0.793096 megajoules/cm²   **53.** $250 \cos(\frac{2\pi}{5}t) + 2500$
55. tw   **57.** 440 Hz   **59.** $5.3 \cos(1.08071t) + 143$
61. (−0.766044, 0.642788)   **63.** (0.809017, −0.587785)

65. $\dfrac{-1 - \sqrt{3}}{2\sqrt{2}}$ **67.** Answers vary.   **69.** Answers vary.

71. tw   **73.** 261.626   **75.** 24 notes above A above
middle C, or the third A above middle C
77. 28 notes above A above middle C

79. **(a)**

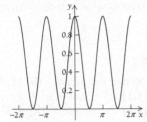

(b) $\frac{1}{2}\cos 2t + \frac{1}{2}$ **(c)** tw
81. **(a)** in red,   **(b)** in blue

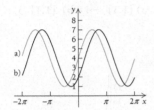

83.   **85.**

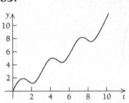   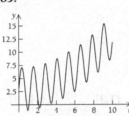

CHAPTER 1 REVIEW

1. a) [1.1] 100 per 1000   **b)** [1.1] 20 and 30
2. [1.2] 13   **3.** [1.2] $2h^2 + 3h + 4$ **4.** [1.2] 3
5. [1.2] 36   **6.** [1.2] $(h-1)^2$ **7.** [1.2] 9
8. [1.2] $f(x) = 2x^2 + 3x - 1$

9. [1.2] $f(x) = 3x^2 - 6x + 1$

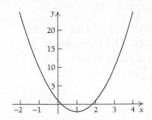

10. [1.3] $f(x) = |x + 1|$

11. [1.2] $f(x) = (x - 2)^2$

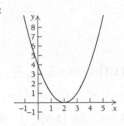

12. [1.3] $f(x) = \dfrac{x^2 - 16}{x + 4}$

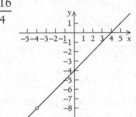

13. (a) [1.1] 1.2   **(b)** [1.1] -3
14. [1.1] $x = -2$

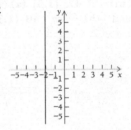

15. [1.1] $y = 4 - 2x$

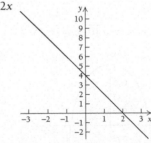

16. [1.1] $y = -\frac{7}{11}x + \frac{6}{11}$ **17.** [1.1] $y = 8x + 7$
18. [1.1] slope: $-\frac{1}{6}$, y-intercept: $(0, 3)$
19. [1.2] $x = -1, x = -4$ **20.** [1.2] $x = 3, x = 4$
21. [1.2] $x = 2, x = -4$ **22.** [1.2] $-3 \pm \sqrt{29}$
23. [1.2] $x = -3, x = -1, x = 1$ **24.** [1.2] $x = 1,$
$x = -2$ **25.** [1.1] 75 pages per day
26. [1.1] $-\frac{20}{3}$ meters per second   **27.** [1.3] 84 lb

28. [1.2] $\dfrac{1}{10} \pm \dfrac{\sqrt{141}}{10}$ **29.** [1.3] $\sqrt[6]{y}$ **30.** [1.3] $x^{3/20}$

31. [1.3] 9 **32.** [1.3]

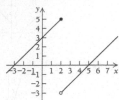

33. (a) [1.1] $G(x) = \dfrac{9}{7}x + \dfrac{437}{7}$

(b) $G(18) = 85.6$, $G(25) = 94.6$ **34.** [1.4] $\dfrac{\sqrt{3}}{2}$

35. [1.5] -1 **36.** [1.5] -1 **37.** [1.4] 119.341

38. [1.5] $\dfrac{\pi}{2} + 2\pi n$ **39.** [1.5] $\dfrac{\pi}{3} + \pi n$

40. [1.5] No solution

41. [1.5] $\dfrac{5\pi}{24} + \dfrac{\pi}{2}n, \dfrac{\pi}{24} + \dfrac{\pi}{2}n$

42. [1.5] $\dfrac{\pi}{6} + 2\pi n, \dfrac{5\pi}{6} + 2\pi n$

43. [1.5] amplitude: 2; period: 6π; mid-line: $y = -4$; maximum: -2; minimum: -6

44. [1.5] amplitude: $\dfrac{1}{2}$, period: 1, mid-line: $y = 3$, maximum: $\dfrac{7}{2}$, minimum: $\dfrac{5}{2}$ **45.** [1.5] $2\sin(2t) + 3$

46. [1.5] $3\cos(\pi t) - 2$ **47.** [1.5] **(a)** $67\cos(4\pi t) + 68$ m above ground **(b)** $\dfrac{203}{2}$ m **48.** [1.3] $\dfrac{1}{32}$

49. [1.2] $x = 0, x = 2, x = -2$

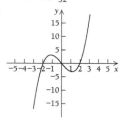

50. [1.3] $x = \pm\sqrt{10}, x = \pm2\sqrt{2}$

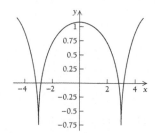

51. [1.3] $(-1.89808, 0.75409), (-0.2737, 1.0743), (2.07926, 0.67227)$

52. (a) [1.2] $G(x) = 0.6255x + 75.4766$

(b) [1.1] $G(18) = 86.7356, G(25) = 91.1141$ **(c)** tw

53. (a) [1.2] $w(h) = 0.00397h^2 + 3.269h - 76.429$

(b) [1.1] 160.415 lb

CHAPTER 1 TEST

1. (a) [1.1] About 1150 min per month **(b)** [1.1] 62

2. (a) [1.1] 11 **(b)** [1.1] $x^2 + 2xh + h^2 + 2$

3. (a) [1.1] 11 **(b)** [1.1] $2x^2 + 4xh + 2h^2 + 3$

4. [1.1] slope is -3, y-intercept is $(0, 2)$

5. [1.1] $y = \dfrac{1}{4}x - 7$ **6.** [1.1] -3

7. [1.1] $-\$700$ dollars per year **8.** [1.1] $\dfrac{1}{2}$ lbs per bag

9. [1.1] $F(W) = \dfrac{2W}{3}$ **10. a)** [1.1] -4 **b)** [1.1] $3, -3$

11. [1.2] $-2 \pm \sqrt{6}$

12. [1.3] $f(x) = \dfrac{4}{x}$

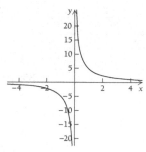

13. [1.3] $t^{-1/2}$ **14.** [1.3] $\dfrac{1}{\sqrt[5]{t^3}}$

15. [1.3] $f(x) = \dfrac{x^2 - 1}{x + 1}$

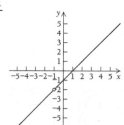

16. [1.5] $-\dfrac{1}{2}$ **17.** [1.5] $-\dfrac{1}{\sqrt{2}}$ **18.** [1.5] 0

19. [1.4] 3.90895 **20.** [1.5] $\dfrac{\pi}{3} + n\pi, -\dfrac{\pi}{3} + n\pi$

21. [1.5] No solution

22. [1.5] $n\dfrac{\pi}{2}, -\dfrac{\pi}{12} + n\pi, \dfrac{7\pi}{12} + n\pi$

23. [1.5] amplitude: 4; period: π; mid-line: $y = 4$; maximum: 8; minimum: 0 **24.** [1.5] amplitude: 6, period: 6π, mid-line: $y = -10$, maximum: -4, minimum: -16 **25.** [1.5] $\dfrac{1}{2}\cos 3t - 1$

26. [1.5] $\dfrac{3}{2}\sin 2\pi t + \dfrac{5}{2}$

27. [1.3]

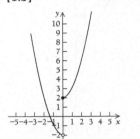

28. [1.3]

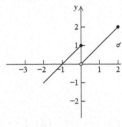

29. (a) [1.1] $M(r) = \frac{1}{5}r + 160$ **(b) [1.1]** 172.4, 175

30. [1.2] $\frac{1}{6} \pm \frac{i\sqrt{95}}{6}$

31. [1.3] $x = -1.25432$ **32. [1.3]** No solution

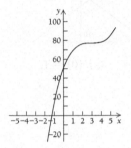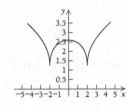

33. [1.3] $(-1.21034, 2.36346)$ **34. (a) [1.2]**
$M(r) = 0.2r + 160$ **(b) [1.1]** 172.4, 175   **(c) tw**
35. (a) [1.2] $291.177 + 37.5761x$,
$-117.725 + 74.6068x - 0.592458x^2$,
$-439.648 + 125.714x - 2.60421x^2 + 0.0220342x^3$,
$507.839 - 88.5121x + 11.4571x^2 - 0.323988x^3$
$+ 0.0028363x^4$

(b)

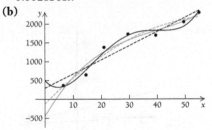

(c) tw   (d) tw

SECTION 2.1

1. Not continuous   **3.** Continuous
5. (a) $-1, 2$, does not exist   **(b)** -1 **(c)** No   **(d)** 3
(e) 3   **(f)** Yes   **7. (a)** 2   **(b)** 2   **(c)** Yes   **(d)** 0
(e) 0   **(f)** Yes   **9. (a)** 3   **(b)** 3   **(c)** 3   **(d)** 2   **(e)** No
(f) Yes   **11. (a)** T   **(b)** T   **(c)** T   **(d)** F   **(e)** T
(f) F   **(g)** T   **(h)** F   **13. (a)** F   **(b)** T   **(c)** F   **(d)** F

(e) F   **(f)** T   **(g)** T   **(h)** F   **(i)** F   **(j)** T   **15. (a)** T
(b) F   **(c)** F   **(d)** T   **(e)** F   **(f)** T   **(g)** F   **(h)** T
17. No, yes, no, yes   **19.** 0.37, 0.60, does not exist
21. 0.83, 0.83, 0.83   **23.** 1.06   **25.** Yes, yes, no, no
27. \$2.30, \$2.60, does not exist   **29.** \$2.90
31. $t = 0.5, 0.75, 1.25, 1.5, 1.75$ **33.** 12
35. $t = 0.1, 0.3, 0.4, 0.5, 0.6, 0.8$ **37.** 35   **39. tw**
41. 100, 30, does not exist   **43.** No, yes
45. tw   47. Justification left to student. 0
49. Justification left to student. 1

51. Justification left to student. $\dfrac{1}{\sqrt{3}}$

53. Justification left to student. $\sqrt{2}$

55. Justification left to student. $\dfrac{\sqrt[4]{3}}{\sqrt{2}}$

57. Justification left to student. $-\frac{1}{2}$ **59.** 0   **61.** 1
63. Does not exist

SECTION 2.2

1. -2 **3.** Does not exist   **5.** 11   **7.** -10 **9.** $-\frac{5}{2}$

11. 5   **13.** 2   **15.** $\dfrac{1}{\sqrt{2}} + \dfrac{\pi}{4}$ **17.** 1

19. Does not exist   **21.** $\frac{2}{7}$ **23.** $\frac{5}{4}$ **25.** $6x^2$ **27.** $-\dfrac{2}{x^3}$

29. 1   **31.** $\sin x$ **33.** Does not exist   **35.** 0   **37.** $\frac{1}{3}$

39. 3   **41.** 6   **43.** $-\dfrac{1}{2\sqrt{3}}$ **45.** $\frac{3}{4}$ **47.** $\frac{1}{4}$

SECTION 2.3

1. (a) $7(2x + h)$ **(b)** 70, 63, 56.7, 56.07
3. (a) $-7(2x + h)$ **(b)** $-70, -63, -56.7, -56.07$
5. (a) $7(3x^2 + 3xh + h^2)$ **(b)** 532, 427, 344.47, 336.841

7. (a) $\dfrac{-5}{x(x + h)}$ **(b)** $-0.2083, -0.25, -0.3049, -0.3117$

9. (a) -2 **(b)** $-2, -2, -2, -2$ **11. (a)** $2x + h - 1$
(b) 9, 8, 7.1, 7.01   **13. (a)** 1.21 lb/mo   **(b)** 0.45 lb/mo
(c) 0.31 lb/mo   **(d)** 0.66 lb/mo   **(e)** At birth
15. (a) 0.58 lb/mo   **(b)** 0.7 lb/mo   **(c)** 0.8 lb/mo
(d) tw   17. (a) 1.25 words/min, 1.25 words/min,
0.625 words/min, 0 words/min, 0 words/min   **(b) tw**
19. (a) 144 ft   **(b)** 400 ft   **(c)** 128 ft/sec
21. (a) 125,000 people/yr for both   **(b) tw**
(c) Population A: 290,000 people/yr,
$-40,000$ people/yr, $-50,000$ people/yr,
300,000 people/yr; Population B: 125,000 people/yr
for each year   **(d) tw   23. tw   25.** $b + a(h + 2x)$

27. $\dfrac{1}{\sqrt{x + h} + \sqrt{x}}$ **29.** $-\dfrac{h + 2x}{x^2(h + x)^2}$

31. $\dfrac{1}{(x+1)(h+x+1)}$

33. $-\dfrac{1}{\sqrt{x}\sqrt{x+h}\,(\sqrt{x}+\sqrt{x+h})}$

SECTION 2.4
1. (a) and **(b)** $f(x) = 5x^2$

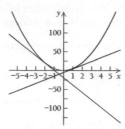

(c) $f'(x) = 10x$ **(d)** $f'(-2) = -20, f'(0) = 0,$
$f'(1) = 10$
3. (a) and **(b)** $f(x) = -5x^2$

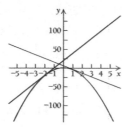

(c) $f'(x) = -10x$ **(d)** $f'(-2) = 20, f'(0) = 0,$
$f'(1) = -10$
5. (a) and **(b)** $f(x) = x^3$

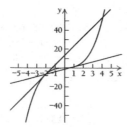

(c) $f'(x) = 3x^2$ **(d)** $f'(-2) = 12, f'(0) = 0,$
$f'(1) = 3$
7. (a) and **(b)** $f(x) = 2x + 3$

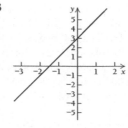

(c) $f'(x) = 2$ **(d)** $f'(-2) = 2, f'(0) = 2, f'(1) = 2$

9. (a) and **(b)** $f(x) = -4x$

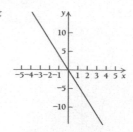

(c) $f'(x) = -4$ **(d)** $f'(-2) = -4, f'(0) = -4,$
$f'(1) = -4$
11. (a) and **(b)** $f(x) = x^2 + x$

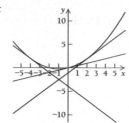

(c) $f'(x) = 2x + 1$ **(d)** $f'(-2) = -3, f'(0) = 1,$
$f'(1) = 3$
13. (a) and **(b)** $f(x) = 2x^2 + 3x - 2$

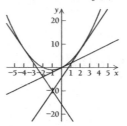

(c) $f'(x) = 4x + 3$ **(d)** $f'(-2) = -5, f'(0) = 3,$
$f'(1) = 7$

15. (a) and **(b)** $f(x) = \dfrac{1}{x}$

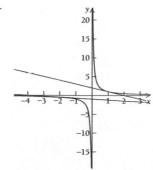

(c) $f'(x) = -\dfrac{1}{x^2}$ **(d)** $f'(-2) = -\frac{1}{4}, f'(0)$ does not

exist, $f'(1) = -1$ **17.** $f'(x) = m$ **19.** $y = 6x - 9,$
$y = -2x - 1, y = 20x - 100$ **21.** $y = 10 - 5x,$
$y = -5x - 10, y = \dfrac{1}{10} - \dfrac{x}{2000}$ **23.** $y = 2x + 5, y = 4,$

$y = 29 - 10x$ **25.** $x_0, x_3, x_4, x_6, x_{12}$
27. 1, 2, 3, 4, 5, 6, 7, 8, 9, 10, 11, 12, . . .
29. Differentiable for all x in the domain
31. Differentiable for all x in the domain

33. tw   **35.** $f'(x) = -\dfrac{2}{x^3}$ **37.** $f'(x) = \dfrac{1}{(x+1)^2}$

39. $f'(x) = -\dfrac{1}{2x^{3/2}}$ **41.** $x = -3$

43, 45, 47. Left to the student

SECTION 2.5

1. $\dfrac{dy}{dx} = 7x^6$ **3.** $\dfrac{dy}{dx} = 6x$ **5.** $\dfrac{dy}{dx} = 12x^2$

7. $\dfrac{dy}{dx} = \dfrac{2}{\sqrt[3]{x}}$ **9.** $\dfrac{dy}{dx} = \dfrac{3}{4\sqrt[4]{x}}$ **11.** $\dfrac{dy}{dx} = 4\cos x$

13. $\dfrac{dy}{dx} = \cos x - \dfrac{1}{x^2}$ **15.** $\dfrac{dy}{dx} = 4(2x+1)$

17. $f'(x) = 0.8x^{2.2}$ **19.** $f'(x) = 10\cos x + 12\sin x$

21. $f'(x) = -\sqrt[3]{9}\sin x$ **23.** $f'(x) = \dfrac{1}{5} - \dfrac{5}{x^2}$

25. $f'(x) = -\dfrac{1}{\sqrt{x}} - \dfrac{1}{x^{3/4}} - \dfrac{1}{2x^{5/4}} + \dfrac{7}{2x^{3/2}}$

27. $f'(x) = \dfrac{18x^{13/5}}{5} - \dfrac{13x^{8/5}}{5}$ **29.** $f'(x) = \dfrac{-3}{x^2} + \dfrac{8}{x^3}$

31. $4 + \dfrac{2}{x^2}$ **33.** $3\left(\dfrac{6}{x^4} - \dfrac{8}{x^5}\right)$

35. $\sqrt{2}(-2\cos x - 3\sin x)$

37. $\dfrac{8}{3}\sqrt[5]{5}\cos x + \dfrac{8}{7}\sqrt[5]{5}\sin x$

39. $\cos x - \dfrac{1}{\sqrt{x}} - \dfrac{3}{2x^{3/2}}$ **41.** $y = 4$ **43.** $y = \dfrac{17x}{6} - \dfrac{8}{3}$

45. $\dfrac{dy}{dx} = -\dfrac{5}{18x^{23/18}}$ **47.** $\dfrac{dy}{dx} = -3\sin x$

49. $\dfrac{dy}{dx} = -\dfrac{\cos x}{2} - \dfrac{1}{2}\sqrt{3}\sin x$ **51.** $y = 0$

53. Left to the student   **55. tw**
57. -0.6922 and 0.6922 **59.** -0.3456 and 1.929

SECTION 2.6

1. (a) $v(t) = 3t^2 + 1$ **(b)** $a(t) = 6t$
(c) $v(4) = 49$ ft/sec, $a(4) = 24$ ft/sec^2
3. (a) $v(t) = 2 - 20t$ **(b)** $a(t) = -20$
(c) $v(1) = -18$ m/sec, $a(1) = -20$ m/sec^2
(d) $t = 0.05$ sec   **5. (a)** $v(t) = 2\cos t + 5$
(b) $a(t) = -2\sin t$ **(c)** $v\left(\dfrac{\pi}{4}\right) = 5 + \sqrt{2}$ m/sec,
$a\left(\dfrac{\pi}{4}\right) = -\sqrt{2}$ m/sec^2 **(d)** $\pi + 2\pi n$ sec
7. (a) $N'(a) = -2a + 300$ **(b)** 2906

(c) 280 units per thousand dollars   **(d) tw**
9. (a) $w'(t) = 1.61 - 0.0968t + 0.0018t^2$
(b) 7.6 lb   **(c)** 1.61 lb/mo   **(d)** 20.9872 lb
(e) 0.7076 lb/mo   **(f)** 1.1156 lb/mo   **(g)** 5.71 mo
11. (a) 2π cm per cm   **(b) tw**
13. (a) $T(t) = 1.2 - 0.2t$ **(b)** 100.175 degrees
(c) 0.9 degrees/day   **(d) tw**

15. (a) $\dfrac{dT}{dW} = 1.31W^{0.31}$ **(b) tw**

17. (a) $\dfrac{dR}{dQ} = kQ - Q^2$ **(b) tw**

19. (a) $\dfrac{dA}{dt} = 0.08$ years per year   **(b) tw**   **21. tw**

23. Position graph: $s(t) = 3t^2 - 2t + 7$
Velocity graph: $v(t) = 6t - 2$
Acceleration graph: $a(t) = 6$

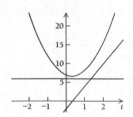

25. Position graph:
$s(t) = 23.7 - 15.8t + 0.023t^2 + 0.0046t^3$
Velocity graph: $v(t) = -15.8 + 0.046t + 0.0138t^2$
Acceleration graph: $a(t) = 0.046 + 0.0276t$

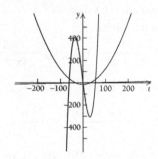

27. (a) Position graph: $s(t) = 2\cos t + 3\sin t$
Velocity graph: $v(t) = 3\cos t - 2\sin t$
Acceleration graph: $a(t) = -2\cos t - 3\sin t$ **(b) tw**

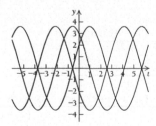

SECTION 2.7

1. $11x^{10}$ **3.** $\dfrac{3\sqrt{x}}{2}$ **5.** $3x^2$ **7.** $2x$

9. $(3x^4 - 3\sqrt{x})(40x^4 - 9x^2) +$

$\left(12x^3 - \dfrac{3}{2\sqrt{x}}\right)(8x^5 - 3x^3 + 2)$

11. $2(\sqrt{x} - \sqrt[3]{x}) + \left(\dfrac{1}{2\sqrt{x}} - \dfrac{1}{3x^{2/3}}\right)(2x + 3)$

13. $\sqrt{x}\,\sec^2 x + \dfrac{\tan x}{2\sqrt{x}}$ **15.** $4(2t + 3)$

17. $(0.04x + 1.3)(4.1x + 11.3) +$
$4.1(0.02x^2 + 1.3x - 11.7)$

19. $\sec^2 x - \csc^2 x$ **21.** $\dfrac{1}{\cos x + 1}$ **23.** $2\sec^2 t \tan t$

25. $2x\left(x + \dfrac{2}{x}\right) + \left(1 - \dfrac{2}{x^2}\right)(x^2 - 3)$

27. $\dfrac{6(x - 1)}{5}$ **29.** $\dfrac{2x^2 - 2x + 5}{(2x - 1)^2}$

31. $-\dfrac{(t + 2)(3t - 8)}{(t^2 - 2t + 6)^2}$ **33.** $\dfrac{x(x + 2)}{(x + 1)^2}$

35. $\dfrac{\sec t}{1 + \sec t}$

37. $\dfrac{2x^2 \cos x + 2x \sec^2 x + x \sin x - \tan x}{2x^{3/2}}$

39. $\dfrac{1}{(\sqrt{t} - 1)^2 \sqrt{t}}$

41. $t \sin t + \tan t \sin t + t \sec t \tan t$

43.–84. Left to the student

85. (a) $f'(x) = \dfrac{1}{(x + 1)^2}$ (b) $g'(x) = \dfrac{1}{(x + 1)^2}$

(c) **tw**   **87.** (a) 0 (b) **tw**   **89.** $y = 2,\ y = \dfrac{x + 4}{2}$

91. $y = \tfrac{1}{2}$ **93.** $y = \dfrac{(\pi + 4)\sqrt{2}}{8}x - \dfrac{\pi^2}{16\sqrt{2}}$

95. (a) $T't = -\dfrac{4(t^2 - 1)}{(t^2 + 1)^2}$ (b) 100.2 degrees

(c) -0.48 degrees/hr

97. (a) $100 \tan t$ (b) $100 \sec^2 t$

(c) $\dfrac{\pi}{4} + \dfrac{\pi}{2} n$

99. $\dfrac{\sec^2 x(x^4 - 1) - 4x \tan x}{(x^2 - 1)^2}$

101. $\dfrac{\sec t(t \sec^2 t + t \tan^2 t - \tan t)}{t^2}$

103. $\dfrac{x \cos^3 x + (x^2 + \sin x) \cos^2 x - x \sec^2 x - x^2 \sin^2 x - \sec x - \sin x \tan x + \tan x}{(x + \cos x)^2}$

105. Left to the student

107.

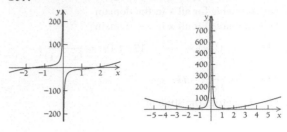

No horizontal tangent lines

109.

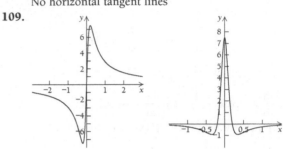

$(0.2, 0.75),\ (-0.2, -0.75)$

111.

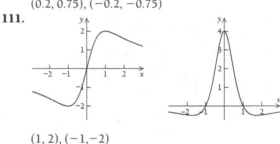

$(1, 2),\ (-1, -2)$

SECTION 2.8

1. $4(2x + 1)$ **3.** $-110(1 - 2x)^{54}$ **5.** $2\sec^2 x \tan x$

7. $-\dfrac{3}{2\sqrt{1 - 3x}}$ **9.** $-\dfrac{12x}{(3x^2 + 1)^2}$

11. $\dfrac{t}{\sqrt{2t + 3}} + \sqrt{2t + 3}$

13. $\dfrac{1}{6} \pi \cos\left(\dfrac{\pi t}{6} + \dfrac{\pi}{3}\right)$

15. $9x^2(x^3 + 1)^2 - 12x^2(x^3 + 1)^3$

17. $\dfrac{\cot x \csc x}{2\sqrt{1 - \csc x}}$ **19.** $\dfrac{2}{3(2x - 1)^{2/3}} - 2(4 - x)$

21. $x^3 \cos x + 3x^2 \sin x - 5x \sin x + 5 \cos x + 4 \sec x \tan x$

23. $\dfrac{2x + 3x^2}{2\sqrt{x^2 + x^3}}(2x^2 + 3x + 5) + \sqrt{x^2 + x^3}(4x + 3)$

25. $-\dfrac{\sin(\sqrt{t})}{2\sqrt{t}}$ **27.** $\dfrac{3x + 2}{\sqrt{2x + 5}} + 3\sqrt{2x + 5}$

29. $-\cos(\cos x)\sin x$ **31.** $-\dfrac{2 \sin 4t}{\sqrt{\cos 4t}}$

33.
$$\dfrac{(x^3 + 2x^2 + 3x - 1)^2(2x^6 - 8x^5 - 30x^4 + 16x^3 + 9x^2 + 12x + 9)}{(2x^4 + 1)^3}$$

35. $\dfrac{29}{2\sqrt{3x - 4}(5x + 3)^{3/2}}$

37. $(0.01x^2 + 2.391x - 8.51)^5 + 5x(0.02x + 2.391)(0.01x^2 + 2.391x - 8.51)^4$

39. $\dfrac{\sin 5x - \csc^2 5x}{(\cot 5x - \cos 5x)^{4/5}}$

41. $8x \cos(\sec^4(x^2)) \sec^4(x^2) \tan(x^2)$

43. $\dfrac{x}{6(x^2 + 2)^{3/4}(\sqrt[4]{x^2 + 2} + 1)^{2/3}}$

45. $\dfrac{\sin^2 x(3x^2 \cos x - 2x \sin x + 15 \cos x)}{(x^2 + 5)^2}$

47. $-3 \cot^2(x \sin(2x + 4)) \csc^2(x \sin(2x + 4)) \times [2x \cos(2x + 4) + \sin(2x + 4)]$

49. $\dfrac{4 \sec^4 x \tan x + 1}{2\sqrt{\sec^4 x + x}}$ **51.** $\dfrac{1}{2\sqrt{u}}, 2x, \dfrac{x}{\sqrt{x^2 - 1}}$

53. $50u^{49}, 12x^2 - 4x, 50(12x^2 - 4x)(4x^3 - 2x^2)^{49}$

55. $2u + 1, 3x^2 - 2, 6x^5 - 16x^3 + 3x^2 + 8x - 2$

57. $y = \frac{1}{4}(5x + 3)$ **59.** $y = 4x - 3$

61. $y = \frac{1}{12}(-6\sqrt{3}x - \sqrt{3}\pi + 3)$

63. $\dfrac{(2 - 3x)x}{(x + 1)^6}$ for each method   **65.** -216

67. $\dfrac{-4}{11^{2/3}} \approx -0.8087$ **69. (a)** $3000(i + 1)^2$

(b) tw   **71. (a)** $D(c) = 4.25(c + 25)$, $c(w) = 2.1991w$ **(b)** 4.25 **(c)** 2.1991
(d) 9.3461 **(e) tw**
73. (a) -1.4218 ppmv/yr   **(b)** 5.3847 ppmv/yr

75. $4\left(16x(x^2 + 4)^7 + \dfrac{3}{2\sqrt{x}}\right)((x^2 + 4)^8 + 3\sqrt{x})^3$

77. $\cos x \cos(\sin x) \cos(\sin(\sin x))$

79. $-3 \sec 3x \tan 3x \csc^2(\sec 3x) \sec^2(\cot(\sec 3x))$

81. 0   **83.** Left to the student   **85.** Left to the student
87. $f(x) = 1.68x\sqrt{9.2 - x^2}$

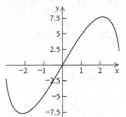

$$f'(x) = \dfrac{15.456 - 3.36x^2}{\sqrt{9.2 - x^2}}$$

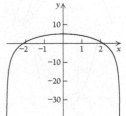

$(2.1448, 7.728)$ and $(-2.1448, -7.728)$

89. $-\dfrac{2(x^2 - 2)}{\sqrt{4 - x^2}}$

SECTION 2.9

1. 0   **3.** $-\dfrac{3}{(x + 1)^3}$ **5.** $-\dfrac{8}{9(2x + 1)^{5/3}}$ **7.** $\dfrac{180}{(4 - 3x)^6}$

9. $-\dfrac{1}{4(x + 1)^{3/2}}$ **11.** $960(2x + 9)^{14}$

13. $9 \sec(3x + 1)[\tan^2(3x + 1) + \sec^2(3x + 1)]$

15. $4(\sec^3(2x + 3) + \sec(2x + 3)\tan^2(2x + 3) + 2)$

17. $2a$ **19.** $\dfrac{3(x^2 + 2)}{4(x^2 + 1)^{5/4}}$

21. $-(4x + 3)\cos x - 8 \sin x$
23. $-a^2 \cos(at + b)$

25. $\left(-\dfrac{8}{(3t^2 + 1)^{5/3}} - \dfrac{1}{7(t^2 + 3)^{3/2}}\right)t^2 + \dfrac{1}{7\sqrt{t^2 + 3}} + \dfrac{2}{(3t^2 + 1)^{2/3}}$

27. 24   **29.** $720x$ **31.** $20(x^2 - 5)^8(19x^2 - 5)$

33. $8 \sec(2x + 3) \tan^3(2x + 3) + 40 \sec^3(2x + 3) \tan(2x + 3)$

35. $a(t) = 36 \sin(3t + 2) - 90 \cos(3t + 2) = -9s(t)$

37. $6t + 2$ **39.** $0.004548t - 0.1192$

41. 200,000   **43.** $\dfrac{x-2}{2(x-1)^{3/2}}, \dfrac{4-x}{4(x-1)^{5/2}}, \dfrac{3(x-6)}{8(x-1)^{7/2}}$

45. $\dfrac{2}{(x-1)^3}$ **47. (a)** $\cos x$ **(b)** $-\sin x$ **(c)** $-\cos x$

(d) $\sin x$ **(e)** $\sin x$ **(f)** $-\sin x$ **(g)** $\cos x$

49. tw   **51.** $-\dfrac{5}{(x-2)^2}, \dfrac{10}{(x-2)^3}, -\dfrac{30}{(x-2)^4},$

$\dfrac{120}{(x-2)^5}, -\dfrac{600}{(x-2)^6}$

53.

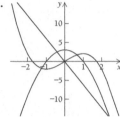

55.

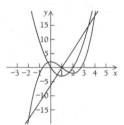

57.

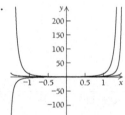

CHAPTER 2 REVIEW

1. (a) [2.1]

x	$f(x)$
-8	-11
-7.5	-10.5
-7.1	-10.1
-7.01	-10.01
-7.001	-10.001
-7.0001	-10.0001
-6	-9
-6.5	-9.5
-6.9	-9.9
-6.99	-9.99
-6.999	-9.999
-6.9999	-9.9999

(b) [2.1] $-10, -10, -10$ **2. [2.1]** -10
3. [2.2] -10 **4. [2.2]** -4 **5. [2.2]** 2
6. [2.2] 17   **7. [2.2]** 5   **8. [2.1]** Not continuous at
$x=-2$ **9. [2.2]** Continuous   **10. [2.1]** -4
11. [2.1] -4 **12. [2.1]** Yes   **13. [2.1]** Does not exist
14. [2.1] -2 **15. [2.1]** No   **16. [2.3]** 2   **17. [2.3]** -3
18. [2.3] $4x+2h$ **19. [2.4]** $y=x-1$
20. [2.5] $(4,5)$ **21. [2.5]** $(5,-108)$

22. [2.5] $20x^4$ **23. [2.5]** $\dfrac{1}{x^{2/3}}$ **24. [2.5]** $\dfrac{64}{x^9}$

25. [2.5] $28\sqrt[3]{x}$ **26. [2.7], [2.8]** $5\sec 5x \tan 5x$
27. [2.7], [2.8] $\cot(x^2) - 2x^2 \csc^2(x^2)$

28. [2.8] $\dfrac{0.46}{\sqrt{0.4x+5.3}} + 0.0017 \cos(0.17x - 0.31)$

29. [2.5] $x^5 + 32x^3 - 5$ **30. [2.5]** $2x$

31. [2.7] $\dfrac{-x^2 + 16x + 8}{(x-8)^2}$ **32. [2.7]** $\dfrac{x \sec^2 x - \tan x}{x^2}$

33. [2.8] $\dfrac{2\sin x \cos x + 1}{2\sqrt{\sin^2 x + x}}$

34. [2.8] $2\sec^2(x\cos x)\tan(x\cos x)(\cos x - x\sin x)$
35. [2.9] $-\csc^2(x - \cos x)(\sin x + 1)$

36. [2.7] $\dfrac{x(11x+6)}{\sqrt[4]{4x+3}}$ **37. [2.9]** $\dfrac{240}{x^6}$

38. [2.9] $20x^3 + \cos x$ **39. (a) [2.6]** $v(t) = 4t^3 + 1$
(b) [2.6] $a(t) = 12t^2$ **(c) [2.6]** $v(2) = 33, a(2) = 48$
40. (a) [2.6]

$$v(t) = 0.036t^2 - 1.85 - 0.0004189 \sin\left(\dfrac{\pi t}{15}\right)$$

(b) [2.6] $a(t) = 0.072t - 0.00008773 \cos\left(\dfrac{\pi t}{15}\right)$

(c) [2.6] $v(2.5) = -1.6252, a(2.5) = 0.1799$
41. (a) [2.8] $67\cos(4\pi t) + 68$ ft above the ground
b) [2.8] 34.5 ft   **(c) [2.8]** 729.147 ft/hr
42. (a) [2.6] $100t$ **(b) [2.6]** 30,000 people
(c) [2.6] 2000 people/year
43. [2.8] $(f \circ g)(x) = (1 - 2x)^2 + 5$,
$(g \circ f)(x) = -2x^2 - 9$
44. [2.8] $\dfrac{-9x^4 - 4x^3 + 9x + 2}{2\sqrt{3x+1}(x^3+1)^2}$

45. [2.2] $-\frac{1}{4}$ **46. [2.2]** $\frac{1}{6}$
47. $(-1.714, 37.445), (0,0), (1.714, 37.445)$

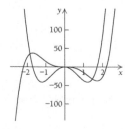

CHAPTER 2 TEST

1. (a) [2.1]

x	$f(x)$
5	11
5.7	11.7
5.9	11.9
5.99	11.99
5.999	11.999
5.9999	11.9999
7	13
6.5	12.5
6.1	12.1
6.01	12.01
6.001	12.001
6.0001	12.0001

(b) [2.1] 12, 12, 12   **2. [2.1]** 12   **3. [2.2]** 12
4. [2.1] Does not exist   **5. [2.1]** 0   **6. [2.1]** Does not exist   **7. [2.1]** 2   **8. [2.1]** 4   **9. [2.1]** 1   **10. [2.1]** 1
11. [2.1] 1   **12. [2.1]** Continuous   **13. [2.1]** Not continuous at $x = 3$ **14. [2.1]** Does not exist
15. [2.1] 1   **16. [2.1]** No   **17. [2.1]** 3   **18. [2.1]** 3
19. [2.1] Yes   **20. [2.2]** 6   **21. [2.2]** $\frac{1}{8}$ **22. [2.2]** Does not exist   **23. [2.3]** $4x + 2h + 3$ **24. [2.4]** $y = \frac{3x}{4} + 2$

25. [2.6] $(0, 0), (2, -4)$ **26. [2.5]** $\frac{5}{\sqrt{x}}$ **27. [2.5]** $\frac{10}{x^2}$

28. [2.5] $\frac{5\sqrt[4]{x}}{4}$ **29. [2.5]** $-x + 0.61$

30. [2.8] $2\sec^2 2x$ **31. [2.7]** $\frac{5}{(x-5)^2}$

32. [2.8] $-\dfrac{5(5x^4 - 12x^2 + 1)}{(x^5 - 4x^3 + x)^6}$

33. [2.8] $\dfrac{2x^2 + 5}{\sqrt{x^2 + 5}}$ **34. [2.8]** $-\dfrac{\sin(\sqrt{x})}{2\sqrt{x}}$

35. [2.8] $\sec 3x(2\sec^2 2x + 3\tan 2x \tan 3x)$

36. [2.8] $\dfrac{\tan x(-2x\sec^2 x + 2\sin^2 x + \tan x + 2)}{(x - \cos^2 x)^2}$

37. [2.8] $-2x\cos(\cos(x^2))\sin(x^2)$

38. [2.8] $2x + \left(\dfrac{x}{\sqrt{x+2}} + 2\sqrt{x+2}\right)\cos(2x\sqrt{x+2})$

39. [2.9] $24x$ **40. (a) [2.6]** $v(t) = \dfrac{2t\cos 2t - \sin 2t}{t^2}$

(b) [2.6] $a(t) = \dfrac{2(1 - 2t^2)\sin 2t - 4t\cos 2t}{t^3}$

(c) [2.6] $\dfrac{3\sqrt{3}}{7\pi}, -\dfrac{6(3\sqrt{3} - 7\pi)}{49\pi^2},$

$-\dfrac{12(-18\sqrt{3} + 42\pi + 49\sqrt{3}\pi^2)}{343\pi^3}$

41. (a) [2.6] $-0.003t^2 + 0.2t$ **(b) [2.6]** 9
(c) [2.6] 1.7
42. [2.8] $(f \circ g)(x) = 4x^6 - 2x^3,$
$(g \circ f)(x) = 2(x - 1)^3 x^3$
43. [2.8] $-2\left(\dfrac{1 + 3x}{1 - 3x}\right)^{1/3} + \left(\dfrac{1 - 3x}{1 + 3x}\right)^{2/3}$
44. [2.2] 27   **45. [2.2]** 2
46. $(1.0836, 25.1029), (2.9503, 8.6247)$

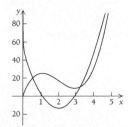

47. [2.1], [2.2] $\frac{1}{2}$

SECTION 3.1

1. Relative minimum at $(2, 1)$

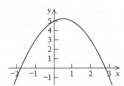

3. Relative maximum at $(1/2, 21/4)$

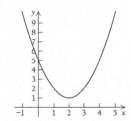

5. Relative minimum at $(-1, -2)$

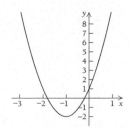

7. Relative maximum at $(-1/3, 59/27)$, relative minimum at $(1, 1)$

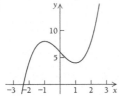

9. Relative maximum at $(-1, 8)$, relative minimum at $(1, 4)$

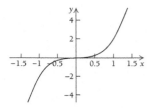

11. No relative minimum or maximum

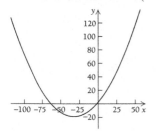

13. Relative minimum at $(-32.5, -18.815)$

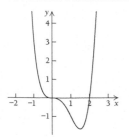

15. Relative minimum at $(3/2, -27/16)$

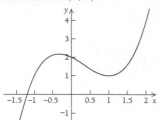

17. Relative minimum at $(-2, -4)$, relative maximum at $(2, 4)$

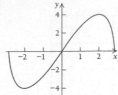

19. Relative maximum at $(0, 1)$

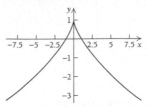

21. Relative minimum at $(0, -8)$

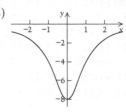

23. Relative minimum at $(-1, -2)$, relative maximum at $(1, 2)$

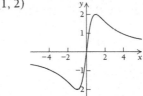

25. No relative maximum or minimum

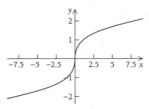

27. Relative minimum at $(-1, 2)$

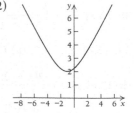

29. Relative maximum at $(0, 1)$

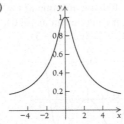

31. Relative maximum at $(\pi/2, 1)$, relative minimum at $(3\pi/2, -1)$

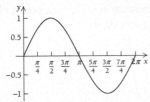

33. Relative maximum at $(3\pi/4, \sqrt{2})$, relative minimum at $(7\pi/4, -\sqrt{2})$

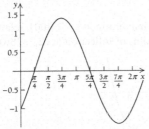

35. Relative maxima at $(0, 1)$, $(\pi, 1)$, and $(2\pi, 1)$; relative minima at $(\pi/2, -1)$ and $(3\pi/2, -1)$

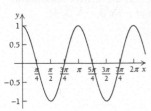

37. Relative maxima at $(\pi/12, 1.128)$ and $(13\pi/12, 4.269)$, relative minima at $(5\pi/12, 0.443)$ and $(17\pi/12, 3.585)$

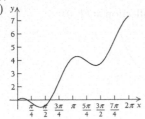

39. Relative maximum at $(\pi/4, 1.128)$, relative minimum at $(5\pi/4, 0.443)$

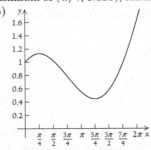

41. Relative maximum at $(\pi/6, \sqrt{3}/3)$, relative minimum at $(5\pi/6, -\sqrt{3}/3)$

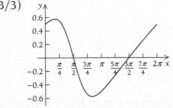

43. Relative maxima at $(\pi/6, 1/4)$ and $(5\pi/6, 1/4)$, relative minima at $(\pi/2, 0)$ and $(3\pi/2, -2)$

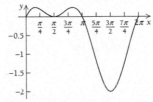

45. Relative maxima at $(\pi/3, 3\sqrt{3})$, $(2\pi/3, 3\sqrt{3})$, and $(3\pi/2, -5)$; relative minima at $(\pi/2, 5)$, $(4\pi/3, -3\sqrt{3})$, and $(5\pi/3, -3\sqrt{3})$

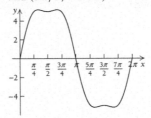

93. 86.6   **95.** 102.2   **97.** tw
99. Relative maximum at $(4, 7)$, relative minimum at $(2, 1)$, critical point at $(1, 3)$
101. Relative minima at $(-5, 425)$ and $(4, -304)$, relative maximum at $(-2, 560)$

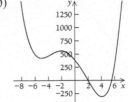

SECTION 3.2

1. Relative maximum at $(0, 2)$, no points of inflection

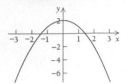

3. Relative maximum at $(-2, 72)$, relative minimum at $(3, -53)$, point of inflection at $(1/2, 19/2)$

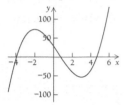

5. Relative maximum at $(-1/2, 1)$, relative minimum at $(1/2, -1/3)$, point of inflection at $(0, 1/3)$

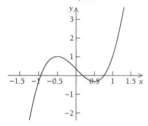

7. Relative minima at $(0, 0)$ and $(3, -27)$, relative maximum at $(1, 5)$, points of inflection approximately at $(0.451, 2.32)$ and $(2.215, -13.358)$

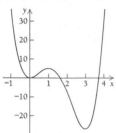

9. Relative minimum and point of inflection at $(-1, 0)$

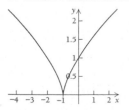

11. Relative minima at $(-\sqrt{3}, -9)$ and $(\sqrt{3}, -9)$, relative maximum at $(0, 0)$, points of inflection at $(-1, -5)$ and $(1, -5)$

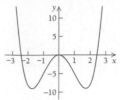

13. Relative minimum at $(-1, -1)$, points of inflection at $(-2/3, -16/27)$ and $(0, 0)$

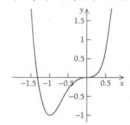

15. Relative maximum at $(-5, 400)$, relative minimum at $(9, -972)$, point of inflection at $(2, -286)$

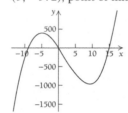

17. Relative minimum at $(-1, -1/2)$, relative maximum at $(1, 1/2)$, points of inflection at $(-\sqrt{3}, -\sqrt{3}/4)$, $(0, 0)$, and $(\sqrt{3}, \sqrt{3}/4)$

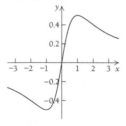

19. Point of inflection at $(1, 0)$

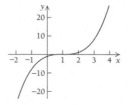

21. Relative minima at $(0, 0)$ and $(1, 0)$, relative maximum at $(1/2, 1/16)$, points of inflection at $((3 - \sqrt{3})/6, 1/36)$ and $((3 + \sqrt{3})/6, 1/36)$

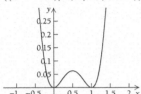

23. Relative minimum at $(-2, -64)$, relative maximum at $(2, 64)$, points of inflection at $(-\sqrt{2}, -28\sqrt{2})$, $(0, 0)$, and $(\sqrt{2}, 28\sqrt{2})$

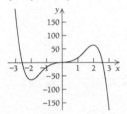

25. Relative minimum at $(-\sqrt{2}, -2)$, relative maximum at $(\sqrt{2}, 2)$, point of inflection at $(0, 0)$

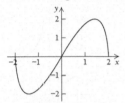

27. Point of inflection at $(1, -1)$

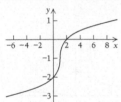

29. Relative maxima at $(\pi/12, 1.128)$ and $(13\pi/12, 4.269)$, relative minima at $(5\pi/12, 0.443)$ and $(17\pi/12, 3.585)$, points of inflection at $(\pi/4, 0.785)$, $(3\pi/4, 2.356)$, $(5\pi/4, 3.927)$, and $(7\pi/4, 5.498)$

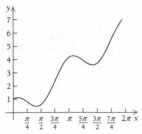

31. Relative minimum at $(\pi/2, -0.342)$, point of inflection at $(3\pi/2, \pi/2)$

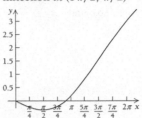

33. Relative maximum at $(\pi/4, \sqrt{2})$, relative minimum at $(5\pi/4, -\sqrt{2})$, points of inflection at $(3\pi/4, 0)$ and $(7\pi/4, 0)$

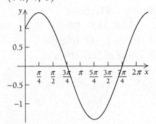

35. Relative maximum at $(\pi/3, 2)$, relative minimum at $(4\pi/3, -2)$, points of inflection at $(5\pi/6, 0)$ and $(11\pi/6, 0)$

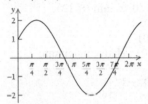

37. Relative maximum at $(\pi/3, \sqrt{3}/3)$, relative minimum at $(5\pi/3, -\sqrt{3}/3)$, points of inflection at $(0, 0)$, $(\pi, 0)$, and $(2\pi, 0)$

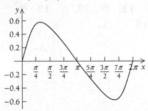

39. Relative maxima at $(0, 1)$, $(\pi, 1)$, and $(2\pi, 1)$; relative minima at $(\pi/2, 0)$ and $(3\pi/2, 0)$; points of inflection at $(\pi/4, 1/2)$, $(3\pi/4, 1/2)$, $(5\pi/4, 1/2)$, and $(7\pi/4, 1/2)$

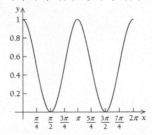

41. Relative maxima at $(0, 1)$, $(\pi, 1)$, and $(2\pi, 1)$; relative minima at $(\pi/2, 0)$ and $(3\pi/2, 0)$; points of inflection at $(\pi/6, 9/16)$, $(5\pi/6, 9/16)$, $(7\pi/6, 9/16)$, and $(11\pi/6, 9/16)$

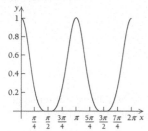

43. 0 **45.** 1/2 **47.** $n\pi$ **49.** $n\pi$ **51.** none
105. $\frac{40}{3}$ **107. (a)** $x = 3.9$, $x = 10.8$ **(b)** tw
109. (a) Relative minimum at $(9.76, 1.44)$, relative maximum at $(56.9, 4.58)$ **(b)** Point of inflection at $(33.33, 3.01)$ **(c)**

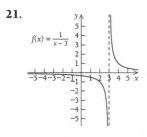

111. tw **113. (a)** Approximately $(2, 2.7)$ **(b)** $(8, 0)$
(c) Approximately $(4, 1)$ **(d)** $(0, 2)$ and $(8, 12)$
(e) $(2, 8)$ **(f)** $(4, 12)$ **(g)** $(0, 4)$
117. Relative minimum at $(0, 0)$
119. Relative maximum at $(0, 0)$, relative minimum at $(0.8, -1.10592)$
121. Relative minimum at $(1/4, -1/4)$

SECTION 3.3

1. $\frac{2}{5}$ **3.** 5 **5.** $\frac{1}{2}$ **7.** $\frac{2}{3}$ **9.** 0 **11.** ∞ **13.** 0 **15.** ∞
17. 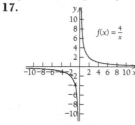 **19.**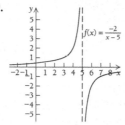

21.

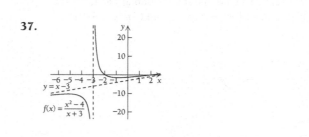

23.

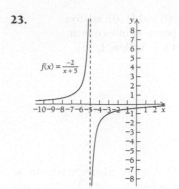

25. 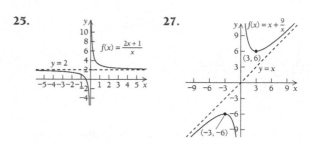 **27.**

29. 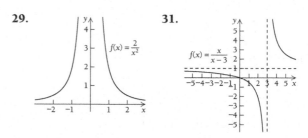 **31.**

33. 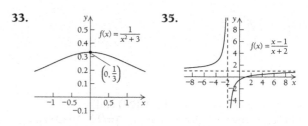 **35.**

37.

39.

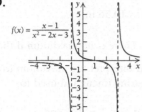

$f(x) = \dfrac{x-1}{x^2 - 2x - 3}$

(c) 101.6   **(d)**

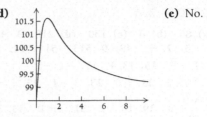

(e) No.

51. (a) 4.00, 4.50, 5.14, 6.00, 7.20, 9.00, 12.00, 18.00, 36.00, 54.00, 108.00;   **(b)** ∞;   **(c)** tw

53. tw   **55.** $\frac{4}{7}$ **57.** $-\infty$ **59.** $\frac{3}{2}$ **61.** $-\infty$

63. undefined   **65.** 0   **67.** ∞   **69.** $-\infty$ **71.** 1

73. $-\infty$ **75.** $-\infty$ **77.** 0

79. $y = \dfrac{x}{\sqrt{x^2 + 1}}$

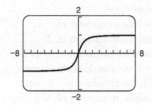

41.

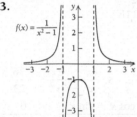

$f(x) = \dfrac{2x^2}{x^2 - 16}$

43.

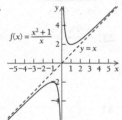

$f(x) = \dfrac{1}{x^2 - 1}$

81. $y = \dfrac{x^3 + 2x^2 - 15x}{x^2 - 5x - 14}$

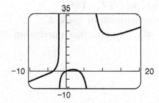

45.

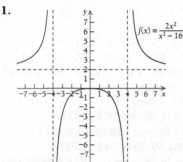

$f(x) = \dfrac{x^2 + 1}{x}$

$y = x$

83. $y = \left| \dfrac{1}{x} - 2 \right|$

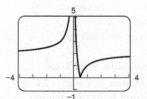

47. (a) \$480, \$600, \$2400, \$4800;   **(b)** ∞;
(c) tw   **(d)**

$C(p) = \dfrac{48,000}{100 - p}$

(e) tw

49. (a) 98.6, 101.6, 101, 99.8, 99.2   **(b)** 98.6

85. (a) $\displaystyle\lim_{x \to \infty} f(x) = 1$, $\displaystyle\lim_{x \to -\infty} f(x) = -1$
(b) undefined   **(c)** $x < -2, -1 < x < 3, x > 3$
(d) Both limits are 0

SECTION 3.4

1. (a) 85 **(b)** 0 **(c)** 150 **(d)** 210 **3.** 4, 1
5. 1, −5 **7.** $\frac{16}{5}$, −48 **9.** 513, −511 **11.** 17, −15
13. 32, −$\frac{27}{16}$ **15.** 13, 4 **17.** −1, −5
19. 20.05, 2 **21.** $\frac{4}{5}$, 0 **23.** 3, −1 **25.** $\frac{1}{2}$, −$\frac{1}{10}$
27. $\frac{3\sqrt{3}}{16}$, 0 **29.** 18, −2 **31.** 2π, 0 **33.** $\frac{1}{3}$, −1
35. $\frac{1}{4}$, 0 **37.** −1 + $\frac{\pi}{2}$, 0 **39.** $\sqrt{2}$, −$\sqrt{2}$
41. Absolute maximum of 1/3, no absolute minimum
43. Absolute maximum of 5700, no absolute minimum
45. No absolute maximum, absolute minimum of 24
47. No absolute maximum, absolute minimum of 108
49. No absolute maximum or minimum
51. No absolute maximum or minimum
53. Absolute maximum of 1, absolute minimum of 0
55. No absolute maximum or minimum
57. No absolute maximum, absolute minimum of −1
59. No absolute maximum, absolute minimum of 2
61. No absolute maximum, absolute minimum of $\sqrt{2}/2$
63. Absolute maximum of −1.46, no absolute minimum
65. Absolute maximum of −$\sqrt{3}$, no absolute minimum
67. No absolute maximum, absolute minimum of $\sqrt{3}$
69. Absolute maximum of −1, no absolute minimum
71. 61.64 **73.** 7.18
75. (a) 0.139443$\sqrt{h}$ − 0.238382$h^{0.5378}$
(b) Absolute maximum of 0.041, absolute minimum
of −0.158. If a person weighs 70 kg, then the two
formulas are at most 0.158 different. **77. tw**
79. (b) $\sqrt{2/3}$ **81.** Absolute maximum of −2$\sqrt[3]{2}$,
absolute minimum of −3$\sqrt[3]{4}$ **83.** No absolute maximum
or minimum

SECTION 3.5

1. 25 and 25; maximum product = 625
3. No; $Q = x(50 − x)$ has no minimum
5. 2 and −2; minimum product = −4
7. $x = \frac{1}{2}$, $y = \sqrt{\frac{1}{2}}$; maximum = $\frac{1}{4}$
9. $x = 3$, $y = 2$; minimum = 30 **11.** $x = 10$, $y = 10$;
minimum = 200 **13.** $x = 2$, $y = \frac{32}{3}$; maximum = $\frac{64}{3}$
15. $x = 1/2$, $y = 1/2$; maximum = $\sqrt{2}$
17. 30 yd by 60 yd; maximum area = 1800 yd^2
19. 13.5 ft by 13.5 ft; 182.25 ft^2
21. 20 in. by 20 in. by 5 in.; maximum = 2000 in^3
23. 5 in. by 5 in. by 2.5 in.; minimum = 75 in^2 **25.** $\frac{\pi}{2}$
27. \$5.75, 72,500 (Will the stadium hold that many?)
29. 25 **31.** 4 ft by 4 ft by 20 ft **33.** 9%
35. (a) 3.1132 **(b)** 39.13
37. 14 in. by 14 in. by 28 in. **39.** $x = y = \dfrac{24}{4 + \pi}$ ft
41. $\sqrt[3]{\dfrac{1}{10}}$

43. Minimum at $x = \dfrac{24\pi}{\pi + 4} \approx 10.56$ in.,

$24 − x = \dfrac{96}{\pi + 4} \approx 13.45$ in. There is no maximum if the
string is to be cut. One would interpret the maximum to
be at the endpoint, with the string uncut and used to
form a circle.
47. Minimum = 6 − 4$\sqrt{2}$ at $x = 2 − \sqrt{2}$ and
$y = −1 + \sqrt{2}$

SECTION 3.6

1. 6x − 9 **3.** −$\frac{x}{16}$ + $\frac{1}{2}$ **5.** 3x − 4 **7.** 1 **9.** x
11. 4.375 **13.** 9.955 **15.** 2.16667 **17.** 9.85 **19.** 0.1
21. −0.04 **23.** 0.381966 **25.** −2.1038 **27.** 2.22677
29. 0.514933 **31.** 0.456625
33. −1.1425, 0.176245, 4.96625
35. −2.8664, −0.566825, 0.408647, 6.02457
37. 0.0545184 cc, 0.152295 cc **39.** 4.3488 mo
41. 1960 **43.** 57.0982, 97.5041 **45.** 1.03272
47. 1.08846 **49. (a)** 11.0856 cm/s **(b)** 0.628697 cm/s
(c) tw **51. (a)** 0 **(b) tw**
53. 259.694, 1925.96, 2950 Hz
55. 797.912, 1290.36, 2840.38 Hz

SECTION 3.7

1. $\dfrac{1 − y}{x + 2}$; −$\dfrac{1}{9}$ **3.** −$\dfrac{x}{y}$; −$\dfrac{1}{\sqrt{3}}$ **5.** $\dfrac{6x^2 − 2xy}{x^2 − 3y^2}$; −$\dfrac{36}{23}$
7. $\dfrac{−2x}{\sin y + \cos y}$, −2 **9.** $\dfrac{x\cos x + \sin x}{1 + \cos y − y\sin y}$, $\dfrac{6}{9 − \pi\sqrt{3}}$
11. −$\dfrac{y}{x}$ **13.** $\dfrac{x}{y}$ **15.** $\dfrac{3x^2}{5y^4}$ **17.** $\dfrac{−3xy^2 − 2y}{3x + 4x^2 y}$
19. −$\sqrt{\dfrac{y}{x}}$ **21.** $\dfrac{2}{3(x + 1)^2 y^2}$ **23.** $\dfrac{−9\sqrt{x}\,y^{1/3}}{4}$
25. $\dfrac{2 − y − 2xy}{x^2 + x − 1}$ **27.** cot x cot y
29. $\dfrac{\cos(\sqrt{y} − x) + 2\sqrt{y} − x}{\cos(\sqrt{y} − x) − 2\sqrt{y} − x}$ **31.** −$\frac{3}{4}$
33. 0.1728π cm^3/day $\approx$ 0.54 cm^3/day
35. (a) 1000$R \cdot \dfrac{dR}{dt}$; **(b)** −0.01125 mm^3/min
37. $\dfrac{−1}{5\sqrt{17}}$ m^2/mo **39.** 65 mph **41.** $\frac{40}{3}$ m/min
43. (a) $\dfrac{20}{\sqrt{91}} \approx$ 2.1 ft/sec **(b)** $\dfrac{3}{5\sqrt{91}}$ rad/sec $\approx$ 3.6°/sec
45. $\dfrac{dy}{dx} = \dfrac{1 + y}{2 − x}$, $\dfrac{d^2y}{dx^2} = \dfrac{2 + 2y}{(2 − x)^2}$
47. $\dfrac{dy}{dx} = \dfrac{x}{y}$, $\dfrac{d^2y}{dx^2} = \dfrac{y^2 − x^2}{y^3}$ **49. tw**

51.

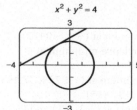

$x^2 + y^2 = 4$

53.

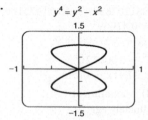

$y^4 = y^2 - x^2$

55.

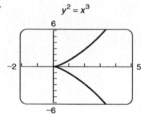

$y^2 = x^3$

CHAPTER 3 REVIEW

1. [3.2] Relative maximum at $(-1, 4)$

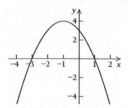

2. [3.2] Relative maximum at $(0, 3)$, relative minima at $(-1, 2)$ and $(1, 2)$, points of inflection at $(1/\sqrt{3}, 22/9)$ and $(-1/\sqrt{3}, 22/9)$

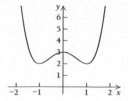

3. [3.2] Relative maximum at $(-1, 4)$, relative minimum at $(1, -4)$, points of inflection at $(-\sqrt{3}, 2\sqrt{3})$, $(0, 0)$, and $(\sqrt{3}, -2\sqrt{3})$

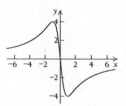

4. [3.2] Point of inflection at $(1, 4)$

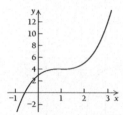

5. [3.2] Relative maximum at $(3\pi/2, 1/5)$, relative minimum at $(\pi/2, 1/9)$, points of inflection at $(7\pi/6, 1/6)$ and $(11\pi/6, 1/6)$

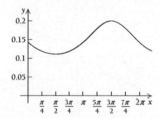

6. Relative minimum and point of inflection at $(0, 0)$

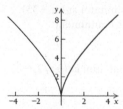

7. [3.2] Relative maximum at $(\pi/2, 3)$, relative minimum at $(3\pi/2, -1)$, points of inflection at $(\pi/6, 5/4)$ and $(5\pi/6, 5/4)$

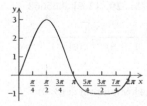

8. [3.2] Points of inflection at $(\pi/4, \pi/2)$, $(3\pi/4, 3\pi/2)$, $(5\pi/4, 5\pi/2)$ and $(7\pi/4, 7\pi/2)$

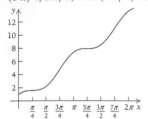

9. [3.3]

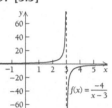

10. [3.3]

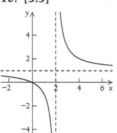

11. [3.3]

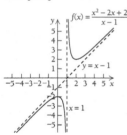

12. [3.3]

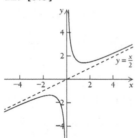

13. [3.4] None
14. [3.4] Maximum at $(-2, 32)$, minimum at $(5, -17)$
15. [3.4] Maximum at $(-1, 19)$, minimum at $(2, -35)$
16. [3.4] Maximum at $(2\pi/3, 0.366)$, minimum at $(4\pi/3, -1.366)$
17. [3.4] Minimum at $(-1, 3)$
18. [3.4] Maximum at $(\pi/2, 7)$, minimum at $(3\pi/2, -1)$
19. [3.4] Minimum at $(\pi/6, 6)$
20. [3.4] Minimum at $(1, 10)$
21. [3.4] Maximum at $(\pi/3, -\sqrt{3})$
22. [3.4] Maximum at $(5/2, 53/4)$ **23. [3.5]** 30, 30
24. $x = -1$, $y = -1$; minimum $= -1$
25. [3.5] 10 ft $\times$ 10 ft $\times$ 25 ft; \$1500   **26. [3.6]** $4 - x$

27. [3.6] $\dfrac{x}{2}$ **28. [3.6]** 7.9375   **29. [3.6]** 0.65662

30. [3.6] 2.47948   **31. [3.7]** $\dfrac{2\sqrt{y}}{1 - 4y^{3/2}\cos(y^2)}$

32. $\dfrac{dy}{dx} = \dfrac{-2x^2 - 3y}{3x + 2y^2}$; $\dfrac{4}{5}$

33. [3.7] $-1\frac{3}{4}$ ft/sec   **34. [3.4]** Minimum at $(3, 0)$

35. [3.7] $\dfrac{3x^5 - 4x^3 - 12xy^2}{12x^2y + 4y^3 - 3y^5}$

36. [3.1] [3.2] Relative minimum at $(-9, -9477)$; relative maximum at $(0, 0)$; relative minimum at $(15, -37{,}125)$
37. [3.1], [3.2] Relative maximum at $(-1.714, 37.445)$; relative minimum at $(1.714, -37.445)$
38. [3.1] [3.2] Relative minima at $(-3, -1)$ and $(3, -1)$; relative maximum at $(0, 1.08)$
39. (a) Linear: $y = 6.9982x - 124.6181$; quadratic: $y = 0.0439274846x^2 + 2.881202838x - 53.51475166$; cubic: $y = -0.003344x^3 + 0.4796x^2 - 11.35931622x + 5.276985809$;

quartic: $y = -0.0000554x^4 + 0.00672x^3 - 0.0997x^2 - 0.841x - 0.246$;

(b) tw ;   **(c) tw** ;   **(d)** Maximum at $(78.966, 466.325)$; approximately age 79.

CHAPTER 3 TEST

1. [3.2] Relative maximum at $(2, -9)$

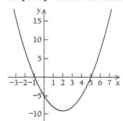

2. [3.2] Relative maximum at $(0, 1)$, relative minima at $(-1, -1)$ and $(1, -1)$, points of inflection at $(-1/\sqrt{3}, -1/9)$ and $(1/\sqrt{3}, -1/9)$

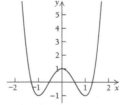

3. [3.2] Relative minimum and point of inflection at $(2, -4)$

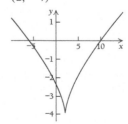

4. [3.2] Relative maximum at $(0, 4)$, points of inflection at $(-2/\sqrt{3}, 3)$ and $(2/\sqrt{3}, 3)$

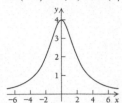

5. [3.2] Relative maximum at $(3\pi/2, 1/9)$, relative minimum at $(\pi/2, -1/5)$, points of inflection at $(\pi/6, -1/12)$ and $(5\pi/6, -1/12)$

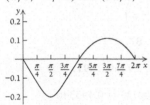

6. [3.2] Relative maximum at $(3\pi/2, 2)$, relative minimum at $(\pi/2, -2)$, points of inflection at $(\pi/6, -1/4)$ and $(5\pi/6, -1/4)$

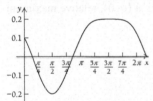

7. [3.2] Point of inflection at $(-2, 0)$

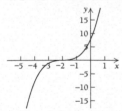

8. [3.2] Relative maximum at $(3/\sqrt{2}, 9/2)$, relative minimum at $(-3/\sqrt{2}, -9/2)$, point of inflection at $(0, 0)$

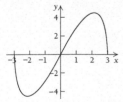

9. [3.3]

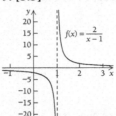

10. [3.3]

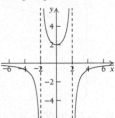

11. [3.3]

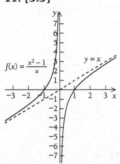

12. [3.3]

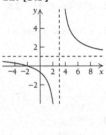

13. [3.4] Maximum at $(3, 9)$
14. [3.4] Maximum at $(-1, 2)$, minimum at $(-2, -1)$
15. [3.4] Maximum at $(7\pi/6, 5/4)$ and $(11\pi/6, 5/4)$, minimum at $(\pi/2, -1)$
16. [3.4] Maximum at $(\pi/3, 3\sqrt{3}/4)$, minimum at $(5\pi/3, -3\sqrt{3}/4)$
17. [3.4] Minimum at $(\pi/6, 16)$
18. [3.4] Maximum at $(0, 0)$
19. [3.4] Minimum at $(4, 48)$
20. [3.5] 4 and -4
21. [3.5] $x = 5$, $y = -5$; minimum $= 50$
22. [3.5] 40 in. by 40 in. by 10 in.; maximum volume $= 16,000$ in^3
23. [3.6] $\dfrac{13x}{3} - \dfrac{32}{3}$ **24. [3.6]** $2x + 1$ **25. [3.6]** 10.2
26. [3.6] -1.15417 **27. [3.6]** 0.456625
28. [3.7] $-\dfrac{x^2}{y^2}, -\dfrac{1}{4}$ **29. [3.7]** -0.96 ft/sec
30. [3.4] Maximum at $(\sqrt[3]{2}, \frac{1}{3}\sqrt[3]{4})$, minimum at $(0, 0)$
31. [3.1], [3.2] Relative maximum at $(1.09, 25.1)$; relative minimum at $(2.97, 8.6)$

SECTION 4.1

1.

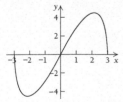

3.

5.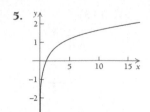

7. 1200, 1440, 2488

77. (a) 0 ppm, 3.68 ppm, 5.41 ppm, 4.48 ppm, 0.05 ppm

(b)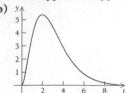

(c) $20te^{-1} - 10t^2 e^{-t}$

9. 292,006,000; 298,138,000; 317,318,000 **11.** $3e^{3x}$

13. $-10e^{-2x}$ **15.** e^{-x} **17.** $-7e^x$ **19.** e^{2x}

21. $e^x x^3(x + 4)$ **23.** $e^x(x^2 + 5x - 6)$

25. $e^x(\cos x + \sin x)$ **27.** $\dfrac{e^x(x-4)}{x^5}$

29. $e^{-x^2+7x}(7 - 2x)$ **31.** $-xe^{-x^2/2}$ **33.** $\dfrac{e^{\sqrt{x-7}}}{2\sqrt{x-7}}$

35. $\dfrac{e^x}{2\sqrt{e^x - 1}}$ **37.** $e^x \sec^2(e^x + 1)$

39. $e^{\tan x} \sec^2 x$

41. $e^{3x+1}(6x + 3\cos x - \sin x + 2)$

43. $e^{-2x} - 2xe^{-2x} - e^{-x} + 3x^2$ **45.** e^{-x}

47. ke^{-kx} **49.** $15e^{3x}(1 + e^{3x})^4$ **51.** $-e^{-t}(1 + 3e^{4t})$

53. $\dfrac{e^x(x-1)^2}{(x^2+1)^2}$ **55.** $\dfrac{1}{2}\left(\sqrt{e^x} + \dfrac{e^{\sqrt{x}}}{\sqrt{x}}\right)$

57. $\dfrac{xe^{x/2}}{2\sqrt{x-1}}$ **59.** $\dfrac{4e^{2x}}{(1+e^{2x})^2}$

61. $f(x) = e^{2x}$

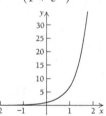

(d) $C \approx 5.4$ ppm at $t = 2$ hr **(e)** **tw**

79. (a) $0.153 \left(e^{0.141T-1.3395} - e^{0.153T-1.8123}\right)$

(b) $0.021573e^{0.141T-1.3395} - 0.023409e^{0.153T-1.8123}$

81. $\frac{1}{2}e^{3x^3+2x-1}\sqrt{x}(18x^3 + 4x + 3)$

83. $\dfrac{1}{2}\left(\dfrac{e^x(x+1)}{\sqrt{xe^x}} + \sqrt{e^x} + \dfrac{1}{\sqrt{x}}\right)$

85. $-e^x \cos(\cos(e^x)) \sin(e^x)$ **87.** $e^{x+e^{1+e^x}+e^x+2}$

89. 2, 2.25, 2.48832, 2.59374, 2.71692 **91.** $f(2) = \dfrac{4}{e^2}$

93. (a) 0 **(b)** 0 **(c)** $f(31.0071) \approx 0.495346$

95. (a) 4.243 per thousand, 3.938 per thousand

(b) 3.92 per thousand **97.** 0.417091

99. -0.0215 per hr **101.** **tw**

103. Left to the student

105. $e^{-x}x^2$; relative minimum at $(0, 0)$, relative maximum at $(2, 0.5413)$

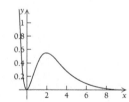

63. $f(x) = e^{-2x}$

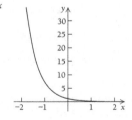

107. e^x; all graphs are identical.

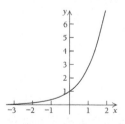

65. $f(x) = 3 - e^{-x}$

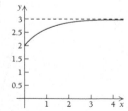

67.–72. Left to the student **73.** $y = x + 1$

75. Left to the student

109. $2e^{0.3x}$, $0.6e^{0.3x}$, $0.18e^{0.3x}$

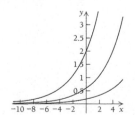

SECTION 4.2

1. $2^3 = 8$ **3.** $8^{1/3} = 2$ **5.** $a^J = K$ **7.** $10^{-p} = h$
9. $\ln b = M$ **11.** $\log_{10} 100 = 2$ **13.** $\log_{10} 0.1 = -1$
15. $\log_M V = p$ **17.** 2.708   **19.** -1.609 **21.** 2.609
23. 2.9957   **25.** -1.3863 **27.** 4   **29.** 8.27563
31. -4.00633 **33.** 8.99962   **35.** 4.6052   **37.** 4.0943

39. 2.3026   **41.** 140.671   **43.** $-\dfrac{6}{x}$

45. $x(4x^2 \ln x + x^2 - 1)$ **47.** $\dfrac{1 - 4 \ln x}{x^5}$ **49.** $\dfrac{1}{x}$

51. $-\tan x$ **53.** $\dfrac{1}{x \ln 4x}$ **55.** $\dfrac{x^2 + 7}{x(x^2 - 7)}$

57. $e^x \left(\dfrac{1}{x} + \ln x \right)$ **59.** $e^x \sec x(\tan x + 1)$

61. $\dfrac{e^x}{1 + e^x}$ **63.** $\dfrac{2 \ln x}{x}$ **65.** $-\dfrac{4}{x \ln^5 x}$ **67.** $\dfrac{15t^2}{t^3 + 1}$

69. $\dfrac{4 \ln^3(x + 5)}{x + 5}$ **71.** $\dfrac{t(5t^3 - 3t + 6)}{t^5 - t^3 + 3t^2 - 3}$ **73.** $\dfrac{24x + 25}{8x^2 + 5x}$

75. $\dfrac{\cos x(1 - \ln(\sin x))}{\sin^2 x}$ **77.** $x^n \ln x$

79. $\dfrac{1}{\sqrt{t^2 + 1}}$ **81.** $\dfrac{\cos(\ln x)}{x}$

83. $\cos x \ln(\tan x) + \sec x$ **85.** $2 \sec 2x$

87. $\dfrac{4e^{2x}}{-1 + e^{4x}}$

89. $5.17x^{1.50}$

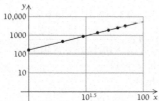

91. $16.5x^{0.136}$

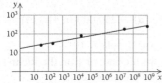

93. $337{,}000{,}000x^{-1.23}$

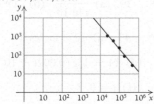

95. (a) 18.1269, 69.8806   **(b)** $20e^{-0.2t}$ **(c)** 11.5 mo
(d) tw   **97. (a)** 68%   **(b)** 35.8%   **(c)** 3.6%

(d) 5.3   **(e)** $\dfrac{-20}{t + 1}$ **(f)** Maximum 68% at 0 mo

(g) tw   **99. (a)** 2.37 ft/sec   **(b)** 3.37 ft/sec   **(c)** $\dfrac{0.37}{p}$

(d) tw   **101.** $\dfrac{3}{x \ln(x)}$ **103.** 1   **105.** $\dfrac{1}{k} \ln \dfrac{P}{P_0}$

107. Left to the student
109. (a) $0.124(e^{0.129T - 1.2255} - e^{0.144T - 1.848})$
(b) $0.015996e^{0.129T - 1.2255} - 0.017856e^{0.144T - 1.848}$

(c) 34.2°C   **111.** $T_U - \dfrac{1}{b - c} \ln \dfrac{b}{c}$

113. (a), (b)   **(c)** tw

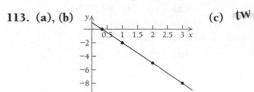

115. (a)

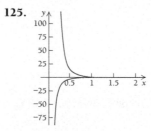

(b) $m = 0.00656, b = -12.3$ **(c)** $N = 10^{(0.00656t - 12.3)}$
117. tw
119. $e^{1/e} \approx 1.44467, e^{1/e} \geq x^{1/x}$ for any $x > 0$ **121.** ∞
123.

125.

127. $f\left(\dfrac{1}{\sqrt{e}} \right) = -\dfrac{1}{2e} \approx -0.1839$ **129.** $21.5x^{0.364}$

131. $132{,}000{,}000x^{-0.958}$ **133.** 203.188   **135.** 204.411
137. 84 and 22,444   **139.** 86 and 9503

SECTION 4.3

1. $Q = ce^{kt}$ **3.** $y = 5e^{2t}$ **5. (a)** $P = 1000e^{0.033t}$
(b) 2691   **(c)** 7243   **(d)** 4.34×10^{23} **(e)** 21 min

7. (a) 1.4748 **(b)** 16692 **(c)** 4.7×10^{17}
(d) 0.7449 hr **9.** 381 **11. (a)** $281e^{0.009t}$ million
(b) 321.6 million **(c)** 77.02 yr
13. (a) $P_0e^{0.065t}$ **(b)** $1067.16, $1138.83
(c) 10.66 yr **15.** 6.93% **17.** 19.8 yr
19. 10.001% **21.** 24.81 yr **23.** 20.4768%
25. (a) 2.2279% **(b)** tw **27. (a)** $47432e^{0.0552846}$
(b) 752,596 **(c)** 2015
(d) 12.54 yr **(e)** Left to the student
29. (a) $85e^{0.062174t}$ **(b)** 116 **(c)** 39.65 hr
(d) 11.15 hr **(e)** Left to the student
31. (a) $14.88 \cdot 0.999960^t$ **(b)** $14.77e^{-0.000040t}$
(c) $14.77e^{-0.000039t}$ **(d)** $-0.00059584e^{-0.000040t}$

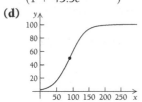

(e) tw **(f)** tw
33. (a) 61.82%, 93.13%, 99.99%
(b) $\dfrac{184.025e^{-0.0425x}}{(1 + 43.3e^{-0.0425x})^2}$ **(c)** (88.662, 50)
(d)

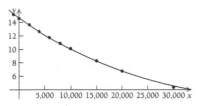

35. (a) 0.17 cm^2, 1.27 cm^2, 8.72 cm^2
(b) $\dfrac{138,180e^{-0.04x}}{(1 + 32,900e^{-0.04x})^2}$ **(c)** (260.031, 52.5)
(d)

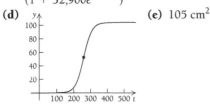

(e) 105 cm^2

37. (a) 0%, 32.97%, 55.07%, 69.88%, 86.47%, 99.18%,
99.83% **(b)** $40e^{-0.4t}$
(c)

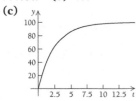

39. 7.57% **41.** 8.84% **43.** $T_3 = \dfrac{\ln 3}{k}$

45. Answers vary. **47.** 0.0289 per hr **49.** tw
51. tw

SECTION 4.4

1. d) **3.** a) **5.** c) **7.** 3.15% per yr
9. 25.7% per hr **11.** 0.00286% per yr **13.** 11.43 g
15. 19,101 yr **17.** 7601 yr **19.** 70.04%
21. 4.55×10^9 yr **23.** 7.99×10^8 yr
25. (a) $A = A_0e^{-kt}$ **(b)** 9 hr **27. (a)** 0.797%
(b) 78.7% **29.** 24.66%, 6.08%, 1.5%
31. (a) 25°C **(b)** 0.0511 per min
(c) 84°C **(d)** 31.5 min **(e)** tw
33. Approximately 7:04 P.M.
35. (a) 1.026%, $p = 51.9e^{-0.01026t}$, $t =$ years since 1995
(b) 42.3 million **(c)** 1731 yr **37. (a)** 11.2 watts
(b) 173.29 days **(c)** 402.36 days **(d)** 50 watts
(e) tw **39. (a)** $84.9 - 0.541 \ln x$
(b) 83.8%, 83.7%, 83.2%, 83% **(c)** 230 mos **(d)** tw
41. (a) 0.4%, $y = 52e^{-0.004t}$, $t =$ years since 1980,
$y =$ lb of beef consumed **(b)** 46.1 lb **(c)** 2392

SECTION 4.5

1. $e^{4 \ln 5}$ **3.** $e^{10 \ln 3.4}$ **5.** $e^{k \ln 4}$ **7.** $e^{kT \ln 8}$
9. $6^x \ln 6$ **11.** $10^x \ln 10$ **13.** $(6.2)^x (x \ln 6.2 + 1)$
15. $10^x x^2(x \ln 10 + 3)$ **17.** $\dfrac{1}{x \ln 4}$ **19.** $\dfrac{2}{x \ln 10}$
21. $\dfrac{1}{x \ln 10}$ **23.** $\dfrac{x^2(3 \ln x + 1)}{\ln 8}$
25. $\dfrac{\csc x(1 - x \cot x \ln x)}{x \ln 2}$ **27.** $\dfrac{\cot x}{\ln 10}$
29. $\dfrac{-\ln 3}{x \ln^2 x}$ **31.** 0 **33. (a)** $-250,000(\ln 4)(1/4)^t$
(b) tw **35.** 5 **37. (a)** $I = I_0 \times 10^7$
(b) $I = I_0 \times 10^8$ **(c)** An increase of 1 on the Richter
increases the intensity by a factor of 10.
(d) $I_0 10^R \ln 10$ **(e)** tw **39. (a)** $\dfrac{1}{I \ln 10}$ **(b)** tw
41. (a) $\dfrac{m}{x \ln 10}$ **(b)** tw **43. (a)** 10^{-7} moles/l
(b) 10^{-7} moles/l **(c)** 7 **(d)** $14 + \log(0.002t + 10^{-7})$
(e) $\dfrac{0.002}{(0.002t + 10^{-7}) \ln 10}$ **(f)** $t = 0, 7$
45. $x^x(\ln x + 1)$ **47.** $x^{e^x}\left(e^x \ln x + \dfrac{e^x}{x}\right)$
49. $\dfrac{f'(x)}{f(x) \ln a}$ **51.** tw

CHAPTER 4 REVIEW

1. [4.2] $\dfrac{1}{x}$ 2. [4.1] e^x 3. [4.2] $\dfrac{4x^3}{x^4 + 5}$

4. [4.1] $\dfrac{e^{2\sqrt{x}}}{\sqrt{x}}$ 5. [4.2] $\dfrac{\cos x + 1}{x + \sin x}$

6. [4.1] $4(x^3 + e^{4x})$ 7. [4.2] $\dfrac{\cot x}{x} - \csc^2 x \ln x$

8. [4.1] $\dfrac{e^{x^2}(2x^2 \ln 4x + 1)}{x}$ 9. [4.1–2] $4e^{4x} - \dfrac{1}{x}$

10. [4.2] $\dfrac{8(x^8 - 1)}{x}$ 11. [4.2] $e^{-x}(1 - x)$

12. [4.2] 6.93 13. [4.2] −3.2698 14. [4.2] 8.7601
15. [4.2] 3.2698 16. [4.2] 2.54995 17. [4.2]
−3.6602 18. [4.3] $Q = Q_0 e^{kt}$ 19. [4.3] 4.3%
20. [4.3] 10.2 yr 21. (a) [4.3] $C = 4.65e^{0.0395t}$
(b) [4.3] $37.66, $45.88 22. (a) [4.3] $N = 60e^{0.12t}$
(b) [4.3] 199 (c) [4.3] 5.8 yr 23. [4.4] 5.3 yr
24. [4.4] 18.2% per day 25. (a) [4.4] $800e^{-0.07t}$
(b) [4.4] 197 g (c) [4.4] 9.9 days
26. [4.3] 0.5, 0.75, 0.97, 0.9991, 0.99994

(b) $0.7e^{-0.7t}$ (c) **tw** (d)

Graph: curve in region with y-axis from 0.6 to 0.9, x-axis 2, 4, 6, 8, 10, 12.

27. [4.5] $3^x \ln 3$ 28. [4.5] $\dfrac{1}{x \ln 15}$

29. [4.1] $-\dfrac{8e^{4x}}{(e^{4x} - 1)^2}$

30. [4.2] $f\!\left(\dfrac{1}{4e^{1/4}}\right) = -\dfrac{1}{1024e}$

31. [4.1]   32. [4.1] 0

Graph: curve with y-axis marks 0.05, 0.1, 0.15, 0.2, 0.25 and x-axis −4, −2, 2, 4.

33. a) [4.2] $984(1.039)^t = 984e^{0.0387t}$, $k = 3.87\%$ per day
b) 47,110; 2,254,840 c) 17.9 days
d) 404.7 days e) **tw**

CHAPTER 4 TEST

1. [4.1] e^x 2. [4.2] $\ln x \sec^2 x + \dfrac{\tan x}{x}$

3. [4.1] $-2xe^{-x^2}$ 4. [4.2] $\dfrac{1}{x}$ 5. [4.1] $e^x - 15x^2$

6. [4.1] $\dfrac{3e^x(x \ln x + 1)}{x}$

7. [4.1–4.2] $\dfrac{e^x - \cos x}{e^x - \sin x}$

8. [4.1–4.2] $\dfrac{e^{-x}(1 - x \ln x)}{x}$ 9. [4.2] 1.0674

10. [4.2] 0.5554 11. [4.2] 0.4057
12. [4.3] $M = M_0 e^{kt}$ 13. [4.3] 17.3% per hr
14. [4.3] 10 yr 15. (a) [4.3] $k = 3.31\%$ per yr,
$C = 0.748e^{0.0331t}$ (b) $1.31, $1.83
(c) Answers vary. 16. (a) [4.3] $3e^{-0.1t}$
(b) 1.1 cc (c) 6.93 hrs
17. [4.4] 83.5 days 18. [4.4] 0.00003% per yr
19. (a) [4.4] 4% (b) 5.22%, 14.45%, 40.66%, 73.54%,
91.85%, 99.46%, 99.87%

(c) $\dfrac{672e^{-0.28t}}{(24 + e^{-0.28t})^2}$ (d) **tw**

(e)

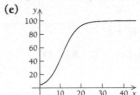

20. [4.5] $20^x \ln 20$ 21. [4.5] $\dfrac{1}{x \ln 20}$
22. [4.2] $\ln^2(x)$
23. [4.2] Maximum: $f(4) = 256e^{-4} \approx 4.6888$;
minimum: $f(0) = 0$
24. [4.1]

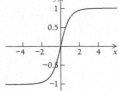

25. [4.1] 0
26. (a) [4.3] $y = 0.027(19.2)^t = 0.027e^{2.96t}$
(b) 1352605, 25970025 (c) 1.3 hrs (d) 0.0354 hrs

SECTION 5.1

1. $\dfrac{x^7}{7} + C$ 3. $2x + C$ 5. $\dfrac{4x^{5/4}}{5} + C$

7. $\dfrac{x^3}{3} + \dfrac{x^2}{2} - x + C$ 9. $\dfrac{t^3}{3} - t^2 + 3t + C$

11. $\dfrac{5e^{8x}}{8} + C$ 13. $\dfrac{w^4}{4} - \dfrac{7w^{15/7}}{15} + C$

15. $1000 \ln |r| + C$ 17. $-\dfrac{1}{x} + C$ 19. $\dfrac{2s^{3/2}}{3} + C$

21. $-18x^{1/3} + C$ 23. $-4e^{-2x} + C$

25. $\dfrac{x^3}{3} - x^{3/2} - \dfrac{3}{\sqrt[3]{x}} + C$ 27. $-\dfrac{5 \cos(2\pi\theta)}{2\pi} + C$

29. $-\cos 5x - 2\sin 2x + C$ **31.** $\tan 3x + C$

33. $3\sec\dfrac{x}{9} + C$ **35.** $\sec x + \tan x + C$

37. $\ln|t| + e^{-t} - \dfrac{1}{t} + C$ **39.** $\dfrac{x^2}{2} - 3x + 13$

41. $\dfrac{x^3}{3} - 4x + 7$ **43.** $\dfrac{2}{3}\sin(3x) + 1$ **45.** $\dfrac{5}{2}(e^{2x} - 5)$

47. $158{,}553$ **49.** $t^3 + 4$ **51.** $2t^2 + 20$

53. $-\dfrac{t^3}{3} + 3t^2 + 6t + 10$ **55.** $s(t) = -16t^2 + v_0t + s_0$

57. 0.25 mi   **59.** 127.18 ft   **61. (a)** $0.1t^2 - 0.001t^3$
(b) 6 **63. (a)** $1016e^{0.0376t} + 606$ **(b)** 3745

(c) $10{,}303$ **65.** $\dfrac{t^{\sqrt{3}+1}}{\sqrt{3}+1} + 8$ **67.** $\dfrac{x^6}{6} - \dfrac{2x^5}{5} + \dfrac{x^4}{4} + C$

69. $\dfrac{2}{5}\sqrt{t}(t^2 + 10t + 45) + C$

71. $\dfrac{t}{4} + t^3 + \dfrac{3t^2}{2} + t + C$ **73.** $\dfrac{be^{ax}}{a} + C$

75. $\dfrac{12}{7}x^{7/3} + C$ **77.** $\dfrac{t^3}{3} - t^2 + 4t + C$

79. $\sin(x) + C$ **81.** $-x - \dfrac{1}{2}\cot(2x) + C$ **83. tw**

SECTION 5.2

1. (a) 0.5118 **(b)** 0.652 **3.** Left to the student
5. Left to the student   **7.** Left to the student
9. Left to the student   **11.** 3.75 **13.** 4.5833
15. 1.8961 **17. tw**   **19.** 146 **21.** $-80{,}000$
23. 190 **25.** 2 **27.** 1 **29.** 64 **31.** $\dfrac{26}{3}$ **33.** $\dfrac{14}{3}$
35. 1 **37.** 9 **39.** 1 **41.** $\dfrac{255}{4}$

43. 2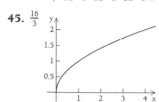

45. $\dfrac{16}{3}$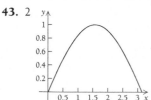

47. $-2 + \ln 64 \approx 2.158883$

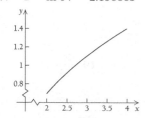

SECTION 5.3

1. 8 **3.** 8 **5.** $\dfrac{125}{3}$ **7.** $\dfrac{1}{4}$ **9.** $\dfrac{32}{3}$ **11.** $e^3 - 1$
13. $\ln(216)$ **15.** 0 **17.** -1
19.–35. Left to the student
37. $e^b - e^a$ **39.** $b^3 - a^3$ **41.** $\dfrac{1}{2}(1 + e^2)$ **43.** $\dfrac{5}{8}$

45. $-\dfrac{75}{2}$ **47.** 4 **49.** $\dfrac{59}{6}$ **51.** 12 **53.** $\dfrac{-1 + e^6}{e}$ **55.** $\dfrac{7}{2}$

57. $\dfrac{5392}{15}$ **59.** $\dfrac{27}{4}$ **61.** $\dfrac{307}{6}$ **63.** $\dfrac{15}{4}$ **65.** 8 **67.** 12

69. 0 **71.** $e - 1$ **73.** $\dfrac{4}{3}$ **75.** 13 **77.** $\dfrac{1}{n + 1}$

79. 148 **81.** $200\left(1 - \dfrac{1}{e^2}\right)$ **83.** $-75{,}000$ **85.** 3927.71

87. (a) $42.03\ \mu g/ml$ **(b)** $22.44\ \mu g/ml$ **89. (a)** Left to the student   **(b) tw**   **91. tw**   **93.** 6.25
95. 4068.79 **97.** 0.4292 **99.** 1250 **101.** 9.5244
103. 10.9872

SECTION 5.4

1. $\dfrac{1}{4}$ **3.** $\dfrac{9}{2}$ **5.** $\dfrac{125}{6}$ **7.** $\dfrac{9}{2}$ **9.** $\dfrac{1}{6}$ **11.** $\dfrac{125}{3}$ **13.** $\dfrac{32}{3}$

15. 3 **17.** $\dfrac{\pi^2}{8} + \dfrac{3\pi}{2} - 2$ **19.** $1 - \dfrac{1}{\sqrt{2}}$

21. $\dfrac{256}{3}$ **23.** 128 **25.** 9 **27.** $\dfrac{52}{3}$ **29.** $\dfrac{1}{e} + e - 2$

31. $\dfrac{2}{3}$ **33.** 726.2 N $\cdot$ mm   **35.** 17.23 degree days

37. 96 **39.** $\dfrac{p\pi R^4}{8Lv}$ **41.** 24.961 **43.** 16.708

45. (a)   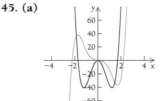   **(b)** $-1.8623, 0, 1.45939$
(c) 64.5239
(d) 17.683

SECTION 5.5

1. $\ln|x^3 + 7| + C$ **3.** $\dfrac{e^{4x}}{4} + C$ **5.** $2e^{x/2} + C$ **7.** $\dfrac{e^{x^4}}{4} + C$

9. $-\dfrac{e^{-t^3}}{3} + C$ **11.** $\dfrac{1}{2}\ln^2(4x) + C$ **13.** $\ln|x + 1| + C$

15. $-\dfrac{3}{4}\cos(4x + 2) + C$ **17.** $-\dfrac{1}{2}\csc(2x + 3) + C$

19. $-\ln|4 - x| + C$ **21.** $(t^3 - 1)^8/24 + C$

23. $-\dfrac{1}{2}\cos(x^2) + C$ **25.** $\tan(x^2 + 2x + 3)/2 + C$

27. $\ln(4 + e^x) + C$ **29.** $\dfrac{1}{4}\ln^2(x^2) + C$

31. $\ln|\ln x| + C$ **33.** $\dfrac{2(b + ax)^{3/2}}{3a} + C$ **35.** $\dfrac{be^{ax}}{a} + C$

37. $-\dfrac{a}{b}\cos(bx + c) + C$ **39.** $-\dfrac{1}{4(x^3 + 1)^4} + C$

41. $\ln|\sin x| + C$ **43.** $-\dfrac{5}{12}(1 - 4t^2)^{3/2} + C$

45. $2e^{\sqrt{w}} + C$ **47.** $\dfrac{1}{3}\sin^3 x + C$

49. $-e^{-\sin t} + C$ **51.** $-\frac{1}{9}\cos(3r^3 + 7) + C$

53. $-1 + e$ **55.** $\frac{21}{4}$ **57.** $\ln 2$ **59.** $\ln 19$

61. $1 - e^{-b}$ **63.** $1 - e^{-bm}$ **65.** $\frac{208}{3}$ **67.** $\frac{\sqrt{3}}{2\pi}$

69. $\frac{1640}{6561}$ **71.** $\frac{315}{8}$ **73.** (a) $\frac{2}{5}KH^{5/2}$

(b) $\frac{1}{20}(8 - \sqrt{2})\, KH^{5/2}$ **(c)** $\frac{1}{8}(8 - \sqrt{2}) \approx 0.8232$

(d) $\frac{1}{4\sqrt{2}} \approx 0.1768$ **75.** $\frac{16}{3}$ **77.** $t + \frac{1}{t + 1} + C$

79. $x + 2\ln|x + 1| + C$ **81.** $\frac{\ln^{1-n}(x)}{1 - n} + C$

83. $\ln|\ln(\ln x)| + C$ **85.** $\ln|\sec x + \tan x| + C$

87. tw

SECTION 5.6

1. $e^{5x}(x - \frac{1}{5}) + C$ **3.** $\sin x - x\cos x + C$

5. $e^{2x}\left(\dfrac{x}{2} - \dfrac{1}{4}\right) + C$ **7.** $e^{-2x}\left(-\dfrac{x}{2} - \dfrac{1}{4}\right) + C$

9. $\dfrac{1}{3}x^3 \ln x - \dfrac{x^3}{9} + C$ **11.** $\dfrac{1}{2}x^2 \ln(x^2) - \dfrac{x^2}{2} + C$

13. $(x + 3)\ln(x + 3) - x + C$

15. $\dfrac{1}{2}x^2 \ln x + 2x\ln x - \dfrac{x^2}{4} - 2x$

17. $-x\cos x + \cos x + \sin x + C$

19. $\frac{2}{15}(x + 2)^{3/2}(3x - 4) + C =$
$\frac{2}{3}x(x + 2)^{3/2} - \frac{4}{15}(x + 2)^{5/2} + C$

21. $\dfrac{1}{4}x^4 \ln(2x) - \dfrac{x^4}{16} + C$ **23.** $e^x(x^2 - 2x + 2) + C$

25. $\frac{1}{2}x\sin(2x) - \frac{1}{4}(2x^2 - 1)\cos(2x) + C$

27. $-\frac{1}{8}e^{-2x}(4x^3 + 6x^2 + 6x + 3) + C$

29. $\ln|\cos x| + x\tan x + C$ **31.** $\frac{1}{9}(-7 + 8\ln 8)$

33. $-4 - 5\ln 5 + 9\ln 9$ **35.** 1

37. $\dfrac{5\pi}{4} - 3 - \dfrac{\sqrt[3]{3}}{2}$ **39.** 152,324 g

41. (a) $10(1 - e^{-T}(T + 1))$

(b) $10 - \dfrac{50}{e^4} \approx 9.08 \text{ kW} \cdot \text{h}$

43. Left to the student   **45.** $\frac{2}{9}x^{3/2}(3\ln x - 2) + C$

47. $\dfrac{e^t}{t + 1} + C$ **49.** $2\sqrt{x}(\ln x - 2) + C$

51. $\dfrac{5(4t + 7)^{4/5}(3744t^2 - 7280t - 32,459)}{16,128} + C$

53. Left to the student   **55.** Left to the student

57. tw   **59.** 355,986

SECTION 5.7

1. $e^{-3x}\left(-\dfrac{x}{3} - \dfrac{1}{9}\right) + C$ **3.** $\dfrac{5^x}{\ln(5)} + C$

5. $\dfrac{1}{8}\ln\left|\dfrac{4 + x}{4 - x}\right| + C$ **7.** $-x - 5\ln|5 - x| + C$

9. $\dfrac{1}{5(5 - x)} + \dfrac{1}{25}\ln\left|\dfrac{x}{5 - x}\right| + C$

11. $x\ln(3x) - x + C$

13. $\dfrac{e^{5x}(625x^4 - 500x^3 + 300x^2 - 120x + 24)}{3125} + C$

15. $3(x^2 - 2)\sin x - x(x^2 - 6)\cos x + C$

17. $\frac{1}{2}\ln|\sec 2x + \tan 2x| + C$

19. $-\ln|\cos(2x + 1)| + C$ **21.** $\ln(x + \sqrt{x^2 + 7}) + C$

23. $\dfrac{2}{5 - 7x} + \dfrac{2}{5}\ln\left|\dfrac{x}{5 - 7x}\right| + C$

25. $\dfrac{5}{4}\ln\left|\dfrac{2x + 1}{2x - 1}\right| + C$

27. $m\sqrt{m^2 + 4} + 4\ln(m + \sqrt{m^2 + 4}) + C$

29. $\dfrac{5\ln x}{2x^2} + \dfrac{5}{4x^2} + C$ **31.** $e^x(x^3 - 3x^2 + 6x - 6) + C$

33. $\frac{1}{15}(2x + 1)^{3/2}(3x - 1) + C$

35. $x\ln^4 x - 4x\ln^3 x + 12x\ln^2 x - 24x\ln x + 24x + C$

37. $\frac{1}{2}\ln|x^2 - 1| - \ln|x| + C$

39. $\frac{4}{27}x\, 3x^2 - 2\sin 3x -$
$\frac{1}{81}(27x^4 - 36x^2 + 8)\cos(3x) + C$

41. $\frac{1}{13}e^{2x}(2\sin 3x - 3\cos 3x) + C$

43. (a) 0.74298   (b) 0.74686   **45.** (a) 0.44998

(b) 0.44714   **47.** (a) 1.50358   (b) 1.50547

49. (a) 0.27096   (b) 0.27092

51. $\dfrac{1}{4}\left(\ln(t) - \ln(t + 2) + \dfrac{2}{t + 2}\right) + 0.87499$

53. 34.7944   **55.** $\dfrac{1}{4}\ln|2x - 3| + \dfrac{3}{4(3 - 2x)} + C$

57. $\frac{1}{2}(e^x\sqrt{e^{2x} + 1} + \ln(e^x + \sqrt{e^{2x} + 1})) + C$

59. $\frac{1}{4}(\ln x\sqrt{(\ln x)^2 + 49} +$
$49\ln(\ln x + \sqrt{(\ln x)^2 + 49})) + C$

61. Left to the student

SECTION 5.8

1. $\dfrac{\pi}{3}$ **3.** π **5.** $\dfrac{(e^{14} - 1)\pi}{2e^4}$ **7.** $\dfrac{2\pi}{3}$ **9.** $\pi\ln 3$

11. 32π **13.** $\dfrac{32\pi}{5}$ **15.** $\dfrac{\pi^2}{4}$ **17.** $\pi - \dfrac{\pi^2}{4}$

19. 56π **21.** $\dfrac{32\pi}{3}$ **23.** 36   **25.** $\dfrac{243\sqrt{3}}{2}$ **27.** $\frac{1024}{5}$

29. $\frac{250}{3}$ **31.** $\frac{800}{3}$ **33.** 212.058 ft^3 **35.** $2e^3\pi$

37. (a) Cone, base radius $= r$, height $= h$ (b) $\frac{1}{3}\pi r^2 h$

SECTION 5.9

1. $\frac{1}{2}$ 3. Divergent 5. 1 7. 1 9. Divergent

11. Divergent 13. 0 15. 0 17. $\frac{1}{2}$ 19. $\frac{4e^3}{9}$

21. Divergent 23. 1 25. $\ln 2 - \frac{1}{2}$

27. Divergent 29. Divergent 31. $-7\sqrt{2\pi}$ 33. 1

35. $\frac{1}{2}$ 37. 1 39. (a) 420.963% per yr

(b) 0.7029 rem (c) 2.3755 rem 41. 4,965 lb

43. Converges for $r < -1$, otherwise diverges

45. $\frac{1}{k^2}$, the total dose of the drug 47. tw

49. 3.1416

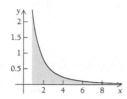

CHAPTER 5 REVIEW

1. [5.1] $\frac{8x^5}{5} + C$ 2. [5.1] $x^3 - 2\cos x + 3x$

3. [5.1] $t^3 + \frac{7t^2}{2} + \ln|t| + C$ 4. [5.3] 9

5. [5.3] $\frac{1}{\sqrt{2}}$ 6. [5.3] Left to the student

7. [5.3] Left to the student 8. [5.3] $\frac{b^6}{6} - \frac{a^6}{6}$

9. [5.3] $-\frac{2}{5}$ 10. [5.3] $e - \frac{1}{2}$

11. [5.3] $3\ln 3$ 12. [5.3] $\frac{1}{\sqrt{3}}$ 13. [5.2] 0

14. [5.2] Negative 15. [5.2] Positive 16. [5.4] $\frac{27}{2}$

17. [5.5] $\frac{e^{x^4}}{4} + C$ 18. [5.5] $\ln(4t^6 + 3) + C$

19. [5.5] $\frac{1}{4}\ln^2(4x) + C$ 20. [5.5] $-\frac{2}{3}e^{-3x} + C$

21. [5.6] $-\frac{1}{3}\cot(x^3) + C$ 22. [5.6] $e^{3x}(x - 1/3) + C$

23. [5.6] $x\ln(x^7) - 7x + C$

24. [5.6] $x^3\ln x - x^3/3 + C$

25. [5.6] $4x\sin x - (2x^2 - 4)\cos x + C$

26. [5.7] $\frac{1}{14}\ln\left|\frac{7+x}{7-x}\right| + C$

27. [5.7] $\frac{1}{125}e^{5x}(25x^2 - 10x + 2) + C$

28. [5.7] $\frac{x}{7} - \frac{1}{49}\ln|7x + 1| + C$

29. [5.7] $\ln|x + \sqrt{x^2 - 36}| + C$

30. [5.7] $\frac{1}{7}x^7\ln x - \frac{x^7}{49} + C$

31. [5.7] $e^{8x}\left(\frac{x}{8} - \frac{1}{64}\right) + C$

32. [5.7] $5(x^4 - 12x^2 + 24)\cos x$
$+ (x^5 - 20x^3 + 120x)\sin x + C$

33. [5.2] 2.167 34. [5.3] $\frac{1}{2}\left(11 - \frac{1}{e^2}\right)$

35. [5.3] 80 km 36. [5.3] 4.8653 in.

37. [5.6] $e^{x/10}(10x^3 - 300x^2 + 6000x - 60,000) + C$

38. [5.5] $\ln|4t^3 + 7| + C$

39. [5.7] $\frac{2}{75}(5x - 8)\sqrt{5x + 4} + C$ 40. [5.5] $e^{x^5} + C$

41. [5.5] $\ln|x + 9| + C$ 42. [5.5] $(t^8 + 3)^{12}/96 + C$

43. [5.6] $x\ln(7x) - x + C$

44. [5.6] $\frac{1}{2}x^2\ln(8x) - \frac{x^2}{4} + C$

45. [5.5] $\frac{1}{5}\sec(t^5 + 1) + C$

46. [5.5] $-\frac{1}{3}\ln|\cos(3x + 7)| + C$ 47. [5.9] 1

48. [5.9] Divergent 49. [5.9] $\frac{1}{4}$ 50. [5.9] 300

51. [5.8] $\frac{127\pi}{7}$ 52. [5.8] $\frac{\pi}{6}$

53. [5.8] $\frac{80,000}{3} \approx 26,666.7$ ft^3

54. [5.8] $3\sqrt{3} \approx 5.1962$ m^3

55. [5.5] [5.7] $\frac{1}{10}\ln^2(t^5 + 3) + C$

56. [5.5] $-\frac{1}{2}\ln(1 + 2e^{-x}) + C$

57. [5.5] [5.6] [5.7] $\frac{\ln^2(x)}{4} + C$

58. [5.6] [5.7] $\frac{1}{92}x^{92}\ln x - \frac{x^{92}}{8464} + C$

59. [5.6]

$4\ln|x - 4| - 3\ln|x - 3| + x\ln\left(\frac{x-3}{x-4}\right) + C$

60. [5.5] $-\frac{1}{3\ln^3 x} + C$ 61. [5.4] $\frac{4}{3}$

62. [5.7] 102.045 63. [5.7] 100.511

CHAPTER 5 TEST

1. [5.1] $x + C$ 2. [5.1] $200x^5 + C$

3. [5.1] $e^x + \ln|x| + \frac{8}{11}x^{11/8} + C$ 4. [5.3] $\frac{1}{6}$

5. [5.3] $4\ln 3$ 6. [5.3] Left to the student

7. [5.3] 12 8. [5.3] $\frac{1}{2} - \frac{1}{2e^2}$ 9. [5.3] $\ln|b/a|$

10. [5.2] Positive 11. [5.5] $\ln|x + 8| + C$

12. [5.5] $-2e^{-0.5x} + C$ 13. [5.5] $-\cos(e^x) + C$

14. [5.6] $\frac{1}{25}\sin(5x) - \frac{1}{5}x\cos(5x) + C$

15. [5.6] $\frac{1}{4}x^4\ln(x^4) - \frac{x^4}{4} + C$ 16. [5.7] $\frac{2^x}{\ln 2} + C$

17. [5.7] $\dfrac{1}{7}\ln\left|\dfrac{x}{7-x}\right| + C$ **18.** [5.3] 6   **19.** [5.4] $\dfrac{1}{3}$

20. [5.2] 70   **21.** [5.3] 42 in.

22. [5.3] 94 words   **23.** [5.7] $\dfrac{1}{10}\ln\left|\dfrac{x}{10-x}\right| + C$

24. [5.6] $e^x(x^5 - 5x^4 + 20x^3 - 60x^2 + 120x - 120) + C$ **25.** [5.5] $\dfrac{e^{x^6}}{6} + C$

26. [5.6] $\dfrac{2}{9}x^{3/2}(3\ln x - 2) + C$

27. [5.6] [5.7] $\dfrac{1}{15}(x^2 + 4)^{3/2}(3x^2 - 8) + C$

28. [5.7] $\dfrac{1}{16}\ln\left|\dfrac{8+x}{8-x}\right| + C$

29. [5.5] $e^{0.1x}(10x^4 - 400x^3 + 12{,}000x^2 - 240{,}000x + 2{,}400{,}000)$ **30.** [5.6] $2x\cos x + (x^2 - 2)\sin x + C$

31. [5.8] $\pi\ln 5$ **32.** [5.8] $\dfrac{5\pi}{2}$

33. [5.8] $\dfrac{112}{3} \approx 37.3333\ \text{ft}^3$ **34.** [5.9] $\dfrac{1}{2}$

35. [5.9] Divergent

36. [5.5] $\dfrac{\ln^4(x)}{4} - \dfrac{4\ln^3(x)}{3} + 5\ln x + C$

37. [5.5] $3\ln|x + 3| + x\ln\left(\dfrac{x+3}{x+5}\right)$
$- 5\ln|x + 5| + C$

38. [5.7]
$$\dfrac{3(5x - 4)^{2/3}(5000x^3 + 4500x^2 + 4320x + 39{,}559)}{34{,}375} + C$$

39. [5.4] 8   **40.** (a) [5.7] 0.881416
(b) [5.7] 0.882032

SECTION 6.1

1. $\begin{bmatrix} 7 & 8 \\ 9 & -11 \end{bmatrix}$ **3.** $\begin{bmatrix} 11 & 12 \\ 2 & -5 \end{bmatrix}$ **5.** Undefined

7. $\begin{bmatrix} 6 & 18 \\ 4 & -4 \end{bmatrix}$ **9.** $\begin{bmatrix} 1 & -10 \\ 5 & -7 \end{bmatrix}$ **11.** $\begin{bmatrix} 10 & -27 & 5 \\ -55 & -69 & 16 \end{bmatrix}$

13. $\begin{bmatrix} 304 \\ 2649 \end{bmatrix}$ **15.** $\begin{bmatrix} -477 \\ 178 \end{bmatrix}$ **17.** $\begin{bmatrix} -11 & 6 \\ -74 & 38 \end{bmatrix}$

19. $\begin{bmatrix} -378 & 213 & 303 \\ 12 & -10 & -46 \end{bmatrix}$ **21.** $\begin{bmatrix} 42 & 53 \\ 59 & 129 \end{bmatrix}$

23. $\begin{bmatrix} 18 & 0 & -22 \\ -16 & 14 & -3 \\ -15 & -10 & -14 \end{bmatrix}, \begin{bmatrix} 16 & -14 & 6 \\ 5 & 21 & 5 \\ 14 & 35 & -19 \end{bmatrix}$

25. $\begin{bmatrix} 28 & -28 & -58 \\ -35 & 1 & 19 \\ -4 & -34 & -8 \end{bmatrix}$

27. (a) $\begin{bmatrix} 16 \\ 5 \\ 14 \end{bmatrix}$ (b) $\begin{bmatrix} -14 \\ 21 \\ 35 \end{bmatrix}$ (c) $\begin{bmatrix} 6 \\ 5 \\ -19 \end{bmatrix}$ (d) tw

29. $\begin{bmatrix} -4 & 4 & -2 \\ 5 & -4 & -3 \\ 6 & 2 & -3 \end{bmatrix}$ **31.** (a) Left to the student

(b) 114 hatchlings   (c) 225 hatchlings
   36 adults   72 adults

33. (a) $\begin{bmatrix} 0.8 & 2 \\ 0.5 & 0.4 \end{bmatrix}$ (b) 4000 hatchlings
   1208 adults

(c) 5616 hatchlings   **35.** (a) Left to the student
   2483 adults

(b) $\begin{bmatrix} 15 & 7.5 \\ 0.8 & 0 \end{bmatrix}$ (c) 300 hatchlings
   16 adults

(d) 4620 hatchlings
   240 adults

37. (a) $\begin{bmatrix} 1 & -16 \\ 32 & 17 \end{bmatrix}$ (b) $\begin{bmatrix} 11 & -35 \\ -5 & 16 \end{bmatrix}$

(c) $\begin{bmatrix} 26 & 15 \\ 15 & 29 \end{bmatrix}$ (d) $\begin{bmatrix} -4 & 3 \\ 21 & 59 \end{bmatrix}$ (e) tw

(f) tw   **39.** (a) $n \times 1, 1 \times n$

(b) tw   **41.** tw   **43.** tw

45.–49. See Solutions for Exercises 23–27.

51. $\begin{bmatrix} 200{,}502 \\ 63{,}757.2 \end{bmatrix}$ **53.** $\begin{bmatrix} 3.414 \times 10^7 \\ 1.399 \times 10^7 \end{bmatrix}$

55. $\begin{bmatrix} 7.644 \times 10^9 \\ 1.021 \times 10^9 \\ 1.568 \times 10^8 \end{bmatrix}$

SECTION 6.2

1. $x = 1, y = 2$ **3.** $w = 4, z = 6$ **5.** No solution
7. $(x, x - 7)$ **9.–15.** See solutions for Exercises 1–7.
17. $x = 68.2, y = 1.6, z = -5.2$
19. $x = 9, y = 12.5, z = 5.5$
21. $x = 14, y = 29, z = 11$
23.–27. See solutions for Exercises 17–21.
29. $x = -17, y = -8, z = 15$
31. $((17z + 25)/3, -(11z + 13)/3, z)$ **33.** No solution
35. $x = 15, y = 25, z = -7$
37. $x = 1, y = -1, z = 2, w = 3$
39. $x = 23, y = 21, z = 20, w = 51$
41. $B = 10\ \text{mg/m}^3, F = 5\ \text{mg/m}^3$
43. $H = 400, S = 299, A = 370$ **45.** tw

47. (a) $\begin{bmatrix} 1 & 2 & 3 \\ 16 & 20 & 24 \\ 7 & 8 & 9 \end{bmatrix}$ **(b)** $\begin{bmatrix} 4 & 2 & 7 \\ 4 & 4 & 12 \\ -4 & 3 & 6 \end{bmatrix}$

(c) tw **49.** Left to the student **50.–55.** See solutions for Exercises 35–40.

SECTION 6.3

1. $\begin{bmatrix} 0 & -1 \\ 1 & 1 \end{bmatrix}$ **3.** $\begin{bmatrix} 0 & 1 \\ 1 & 0 \end{bmatrix}$ **5.** $\begin{bmatrix} 7 & -4 \\ -5 & 3 \end{bmatrix}$

7. $\begin{bmatrix} \frac{1}{31} & \frac{7}{62} \\ \frac{4}{31} & -\frac{3}{62} \end{bmatrix}$ **9.** $\begin{bmatrix} 0 & 1 & 2 \\ 0 & 0 & -1 \\ 1 & 2 & 6 \end{bmatrix}$ **11.** $\begin{bmatrix} 2 & 1 & -2 \\ 2 & 0 & 1 \\ -1 & 0 & 0 \end{bmatrix}$

13. $\begin{bmatrix} 8 & 2 & -1 \\ 8 & 3 & -3 \\ -3 & -1 & 1 \end{bmatrix}$ **15.** $\begin{bmatrix} -20 & -27 & -14 \\ 7 & 10 & 5 \\ 10 & 13 & 7 \end{bmatrix}$

17. $\begin{bmatrix} 0 & -\frac{3}{2} & -1 \\ -1 & -\frac{3}{2} & -\frac{5}{2} \\ 0 & -1 & -\frac{1}{2} \end{bmatrix}$ **19.** $\begin{bmatrix} -7 & -5 & -6 \\ 1 & 1 & 2 \\ \frac{13}{2} & \frac{9}{2} & \frac{9}{2} \end{bmatrix}$

21. $\begin{bmatrix} \frac{8}{3} & \frac{4}{3} & -\frac{3}{2} \\ \frac{7}{3} & \frac{5}{3} & -\frac{3}{2} \\ \frac{19}{3} & \frac{8}{3} & -\frac{7}{2} \end{bmatrix}$ **23.** $\begin{bmatrix} 7 & 7 & 37 & 8 \\ 10 & 10 & 53 & 10 \\ -3 & -3 & -16 & -3 \\ 2 & 3 & 13 & 4 \end{bmatrix}$

25. 9, invertible **27.** 37, invertible **29.** 0, not invertible **31.** -219, invertible **33.** 0, not invertible **35.** 0, not invertible **37.** -120, invertible **39.** -96, invertible **41.** 12, invertible

43. $\begin{bmatrix} 26 \\ 114 \end{bmatrix}$ **45. (a)** $\begin{bmatrix} 0 & 1.25 \\ 0.1333 & -2.5 \end{bmatrix}$ **(b)** $\begin{bmatrix} 3825 \\ 192 \end{bmatrix}$

(c) $\begin{bmatrix} 240 \\ 30 \end{bmatrix}$ **47.** Left to the student **49. (a)** -78

(b) 1032 **51–57.** See solutions to Exercises 21–23 and 39–41.

SECTION 6.4

1. Yes, 2 **3.** No **5.** Yes, -2 **7.** No **9.** Yes, 3 **11.** No **13.** $\mathbf{v} = 3\mathbf{w} + 4\mathbf{u}$ **15.** $\mathbf{v} = 2\mathbf{w} + \mathbf{u}$

17. $\mathbf{v} = -\mathbf{w} + 3\mathbf{u}$ **19.** $2, t\begin{bmatrix} 0 \\ 1 \end{bmatrix}; 1, t\begin{bmatrix} 1 \\ 1 \end{bmatrix}$ $(t \neq 0)$

21. $-3, t\begin{bmatrix} -1 \\ 4 \end{bmatrix}; -1, t\begin{bmatrix} -1 \\ 3 \end{bmatrix}$ $(t \neq 0)$ **23.** $3,$

$t\begin{bmatrix} -3 \\ 2 \end{bmatrix}; -\frac{3}{2}, t\begin{bmatrix} -7 \\ 4 \end{bmatrix}$ $(t \neq 0)$ **25.** $2, t\begin{bmatrix} 3 \\ 5 \end{bmatrix}; \frac{1}{2},$

$t\begin{bmatrix} 1 \\ 2 \end{bmatrix}$ $(t \neq 0)$ **27.** $2, t\begin{bmatrix} 1 \\ 2 \\ 0 \end{bmatrix}; 1, t\begin{bmatrix} -3 \\ -3 \\ 1 \end{bmatrix}; 0,$

$t\begin{bmatrix} -5 \\ -5 \\ 2 \end{bmatrix}$ $(t \neq 0)$ **29.** $2, t\begin{bmatrix} 0 \\ -1 \\ 4 \end{bmatrix}; 1, t\begin{bmatrix} -1 \\ -3 \\ 9 \end{bmatrix}; 0,$

$t\begin{bmatrix} -2 \\ -5 \\ 15 \end{bmatrix}$ $(t \neq 0)$ **31.** $-2, t\begin{bmatrix} -1 \\ 1 \\ 2 \end{bmatrix}; -1, t\begin{bmatrix} -1 \\ 2 \\ 4 \end{bmatrix}; 1,$

$t\begin{bmatrix} 0 \\ 1 \\ 1 \end{bmatrix}$ $(t \neq 0)$ **33.** $-4, t\begin{bmatrix} 1 \\ 1 \\ 1 \end{bmatrix}; 4, t\begin{bmatrix} 2 \\ 1 \\ 0 \end{bmatrix}; 2, t\begin{bmatrix} 0 \\ 0 \\ 1 \end{bmatrix}$ $(t \neq 0)$

35. $2, t\begin{bmatrix} -2 \\ -2 \\ 1 \end{bmatrix}; -1, t\begin{bmatrix} -1 \\ -1 \\ 1 \end{bmatrix}; 1, t\begin{bmatrix} 2 \\ 1 \\ 0 \end{bmatrix}$ $(t \neq 0)$

37. $\begin{bmatrix} 2048 \\ 3 \end{bmatrix}$ **39.** $\begin{bmatrix} 2048 \\ 2048 \end{bmatrix}$ **41.** 1.5, 50% **43.** 2.1,

110% **45.** 1.5 **47.** 15.39 **49. (a)** $\begin{bmatrix} 1 \\ 1 \end{bmatrix}$ **(b) tw**

51. (a) $\begin{bmatrix} 6 & 10 \\ -2 & -3 \end{bmatrix}$ **(b)** $r^2 - 3r + 2$ **(c) tw**

53. (a) Left to the student **(b)** $r_1 r_2$ **55. (a)** Left to the student **(b)** $\begin{bmatrix} 0 & 0 \\ 0 & 0 \end{bmatrix}$ **(c)** Answers vary

SECTION 6.5

1. Not linear **3.** Homogeneous linear
5. Homogeneous linear **7.** $x_n = c_1(-1)^n + c_2 3^n$
9. $x_n = c_1(-3)^n + c_2 2^n$ **11.** $x_n = c_1(-1)^n + c_2 2^n$
13. $x_n = (-1)^n + 2^n$ **15.** $x_n = 2^{-n} - 2^n$
17. $x_n = (-2)^n$ **19.** $x_n = 3(1/2)^n + 2(-2)^n$
21. $x_n = c_1(2)^n + c_2(3)^n + 4$
23. $x_n = c_1(4)^n + c_2(5)^n + 1$
25. $x_n = -(3)^n + (4)^n + 2$
27. $x_n = 3(-4)^n - (2)^n + 9$
29. $x_n = 2(3)^n + 3(6)^n + 2$ **31.** Decreases exponentially to 0 **33.** Increases exponentially
35. Increases exponentially
37. (a) $x_n = -41.57(-0.1413)^n + 41.57(1.0613)^n$
(b) tw **39.** $x_{n+1} = ax_n - abx_{n-1} + abM$
41. (a) $x_n = c_1(0.1136)^n + c_2(0.8364)^n + 655.2$
(b) $x_n = 26.28(0.1136)^n - 631.45(0.8364)^n + 655.2$
(c) tw **43. (a)** $x_{n+1} = x_{n-1}$ **(b)** $x_n = 3 - (-1)^n$
(c) $3 - (-1)^k$ **45.** $x_n = c_1(2)^n + c_2 n(2)^n$
47. $x_n = c_1(-1)^n + c_2 n(-1)^n + 3$ **49.** Left to the student **51.** Left to the student **53. tw**
55. (a) 1.2987 g, 1.6861 g, 2.1883 g, 2.8385 g, 3.6797 g

(b) tw   **57. (a)**

CHAPTER 6 REVIEW

1. [6.1] $\begin{bmatrix} 3 & -9 & 6 \\ -12 & 21 & 15 \end{bmatrix}$ **2. [6.1]** Undefined

3. [6.1] $\begin{bmatrix} 22 & 13 \\ 13 & -5 \end{bmatrix}$ **4. [6.1]** Undefined

5. [6.1] Undefined   **6. [6.1]** $\begin{bmatrix} -30 & 50 & 44 \\ -6 & 13 & 1 \\ -9 & 7 & 34 \end{bmatrix}$

7. [6.1] $\begin{bmatrix} 70 & 31 \\ 53 & -9 \end{bmatrix}$ **8. [6.1]** $\begin{bmatrix} 34 & -9 & 8 \\ -78 & -18 & 33 \end{bmatrix}$

9. [6.1] $\begin{bmatrix} 380 & 54 & -116 \\ 4 & -117 & 155 \end{bmatrix}$ **10. [6.1]** $\begin{bmatrix} -14 \\ 36 \end{bmatrix}$

11. [6.1] $\begin{bmatrix} -384 \\ -1,174 \end{bmatrix}$ **12. [6.1]** $\begin{bmatrix} -2300 \\ -2529 \\ 2663 \end{bmatrix}$

13. [6.2] $x = -5, y = 7$ **14. [6.2]** $x = 1, y = 1, z = 2$
15. [6.2] $x = 2, y = -4, z = 3$ **16. [6.3]** -3,
invertible   **17. [6.3]** 0, not invertible   **18. [6.3]** 24,
invertible   **19. [6.3]** 0, not invertible   **20. [6.3]** 25,
invertible   **21. [6.3]** 0, not invertible

22. [6.3] $\begin{bmatrix} \frac{1}{3} & 0 \\ \frac{1}{3} & -1 \end{bmatrix}$ **23. [6.3]** $\begin{bmatrix} -\frac{1}{5} & -\frac{3}{5} \\ \frac{2}{5} & \frac{1}{5} \end{bmatrix}$

24. [6.3] $\begin{bmatrix} -1 & \frac{3}{2} & -2 \\ -\frac{1}{2} & \frac{1}{2} & \frac{1}{2} \\ -\frac{1}{2} & 1 & -\frac{3}{2} \end{bmatrix}$ **25. [6.3]** $\begin{bmatrix} -5 & -1 & 9 \\ -7 & 0 & 10 \\ -5 & -2 & 11 \end{bmatrix}$

26. [6.4] $4, t\begin{bmatrix} 0 \\ 1 \end{bmatrix}; 3, t\begin{bmatrix} -1 \\ 2 \end{bmatrix} (t \neq 0)$

27. [6.4] $4, t\begin{bmatrix} 1 \\ 0 \end{bmatrix}; -1, t\begin{bmatrix} -2 \\ 5 \end{bmatrix} (t \neq 0)$

28. [6.4] $3, t\begin{bmatrix} -7 \\ 3 \end{bmatrix}; 1, t\begin{bmatrix} -2 \\ 1 \end{bmatrix} (t \neq 0)$

29. [6.4] $3, t\begin{bmatrix} 7 \\ 2 \end{bmatrix}; 1, t\begin{bmatrix} 3 \\ 1 \end{bmatrix} (t \neq 0)$

30. [6.4] $5, t\begin{bmatrix} 1 \\ 0 \\ 0 \end{bmatrix}; -3, t\begin{bmatrix} 0 \\ -4 \\ 1 \end{bmatrix}; -2,$

$t\begin{bmatrix} 0 \\ 1 \\ 0 \end{bmatrix} (t \neq 0)$ **31. [6.4]** $-7, t\begin{bmatrix} -3 \\ 1 \\ 1 \end{bmatrix}; 2,$

$t\begin{bmatrix} 0 \\ 1 \\ 2 \end{bmatrix}; -1, t\begin{bmatrix} 0 \\ 0 \\ 1 \end{bmatrix} (t \neq 0)$

32. [6.4] $-3, t\begin{bmatrix} 0 \\ 0 \\ 1 \end{bmatrix}; 3, t\begin{bmatrix} -2 \\ 1 \\ 0 \end{bmatrix}; 2,$

$t\begin{bmatrix} 1 \\ -1 \\ 1 \end{bmatrix} (t \neq 0)$ **33. [6.4]** $-1, t\begin{bmatrix} -1 \\ 2 \\ 2 \end{bmatrix}; 1,$

$t\begin{bmatrix} -1 \\ 1 \\ 1 \end{bmatrix}; 0, t\begin{bmatrix} 0 \\ 1 \\ 0 \end{bmatrix} (t \neq 0)$

34. [6.4] $7, t\begin{bmatrix} -1 \\ 3 \\ 6 \end{bmatrix}; 6, t\begin{bmatrix} -1 \\ 2 \\ 4 \end{bmatrix}; 0,$

$t\begin{bmatrix} 0 \\ 1 \\ 1 \end{bmatrix} (t \neq 0)$ **35. [6.5]** $x_n = 2(2)^n + (-4)^n$

36. [6.5] $-\frac{3}{5}(-2)^n - \frac{2}{5}(1/2)^n + 1$

37. [6.2] $B = \frac{2}{3}$ mg, $L = \frac{4}{3}$ mg

38. (a) [6.1] $\begin{bmatrix} 0.7 & 0.4 \\ 0.6 & 1 \end{bmatrix}$ **(b) [6.1]** $\begin{bmatrix} 58 \\ 76 \end{bmatrix}$

(c) [6.1] $\begin{bmatrix} 71 \\ 111 \end{bmatrix}$ **(d) [6.4]** 1.362

39. (a) [6.1] $\begin{bmatrix} 200 \\ 74 \end{bmatrix}$ **(b) [6.1]** $\begin{bmatrix} 308 \\ 130 \end{bmatrix}$

(c) [6.4] 1.62

40. [6.2] tw   **41. [6.4] (a)** $1, 0.5$ **(b)** $1, t\begin{bmatrix} 1 \\ 6 \end{bmatrix}; 0.5,$

$t\begin{bmatrix} 0 \\ 1 \end{bmatrix} (t \neq 0)$ **(c)** $\begin{bmatrix} 1 & 0 \\ 6 & 7.889 \times 10^{-31} \end{bmatrix}$

42. [6.4] (a) $1, 0.9$ **(b)** $1, t\begin{bmatrix} 1 \\ 40 \end{bmatrix}; 0.9, t\begin{bmatrix} 0 \\ 1 \end{bmatrix} (t \neq 0)$

(c) $\begin{bmatrix} 1 & 0 \\ 39.9989 & 2.656 \times 10^{-5} \end{bmatrix}$

CHAPTER 6 TEST

1. [6.1] $\begin{bmatrix} -3 & 44 & -29 \\ 20 & 27 & -3 \end{bmatrix}$

2. [6.1] $\begin{bmatrix} -42 & 56 & -32 \\ 0 & 7 & -40 \\ -39 & 44 & 16 \end{bmatrix}$ **3. [6.1]** $\begin{bmatrix} -6 & -38 \\ 71 & -10 \end{bmatrix}$

4. [6.2] $x = 4, y = -3$ **5. [6.2]** $x = 2, y = -3, z = 4$
6. [6.3] 0, not invertible   **7. [6.3]** 0, not invertible
8. [6.3] -41.283, invertible   **9. [6.3]** 44, invertible

10. [6.3] $\begin{bmatrix} -1 & 4 \\ 1 & -3 \end{bmatrix}$ **11. [6.3]** $\begin{bmatrix} 1 & 1 & 0 \\ 1 & 0 & 2 \\ -1 & 0 & -1 \end{bmatrix}$

12. [6.4] $-1, t\begin{bmatrix} -5 \\ 3 \end{bmatrix}; 1, t\begin{bmatrix} -2 \\ 1 \end{bmatrix} (t \neq 0)$

13. [6.4] $3, t\begin{bmatrix} -1 \\ 3 \end{bmatrix}; 1, t\begin{bmatrix} -2 \\ 7 \end{bmatrix} (t \neq 0)$

14. [6.4] $-3, t\begin{bmatrix} -2 \\ -1 \\ 1 \end{bmatrix}; 3, t\begin{bmatrix} 0 \\ -2 \\ 1 \end{bmatrix}; -1,$

$t\begin{bmatrix} -1 \\ -1 \\ 1 \end{bmatrix} (t \neq 0)$ **15. [6.4]** $-7, t\begin{bmatrix} 0 \\ 1 \\ 0 \end{bmatrix}; 7,$

$t\begin{bmatrix} -1 \\ 1 \\ 1 \end{bmatrix}; 3, t\begin{bmatrix} -2 \\ 2 \\ 1 \end{bmatrix} (t \neq 0)$

16. [6.4] $-6, t\begin{bmatrix} 1 \\ 0 \\ 0 \end{bmatrix}; 6, t\begin{bmatrix} 0 \\ -2 \\ 1 \end{bmatrix}; 5, t\begin{bmatrix} -1 \\ -1 \\ 1 \end{bmatrix} (t \neq 0)$

17. [6.5] $x_n = 2(5)^n - 3(-2)^n$
18. [6.5] $x_n = -(-2)^n - (4)^n + 3$
19. [6.2] $B = 6$ mg/m^3, $F = \frac{12}{5}$ mg/m^3

20. (a) [6.1] $\begin{bmatrix} 660 \\ 116 \end{bmatrix}$ **(b) [6.1]** $\begin{bmatrix} 744 \\ 287 \end{bmatrix}$

21. [6.4] 1.1   **22. (a) [6.3]** $\begin{bmatrix} -\frac{1}{29} & \frac{6}{29} \\ \frac{5}{29} & -\frac{1}{29} \end{bmatrix}$

(b) [6.3] $\begin{bmatrix} -\frac{1}{11} & \frac{3}{11} \\ \frac{4}{11} & -\frac{1}{11} \end{bmatrix}$ **(c) [6.3]** $\begin{bmatrix} \frac{25}{319} & -\frac{9}{319} \\ -\frac{9}{319} & \frac{16}{319} \end{bmatrix}$

(d) [6.3] $\begin{bmatrix} \frac{16}{319} & -\frac{9}{319} \\ -\frac{9}{319} & \frac{25}{319} \end{bmatrix}$ **(e) [6.3]** $\begin{bmatrix} \frac{25}{319} & -\frac{9}{319} \\ -\frac{9}{319} & \frac{16}{319} \end{bmatrix}$

(f) [6.3] tw

23. (a) [6.1] $\begin{bmatrix} 439 \\ 132 \end{bmatrix}$ **(b) [6.1]** $\begin{bmatrix} 563 \\ 303 \end{bmatrix}$

(c) [6.1] $\begin{bmatrix} 3,174,909 \\ 1,407,942 \end{bmatrix}$ **(d) [6.4]** 1.653

SECTION 7.1

1. $0, -8, 200$ **3.** $1, -\frac{125}{9}, 23$ **5.** $0, -\sqrt{\frac{3}{2}}, \frac{1}{2}$ **7.** 6, 12
9. 1.91485 m^2 **11. (a)** 3466 g
(b) 3868 g   **13.** 244.7 mi/hr   **15.** 0.025   **17.** tw
19. -22 **21.–25.** Left to the student

SECTION 7.2

1. $2 - 3y, -3x, 11, 0$ **3.** $6x - 2y, 1 - 2x, -6, 1$

5. $2, -3, 2, -3$ **7.** $\dfrac{x}{\sqrt{x^2 + y^2}}, \dfrac{y}{\sqrt{x^2 + y^2}}, \dfrac{-2}{\sqrt{5}}, \dfrac{-2}{\sqrt{13}}$

9. $2, -3$ **11.** $\dfrac{1}{2\sqrt{x}} + y\cos(xy), x\cos(xy)$ **13.** $\ln y, \dfrac{x}{y}$

15. $3x^2 - 4y, -4x + 2y$
17. $8x(x^2 + 2y + 2)^3, 8(x^2 + 2y + 2)^3$
19. $e^{x+y}\cos(e^{x+y}), e^{x+y}\cos(e^{x+y})$
21. $\dfrac{e^x}{y^2 + 1}, \dfrac{-2e^x y}{(y^2 + 1)^2}$
23. $20x^4(x^5 + \tan(y^2))^3, 8y\sec^2(y^2)(x^5 + \tan(y^2))^3$
25. $\dfrac{y}{x(y^3 - 1)}, -\dfrac{(2y^3 + 1)\ln x}{(y^3 - 1)^2}$
27. $6b + 12m - 30, 12b + 28m - 64$
29. $\dfrac{-4t^2 x}{(x^2 - t^2)^2}, \dfrac{4tx^2}{(x^2 - t^2)^2}$
31. $\dfrac{1}{(1 + 2\sqrt{t})\sqrt{x}}, \dfrac{-1 - 2\sqrt{x}}{(1 + 2\sqrt{t})^2\sqrt{t}}$ **33.** $\dfrac{3t^{5/4}}{4x^{1/4}}, \dfrac{5t^{1/4}x^{3/4}}{4}$

35. $\begin{bmatrix} 1 & 3 \\ 1 & -2 \end{bmatrix}$ **37.** $\begin{bmatrix} \dfrac{1}{2\sqrt{x + 3y}} & \dfrac{3}{2\sqrt{x + 3y}} \\ -e^{-x-y} & -e^{-x-y} \end{bmatrix}$

39. $0, 0, 0, 0$

41. $\dfrac{-1}{4x^{3/2}} - y^2\sin(xy), \cos(xy) - xy\sin(xy), \cos(xy)$
$\qquad - xy\sin(xy), -x^2\sin(xy)$

43. $0, \dfrac{1}{y}, \dfrac{1}{y}, -\dfrac{x}{y^2}$ **45.** $6x, -4, -4, 2$
47. $2xy^3 z^4, 3x^2 y^2 z^4, 4x^2 y^3 z^3$
49. $e^{x+y^2+z^3}, 2ye^{x+y^2+z^3}, 3z^2 e^{x+y^2+z^3}$ **51.** 18.33
53. 0.02   **55. (a)** 0.881917 m^2 **(b)** 0.922233 m^2
57. (a) 3440 g   **(b)** 3392 g   **59. (a)** 0.467135
(b) 0.51688   **61.** 121.3   **63.** tw   **65.** 78.244
67. -0.846 **69.** Left to the student   **71. (a)** $-y$
(b) x **(c)** $1, -1$

SECTION 7.3

1. Relative minimum at $x = -\frac{1}{3}, y = \frac{2}{3}$
3. Saddle point at $x = 0, y = 0$; relative maximum at $x = \frac{2}{3}, y = \frac{2}{3}$ **5.** Saddle point at $x = 0, y = 0$; relative minimum at $x = 1, y = 1$ **7.** Relative minimum at $x = 1, y = -2$ **9.** Relative minimum at $x = -1, y = 2$
11. Saddle point at $x = 0, y = 0$ **13.** Relative minimum at $x = 0, y = 0$ **15.** Relative maximum at $x = 415.985, y = 0.561253$ **17.** Maximum of 418 at $a = 5, n = 1$ **19.** Minimum of $-18°$ at $x = 4, y = -1$
21. Saddle point at $x = 0, y = 0$ **23.** Relative minimum at $x = 0, y = 0$; saddle point at $x = -2, y = 1$; saddle point at $x = 2, y = 1$ **25.** tw
27. (a) $e^{-1.236+1.35\cos l \cos s - 1.707 \sin l \sin s} \times$
$\qquad (-1.35 \cos s \sin l - 1.707 \cos l \sin s)$
(b) Left to the student **(c)** Left to the student
(d) tw **29.** Relative minimum at $x = 0, y = 0$

SECTION 7.4

1. (a) $33.1775 + 2.52464x$ **(b)** $73.57, $86.20
3. (a) $71.3524 + 0.176571x$ **(b)** $81.9, 82.8$
5. (a) $85.54 - 1.63x$ **(b)** 62.72 per thousand
7. (a) $-1.23684 + 1.06842x$ **(b)** 85.3 **9.** tw
11. (a) Left to the student **(b)** $1.21709 + 0.135682X$
(c) $16.4849x^{0.135682}$ **(d)** 107
13. (a) $15.5719 - 0.00593796x$
(b) $3{:}38.20, 3{:}36.42$ **(c)** $3{:}42.12$

SECTION 7.5

1. 1 **3.** 0 **5.** 6 **7.** $\frac{3}{20}$ **9.** $\frac{1}{2}$ **11.** 4 **13.** $\frac{4}{15}$
15. $\frac{4}{3}(e - 1)$ **17.** $\frac{81}{5}$ **19.** $\frac{1}{60}$ **21.** $\frac{1}{6}$ **23. (a)** 54
(b) 54 **(c)** tw **25.** 39 **27.** $\frac{13}{240}$ **29.** tw
31. 2.9574 **33.** 0.3354

CHAPTER 7 REVIEW

1. [7.1] 1 **2. [7.2]** $3y^3$ **3. [7.2]** $2 + e^y + 9xy^2$
4. [7.2] $9y^2$ **5. [7.2]** $9y^2$ **6. [7.2]** 0

7. [7.2] $e^y + 18xy$ **8. [7.2]** $\dfrac{6x^2}{2x^3 + y} + y^2 \cos(xy^2)$

9. [7.2] $\dfrac{1}{2x^3 + y} + 2xy \cos(xy^2)$

10. [7.2] $\dfrac{-6x^2}{(2x^3 + y)^2} + 2y \cos(xy^2) - 2xy^3 \sin(xy^2)$

11. [7.2] $\dfrac{12x(y - x^3)}{(2x^3 + y)^2} - y^4 \sin(xy^2)$

12. [7.2] $-(2x^3 + y)^{-2} + 2x \cos(xy^2) - 4x^2y^2 \sin(xy^2)$

13. [7.2] $\begin{bmatrix} 1 + 2xy^3 & 3x^2y^2 \\ e^{x-5y} & -5e^{x-5y} \end{bmatrix}$ **14. [7.2]** 1.03

15. [7.2] 4.05

16. [7.3] Saddle point at $x = 1, y = \frac{3}{2}$, Relative minimum at $x = 5, y = \frac{27}{2}$
17. [7.3] Relative minimum at $x = 0, y = -2$
18. [7.3] Relative maximum at $x = \frac{3}{2}, y = -3$
19. [7.3] Relative minimum at $x = -1, y = 2$
20. [7.4] (a) $0.442 + 0.0705455x$
(b) $1.57 million, $1.78 million **21. [7.4] (a)** $\frac{3}{5}x + \frac{20}{3}$
(b) 9.1 million **22. [7.5]** $\frac{97}{30}$ **23. [7.5]** $\frac{1}{60}$
24. [7.5] 0 **25. [7.1]** Left to the student

CHAPTER 7 TEST

1. [7.1] $\cos(e^4)$ **2. [7.2]** $-2xe^{2y}\sin(e^{2y}x^2)$
3. [7.2] $-2x^2e^{2y}\sin(e^{2y}x^2)$
4. [7.2] $-4x^2e^{4y}\cos(e^{2y}x^2) - 2e^{2y}\sin(e^{2y}x^2)$
5. [7.2] $-4x^3e^{4y}\cos(e^{2y}x^2) - 4xe^{2y}\sin(e^{2y}x^2)$
6. [7.2] $-4x^4e^{4y}\cos(e^{2y}x^2) - 4x^2e^{2y}\sin(e^{2y}x^2)$
7. [7.2] 2.055 **8. [7.2]** 8.985

9. [7.2] $\begin{bmatrix} \dfrac{1}{y^2 + 1} & \dfrac{-2xy}{(y^2 + 1)^2} \\ -e^xy \sin(e^xy) & -e^x \sin(e^xy) \end{bmatrix}$

10. [7.3] Saddle point at $x = \frac{1}{3}, y = -\frac{1}{3}$, Relative minimum at $x = \frac{3}{4}, y = \frac{1}{2}$
11. [7.3] Saddle point at $x = 0, y = 0$
12. [7.4] (a) $88.2667 - 9.45x$ **(b)** 41.0 per 100,000
13. [7.5] 1 **14. [7.5]** $\frac{7}{8}$ **15. [7.1]** Left to the student

SECTION 8.1

1. $y = x^4 + C$ **3.** $y = \dfrac{x^6}{6} - \dfrac{x^3}{3} + 3 \ln x + C$

5. $y = \dfrac{4e^{3x}}{3} + \dfrac{2x^{3/2}}{3} + C$ **7.** $y = \dfrac{2(3x^3 - 5)^{3/2}}{27} + C$

9. $y = \dfrac{1}{4(4 + \cos 2x)^2} + C$

11. $y = \dfrac{1}{2} \ln \left| \dfrac{x + 1}{x - 1} \right| + C$

13. $y = \dfrac{x^3}{3} + x^2 - 3x + 4$ **15.** $y = \dfrac{e^{3x}}{3} + x + \dfrac{5}{3}$

17. $y = \dfrac{6x^{5/3} - 5x^2 - 61}{10}$ **19.** $y = \dfrac{(x^2 + 1)^{3/2} + 8}{3}$

21. $y = -\dfrac{1}{4(x^4 + 1)} - \dfrac{15}{8}$ **23.** $y = xe^x - x + 3$

25. $y = x \ln x - x + 3$ **27.** $y = \dfrac{7 - \cos(x^2)}{2}$

29. $y = x^2 + 4x + 3$ **31.** $y = \dfrac{x^3}{6} - \dfrac{15x}{8} + \dfrac{19}{6} + \dfrac{1}{2x}$

33. $y = -\dfrac{10x}{3} - \dfrac{\sin 3x}{9} - 2 + \dfrac{10\pi}{3}$

35. 3

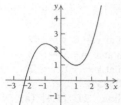

37. $-\frac{1}{3}$

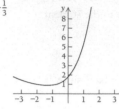

39.–45. Left to the student

47. $R = k \ln(S/S_0)$

49. (a)   (b)   (c)

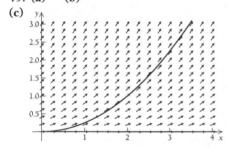

(d) tW   (e) $\dfrac{1}{2\sqrt{y}}$ **(f) tW   51. (a)** $x = 1$ and
$x = -1$ **(b) tW   (c) tW   53. (a)** $y = -2x/3$
(b) tW   (c) tW

SECTION 8.2

1. All real numbers   **3.** $(-\pi/2, \pi/2)$ **5.** $(2, \infty)$
7. $y = Ce^{3x}$ **9.** $y = 1 + Ce^{-\frac{1}{2}\sin 2x}$
11. $y = -1 + Ce^{t^2}$ **13.** $y = (t + C)e^{t}$
15. $y = \dfrac{1}{2}x^4 + 3x^4 \ln|x| + \dfrac{6}{5x} + Cx^4$
17. $y = \dfrac{C}{x^3} - \dfrac{x}{16} + \dfrac{x \ln x}{4}$ **19.** $y = (t \ln t - t + C)e^{-t}$
21. $y = \dfrac{3}{2} + \dfrac{1}{2}e^{-4x}$; all real numbers
23. $y = \dfrac{\sin^2 x + 2 \sin x - 7}{2(1 + \sin x)}$; $-\pi/2 < x < 3\pi/2$
25. $y = \dfrac{t^3}{4} + \dfrac{6}{t}$; $t > 0$ **27.** $y = \dfrac{31}{25}e^{-5x} - \dfrac{x}{5} - \dfrac{6}{25}$; all
real numbers   **29.** $y = \dfrac{7 - e^{-3x} + 3e^{-2x}}{6 - 3e^x}$; $x < \ln 2$
31. $y = \dfrac{17}{6}e^{-t^3} + \dfrac{1}{6}e^{t^3}$; all real numbers
33. $y = \dfrac{1}{4} + \dfrac{11}{4}e^{-2t^2}$; all real numbers
35. $P(t) = 15 + Ce^{-t/5}$ **37.** $Q(t) = -50 + Ce^{-t/10}$
39. $P = P_0 e^{kt}$ **41.** 415.8 lb   **43. (a)** $C + (T_0 - C)e^{-kt}$
(b) 88.2 min   **45. (a)** $x > 0$ **(b)** x^3 **(c) tW**
(d) tW   47. 79.37 lb

49. $P(t) = 1 - 1.2te^{-1.2t} - e^{-1.2t}$
51. (a)   (b) $\dfrac{a}{b}(1 - e^{-bt})$ **(c)** $\dfrac{a}{b}$

SECTION 8.3

1. (a) $y = 2$ **(b)** Asymptotically stable   **(c)** No points
of inflection
(d)

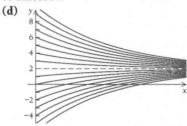

3. (a) $y = 1, 4$ **(b)** Asymptotically stable, unstable
(c) $y = \dfrac{5}{2}$
(d)

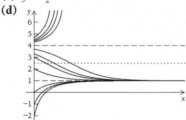

5. (a) $y = 0, 2$ **(b)** Semistable, unstable   **(c)** $y = \dfrac{4}{3}$
(d)

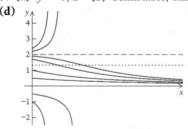

7. (a) $y = -5, -3, 0$ **(b)** Unstable, asymptotically
stable, unstable   **(c)** $y = \dfrac{-8 \pm \sqrt{19}}{3}$
(d)

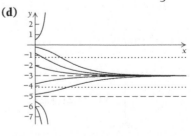

9. (a) $y = 0$ **(b)** Unstable   **(c)** $y = -\ln 2$
(d)

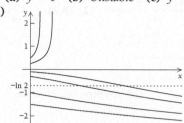

11. (a) 146.4 million   **(b)** 11.70 hr
13. (a) Left to the student   **(b)** $k(1 - [y/L]^\theta)$ **(c)** 0 is unstable; L is asymptotically stable.
(d)

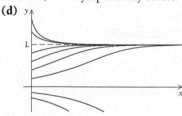

15. (a) 0 is unstable; P is asymptotically stable.
(b)

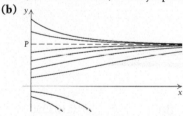

(c) The entire population becomes infected.
17. (a) $y = 0$ is unstable; $y = c$ is asymptotically stable.
(b) $Ls - \dfrac{Ls^2}{k}$ **(c)** $\dfrac{k}{2}$ **19.** 0   **21.** tw   **23.** Left to the student   **25.** $\dfrac{L + T \pm \sqrt{L^2 - LT + T^2}}{3}$ **27.** tw
29. Left to the student   **31.** $\left(\dfrac{1}{n}\right)^{1/(n-1)}$

33. (a)

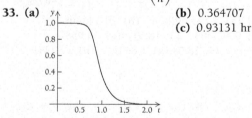

(b) 0.364707
(c) 0.93131 hr

SECTION 8.4

1. $y = Ce^{x^4}$ **3.** $y = \pm\sqrt{\dfrac{x^2 + C}{2}}$
5. $y = \pm\sqrt{\left(\dfrac{1}{2}x^2 + C\right)^2 - 1}$

7. $\sin y + \dfrac{1}{6(x^3 + 1)^2} = C$ **9.** $y = \sqrt[3]{Ce^{3t} - 1}$
11. $\tan y - x^2/2 = C$ and $y = (2n + 1)\pi/2$
13. $\frac{2}{3}x^{3/2} + \cos y - \sin y = C$
15. $\frac{1}{5}y^5 + y - \frac{1}{2}\ln(t^2 + 1) = C$
17. $y = 2 + \dfrac{1}{C - x^3}$ and $y = 2$ **19.** $y = \sqrt[3]{x^2 + 121}$
21. $e^{2t} + 2\cot y = 3$ **23.** $y = \dfrac{-4}{1 + 2\ln^2 t}$
25. $y = \sqrt[3]{\dfrac{10x^2 - 2}{6 - 5x^2}}$ **27.** $2 + \sin y = \dfrac{15}{1 - 5t^3}$
29. $\dfrac{e^{4y}}{4} + \dfrac{e^{5y}}{5} = \dfrac{2}{3}t^{3/2} - \dfrac{13}{60}$ **31.** $y = \sqrt{x^2 - 4}$
33. (a) $y = y_0 e^{kt}$ **(b)** tw   **35.** 1.154% per day
37. Left to the student   **39. (a)** Left to the student
(b) 8.06 billion; 9.03 billion   **(c)** 2019
41. Left to the student   **43. (a)** $\dfrac{LR_0}{R_0 + (L - R_0)e^{-kt}}$
(b) Left to the student

SECTION 8.5

1. (a) $y(1) \approx -1.2$ **(b)** $y(1) \approx -1$ **(c)** $x^2 - 2$
(d) $y(1) = -1$ **3. (a)** $y(2) \approx 18.2355$
(b) $y(2) \approx 40.0675$ **(c)** $2e^{x^2-1}$ **(d)** $y(2) \approx 40.1711$
5. (a) $y(2) \approx 28.8988$ **(b)** $y(2) \approx 34.171$
(c) $2(e^{x+1} - x - 1)$ **(d)** $y(2) \approx 34.1711$
7. (a) $y(2) \approx 9.30409$ **(b)** $y(2) \approx 14.3902$ **(c)** $e^{x^3/3}$
(d) $y(2) \approx 14.3919$ **9. (a)** 101.782, 404.558
(b) 108.307, 436.449   **(c)** 108.307, 436.449
11. (a) 0.697503, 0.958943   **(b)** 0.691557, 0.952266
(c) 0.691559, 0.952267   **13. (a)** 3.07458   **(b)** 3.15527
15. (a) No   **(b)** No   **(c)** Yes   **(d)** $\Delta x = 1/n$ for some integer n **(e)** $\Delta x = a/n$ for some integer n
17. (a) 2.47523873   **(b)** 2.47522970   **(c)** 2.47522913
(d) 9.6×10^{-6}, 5.8×10^{-7} **(e)** 16

CHAPTER 8 REVIEW

1. [8.1] $y = 2x^3 + C$ **2.** [8.1] $y = \dfrac{\ln^2 x}{2} + C$
3. [8.1] $y = xe^x - e^x + C$ **4.** [8.2] $y = \frac{5}{6} + Ce^{-2x^3}$
5. [8.2] $y = \dfrac{C}{x^2} + \dfrac{x}{3} + \dfrac{x^3}{5}$ **6.** [8.2] $y = Ce^{-x^3/3} + \frac{1}{4}e^{x^3}$
7. [8.4] $y = \pm\sqrt{C - x^2}$ **8.** [8.4] $x^2 + 2\cos y = C$
9. [8.4] $6y^2 + 8y^{3/2} = 3x^4 - 18x^2 + 12x + C$
10. [8.1] $y = 2x^2 - 5x - 4$
11. [8.1] $y = -\frac{2}{9}(4 - 3\sin x)^{3/2} - \frac{7}{9}$
12. [8.2] $y = x - \frac{1}{5} + \frac{11}{5}e^{-5x}$ **13.** [8.2] $y = 2 + \dfrac{2}{x^4}$
14. [8.4] $2y^8 + 4y = x^4 - 18$
15. [8.4] $y = -\sqrt{2x^3 + 2}$ **16.** [8.3] **(a)** $y = -4, 6$
(b) Asymptotically stable, unstable   **(c)** $y = 1$

(d)

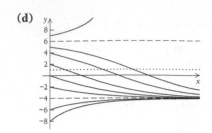

17. [8.3] (a) 0, 3 **(b)** Unstable, semistable **(c)** $y = 1$
(d)

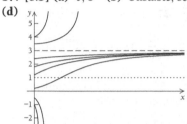

18. [8.3] (a) 82.3 million **(b)** 7.06 hrs
19. [8.2] (a) $1 - e^{-kt}$ **(b)** 0.173287 **(c)** 13.3 months
20. [8.2] (a)   **(b)** Left to the student
                        (c) $C(t) = (C_0 - b/a)e^{-at} + b/a$

(d) b/a **21. [8.2]** **(a)** $Ce^{-t/(cr)} + cdp + x_0$
(b) $(cdp + x_0)(1 - e^{-t/(cr)})$ **22. [8.2]**
(a) $Ce^{-t/(cr)} + x_0$ **(b)** $(a - x_0)e^{(T-t)/(cr)} + x_0$
23. [8.5] **(a)** 1.4295 **(b)** 1.54136 **24. [8.5]**
(a) 1.36493 **(b)** 1.36048

CHAPTER 8 TEST

1. [8.1] $y = \dfrac{2x^3}{3} - 4x + C$

2. [8.1] $y = \frac{1}{10}(e^{2x} - 2)^5 + C$

3. [8.1] $y = \sin x - x \cos x + C$

4. [8.2] $y = \frac{3}{2} + Ce^{-4x}$ **5. [8.2]** $y = \dfrac{\ln |x| + C}{x^2}$

6. [8.2] $y = 3 + Ce^{-x^2/2}$

7. [8.4] $y = \dfrac{1}{2} \ln\left(\dfrac{2}{3}e^{3x} + C\right)$

8. [8.4] $y = \pm \dfrac{1}{\sqrt{C - x^2}}$ and $y = 0$

9. [8.4] $3y^4 - 12y^2 = 4x^3 + C$

10. [8.1] $y = \dfrac{x^3}{3} + \dfrac{3x^2}{2} + 5x + \dfrac{1}{6}$

11. [8.1] $y = \dfrac{-\cos(x^4 - \pi) - 5}{4}$ **12. [8.2]** $y = \dfrac{64}{x^3}$

13. [8.2] $y = \tan x + \dfrac{\sqrt{2}}{2} \sec x$

14. [8.4] $y = \sqrt{x^2 + 2x + 8}$
15. [8.4] $e^y + y = x - 1 + 1/e$
16. [8.3] (a) $y = -3, 0$ **(b)** Asymptotically stable,
unstable **(c)** $y = -\frac{3}{2}$
(d)

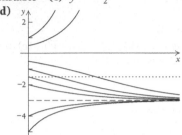

17. [8.3] (a) $y = -3, 3$ **(b)** Asymptotically stable,
unstable **(c)** $y = 0$
(d)

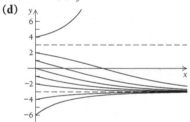

18. (a) $P' + kP = k$ **(b)** $1 - e^{-kt}$ **(c)** 0.183258
(d) 0.84 **(e)** 12.5647. Approximately 13 trials are
required **19. [8.2]** 415.8 lb
20. [8.2] (a)

(b) Left to the student
(c) $D(t) = D_0 e^{-0.226t} - (1.327 \times 10^{-7})(1 - e^{-0.226t})$
21. [8.3] (a) 3.295 million **(b)** 20.3 hours
22. [8.3] 0.493303 **23. [8.5]** **(a)** 28.9587
(b) 34.9444 **24. [8.5] (a)** 1.84211 **(b)** 1.82241

SECTION 9.1

1. $C_1 e^x + C_2 e^{5x}$ **3.** $C_1 e^{-x} + C_2 e^{2x}$ **5.** $C_1 e^{-2x} + C_2 e^{-x}$
7. $C_1 e^{x/2} + C_2 e^{2x}$ **9.** $C_1 e^{-3x} + C_2 e^{3x}$
11. $C_1 e^{-5x} + C_2 x e^{-5x}$ **13.** $C_1 e^{-3x/2} + C_2 x e^{-3x/2}$
15. $C_1 e^{-x} + C_2 \sin 2x + C_3 \cos 2x$
17. $C_1 e^{-2x} + C_2 x e^{-2x} + C_3 x^2 e^{-2x}$
19. $C_1 e^{6x} + C_2 \sin(x\sqrt{3}) + C_3 \cos(x\sqrt{3})$
21. $C_1 + C_2 x + C_3 e^x + C_4 e^{4x}$
23. $C_1 \sin 6x + C_2 \cos 6x$
25. $C_1 e^{-4x} \sin 5x + C_2 e^{-4x} \cos 5x$

27. $C_1 e^{-x}\sin 2x + C_2 e^{-x}\cos 2x + C_3$

29. $C_1 e^x + C_2 e^{-x/2}\sin\dfrac{x\sqrt3}{2} + C_3 e^{-x/2}\cos\dfrac{x\sqrt3}{2}$

31. $C_1 e^{x(-1-\sqrt{11})/2} + C_2 e^{x(-1+\sqrt{11})/2}$

33. $C_1 e^{x/3}\sin\dfrac{x\sqrt{29}}{3} + C_2 e^{x/3}\cos\dfrac{x\sqrt{29}}{3}$

35. $1 - e^x$ **37.** $2e^x - e^{-x}$ **39.** $3e^{-2x} + 8xe^{-2x}$

41. $5 + 4\cos x - 8\sin x$ **43.** $29{,}537$

45. $C_1 e^{-6.6976x} + C_2 e^{-0.0576x}$ **47.** $C_1 e^{-2.46x} + C_2 e^{-0.51x}$

49. $C_1 e^{2x}\sin 4x + C_2 e^{2x}\cos 4x +$
$C_3 xe^{2x}\sin 4x + C_4 xe^{2x}\cos 4x + C_5 e^{7x} + C_6 e^{9x}$

51. Left to the student **53.** Left to the student

SECTION 9.2

1. $C_1\sin x + C_2\cos x + 7$ **3.** $C_1 e^x + C_2 xe^x + 3$

5. $C_1 e^{-2x} + C_2 xe^{-2x} - 3x + 5$

7. $C_1 e^{-3x} + C_2 e^{-x} + 2x^2 - \dfrac{16x}{3} + \dfrac{40}{9}$

9. $C_1 e^{-x} + C_2 e^{2x} - \dfrac{x^3}{2} + \dfrac{3x^2}{4} - \dfrac{9x}{4} + \dfrac{19}{8}$

11. $C_1 + C_2 e^{3x} - \dfrac{4x}{3}$ **13.** $C_1 e^{-x} + C_2 + C_3 x - x^2$

15. $C_1 e^{-2x}\sin 4x + C_2 e^{-2x}\cos 4x + C_3 + x^2 - x$

17. $3e^{-x} + 2e^{2x} - x + 1$

19. $-5e^{-x} - xe^{-x} + x^2 - 4x + 6$

21. $\dfrac{1}{2}e^{-4x} + 2x^2 - x + \dfrac{3}{2}$

23. $e^{-x}\cos x + 2e^{-x}\sin x + 1$

25. $-\dfrac{7}{5}e^{-2x}\sin x - \dfrac{16}{5}e^{-2x}\cos x + \dfrac{5x^2}{2} - 5x + \dfrac{21}{5}$

27. $\dfrac{1 - \cos(4t)}{16}$ **29.** $2 - 2e^{-t}\cos 2t - e^{-t}\sin 2t$

31. **(a)** $C_1\sin\dfrac{t}{\sqrt{cm}} + C_2\cos\dfrac{t}{\sqrt{cm}} + cdp + x_0.$

(b) $(cdp + x_0)\left(1 - \cos\dfrac{t}{\sqrt{cm}}\right)$ **33.** $1 - e^{-t/20}$

35. $C_1 e^{-2x} + \dfrac{x^2}{2} - \dfrac{x}{2} + \dfrac{1}{4}$

SECTION 9.3

1. $x(t) = C_1 e^t + C_2 e^{2t},\ y(t) = C_1 e^t + 2C_2 e^{2t}$

3. $x(t) = C_1 e^t + C_2 e^{5t},\ y(t) = -C_1 e^t + 3C_2 e^{5t}$

5. $x(t) = C_1 e^{-t} + C_2 e^{t/2},\ y(t) = -\dfrac{1}{2}C_1 e^{-t} + C_2 e^{t/2}$

7. $x(t) = C_1 e^{3t} + C_2 te^{3t},\ y(t) = (C_2 - C_1)e^{3t} - C_2 te^{3t}$

9. $x(t) = 3C_1\cos 6t + C_2\sin 6t,\ y(t) =$
$-9C_1\sin 6t + 3C_2\cos 6t$

11. $x(t) = C_1 e^t\sin t + C_2 e^t\cos t,$
$y(t) = \dfrac{1}{5}(2C_1 + C_2)e^t\sin t + \dfrac{1}{5}(2C_2 - C_1)e^t\cos t$

13. $x(t) = C_1 e^{3t} + C_2 e^{4t} + \dfrac{2}{3},$
$y(t) = 2C_1 e^{3t} + C_2 e^{4t} - \dfrac{5}{3}$

15. $x(t) = C_1\cos t + C_2\sin t + 4t,\ y(t) =$
$-C_1\sin t + C_2\cos t - 2t + 1$

17. $x(t) = e^{3t} + 2e^{4t},\ y(t) = -e^{3t} - 4e^{4t}$

19. $x(t) = e^{5t},\ y(t) = e^{5t}$

21. $x(t) = \cos 2t + \sin 2t,\ y(t) = 2\cos 2t - 2\sin 2t$

23. $x(t) = 2e^{-3t} + 6e^{3t} - 2,\ y(t) = -10e^{-3t} + 6e^{3t} + 1$

25. $P(t) = C_1 + C_2 e^{-5t},\ Q(t) = \dfrac{3}{2}C_1 - C_2 e^{-5t}$

27. $x(t) = C_1 + C_2 e^{-4t} - 5t,$
$y(t) = C_1 - C_2 e^{-4t} - 5t + \dfrac{5}{2}$

29. **(a)** $A(t)/2$ per hr, $B(t)/2$ per hr

(b)   **(c)** Left to the student

(d) $A(t) = 1500 + 500e^{-t},\ B(t) = 1500 - 500e^{-t}$

(e)   **(f)** $1500, 1500$

31. **(a)**

(b) $P' = -5.24P + 0.6L,\ L' = 3P - 0.6L$

(c) $P(t) = 1.79851e^{-0.24t} - 1.79851e^{-5.6t},$
$L(t) = 14.9876e^{-0.24t} + 1.0791e^{-5.6t}$

33. **(a)** Left to the student **(b)** Left to the student

(c)

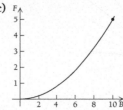

35. $L' = -(a + m)L + cM + p, M' = mL - (a + c)M$
37. Left to the student   **39.** Left to the student
41.

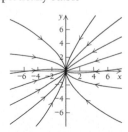

43. Left to the student   **45.** Left to the student

SECTION 9.4

1. $\begin{bmatrix} x \\ y \end{bmatrix}' = \begin{bmatrix} 1 & -1 \\ 3 & 2 \end{bmatrix} \begin{bmatrix} x \\ y \end{bmatrix}$ **3.** $\begin{bmatrix} x \\ y \end{bmatrix}' = \begin{bmatrix} 4 & 2 \\ 0 & 1 \end{bmatrix} \begin{bmatrix} x \\ y \end{bmatrix}$

5. $x' = x + 3y, y' = 5x + 7y$ **7.** $x' = 3y, y' = x - 2y$
9. Left to the student   **11.** Left to the student
13. $x(t) = -C_1 e^{2t} - 2C_2 e^t, y(t) = C_1 e^{2t} + C_2 e^t$
15. $x(t) = -2C_1 e^{-4t} + 2C_2 e^{4t}, y(t) = 3C_1 e^{-4t} + C_2 e^{4t}$
17. $x(t) = C_1 e^{-6t} + 4C_2 e^{-3t}, y(t) = -C_1 e^{-6t} - C_2 e^{-3t}$
19. $x(t) = C_1 e^t \sin 2t + C_2 e^t \cos 2t,$
$y(t) = \frac{1}{5}(3C_1 - C_2)e^t \sin 2t + \frac{1}{5}(C_1 + 3C_2)e^t \cos 2t$
21. $x(t) = C_1 e^t + C_2 t e^t, y(t) = -(C_1 + C_2)e^t - C_2 t e^t$
23. Saddle point, unstable

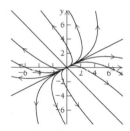

25. Node, unstable

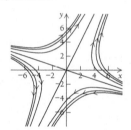

27. Node, asymptotically stable

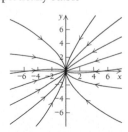

29. Node, unstable

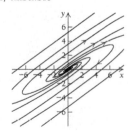

31. Saddle point, unstable

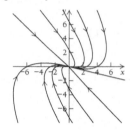

33. Node, asymptotically stable

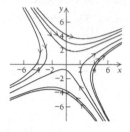

35. Spiral point, unstable

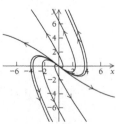

37. Improper node, unstable

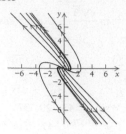

39. (a) $\begin{bmatrix} 1 \\ 1 \end{bmatrix}$ (b) tw
41. Left to the student   **43.** (a) Left to the student

(b) $x(t) = 3C_1 + C_2e^{-8t}$, $y(t) = C_1 - C_2e^{-8t}$ **(c)** Left to the student   **(d)** Left to the student
(e)

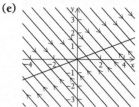

45. $x(t) = C_1e^{4t} - C_2e^{-2t} - 3C_3e^{2t}$,
$y(t) = -C_1e^{4t} - C_2e^{-2t} + C_3e^{2t}$,
$z(t) = C_1e^{4t} + C_2e^{-2t} + C_3e^{2t}$
47. $x(t) = -2C_1e^{2t} - C_3e^t$, $y(t) = C_2e^{2t} + C_3e^t$,
$z(t) = C_1e^{2t} + C_3e^t$
49. $x(t) = -3C_1e^{-2t} + C_2e^{2t} - 2C_3e^{-t}$,
$y(t) = C_1e^{-2t} + C_2e^{2t} + C_3e^{-t}$, $z(t) = 2C_1e^{-2t} + C_3e^{-t}$
51. $S' = -aS, B' = aS - (b + c)B + dH, H' = cB - dH$
53. Asymptotically stable
55. $S(t) = e^{-5t}$, $B(t) = -\frac{14}{9}e^{-8t} + \frac{40}{27}e^{-5t} + \frac{2}{27}e^{-t/2}$,
$H(t) = \frac{7}{9}e^{-8t} - \frac{35}{27}e^{-5t} + \frac{14}{27}e^{-t/2}$

SECTION 9.5

1. $(-4, 0)$, node, asymptotically stable; $(-3, -1)$, saddle point, unstable; $(0, 0)$, node, unstable; $(0, \frac{1}{2})$, saddle point, unstable   **3.** $(3, 4)$, saddle point, unstable; $(4, 3)$, node, unstable   **5.** $(2, 4)$, saddle point, unstable
7. $(1, -1)$, node, asymptotically stable; $(1, 1)$, saddle point, unstable   **9.** $(1, 0)$, spiral point, unstable
11. (a) $(0, 0)$, node, unstable; $(0, 25)$, node, asymptotically stable; $(10, 0)$, saddle point, unstable   **(b)** Only the second species survives.   **13. (a)** $(0, 0)$, node, unstable; $(0, 25)$, saddle point, unstable; $(12.5, 18.75)$, node, asymptotically stable; $(20, 0)$, saddle point, unstable
(b) Coexistence   **15. (a)** $(0, 0)$, node, unstable; $(0, 20)$, saddle point, unstable; $(100, 0)$, node, asymptotically stable   **(b)** Only the first species survives.   **17.** $(8, 4)$
19. (a) $(0, 0)$, node, unstable; $(2, 0)$, saddle point, unstable; $(0, 10)$, node, asymptotically stable   **(b)** $(0, 10)$
(c) Excessive fishing eliminates the first species.
21. $(0, 0)$, saddle point; $(5, 2)$, center   **23.** $(0, 0)$, saddle point; $(4, 2.5)$, center   **25.** $(0, 0)$, saddle point, unstable; $(20, 0)$, saddle point, unstable; $(5, 1.5)$, spiral point, asymptotically stable   **27.** $(0, 0)$, saddle point, unstable; $(10, 0)$, saddle point, unstable; $(4, 1.5)$, spiral point, asymptotically stable   **29. (a)** Left to the student
(b) $(0.671855, 0.895806)$
(c) Spiral point, unstable.

31. (a) -0.1, $\begin{bmatrix} 0 \\ 1 \end{bmatrix}$; 0.1, $\begin{bmatrix} 4 \\ -1 \end{bmatrix}$. Near $(0, 5)$, trajectories approach parallel to $\begin{bmatrix} 0 \\ 1 \end{bmatrix}$ and leave parallel to $\begin{bmatrix} 4 \\ -1 \end{bmatrix}$.

(b) -0.2, $\begin{bmatrix} 1 \\ 0 \end{bmatrix}$; 0.06, $\begin{bmatrix} -4 \\ 13 \end{bmatrix}$. Near $(4, 0)$, trajectories

approach parallel to $\begin{bmatrix} 1 \\ 0 \end{bmatrix}$ and leave parallel to $\begin{bmatrix} -4 \\ 13 \end{bmatrix}$.

(c) -0.15, $\begin{bmatrix} 2 \\ 1 \end{bmatrix}$; -0.05, $\begin{bmatrix} -2 \\ 3 \end{bmatrix}$. Most trajectories approach

$(2.5, 3.75)$ parallel to $\begin{bmatrix} -2 \\ 3 \end{bmatrix}$.   **33. (a)** Exercises 11, 13, 15,

and 16, a_1a_2; Exercises 12 and 14, b_1b_2 **(b)** tw
35. Left to the student   **37.** $(2, 5.7)$, spiral point, unstable   **39. (a)** Left to the student   **(b)** Left to the student   **(c)** Left to the student   **(d)** One population assumes its extreme values when the other population agrees with the coexistence equilibrium value.
41. $2 \ln y - y + \ln x - 0.4x = \ln 5 - 3$
43. $1.01594, 5$ **45.** $1.63126, 3.63308$

SECTION 9.6

1. $x(1) \approx 2.42368, y(1) \approx 1.048$ **3.** $x(2) \approx 38.2193$,
$y(2) \approx 83.4785$ **5.** $x(3) \approx 75.1539, y(3) \approx 212.322$
7. (a)

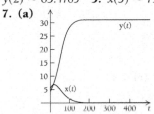

(b)

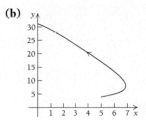

9. (a)

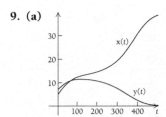

(b)

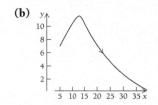

11. (a)

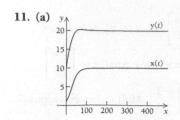

(b)

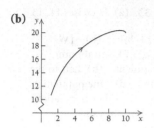

13. (a)

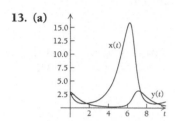

(b)   **(c)** 7.4

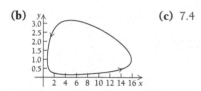

15. (a)

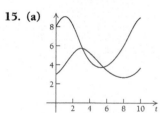

(b)   **(c)** 9.2

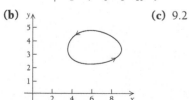

17.

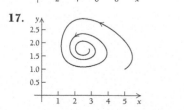

19. (a) $W(20) \approx 0.00664039$, $E(20) \approx 0.806201$, $U(20) \approx 0.190159$

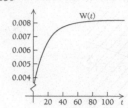

(b)

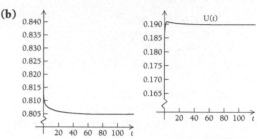

(c) Approximately 0.0082, 0.8050, and 0.1899
21. $(0, 0)$, node, unstable; $(0, 31.25)$, node, asymptotically stable; $(10, 0)$, saddle point, unstable
23. $(0, 0)$, node, unstable; $(0, 25)$, saddle point, unstable; $(10, 20)$, node, asymptotically stable; $(50/3, 0)$, saddle point, unstable   **25.** $(0, 0)$, saddle point, unstable; $(6, 4)$, center, stable   **27.** Left to the student

CHAPTER 9 REVIEW

1. [9.1] $C_1e^{2x} + C_2e^{5x}$ **2. [9.1]** $C_1e^{3x} + C_2xe^{3x}$
3. [9.1] $C_1 + C_2 \sin 2x + C_3 \cos 2x$
4. [9.2] $C_1e^{-x} \sin 5x + C_2e^{-x} \cos 5x + \frac{50}{13}$
5. [9.2] $C_1e^{-4x} + C_2e^{5x} - 2x^2 + x - 1$
6. [9.1] $3e^{-2x} - e^{-3x}$ **7. [9.1]** $3e^{-2x} + 12xe^{-2x}$
8. [9.1] $-4e^{-4x} \cos 2x - \frac{15}{2}e^{-4x} \sin 2x$
9. [9.2] $\frac{41}{10}e^{5t} - \frac{43}{6}e^{3t} + 2t + \frac{76}{15}$
10. [9.2] $-3e^x - 11xe^x + x^2 + 4x + 6$
11. [9.3], [9.4] $x(t) = C_1e^{-4t} + 3C_2e^{6t}$,
$y(t) = -C_1e^{-4t} + 2C_2e^{6t}$
12. [9.3], [9.4] $x(t) = C_1e^{3t} + C_2te^{3t}$,
$y(t) = (-C_1 - C_2)e^{3t} - C_2te^{3t}$
13. [9.3], [9.4] $x(t) = -2C_1e^{3t} \sin 6t + 2C_2e^{3t} \cos 6t$,
$y(t) = (C_1 + 3C_2)e^{3t} \sin 6t + (3C_1 - C_2)e^{3t} \cos 6t$
14. [9.3], [9.4] $x(t) = C_1e^{-2t} + C_2e^{4t}$,
$y(t) = -3C_1e^{-2t} - C_2e^{4t}$
15. [9.4] Saddle point, unstable

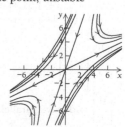

16. [9.4] Node, unstable

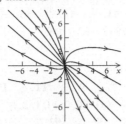

17. [9.4] Spiral point, asymptotically stable. Precise trajectories may vary.

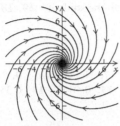

18. [9.4] Center. Precise trajectories may vary.

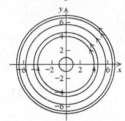

19. [9.3], [9.4] $x(t) = e^{2t} + e^{4t}$, $y(t) = -e^{2t} + e^{4t}$
20. [9.3], [9.4] $x(t) = 3e^{3t} + 4e^{3t}t$, $y(t) = e^{3t} - 4e^{3t}t$
21. [9.5] $(-3, -2)$, spiral point, unstable; $(0, -5)$, saddle point, unstable **22. [9.3], [9.4]**
$x(t) = 2C_1e^{-3t} + C_2e^{-t/2}$, $y(t) = -C_1e^{-3t} + 2C_2e^{-t/2}$
23. [9.3], [9.4] (a)

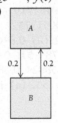

(b) $A(t) = 450 - 50e^{-2t/5}$, $B(t) = 450 + 50e^{-2t/5}$
(c) 450 lb, 450 lb
24. [9.5] $(0, 0)$, node, unstable; $(0, 15.625)$, saddle point, unstable; $(40, 0)$, node, asymptotically stable
25. [9.5] $(0, 0)$, saddle point, unstable; $(4, 1)$, center
26. [9.6]

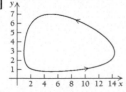

CHAPTER 9 TEST

1. [9.1] $C_1e^{3x} + C_2e^{5x}$ **2. [9.1]** $C_1e^{6x} + C_2xe^{6x}$
3. [9.2] $C_1 \sin 6x + C_2 \cos 6x + 2$
4. [9.2] $C_1e^{-4x} + C_2e^{3x} - \dfrac{x}{3} + \dfrac{2}{9}$
5. [9.1] $\dfrac{5}{4}e^x - \dfrac{1}{4}e^{5x}$ **6. [9.1]** $2e^{-2x}\cos x + 2e^{-2x}\sin x$
7. [9.2] $\dfrac{4}{3} - \dfrac{4}{3}e^{-3x} - xe^{-3x}$
8. [9.2] $\dfrac{16}{5}\cos 5x - \dfrac{8}{25}\sin 5x + \dfrac{3x}{5} - \dfrac{1}{5}$
9. [9.3], [9.4] $x(t) = C_1e^{-3t} + 5C_2e^{5t}$,
$y(t) = -C_1e^{-3t} + 3C_2e^{5t}$
10. [9.3], [9.4] $x(t) = 3C_1e^t + C_2e^{2t}$,
$y(t) = -2C_1e^t - C_2e^{2t}$
11. [9.3], [9.4] $x(t) = C_1e^{2t} + C_2e^{5t}$,
$y(t) = C_1e^{2t} - 2C_2e^{5t}$
12. [9.3], [9.4] $x(t) = C_1e^t \sin 8t + C_2e^t \cos 8t$,
$y(t) = \frac{1}{10}(4C_2 - 3C_1)e^t \sin 8t - \frac{1}{10}(4C_1 + 3C_2)e^t \cos 8t$
13. [9.4] Spiral point, asymptotically stable. Precise trajectories may vary.

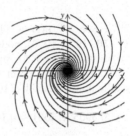

14. [9.4] Saddle point, unstable

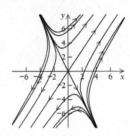

15. [9.4] Node, asymptotically stable

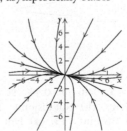

16. [9.4] Center. Precise trajectories may vary.

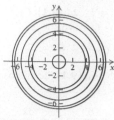

17. [9.5] $(-3, 2)$, node, asymptotically stable; $(-2, 3)$, saddle point, unstable
18. [9.3], [9.4] $x(t) = 2C_1 + C_2e^{-3t/2}$
$y(t) = C_1 - C_2e^{-3t/2}$
19. [9.3], [9.4]
(a)

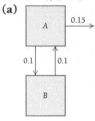

(b) $A(t) = 20e^{-3t/10} + 80e^{-t/20}$, $B(t) = -10e^{-3t/10} + 160e^{-t/20}$ **(c)** 0, 0   **20. [9.5]** $(0, 0)$, node, unstable; $(0, 20)$, saddle point, unstable; $(22.5, 5)$, node, asymptotically stable; $(25, 0)$, saddle point, unstable
21. [9.5] $(0, 0)$, saddle point, unstable; $(10, 2)$, center, stable
22. [9.6] $x(500) \approx 39.3905$, $y(500) \approx 0.106101$

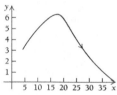

23. [9.6] $x(500) \approx 9 \times 10^{-11}$, $y(500) \approx 49.612$

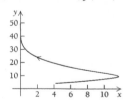

SECTION 10.1

1. 0.06   **3.** 0.28   **5.** 0.46   **7.** 0.48   **9.** 0.72
11. (a) Independent   **(b)** Not disjoint   **13. (a)** Not independent   **(b)** Disjoint   **15. (a)** Not independent
(b) Not disjoint   **17. (a)** Not independent
(b) Disjoint   **19.** 0.2   **21.** 0.8   **23.** 0.4   **25.** 0.05
27. 0.15   **29.** 0.04   **31.** 0.16   **33.** 0.4   **35.** 0.9

37. 0.7   **39.** $\frac{1}{45}$ **41.** $\frac{1}{18}$ **43.** 0.04   **45.** 0.06
47. $\frac{1}{216}$ **49.** $\frac{215}{216}$ **51.** $\frac{1}{32}$ **53.** $\frac{31}{32}$ **55.** $\frac{1}{1296}$ **57.** $\frac{5}{18}$
59. $\frac{11}{36}$ **61.** 0.147813   **63.** tw

SECTION 10.2

1. $\frac{1}{3}$ **3.** $\frac{1}{3}$ **5.** $\frac{1}{3}$ **7.** $\frac{1}{3}$ **9.** $\frac{1}{16}$ **11.** $\frac{1}{8}$ **13.** $\frac{1}{4}$
15. 0.382353   **17. (a)** FF, 0; Ff, $1/2$; ff, $1/2$ **(b)** tw
19. (a) $\frac{9}{16}$ **(b)** $\frac{3}{16}$ **(c)** $\frac{3}{16}$ **(d)** $\frac{1}{16}$
21. (a) 0.880952   **(b)** 0.963636   **(c)** 0.030
23. (a) 0.236749   **(b)** 0.736264   **25.** Left to the student   **27.** Left to the student   **29.** DD, 0.312; Dd, 0.623; dd, 0.065

SECTION 10.3

1. 6   **3.** 1   **5.** 1   **7.** Not defined   **9.** Not defined
11. 8008   **13.** $\frac{2}{9}$ **15.** 0.4096   **17.** 0.354294
19. 0.117142   **21.** 0

23.

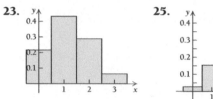

25.

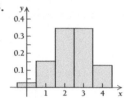

27.

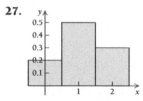

29. $\frac{7}{32}$ **31.** $\frac{7}{64}$

33. 0.160751   **35.** 0.838494
37. 0.0000938707   **39.** 0.0181719   **41.** 0.384
43. (a) 0.158095   **(b)** 0.0051375
45. CC, 0.0064; CG, 0.1472; GG, 0.8464
47. 0.0392   **49.** 0.00332778   **51.** 0.31744
53. (a) $\{0, 1, 2\}$ **(b)** $\{0.558824, 0.382353, 0.0588235\}$
(c)

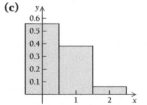

55. (a) $\{1, 2\}$ **(b)** $\frac{2}{3}, \frac{1}{3}$

(c)

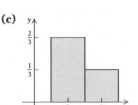

57. Left to the student

59. Left to the student **61.** 56 **63.** 1
65. tw **67. (a)** $SSSS$, $SSSF$, $SSFS$, $SSFF$, $SFSS$, $SFSF$, $SFFS$, $SFFF$, $FSSS$, $FSSF$, $FSFS$, $FSFF$, $FFSS$, $FFSF$, $FFFS$, $FFFF$ **(b)** 6 **(c)** 4 **(d)** tw **(e)** tw
69. 0.0796118 **71. (a)** $2pq, q^2$; 1/2, 1/2, 0, 1
(b) $FFSF$, $FFFS$ **(c)** tw **(d)** tw

SECTION 10.4

1. 1.3 **3.** 0.61, 0.781025 **5.** $\frac{1}{2}$ **7.** $\frac{3}{2}$ **9.** 0.612372
11. 0.866025 **13.** 1.2, 0.979796 **15.** 2, 1.34164
17. 20, 3.4641 **19.** 2 **21.** $-\frac{5}{3}$ **23.** 2 **25. (a)** 8.76,
1.53792 **(b)** 0.21215 **27.** 7.656, 1.66477 **29.** tw

SECTION 10.5

1. Left to the student **3.** Left to the student **5.** $\frac{5}{16}$
7. 0.632813 **9.** 0.42265 **11.** 0.377541 **13.** $\frac{1}{16}$
15. 0.0333733 **17.** 0.0588235 **19.** 20 **21.** 2.99573
23. 6.48074 **25.** $\frac{1}{4}$ **27.** $\frac{3}{2}$ **29.** Not a pdf

31. (a) $\frac{1}{4}, \frac{(2-x)^3}{4}$ **(b)** $\frac{15}{16}$ **33. (a)** $\frac{2\sqrt{3}}{3}, \frac{2\sqrt{3}}{3}\sin x$

(b) 0.577350 **35. (a)** $\frac{1}{\ln 11}, \frac{2x+3}{(\ln 11)(x^2+3x+4)}$

(b) 0.344752 **37. (a)** $\frac{1}{2\pi}, \frac{x\sin x}{2\pi}$ **(b)** 0.659155

39. $\frac{1}{2}$ **41.** $\frac{3}{8}$
43. $E(X) = 2$, $\text{Var}(X) = 0.5$, $\text{SD}(X) = 0.707107$
45. $E(X) = 1.4427$, $\text{Var}(X) = 0.0826736$,
$\text{SD}(X) = 0.28753$
47. $E(X) = 1.6254$, $\text{Var}(X) = 0.0723709$,
$\text{SD}(X) = 0.269018$

49. $E(X) = \frac{\pi}{2}$, $\text{Var}(X) = 0.467401$,
$\text{SD}(X) = 0.683667$
51. $E(X) = 1.65714$, $\text{Var}(X) = 0.715102$,
$\text{SD}(X) = 0.845637$
53. $E(X) = 1/3$, $\text{Var}(X) = 1/9$, $\text{SD}(X) = 1/3$
55. $E(X) = 6$, $\text{Var}(X) = 3$, $\text{SD}(X) = 1.73205$
57. $E(X) = 15$, $\text{Var}(X) = \frac{25}{3}$, $\text{SD}(X) = 2.88675$
59. $\frac{3}{5}$ **61. (a)** $\frac{1}{2}$ **(b)** $\frac{1}{5}$ **63.** $\sqrt{2}$ **65. (a)**
(b) $\frac{1}{3}(a^2 + ab + b^2)$ **(c)**   **67. (a)** 5.69333×10^{-7}
(b) 66.7012 **(c)** 16.7748 **69. (a)** tw **(b)** 0.82347
71. 0.0285955 **73.** 0.0189968 **75.** 0.00470122
77. 0.265026 **79.** 9.48773 **81.** 12.5916

SECTION 10.6

1. 0.110803 **3.** 0.354291 **5.** 0.0595404 **7.** 0.42319
9. 0.191153 **11.** 0.119013 **13.** 2.3 **15.** 2
17. 0.864665 **19.** 0.997521 **21.** 0.0183156
23. 0.713495 **25.** $\frac{1}{4}$ **27.** $\frac{1}{4}$
29.–47. See Exercises 1–11 and 17–23.
49. $E(X) = 4$, $\text{Var}(X) = 4$, $\text{SD}(X) = 2$

51. $E(X) = 1/3$, $\text{Var}(X) = 1/9$, $\text{SD}(X) = 1/3$
53. 0.59711 **55. (a)** 0.0148725 **(b)** 0.0607269
(c) 0.0497871 **57. (a)** 0.736403 **(b)** 0.513417
(c) 0.716531 **(d)** 0.238844 **(e)** 0.0446249
59. (a) $E(X) = 6$, $\text{SD}(X) = 2.44949$ **(b)** 0.00247875
61. Left to the student **63.** $\sqrt{2}$ **65.** tw
67. (a) $\ln 2$ **(b)** tw **69.** Left to the student
71. (a) 0.168284 **(b)** 0.168031 **73. (a)** 0.256561
(b) 0.256516 **75. (a)** 0.183949 **(b)** 0.18394

SECTION 10.7

1. 0.9545 **3.** 0.3606 **5.** 0.0275 **7.** 0.3452 **9.** 0.5
11. 0.121 **13. (a)** 2 **(b)** 5 **(c)** 0.3446 **(d)** 0.5793
(e) 0.8186 **15. (a)** 0.9772 **(b)** 0.3085 **(c)** 0.6041
17.–31. See Exercises 1–15. **33.** 0.734 **35.** 0.266
37. 0.397 **39. (a)** 0.1587 **(b)** 0.9772 **(c)** 0.0013
41. 0.2317 **43.** 0.7095 **45.** 0.4419 **47. (a)** 0.0447
(b) tw **49. (a)** 464 **(b)** 507 **(c)** 663 **51.** 0.1762
53. 0.0099 **55.** 0.1075 **57.** 2.33 **59.** −1.64
61. −2.58 **63.** 0.413223 **65.** 56.25 **67.** −1.46
69. 2.8, 6 **71.** −0.85, 2.20588 **73.** tw
75. (a) 0.4016 **(b)** 0.3944 **(c)** 0.0072 **77.** tw
79. 0.341529, 0.341355, 0.341347, 0.341345

CHAPTER 10 REVIEW

1. [10.1] $\frac{1}{4}$ **2.** [10.1] $\frac{3}{4}$ **3.** [10.1] $\frac{1}{2}$ **4.** [10.1] $\frac{1}{16}$
5. [10.3] 0.288 **6.** [10.3] 0.059535 **7.** [10.5] $\frac{2}{5}$
8. [10.5] $\frac{5}{7}$ **9.** [10.7] 0.5404 **10.** [10.7] 0.822
11. [10.6] 0.249236 **12.** [10.6] 0.393469
13. [10.6] 0.213763 **14.** [10.6] 0.238103
15. [10.4] 0.72, 0.6336, 0.79599 **16.** [10.4] 11.2,
2.24, 1.49666 **17.** [10.5] 28, $\frac{64}{3}$, 4.6188
18. [10.5] 1, $\frac{16}{3}$, 2.3094 **19.** [10.7] 0, 1, 1
20. [10.7] −5, 16, 4 **21.** [10.6] 0.25, 0.0625, 0.25
22. [10.6] 2.5, 6.25, 2.5 **23.** [10.6] 4, 4, 2
24. [10.6] 3, 3, $\sqrt{3}$ **25.** [10.5] 0.729167
26. [10.5] 0.367544 **27.** [10.5] 0.002
28. [10.5] 0.748933 **29.** [10.5] $\frac{1}{76}$
30. [10.5] $\frac{1}{\ln 10}$ **31.** [10.5] 3.30667, 0.265958,
0.515709 **32.** [10.2] $\frac{1}{4}$ **33.** [10.2] 0.845361
34. [10.3] 0.160751 **35.** [10.7] 0.8623
36. [10.5] 0.25 **37.** [10.7] **(a)** 0.0304
(b) 0.2356 **(c)** 0.0304 **38.** [10.7] 0.927
39. [10.6] 0.550671 **40.** [10.6] 0.503415 **41.** [10.3,
10.7] Exactly 50 times **42.** [10.7] **(a)** 0.4986501
(b) 0.0013182 **(c)** 0.0000314 **(d)** 3×10^{-7}

CHAPTER 10 TEST

1. [10.1] $\frac{1}{4}$ **2.** [10.1] $\frac{12}{13}$ **3.** [10.1] $\frac{1}{2}$
4. [10.1] 0.0637255 **5.** [10.3] 0.233474
6. [10.6] 0.0470665 **7.** [10.5] 0.2

8. [10.6] 0.285794 **9. [10.7]** 0.6687
10. [10.3] 3.6, 2.16, 1.46969 **11. [10.6]** 2.4, 2.4, 1.54919 **12. [10.5]** 30, $\frac{100}{3}$, 5.7735
13. [10.6] 0.2, 0.04, 0.2 **14. [10.7]** 3.2, 0.25, 0.5
15. [10.5] 0.875 **16. [10.5]** 0.056 **17. [10.5]** 3
18. [10.5] $\sqrt{3}$ **19. [10.5]** 0.015625
20. [10.5] 5.42884 **21. [10.2]** 0.678899
22. [10.3] 0.15625 **23. [10.7]** 0.0772 **24. [10.5]** $\frac{1}{4}$
25. [10.6] 0.527633 **26. [10.7]** 0.0478
27. [10.5] 0.287297 **28. [10.5]** $\frac{2}{5}$
29. [10.5] 0.282843 **30. [10.7]** **(a)** 0.1504
(b) 0.2406

APPENDIX A

Exercise Set A, p. 737

1. $5 \cdot 5 \cdot 5$, or 125 **3.** $(-7)(-7)$, or 49 **5.** 1.0201

7. $\frac{1}{16}$ **9.** 1 **11.** t **13.** 1 **15.** $\frac{1}{3^2}$, or $\frac{1}{9}$ **17.** 8

19. 0.1 **21.** $\frac{1}{e^b}$ **23.** $\frac{1}{b}$ **25.** x^5 **27.** x^{-6}, or $\frac{1}{x^6}$

29. $35x^5$ **31.** x^4 **33.** 1 **35.** x^3 **37.** x^{-3}, or $\frac{1}{x^3}$

39. 1 **41.** e^{t-4} **43.** t^{14} **45.** t^2 **47.** a^6b^5 **49.** t^{-6}, or $\frac{1}{t^6}$ **51.** e^{4x} **53.** $8x^6y^{12}$ **55.** $\frac{1}{81}x^8y^{20}z^{-16}$, or $\frac{x^8y^{20}}{81z^{16}}$

57. $9x^{-16}y^{14}z^4$, or $\frac{9y^{14}z^4}{x^{16}}$ **59.** $\frac{c^4d^{12}}{16q^8}$ **61.** $5x - 35$

63. $x^2 - 7x + 10$ **65.** $a^3 - b^3$ **67.** $2x^2 + 3x - 5$
69. $a^2 - 4$ **71.** $25x^2 - 4$ **73.** $a^2 - 2ah + h^2$
75. $25x^2 + 10xt + t^2$ **77.** $5x^5 + 30x^3 + 45x$
79. $a^3 + 3a^2b + 3ab^2 + b^3$
81. $x^3 - 15x^2 + 75x - 125$ **83.** $x(1 - t)$
85. $(x + 3y)^2$ **87.** $(x - 5)(x + 3)$ **89.** $(x - 5)(x + 4)$
91. $(7x - t)(7x + t)$ **93.** $4(3t - 2m)(3t + 2m)$
95. $ab(a + 4b)(a - 4b)$
97. $(a^4 + b^4)(a^2 + b^2)(a + b)(a - b)$
99. $10x(a + 2b)(a - 2b)$
101. $2(1 + 4x^2)(1 + 2x)(1 - 2x)$
103. $(9x - 1)(x + 2)$ **105.** $(x + 2)(x^2 - 2x + 4)$
107. $(y - 4t)(y^2 + 4yt + 16t^2)$ **109.** $(3x^2 - 1)(x - 2)$
111. $(x - 3)(x + 3)(x - 5)$ **113.** $\frac{7}{4}$ **115.** -8
117. 120 **119.** 200 **121.** $0, -3, \frac{4}{5}$ **123.** 0, 2
125. 0, 3 **127.** 0, 7 **129.** $0, \frac{1}{3}, -\frac{1}{3}$ **131.** 1
133. No solution **135.** -23 **137.** 5 **139.** $x \geq -\frac{4}{5}$
141. $x > -\frac{1}{12}$ **143.** $x > -\frac{4}{7}$ **145.** $x \leq -3$
147. $x > \frac{2}{3}$ **149.** $x < -\frac{2}{5}$ **151.** $2 < x < 4$
153. $\frac{3}{2} \leq x \leq \frac{11}{2}$ **155.** $-1 \leq x \leq \frac{14}{5}$ **157.** $650
159. More than 7000 units **161.** 480 lb
163. 810,000 **165.** $60\% \leq x < 100\%$

APPENDIX B

Exercise Set B, p. 745

1.
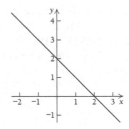

3. See graph. **5.** See graph. **7.** **(a)** 10 **(b)** 25
(c) 13 **(d)** 45 **(e)** 25 **(f)** $x^2 + 4x + 13$
(g) $x^2 + 2xh + h^2 + 2x + 2h + 10$
9. $y + x = 2$

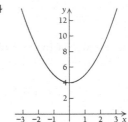

11. $y - x^2 = 4$

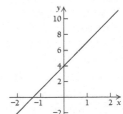

13. **(a)** Function **(b)** Not a function **(c)** Not a function **(d)** Function **15.** **(a)** Not a function **(b)** Function **(c)** Not a function **(d)** Function
17. $f(x) = 3x + 4$

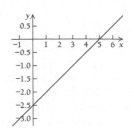

19. $g(x) = 0.5x - 2.5$

21. $f(x) = x^3$

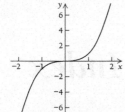

23. $r(x) = x^2 + x + 1$

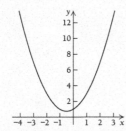

25. Domain: all real numbers except $x = -3$; range: all real numbers except $y = 0$ **27.** Domain: $x \geq -4$; range: $y \geq 0$ **29.** All real numbers   **31.** $x \geq -1$ **33.** $x \neq -2$ **35.** $x > -1$ **37.** $x > 1$

Index